Methods in Enzymology

Volume 204
BACTERIAL GENETIC SYSTEMS

METHODS IN ENZYMOLOGY

EDITORS-IN-CHIEF

John N. Abelson   Melvin I. Simon

DIVISION OF BIOLOGY
CALIFORNIA INSTITUTE OF TECHNOLOGY
PASADENA, CALIFORNIA

FOUNDING EDITORS

Sidney P. Colowick and Nathan O. Kaplan

Methods in Enzymology

Volume 204

Bacterial Genetic Systems

EDITED BY

Jeffrey H. Miller

DEPARTMENT OF BIOLOGY
UNIVERSITY OF CALIFORNIA, LOS ANGELES
LOS ANGELES, CALIFORNIA

ACADEMIC PRESS, INC.
Harcourt Brace Jovanovich, Publishers
San Diego   New York   Boston
London   Sydney   Tokyo   Toronto

Academic Press, Inc.
San Diego, California 92101

United Kingdom Edition published by
·ACADEMIC PRESS LIMITED
24-28 Oval Road, London NW1 7DX

Library of Congress Catalog Card Number: 54-9110

ISBN   0-12-182105-6   (alk. paper)

PRINTED IN THE UNITED STATES OF AMERICA
91 92 93 94   9 8 7 6 5 4 3 2 1

Table of Contents

Section II. Other Bacterial Systems

Contributors to Volume 204

Article numbers are in parentheses following the names of contributors.
Affiliations listed are current.

GERARD J. BARCAK (14), *Department of Biological Chemistry, University of Maryland School of Medicine, Baltimore, Maryland 21201*

RAÚL G. BARLETTA (25), *Department of Veterinary Science, University of Nebraska-Lincoln, Lincoln, Nebraska 68583*

KATARZYNA BEBENEK (6), *Laboratory of Molecular Genetics, National Institute of Environmental Health Sciences, National Institutes of Health, Research Triangle Park, North Carolina 27709*

JON BECKWITH (1), *Department of Microbiology and Molecular Genetics, Harvard Medical School, Boston, Massachusetts 02115*

JUDITH BENDER (7), *Whitehead Institute for Biomedical Research, Cambridge, Massachusetts 02142*

A. BERRY (23), *Genencor International, Rochester, New York 14652*

ELAINE A. BEST (18), *Calgene, Inc., Davis, California 95616*

BARRY R. BLOOM (25), *Department of Microbiology and Immunology, Howard Hughes Medical Institute, Albert Einstein College of Medicine, Bronx, New York 10461*

FREDRIC R. BLOOM (4), *Molecular Biology Research and Development, Life Technologies, Inc., Gaithersburg, Maryland 20877*

RICHARD CALENDAR (11), *Department of Molecular and Cell Biology, University of California, Berkeley, Berkeley, California 94720*

GERARD A. CANGELOSI (18), *MicroProbe Corp., Bothell, Washington 98021*

MICHAEL G. CAPARON (26), *Department of Molecular Microbiology, Washington University School of Medicine, St. Louis, Missouri 63017*

A. M. CHAKRABARTY (23), *Department of Microbiology and Immunology, University of Illinois College of Medicine, Chicago, Illinois 60612*

MARK S. CHANDLER (14), *Department of Molecular Biology and Genetics, Johns Hopkins University, School of Medicine, Baltimore, Maryland 21205*

JEFFREY D. CIRILLO (25), *Department of Microbiology and Immunology, Howard Hughes Medical Institute, Albert Einstein College of Medicine, Bronx, New York 10461*

A. DARZINS (23), *Department of Microbiology, Ohio State University, Columbus, Ohio 43210*

GIANNI DEHÒ (11), *Dipartimento di Genetica e di Biologia dei Microorganismi, Università degli Studi di Milano, 20133 Milan, Italy*

TIMOTHY J. DONOHUE (22), *Department of Bacteriology, University of Wisconsin, Madison, Madison, Wisconsin 53706*

BERT ELY (17), *Department of Biological Sciences, The University of South Carolina, Columbia, South Carolina 29208*

PATRICIA L. FOSTER (5), *Boston University School of Public Health, Boston University School of Medicine, Boston, Massachusetts 02174*

JANE GLAZEBROOK (19), *Department of Biology, Massachusetts Institute of Technology, Cambridge, Massachusetts 02139*

SUSAN GOTTESMAN (7), *Laboratory of Molecular Biology, National Cancer Institute, National Institutes of Health, Bethesda, Maryland 20892*

EDUARDO A. GROISMAN (8), *Department of Molecular Microbiology, Washington*

University School of Medicine, St. Louis, Missouri 63110

DOUGLAS HANAHAN (4), Department of Biochemistry and Biophysics, Hormone Research Institute, University of California, San Francisco, San Francisco, California 94143

ROBERT HASELKORN (20), Department of Molecular Genetics and Cell Biology, The University of Chicago, Chicago, Illinois 60637

JAMES A. HOCH (13), Division of Cellular Biology, Scripps Research Institute, La Jolla, California 92037

DAVID A. HOPWOOD (21), Department of Genetics, John Innes Institute, Norwich NR4 7UH, England

WILLIAM R. JACOBS, JR. (25), Department of Microbiology and Immunology, Howard Hughes Medical Institute, Albert Einstein College of Medicine, Bronx, New York 10461

JOEL JESSEE (4), Molecular Biology Research and Development, Life Technologies, Inc., Gaithersburg, Maryland 20877

WILBUR JONES (25), Mycobacteriology Laboratories, Centers for Disease Control, Atlanta, Georgia 30333

MICHAEL L. KAHN (11), Department of Microbiology, Institute of Biological Chemistry, Washington State University, Pullman, Washington 99164

DALE KAISER (16), Departments of Biochemistry and Developmental Biology, Stanford University, Stanford, California 94305

GANJAM V. KALPANA (25), Department of Microbiology and Immunology, Howard Hughes Medical Institute, Albert Einstein College of Medicine, Bronx, New York 10461

SAMUEL KAPLAN (22), Department of Microbiology, The University of Texas Medical School at Houston, Houston, Texas 77225

TOBIAS KIESER (21), Department of Genetics, John Innes Institute, Norwich NR4 7UH, England

NANCY KLECKNER (7), Department of Biochemistry and Molecular Biology, Harvard University, Cambridge, Massachusetts 02138

THOMAS A. KUNKEL (6), Laboratory of Molecular Genetics, National Institute of Environmental Health Sciences, National Institutes of Health, Research Triangle Park, North Carolina 27709

K. BROOKS LOW (3), Radiobiology Laboratories, Yale University School of Medicine, New Haven, Connecticut 06510

GLADYS MARTINETTI (18), Department of Medical Microbiology, University of Zurich, CH-8082 Zurich, Switzerland

RUSSELL MAURER (2), Department of Molecular Biology and Microbiology, Case Western Reserve University School of Medicine, Cleveland, Ohio 44106

LINDA McCARTER (24), Microbial Genetics, The Agouron Institute, La Jolla, California 92037

JOHN McCLARY (6), Codon, South San Francisco, California 94080

NOREEN E. MURRAY (12), Institute of Cell and Molecular Biology, University of Edinburgh, Edinburgh EH9 3JR, Scotland

EUGENE W. NESTER (18), Department of Microbiology, University of Washington, Seattle, Washington 98195

RICHARD P. NOVICK (27), Department of Plasmid Biology, Public Health Research Institute, New York, New York 10016

DAVID W. OW (11), Plant Gene Expression Center, United States Department of Agriculture, Albany, California 94710

LISA PASCOPELLA (25), Department of Microbiology and Immunology, Howard Hughes Medical Institute, Albert Einstein College of Medicine, Bronx, New York 10461

ROSEMARY J. REDFIELD (14), Department of Biochemistry, University of British Columbia, Vancouver, British Columbia V6T 1W5, Canada

R. K. ROTHMEL (23), Envirogen, Inc., Lawrenceville, New Jersey 08648

KENNETH E. SANDERSON (10), Department of Biological Sciences, Salmonella Genetic Stock Centre, University of Calgary, Calgary, Alberta T2N 1N4, Canada

JUNE R. SCOTT (26), Department of Microbiology and Immunology, Emory University Health Sciences Center, Atlanta, Georgia 30322

H. STEVEN SEIFERT (15), Department of Microbiology-Immunology, Northwestern University Medical School, Chicago, Illinois 60611

RICHARD SHOWALTER (24), Microbial Genetics, The Agouron Institute, La Jolla, California 92037

THOMAS J. SILHAVY (9), Department of Molecular Biology, Princeton University, Princeton, New Jersey 08544

MICHAEL SILVERMAN (24), Microbial Genetics, The Agouron Institute, La Jolla, California 92037

JAMES M. SLAUCH (9), Department of Microbiology and Molecular Genetics, Harvard Medical School, Boston, Massachusetts 02115

SCOTT B. SNAPPER (25), Department of Microbiology and Immunology, Howard Hughes Medical Institute, Albert Einstein College of Medicine, Bronx, New York 10461

MAGDALENE SO (15), Department of Micro-

biology-Immunology, Oregon Health Sciences University, Portland, Oregon 97201

NAT L. STERNBERG (2), Du Pont-Merck Pharmaceutical, Wilmington, Delaware 19880

MELVIN G. SUNSHINE (11), Department of Microbiology, School of Medicine, University of Iowa, Iowa City, Iowa 52242

JEAN-FRANÇOIS TOMB (14), Department of Molecular Biology and Genetics, Johns Hopkins University, School of Medicine, Baltimore, Maryland 21205

RUPA A. UDANI (25), Department of Microbiology and Immunology, Howard Hughes Medical Institute, Albert Einstein College of Medicine, Bronx, New York 10461

GRAHAM C. WALKER (19), Department of Biology, Massachusetts Institute of Technology, Cambridge, Massachusetts 02139

DANIEL R. ZEIGLER (10), Department of Biochemistry, Bacillus Genetic Stock Center, Ohio State University, Columbus, Ohio 43210

RAINER ZIERMANN (11), Department of Molecular and Cell Biology, University of California, Berkeley, Berkeley, California 94720

Preface

This volume contains contributions which offer an extraordinary overview of the state of bacterial genetics. The articles provide not only enjoyable reading, but furnish a useful manual for those performing experiments in these areas.

The book is divided into two parts. First, the arsenal of genetic methods available in *Escherichia coli* and *Salmonella typhimurium* is covered. Separate articles are devoted to methods for storing and maintaining bacterial strains, use of different mutagens *in vivo, in vitro* mutagenesis, devising selections for mutants, gene transfer by transduction, gene transfer by transformation, gene transfer by conjugation, utilization of lambda derivatives, genetic engineering with transposable elements, use of Mu phage for genetic engineering, employment of phages P2 and P4, and for the use of gene fusions.

The second portion of the book reviews the genetic systems available in other bacteria. Separate articles detail the state of the art in *Agrobacterium,* Rhodospirillacae *Staphylococcus,* pathogenic *Neisseria, Bacillus subtilis, Vibrio,* mycobacteria, myxobacteria, *Haemophilus influenzae,* pathogenic *Streptococcus, Streptomyces,* cyanobacteria, and *Pseudomonas.*

The complete set of reviews places the reader at the forefront of bacterial genetic methodology, and should aid many laboratories in carrying out genetic procedures in the organisms described.

JEFFREY H. MILLER

METHODS IN ENZYMOLOGY

VOLUME XXVII. Enzyme Structure (Part D)
Edited by C. H. W. HIRS AND SERGE N. TIMASHEFF

VOLUME XXVIII. Complex Carbohydrates (Part B)
Edited by VICTOR GINSBURG

VOLUME XXIX. Nucleic Acids and Protein Synthesis (Part E)
Edited by LAWRENCE GROSSMAN AND KIVIE MOLDAVE

VOLUME XXX. Nucleic Acids and Protein Synthesis (Part F)
Edited by KIVIE MOLDAVE AND LAWRENCE GROSSMAN

VOLUME XXXI. Biomembranes (Part A)
Edited by SIDNEY FLEISCHER AND LESTER PACKER

VOLUME XXXII. Biomembranes (Part B)
Edited by SIDNEY FLEISCHER AND LESTER PACKER

VOLUME XXXIII. Cumulative Subject Index Volumes I–XXX
Edited by MARTHA G. DENNIS AND EDWARD A. DENNIS

VOLUME XXXIV. Affinity Techniques (Enzyme Purification: Part B)
Edited by WILLIAM B. JAKOBY AND MEIR WILCHEK

VOLUME XXXV. Lipids (Part B)
Edited by JOHN M. LOWENSTEIN

VOLUME XXXVI. Hormone Action (Part A: Steroid Hormones)
Edited by BERT W. O'MALLEY AND JOEL G. HARDMAN

VOLUME XXXVII. Hormone Action (Part B: Peptide Hormones)
Edited by BERT W. O'MALLEY AND JOEL G. HARDMAN

VOLUME XXXVIII. Hormone Action (Part C: Cyclic Nucleotides)
Edited by JOEL G. HARDMAN AND BERT W. O'MALLEY

VOLUME XXXIX. Hormone Action (Part D: Isolated Cells, Tissues, and Organ Systems)
Edited by JOEL G. HARDMAN AND BERT W. O'MALLEY

VOLUME 80. Proteolytic Enzymes (Part C)
Edited by LASZLO LORAND

VOLUME 81. Biomembranes (Part H: Visual Pigments and Purple Membranes, I)
Edited by LESTER PACKER

VOLUME 82. Structural and Contractile Proteins (Part A: Extracellular Matrix)
Edited by LEON W. CUNNINGHAM AND DIXIE W. FREDERIKSEN

VOLUME 83. Complex Carbohydrates (Part D)
Edited by VICTOR GINSBURG

VOLUME 84. Immunochemical Techniques (Part D: Selected Immunoassays)
Edited by JOHN J. LANGONE AND HELEN VAN VUNAKIS

VOLUME 85. Structural and Contractile Proteins (Part B: The Contractile Apparatus and the Cytoskeleton)
Edited by DIXIE W. FREDERIKSEN AND LEON W. CUNNINGHAM

VOLUME 86. Prostaglandins and Arachidonate Metabolites
Edited by WILLIAM E. M. LANDS AND WILLIAM L. SMITH

VOLUME 87. Enzyme Kinetics and Mechanism (Part C: Intermediates, Stereochemistry, and Rate Studies)
Edited by DANIEL L. PURICH

VOLUME 88. Biomembranes (Part I: Visual Pigments and Purple Membranes, II)
Edited by LESTER PACKER

VOLUME 89. Carbohydrate Metabolism (Part D)
Edited by WILLIS A. WOOD

VOLUME 90. Carbohydrate Metabolism (Part E)
Edited by WILLIS A. WOOD

VOLUME 91. Enzyme Structure (Part I)
Edited by C. H. W. HIRS AND SERGE N. TIMASHEFF

Section I

Escherichia coli and *Salmonella typhimurium*

[1] Strategies for Finding Mutants

By Jon Beckwith

Introduction

The heights of accomplishment in bacterial or bacteriophage genetics have usually been ingenious genetic schemes for isolating mutants or for determining the basis of some biological phenomenon. One of the most striking examples of exploits in this field is the use of frameshift mutations to determine the triplet nature of the genetic code.[1] In this case, a problem that would appear to be one amenable only to biochemical approaches yielded to an indirect but convincing genetic approach.

The strategy of the geneticist is to look for mutants that affect a particular biological process in order to use them to understand that process. The geneticist starts with the premise that genetics will be necessary either to show that some biochemical finding has a reality in the living cell or to determine the nature of the biochemical process in the first place. The current ability to do reverse genetics has heightened awareness of the importance of the genetic approach. Enough cases have now appeared where a particular biochemical basis for a phenomenon has been assumed, only for reverse genetics to show that the link does not exist, to emphasize what geneticists have held as self-evident for many years: mutants are necessary to establish a link between *in vitro* and *in vivo* phenomena.

In many cases, mutant hunts are straightforward. For example, if one wants to study the basis of antibiotic action in bacteria, one can often, by simply selecting antibiotic-resistant mutants and defining the mutated gene and its product, establish where the agent acts in the cell. Determining the steps in a biosynthetic pathway requires obtaining auxotrophic mutants for that pathway and using them to seek out the missing enzymatic step. Mutagenesis can be used to enrich for mutants, making the hunt for auxotrophs much easier. There are mutagenic techniques which will yield a variety of types of mutations. These include ultraviolet irradiation which gives a spectrum of classes of mutations, including frameshifts, deletions, and base substitutions. The *mutD* mutator strains favor transitions but also give a variety of other base changes. There also exist three mutator

---

[1] F. H. C. Crick, L. Barnett, S. Brenner, and R. J. Watt-Tobin, *Nature (London)* **192**, 1227 (1961).

METHODS IN ENZYMOLOGY, VOL. 204

genes, *mutT*,[2] *mutY*,[3] and *mutM*,[4] which give specific transversion changes. After mutagenesis, either replica plating or penicillin selection for enrichment of mutants makes detecting such mutants relatively easy. Penicillin selection may enrich the mutant population by 100-fold.

However, there are often important reasons to develop genetic selections in order to obtain spontaneous mutants. Spontaneous mutations include insertions of insertion sequence (IS) elements, frameshifts, all possible base substitutions, and deletions. All mutagens or mutator genes are limited in the classes of mutations they induce. Some of the most powerful mutagens (e.g., *N*-methyl-*N'*-nitronitrosoguanidine) induce only G–C to A–T transitions. Such transitions are the class of mutations least likely to affect function of a protein, and they are likely to yield a high proportion of nonsense mutations. Even ultraviolet light or the mutator *mutD* still favor certain kinds of changes over others. When one is looking for a wide range of mutations affecting a particular gene or is looking for a very rare or special kind of mutation (e.g., altering substrate specificity of a protein), mutagenic approaches which cause a limited range of DNA alterations may not be appropriate. Thus, it is useful, in many cases, to develop techniques for obtaining spontaneous mutants. Since the spontaneous forward mutation frequency for any individual gene is on the order of 10^{-5}, the use of replica plating (in which perhaps 200 colonies can be tested per plate) is impractical and penicillin enrichment not strong enough. The result of this dilemma is that geneticists have devoted considerable energy to devising inventive schemes for selecting spontaneous mutations in various biological processes. It should be pointed out that with the increasing number of mutator genes and mutagens available, a combination of these mutagenic methods begins to approach the range of classes of mutations obtainable by the use of spontaneous selections.

In this chapter, I cover a variety of strategies for obtaining mutants in the bacteria *Escherichia coli* and *Salmonella typhimurium*. Some of the greatest ingenuity can be found in the schemes developed for studying bacteriophage λ genetics. Since these schemes are usually rather specific to λ, they are not covered in this chapter. Examples of the approaches are covered in volumes that review the biology of the phage.[5]

[2] C. Yanofsky, E. C. Cox, and V. Horn, *Proc. Natl. Acad. Sci. U.S.A.* **55,** 274 (1966).

[3] T. Nghiem, M. Cabrera, C. G. Cupples, and J. H. Miller, *Proc. Natl. Acad. Sci. U.S.A.* **85,** 2709 (1988).

[4] M. Cabrera, Y. Nghiem, and J. H. Miller, *J. Bacteriol.* **170,** 5405 (1988).

[5] A. D. Hershey (ed.), "The Bacteriophage Lambda." Cold Spring Harbor Laboratory, Cold Spring Harbor, New York, 1971.

Mutations in Catabolic Pathways

Suicide Compounds

It is instructive to review the strategies used for obtaining various types of mutants of the *lac* system for two reasons. First, the approaches used represent a good fraction of the available types of strategies. Second, *lac* fusions have been widely used for studying a variety of genetic systems. The availability of these fusions allows the use of the strategies employed in *lac* genetics for studying other systems. These methods and their use with fusions are described in [9] in this volume. However, some generalizations about the genetic approaches to the *lac* system are worth noting. First, the concept of a "suicide" compound is one that may be applicable to a wide variety of systems. In the case of the *lac* operon, the lactose analog thio-*o*-nitrophenyl-β-D-galactoside causes bacterial stasis when transported into the cell by the *lac* permease. Selection for resistance to this compound is a direct selection for forward mutations in the *lac* operon.[6,7] It also can be used to select for revertants of *lacO*^{c} or *lacI*$^-$ mutations.

A suicide compound for two other pathways is phosphonomycin.[8] This antibiotic can be transported into the cell by either the β-glycerol phosphate transport system (product of the *glpD* gene) or the hexose phosphate transport system (product of the *uhpT* gene). If either transport system is functional, phosphonomycin will kill the cell. Selection for resistance to phosphonomycin is a selection for inactivation of both genes. Since double mutations are rare, the only way in which resistance can occur in a single step is by inactivation of the CAP–cyclic AMP system which is required for expression of both *glpD* and *uhpT*. Thus, phosphonomycin-resistant mutants carry either *cya* or *crp* mutations. This example also illustrates the utility of developing selections for the loss of two functions at the same time. Numerous other examples of the use of analogs of normal metabolites to select mutants altered in either transport or some other step of a metabolic pathway can be found in a review by Vinopal.[9]

Selection for Resistance to Substrate in Substrate-Sensitive Mutants

A second approach to obtaining forward mutations in a catabolic pathway derives from the finding that mutants in certain of these pathways

[6] T. F. Smith and J. R. Sadler, *J. Mol. Biol.* **59**, 273 (1971).
[7] J. Hopkins, *J. Mol. Biol.* **87**, 715 (1974).
[8] M. Alper and B. Ames, *J. Bacteriol.* **133**, 149 (1978).
[9] R. T. Vinopal, *in* "*Escherichia coli* and *Salmonella typhimurium*" (J. L. Ingraham, *et al.*, eds.), p. 990. American Society for Microbiology, Washington, D.C., 1987.

become sensitive to the substrate. Examples include the *galE* mutant which eliminates the UDP-gal epimerase of the galactose catabolic pathway and makes cells sensitive to galactose;[10] the *araD* mutant which eliminates the ribulose-5-phosphate epimerase of the arabinose pathway and makes cells sensitive to arabinose;[11] and the *glpD* mutation which abolishes α-glycerol phosphate dehydrogenase required for glycerol metabolism and makes cells sensitive to glycerol.[12] In fact, it appears that for most catabolic pathways which include phosphorylated intermediates sensitivity to substrate is manifested in certain defective mutants. These mutants can be used to select for mutants in steps of the pathway preceding the one inactivated by the mutation. For instance, selection of galactose-resistant mutations in a *galE* mutant yields strains defective in the genes *galK* and *galT*, whose products act at earlier steps in the pathway.

This selection, employing sensitivity to substrate, can be extended to other pathways by modifications of the approach. For instance, since lactose is cleaved to glucose and galactose, a Lac$^+$ strain carrying a *galE* mutation is killed by the galactose released by β-galactosidase hydrolysis of lactose. Therefore, such a strain can be used to select for spontaneous Lac$^-$ mutants.[13] Similarly, β-glycerol phosphate is cleaved by alkaline phosphatase to yield glycerol. In a *glpD* background, cells are sensitive to β-glycerol phosphate if alkaline phosphatase is present. Selection for resistance to β-glycerol phosphate is a direct selection for mutants lacking alkaline phosphatase (or some components of the glycerol pathway).[14]

Selections for Constitutive or Up-Expression Mutants

Gene fusions provide a means of easily selecting for mutations which increase the expression of a gene. In particular, with a fusion of the *lac* operon to a gene which is expressed at low level, one may be able to select for Lac$^+$ derivatives directly. This was done with a *lac* fusion to one of the *mal* operons. Selection for Lac$^+$ yielded constitutive mutations in the regulatory gene for the system, *malT*.[15] In many cases, the level of expression of the gene to which *lac* is fused may be high enough so that a direct selection for Lac$^+$ is not possible. However, this problem can be overcome through the use of an inhibitor of β-galactosidase, thiophenyl-

[10] T. Fukusaws and H. Nikkaido, *Biochim. Biophys. Acta* **48,** 479 (1961).

[11] E. R. Englesberg, R. L. Anderson, R. Neinberg, N. Lee, D. Hoffee, G. Huttenhauer, and H. Boyer, *J. Bacteriol.* **84,** 137 (1962).

[12] N. R. Cozzarelli, W. B. Freedberg, and E. C. C. Lin, *J. Mol. Biol.* **31,** 371 (1968).

[13] M. Malamy, *Cold Spring Harbor Symp. Quant. Biol.* **31,** 189 (1966).

[14] A. Sarthy, S. Michaelis, and J. Beckwith, *J. Bacteriol.* **145,** 288 (1981).

[15] M. Debarbouille, H. A. Shuman, T. J. Silhavy, and M. Schwartz, *J. Mol. Biol.* **124,** 359 (1978).

ethyl-β-D-galactoside (TPEG). The inhibitor damps down the activity of the enzyme enough to restore a Lac⁻ phenotype, permitting direct selection for Lac⁺ derivatives.[16]

Mutations in Biosynthetic Pathways

For biosynthetic pathways, direct selections for defective mutants do not usually exist. Analogs of end products of a pathway have been used to select constitutive mutants. For example, resistance to 5-methyltryptophan yields *trpR* mutants constitutive for the tryptophan biosynthetic pathway. The review by Vinopal[9] lists numerous selections of this sort. In addition, he describes selections using analogs that give mutations in a few genes of biosynthetic pathways. As described above, *lac* fusions also provide a means of selecting constitutive mutants for a pathway or gene.

Brute Force Genetics

For many researchers working in bacterial genetics, much of the original appeal of the field was the degree to which it required constant thinking in order to come up with ever more creative schemes for mutant hunts. It was, then, something of a shock, when a "novel," less elegant way of obtaining important mutants was presented in the late 1960s. This "brute force" approach was first described in a paper by Ray Gesteland who sought a mutant lacking RNase I activity.[17] The problem confronting the "ingenious" strategy here is that the function of the enzyme and whether or not it was essential were not known. Gesteland's method was to mutagenize bacteria with the potent mutagen nitrosoguanidine, grow up cultures from hundreds of individual colonies formed after the mutagenic treatment, and assay extracts of these cultures for the enzymatic activity. Using this approach, he found two mutants missing RNase I activity.

Similar methods to that employed by Gesteland have been used in several other cases. The most noteworthy of these is the isolation of a mutant defective in DNA polymerase I activity in *E. coli*.[18] At the time DeLucia and Cairns did these studies, it was still not clear whether this enzyme played a central role in the polymerization of bases during DNA replication. Not knowing the role of the enzyme, no selection strategy could be devised. Bacteria were mutagenized and individual colonies as-

[16] M. N. Hall, M. Gabay, M. Debarbouille, and M. Schwartz, *Nature (London)* **295**, 616 (1982).

[17] R. F. Gesteland, *J. Mol. Biol.* **16**, 67 (1966).

[18] P. DeLucia and J. Cairns, *Nature (London)* **224**, 1164 (1969).

sayed for the enzymatic activity. Since it was possible that the enzyme was essential for replication and, therefore, for growth, it was necessary to assume that a mutant missing the activity would be unable to grow. To circumvent this problem, DeLucia and Cairns sought a conditional mutation, one that would code for an enzyme that had activity at 30° and not at 42°. Up until the point of assay, cells were grown at 30°; extracts were then incubated and assayed at 42°. This was an unnecessary precaution, as the mutant had very low DNA polymerase activity at both temperatures.

Detecting Mutations in Essential Genes

In designing screening or selection procedures for mutants, one of the important considerations to keep in mind is whether the process being studied is essential for cell growth. If so, several different approaches are used to detect mutations in essential genes.

General Screens for Conditional Lethal Mutants

The conditional lethal phenotype provides a means of amassing a collection of mutants which could affect almost any essential gene in an organism. Mutagenized bacteria or phage are tested individually for their ability to grow under one condition, but not another. The most common conditions used are low and high temperature. With a large collection of temperature-sensitive (*ts*) mutants, one can assay each one after a shift to the nonpermissive temperature for any enzymatic or biochemical activity or for other phenotypes. Many mutations in DNA synthesis have been detected in this way. In the late 1950s and early 1960s, conditional lethal mutations to analyze bacterial and bacteriophage genomes came into use. This generalized approach was first popularized by the work on bacteriophage T4 from the group of Bob Edgar at the California Institute of Technology.[19]

Although collections of temperature-sensitive lethal mutations of *E. coli* had been described by Horowitz and Leupold[20] in 1951, their utility for genetic analysis was not recognized until much later. The discovery of suppressible nonsense mutations provided a second class of conditional lethals for bacteriophage. The temperature-sensitive mutants could be used in either bacteria or bacteriophage, but suppressible nonsense mutants were limited to bacteriophage at that time. This was because one could maintain the phage by growing them on suppressor-containing hosts

---

[19] R. H. Epstein, A. Bolle, C. M. Steinberg, E. Kellenberger, E. Boy de la Tour, and R. Chevalley, *Cold Spring Harbor Symp.* **28,** 375 (1963).

[20] N. H. Horowitz and U. Leupold, *Cold Spring Harbor Symp.* **16,** 65 (1951).

and testing them on *sup*⁻ hosts. With bacteria, a nonsense mutation in an essential gene could not be maintained in the absence of the suppressor and could not be detected in the presence of a suppressor. It was only with the later construction of suppressor genes whose activity could be controlled[21,22] and the development of other novel techniques (see below) that general approaches to obtaining nonsense mutations in *E. coli* essential genes were made possible.

The general conditional lethal strategy is not foolproof. There are a number of reasons why mutations in certain genes will be missed. First, *ts* mutations may vary widely in their frequency from gene to gene. It has been suggested that for ribosomal proteins cold-sensitive (*cs*) mutation will be much more common than *ts* mutation.[23] To enhance chances of detecting mutations in a gene, it may be advantageous to screen for both *cs* and *ts* mutations. Second, the existence of mutations termed temperature-sensitive synthesis (*tss*) creates a problem for analyzing conditional lethals in certain pathways. The *tss* mutants are ones in which the protein, if assembled at the permissive temperature, retains its activity at the nonpermissive temperature. It is only its assembly at high temperature that is defective. This type of mutant contrasts with the more standard *ts* mutant in which the protein itself is temperature-sensitive for its activity at high temperature. Since the strategy is to take temperature-sensitive mutations and look for inactivation of an activity or a change in phenotype shortly after a shift to high temperature, temperature-sensitive synthesis mutations will be missed. This appears to be the reason that *ts* mutations in RNA polymerase were never detected among the general *ts* lethal collection.[24] It has been suggested that among proteins that assemble into complexes (such as RNA polymerase) *tss* mutants will be the common class.

Direct Screen for Conditional Lethals with Specific Phenotype at Nonpermissive Temperature

Screens for conditional lethal mutations of the type described above yield mutants with a wide variety of phenotypes as there is no way to demand mutations affecting a particular process. One only screens for a

[21] D. Beckman and S. Cooper, *J. Bacteriol.* **116,** 1336 (1973).

[22] H. Inoko, K. Shigesada, and M. Imai, *Proc. Natl. Acad. Sci. U.S.A.* **74,** 1162 (1977).

[23] C. Guthrie, H. Nashimoto, and M. Nomura, *Proc. Natl. Acad. Sci. U.S.A.* **63,** 384 (1969).

[24] J. Miller, I. V. Claeys, J. B. Kirschbaum, S. Nasi, S. Van der Elsacker, B. Molholt, G. Gross, D. A. Fields, and E. K. F. Bautz, *in* "RNA Polymerase" (R. Losick and M. Chamberlin, eds.), p. 519. Cold Spring Harbor Laboratory, Cold Spring Harbor, New York, 1976.

phenotype or the lack of a particular activity after taking a collection of conditional lethal mutations and shifting each mutant to the nonpermissive temperature. A little used but quite valuable approach has been developed for screening mutagenized colonies at a nonpermissive temperature for a desired phenotype.[25] This is done by growing the bacteria into colonies first at a permissive temperature and then switching the plates to a nonpermissive temperature on media where the desired phenotype can be observed. The strategy described by Schedl and Primakoff was designed to yield mutants affecting tRNA synthesis and maturation. The principles of the approach were (1) to use a phenotype that can be detected on an easily scored indicator medium and (2) to introduce the tRNA gene into cells after they have been switched to the nonpermissive temperature so that the tRNA accumulated before the switch does not interfere with the screen. This was accomplished by using a strain carrying a *lacZ* nonsense mutation and a ϕ80 transducing phage carrying a mutant tRNA which could suppress the amber mutation.

Bacteria were mutagenized and spread for single colonies on lactose MacConkey agar at 30°. After colonies had formed, the plates were put at 42° and sprayed with a lysate containing the ϕ80^{sup+}. After incubation for several hours longer, bacteria which could express the tRNA gene from the phage would suppress the *lacZ* amber and the colonies would turn red. Bacteria unable to express the tRNA activity formed white colonies. In this way, a number of mutants were detected, some of which were shown to be defective in the enzymes which process tRNA.

The Schedl–Primakoff approach has also been used with a ϕ80*phoA* phage to detect mutants defective in expression or secretion of alkaline phosphatase.[26] In this case, a mutation in a gene which was later characterized as part of a two-component regulatory system was found. This method could be used for any cellular process where the phenotype can be detected on indicator plates and a transducing phage carrying the gene exists. Since *lac* fusions can be obtained to any bacterial gene, it is possible to use this approach for a wide variety of genes.

Conditional Lethal or Null Mutations in Already Defined Genes

The approaches described in the previous section are designed to detect mutations in genes which had not previously been characterized. These approaches are important in the initial characterization of a biological process. Once one has found the genes involved, one usually needs to

[25] P. Schedl and P. Primakoff, *Proc. Natl. Acad. Sci. U.S.A.* **70**, 2091 (1973).
[26] B. Wanner, A. Sarthy, and J. Beckwith, *J. Bacteriol.* **140**, 229 (1979).

obtain additional mutations in the gene, either for structure–function analysis or to obtain true null mutations for the study of essentiality and function. At least two general *in vivo* approaches can be used. One approach utilizes diploids for the gene under study and the other employs localized mutagenesis of the region of the chromosome containing the gene. Obviously, *in vitro* approaches for mutagenizing plasmids containing the gene are an alternative. The advantages of the *in vivo* approach are (1) in some cases spontaneous mutants can be selected and (2), when not, the range of specificities of mutagenic agents and mutator genes allows the isolation of mutants with a wide variety of base alterations.

Selecting or Screening for Mutants in Diploids

Selecting for mutants in diploids was first used to obtain mutations in the β and β' subunits of RNA polymerase.[27] Mutations of *E. coli* resistant to the antibiotic rifamycin occur in the gene for the β subunit, *rpoB*. A *rif*^{s}/*rif*^{r} diploid strain is Rif^s, showing that the *rif*^{r} mutation is recessive. In such a strain, selection for Rif^r gives mutations which inactivate the *rif*^{s} copy of the *rpoB* gene. Since, in these mutant diploids, the only copy of *rpoB* that is expressed is the *rif*^{r} one, the cells are resistant to the antibiotic. To avoid selection for *rif*^{r} homogenotes, the starting strains are made *recA*$^-$. Using this approach, nonsense and temperature-sensitive mutants of *rpoB* were obtained.

Selection or screening procedures of this sort can be done with any gene where recessive mutations exist and there is a phenotype which can be selected or screened for. In the *rpoB* example, an F' episome was used to construct the diploid. A plasmid carrying the gene could be used as well. However, with a multicopy plasmid, it should be noted that it may be much more difficult to obtain mutations other than null mutations given the high copy number of the gene product in the cell.

Localized Mutagenesis

Another approach to obtaining a collection of mutations in a specific gene is to mutagenize only the region in which the gene lies. This can be done in several ways. One approach involves mutagenesis of a strain followed by transfer of the region of interest to another strain background. This can be done by making generalized transducing lysates of the mutagenized strain and transducing an unmutagenized recipient for a marker which is closely linked to the gene of interest.[28] The transductants are then

[27] S. Austin and J. Scaife, *J. Mol. Biol.* **49,** 263 (1970).
[28] E. Hawrot and E. P. Kennedy, *Proc. Natl. Acad. Sci. U.S.A.* **72,** 1112 (1975).

screened either for a phenotype or to obtain, for example, a collection of temperature-sensitive mutations. In the latter case, any *ts* mutants are further screened to see if they are complemented by a plasmid carrying the gene under study. Because of the way in which this experiment is done, the only region of the recipient strain to have seen a mutagen in its history is the region of interest. Therefore, most of the mutations detected should be in this region.

There are alternative ways of doing localized mutagenesis. Rather than mutagenizing the donor strain, one can treat the transducing lysate with mutagen.[29] This approach is limited by the types of mutagens which will work on phage lysates and the limited spectrum of mutations they induce. Also, one can mutagenize a donor strain carrying a plasmid and then transform the mutagenized DNA into the recipient strain.

A general approach to obtaining mutations in an already defined essential gene of *Salmonella typhimurium* has been described.[30] Using λ phages carrying the gene under study, a color test for complementation in phage plaques was used to detect phage carrying mutant genes. For instance, a λ phage carrying the *dnaC* gene is mutagenized and allowed to make plaques on a first nonmutant indicator strain; then this indicator is killed by exposure to chlorofirm. Subsequently, the plates are overlayed with soft agar containing a strain with a *dnaC* temperature-sensitive mutation and maltose plus tetrazolium dye to allow red color formation in plaques that are complementing. The plates are incubated at 42°. The *dnaC*+ phage make red plaques because they allow growth of lysogens in the plaque, while any *dnaC*− phage make white plaques since they do not allow such growth. In this way, a collection of *dnaC* mutants can be obtained. The approach is applicable to other essential genes.

Nonsense Mutations in Essential Genes

It is often important to obtain nonsense mutations in an essential gene. They can be used to determine the essentiality of the gene and potentially to study the physiology of a null mutation. In addition to obtaining the mutations on a plasmid by techniques described above, it is also possible to detect chromosomally located nonsense mutations in essential genes. Two approaches are used. In one, the localized mutagenesis technique is applied to a strain which carries an amber suppressor tRNA which can be turned on and off. The amber suppressors exist either as *ts* tRNA mutants of the suppressor[21,22] or as suppressor genes regulated by the *lac* pro-

---

[29] J. Hong and B. N. Ames, *Proc. Natl. Acad. Sci. U.S.A.* **68,** 3158 (1971).
[30] R. Maurer, B. C. Osmond, E. Shekhtman, A. Wong, and D. Botstein, *Genetics* **108,** 1 (1984).

moter.[31] With such strains, one can look after localized mutagenesis for mutants dependent on the expression of the suppressor.

A second approach to obtaining amber mutations in essential genes uses what is known as the "blue ghost" technique.[32] The basis of this approach is a replica plating technique after P1 transduction which detects colonies which are dependent for their survival on the presence of the amber suppressor. Amber mutations in the *rho*, *secA*, and *dnaA* genes have been detected in this way.

Subtractive Mutagenesis

The localized mutagenesis approaches described above are designed to ensure that mutations will be limited to the chromosomal region of the gene of interest. However, often just the opposite is wanted. If one is looking for unlinked mutations that affect the expression of a gene, one may want to eliminate the possibility of obtaining mutations in that gene. This can be done by mutagenizing a strain that is deleted for the particular gene and then bringing in a copy of the gene from an unmutagenized background. Beyond this step, the procedures are the same as for localized mutagenesis, except that the recipient strain cannot contain a mutator gene. If it did, the introduced gene would also be subject to mutagenesis.

Use of Suppressor Mutations to Define New Genes in Biological Processes

Suppressor mutations have played an important role in molecular biology. Suppressors can occur by a number of different mechanisms. These include (1) alterations of the translational machinery that restore synthesis of a functional protein, (2) bypass mutations that generate a substitute pathway for the defective one, and (3) alteration of one component of a pathway compensating for a defect in another component. In the case of the last class (3), such suppressors may act either by promoting an improved interaction between the two mutated components or by allowing the second altered component to compensate in some way for the defective function. Mutations of the third class can allow one to define previously undiscovered genes of a pathway or to obtain evidence for protein–protein interactions in a pathway.

One example of the use of suppressors in the discovery of a new gene involved the study of signal sequence mutations of the *lamB* gene, coding

---

[31] M. L. Berman has constructed such a regulatable derivative of the *supF* amber suppressor gene.

[32] S. Brown, E. R. Brickman, and J. Beckwith, *J. Bacteriol.* **146**, 422 (1981).

for bacteriophage receptor.[33] The signal sequence mutation resulted in the inability to export LamB protein to the outer membrane. This defect resulted in the inability to utilize maltodextrin as a carbon source, since the LamB protein also can act as a pore in the membrane for maltodextrins. The main class of suppressor mutations, restoring export of LamB, mapped in a gene which was named *prlA*. The *prlA* gene has since been shown to code for a cytoplasmic membrane protein central to the process of protein export.[34]

Brown has isolated suppressor mutations which restore viability to a cell which is underproducing an essential cell component.[35] He put the *ffs* gene (coding for 4.5 S RNA) under a regulatable promoter on a plasmid in a strain deleted for the chromosomal copy of the gene. At low levels of expression, the strain was nonviable. However, mutations in genes unlinked to the plasmid restored growth. One of the mutations mapped to the gene for EF-G, a protein synthesis factor, a result that suggested an interaction between the two components.

A general way to use suppressor mutations to detect protein–protein interactions has been proposed by Jarvik and Botstein.[36] Revertants of temperature-sensitive mutations in one gene are screened for new cold-sensitive phenotypes. Those cold-sensitive mutations which are unlinked are presumed to be in a gene whose product interacts with the product of the first gene. This approach appeared to give the desired results when used to study the morphogenic genes of phage P22 of *Salmonella typhimurium*. This approach has the potential of revealing new genes in a pathway. Also, if carefully done, it can indicate those genes whose products interact. However, in some cases, *ts* mutations appear to give suppressor mutations in genes coding for products in a pathway not directly related to the original gene. For instance, mutations in a gene required for protein secretion could be suppressed by mutations in a host of genes involved in protein synthesis.[37] Furthermore, it is important to do careful allele specificity studies to establish significant interactions.[38]

The red plaque assay of Maurer and co-workers described above has also been used to look for suppressor mutations. For instance, using a λ phage carrying *dnaN* or *Salmonella,* Engstrom *et al.* showed that mutations in the *dnaN* gene could suppress temperature-sensitive mutations in

[33] S. D. Emr, S. Hanley-Way, and T. J. Silhavy, *Cell* (*Cambridge, Mass.*) **23,** 79 (1981).
[34] J. P. Fandl and P. C. Tai, *Proc. Natl. Acad. Sci. U.S.A.* **84,** 7448 (1987).
[35] S. Brown, *Cell* (*Cambridge, Mass.*) **49,** 825 (1987).
[36] J. Jarvik and D. Botstein, *Proc. Natl. Acad. Sci. U.S.A.* **72,** 2738 (1975).
[37] C. Lee and J. Beckwith, *J. Bacteriol.* **166,** 878 (1986).
[38] N. A. Treptow and H. A. Shuman, *J. Mol. Biol.* **202,** 809 (1988).

the *dnaZ* gene.[39] A mutagenized λ*dnaN* gave rise to some red plaques when plated on a *dnaZts* host.

Mutations in Complex Pathways

When studying a cellular process where no knowledge of the pathway exists, it often becomes crucial to develop novel genetic approaches to reveal the genes involved. Two examples are presented below to illustrate the kinds of strategies that can be developed.

Mutations Affecting Protein Export Pathways

Protein secretion in both prokaryotes and eukaryotes requires a complex cellular machinery. One genetic approach to the analysis of this machinery, namely, the isolation of suppressors of signal sequence mutations, has been described. A more general strategy for obtaining mutants in the machinery would require a selection or screening procedure for mutants that cause the internalization of normally exported proteins. Such a selection was made possible through the construction of hybrid proteins in which a signal sequence was placed in front of the enzyme β-galactosidase, which is normally located in the cytoplasm. These hybrid proteins are found embedded in the cytoplasmic membrane, led there by the amino-terminal signal. In this membrane location, β-galactosidase is unable to assemble into active enzyme, and cells are phenotypically Lac⁻. Thus, selection for Lac⁺ derivatives of strains making these hybrid proteins is a direct selection for internalization of the protein.[40] In other words, the gene fusion strains provide a direct selection for mutations (*sec⁻*) causing a defect in the secretion machinery.

The selection for *sec* mutants has several important features to it. Because secretion is apparently an essential process for the viability of *E. coli*, mutations which caused severe defects in the machinery are lethal. However, because the selection for Lac⁺ only requires a small amount of β-galactosidase to be internalized, mutations with only slight defects in the machinery can be detected in the selection. For further studies on *sec* mutants, it is preferable to have conditional lethal mutations. This makes it easier to map the mutation and to study its phenotype under conditions where the protein is totally nonfunctional. In order to obtain conditional lethals in this procedure, selections for Lac⁺ were carried out at 30° and mutants scored for a temperature-sensitive phenotype. In this way, conditional lethals were found in the *secA* and *secD* genes.

---

[39] J. Engstrom, A. Wong, and R. Maurer, *Genetics* **113**, 499 (1986).
[40] D. Oliver and J. Beckwith, *Cell* (*Cambridge, Mass.*) **25**, 765 (1981).

This scheme has been adapted for the same purpose in the yeast *Saccharomyces cerevisiae.*[41] In that case, a signal sequence was fused to the protein histidinol dehydrogenase. The enzyme was translocated into the rough endoplasmic reticulum where it was separated from any substrate that might be present in the cytoplasm. When an otherwise HIS⁻ strain carrying this gene fusion was incubated in the presence of histidinol, it was unable to utilize the histidinol as a source of histidine and remained auxotrophic. Selection for histidinol⁺ derivatives of the strain yielded mutations in *SEC* genes analogous to those identified in *E. coli.*

Chromosome Partition

Chromosomes in all cells must be accurately partitioned between daughter cells during the cell division process. There has been essentially nothing known about the molecular mechanism of this process until quite recently. In addition to chromosomes, certain plasmids in bacteria also partition precisely. Such plasmids provide a means of studying the mechanism, in general. Austin and Abeles[42] devised a scheme for isolating mutants of the P1 plasmid which were defective in partitioning (*par* mutants). The scheme involved incorporating the portion of the P1 genome required for partitioning into a λ phage. This λ derivative was now accurately partitioned at cell division and, therefore, gave rise to stable lysogens. The background of λ genetics made genetic manipulation of the system much easier. In this case, one could look for mutants defective in partitioning by looking for λ that were unable to form stable lysogens. These derivatives could be detected on EMBO (eosin methylene blue) agar. When plaques from a mutagenized lysate of the λ*par* phage are plated on a sensitive indicator strain in the presence or absence of a phage, lambda *c*I⁻, which will kill nonlysogens, and then restabbed onto the same indicator, *par*⁺ and *par*⁻ derivatives can be distinguished.

Special Techniques

Papillation

It is possible to examine single colonies on certain indicator plates and distinguish those that show an increased or decreased rate of certain cellular processes. These processes include recombination, mutation, and transposition. The media used include such sugar indicator plates as tetrazolium and MacConkey agar. The approach was first described by Monod

[41] R. J. Deshaies and R. Schekman, *J. Cell Biol.* **105,** 633 (1987).
[42] S. Austin and A. Abeles, *J. Mol. Biol.* **169,** 373 (1983).

and Audureau.[43] An example of the use follows. If Lac⁻ bacteria are spread on lactose–tetrazolium agar, they form red colonies. If the mutation causing the Lac⁻ phenotype is revertible, then, after a few days, papillae will appear on some of the red colonies. These papillae are microcolonies of Lac⁺ revertants which outgrow the rest of the colony. Apparently, the concentration of carbon sources in the vicinity of the colony is reduced, so that those bacteria within the colony capable of using lactose have a selective advantage.

One can employ this phenomenon to detect mutator genes of *E. coli*. A culture of a Lac⁻ mutant is mutagenized and then spread on the indicator media.[3] A small proportion of the bacteria plated should now have mutator mutations, which increase the frequency of reversion of the *lac*⁻ mutation. If the plates are incubated for several days, the colonies that arise from a mutator derivative should have a much greater number of Lac⁺ papillas than their neighbors. A similar strategy has been used to detect mutations which cause an increase in recombination frequency (hyperrec mutants).[44] In this case, the starting strain carries two copies of the *lac* region, each with a different deletion mutation. The deletions can recombine at low frequency to give Lac⁺ derivatives. When such a strain is mutagenized and then examined on lactose–tetrazolium agar, papillating colonies are ones in which an increase in recombination frequency leads to an increase in the number of Lac⁺ bacteria within a colony.

The *bgl* operon of *E. coli* codes for proteins required for β-glucoside utilization. Ordinarily, however, this operon is cryptic. It can be turned on by the insertion of IS elements in sites near the promoter region. When a wild-type (*bgl*-cryptic) strain is spread on MacConkey agar containing a β-glucoside, colonies form which eventually show Bgl⁺ papillas. Mutagenesis of such a strain and the detection of colonies that fail to yield papillas reveal mutants which are defective in the transposition of IS elements and other transposons.[45]

Halos

A specialized technique allows detection of strains that are exporting proteins beyond the outer membrane. A screen for mutants of *E. coli* which have leaky outer membranes provides one example of this class of genetic screening procedures.[46] Mutants were sought which excreted a normally periplasmic ribonuclease into the media. Included in the agar

[43] J. Monod and A. Audureau, *Ann. Inst. Pasteur (Paris)* **72,** 871 (1946).
[44] E. B. Konrad and I. R. Lehman, *Proc. Natl. Acad. Sci. U.S.A.* **71,** 2048 (1974).
[45] M. Syvanen, J. D. Hopkins, and M. Clements, *J. Mol. Biol.* **158,** 203 (1982).
[46] J. Lopes, S. Gottfried, and L. Rothfield, *J. Bacteriol.* **109,** 520 (1972).

growth medium was a concentration of RNA sufficient to lend a cloudy appearance to the plates. Mutant colonies which export the ribonuclease will degrade the RNA in the surrounding medium, giving a "halo" of clearing around the colonies. Mutants which do not make the enzyme or fail to export it will have no halo.

This approach has been extended by using a leaky outer membrane mutant in another genetic screening procedure. In these mutant cells, an unstable hybrid membrane protein composed of the Tsr protein and alkaline phosphatase leaks the alkaline phosphatase portion into the growth medium. The enzyme in the medium cleaves an indicator dye, 5-bromo-3-chloroindolylgalactoside, to generate the blue dye indigo. These bacteria form colonies with a blue halo. Mutants that fail to cleave the hybrid protein could be detected because they do not form blue halos. One class of such mutants was defective in a periplasmic protease, DegP.[47]

[47] K. Strauch and J. Beckwith, *Proc. Natl. Acad. Sci. U.S.A.* **85,** 1576 (1988).

[2] Bacteriophage-Mediated Generalized Transduction in *Escherichia coli* and *Salmonella typhimurium*

By NAT L. STERNBERG and RUSSELL MAURER

Introduction

Phage-mediated transduction is one of three ways, along with conjugation and transformation, by which DNA is transferred from one bacterium to another. In transduction, bacterial DNA is encapsidated into phage particles during lytic growth of the phage in the donor cell and is transferred with near unit efficiency to the recipient cell by infection. There it either undergoes recombination with the host chromosome to produce a stable transductant or remains extrachromosomal and gives rise to an abortive transductant. In the latter case, the DNA fails to replicate and is passaged to only one of the daughter cells at cell division.

Based on the mode of production of transducing DNA and the means by which it is incorporated into the recipient cell chromosome, transduction can be classified into one of two types, specialized or generalized. In specialized transduction (see [12], this volume) the DNA present in phage is produced by aberrant excision of prophage DNA and, therefore, contains chromosomal sequences adjacent to the prophage DNA as well as prophage sequences. When this DNA is injected into the recipient host, it is established and maintained as a prophage independent of host general homologous recombination. The transductants are thus diploid for the transduced DNA. In contrast, generalized transducing particles contain only host chromosomal sequences that are packaged from random or specific sites in host chromosomes. Following injection of that DNA into the recipient cell, stable transductants are generated by the replacement of the homologous cell DNA with the transducing DNA. That process is dependent on the general recombination system of the host.

In this chapter we discuss generalized transduction mediated by phages P1 and λ in *Escherichia coli* and phage P22 in *Salmonella typhimurium*. We review what we know about the mechanisms that produce transducing particles in donor cells and that incorporate transducing DNA in the chromosome of the recipient cells. The significance of phage- and host-encoded gene products in the transduction process is emphasized. Finally, we describe in detail the procedures used to generate transducing lysates and stable transductants. It should be noted that phages other than those described here, most notably T4[1] and Mu,[2] have been used to promote generalized transduction in *E. coli* or *S. typhimurium*. Although we do not describe methods for using these phages in transduction here, the procedures and principles are largely the same as for the phages described.

Bacteriophage P1

General Principles of Phage Packaging

Bacteriophage P1 packages its DNA by a processive headful mechanism.[3] The normal DNA substrate for this process in the viral life cycle is a concatemer consisting of tandemly repeated units of the phage genome arranged in a head-to-tail configuration. Packaging is initiated when a unique 162-base pair (bp) sequence on P1 DNA, called *pac*, is recognized and cleaved by phage-encoded pacase proteins.[4] One of the cleaved *pac* ends is then brought into an empty phage prohead and the head filled until no more DNA can be inserted. The packaging process is completed when the packaged DNA is cleaved away from the rest of the concatemer by a headful cutting reaction that appears not to recognize any specific DNA sequence. A second round of packaging is initiated from the unpackaged DNA end produced by the headful cut. In this way a processive series of

[1] G. C. Wilson, K. Y. Y. Young, and G. J. Edlin, *Nature (London)* **280,** 80 (1979).
[2] M. Howe, *Virology* **55,** 108 (1973).
[3] B. Bächi and W. Arber, *Mol. Gen. Genet.* **153,** 311 (1977).
[4] N. Sternberg and J. Coulby, *J. Mol. Biol.* **194,** 469 (1987).

DNA headfuls are packaged unidirectionally from a single *pac* site on DNA.

The P1 processive headful packaging process generates a population of viral DNAs that are terminally redundant and cyclically permuted. Since the P1 genome is about 100 kilobase pairs (kbp) and a P1 headful is about 110–115 kbp,[5] each packaged DNA molecule will have the same 10 to 15-kbp sequence at both ends. When that DNA is injected into a recombination-proficient host, homologous recombination betweeen the terminally redundant sequences cyclizes the linear viral DNA, protecting it from cellular exonucleases and permitting the onset of the viral life cycle.[6] The packaged DNA is cyclically permuted because more than one headful is packaged from each P1 concatemer. By measuring the molarity of the *pac* end in virion DNA, Bächi and Arber[3] concluded that on average 3–4 headfuls are produced per packaging series. In contrast, the packaging of bacterial DNA from a P1 *pac* site inserted in the bacterial chromosome indicates that P1 packaging can be processive for about 5–10 headfuls.[7]

Why is P1 DNA preferentially packaged into phage proheads? Specificity is presumably provided by phage-encoded proteins that either cleave *pac* or make the headful cut, remain associated with that DNA, and then interact with phage proheads. The juxtaposition of the cut DNA and the phage head presumably facilitates the encapsidation process. The P1 genes whose products carry out these reactions are located on the phage genome between *pac* at map coordinate 96 and the P1 *c*1 repressor gene at map coordinate 99.[7,8]

A central element of the P1 packaging system is *pac* (Fig. 1). It is a 162-bp sequence that contains four hexanucleotide sequences (5' TGATCA/G) at one end of the site and three at the other. In Fig. 1, *pac* cleavage events (indicated by the arrows) are localized to about a turn of the DNA helix in a 90-bp region that separates the hexanucleotide domains. Each hexanucleotide element contains a DNA adenine methylation (*dam*) sequence (5' GATC), and Dam methylation is necessary for *pac* cleavage.[9] These results suggest that P1 should not produce virus particles in an *E. coli dam*⁻ host. In fact, while *pac* cleavage and particle production are delayed in that host the final yield of phage is nearly the same as it is in a *dam*⁺ host. Moreover, the DNA in particles produced in a *dam*⁻ host contain *pac* ends. These results can be explained by the recent demonstration that P1 produces its own DNA adenine methylase late in the viral infection

[5] N. Sternberg, *Proc. Natl. Acad. Sci. U.S.A.* **87,** 103 (1990).
[6] N. Sternberg, B. Sauer, R. Hoess, and K. Abremski, *J. Mol. Biol.* **187,** 197 (1986).
[7] N. Sternberg and J. Coulby, *J. Mol. Biol.* **194,** 453 (1986).
[8] M. Yarmolinsky, *Genet. Maps* **4,** 38 (1987).
[9] J. Coulby and N. Sternberg, unpublished.

5' CA TGATCA T TGATCA CTCTAA TGATCA ACATGCAAGG TGATCA CATTGCGG

CTGAAATAGCGGAAAAACAAAGAGTTAATGCCGTTGTCAGTGCCGCAGTCGAGAA
                                   CAACAGTCACGG

TGCGAAGCGCCAAAATAAGCGCATAAA TGATCG TTCAGA TGATCA TGACG TGATCA C

FIG. 1. The P1 packaging site. The sequence shown is the minimal functioning P1 *pac* site. It is located at P1 map coordinate 96 on the circular P1 map.[8] Boxed sequences are the hexanucleotide elements whose methylation is essential for *pac* cleavage. The positions of the *pac* cleavage events are indicated by arrows, with the large arrows representing frequent cleavage sites and the small arrows infrequent cleavage sites.

cycle.[9] P1 *dam*⁻ mutants fail to cleave *pac* in a *dam*⁻ *E. coli* host and produce only about 5–10 phage per cell in that host. These phage do not contain *pac* ends. P1 *dam*⁻ pseudorevertants have been isolated that produce 3–6 times more phage in a *dam*⁻ host.[9] They also produce more P1 transducing particles per cell.[10] The significance of these results are discussed in the next section.

General Principles of Transduction

Production of P1 Transducing Particles. The density shift experiments of Ikeda and Tomizawa indicate that P1 generalized transducing particles lack significant amounts of phage DNA (<5%) and represent about 0.3% of total phage particles in any P1 lysate.[11] More recent estimates are as high as 6%.[12] Measurements of the density of DNA in transducing and infectious particles indicate that the former contain protein covalently linked to the DNA that is not present in the latter.[13]

Since P1 encapsidates 110–115 kbp of DNA, about 2.3% of the *E. coli* chromosome can be transferred from cell to cell. Taylor and Trotter[14] noted that the longest distance for cotransduction of two markers is about 2.2 min of the *E. coli* map, at a frequency of 0.3%. The frequencies with

[10] G. Lucey and N. Sternberg, unpublished (1989).
[11] H. Ikeda and J. I. Tomizawa, *J. Mol. Biol.* **14**, 85 (1965).
[12] M. C. Hanks, B. Newman, I. R. Oliver, and M. Masters, *Mol. Gen. Genet.* **214**, 523 (1988).
[13] H. Ikeda and J. I. Tomizawa, *J. Mol. Biol.* **14**, 110 (1965).
[14] A. L. Taylor and C. D. Trotter, *Bacteriol. Rev.* **31**, 332 (1967).

TABLE I
PROPERTIES OF TRANSDUCING PHAGE

Phage	DNA content of transducing particle (kbp)	Special conditions required for production	Stable transductants per infected donor cell
P1	100–115	None	3×10^{-4} to 1×10^{-2}
P22	42	None	5×10^{-7} to 5×10^{-4}
λ	50	Treatment of lysate with DNase, tail addition *in vitro*	7×10^{-7} to 1.5×10^{-5}
T4	165	Phage must be mutant in genes that destroy host DNA	0 to 10^{-2}
Mu	45	None	10^{-7} to 10^{-5}

which markers are transduced can vary by as much as 30-fold, ranging from 3×10^{-4} to 10^{-5} transductants per infected donor cell (Table I). These differences are due primarily to variation in the efficiency of incorporation of transducing DNA into the chromosome of the recipient cell. A small contributing factor is also gene dosage differences in the donor cell.

If P1 lysates are treated with UV light to introduce damage in the transducing DNA in order to stimulate the insertion of that DNA into the chromosome of recipient cells, much of the transductional variability between markers is eliminated.[15] Direct measurements by Southern blot hybridization of the relative abundance of specific bacterial DNA in transducing particles show that all sequences except for a group of markers near 2 min and 90 min on the *E. coli* map are present in approximately equal amounts.[15] Markers near 2 min and 90 min are packaged at levels 2 to 3-fold higher than are other markers. These results show that the packaging of chromosomal DNA into P1 phage is nonselective. Because the insertion of a normal P1 *pac* site in the chromosome can increase the transduction, and presumably the packaging, of adjacent chromosomal markers about 80-fold,[7] it is unlikely that *pac* or pseudo *pac* sites are used to initiate the packaging of transducing DNA. This is supported by the failure to demonstrate cross-hybridization between P1 *pac* and any chromosomal sequence.[12] Rather, the most likely hypothesis is that chromosomal DNA is packaged into transducing particles from ends or nicks in the DNA. These results contrast with those for phage P22, where differences in transduction frequencies for different chromosomal markers are much greater than they are for P1 (see below), and cannot be explained

[15] B. J. Newman and M. Masters, *Mol. Gen. Genet.* **180,** 585 (1980).

by either gene dosage effects or DNA insertion differences in the recipient cell.

Like the packaging of phage DNA, the packaging of host DNA into P1 particles is also processive. Using prophages as chromosomal markers, Harriman[16] concluded that transducing particles were generated in a processive series which packages about 20% of the *E. coli* chromosome and that cells producing transducers are special cells in that they are prone to initiate more than one packaging series per infection, whereas the majority of cells yield no transducers.

High-frequency transducing mutants of P1 have been isolated and fall into three classes. Mutants of the first class[17] transduce all markers about 5–10 times more efficiently than does P1 wild type and produce about 10 times more transducing particles than does parental P1. None of these mutants has been mapped. A second class of mutant, called *gta* or *teu*, is a deletion of DNA encompassing the region between P1 map coordinates 25 and 30. Transduction with this mutant is also insensitive to stimulation by UV light. The high-transducing mutant of Yamamoto,[18] *sup50*, has the same properties as *gta-teu*. The third class of mutants consists of the P1 *dam⁻* pseudorevertants previously discussed. The mutants in this class produce transducing particles 10–20 times more efficiently than does P1 wild type (Table II), and at least one of these mutants is probably located in a pacase gene since it cannot cleave *pac* even in a *dam⁺* host.

Fate of Transducing DNA in Recipient. Sandri and Berger[19] showed that only about 10% of the transducing DNA injected into a recipient cell is ever stably incorporated into the chromosome. This process occurs within the first hour after infection and is dependent on the *recA* protein of the host. Based on the density of transducing DNA released from bacterial chromosomes 1 hr after P1 infection, it was concluded that both strands of the infecting DNA were simultaneously incorporated into chromosomes. This result contrasts with physiological experiments of Hanks and Masters[20] that are best explained by assuming that stable transductants result from single-strand replacement.

As previously noted, the frequency of P1 transduction can vary by about 30-fold. Much of this effect is due to recombination differences in the recipient cell since treatment of the phage lysate with UV light, or use of a recipient cell in which the only recombination pathway is the *recF* pathway, eliminates the frequency differences for non-gene dosage-depen-

[16] P. Harriman, *Virology* **48**, 595 (1972).

[17] J. D. Wall and P. Harriman, *Virology* **59**, 532 (1974).

[18] Y. Yamamoto, *Virology* **118**, 329 (1982).

[19] R. M. Sandri and H. Berger, *Virology* **106**, 14 (1980).

[20] M. C. Hanks and M. Masters, *Mol. Gen. Genet.* **210**, 288 (1987).

TABLE II
ELEVATED TRANSDUCTION USING UV-IRRADIATED
PHAGE, P1 *dam⁻* MUTANTS, AND *dam⁻*
PSEUDOREVERTANTS[a]

P1 phage	*argH* transductants produced per infected donor cell
P1	4,7
P1-UVa	19,38
P1-UVb	22,48
P1 *dam*::Kan	12,20
P2 *dam*::Kan *rev7*	40,56
P1 *dam*ΔMB	13,23
P1 *dam*ΔMB*rev6*	64,70

[a] Transduction values should be multiplied by 10^{-4}. The two values for each phage represent results for separate lysates preparerd by the plate stock method. P1-UVa was irradiated to inactivate plaque forming units (pfu) 10-fold. P1-UVb was irradiated to inactivate pfu 100-fold.

dent markers. Transductional equivalence is obtained by increasing the frequencies of the poorly transduced markers. The most likely explanation for these results is that DNA damage or the *recF* pathway renders poor recombination substrates (e.g., those with few if any χ sites) recombinogenic.

Genetic studies with P1 (as well as P22) indicate that the majority of transductants are abortive, that is, the transducing DNA is not stably integrated into the recipient cell genome but rather remains extrachromosomal and is transferred to only one daughter cell at cell division. Sandri and Berger[21] showed that 75% of total P1 transducing DNA injected into a *recA⁺* host remains unassociated with chromosomes and does not replicate for as long as 5 hr after infection. Since the number of stable transductants does not increase after the first hour of infection, abortively transduced DNA must only rarely generate stable transductants. Further analysis of the state of abortively transduced DNA indicated that it was circular and held in that form by a protein. The obvious candidate for that protein is the protein detected by Ikeda and Tomizawa[11] which is specifically associated with transducing DNA. Presumably, when transducing DNA is injected into a cell this protein, possibly associated with one of the DNA ends, associates with the other to cyclize DNA. Less

[21] R. M. Sandri and H. Berger, *Virology* **106**, 30 (1980).

frequently this does not occur, and the DNA either is stably inserted into the chromosome or is rapidly degraded. It is noteworthy that the elevated stable transduction frequency associated with the *sup50* mutation is associated with a decrease in abortive transduction.

Several investigators have shown that there is a lag of about six generations after P1 infection before the number of stable transductants increases. Moreover, Bender and Sambucetti[22] showed that stable transductants were 1000 times more abundant among filamentous cells generated after P1 infection than they were among normal cells. They interpreted these results to mean that transductants initially grow as long, nondividing filaments and that the effect of integration of transducing DNA is to suppress cell division. In contrast, Hanks and Masters[20] showed that the transduction of markers located at different positions on the chromosome increases in reverse order to that in which they are replicated. They excluded transduction-induced filamentation as a cause of the initial lag and suggested that the lag might result from the way donor DNA is inherited.

Materials and Methods for Bacteriophage P1

Phage

The phage strain traditionally used to prepare transducing lysates is P1vir^S. This phage is virulent, it will grow in a P1 lysogen, and it cannot lysogenize bacteria because it constitutively produces a phage antirepressor[23,24] that interferes with P1 *c1* repressor function. Another P1 phage commonly used to generate transducing lysates is P1 cm *c1.100*. This phage contains the temperature-sensitive *c1.100* repressor mutation and a Tn9 transposon (cm) at P1 map coordinate 25. It can lysogenize bacteria at temperatures below 33°, but not at temperatures above 37°. Lysogens are detectable as chloramphenicol-resistant cells. P1 cm *c1.100* $r^- m^-$ *dam*::Kan (P1 *dam*::Kan) and its large plaque revertant (P1 *dam*::Kan *rev7*) as well as P1 cm *c1.100* $r^- m^-$ *dam*ΔMB (P1 Δ*dam*) and its large plaque revertant (P1 Δ*dam rev6*) are recently isolated phage which show elevated transduction. Besides the mutations described above these phage contain a mutation that inactivates the viral restriction and modification enzymes ($r^- m^-$),[25] mutations that inactivate the viral DNA adenine meth-

[22] R. A. Bender and L. C. Sambucetti, *Mol. Gen. Genet.* **189**, 236 (1983).
[23] C. Wandersman and M. Yarmolinsky, *Virology* **77**, 386 (1977).
[24] J. R. Scott, B. W. West, and J. L. Laping, *Virology* **85**, 587 (1978).
[25] S. W. Glover, J. Schell, W. Symonds, and K. A. Stacey, *Genet. Res.* **13**, 227 (1963).

ylase (*dam*) gene,[7,10] and mutations that permit P1 *dam*⁻ mutants to grow better in a *dam*⁻ host (*rev6* and *rev7*). The *dam*::Kan strain contains an insertion of the Tn*903 kan*ʳ gene in *dam*, and ΔMB contains a 320-bp deletion in *dam*.

Bacterial Strains

P1 lysates can be made on most strains of *E. coli* K12 that are *recA*⁺ although the phage yield (25–150 infectious phage per cell) may vary significantly. Overall, *recA*⁻ strains are poor hosts for viral growth, generally yielding less than 5 infectious phage per cell and failing to permit P1 plaque formation. In contrast, the yield of transducing phage is only reduced severalfold in a *recA* host compared to a *recA*⁺ host. The frequency of stable transductants per infectious particle is usually about 10 times higher in a *recA*⁻ lysate than in a *recA*⁺ lysate.

Preparation of Phage Stocks

Phage Infection. Plate lysates. Prepare an agar plate containing P1 plaques as follows. An aliquot of a P1 lysate containing about 100 plaque-forming units (pfu) is added to 50 μl of a fresh overnight culture of bacteria (~10⁸ cells) and the mixture incubated for 5 min at 37°. To this phage–cell mix is added 2 ml of L (Luria) broth containing 10 mM CaCl$_2$ followed by 2 ml of L top agar, and the suspension is pipetted onto an LCA agar plate (see Table IV). The agar is allowed to solidify, and the plate is incubated overnight at 37° in an upright position. The next day a single plaque (~10⁶ phage) is picked with a microcapillary tube and mixed with 50 μl of an overnight bacterial culture. The mixture is incubated for 5 min at 37° to permit phage adsorption and then plated as described above onto an LCA plate that is prewarmed to 38°. The plate is immediately placed at 38° to prevent complete solidification of the top agar and is incubated for 5–6 hr at that temperature. It is convenient to include a plate with cells, but without phage, as a control to evaluate phage-induced cell lysis.

After about 5 hr the plate containing phage should be clear while the control should be turbid. At this time the top agar suspension (~2 ml) is pipetted into a tube containing 100 μl of 0.1 M CaCl$_2$ and 50 μl chloroform, and the suspension is vigorously vortexed. If the top agar is difficult to remove from the plate it can be loosened by pipetting 1.5 ml of L broth onto the plate and mixing it with the top agar. Then proceed as described above. The phage suspension is centrifuged (e.g., 6000 rpm for 5 min in a Sorvall SS34 rotor at 4°) in a chloroform-resistant tube to remove lysed bacterial debris and any excess chloroform, and the supernatant is tested for bacterial contamination by spreading an aliquot onto L agar plates.

If free of contaminating cells the lysate is stored at 4°. If the lysate is contaminated several additional drops of chloroform are added, the lysate vortexed and incubated 5 min at 37°, and centrifuged as described above. This procedure yields phage lysates with titers ranging from 1 to 5 × 10^{10} phage/ml.

Liquid lysates. Several P1 plaques (~10^6 phage) are added to 50 μl of a fresh overnight culture (~10^8 cells) and the mixture incubated for 5 min at 38°. It is then diluted 50-fold with L broth containing 5 mM $CaCl_2$ and incubation continued at 38° with vigorous aeration until visible lysis is detected (usually 3–5 hr). Several drops of chloroform are added, the solution vortexed, and debris removed as described above. Phage titers achieved by this procedure are usually 3 × 10^9 to 2 × 10^{10} phage/ml.

Lysogen Induction. Lysogens containing a P1 cm c1.100 prophage are grown on L broth with 25 μg/ml chloramphenicol in a water bath at 32° to 4 × 10^8 cells/ml (OD_{630} = 0.6). The culture is then diluted with an equal volume of L broth containing 10 mM $CaCl_2$ that was preheated to 50°, and the culture is vigorously aerated at 42° for 15 min. The temperature of the water bath is then reduced to 39° and aeration continued until visible cell lysis is detected, about 50–60 min after temperature shift-up. The lysate is treated with chloroform and debris removed by centrifugation. Lysate titers of 5 × 10^9 to 2 × 10^{10} phage/ml are usually obtained by this procedure. This procedure is the preferred method of preparing phage stocks in a *recA⁻* host.

Transduction Assay

Preparing Recipient Culture. Individual colonies of the recipient bacterial strain are separately incubated overnight in 5 ml of L broth. The cells are pelleted by centrifugation (6000 rpm in a Sorvall SS34 rotor for 10 min at 4°) and resuspended in an equal volume of sterile 10 mM $MgSO_4$. The centrifugation step is repeated. The reversion frequency of each of the cultures is measured by plating 0.1 ml of the cells on selection plates and incubating the plates overnight at 38°. For amino acid markers we use M63–glucose–B_1 minimal agar plates (see Table IV) lacking the amino acid being assayed. The culture containing the lowest marker reversion frequency is used to measure transduction.

Assay. We use here the transduction of the *argH* defect in *E. coli* strain AB1157 (*his⁻, leu⁻, argH⁻, proA⁻, thr⁻*) to illustrate the transduction assay. Obviously the P1 transducing lysates should be prepared on an *argH⁺* donor strain. Add 2–20 μl of the P1 lysate (10^7–10^8 infectious phage) to 50 μl of a fresh overnight culture of AB1157 in 10 mM $MgSO_4$ (1 × 10^9 cells/ml) and incubate the mixture for 10 min at 38° to permit the phage to adsorb to the cells. Now add 1 ml of L broth containing 10 mM

sodium citrate and incubate the infected cells at 38° for 1 hr with vigorous aeration. The cells are then poured into a 1.5-ml sterile Eppendorf centrifuge tube and pelleted at 12,000 rpm for 2 min in an Eppendorf tabletop minifuge at room temperature. They are washed with 1 ml of 10 mM MgSO$_4$, resuspended in 50 μl of 10 mM MgSO$_4$, and spread on M63–glucose–B$_1$ minimal agar plates containing 10 mM sodium citrate and 10 μg/ml of leucine, histidine, threonine, and proline. The plates are incubated overnight at 38° and colonies scored. A control experiment without P1 measures the reversion frequency of the marker.

It is often convenient to prepare minimal agar plates without amino acids and to supplement the plates as needed. To do this we usually add 50 μl of a sterile 10 mg/ml solution of each needed amino acid to a plate containing about 40 ml of agar and spread the solution evenly on the agar surface. In these experiments sodium citrate inhibits phage infection of sensitive cells after the initial adsorption period, thereby preventing the loss of transductants owing to infectious phage in the lysate. Typical *argH* transduction frequencies using AB1157 as a recipient are shown in Table II.

Enhanced Transduction

UV Treatment of P1 Lysate. The P1 transducing lysate is diluted 10-fold into a sterile solution of 10 mM Tris-HCl, pH 8.0, 10 mM MgSO$_4$ and exposed to a shortwave UV light source (260 nm) until the phage titer is reduced about 10-fold. We find it convenient to treat 1 ml of the diluted lysate in a 35-mm petri dish with a UVG-54 shortwave lamp to achieve the desired UV dose. The UV-irradiated lysate produces about 5–20 times more transductants (depending on the marker examined) per unirradiated plaque-forming unit than does the untreated lysate. Data shown in Table II illustrate our results for the *argH* marker in AB1157. Treatment of the lysate with UV light has the added advantage of reducing the loss of transductants owing to superinfection with infectious phage.

Use of P1 dam$^-$ Revertants. Results shown in Table II illustrate that P1 *dam$^-$* phage and their revertants (*rev6* and *rev7*) produce 5–20 times more *argH* transductants per infected donor cell than does the *dam$^+$* parent phage. The same results were obtained for the *proA* marker in AB1157 and the *trpE* marker in *E. coli* strain NS440. These effects are even more pronounced for *rev6* when transductants are standardized relative to plaque forming units in the lysate since *rev6* is defective for infectious particle production (it cleaves *pac* poorly).

Cotransduction

Because the size of P1 DNA is about 2.3% that of the *E. coli* chromosome, it should be possible to cotransduce markers that are less than 2.3

min apart. Moreover, if one assumes that all segments of chromosomal DNA are packaged with equal frequency and that they recombine with the chromosome of the recipient cell by crossover events that occur at random, it is possible to predict the expected marker distances from their cotransduction frequencies. A graph showing the relationship between these two parameters is given by Masters.[26] It should be noted, however, that unless the two adjacent markers are transduced with the same frequency and the reciprocal cotransduction frequencies are the same, cotransductional linkage may not be a good measure of physical distance.

Bacteriophage λ

General Principles of Packaging

Bacteriophage λ packages its DNA from a concatemer by a site-specific packaging mechanism. The packaging site, *cos*, is a 100 to 150-bp sequence that is cleaved at the initiation and termination of the packaging process. The cleavage reaction generates staggered nicks that are separated by 12 bp. Thus, the two ends of packaged λ DNA contain complementary 12-base single strands which can anneal to form circles after that DNA is injected into the recipient cell. Since the termination of λ packaging is not dictated by the filling of the phage head, as it is for P1, but rather by the distance between *cos* sites, λ can package a variable amount of DNA that is limited to a narrow range between 75 and 105% that of wild-type DNA content (37.5–52.5 kbp).[27] Presumably, if the head is less than three-quarters full the terminal *cos* cleavage reaction is inhibited. The 105% upper packaging limit probably reflects the headful size of λ.

General Principles of Transduction

Phage λ packages bacterial DNA into transducing particles much as does P1 except for one important difference. As λ cannot perform the headful cut when the head has been filled with DNA, the maturation of λ transducing particles must occur *in vitro* after the infected cells have lysed.[28] This is achieved by treating the lysed cells with DNase to degrade the DNA protruding from the phage head so that phage tails can attach to

[26] M. Masters, *Genet. Bact. 1985,* Chap. 11, 197 (1985).
[27] M. Feiss and D. A. Siegele, *Virology* **92,** 190 (1979).
[28] N. Sternberg and R. Weisberg, *Nature (London)* **256,** 97 (1975).

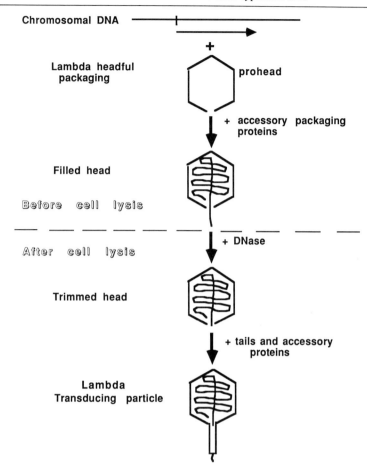

FIG. 2. Generation of λ transducing particles. The steps are described in the text.

generate infectious particles (Fig. 2). As predicted by this model, the DNA packaged into λ transducing particles is of uniform size and approximately 105–109% that of λ wild type, presumably a λ headful.

Several λ mutations affect the yield of transducing particles.[29] First, the λ phage used to generate the particles must contain a mutation in the phage S gene,[30] which inactivates the product of that gene and delays the lysis of infected cells but still permits the progression of the viral lytic cycle beyond the time that infected cells would normally lyse (50–60 min after infection). The S⁻ phenotype is essential because transducing

[29] N. Sternberg, *Gene* **50,** 69 (1986).
[30] A. R. Goldberg and M. Howe, *Virology* **38,** 200 (1969).

particles are not produced until very late in the viral life cycle. They are barely detectable by 60 min after infection and only reach maximal levels by 180 min after infection. Possibly the need for packaging proteins is greater for transducers than it is for infectious particles. Alternatively, the DNA substrate for the initiation of packaging is not produced until late in the phage infection.

A second λ mutation critical for the production of transducers is one that inactivates the phage exonuclease gene (the *red* exo gene).[31] λ *exo⁻* mutations produce 10–30 times more transducing particle than *exo⁺* phage. Since this effect is not seen with *redβ* mutations, we presume the result is due to the destruction of the DNA normally packaged into transducing particles by the λ exonuclease. A third factor affecting the yield of transducers is gene dosage. If transducers are produced by the induction of a λ prophage, markers on either side of the prophage show enhanced transduction relative to that seen when transducing lysates are prepared by infection. This effect can be as much as 200-fold and reflects the replication of the adjacent chromosomal sequences by the induced λ replicon. Indeed, it is this latter property that represents a unique feature of the λ system in that it permits one to tailor elevated transduction to those regions of the bacterial chromosome adjacent to the integration site of λ prophage used to generate the transducers.

If one eliminates gene dosage effects, λ transduction frequencies vary from 1.5×10^{-5} to 7×10^{-7} per infected donor cell. This variation is much the same as that seen for P1 in the absence of UV irradiation. Whether irradiation can equalize the frequency of λ transduction for different markers, as it does for P1 transduction, has not been tested. Although the frequency of marker transduction with λ appears to be about 100-fold lower than it is with P1, this is probably not due to packaging differences between the two phages, but rather reflects inefficient injection of DNA into recipient cells by λ particles matured *in vitro* after headful packaging. Although λ infectious particles inject their DNA into cells with near unit efficiency, λ particles matured *in vitro* inject DNA into cells with an efficiency no greater than 5%.

Materials and Methods for Bacteriophage λ

Phage

The λ phage used to produce a transducing lysate are λ*b2 red3 c*I857 *Sam7* and λ*int6 red3 c*I857 *Sam7*. Both phages are integration defective, the former because it carries a deletion (b2) of the phage attachment (*att*)

---

[31] E. Signer, H. Echols, J. Weil, C. Radding, M. Shulman, L. Moore, and K. Manley, *Cold Spring Harbor Symp. Quant. Biol.* **33,** 711 (1968).

site and the latter because it contains a mutation in the phage integrase (*int*) gene.[32] The *red3* mutation in both phages inactivates the phage exonuclease gene, and the *Sam7* mutation in both phages inactivates the phage *S* gene function in a *sup°* host and results in delayed cell lysis after infection. The *cI857* mutation is a temperature-sensitive *cI* repressor mutation. λ *cI857* lysogens are grown at 32° and can be induced to undergo lytic growth when the temperature of the culture is shifted to 40°. While it is critical that the λ phage used to generate a transducing lysate exhibits a delayed lysis phenotype and an exo⁻ phenotype, the integration defect is optional (see below).

Clear phage used to select λ lysogens are λh$_\lambda$Δ*attc*I⁻ (W30) and λh$_{80}$Δ*attc*I⁻ (W248). Neither phage can integrate into the bacterial chromosome, and both are *cI* repressor defective so they cannot produce lysogens. The use of phage containing both h$_\lambda$ and h$_{80}$ host ranges prevents the survival of cells that are resistant to only one host range. Colonies on agar plates containing both of these selector phage are almost always λ lysogens.[33]

Bacterial Strains

Transducing lysates can be prepared in *recA*⁻ or *recA*⁺ strains of *E. coli* K12. In general, the yield of transducers per infected cell or induced lysogen is about 2–3 times higher in a *recA*⁻ host than it is in a *recA*⁺ host.[29] λ*b2 red3 cI857 Sam7* lysogens are produced as follows. First, prepare a lysogen selector plate by spreading evenly 10⁸ infectious W30 and W248 phage onto an EMBO (eosin methylene blue) plate that is then allowed to dry. Next, 50 μl of a fresh overnight culture of a bacterial strain (~10⁸ cells) is mixed with 10⁸ infectious λ*b2 red3 cI857 Sam7* phage and incubate the mixture for 5 min at 32° to permit phage adsorption. The infected cells are then streaked onto the EMBO selector plate, and the plate is incubated overnight at 32°. The next day healty pinkish colonies with smooth edges (~10³–10⁴ per plate) are purified by restreaking on L agar plates. These colonies are used to inoculate overnight cultures which are then screened to ensure that they are lysogens. λ *cI857* lysogens are immune to λ superinfection at 32° and die at 42°. The lysogen we frequently use in preparing a transducing lysate is NS490 (=N205; λ*b2 red3 cI857 Sam7*). In principle, any λ*red3 cI857 Sam7* lysogen could be used.

[32] R. A. Weisberg and A. Landy, *in* "Lambda II" (R. Hendrix, J. Roberts, F. Stahl, and R. Weisberg, eds.), p. 211. Cold Spring Harbor Laboratory, Cold Spring Harbor, New York, 1983.

[33] N. Sternberg, *Virology* **71**, 568 (1976).

Preparation of λ Transducing Lysates

The preparation of standard λ lysates is described in [12] in this volume and is not described here. Transducing lysates can be prepared either by phage infection or prophage induction.

Phage Infection. An overnight culture of *E. coli* strain NS205 (*recA, sup°*) is diluted 100-fold into 20 ml of L broth and grown to a density of 2 × 10⁸ cells/ml (OD₆₃₀ = 0.4) at 37°. The cells are then pelleted by centrifugation (6000 rpm in a Sorvall SS34 rotor for 10 min at 4°) and resuspended in 2 ml of sterile 10 mM MgSo₄. A lysate of λ*int6 red3* c*I857 Sam7* is added to the cells at a multiplicity of 5 phage/cell (2 × 10¹⁰ phage). The mixture is incubated for 5 min at 37°, then diluted 10-fold into L broth and shaken vigorously for 3 hr at that temperature. The unsuppressed *Sam7* mutation prevents cell lysis. The cells are then chilled to 4°, pelleted as described above, and resuspended in 1/50 volume (0.4 ml) of TMG buffer (see Table IV) containing 10 μg/ml pancreatic DNase I. Chloroform (20 μl) is added, the suspension of cells is vigorously vortexed for 5 sec, and the cells are incubated for 30 min at 37°. During this period the cells are lysed by the chloroform, DNA not already encapsidated or protruding from filled heads is degraded, and phage tails are added to filled heads whose protruding DNA has been removed (Fig. 2). The resulting lysate can be stored in this form at 4° for months without loss of phage titer. These lysates contain about 1–5 × 10¹¹ infectious phage/ml. Infected cells should be concentrated at least 50-fold before lysis to ensure efficient tail addition *in vitro*.

Prophage Induction. Twenty milliliters of strain NS490 is grown to a density of 2 × 10⁸ cells/ml at 32° in L broth. The cells are pelleted by centrifugation as described above, resuspended in 2 ml of L broth, and then diluted 10-fold into L broth that has been prewarmed to 42°. The culture is grown with vigorous aeration at this temperature for 15 min to induce the prophage and then for an additional 165 min at 39°. The induced cells are harvested and treated as described above for infected cells. As the prophage in NS490 cannot excise from the chromosome after the phage lytic cycle is induced, the yield of infectious phage in these lysates is very low, usually less than 10⁷ phage/ml.

Transduction Assay

The preparation of recipient cells and the assay itself are carried out much the same as for P1 transduction. Usually we add 5–10 μl of transducing lysate to 50 μl of a fresh overnight culture of cells in 10 mM MgSO₄ (10⁸ cells). After adsorption for 5 min the cells are diluted with 1 ml of broth, grown for 1 hr at 38°, and then spread on selection plates. If the lysate used is generated by phage infection it is desirable to use a recipient

TABLE III

λ Transduction Frequencies[a]

Marker tested	E. coli map position	Transduction frequencies with lysates prepared by		
		Infection of N205	Induction of NS490(a)	Induction of NS1591(b)
thr	0.1	0.1	0.1	
leu	1.5	0.7	0.8	
proA	6	0.8	4	
purE	11.5	1.5	13.5	
gltA	16	1.0	50	1.0
nadA	16.5	0.07	18	
attB	17			
aroA	20	0.8	104	
pyrD	21	0.2	9	
pyrC	23.5	0.07	4.2	
trpE	27.5	0.5	4.5	
pyrT	28	0.15	0.45	
his	44	0.08	0.05	
thyA	60.5	0.22		8
lysA	61	0.22		9
serA	62	0.12		11
metC	64.5	0.2		24
loxB	66			
argG	68.5	0.2		20
argH	89	0.5	0.3	0.4

[a] All tranduction values are standardized relative to one value of 10^{-5} transducers per infected or induced cell equals 1. The λ prophage in NS490 is located at *attB*. The λ prophage in NS1591 is located at *loxB*.

cell line that is a λ lysogen. This prevents cell killing by the large number of infectious particles added along with the transducers.

Transduction results with lysates prepared by infection or by induction for 19 bacterial markers distributed about the *E. coli* chromosome are shown in Table III. The major difference between the two classes of lysate is the elevated transduction frequencies in the lysates prepared by induction for markers present on both sides of the resident prophage. The effect extends for about 10 min from the prophage in either direction and was observed for lysogens in which the prophage was at the normal bacterial *attB* site (16 min on the *E. coli* map) or at the bacterial *loxB* site (68 min on the *E. coli* map). As noted previously we believe that this effect is due to λ-mediated replication of adjacent chromosomal sequences following prophage induction.

Salmonella Phage P22

In a number of respects, the biology of P22 (reviewed by Susskind and Botstein[34]) is similar to that of P1. P22 packages its own DNA by processive headful packaging initiating at (or near) a *pac* site to give terminally redundant linear DNA molecules exhibiting limited circular permutation within a population. P22 DNA circularizes by homologous recombination early in infection. Transducing particles contain predominantly bacterial DNA synthesized before infection, but they may also have a variable small component (<10%) of DNA synthesized after infection, the origin of which (whether phage or bacterial) has been a matter of dispute.[35-37] After entering a recipient cell, transducing DNA gives rise to abortive or stable transductants as described above for P1.

An obvious difference between P1 and P22 is the much smaller size of P22 virion DNA, 43.4 ± 0.75 kb.[38] Consequently, the distance over which cotransduction of two markers can be detected with P22 is about 1 min of the chromosomal map, less than half the distance cotransducible by P1.

General Principles of Packaging and Transduction

The variation in the normalized rates of transduction of different markers is much greater for P22 than for P1 (Table I) and has a different underlying cause. As mentioned above, for P1 these rates can vary over a 30-fold range, and most of this variation results from differences in recombination efficiency in the recipient cell. With P22, transduction rates can vary over a 1000-fold range, and much of this variation is accounted for by differences in the efficiency with which different chromosomal sequences are packaged into transducing particles.[39,40]

An important insight into the packaging of chromosomal DNA was derived from the experiment of Chelala and Margolin[41] in which some deletions in donor strains strikingly altered the cotransduction frequency of marker pairs located to one side of the deletion. These results were interpreted to mean that most (if not all) packaging of chromosomal DNA from the two regions studied initiated at a small number of defined sites and was processive. By altering the distance between a hypothesized

[34] M. M. Susskind and D. Botstein, *Microbiol. Rev.* **42**, 385 (1978).

[35] H. Schmieger, *Mol. Gen. Genet.* **109**, 323 (1970).

[36] H. Schmieger and H. Backhaus, *Mol. Gen. Genet.* **120**, 181 (1973).

[37] J. Ebel-Tsipis, D. Botstein, and M. S. Fox, *J. Mol. Biol.* **71**, 433 (1972).

[38] S. Casjens and M. Hayden, *J. Mol. Biol.* **19**, 467 (1988).

[39] H. Schmieger, *Mol. Gen. Genet.* **119**, 75 (1972).

[40] H. Schmieger, *Mol. Gen. Genet.* **187**, 516 (1982).

[41] C. A. Chelala and P. Margolin, *Mol. Gen. Genet.* **131**, 97 (1974).

packaging initiation site and the marker genes, the deletion would alter the placement of the two genes relative to the likely locations of headful packaging cuts. Consequently, the number of fragments carrying both genes and having a potential to produce a cotransductant would change relative to the number of fragments carrying one gene but not the other.

The effect seen by Chelala and Margolin was also observed in other parts of the *Salmonella* chromosome and could be caused by insertions, too.[42] Alterations in cotransduction frequency were detectable in some cases as far away as 5–10% of the chromosome length from the site of the insertion or deletion, suggesting that the number of sequential headful events from a single initiation site could be as many as 10. Thus, packaging of chromosomal DNA exhibited two hallmarks of phage DNA packaging: initiation at a specific site and processive headful packaging. The properties of P22 mutants with high transducing frequencies (HT mutants; see below) argue that the chromosomal specific initiation sites are selected by P22-encoded proteins. Production of transducing particles by P22 is therefore thought to reflect erroneous choice of substrate DNA by the P22 packaging mechanism. However, a rather different, recombination-dependent mechanism has been suggested by Schmieger for a fraction of transducing DNA (see below).

The hypothesis that transducing fragments are produced by limited processive headful packaging from specific initiation sites provides an explanation for the wide variation in normalized transduction frequencies observed for single markers. Markers located near an initiation site would likely be included in every packaging series and therefore would exhibit a high transduction frequency; markers located far from such a site would be omitted from most packaging series and would exhibit a low transduction frequency. Schmieger,[40] by equating isolated high-frequency markers with initiation sites, concluded that there were at least 5–6 such sites in the *Salmonella* chromosome; Chelala and Margolin[41] hypothesized 10–15 such sites.

Phage pac Site and Search for Bacterial pac Sites

Plasmids such as pBR322 are not transduced by wild-type P22 unless they carry a fragment of P22 DNA.[43] This transduction is dependent on homologous recombination in the donor (see below) unless the cloned segment carries *pac*.[44] This selectivity provided an operational definition

[42] K. Krajewska-Grynkiewicz and T. Klopotowski, *Mol. Gen. Genet.* **176**, 87 (1979).

[43] M. H. Orbach and E. N. Jackson, *J. Bacteriol.* **149**, 985 (1982).

[44] C. Schmidt and H. Schmieger, *Mol. Gen. Genet.* **196**, 123 (1984).

for P22 *pac*. H. Backhaus (unpublished data cited in Backhaus[45]) has delimited P22 *pac* to a region of 114 bp located within the coding sequence of gene 3. The initiating packaging cuts (end sites) are made in six distinct areas contained within a region of some 120 bp that partially overlaps the 114-bp *pac*.[45,46] Although the distribution of end sites has been determined in detail at the nucleotide resolution level, the mechanism generating the complex pattern found remains a mystery. The defining sequence of *pac* also is not known with certainty, but Casjens *et al.*[47] have proposed that a short (10 bp) sequence located near the most frequently used end region is an important element of *pac*. Unlike P1 *pac*, the 114-bp P22 *pac* has no *dam* methylation sites, making it unlikely that adenine methylation plays any role in regulating packaging of P22.

Vogel and Schmieger[48] took advantage of the inability of P22 to transduce pBR322 to clone selectively bacterial DNA fragments that behave like *pac*. Seven distinct such fragments were found in a library representative of the entire chromosome of *Salmonella*, in promising agreement with the prior estimates of the number of bacterial *pac*-like sites. As yet, no details have emerged regarding the sequence of these fragments, their similarity to P22 *pac*, or their map position on the *Salmonella* chromosome.

Recombination-Dependent Transduction

It is well established that the insertion of a *pac* site into another piece of DNA will make that DNA a substrate for packaging. For example, an integrated P22 prophage promotes the efficient packaging of adjacent chromosomal regions, a fact elegantly exploited by Youderian *et al.*[49] to produce transducing lysates highly enriched for specific chromosomal segments. In the experiments of Orbach and Jackson,[43] a short piece of P22 DNA inserted in a nontransducible plasmid renders the plasmid transducible. Such transduction depends on insertion of P22 itself into the plasmid, by homologous recombination with the cloned P22 DNA. A similar mechanism could be operating to produce chromosomal transducing particles if there are regions homologous to P22 in the *Salmonella* chromosome. Even though they could not detect P22 homologous regions by Southern analysis of *Salmonella* DNA, Vogel and Schmieger[48] cloned

---

[45] H. Backhaus, *J. Virol.* **55**, 458 (1985).
[46] S. Casjens and W. M. Huang, *J. Mol. Biol.* **157**, 287 (1982).
[47] S. Casjens, W. M. Huang, M. Hayden, and R. Parr, *J. Mol. Biol.* **194**, 411 (1987).
[48] W. Vogel and H. Schmieger, *Mol. Gen. Genet.* **2205**, 563 (1986).
[49] P. Youderian, P. Sugiono, K. L. Brewer, N. P. Higgins, and T. Elliot, *Genetics* **118**, 581 (1988).

three distinct pieces of *Salmonella* DNA that render a plasmid transducible by P22 in donor-recombination-dependent fashion. If chromosomal transducing fragments are generated by a P22 integration mechanism, some of these fragments (i.e., the first one in each processive series) would have to have a bit of P22 DNA at one end. Schmieger[35,36] argued for such a structure for transducing DNA based on the results of density shift experiments. However, such a mechanism could account, at most, for only a small fraction of total transducing particles in a lysate since the yield of transducing particles is normal in the simultaneous absence of phage and host homologous recombination functions.[37]

High-Transducing Mutants of P22

Schmieger[39] isolated P22 mutants with altered generalized transduction efficiencies. In general, high-transducing (HT) mutants transduce all markers with similar frequencies (within a factor of 10). The efficiency is increased for all markers, but more dramatically so for markers that are poorly transduced by wild-type P22. Transducing particles comprise a much higher fraction of the total phage lysate in HT mutants (up to 50%) than in wild type. HT mutants, or derivatives of them containing an *int* mutation (see below), have become the phage of choice for routine transductions.

All HT mutants studied to date map to the vicinity of gene 3 of P22.[50] HT 12/4, which changes glutamate to lysine at position 83 of the gene 3 product,[47] has been analyzed in some detail to try to understand the basis for its increased transduction. The mutation has dramatic, trans-acting effects on both phage and host DNA packaging.[51] Tye[52] found that the circular permutation of DNA in phage lysates, normally restricted to about 20% of the genome owing to the limited length of processive packaging series and initiation at a single site, was unrestricted in the mutant. Tye suggested that HT 12/4 was initiating packaging at random. A more refined model emerged from analysis (by restriction digestion) of the location of the ends of mature phage DNA generated by the initiating packaging cut. Randomization of the ends anticipated from the results of Tye was not observed; rather the ends were localized to a specific region different from the region normally used.[51] However, Jackson *et al.*[51] misinterpreted a second submolar fragment in restriction digests of HT 12/4 DNA as originating from the first headful packaging cut of a series. Casjens *et al.*[47] convincingly showed that this additional submolar fragment reflected

[50] A. S. Raj, A. Y. Raj, and H. Schmieger, *Mol. Gen. Genet.* **135,** 175 (1974).
[51] E. N. Jackson, F. Laski, and C. Andres, *J. Mol. Biol.* **154,** 551 (1982).
[52] B.-K. Tye, *J. Mol. Biol.* **100,** 421 (1976).

packaging initiation at a second, distinct, novel site. The two sites were mapped on P22 DNA to locations separated by about 8 kb (~20% of P22 chromosome length) and were found to direct packaging in opposite directions. Short packaging series (~3–5) initiating at these two sites are able to account for the results of Tye.

The greatly increased transduction and more uniform marker-specific frequencies observed with HT 12/4 initially suggested a loss of specificity for packaging initiation on the bacterial chromosome. But with three HT mutants, as with wild-type P22, Chelala and Margolin[53] found that neighboring deletions altered cotransduction frequencies, leading them to suggest that each HT mutant was initiating packaging at a limited number of sites. Moreover, the site preference of each HT mutant was different since the pattern of cotransduction behavior for each HT mutant was distinct.

These studies of P22 and chromosomal DNA packaging by the HT 12/4 mutant lead to a consistent interpretation that the mutant exhibits a strong preference for initiating packaging at defined sites different from those used by wild type. This observation alone could account for the higher and more uniform transduction by HT 12/4 if its preferred sequence occurs often on the bacterial chromosome. But this explanation is suspect because it lacks generality. Of five HT mutants studied by Jackson et al.[51], HT 12/4 was the only one that used novel end sites. Two others showed reduced use of the normal end region, and the remaining two were indistinguishable from wild type. Curiously, one of the latter (HT 13/4) was deduced by Chelala and Margolin[53] to have altered packaging specificity on chromosomal DNA. One important area which has not been explored is the possibility that HT mutants improve the potential for low-efficiency initiations independent of sequence, with the alteration of preferred-site specificity being an incidental effect.

Transduction of Plasmids

Bauerle[54] found that HT mutants could transduce plasmids that were not transducible by wild-type P22 (e.g., pBR322). This technique was quickly adopted in many laboratories working with *Salmonella* because of its convenience even though many details of the mechanism were not and still are not known. For example, it is not known whether transduction depends on the presence of any specific sequence in the plasmid. Transduction can be highly efficient (up to 10^{-3} per pfu) and seems to work best when both donor and recipient strains are competent for homologous recombination. Sometimes, the transduced plasmid is found to be

---

[53] C. A. Chelala and P. Margolin, *Genet. Res.* **27**, 315 (1976).
[54] Bauerle, *Genetics* **104** (Suppl.), 55 (1983).

multimeric. A working model for this process is that the packaging substrate is multimeric rolling circle DNA, and the plasmid is regenerated in the recipient by recombination.

Lysogeny

It is usually desirable to avoid lysogenization of transductants. Since P22 lysogeny involves Campbell-type insertion of phage DNA into the chromosome (in contrast to P1), use of an integration-defective phage mutant solves this problem. A number of HT mutants have been crossed to *int* mutants to derive the HT, *int* double mutants now in wide use.[55]

Use of P1 in Salmonella

Certain mutants of *Salmonella* (notably, *galE*) are susceptible to infection by P1, and P1 will complete its life cycle in *Salmonella*. Therefore, *Salmonella* can serve as donor or recipient in P1 transductions. This has an obvious, albeit limited, utility for measuring cotransduction of markers too far apart for P22. More importantly, it provides a general mechanism to carry out intergeneric crosses between *E. coli* and *S. typhimurium*. Typically, such crosses yield few if any transductants because of significant dissimilarities between the two genomes. Recently, however, it has been learned that the use of recipients defective in methyl-directed mismatch repair dramatically increases the yield in such crosses.[56] The potential for construction of useful hybrid strains by this method has scarcely begun to be exploited.

Methods for Phage P22

Preparation of Phage Stock

The following is based on the method of Davis, Botstein, and Roth.[57] Prepare an overnight culture of the donor strain in L broth. To 1 ml of this culture add 4 ml of P22 broth (see Table IV[57–59]). Incubate this mix in a culture tube (20 × 150 mm or similar) with aeration for 5–18 hr (i.e., during the day or overnight). Routinely, this would be at 37°, but other

[55] R. P. Anderson and J. R. Roth, *J. Mol. Biol.* **119**, 147 (1978).
[56] C. Rayssiguier, D. Thaler, and M. Radman, *Nature (London)* **342**, 396 (1989).
[57] R. W. Davis, D. Botstein, and J. R. Roth, "Advanced Bacterial Genetics, A Manual for Genetic Engineering." Cold Spring Harbor, Laboratory, Cold Spring Harbor, New York (1980).
[58] M. B. Schmid, personal communication (1990).
[59] M. Levine, *Virology* **3**, 22 (1957).

TABLE IV
PREPARATION OF MEDIA

Type	Preparation
L broth	10 g Difco (Detroit, MI) tryptone, 5 g Difco yeast extract, 5 g NaCl, 1 liter deionized water. Adjust pH to 7.4 with 10 N NaOH (usually 0.2–0.4 ml per liter) and sterilize by autoclaving.
L agar plates	Prepare L broth but include 15 g Difco agar per liter and pour the autoclaved medium into standard petri plates. Fill the plates about half full and allow to solidify.
L top agar	Prepare L broth but include 7 g agar per liter. Autoclave and distribute into 100-ml bottles. Melt agar prior to use and keep at 55°.
LCA agar plates	Same as L agar plates except add 5 ml of 0.5 M CaCl$_2$ per liter of media after autoclaving.
TMG buffer	Prepare a solution of 10 mM Tris-HCl, pH 8.0, 10 mM MgSO$_4$, and 0.1% gelatin. Autoclave and distribute into 100-ml bottles.
M63–glucose–B$_1$	Dissolve 3.0 g KH$_2$PO$_4$, 7.0 g K$_2$HPO$_4$, and 2.0 g (NH$_4$)$_2$SO$_4$ in 1 liter of deionized water. Add 0.5 ml of a 1 mg/ml solution of FeSO$_4$ and autoclave. When cool, add 10 ml of a sterile 20% (w/v) glucose solution, 2 ml of a sterile 50 mg/ml thiamin solution, and 1 ml of a sterile 1 M MgSO$_4$ solution. It is often convenient to prepare a solution of 10× M63 [30 g KH$_2$PO$_4$, 70 g K$_2$HPO$_4$, 20 g (NH$_4$)$_2$SO$_4$, 5 ml of 1 mg/ml FeSO$_4$ per liter] and to dilute this 10-fold into deionized water prior to autoclaving.
M63 agar plates	Same as for M63–glucose–B$_1$ medium except add 15 g Bacto-agar (Difco) to the M63 medium before autoclaving. Pour 40 ml per sterile plastic dish.
EMB plates	Add 27.5 g EMB (eosin methylene blue) agar base (Difco) to 1 liter of water and autoclave. Add 50 ml of a sterile 20% solution of the desired sugar.
P22 broth	L broth containing P22 lysate at 5 × 10^6 pfu/ml and other ingredients. Davis et al.[57] also add 1× E salts and 0.2% (w/v) glucose. The author (R.M.) obtains adequate results omitting the salts and glucose and including 0.05% galactose. The latter permits propagation of P22 on $galE$ mutants without interfering with P22 growth on other strains. The titer of P22 in P22 broth is stable for months under refrigeration.
EGTA	A 1 M solution can be made by slowly adding concentrated NaOH or NaOH pellets just until the EGTA goes into solution. The final pH will be about 8. The solution should be sterilized by filtration, and it is added to agar just before pouring.
Green plates	8 g Bacto-tryptone, 1 g Bacto yeast extract, 15 g NaCl, and 15 g Bacto-agar per liter. After autoclaving, the following additions are made: 16.8 ml of 40% glucose, 25 ml of 2.5% Alizarin Yellow G,GG (EM Science;, Cherry Hill, NJ), and 3.3 ml of 2% Aniline Blue (Fisher, Fairlawn, NJ). Alizarin Yellow is insoluble at room temperature and therefore must be dissolved by heating just before each use, or it can be added as a solid to hot agar. The dyes can also be autoclaved with the agar. This recipe is a modification[58] of the Levine[59] formulation.
P22 phage buffer	P22 is stable in many solutions. We have used either a phosphate-buffered saline solution (per liter, 8.5 g NaCl, 5.8 g anhydrous Na$_2$HPO$_4$, 3 g KH$_2$PO$_4$) or "SMO$^+$" (50 mM Tris-HCl, pH 7.5, 100 mM NaCl, 10 mM MgSO$_4$, and 0.01% (w/v) gelatin, the last two ingredients being added after autoclaving).

temperatures can be used if the donor strain requires it. P22 gives poor yields of plaque-forming units and transducing particles above 40°.

Add a few drops of chloroform to the lysate (whether the culture has cleared or not) and continue incubation at 37° for 5 min. Then remove cells and debris by low-speed centrifugation (e.g., 10 min at 10,000 rpm in a Sorvall SS34 rotor at 4°). Remove the clarified lysate to a sterile tube for storage at 4°. As an alternative to low-speed centrifugation, the lysate can be diluted 100× into sterile phage buffer. The dilution, at about 10^8 pfu/ml, has enough phage for many transductions with HT phage.

Transduction

The following low-multiplicity protocol is appropriate for routine strain constructions and linkage analysis. An overnight culture of the recipient strain should be diluted 50 to 100-fold into LB broth and regrown to log phase to about 10^8 cells/ml (the exact cell concentration is not critical). In the next step, phage and cells are going to be mixed; their proportions should be designed with several considerations in mind. First, the yield of transductants for HT 12/4 will be about 1 per 10^4 pfu. Second, it is desirable to have cells in excess over phage pfu by a factor of at least 10 so as to minimize loss of transductants through coinfection by a phage particle. Third, an outgrowth period for expression may be needed prior to selection. Most commonly used drug resistance markers need an outgrowth, but β-lactamase (ampicillin resistance) does not. Fourth, the amount finally plated should be sufficient to give about 100 colonies under selection. A typical experiment would use 1 ml (10^8) cells plus 0.1 ml of lysate at 10^8 pfu/ml; additional mixtures with higher and lower amounts of phage would also be made.

The mixtures are held for 10 min at room temperature or on ice to allow phage adsorption. If no outgrowth is needed, 0.1 ml of each mix can then be plated. We routinely formulate selection plates with 10 mM EGTA [ethylene glycol bis(β-aminoethyl ether)-N,N,N',N'-tetraacetic acid; Sigma, St. Louis, MO] to prevent infection on the plates. Alternatively, if an outgrowth is needed, the mixtures are diluted with an equal volume of L broth containing 20 mM EGTA and then incubated for the length of time desired (seldom more than 1–2 hr) before plating. Transductants apparently do not multiply during such a short outgrowth period.[60] As with P1 transduction, it is always important to perform controls for the sterility of the lysate and reversion of the recipient strain.

Even though P22 HT *int* cannot lysogenize, it is important to ensure

---

[60] J. Ebel-Tsipis, M. S. Fox, and D. Botstein, *J. Mol. Biol.* **71,** 449 (1972).

that the transductant colony is not undergoing active infection. Infection cannot take place in the presence of EGTA, but competent phage are still present on the plate and can initiate infection once the EGTA has been removed, that is, during purification or testing of a transductant. We routinely maintain transductants in the presence of EGTA until they have been single-colony purified at least one time. After this, colonies can be considered phage free and EGTA may be omitted.

In some laboratories, transduction is carried out directly on selective plates by simply spreading together cells and phage. In this method, EGTA must be omitted, and the absence of infection in a transductant colony must be established. It is helpful to pick transductants as soon as they become visible. They should then be purified on "green plates," on which infected colonies turn dark green while uninfected colonies remain pale in color. In the experience of the author (R.M.), the use of EGTA is more convenient. However, certain complex transductional events requiring the participation of two transducing fragments in the same cell can be detected only under high multiplicity (phage in excess) conditions[61,62]; in this case the use of green plates to identify phage-free segregants is preferred.

Because P22 HT12/4 *int* has a turbid plaque phenotype, it is susceptible to overgrowth by clear-plaque mutants if propagated repeatedly in liquid culture. For this reason, it is advisable to prepare master stocks of this phage starting from a single, turbid plaque. However, the inevitable occurrence of clear-plaque mutants in any particular lysate has no significant effect on transduction carried out at low multiplicity as described here.

[61] K. T. Hughes and J. R. Roth, *Genetics* **109**, 263 (1985).
[62] M. B. Schmid and J. R. Roth, *Genetics* **105**, 517 (1983).

[3] Conjugational Methods for Mapping with Hfr and F-Prime Strains

By K. Brooks Low

Introduction

Conjugational mapping methods are particularly useful when the map location of a locus is totally unknown, and there is no clone available for molecular mapping of the locus, or when a clone bearing an antibiotic resistance marker within or near an unmapped locus has been inserted in the chromosome by homologous recombination.[1-5] The procedures described in this chapter have all been developed for *Escherichia coli* K-12, but similar approaches can be used with certain other bacteria such as *Salmonella*,[6] *Proteus*,[7] and *Pseudomonas*.[8]

Use of Hfr Strains: General Considerations

Commonly Used Hfr Strains

The positions of 20 origins of transfer for the most commonly used Hfr strains are shown in Fig. 1.[9,10] The names shown on Fig. 1 are those of the original (ancestral) strains isolated, from which a number of derivatives have been made including the ones listed in Table I.[11] One useful set of derivative Hfr strains constructed by Wanner[12] carries a Tn*10* insertion in each strain, thus providing an easily selectable Hfr marker (Tet^R) at a convenient distance from the origin of transfer. These strains have the prefix BW in Table I. Sets of these Hfr strains, with our without the Tn*10* insertions, are available as indicated in Table I.

Hfr strains with other points of origin have also been isolated.[13] It is also possible to tailor conjugal donor strains using small F or F-prime (F') factors which carry some homology (e.g., a Tn*10* transposon) to a site (e.g., Tn*10*) which has been located on the chromosome.[11] The "chromosome mobilization" (see below) from such a strain results in oriented transfer of donor markers beginning at the site of homology (e.g., the transposon) between the F derivative and the chromosome. Specific chromosomal

[1] M. Russell and P. Model, *J. Bacteriol.* **159**, 1034 (1984).

[2] C. A. Lee, M. J. Fournier, and J. Beckwith, *J. Bacteriol.* **161**, 1156 (1985).

[3] M. Jasin and P. Schimmel, *J. Bacteriol.* **159**, 783 (1984).

[4] S. C. Winans, S. J. Elledge, J. H. Krueger, and G. C. Walker, *J. Bacteriol.* **161**, 1219 (1985).

[5] C. B. Russell, D. S. Thaler, and F. W. Dahlquist, *J. Bacteriol.* **171**, 2609 (1989).

[6] K. E. Sanderson and P. R. MacLachlan, *in* "*Escherichia coli* and *Salmonella typhimurium*: Cellular and Molecular Biology" (F. C. Neidhardt, *et al.*, eds.), p. 1138. American Society for Microbiology, Washington, D.C., 1987.

[7] B. W. Holloway, *Plasmid* **2**, 1 (1979).

[8] B. W. Holloway, *in* "Pseudomonads. The Bacteria, Volume 10" (J. R. Sokatch, ed.), Academic Press, New York, 1985.

[9] K. B. Low, *in* "*Escherichia coli* and *Salmonella typhimurium*: Cellular and Molecular Biology" (F. C. Neidhardt, *et al.*, eds.), p. 1134. American Society for Microbiology, Washington, D.C., 1987.

[10] B. J. Bachmann, *Microbiol. Rev.* **54**, 130 (1990).

[11] F. G. Chumley, R. Menzel, and J. R. Roth, *Genetics* **91**, 639 (1979).

[12] B. L. Wanner, *J. Mol. Biol.* **191**, 39 (1983).

[13] B. Low, *Bacteriol Rev.* **36**, 587 (1972).

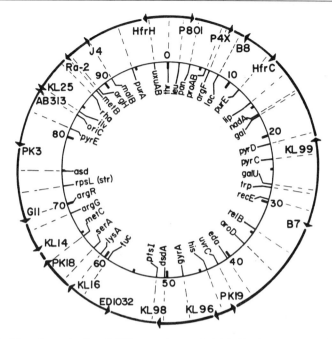

FIG. 1. Map of certain *E. coli* K12 genes (inner circle) indicating map coordinates[10] and relative positions of Hfr points of origin (arrowheads). The sequence of markers transferred from a given Hfr is indicated by the direction of the arrowhead and begins with the region just *behind* the arrowhead; that is, the arrowhead points in the opposite direction from the sequence of markers transferred (example: HfrH transfers the sequence *uxuAB, thr, leu, . . .*).

transposon insertions for such constructions can be chosen from among the many already available.[14,15]

Hfr Stability

During vegetative growth, most Hfr strains occasionally produce daughter cells in which the F factor has recombined out of the chromosome and which are thus no longer Hfr cells. Some Hfr strains are extremely stable (never known to revert) and are marked S in Table I. Certain Hfr strains are extremely unstable (marked VU in Table I), and even newly

[14] M. L. Singer, T. A. Baker, G. Schnitzler, S. M. Deischel, M. Goel, W. Dove, K. J. Jaacks, A. D. Grossman, J. W. Erickson, and C. Gross, *Microbiol. Rev.* **53**, 1 (1989).
[15] C. M. Berg and D. E. Berg, in *"Escherichia coli* and *Salmonella typhimurium:* Cellular and Molecular Biology" (F. C. Neidhardt, *et al.,* eds.), p. 1071. American Society for Microbiology, Washington, D.C., 1987.

TABLE I
USEFUL *Escherichia coli* K-12 Hfr STRAINS

Hfr strain	Ancestral Hfr	PO No.[a]	Earliest marker known	Latest marker known	Site of Tn10[b]	Stability[c]	Comments
3000 (=HfrH *thi⁻* λ⁻)	HfrH	1	*uxuBA*	*valS*	—	SU	*d*
NK6051	HfrH	1	*uxuBA*	*valS*	*purE*	SU	*e*
P801	P801	120	*leu*	*pan*	—	SU	*d*
BW6165	P801	120	*leu*	*pan*	*argE*	SU	*e*
BW113	P4X	3	*argF*	*lac*	—	S	*d*
BW6156	P4X	3	*argF*	*lac*	*zje*	S	*e*
B8	B8	118	*tsx*	*lac*	—	SU	*d*
BW6160	B8	118	*tsx*	*lac*	*zdh*	SU	*e*
KL226	HfrC	2A	*gsk*	*fep*	—	S	*d*
BW7261	HfrC	2A	*gsk*	*fep*	*leu*	S	*e*
KL99	KL99	42	*pyrC*	*pyrD*	—	VU	*d*
BW7623	KL99	42	*pyrC*	*pyrD*	*zed*	VU	*e*
KL208	B7	43	*rac*	*trg*	—	SU	*d*: F deleted from 33–43 kb
BW7620	B7	43	*rac*	*trg*	*purE*	SU	*e*; F deleted from 33–43 kb
PK191	PK19	66	*supD*	*cheC*	—	S	*d*; fertility factor is ColV
BW5660	PK19	66	*supD*	*cheC*	*srlC*	S	*e*; fertility factor is ColV
KL96	KL96	44	*his*	*purF*	—	SU	*d*
BW7622	KL96	44	*his*	*purF*	*trpB*	SU	*e*
KL983	KL98	53	*dsdA*	*supN*	—	SU	*d*
BW5659	KL98	53	*dsdA*	*supN*	*zdh*	SU	*e*
ED1032	ED1032	201	*thy*	*his*	—		*e* Transposition of F42-114; transfers *tra* genes early

Strain	PO					
KL16	45	lysA	serA	—	S	d
BW1663	45	lysA	serA	zed	S	e
PK18	132	metC	argG	—	S	Fertility factor is ColV
KL14	68	cca	tolC	—	S	d
BW6159	68	cca	tolC	ilv	S	e
G11	124	str	argG	—	SU	d
KL800	131	xyl	malA	—	S	d; fertility factor is ColV
BW6175	131	xyl	malA	argE	S	e; fertility factor is ColV
AB313	13	rbs	ilvE	—	SU	d
BW6129	13	rbs	ilvE	argA	SU	e
KL25	46	ilvE	pyrE	—	VU	d
Ra-2	48	metB	rha	—	VU	d
BW6164	48	metB	rha	thr	VU	e
KL209	18 (=P10)	argE	purA	—	SU	d; F factor in malB
BW6166	18 (=P10)	argE	purA	zhf	SU	e; F factor in malB

[a] The PO (point of origin) numbers were arbitrarily assigned to distinguish independent F factor insertion events in the formation of Hfr strains. These numbers do not relate to map position.

[b] The sites of Tn10 insertion in these strains lie within the range of 10 to 30 min from the points of origin. Some of these strains also carry a deletion of the *lac* region, as well as certain other markers. In the *zxy* system for naming the sites of unknown Tn insertions, [11] the second letter (*x*) indicates the appropriate 10-min interval of the map (*a* = 0–10, *b* = 10–20, etc.), and the third letter (*y*) indicates the particular minute within that interval (*a* = 0–1, *b* = 1–2, etc.). All other gene symbols are as given on the *E. coli* K12 map. [10]

[c] S, Extremely stable; SU, somewhat unstable; VU, very unstable.

[d] These strains are included in the Hfr Kit available from the *E. coli* Genetic Stock Center, c/o Dr. Barbara J. Bachmann, Department of Biology, Osborn Laboratory, Yale University, New Haven, CT 06510. References for the Hfr strains are given by Low. [9]

[e] These strains are included in the Hfr::Tn10 Kit, available as above.

isolated clones (see below) contain over 10% F^+ revertants. The remaining Hfr strains listed in Table I, designated SU (somewhat unstable), accumulate F^+ revertants at a low rate during subculture, and they occasionally require repurification (see below) or a new subculture from master stock cultures.

Purification and Storage of Hfr Strains

1. Grow an overnight culture (at least 5 ml) of a polyauxotrophic streptomycin-resistant F^- strain such as AB1157.[16,17] Resuspend in the same volume of minimal medium such as medium 56[16] or buffered saline. This can be stored for 1–2 weeks at 4°.

2. Prepare recombinant-selective plates appropriate for the strain used in Step 1 above. Choose three or more auxotrophic markers well spaced around the map, such as *proA, his,* and *argE* (close to *argH,* Fig. 1) when using AB1157. Omit the selected growth requirement from the medium; include streptomycin in all cases to counterselect the Hfr (streptomycin-sensitive) cells.

3. Streak Hfr strains onto plates and incubate long enough to obtain 20–50 well-separated colonies (e.g., overnight at 30°–37°) on a rich medium such as LB (Luria broth) agar.[16]

4. Pick colonies onto a master plate, using a template such as shown in Fig. 2. A small loop holder with a fine wire which may be quickly steriled in a pilot flame,[16,18] or sterile toothpicks, can be used. A rich medium may be used, but a minimal medium (e.g., medium 56 plus 0.5% (w/v) casamino acids plus 0.3% (w/v) glucose[16]) is preferable if prolonged storage at 4° is desired.

5. Incubate the master plate until patches are clearly grown (8–12 hr at 37° or 12–20 hr at 30°). Store at 4° at this stage.

6. Using a velvet replicator,[16] replica-plate (print) the set of patches from the master grid made in Step 5 onto rich agar such as LB agar. Incubate at 37° until printed patches are clearly visible but not yet thick (4–7 hr, depending on the age of the master grid). Leave at 37° until use.

7. Prepare one recombinant-selective plate for each marker. The

[16] B. Low, *J. Bacteriol.* **113,** 798 (1973). A replicating block is now commercially available (Ann Arbor Plastics, Ann Arbor, MI 48104). This block is improved if the three small feet are removed and the top edge rounded off.

[17] B. J. Bachmann, *in* "*Escherichia coli* and *Salmonella typhimurium:* Cellular and Molecular Biology" (F. C. Neidhardt, *et al.,* eds.), p. 1190. American Society for Microbiology, Washington, D.C., 1987.

[18] 90% Pt/10% Ir wire (0.005 inch) is available (#10IR 5T) from Medwire Corp. (Mount Vernon, NY 10553).

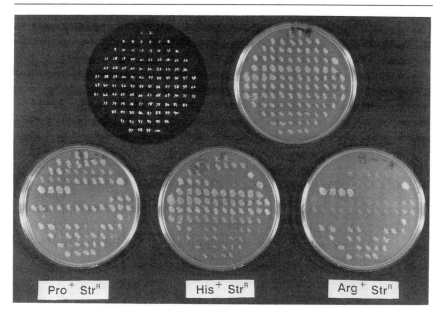

FIG. 2. Hfr purification. A full-sized template can be prepared by photocopying one given in Ref. 16. The agar plate at upper right is the fresh copy of the Hfr patches, ready to replica plate. The other three plates select for either Pro$^+$ [Str^R], His$^+$ [Str^R], or Arg$^+$ [Str^R] recombinants. The groups of Hfr patches shown are as follows (see Table I and Fig. 1 for corresponding points of origin): HfrH (patches #1–10), KL14 (#11–15), KL16 (#16–20), KL226 (#21–25), KL800 (#26–31), PK18 (#31–40), PK191 (#41–45), KL983 (#46–60), B8 (#61–70), and KL25 (#71–100). Hfr strains PK18 and PK191 are deleted for *proAB* and thus cannot produce any Pro$^+$ [Str^R] recombinants in this experiment.

plates should be somewhat dry (very faint wrinkles on the agar surface). Flood the plate with recipient cells (prepared in Step 1 above) by pipetting 2–3 ml of culture onto the surface of the plate. Tip the plate to fully cover with liquid and promptly pour off the excess (onto another recombinant-selective plate, if desired). Use a Kimwipe to soak up the remaining "puddle" at the edge after the culture is poured off. Return the plate promptly to a flat surface to allow the lawn to dry (1–5 min). Repeat with the other recombinant-selective plates.

8. Use the fresh 37° copy of the grid (prepared in Step 6) as a master to replica-plate the newly growing patches onto the flooded recombinant-selective plates. (Inoculate the velvet once and print all of the flooded plates from this one velvet.)

9. Promptly incubate the recombinant selective plates at 37° for 16–20

hr (do not overincubate). At this point there will clearly be strong patches of recombinants corresponding to Hfr clones which transfer a given marker fairly early. For example, as shown in Fig. 2, certain patches from Hfr KL25 (bottom one-third of patches on plate) produce strong patches of Arg$^+$ [Str^R] recombinants, and certain patches (F$^+$ revertants) do not. These same Hfr clones often produce very weak patches of recombinants for a selected marker which is transferred very late (e.g., the Arg$^+$ [Str^R] selective plate for HfrH, the top 10 patches on the plate), because of the gradient of transmission from the Hfr (see below).

10. A record of the strengths of recombinant patches can be simply obtained by laying the recombinant-selective plates on a photocopy machine, covering with a black background to maximize contrast, and "Xeroxing" the image of the plates.

11. To preserve good long-term stocks, scrape up several good Hfr patches for a given strain from the original master grid, resuspend together in 1 ml of 15% glycerol in rich broth or minimal medium, and freeze at $-80°$ without further growth. Lyophilization is also advised, particularly for very unstable Hfr strains.

Growth of Cells Prior to Conjugation

The condition of the donor strain in conjugation is generally much more crucial than that of the recipient strain. For highest transfer efficiency, the donor strain should be exponentially growing. Dilute stock (or overnight) cultures approximately 1 : 50 into fresh medium and grow ($\sim$2 hr) until the OD$_{600}$ reaches 0.3–0.4 (2–3 $\times$ 10^8 cells/ml). At this stage donor and recipient cultures can be kept on ice for several hours, if desired, before starting the cross (except for time-of-entry matings, see below).

Media. Growth in a rich medium such as Luria broth[16] results in much higher efficiencies of conjugation than growth in minimal medium.[19]

Temperature. Conjugation efficiency is optimal at 37° and falls off sharply at temperatures below 30° or above 40°.[19] The temperature at which the donor cells are grown is particularly crucial: donor cells grown at 30° or below are usually infertile. (They produce very few F pili under these circumstances.) At 33°, pregrowth can result in only 10% of the mating efficiency at 37°.[20]

Shaking. For pregrowth of cultures as well as during conjugation, gentle shaking is recommended for uniformity and reproducibility, al-

[19] T. H. Wood, *J. Bacteriol.* **96,** 2077 (1968).
[20] K. B. Low, unpublished (1978).

though slightly higher transfer efficiency has been reported if the cultures are incubated without shaking.[21]

F⁻ Phenocopy: Use of Hfr, F⁺, or F' Strains as Recipients. The presence of an F factor in a strain renders it a very poor conjugational recipient during exponential growth.[22] However, such a strain will be a highly efficient recipient if it is grown to stationary phase in rich medium, with aeration [e.g., 1 ml in a 25 mm (diameter) culture tube, with gentle shaking or rotation) for 1–2 days. Mix this culture directly 1 : 1 with a freshly growing 37° culture (in rich medium) of the donor strain to be used in the cross. Under these conditions the donor to recipient cell ratio is approximately 1 : 10, and the efficiency of recombinant formation per donor can be very high (10–50%).[20,21]

Mating Conditions

Long Matings. A small volume of mating mixture [e.g., 2 ml in a 25 mm (diameter) culture tube] is usually more than enough to enable selection of even rare recombinants. Uninterrupted matings of 90 min–2 hr allow transfer of even late (distal) Hfr markers into recipients at detectable levels.

Time-of-Entry Matings. When kinetics of transfer are being determined (see below) it is important to mix exponentially growing donor and recipient cultures directly without temporary storage at 4°. This prevents unnecessary delays in initiation of transfer. To facilitate the rapid sampling of the culture, it is convenient to use a 10–30 ml mating mixture and shake slowly in a 125 ml flask covered with easily removable foil or a loose cap.

Mating Ratios. The maximum number of recombinants per milliliter of mating mixture is usually obtained by using a 1 : 2 to 1 : 3 donor to recipient ratio.[20] If an optimum number of recombinants per donor cell is desired, a ratio of 1 : 10 (or greater) is needed. For blender (time-of-entry) experiments, an even lower ratio is often used (1 : 20 to 1 : 300) so that individual samples (0.1 ml) of the mating mixture can be taken, treated to stop transfer directly (see below), and plated without dilution.

Selective and Counter-selective Markers

In principle, any two differing donor and recipient selectable phenotypes can be used to select recombinants in conjugational crosses. These include auxotrophic growth requirements, sensitivity or resistance to anti-

[21] R. Curtiss III, L. G. Caro, D. P. Allison, and D. R. Stallions, *J. Bacteriol.* **100,** 1091 (1969).
[22] N. Willetts and R. Skurray, *in* "*Escherichia coli* and *Salmonella typhimurium:* Cellular and Molecular Biology" (F. C. Neidhardt, *et al.,* eds.), p. 1110. American Society for Microbiology, Washington, D.C., 1987.

biotics or other growth inhibitors, sensitivity or resistance to bacterio-phage infection, and temperature- or cold-sensitive growth. Background growth of parental cells can occur with certain pairs of markers owing to cross-feeding or inactivation of an antibiotic or other inhibitor. When this is a possibility, donor and recipient strains can be streaked close to each other on the recombinant-selective medium to be used, to check for cross-feeding.

The best recipient markers to use to counterselect against donor cells, in particular for time-of-entry experiments, are streptomycin resistance ($Str^R = rpsL^-$) or nalidixic acid resistance ($Nal^R = gyrA^-$), in the presence of 100 μg/ml streptomycin or 50 μg/ml nalidixic acid, respectively. A detailed discussion of these and other selective markers is given by Miller.[23]

Approaches to Map Location Using Hfr Crosses

Long Matings and the Transfer Gradient

Dependence on Distance from Transfer Origin. For reasons that re-main obscure 33 years after its discovery, the process of transfer of DNA from Hfr to recipient in conjugation terminates seemingly at random at various points along the chromosome. This results in a continuous logarith-mic decrease (transfer gradient) in the numbers of recipient cells which receive a given donor marker from an Hfr strain, as a function of distance of the marker from the origin of transfer.[19,24,25] Since the chance of incorpo-ration of most Hfr markers into recombinants after transfer is roughly the same, this gradient of transfer also results in an exponentially decreasing number of actual recombinants formed as a function of distance, namely, the gradient of transmission.[26]

Comparison of Known and Unknown Markers. An example of the dependence of recombination frequency on distance is shown in Fig. 3.[27] This type of curve can be used to find the approximate map position of an unmapped marker, provided two (or more, if possible) known points have been determined for the given Hfr, F$^-$ combination. Other examples of this type of experiment have been described.[19,23-25] Although the slopes of

[23] J. H. Miller, "Experiments in Molecular Genetics." Cold Spring Harbor Laboratory, Cold Spring Harbor, New York, 1972.

[24] E. L. Wollman and F. Jacob, *Ann. Inst. Pasteur (Paris)* **95**, 641 (1958).

[25] P. G. deHaan, W. P. M. Hoekstra, C. Verhoef, and H. S. Felix, *Mutat. Res.* **8**, 505 (1969).

[26] F. Jacob and E. L. Wollman, "Sexuality and the Genetics of Bacteria." Academic Press, New York, 1961.

[27] B. Low, *J. Bacteriol.* **93**, 98 (1967).

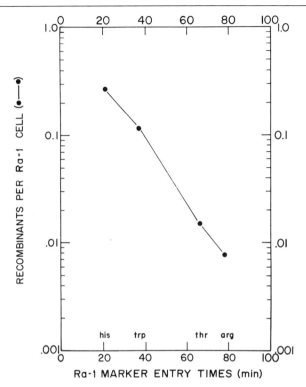

FIG. 3. Gradient of transmission for Hfr Ra-1 [*his*$^+$ *trp*$^+$ *thr*$^+$ *arg*$^+$ *rpsL*$^+$ (Str^S)] × F$^-$ PA309 [*his*$^-$ *trp*$^-$ *thr*$^-$ *arg*$^-$ *rpsL*$^-$ (Str^R)]. Cells were mixed in a 1 : 10 Hfr × F$^-$ ratio and incubated for 90 min before plating onto recombinant selective media with no interruption.[27] The point of origin of Hfr Ra-1 is similar to that of KL16 (Fig. 1 and Table I).

the transfer gradient curves are fairly similar for most Hfr strains,[19] a few Hfr strains such as KL800, PK18, and PK191 (all derived from an integrated ColV fertility factor instead of the F factor; see Table I) produce more shallow transfer gradients.[28]

Exceptions: Markers Very Close to Point of Origin. If an Hfr marker is located less than 1 min or so from the point of origin (i.e., transferred very early by the Hfr), there is less than the usual chance that it will be incorporated into recombinants.[29-31] In some cases this can result in more than a 10-fold decrease in recombination frequency observed compared

[28] K. B. Low and J. H. Miller, unpublished (1984).
[29] B. Low, *Genet. Res. Camb.* **6**, 469 (1965).
[30] N. Glansdorff, *Genetics* **55**, 49 (1966).
[31] J. Pittard and E. M. Walker, *J. Bacteriol.* **94**, 1656 (1967).

to what would be expected from the transfer gradient. An uncertainty of this type can be resolved by obtaining gradient plots using two different Hfr strains.

Rapid Mapping by Replica Plating ("Print Mating")

The phenomenon of the transfer gradient allows the use of replica plating to determine quickly the approximate location of a selectable Hfr marker.[16] In this system patches of a wide variety of Hfr strains are all grown on the same master plate. They are then replica plated onto a lawn of mutant cells on recombinant-selective plates, as in the purification of Hfr strains (Fig. 2, above). By observing which Hfr strains produce the strongest patches of recombinants (and thus are close to the selected marker because of the transfer gradient), the corresponding Hfr origins of transfer closest to the selected marker can be determined.[16,32]

Analysis of Genetic Linkage between Markers

An Hfr × F⁻ cross can also provide information on unknown map locations by an analysis of genetic linkage between a selected and unselected marker in the cross.[26,33–36] This genetic analysis is most conveniently carried out using velvet replica plating methods,[16,18] although other devices have also been reported.[37,38] The mechanisms of recombination involved in incorporation of various continuous (genetically linked) stretches of donor genetic information into recipient chromosomes are not understood, and an exact predictable linkage relationship at this point is not available. However, certain categories of genetic linkage are generally reliable for crosses carried out at 37° and useful in genetic analysis:

1. If an unselected Hfr marker appears with very low frequency (less than 5%) among selected recombinants in an Hfr × F⁻ cross, this is usually due to the transfer gradient between the two markers, that is, the unselected marker lies greatly distal to (is transferred at least 30 min later than) the selected one.

2. If an unselected marker appears with very high frequency (greater

[32] G. N. Godson, *J. Bacteriol.* **113**, 813 (1973).

[33] P. G. deHaan and C. Verhoef, *Mutat. Res.* **3**, 111 (1966).

[34] C. Verhoef and P. G. deHaan, *Mutat. Res.* **3**, 101 (1966).

[35] K. B. Low, *in* "*Escherichia coli* and *Salmonella typhimurium*: Cellular and Molecular Biology" (F. C. Neidhardt, *et al.*, eds.), p. 1184. American Society for Microbiology, Washington, D.C., 1987.

[36] G. R. Smith, *Cell* **64**, 19 (1991).

[37] T. H. Wood and S. K. Mahajan, *Genet. Res. Comb.* **15**, 335 (1970).

[38] W. M. Thwaites, *Appl. Microbiol.* **16**, 956 (1968).

than 95%) among selected recombinants, the two markers probably lie within 1 min of distance on the chromosome. If the linkage is greater than 99% (i.e., less than 1% recombination between the markers), the two markers are probably within the same or neighboring genes.[39–41]

3. For other cases (linkage between 5 and 95%), the unselected marker probably lies either some distance proximal to (i.e., enters earlier than) the selected one or else within 30 min of distance on the distal side (i.e., where the effect of the transfer gradient is not yet too great). Estimates of an "average" number of crossovers per unit of length are sometimes given as 10–20% per minute of chromosome length, but they may actually vary significantly outside this range from case to case.[26,29–31,33–36] The percentage of recombinants which show crossovers between close markers can be increased, roughly in proportion to the distance between the markers, by irradiation of the zygotes soon after the cross.[42]

Time-of-Entry Experiments

Importance of Mating Condition. In order to relate various time-of-entry results to each other, parental cells need to be grown and maintained at 37° to exponential phase in rich medium, then mixed (e.g., in a 1 : 20 to 1 : 300 donor to recipient ratio) without any changes in temperature which could delay the onset of chromosome transfer or alter the speed of transfer.[19] It is convenient to work with 10–30 ml of mating mixture shaking gently in a 125-ml Erlenmeyer flask. Begin a timing clock at the time of mixing the donor and recipient strains.

Mating Interruption. Two methods of mating interruption are dependable and convenient, and the choice may depend on the region of chromosome being mapped. Mechanical agitation is efficiently achieved by using a vibratory shaker,[43] and an updated version of this mechanism is illustrated in Fig. 4. The use of this device is described further by Miller.[23] Either 100 μg/ml streptomycin [in conjunction with a streptomycin-resistant ($rpsL^-$) recipient] or 50 μg/ml nalidixic acid [in conjunction with a nalidixic acid-resistant ($gyrA^-$) recipient] can be used to provide a clean counterselection and also to prevent plate mating after using the vibratory shaker, and the choice may depend on which region of the chromosome is being mapped. (It is best to avoid transfer of the $rpsL^+$ or $gyrA^+$ region by the Hfr prior to the selected Hfr marker, so as not to kill the resulting

[39] C. Willson, D. Perrin, M. Cohn, F. Jacob, and J. Monod, *J. Mol. Biol.* **8**, 582 (1964).
[40] L. C. Norkin, *J. Mol. Biol.* **51**, 633 (1970).
[41] S. I. Feinstein and K. B. Low, *Genetics* **113**, 13 (1986).
[42] M. Wann, S. K. Mahajan, and T. H. Wood, *J. Bacteriol.* **103**, 601 (1970).
[43] B. Low and T. H. Wood, *Genet. Res., Comb.* **6**, 300 (1965).

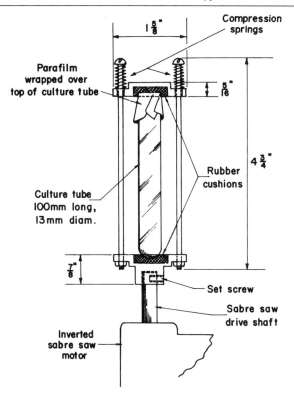

FIG. 4. Apparatus for interruption of mating pairs. Further details of the construction and mounting of this device are given by Miller.[23]

merozygotes with the antibiotic.) In crosses where the nalidixic acid selection can be used, it may be more convenient to interrupt mating by simply adding nalidixic acid to samples of the mating mixture, followed by plating.[44]

Shape of Time-of-Entry Curves; Extrapolation. Even under optimal conditions, not all mating pairs begin chromosome transfer at the same time; thus the numbers of various kinds of recombinant-forming cells per milliliter increase gradually with time. Furthermore, there appears to be a delay of approximately 3 min after mixing donor and recipient cultures before even the earliest donor marker is transferred.[29] Thus, the time-of-entry curve for any one donor marker from a known Hfr strain (e.g., see Figs. 5 and 6A) can give only an approximate estimation of its position on the chromosome. A more accurate measurement of chromosomal distance can only be obtained by comparison of two time-of-entry curves in the

[44] D. Zipkas and M. Riley, *J. Bacteriol.* **126,** 559 (1976).

same cross (Figs. 5 and 6A). As can be seen from Fig. 5, the rate of increase in recombinants for late markers is less than for early markers (as would be expected owing to the transfer gradient). In drawing an initial slope for these curves, therefore, comparative slopes are much steeper for early markers than late ones (Fig. 5). It is important to obtain samples at times prior to the extrapolated "entry time," to make sure the background number is low. In addition, it is generally more difficult to estimate precise entry times for late markers, so that for comparison with the published *E. coli* map,[10] time-of-entry curves in the 5–30 min range are most useful. In this range, entry time differences of 1 min or more can be resolved.

Very Closely Linked Markers: Combined Time-of-Entry and Genetic Analysis. Figure 6A illustrates time-of-entry curves for two nearby loci (*aroC* and *purF*) whose relative position cannot be resolved from the curves alone. However, samples of the two kinds of recombinants which are then scored for the other marker (Fig. 6B) clearly indicate the order of transfer, since recombinants selected for the earlier marker (*aro*+) at its earliest entry times will inherit the later marker (*pur*+) much less often than at later times, when mating interruption takes place after both markers have been transferred and are available for cointegration into recombinants. Thus, in this example, *aroC* is transferred earlier than *purF*.

Use of F' (and R') Factors for Mapping

Availability of F' and R' Strains

Figure 7 illustrates representative F' factors used for dominance studies in *E. coli* and other organisms.[45] Other F' and R' factors have also been reviewed.[13,45]

Isolation of New F' Strains. F' factors, which are formed at very low frequency during the growth of Hfr strains, are conveniently recovered by mating an Hfr strain with an F⁻ recA⁻ strain which carries an auxotrophic mutation corresponding to a chromosomal region near the point of origin of the Hfr.[46] Table II shows the frequencies of F' strains obtained in some crosses of this type.[20] Another convenient selection is for an antibiotic resistance marker (e.g., a transposon insertion) introduced into a desired region of the Hfr and selected for in crosses with an F⁻ recA⁻ strain which is sensitive to the antibiotic. In these types of crosses it is essential to interrupt the matings (see above) before the recA⁺ gene of the normal Hfr

[45] B. Holloway and K. B. Low, *in* "*Escherichia coli* and *Salmonella typhimurium*: Cellular and Molecular Biology" (F. C. Neidhardt, *et al.*, eds.), p. 1145. American Society for Microbiology, Washington, D.C., 1987.

[46] B. Low, *Proc. Natl. Acad. Sci. U.S.A.* **60,** 160 (1968).

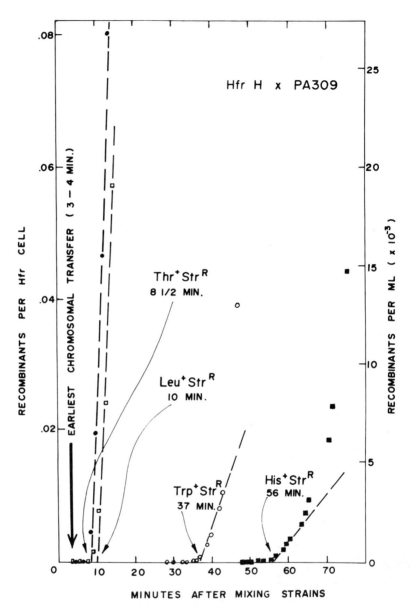

FIG. 5. Time-of-entry curves for a cross between HfrH (Fig. 1, Table I) and F⁻ PA309.[17] One-tenth milliliter of HfrH was mixed with 30 ml of PA309 in a 125-ml flask at time zero. Each sample of 0.1 ml was withdrawn and added to 3 ml of 0.8% agar in a 13 × 100 mm culture tube (kept at 43°–45° before use). This tube was then covered with Parafilm and shaken (Fig. 4) for 7 sec, at which point the elapsed time was noted and marked on the recombinant-selective plate. The tube contents were immediately poured onto the plate and allowed to solidify. Numbers of recombinant colonies were counted after 1½–2 days of incubation at 37°.[20]

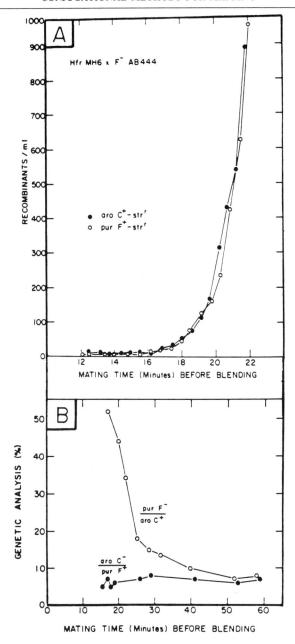

FIG. 6. Combined time-of-entry (A) and genetic analysis (B) (see text). Data of Wann *et al.*[42]

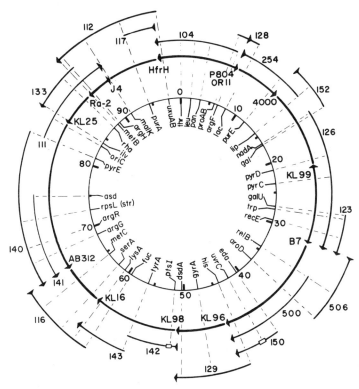

FIG. 7. Map of representative F′ factors.[45] These F′ factors are available separately or as a kit from the *E. coli* Genetic Stock Center (see Table I).

cells in the culture could be transferred and allow the formation of haploid recombinants which would mask the rare F′-bearing clones.[46]

Isolation of Derived F′ Strains. Various genetic approaches are available to recombine known mutations onto existing F′ factors by F′-chromosome recombination, to obtain truncated derivatives of F′s by transduction (transductional shortening), or to obtain chromosomal insertions or fusions of F′s with other genetic elements. These procedures have been reviewed.[13,23,41,45]

Transposon-bearing derivatives of an F′ factor can be readily obtained by mating the F′ out of a host strain which bears a transposon, then selecting for both F′ and transposon markers transferred to the recipient in the cross.[47–49]

[47] T. J. Foster, *Mol. Gen. Genet.* **154,** 305 (1977).

[48] R. Binding, G. Romanski, R. Bitner, and P. Kuempel, *Mol. Gen. Genet.* **183,** 333 (1981).

[49] J. Feutrier, M. Lepelletier, M.-C. Pascal, and M. Chippaux, *Mol. Gen. Genet.* **185,** 518 (1982).

TABLE II

REPRESENTATIVE Hfr × F⁻ $recA^-$ CROSSES TO SEARCH FOR F' FACTORS[a]

Hfr[b]	fr[b]	Selected markers[c] Hfr [F⁻]	"Recombinant" colonies/ml	Fraction $MS-2^{S}$ [d]	F' obtained[e]
HfrH	AB2463	Leu^+ [Str^R]	5×10^3	Not tested	F101 (thr^+ leu^+)
HfrH	AB2463	$ProA^+$ [Str^R]	3×10^3	16/20	F104 (thr^+ leu^+ $proA^+$)
					F139 (thr^+ leu^+ $proA^+$ lac^+)
HfrC	AB2463	$ProA^+$ [Str^R]	3×10^3	0/19	None
JC12	JC1553	$ArgG^+$ [$PurF^+$ $MetB^+$]	10^4	12/18	F102 ($argG^+$)
					F141 ($argG^+$ str^S $malA^+$)

[a] For all crosses, cells were grown in rich medium to $1–2 \times 10^8$ cells/ml, then mixed in a 1 : 10 Hfr to F⁻ ratio (total volume 2 ml). After 60 min of gentle mixing at 37°, samples were shaken in a vibratory blender (see text) and plated.

[b] Hfr points of origin are shown in Fig. 1. Hfr JC12 (derived from AB312) has a point of origin similar to that of KL14. Strain pedigrees can be found in Ref. 17.

[c] All symbols are described by Bachmann.[10]

[d] As explained in Ref. 46, not all "recombinants" from Hfr × F⁻ $recA^-$ crosses carry F' factors with the F element intact. In general, those which confer sensitivity to male-specific phage MS-2 (i.e., still produce normal F pili[22]) do carry transferable F' factors.

[e] F' factors are referred to in Refs. 13 and 45 and Fig. 7.

*Use of an F′ to Convert F⁻ Strains into Donors with Specific Origins
of Transfer: Chromosome Mobilization*

An F′ which carries homology with the chromosome is able to recombine with it, and many cells in such a culture can thus transfer the chromosome conjugationally linked to the origin of transfer of the F′.[50] The general notion of recombination between a conjugative plasmid and the chromosome, followed by chromosome transfer, has been termed "chromosome mobilization" by Signer and Beckwith[51] and by many other authors since. A chromosome mobilization system allows the use of the Hfr techniques described above, with the modification that the chromosome transfer frequencies of F′ chromosome mobilization systems are usually only 10% or less than the corresponding transfer frequencies for a similarly located Hfr point of origin.

One particularly useful F′ chromosome mobilization system was developed by Chumley *et al.*,[11] who constructed two Tn*10* derivatives of a temperature-sensitive (*ts*) F-*lac* F′ (with the orientation of Tn*10* differing in the two derivatives). In any strain which is deleted for *lac* but carries a chromosomal Tn*10* insertion, either of these F_{ts}-*lac* (Tn*10*) factors can be introduced and maintained in an integrated state at the site of the Tn*10* homology between F′ and chromosome, simply by growing the strain on lactose at high temperatures (which prevents autonomous replication of the F′). The strain is thus converted to a pseudostable Hfr-like state with a predetermined point of origin and [depending on which of the two F_{ts}-*lac* (Tn*10*) derivatives is used] direction of transfer. A similar use of these F′ factors to mobilize the chromosome through *lac* homology (e.g., at the site of a *lac* fusion on the chromosome) has been described.[52]

Other Uses of F′ Factors for Mapping

F′ factors are sometimes advantageous to locate and analyze chromosomal mutations by complementation or recombination after transfer.[13,45] In these cases chromosome mobilization is to be avoided, and this can be accomplished by the use of *recA*⁻ F′-bearing donor strains. F′ strains can also provide preliminary mapping information owing to their specific increased gene dosage.[53]

[50] J. Pittard and E. A. Adelberg, *Genetics* **49**, 995 (1964).
[51] E. R. Signer and J. R. Beckwith, *J. Mol. Biol.* **22**, 33 (1966).
[52] A. Middendorf, H. Schweizer, J. Vreemann, and W. Boos, *Mol. Gen. Genet.* **197**, 175 (1984).
[53] T. Ikemura and H. Ozeki, *J. Mol. Biol.* **117**, 419 (1977).

[4] Plasmid Transformation of *Escherichia coli* and Other Bacteria

By Douglas Hanahan, Joel Jessee, and Fredric R. Bloom

I. Introduction

The discoveries that phage[1] and plasmid[2] DNAs could be transferred directly into bacterial cells following a competence induction process involving treatment with calcium chloride at low temperatures set the stage for molecular cloning of recombinant DNAs. The technique of DNA transformation has become important in virtually all aspects of molecular genetics. *Escherichia coli* has developed into a universal host organism both for molecular cloning of DNA and for a diverse set of assays involving clones genes. Transformation of other bacteria is also being applied for more specialized purposes, in addition to genetic studies of these organisms themselves, as the chapters throughout this volume testify.

This chapter presents the major techniques and parameters that affect transformation of bacteria. The principal focus is on *E. coli*, although the basic principles for effectively transforming other gram-negative and gram-positive bacteria are discussed as well. There are two major parameters involved in efficiently transforming a bacterial organism. The first is the method used to induce competence for transformation. There are two primary technical variations: chemical induction of competence and high-voltage electroshock treatment (electroporation). Both the characteristic of the cells being transformed and the purpose of the transformation will affect the choice of method. The second major parameter is the genetic constitution of the host strain of the organism being transformed, as there is clear evidence that a variety of genes can dramatically influence the outcome of transformation experiments. Since these genetic factors are increasingly well understood, they are discussed in some depth below. Methods for inducing competence by chemical treatments and electroshock are described, along with other relevant parameters, such as handling of bacteria for transformation purposes. Finally, application of these principles (especially electroshock treatment) to other bacteria is discussed. The *E. coli* transformation protocols to be presented are summarized in Table I.

---

[1] M. Mandel and A. Higa, *J. Mol. Biol.* **53**, 159 (1970).
[2] S. Cohen, A. Chang, and L. Hsu, *Proc. Natl. Acad. Sci. U.S.A.* **69**, 2110 (1972).

TABLE I
Escherichia coli TRANSFORMATION METHODS[a]

Protocol	Method	Typical transformation efficiency (*XFE*)	Typical competence (*F*$_c$) (%)	Best for strains	Recommended applications
5	Calcium manganese-based (CCMB)	$1–10 \times 10^8$	1–5	MC1061, DH10B	cDNA cloning, mutagenesis high-efficiency cloning
6	TFB-based high efficiency	$0.4–2 \times 10^9$	2–10	DH1, DH5, HB101	cDNA cloning, mutagenesis high-efficiency cloning
7	FSB-based frozen competent	$1–5 \times 10^8$	1–4	DH1, DH5, HB101	cDNA cloning, mutagenesis, high-efficiency cloning
8	PEG/DMSO one step	$10^6–10^7$	0.01–0.1	Most common cloning hosts	Subcloning, plasmid constructions
9	Rapid colony transformation	$10^3–10^4$	<0.001	Most common cloning hosts	Reintroduction of cloned plasmids into *E. coli*
10	Electroshock	$0.5–5 \times 10^{10}$	80	Most common strains especially MC1061 and DH10B	High complexity cDNA and genomic cloning

[a] Transformation efficiency (*XFE*) is given here in transformants (colonies formed) per microgram of the plasmid pBR322. F_c is the fraction of viable cells that are competent for plasmid transformation. See Section V for discussion of these parameters.

II. General Handling of Bacteria for Transformation Experiments

A. *Storage of Strains*

Despite the resilience of *E. coli*, the conditions under which cells are stored has been found to influence their subsequent ability to be rendered competent for tranformation. *Escherichia coli* maintained either in a suspension in glycerol at $-20°$ or on plates at $4°$ both produce erratic competence when cells stored in these conditions are used to initiate cultures for preparing competent cells. Moreover, cells grown to saturation and nutrient limitation (e.g., the classic overnight culture) also produce erratic competence. There is no information about the physiological basis for this "memory," which is also not obvious, since in every case the stored cells are substantially expanded through growth at 37° and multiple cell divisions prior to the competence inducing procedures. Yet none of these conditions produce reliable sources for initiating the preparation of competent cells. Rather, it is our recommendation that the procedures described below be initiated from growing colonies on a freshly streaked agar plate, and that this plate be derived from either a frozen stock stored at −80° or from cells maintained in slow anaerobic growth in agar stab bottles stored in the dark at room temperature. These storage conditions produce reliably competent *E. coli* cells, and protocols for each are presented below. There is no information on storage of other bacteria prior to competence induction, but investigators using them for plasmid transformation are advised to consider these conditions as well.

Protocol 1. Frozen Stocks at −80°

1. Streak single cells on an SOB agar plate and incubate to allow colonies of 2–3 mm diameter to form. Pick several colonies and inoculate 5–10 ml of SOB medium (see Section II, C,2 for formulations).
2. Incubate the culture at 37° with agitation until the culture is midway into the logarithmic phase of growth (1–2 × 10⁸ cells/ml; at this density the cells are visible as a cloudy suspension).
3. Dilute the culture 1 : 1 with a solution that is 60% SOB medium/40% glycerol (v/v), to give a final concentration of 20% glycerol.
4. Aliquot 0.5 to 1-ml volumes of the cell suspension into a series of 2-ml screw-cap polypropylene tubes or 1.5-ml microcentrifuge tubes. Chill the tubes on ice and flash freeze them in a dry ice/ethanol bath or in liquid nitrogen. Place the tubes at −80°.
5. To retrieve cells, remove a tube from the freezer and, if possible, place on dry ice. Quickly scrape a clump of frozen cells from the surface of the frozen suspension, using, for example, a sterile yellow micropipette tip or a flamed tungsten loop. Immediately return the tube to the freezer, to minimize thawing of the contents.
6. Touch the frozen clump of cells (now thawing) onto an SOB agar plate, where it will immediately thaw. Spread the puddle of medium to disperse the cells, then incubate the plate at 37° to establish colonies.

Notes. Each tube of frozen cells can be repeatedly used as a source of cells. However, the warming during removal from the freezer will affect cell viability over time. Thus, it is recommended that a serially numbered set of tubes be prepared initially, with each tube used for streaking cells 10–20 times, after which the next in order is used. However, if the titer of viable cells in a clump declines precipitously (i.e., only a few colonies develop) before this number of retrievals is reached, the tube should nevertheless be discarded and another aliquot used in its place. The storage lifetime of cells under these conditions is essentially indefinite, and is at least 5–10 years, provided that the cells do not experience any freeze–thaw cycles.

Protocol 2. Storage in Agar Stab Bottles

1. Prepare a solution of SOB medium plus 0.8% Bacto-agar. The components can be added and the suspension heated briefly in a microwave oven to dissolve the agar. (The formulation of SOB is given in Section II,C,2.)

2. Aliquot 4-ml volumes of the hot medium into 5-ml glass Bijou bottles (or other screw-cap bottles that can be tightly sealed). Autoclave to sterilize the bottles and their contents. Leave the bottles loosely capped both during cooling and overnight at room temperature to allow evaporation of condensed water. Then seal tightly and store in the dark.

3. To prepare stabs, streak cells on an SOB/agar plate and incubate to develop fresh 2–3 mm diameter colonies. Pick several colonies with a flame-sterilized tungsten inoculating loop. Flame the cap of the stab bottle. Remove the cap, and stab the loop of cells into the agar. Quickly remove the loop and replace the cap, screwing it tightly to seal the bottle.

4. Store the stab bottle in the dark at room temperature. Do not store stabs of *E. coli* at 4°. This temperature reduces long-term viability and affects the reliability of transformations originating from cells stored under such conditions.

5. To retrieve cells, flame the bottle cap briefly and flame sterilize a tungsten inoculating loop. Remove the cap, stab the loop into the agar, remove the loop, and replace the cap. Streak the loop of cells on a SOB agar plate to disperse the cells, then incubate at 37° to establish colonies.

Notes. Escherichia coli stored in the dark at room temperature in Bijou bottles with 4 ml of medium have a remarkably long lifetime, up to 5 years with many strains. For transformation experiments, it is recommended that fresh stabs be prepared every year. It is always advised that passaging from stab to stab be separated by streaking on agar plates, such that pure colonies are used to inoculate the fresh stab. This prevents contaminants from overgrowing the culture. We recommend Bijou bottles which are made by United Glass Company (Stock Number 528), they are available from Johns Scientific (175 Hanson Street, Toronto, Ontario M4C 1A7, Canada).

B. Plating and Storage of Transformed Cells

1. Plating. *Escherichia coli* cells are somewhat fragile following the transformation procedures, and it is advisable to treat them gently immediately following the heat or electroshock and 37° recovery periods. In particular the vigorous spreading of cells on a dry agar plate can significantly reduce the number of transformed cells that establish colonies following a transformation. Therefore, the following procedure, or variations thereon, is recommended.

Protocol 3. Plating Transformed Cells

1. Spot 150–200 μl of growth medium (e.g., SOB) in the middle of a 10-cm agar plate. Aliquot the desired volume of the transformed cell suspension into the middle of the puddle.
2. Quickly spread the puddle evenly over the plate using an L-shaped Pasteur pipette (bent in a Bunsen burner). The spreading should be completed before the medium dries into the plate, and it can be visualized by an even surface reflection off a ceiling light (or hood light if plating in a biosafety hood). A plating wheel is useful for spreading the cell suspension evenly.
3. Leave the plates at room temperature for 5–10 min to allow the cell suspension to dry. Then invert the plates and incubate at 37° to develop colonies of transformed cells.

Notes. If an entire transformation is to be spread on a single 10 cm plate, the cells should first be concentrated by gentle centrifugation (5 min at 600-800 g, or 1500-2000 rpm in a table-top centrifuge at 4°). After decanting the supernatant, gently resuspend the pellet of cells with 250 μl of SOB/SOC medium, and plate as described above, omitting the initial puddle of medium.

2. Storage of Transformed Cells. Chemically transformed cells in their mixture of transformation buffer (20%) and SOC medium (80%) can be stored overnight at 4° with some loss of viability, which varies with strain and transformation procedure. However, it is advisable to wash the cells and remove the transformation chemicals, as exemplified by the following protocol. (This procedure is unnecessary for cells transformed by electroshock treatment.)

Protocol 4. Storage of Transformed Cells

1. Pellet the transformed cells in a clinical centrifuge (5 min, 1500–2000 rpm, 600–800 g at 4°). Decant the supernatant carefully, so as not to dislodge the cell pellet.
2a. For short-term storage, resuspend the cells in SOC medium by gently vortexing or pipetting. Place the cells at 4° for up to several days (the viability of the transformed cells should remain >90%).
2b. For long-term storage, resuspend the cell pellet in a mixture of 80% SOB medium/20% redistilled glycerol, transfer the cell suspension to a 1.5 to 2-ml polypropylene tube (preferably screw-capped), chill on ice, flash freeze in a dry ice/ethanol bath, and place at −80°.

C. Reagents and Growth Media

Both chemical and electroshock methods of transformation show some sensitivity to the quality of reagents used in the transformation buffers. This is particularly true for the high-efficiency protocols which employ dimethyl sulfoxide (DMSO), where it appears that organic contaminants, including oxidation products of DMSO itself, are highly inhibitory to the induction of competence for plasmid transformation. With regard to growth medium, there has also been variability noted in lots of digested casein (tryptone) and extracts of yeast with regard to competence induction. Besides possible effects of reagent quality, the transformation methods differ in the benefits of elevated levels (20 mM) of magnesium during cell growth prior to competence induction. The protocols using standard transformation buffer (TFB) or frozen storage buffer (FSB) and DMSO benefit from added magnesium, whereas the chemical protocol optimized for *E. coli* MC1061 derivatives and the electrocompetence protocol require growth without added magnesium.

1. Quality and Suppliers. As a general rule ultrapure or reagent-grade chemicals should be employed, and most major suppliers are satisfactory (including Fluka, Ronkonkoma, NY; Mallinkrodt, St. Louis, MO; and Fisher, Fairlawn, NJ). The quality of the water is very important for the procedures using DMSO and can impact on all of the methods. In general we recommend the use of water purified by reverse osmosis, as this procedure effectively removes not only salts but also organic contaminants, including those with similar volatility to water, which are not removed by distillation. However, it is important that the carbon filters which remove organics be replaced regularly, as there is no gauge to measure their saturation (except drastically reduced competence when it occurs). Distilled or deionized water can also be further purified in the laboratory with carbon, as described by Hanahan.[3]

Dimethyl sulfoxide (DMSO) is subject to oxidation, and its oxidation products can be a major source of variability in the levels of competence induced by procedures which employ it. Note also that DMSO dissolves polystyrene, and therefore tubes composed of this material cannot be used. We recommend that DMSO be purchased in the smallest possible amounts (e.g., 100-ml bottles) at spectroscopy grade, aliquoted to fill small (0.5 ml) polypropylene tubes completely, and then stored at $-80°$. A tube of DMSO is thawed, used for that day, and then discarded. Among the adequate suppliers of DMSO are Fluka, Mallinkrodt, and MCB. Similarly, the dithiothreitol (DTT) used in the TFB method (Protocol 6) is also subject to impurities which inhibit competence. One of the most reliable suppliers of DTT is Calbiochem (La Jolla, CA). Impurities in glycerol products can affect all of the competence and storage protocols used here, and it is recommended that redistilled or spectroscopy grades of glycerol be used. Among the recommended suppliers are Fluka and BRL (Gaithersburg, MD).

Other variables which can influence competence induction are in the flasks and tubes used for culturing and treating the cells during transformation. Of particular concern are soap residues in glassware and surfactants used in manufacturing some brands of polypropylene tubes. It is recommended that glassware used in transformation procedures be thoroughly rinsed with distilled water, and autoclaved half full of water, to prevent deposition of surface residues (including soap). For *E. coli* we do not use glassware washing services but rather wash by hand without soap. Among the reliable suppliers of polypropylene tubes for transformation procedures are Falcon (Oxnard, CA) and Corning (Corning, NY).

The components of the growth media can also influence transformation. We routinely use Bacto-tryptone and Bacto yeast extract. Alternative suppliers should be carefully compared, as lot variations have been reported with regard to transformation experiments. Again, reagent-grade chemicals and the purest available water are recommended.

2. Formulations of Media Used for Cell Growth and Recovery in Transformation Procedures. All the transformation protocols described herein use a common set of growth media, with the only major distinction

---

[3] D. Hanahan, *in* "DNA Cloning Techniques" (D. Glover, ed.), p. 109. IRL Press, London, 1985.

being the presence or absence of elevated levels of magnesium. These differ from many traditional media in that the levels of sodium are low. The recovery of cells from all of these procedures is enhanced in SOC medium, which contains glucose. We recommend that these media be filter sterilized just prior to use, to remove particulate matter and any slow-growing contaminating microorganisms. Each medium is identical to or derived from a stock of SOB–Mg medium (SOB without magnesium), and their formulations are as follows.

Compound	Amount/liter	Final concentration
Growth medium SOB–Mg		
Bacto-tryptone	20 g	2.0%
Bacto yeast extract	5 g	0.5%
NaCl	0.58 g	10.0 mM
KCl	0.19 g	2.5 mM
Growth medium SOB		
SOB–Mg medium	1 liter	99%
MgCl$_2$ and MgSO$_4$	10 ml of a 2 M stock	20 mM
Recovery medium SOC		
SOB–Mg medium	1 liter	98%
MgCl$_2$ and MgSO$_4$	10 ml of a 2 M stock	20 mM
Glucose	10 ml of a 2 M stock	20 mM

Preparation. Combine tryptone, yeast extract, NaCl, and KCl in the purest available water. Aliquot into prerinsed glass flasks and autoclave for 30–40 min to generate SOB–Mg medium. Sterilize by filtration just prior to use for protocols requiring this medium itself. Prepare a 2 M stock solution of Mg^{2+} by combining 203 g MgCl$_2$ · 6H$_2$O and 247 g MgSO$_4$ · 7H$_2$O per liter of purest available water and sterilize by filtration through a prerinsed 0.2-μm filter unit. Similarly prepare a 2 M stock of glucose (360 g/liter with purest water), sterilize by filtration, and store frozen in aliquots. For SOB and SOC, combine the magnesium and glucose as specified with autoclaved SOB–Mg medium, then sterilize by filtration through a 0.2-μm filter unit just prior to use. It is advisable to use detergent-free filter units for all of these purposes. The final pH of the media should be 6.8 to 7.2.

TB Medium for Plasmid Preparations and Stabilizing Unstable Inserts. TB medium[4] has proved to be beneficial for maximizing the yield of plasmid DNA from *E. coli* transformed with plasmids based on the high copy number pUC vectors. In addition, this medium facilitates the stabilization

[4] R. Tartof and C. Hobbs, *BRL Focus* **9**, 2 (1987).

of inserts that are unstable and prone to rearrangement, as discussed in Section VI,B,4.

PLASMID PREPARATION MEDIUM TB

Compound	Amount/liter	Final concentration
Solution a		
Bacto-tryptone	12 g	1.2%
Bacto yeast extract	24 g	2.4%
Redistilled glycerol	4 g	0.4%
Combine in 900 ml water and autoclave		
Solution b		
KH_2PO_4	2.3 g	17 mM
K_2HPO_4	12.5 g	72 mM
Dissolve the two potassium phosphates in 100 ml water and autoclave		

When solutions (a) and (b) are cool, combine them (1 : 1) to give complete TB medium

III. Protocols for Preparing Competent *Escherichia coli* Using Chemical Treatments

The original discovery by Mandel and Higa[1] that incubation of *E. coli* at low temperatures in the presence of calcium chloride induced a state of competence for DNA uptake (phase transfection) has in subsequent years been extended to include a variety of different chemicals and protocols for inducing competence for stable plasmid transformation. There have proved to be two major alternative chemical treatment regimens for inducing high degrees of competence. Each seems to be favored by different strains of *E. coli*; it is likely that genetic differences in the cell envelope determine susceptibility to transformation by one or the other of these conditions (J. Jessee and F. Bloom, unpublished results). The first method extends the simple Ca^{2+}-based procedure through the utilization of manganese and potassium in addition to calcium. The second method originated from a series of experiments into the mechanisms of transformation, conducted primarily with an *E. coli* strain called DH1.[5] This protocol produced very high transformation efficiencies using DH1 and many other strains. It is considerably more complex in its conditions, in that DMSO, DTT, and hexamine cobalt trichloride are employed in addition to those used in the simple Mn^{2+}/Ca^{2+}-based version. Each method is suitable for either immediate transformation or for frozen storage of competent cells, and

[5] D. Hanahan, *J. Mol. Biol.* **166,** 557 (1983).

variations for each purpose are described below. In addition, two convenient, rapid transformation protocols for routine use are described. The application of the various protocols for different strains and purposes is then considered.

One significant new development in the methodology for preparing frozen competent cells is the observation that growth of the cells at lower temperatures than 37° can improve competence following a freeze–thaw cycle.[5a,5b] We have found that growth of *E. coli* at a temperature of 30° is preferable to 37° for most of the strains described below, using either protocols 5 or 7.[5a,5c] Following the submission of this review, Inoue *et al.*[5b] reported that growth at 18° is optimal for a new procedure which they have established (at press time we have not been able to evaluate this method). Thus, cell growth at lower temperatures than 37° can be added to the incubation in transformation buffers at 0° and the 42° heat pulse (or heat shock) as significant conditions for the induction of competence for transformation.

A. Moderate Efficiency Methods Based on Mn^{2+}/Ca^{2+} Treatment

MC1061[6] and its derivative, DH10B,[7] are preferentially transformed by a modification of the Mandel and Higa method.[1] These strains differ significantly from those derived from Hoffman Berling strain 1100[8] (e.g., MM294, DH1, DH5, DH5α, DH5αMCR) as well as from many other strains (e.g., HB101, C600) in that Mg^{2+} is *not* beneficial in the growth medium, and the addition of either DMSO or DTT to the transformation buffer reduces competence. The following protocol works very well for MC1061 derivatives and may be preferable for certain other *E. coli* strains as well.

Protocol 5: Calcium/Manganese-Based (CCMB) Transformation for MC1061 Derived Strains

1. Pick several 2–3 mm diameter colonies from a freshly streaked plate into 1 ml of SOB–Mg growth medium (SOB without Mg). Avoid collecting agar fragments. Vortex gently to disperse cells.

[5a] J. Jessee and F. Bloom, U.S. Patent #4,981,797 (1991).
[5b] H. Inoue, H. Nojima, and H. Okayama, *Gene* **96**, 23 (1990).
[5c] J. Jessee and F. Bloom, unpublished observations.
[6] M. Casadaban and S. Cohen, *J. Mol. Biol.* **138**, 179 (1980).
[7] S. G. N. Grant, J. Jessee, F. R. Bloom, and D. Hanahan, *Proc. Natl. Acad. Sci. U.S.A.* **87**, 4654 (1990).
[8] M. Meselson and R. Yuan, *Nature (London)* **217**, 1110 (1968).

2. Inoculate the dispersed colonies into a shake flask containing SOB–Mg medium (50 ml medium in a 500-ml nonbaffled Erlenmeyer flask, or 200 ml in a 2.8-liter Fernbach flask).

3. Incubate at 275 rpm, 30° until the OD_{550} reaches 0.3, which corresponds to 5×10^7 cells/ml for MC1061.

4. Collect the cell suspension into sterile 50-ml polypropylene centrifuge tubes and chill on ice for 10 min. (Take a 10-μl aliquot to determine the density of viable cells by plating a 10^6 dilution onto a SOB agar plate.)

5. Pellet the cells at 750–1000 g (2000–3000 rpm in a clinical centrifuge) for 10–15 min at 4°. Decant the supernatant and invert the tubes to remove excess culture medium.

6. Disperse the cells in 1/3 volume of CCMB 80 (see below) by gentle vortexing or rapping of the centrifuge tube.

7. Incubate on ice for 20 min.

8. Centrifuge for 10 min at 750–1000 g at 4°. Decant as in Step 5.

9. Resuspend the cells in CCMB 80 at 1/12 of original volume.

10. The competent cells can be used immediately or aliquoted in 1-ml volumes into chilled 2-ml screw-capped polypropylene tubes, flash frozen in a dry ice/ethanol bath, and then stored at −80°.

For immediate transformation

11. Aliquot 200-μl volumes of the competent cell suspension into chilled Falcon 2059 polypropylene tubes. Incubate on ice for 10 min.

12. Add DNA in less than 20 μl of 0.5× TE buffer. Incubate on ice for 30 min (0.5× TE is 5 mM Tris, 0.2 mM EDTA, pH 7.4).

13. Heat-shock in a water bath at 42° for 90 sec. Place on ice for 2 min.

14. Add 800 μl of SOC medium. Incubate at 37° with mild agitation for 60 min. Then plate onto SOB agar plates with appropriate drug selection and incubate at 37° to develop colonies of the transformed cells.

For frozen competent cells

10a. Remove frozen competent cells from the freezer. Thaw on ice. Then follow steps 11–14 above.

Materials

TRANSFORMATION BUFFER CCMB 80

Compound	Amount/liter	Final concentration
$CaCl_2 \cdot 2H_2O$	11.8 g	80 mM
$MnCl_2 \cdot 4H_2O$	4.0 g	20 mM
$MgCl_2 \cdot 6H_2O$	2.0 g	10 mM
Potassium acetate	10 ml of a 1 M stock (pH 7.0)	10 mM
Redistilled glycerol	100 ml	10% (v/v)

Preparation. First prepare a 1 M solution of potassium acetate and adjust to pH 7.0 using KOH, then sterilize by filtration through a prerinsed 0.2-μm membrane and store frozen. Then prepare a solution of 10 mM potassium acetate, 10% glycerol (v/v) using these reagents and the purest available water. Add salts as solids and allow each to enter into solution before adding the next. Adjust pH to 6.4 with 0.1 N HCl. Do not adjust pH upward with base. Sterilize the solution by filtration through a prerinsed 0.22-μ filter and store at 4°.

The formulation of SOB–Mg is given in Section II,C,2 above.

B. High-Efficiency Transformation Methods Utilizing Complex Conditions (TFB/FSB)

The high-efficiency transformation procedures described below are effective with many *E. coli* strains, including several which have wide applicability in molecular cloning experiments (DH5, DH5α, HB101). The method and its evaluation has been described in considerable depth previously.[3,5] Moreover, detailed descriptions of the evaluation and optimization of this method, as well as troubleshooting strategies, important parameters, and variations (e.g., for the *E. coli* strain χ1776), can be found in these papers. The critical parameters for success with this method are as follows: (1) growth of the cells in medium containing magnesium (20 mM); (2) collection of the cells during mid log-phase growth following an "appropriate" inoculation regimen; (3) use of ultrapure water lacking organic contaminants; and (4) use of DMSO free of significant oxidation. When this method is used with these parameters satisfied, very high transformation frequencies can routinely be achieved (2–3% of the plasmid molecules affecting a transformed cell; 5–10% of the cells rendered competent for transformation). On average, the immediate transformation procedure (Protocol 6) gives higher frequencies than does the frozen competent version (Protocol 7, which of necessity omits DTT). Interestingly, some commercial preparations of competent cells often equal or exceed those obtained with these protocols, and they can be considered as a valid option to preparing high-efficiency competent cells in the laboratory.

Protocol 6. TFB-Based Chemical Transformation Protocol

1. Pick several 2–3 mm diameter colonies from a freshly streaked SOB agar plate and disperse in 1 ml of SOB medium by vortexing. Use one colony per 10 ml of culture medium. The cells are best streaked from a frozen stock or fresh stab about 16–20 hr prior to initiating liquid growth.

2. Inoculate the cells into an Erlenmeyer flask containing SOB medium. Use a culture volume to flask volume ratio between 1 : 10 and 1 : 30 (e.g., 30–100 ml in a 1-liter flask).

3. Incubate at 37° with moderate agitation until the cell density is 4–7 × 10^7 viable cells/ml (OD_{550} 0.4 for DH5, 0.5 for DH5α and DH5αF').

4. Collect the culture into 50-ml polypropylene centrifuge tubes (such as Falcon 2070 tubes) and chill on ice for 10–15 min. (Take a 10-μl aliquot of cells to determine the density of viable cells by plating a 10^6 dilution onto an SOB/agar plate.)

5. Pellet the cells by centrifugation at 750–1000 g (2000–3000 rpm in a clinical centrifuge) for 12–15 min at 4°. Drain the pelleted cells thoroughly, by inverting the tubes on paper towels and rapping sharply to remove any liquid. A micropipette can be used to draw off recalcitrant drops.

6. Resuspend the cells in 1/3 of the culture volume of TFB (see below) by vortexing moderately. Incubate on ice for 10–15 min.

7. Pellet the cells and drain thoroughly as in Step 5.

8. Resuspend the cells in TFB to 1/12.5 of the original volume. This represents a concentration from each 2.5 ml of culture into 200 μl of TFB.

9. Add DMSO and DTT solution (DnD, see below) to 3.5% (v/v) (7 μl per 200 μl of cell suspension). Squirt the DnD into the center of the cell suspension and immediately swirl the tube for several seconds. Incubate the tubes on ice for 10 min.

10. Add a second, equal aliquot of DnD as in Step 9 to give a 7% final concentration. Incubate the tubes on ice for 10–20 min.

11. Pipette 210-μl aliquots into chilled 17 × 100 mm polypropylene tubes (Falcon 2059 or equivalent).

12. Add the DNA solution in a volume of less than 20 μl, swirling to mix. Incubate the tubes on ice for 20–40 min. (Ligations should be diluted or precipitated in ethanol.)

13. Heat-shock the cells by placing the tubes in a 42° water bath for 90 sec. Return the tubes to ice to quench the heat shock, allowing 2 min for cooling.
14. Add 800 μl of SOC medium to each tube. Incubate at 37° with moderate agitation for 30–60 min. Spread the cells on agar plates containing appropriate antibiotics (or other conditions) to select for transformants. The incubation period should be omitted for M13 transfections.

Materials

STANDARD TRANSFORMATION BUFFER (TFB)

Compound	Amount/liter	Final concentration
KCl (ultrapure)	7.4 g	100 mM
MnCl · 4H$_2$O	8.9 g	45 mM
CaCl$_2$ · 2H$_2$O	1.5 g	10 mM
HACoCl$_3$	0.8 g	3 mM
Potassium MES (Final pH 6.20 ± 0.10)	20 ml of 0.5 M stock (pH 6.3)	10 mM

Preparation. Equilibrate a 0.5 M solution of MES [2(N-morpholino) ethane sulfonic acid] to pH 6.3 using concentrated KOH, then sterilize by filtration through a 0.2-μm membrane and store in aliquots at $-20°$. Make a solution of 10 mM potassium MES, using the 0.5 M MES stock and the purest available water. Add the salts as solids, then filter the solution through a 0.22-μm prerinsed membrane. HACoCl$_3$ is hexamminecobalt trichloride. Aliquot into sterile flasks and store at 4°. (TFB is stable for over 1 year.)

DMSO AND DTT SOLUTION (DnD)

Compound	Amount/10 ml final volume	Final concentration
DTT	1.53 g	1 M
DMSO (spectroscopy grade)	9 ml	90% (v/v)
Potassium acetate	100 μl of a 1 M stock (pH 7.5)	10 mM

The formulation of SOB medium is given in Section II,C,2.

Protocol 7. FSB-Based Frozen Storage of Competent Cells

1. Pick several 2–3 mm diameter colonies from a freshly streaked SOB agar plate and disperse in 1 ml of SOB medium by vortexing.

Use one colony per 10 ml of culture medium. The cells are best streaked from a frozen stock or fresh stab about 16–20 hr prior to initiating liquid growth.

2. Inoculate the cells into an Erlenmeyer flask containing SOB medium. Use a culture volume to flask volume ratio between 1 : 10 and 1 : 30 (e.g., 30–100 ml in a 1-liter flask).

3. Incubate at 30° with moderate agitation until the cell density is 6–9 × 10^7 viable cells/ml (OD_{550} 0.4 for DH5, 0.5 for DH5α and DH5αF').

4. Collect the culture into 50-ml polypropylene centrifuge tubes (Falcon 2070 tubes) and chill on ice for 10–15 min. (Take an aliquot of cells to determine viable cell density by plating a 10^6 dilution onto an SOB agar plate.)

5. Pellet the cells by centrifugation at 750–1000 g (2000–3000 rpm in a clinical centrifuge) for 12–15 min at 4°. Drain the pelleted cells thoroughly, by inverting the tubes on paper towels and rapping sharply to remove any liquid. A micropipette can be used to draw off recalcitrant drops.

6. Resuspend the cells in 1/3 of the culture volume of frozen storage buffer (FSB, see below) by vortexing moderately. Incubate on ice for 10–15 min.

7. Pellet the cells as before and drain thoroughly as in Step 5.

8. Resuspend the cells in FSB to 1/12.5 of the original volume. (Each 2.5 ml of culture is concentrated into 200 μl of FSB.)

9. Add DMSO to 3.5% (v/v) (7 μl per 200 μl of cell suspension). Squirt the DMSO into the center of the cell suspension and immediately swirl the tube for several seconds. Incubate the tube on ice for 5 min. (DTT is not used in this protocol.)

10. Add a second, equal aliquot of DMSO, as before, giving a 7% final concentration. Incubate on ice for 10–15 min.

11. Pipette 210-μl aliquots into chilled screw-cap polypropylene tubes, 1.5-ml microcentrifuge tubes, or snap-cap polypropylene tubes.

12. Flash freeze by placing the tubes in a dry ice/ethanol bath or into liquid nitrogen for several minutes. (Be careful that the ethanol does not get inside the tubes. An option is to avoid total immersion but instead just set the bottom half of the tubes into the bath.)

13. Transfer tubes to a −80° freezer.

Use of frozen competent cells

1. Remove a tube(s) from the freezer and thaw on ice. If small vials were used, it is recommended that the cells be transferred in 200-μl aliquots to 12-ml polypropylene tubes (e.g., Falcon 2059).

2. Add the DNA solution in a volume of less than 20 μl. Swirl the tube to mix the DNA evenly with the cells.
3. Incubate the tube(s) on ice for 10–30 min.
4. Heat-shock the cells by placing the tubes in a 42° water bath for 90 sec, and then chill by returning the tubes immediately to 0° (crushed ice).
5. Add 800 μl of SOC medium and incubate at 37° with moderate agitation for 30–60 min.

Materials

FROZEN STORAGE BUFFER (FSB)

Compound	Amount/liter	Final concentration
KCl	7.4 g	100 mM
MnCl$_2$ · 4H$_2$O	8.9 g	45 mM
CaCl$_2$ · 2H$_2$O	1.5 g	10 mM
HACoCl$_3$	0.8 g	3 mM
Potassium acetate	10 ml of a 1 M stock (pH 7.5)	10 mM
Redistilled glycerol (Final pH 6.20 ± 0.10)	100 g	10% (w/v)

Preparation. Equilibrate a 1 M solution of potassium acetate to pH 7.5 using KOH, then sterilize by filtration through a 0.2-μm membrane and store frozen. Prepare a 10 mM potassium acetate, 10% glycerol solution using this stock and the purest available water. Add the salts as solids, and adjust the pH (if necessary) to 6.4 using 0.1 N HCl. Do not adjust the pH upward with base. (The pH may drift for 1–2 days before settling at 6.1–6.2.) Sterilize the solution by filtration through a prerinsed 0.2-μm filter and store at 4°. HACoCl$_3$ is hexamminecobalt trichloride. SOB and DMSO are described in Section II,C,2.

C. PEG/DMSO-Mediated Transformation

Competence induction in *E. coli* by treatment with polyethylene glycol (PEG) and DMSO is a convenient procedure which produces moderate transformation frequencies.[9] This method may also be useful for transformation of other gram-negative microorganisms. Transformation efficiencies of 5×10^6 to 10^8 transformants/μg can be obtained with many *E. coli* strains. No heat shock is required for this procedure.

[9] C. Chung, S. Neimela, and R. Miller, *Proc. Natl. Acad. Sci. U.S.A.* **86**, 2172 (1989).

Protocol 8. PEG/DMSO One-Step Transformation Procedure

1. Incubate *E. coli* at 37° in LB broth to OD_{550} 0.4–0.5, corresponding to a cell density of 5 × 10^7 cells/ml.
2. Pellet the cells by centrifugation at 1000 *g* for 10 min at 4°. Resuspend in 1/10 volume cold transformation and storage solution (TSS, see below).
3. Incubate the cells on ice for 20 min.
4. Transfer 100-μl aliquots of the cell suspension into chilled polypropylene tubes and mix with 100 pg–1 ng DNA. Incubate on ice for 30 min. (Any remaining cells can be frozen in dry ice/ethanol and stored at −80° for subsequent use.)
5. Add 0.9 ml LB medium containing 20 mM glucose (or use SOC) and incubate at 37° with moderate agitation (225 rpm) for 1 hr.
6. Plate cells, diluting into LB or SOC if necessary.

Materials

TSS COMPONENTS

Compound	Amount/liter	Final concentration in TSS
a. LB broth		
Bacto-tryptone	10 g	~0.85%
Bacto yeast extract	5 g	~0.4%
NaCl	5 g	~8 mM
Dissolve in 1 liter ultrapure water and autoclave		
b. 2.0 M Mg^{2+} solution		
1 M MgCl$_2$ · 6H$_2$O	203 g	10 mM
1 M MgSO$_4$ · 7H$_2$O	247 g	10 mM
Dissolve in 1 liter ultrapure water and filter sterilize		
c. 2 M glucose solution	360 g	20 mM
Prepare in 1 liter ultrapure water and sterilize by filtration		
d. Polyethylene glycol (PEG)	Use either molecular weight 3350 or 8000; add as a solid directly into TSS	10%
e. Dimethyl sulfoxide (DMSO)	Ultrapure spectroscopy grade	5%

Preparation of Transformation and Storage Solution (TSS)

1. Add solid PEG to LB to make a 10% (w/v) solution.
2. Add an aliquot of the 2.0 M Mg^{2+} solution to achieve a final concentration of 20 mM.
3. Measure the pH, which should be about 6.8. If the pH is higher, the solution can be titrated with 1.0 N HCl. (The final pH of the TSS should be between 6.5 and 6.8.)
4. Filter sterilize the solution through a standard 0.45-μm filter unit.
5. Add DMSO to the filtered solution to a final concentration of 5% (v/v). Store TSS at 4° or on ice until ready to use. Alternatively, the PEG, Mg^{2+}, and DMSO can be added to the LB, the pH adjusted, and the solution sterile filtered through a 0.2-μm nylon filter unit.

D. Rapid Colony Transformation

It is possible to collect a few colonies of cells from an agar plate, disperse them in transformation buffer, chill on ice to induce competence, and add DNA to effect transformation. This procedure is extremely fast and easy. It is limited in efficiency, however, and typically gives transformation efficiencies (XFE values) of $10^4/\mu$g pBR322. It is not recommended for primary cloning experiments but is very useful for reintroducing cloned plasmid DNA back into *E. coli* prior to growing up large-scale cultures for plasmid DNA purification and other purposes.

Protocol 9. Rapid Colony Transformation

1. Pick several colonies (or a clump of cells) from a plate using a tungsten inoculating loop or a wooden applicator stick, being careful to take no agar along with the cells.
2. Disperse the colonies in 200 μl of chilled standard transformation buffer (TFB) by vigorous vortexing or by repeated pipetting (CCMB 80 or FSB can be used instead of TFB).
3. Incubate the cells on ice for 10 min.
4. Add DNA solution (10–1000 ng) in less than 20 μl, swirl to mix, and incubate on ice for 10 min.
5. Heat-shock the cells at 37–42° for 90 sec. (This step is optional if >100 ng of DNA is used.)
6. Add 400–800 μl SOC medium and incubate 20–60 min at 37°. (This step is also optional and unnecessary if >100 ng of DNA is used in the transformation.)
7. Plate several fractions (e.g., 1%, 5%, 25%) on appropriate selective media. Incubate to establish colonies.

E. Choice of Transformation Method

Bacterial transformations can be divided into three classes with regard to purpose and consequent requirements for transformation efficiency: (1) primary cloning of high-complexity plasmid populations, for example, cDNAs representing an mRNA population or fragments of the genome of an organism, (2) cloning of more restricted populations of plasmids, for example, following *in vitro* manipulations such as oligonucleotide mutagenesis, or subcloning fragments of previously cloned DNAs, and (3) reintroduction of a pure plasmid population (a single clone) for purposes such as the preparation of large quantities of plasmid DNA. The latter purpose does not require cells with high competence, and various methods will suffice. Among these are the rapid colony transformation (Protocol 9), the PEG/DMSO one-step procedure (Protocol 8), or methods employing frozen competent cells (Protocols 5 and 7). If frozen competent cells are used, we suggest older lots which have decayed in competence, since these are nevertheless more than adequate for the purpose. Alternatively, a small clump of highly competent frozen cells can be quickly scraped off the surface of a frozen aliquot, which is returned immediately to the freezer without thawing. A 5- to 10-μl clump of high-competence cells is sufficient for retransforming a cloned plasmid into *E. coli*.

The intermediate purpose of manipulating and modifying previously cloned DNA usually requires transformation efficiencies in the range of 10^7 to 10^8 transformants/μg. These values are achieved with the immediate and frozen storage versions of the high-efficiency transformation methods (Protocols 5–7), and to some extent with the PEG/DMSO one-step procedure (Protocol 8). Experience will to some extent dictate the choice of method. If insufficient numbers of transformants are being generated, the options are to increase either the number of discrete transformations performed or the amount of DNA being used, or to change the type of transformation procedure (to improve the competence). For some types of *in vitro* modifications, very high efficiencies may be necessary to obtain the desired number of transformants, in which case the highest efficiency methods should be used.

Transformations which are intended to represent a high-complexity population or transformations with very small amounts of DNA (e.g., plasmid rescue from total genomic DNA) require very high efficiency transformation protocols. It is increasingly clear that the electroshock transformation method described below is the most efficient, and therefore it is the method of choice for such purposes. The method suffers from the requirement for a high-voltage electroporation device, and from the large volume of cells that must be grown during the preparation of electrocompetent cells. The latter problem is being obviated by the availability of commercial preparations of concentrated cells for electroshock transformation. A protocol for electroshock transformation is given in Section IV,A (Protocol 10).

An alternative to electroshock transformation is the use of the high-efficiency chemical transformation procedures under conditions that produce optimal competence. Both Protocol 7 with DH5α and Protocol 5 with DH10B will produce transformation efficiencies in excess of 10^9 transformants/µg pBR322, provided that all the important parameters of the protocols are carefully satisfied. Given careful attention to the transformation protocols, it is routinely possible to achieve high-efficiency competent cell preparations. It is noted in addition that frozen competent cells in all four competence ranges are commercially available, including maximally efficient chemically competent cells and electrocompetent cells. This alternative may prove reasonable for investigators who prefer to avoid the rigors of achieving routine high-efficiency transformations. However, all of the procedures presented herein will provide reliable transformations, given commitment to the details of the protocols.

F. Optimization of Competence for New Escherichia coli Strains

If the ultimate goal of a transformation experiment requires the use of a strain of *E. coli* that has not been described here, nor in previous reports by Hanahan,[3,5] it may be necessary to determine which transformation protocol is most suited for the strain. It is first recommended that trial transformations be performed on the new strain using Protocols 5 (CCMB based) and 6 (TFB based). For the CCMB protocol cells should be grown without added magnesium (in SOB–Mg), whereas for the TFB-based version complete SOB medium is recommended. In addition, with the TFB version, the procedure can be tried without DMSO, without DTT, or without both. Both the density at which the cells are collected following their growth in SOB medium ($+/- Mg^{2+}$) and the temperature at which the cells are grown can be varied. Regarding growth temperature, based on recent results one could consider testing new strains over a range from 18° to 37°.[5a,5b] If neither of these protocols gives adequate competence, the PEG/DMSO one-step procedure (Protocol 8) could be tried, as well as other variations on transformation buffers given by Hanahan.[3] Finally, electroshock transformation can be assessed, using the procedure given below (Protocol 10) or the variations suggested in Section V,C. It is our experience that most *E. coli* strains can be rendered competent to at least 10^7 transformants/µg, given a serious optimization process. However, unless the intended purpose specifically requires a specific new strain, the strains recommended in Table II should be considered, as they have proved to be reliable hosts for plasmid tranformation.

IV. Induction of Competence with Electroshock Treatment

The treatment of cells with a brief pulse of high-voltage electricity has been found to permeabilize them toward entry of a variety of macromolecules.[10,11] It is presumed that the discharge of a voltage potential across a field which includes cells transiently depolarizes their membranes and induces pores which can be entry points for macromolecules. Such electroshock treatment has recently been shown to be applicable to competence induction of *E. coli* and other bacteria. For *E. coli*, electroshock transformation is the most efficient method available and approaches the theoretical maximum of 100% cell transformation frequencies. Efficiencies of greater than 10^{10} transformants/μg of pUC18/19 have been reported,[12,13] and greater than 80% of all cells can be transformed by this method. More recent improvements in electronics, *E. coli* strains, and procedures have increased the transformation efficiency up to 5×10^{10}, as is described below.[14] For other bacteria, electroporation is the only method presently available for the efficient introduction of plasmid DNA.

The exposure of a dense suspension of bacterial cells and plasmid DNA to a high-strength electric field of short duration has been found to induce DNA uptake and to elicit competence for stable plasmid transformation. The generation of an electroshock is usually accomplished by the discharge of a high-voltage capacitor through a mixture of bacterial cells and DNA suspended between two electrodes. The pulse length of the capacitor discharge can be varied by increasing the capacitor size or the resistance in the circuit itself, which includes the mixture of cell suspension and DNA. A parallel resistor can also be used to modulate the resistance of the electroshock circuit. The time of the electric current pulse (the shock) is described by a decay time constant τ, which corresponds to the time at which the voltage has dropped to approximately 37% of its original value. The time constant of the electroshock is determined by the product ($\tau = RC$) of the resistance R (both of the cell/DNA mixture and any parallel resistor) and the capacitance C of the circuit through which the electric field is being discharged.

[10] D. Knight, in "Techniques in Cellular Physiology" (P. F. Baker, ed.), pp. 1–20. Elsevier, Amsterdam, 1981.

[11] E. Neumann, M. Schaefer-Ridder, Y. Wang, and P. Hofschneider, *EMBO J.* **7,** 841 (1982).

[12] W. Dower, W. Miller, and C. Ragsdale, *Nucleic Acids Res.* **16,** 6127 (1988).

[13] W. Calvin and P. Hanawalt, *J. Bacteriol.* **170,** 2796 (1988).

[14] M. Smith, J. Jessee, T. Landers, and J. Jordan, *Focus* **12,** 38 (1990).

TABLE II

RECOMMENDED *Escherichia coli* STRAINS FOR GENERATING cDNA/GENOMIC LIBRARIES

Strain	Relevant genetic loci[b]								Maximal competence (XFE, colonies/μg pBR322)		Ref.
	$recA$[a]	$hsdR$	$mcrA$	$mcrB$	mrr	$gyrA$	$deoR$	$lacZ\ \Delta M15$[c]	Chemical treatment[d]	Electroshock[e]	
RR1	+	−	+	−	−	+	+	−	0.5–2×10^8	1–2×10^{10}	f
HB101	−(13)	−	+	−	−	+	+	−	0.5–2×10^8	1–2×10^{10}	g
MM294	+	−	+	+	+	+	+	−	1–3×10^8	3–10×10^9	h
DH1	−	−	+	+	+	−	+	−	1–3×10^8	3–10×10^9	i
DH5	−	−	+	+	+	−	−	−	2–10×10^8	3–10×10^9	j
DH5α	−	−	+	+	−	−	−	+	2–10×10^8	3–10×10^9	j
DH5αMCR	−	Δ	−	Δ	Δ	−	−	+	1–5×10^8	1–5×10^9	j
MC1061	+	−	−	+	+	+	+	−	1–5×10^8	1–5×10^{10}	k
MC1061/p3	+	−	−	−	+	+	−	−	1–5×10^8	1–5×10^{10}	l
DH10	−	Δ	−	Δ	Δ	+	+	+	2–10×10^8	1–5×10^{10}	j
DH10B	−	Δ	−	Δ	Δ	+	−	+	2–10×10^8	1–5×10^{10}	j
DH10B/p3	−	Δ	−	Δ	Δ	+	−	+	2–10×10^8	1–5×10^{10}	m
ER1648	+	Δ	Tn10	Δ	Δ	+	+	−	(n.d.)	(n.d.)	n

[a] Unless indicated all $recA^-$ alleles are $recA1$.

[b] Δ, Complete deletion of noted gene.

[c] Complements $lacZ$ α subunit carried in many vectors to produce active β-Gal protein (+ designates that this deletion is present).

[d] Range of values given for MC1061, DH10, DH10B and their derivatives are based on the CCMB 80 transformation protocol, whereas all other strains are rendered most competent with the TFB/FSB-based protocols.

[e] These values are obtained using electrical field strengths of 16.6 kV/cm. (Some commercial electroporation devices cannot achieve this field strength and therefore cannot produce these transformation efficiencies.)

[f] F. Bolivar, R. L. Rodriguez, P. J. Greene, M. C. Betlach, H. L. Heyneker, H. W. Boyer, J. H. Crosa, and S. Falkow, *Gene* **2**, 95 (1977).

[g] H. W. Boyer and D. Roulland-Dussoix, *J. Mol. Biol.* **41**, 459 (1969).

[h] M. Meselson and R. Yuan, *Nature (London)* **217**, 1110 (1968).

[i] D. Hanahan, *J. Mol. Biol.* **166**, 557 (1983).

[j] S. G. N. Grant, J. Jessee, F. R. Bloom, and D. Hanahan, *Proc. Natl. Acad. Sci. U.S.A.* **87**, 4645 (1990).

[k] M. J. Casadaban and S. J. Cohen, *J. Mol. Biol.* **138**, 179 (1980).

[l] B. Seed, *Nucleic Acids Res.* **11**, 2427 (1983).

[m] Bethesda Research Laboratories, Gaithersburg, MD 20877.

[n] D. M. Woodcock, P. J. Crowther, J. Doherty, S. Jefferson, E. DeCruz, M. Noyer-Weidner, S. S. Smith, M. Z. Michael, and M. W. Graham, *Nucleic Acids Res.* **17**, 3469 (1989).

A. Electroporation of Escherichia coli

Field strengths used for optimal electroshock transformation of *E. coli* range from 12.5 to 16.7 kV/cm.[12,14] The field strength is the maximum voltage applied to the cell suspension divided by the distance between the electrodes in which the cell suspension rests. The optimal combination of capacitor and resistor is different for each of these field strengths, but the decay time τ should be about 5 msec for each.

Several parameters other than the electroshock itself affect electroporation efficiencies. The *E. coli* strain MC1061,[6] and its derivative DH10B[7] give the highest efficiencies of electroporation (Table II). Growth of cells in medium without added Mg produces the highest competence. In addition, cells should be washed extensively to remove all salts, and the final cell slurry should be at a density of 5×10^{10} to 10^{11} cells/ml, with an OD_{500} above 200 units.

Protocol 10. Electroshock Transformation of *Escherichia coli*

1. Pick a single 2–3 mm colony of a freshly streaked SOB–Mg agar plate and disperse it in 1 ml of SOB–Mg by vortexing. The cells are best streaked from a frozen stock or a fresh stab about 16–20 hr prior to initiating liquid growth.
2. Inoculate the cells into a 500-ml Erlenmeyer flask containing 50 ml of SOB–Mg.
3. Incubate 37° at 275 rpm overnight (preferably <12 hr).
4. The following day use 7.5 ml of this fresh overnight culture to inoculate 750 ml of SOB–Mg in a 2.8-liter Fernbach flask (non-baffled).
5. Incubate at 37° with moderate agitation until an optical density of 0.75 OD_{550} units is reached (3–6 $\times$ 10^8 cells/ml).
6. Collect the cell suspension into chilled polypropylene centrifuge tubes.
7. Pellet the cells by centrifugation at 2600 g for 12–15 min at 4°. Carefully decant the supernatant.
8. Resuspend the cell pellet in an equal volume (to the original) of chilled 4° EWB (10% redistilled glycerol in ultrapure water). Resuspension requires vigorous agitation by vortexing or rapping against a solid object.
9. Centrifuge again to pellet the cells, the immediately decant the supernatant. (Cell loss is difficult to avoid but should be minimized.)

10. Resuspend the cell pellet as in Step 8 and centrifuge the cells, again carefully decanting the supernatant.
11. Resuspend the cell pellet with the few drops of excess liquid left in tube, and measure the volume of the cell suspension.
12. Determine the OD_{550} by diluting a small portion (e.g., 1%) of the cell suspension 300 ×. Then adjust the volume of the concentrated cell slurry with cold EWB to produce a final OD_{550} of 200–250 units/ml.
13. Dispense the cell suspension in 120-μl aliquots into cold cryotubes, freeze in a dry ice/ethanol bath, and store at −80°

Electroshock transformation
1. Thaw cells on ice, and aliquot 20 μl volumes into chilled polypropylene tubes. Add 1 μl of DNA solution in 0.5 × TE.
2. Transfer individual aliquots of cells plus DNA into chilled electroporation chambers and electroshock under optimal conditions for the strain. For DH10B, these values are 16.7 kV/cm, a resistance of 4000 ohms, and a capacitance of 2 μF. (An alternative, if 16.7 kV/cm cannot be achieved, is 12.5 KV/cm, 25 μF, and 200 ohms.)
3. Remove electroshocked cells immediately from the chamber, place in a chilled Falcon 2059 tube, and add 1 ml of SOC recovery medium. Incubate for 1 hr at 37°, 225 rpm.
4. Plate on selective medium.

Preparation of DNA for Electroportion. DNA used for electroporation should be free of phenol, ethanol, and detergents, as with the high-efficiency chemical transformation protocols. In addition, it is very important that the DNA solution being used for electroshock transformation has a very low ionic strength, and thus a high resistance. The following protocols achieve these conditions.

**Protocol 11a. Precipitation Purification of DNA for Electroshock
          Transformation**

1. For a typical 20-μl ligation reaction, add 5–10 μg tRNA and adjust the solution to 0.3 M in ammonium acetate using a 5 M ammonium acetate stock solution. The volumes may be scaled up according to the amount of DNA being prepared.
2. Add 2 volumes of ethanol.
3. Centrifuge for 15 min at 4° at greater than 12,000 g. Remove the supernatant with a micropipette.

4. Add 60 μl of 70% ethanol (30% ultrapure water, no added salts or buffers) and centrifuge for 15 min. Remove the supernatant with a micropipette.
5. Dry the pellet.
6. Resuspend the DNA in 0.5× TE to a concentration of 10 ng/μl DNA. Use 1 μl per discrete transformation of 20 μl of cell suspension (0.5× TE is 5 mM Tris, 0.2 mM EDTA, pH 7.4).

Protocol 11b. Microdialysis Purification of DNA for Electroshock Transformation

Ligations and other DNA preparations can be microdialyzed against 0.5× TE as drops on hydrophobic membrane filters to remove salts and thereby reduce conductivity, as follows.[15]

1. Remove the DNA solution (20–50 μl) from its container and place it as a drop on top of a Millipore (Bedford, MA) filter, type VS 0.0025 μm, that is floating on a pool of 0.5× TE (or 10% glycerol), for example, in a small plastic petri dish.
2. Incubate at room temperature for several hours.
3. Withdraw the DNA drop from the filter and place it in a polypropylene microcentrifuge tube. Use 1 μl in a discrete electroshock transformation. This method assumes that the DNA concentration is already appropriate for the transformation.

B. Electroshock Transformation of Other Gram-Negative Microorganisms

A considerable number gram-negative bacteria can be successfully transformed by electroshock treatment, as summarized in Table III. In many previous studies, field strengths in the range of 6.25–12.5 kV/cm, with pulse lengths of about 5 msec, and cell suspensions of high density (5×10^{10} cells/ml) have been used. When attempting to electroporate other bacterial organisms, genetic restriction barriers and other genetic factors potentially capable of affecting transformation should also be considered if high efficiencies are desired, as discussed below. Although no one procedure is optimal for all gram-negative microorganisms, the following protocol can serve as a starting point for developing efficient electroshock transformation.

[15] M. Jacobs, S. Wnendt, and U. Stahl, *Nucleic Acids Res.* **18,** 1653 (1990).

TABLE III
ELECTROSHOCK TRANSFORMATION OF VARIOUS BACTERIA

Species	Field strength (kV/cm)	Transformation efficiency (per μg plasmid DNA)	Ref.
Gram-negative bacteria			
Agrobacterium rhizogenes	12.5	10^6	a
Agrobacterium tumefaciens	6.25–12.5	10^6	a
Campylobacter jejuni	12.5	10^6	b
Citrobacter freundii	6.25	10^3	c
Enterobacter aerogenes	6.25	10^2	c
Erwinia carotovora	6.25	$10^4–10^5$	c
Escherichia coli	12.5, 16	$1–5 \times 10^{10}$	d,e
Klebsiella pneumoniae	6.25	2×10^2	c
Proteus mirabilis	6.25	4×10^2	c
Proteus vulgaris	6.25	8.7×10^2	c
Pseudomonas oxalaticus	6.25	9×10^2	c
Pseudomonas putida	6.25–12.5	10^5	c
Pseudomonas aeruginosa	6.25	10^5	d
Salmonella typhimurium	12.5	10^5	f
Serratia marcescens	6.25	4×10^3	c
Xanthomonas campestris	6.25	10^2	c
Gram-positive bacteria			
Bacillus cereus	6.25–12.5	10^2	g
Bacillus subtilis	16	10^5	d
Clostridium perfringens	12.5	10^4	h
Lactobacillus casei	5	10^5	g
Lactobacillus acidophilus	6.25–12.5	10^4	g
Lactobacillus fermentum	6.25–12.5	10^6	g
Lactobacillus plantarum	6.25	10^4	g
Lactobacillus reuteri	6.25	10^3	g
Lactococcus lactis	6.25	$<10^1$	g
Listeria innocua	6.25	10^2	g
Pediococcus acidilactici	6.25	10^2	g
Proprionibacterium jensenii	6.25	5×10^1	g
Staphylococcus aureus	6.25–12.5	$<10^1$	g
Streptococcus cremoris	6.25–12.5	$10^4–10^5$	i
Streptococcus lactis	6.25–12.5	$10^4–10^5$	i

[a] D. Mattanovich, F. Ruker, A. da Camara Machado, M. L. Aimer, F. Regner, H. Steinkellner, G. Himmler, and H. Katinger, *Nucleic Acids Res.* **17**, 6747 (1989).

[b] J. F. Miller, W. J. Dower, and L. S. Tompkins, *Proc. Natl. Acad. Sci. U.S.A.* **85**, 856 (1988).

[c] R. Wirth, A. Friesenegger, and S. Fieder, *Mol. Gen. Genet.* **216**, 175 (1989).

[d] M. Smith, J. Jessee, T. Landers, and J. Jordan, *Focus* **12**, 38 (1990).

[e] W. J. Dower, J. Miller, and C. W. Ragsdale, *Nucleic Acids Res.* **16**, 6127 (1988).

[f] A. Taketo, *Biochim. Biophys. Acta* **949**, 318 (1988).

[g] J. B. Luchansky, P. M. Muriana, and T. R. Klaenhammer, *Mol. Microbiol.* **2**, 637 (1988).

[h] A. Y. Kim and H. P. Blaschek, *Appl. Environ. Microbiol.* **55**, 360 (1989).

[i] I. B. Powell, M. G. Achen, A. J. Hiller, and B. E. Davidson, *Appl. Environ. Microbiol.* **54**, 655.

**Protocol 12. General Protocol for Electroshock Transformation
of Gram-Negative Microorganisms**

1. Incubate a liquid culture to mid-log phase in sufficient volume to give at least 0.5–1.0 ml of cells at a density of 5×10^{10} to 10^{11} cells/ml. The *E. coli* protocol (Protocol 10) may prove to be a useful example.
2. The growth medium should be rich in nutrients and low in ionic strength (i.e., do not add salts unless necessary for growth).
3. Chill the cells on ice and wash twice in an equal volume of growth medium by pelleting cells with low-speed centrifugation, decanting the supernatant, adding medium, and resuspending the cell pellet by vortexing.
4. Wash the cells twice in an equal volume of chilled electroshock buffer, as in Step 3. Electroshock buffer can be formulated as 10% glycerol in ultrapure water (EWB), as for *E. coli*, or it can be buffered with 0.2 mM phosphate or 1 mM HEPES, both at pH 7.0[12] It is advisable to try both buffer and nonbuffer versions initially.
5. Following the second wash in electroshock buffer, pellet the cells by centrifugation, drain the supernatant, and resuspend the pellet in the residual buffer on the wall of the centrifuge tube. Adjust the concentration to approximately 10^{10}–10^{11} cells/ml by determining the optical density of a dilution of the cell suspension, as is elaborated in the *E. coli* protocol (Protocol 10).
6. Cells can be used for immediate electroshock transformation, or aliquots can be frozen in a dry ice/ethanol bath and stored at $-80°$ for subsequent use.

C. Electroshock Transformation of Gram-Positive Microorganisms

The efficiency of electroshock transformation of gram-positive bacteria is considerably reduced compared to gram-negative microorganisms such as *E. coli*. Presumably the peptidoglycan cell wall structure is a major factor in the reduced electrotransformation frequencies, as well as in the low frequencies of competence following chemical induction regimens. However, electroshock plasmid transformation is feasible for a number of gram-positive bacteria (Table III). It is of note that many of these microorganisms were electroporated at field strengths (6.25 kV/cm) lower than many commercial electroporation devices are now capable of achieving. Thus, the higher field strengths of 12.5–16.7 kV/cm may improve transformation.

Various conditions have been used in attempts to improve transformation of gram-positive microorganisms. These include direct electroshock treatment of intact cells and electroshock following mild digestion of the cell wall[16] or after generation of spheroplasts,[17] which must be regenerated into viable cells following transformation. The following protocol for transformation of *Bacillus subtilis* is exemplary of the gram-positive methods, and variations on it may prove applicable to other gram-positive bacteria.

Protocol 13. Procedure for Electroshock Transformation of *Bacillus subtilis*ˋ

1. Pick a single colony of a freshly streaked agar plate and disperse into 1 ml SOB–Mg.
2. Inoculate the cells into a 500-ml Erlenmyer flask containing 50 ml SOB–Mg.
3. Incubate at 32°, 275 rpm, overnight.
4. The following day inoculate 500 ml of SOB–Mg in a 2-liter Fernbach flask (nonbaffled) with 5 ml of the overnight culture.
5. Incubate at 37°, 275 rpm, until an OD_{550} of 0.75 is reached.
6. Collect the cell pellet in chilled polypropylene centrifuge tubes (e.g., Falcon 2070).
7. Centrifuge at 2600 g for 12–15 min at 4°.
8. Resuspend the cells in an equal volume of cold EWB plus 1 mM EDTA, pH 8.0. Resuspend by vigorous vortexing.
9. Centrifuge again and wash in EWB without EDTA.
10. Resuspend the cells in 2 ml EWB and freeze in a dry ice/ethanol bath in 100-μl aliquots. Cells should be a thick paste of more than 10^{10} cells/ml.
11. Thaw cells on ice and add DNA (in 2–5 μl). Then add 100 μl of sterile 40% PEG 8000 and transfer the suspension to an electroporation cuvette.
12. Electroshock cells at 16 kV/cm, 4000 ohms, 2 μF. Transfer to a chilled polypropylene tube.
13. Add 800 μl of growth medium (e.g., SOC).
14. Incubate at 32° for 1 hr with moderate agitation (250 rpm), then plate on selective medium.

Materials. The formulation of SOB–MG is described in Section II,C,2.

---

[16] I. Powell, M. Achen, A. Hiller, and B. Davidson, *Appl. Environ. Microbiol.* **54,** 655 (1988).
[17] N. Shiverova, W. Foster, H. Jacob, and R. Grigorova, *Z. Allg. Mikrobiol.* **23,** 595 (1983).

EWB is 10% redistilled glycerol in ultrapure water. PEG 8000 is available from Sigma (St. Louis, MO).

D. Optimization of Electroporation

The general approach that we have found useful for optimization of electroporation for new bacteria (or new bacterial strains) is to use very high field strengths with a buffer of low conductivity. Field strengths between 12.5 and 16 kV/cm are good starting points for gram-negative organisms, and 6.5 kV/cm is similarly an appropriate beginning for gram-positive bacteria, although higher voltages may prove to be more effective. Capacitance and resistance should be tuned to give pulse lengths of 2–6 msec, depending on the type of bacteria. Standard conditions for the BRL Cell-porater system are a capacitance of 2 μF and resistance of 4000 ohms. For the Bio-Rad (Richmond, CA) Gene Pulser, the values are 25 μF and 200 ohms. Field strengths are easily reduced by applying lower voltages to the cells. Again, the field strength is affected by the size of the electroporation cuvette, and it is determined by dividing the peak voltage set on the electroporation device by the distance between the electrodes (the path length).

High field strengths maintained over long pulse lengths can cause arcing and should be avoided. Foaming of the cell suspension following electroshock is an indication of excess pulse length. Viability of cells is also an indication of optimal conditions. For bacteria, in contrast to mammalian cells, 50–80% survival rates should be maintained for the highest transformation efficiency. Thus, cell viability following competence induction should be assessed when optimizing the protocol for a particular strain.

V. Evaluation of Bacterial Transformation Parameters and Applications

There are several criteria relevant to assessing the success of a competence induction protocol and its suitability for the intended transformation. The results of a transformation can be considered from the perspective of the plasmid or of the cell. Namely, one can either ask what is the probability that a given plasmid molecule will effect a transformed cell, or what fraction of the viable cells is competent for transformation? The first parameter is often referred to as the transformation efficiency (XFE) and is given in transformed colonies formed per unit mass of DNA (e.g., transformants per microgram of pBR322). It is more properly considered as the molecular transformation frequency, expressed as the probability that a plasmid molecule of a given size will transform a cell. The second parameter, the fraction of competent cells (F_c), illustrates the fact that not every cell becomes competent for transformation, which means that a given aliquot of competent cells can produce only a limited number of transformed cells, no matter how much DNA is added. This latter parameter becomes important when plasmid libraries of maximal number are the intended outcome of a transformation experiment. In most cases the primary parameter is the transformation efficiency, since this measures the amount of DNA which must be transformed to produce the desired number of colonies.

Another parameter of note is the growth state and cell density of the culture that was collected for preparing the competent cells. Cell density significantly influences subsequent induction of competence and therefore can provide feedback for troubleshooting preparations with suboptimal competency. Other factors which can affect a transformation include competition from nontransforming DNA molecules and effects of size and form of the transforming DNA. These parameters are described briefly below.

A. *Determination of Transformation Efficiency*

The transformation efficiency should be measured under conditions of cell excess, such that there is no competition between plasmids for the same competent cell. The traditional standard is the plasmid pBR322, although smaller plasmids based on the pUC vector are also used. There are about 2×10^{11} molecules of pBR322 per microgram and about 3×10^{11} molecules of pUC. The number of viable cells used in an individual transformation ranges from 2×10^8 to 10^{11}, and thus the cells will always be in excess ($>10:1$) if the control transformation uses less than 2×10^7 DNA molecules, which corresponds to 100 pg of pBR322. Values for *XFE* range from 10^4 for rapid colony transformations to over 2×10^{10} transformed colonies per microgram of pBR322 with electrocompetent cells. Thus, with electrocompetent cells, up to 10% of the plasmid molecules are producing transformed cells. The amount of plasmid DNA used as a standard should be adjusted for the expected range of *XFE* values. For high-efficiency transformations, 10 pg is a reliable amount. Following the transformation, different fractions of the transformation should be plated onto drug selection plates: 10 and 1% (and 0.1% for electrocompetent cells). The 1% plating of 10 pg pBR322 will have 10 colonies if the *XFE* is 10^8 and 200 colonies if the *XFE* is 2×10^9. For electrocompetent cells, 100 μl of the 1 ml transformation mix is diluted into 900 μl SOC, and 10 μl of this dilution is plated (into a puddle of medium and spread). This

0.1% dilution of a 10 pg transformation will produce 10 colonies if the *XFE* is 10^9, 100 if the *XFE* is 10^{10}, etc.

For less efficient competent cell preparations, and also for assessing linearity of the response in the highly competent cells, 1 ng of pBR322 is a reasonable amount for a second standard. The 1% plating of 1 ng will give 10 colonies if the *XFE* is 10^6 and 100 if the *XFE* is 10^7. For more efficient cells, perform serial dilutions of 10 μl of transformed cells into 1 ml SOC, plating 1 and 10%, etc. To assess the rapid colony transformation procedure, use 10 ng, plating 10%, which will give about 10 colonies if the *XFE* is 10^4.

It is recommended that a control transformation be included on each occasion that a series of transformations are performed, as the *XFE* gives important information on the state of the competent cells in the experiment. For example, if the experimental DNAs give no colonies, but the control *XFE* is as expected, then the experimental DNA is suspect. On the other hand, if the *XFE* is low, then the failure of the experiment may be more accurately explained by inadequate competence, with a clear remedy being the acquisition of more efficient competent cells. Without this control there is no guidance into the problem.

B. Fraction of Competent Cells

The fraction of viable cells (F_c) that are competent is not unity, and as a result this parameter can influence the colony-forming potential of an individual transformation, which in turn can determine the number of discrete transformations that need be performed in order to generate the desired number of transformants. The fraction of competent cells is traditionally measured with saturating levels of DNA, for example, at cell to plasmid levels of 1 : 10, which is the inverse of that employed to assess the *XFE*. Thus, at least 100 ng and generally 500–1000 ng of pBR322 DNA is typically used to determine F_c values. Following transformation, the cells should be serially diluted by aliquoting 10 μl into 1 ml of SOC and 10 μl of that into another 1 ml to give a 10^4 dilution. For chemically competent cells, where 2–5×10^8 cells are used in a transformation, 10 and 100 μl of this dilution are plated onto rich (no drug) and drug selection plates (10^5 and 10^6 final dilutions, respectively). The ratio of colonies formed on drug versus nondrug plates gives F_c. For electrocompetent cells, it may be useful to use 10 μg of plasmid DNA, and then take an additional $100\times$ dilution, such that the colony number per plate is easily counted.

The fraction of competent cells ranges from around 0.001% for the rapid colony protocol, to 5–10% with the high-efficiency chemical protocols, and up to 80% with the electroshock method. Moreover, when cells are analyzed below saturation, the *XFE* shows that electrocompetent cells are more efficient at producing transformation of a cell by a single plasmid molecule. Thus, both *XFE* and F_c are important for maximizing the number of transformants actually produced in an experimental transformation. When crucial transformations are envisioned, such as for cDNA or genomic DNA library construction, both parameters should be assessed. Because electrocompetent cells have a 10-fold higher *XFE* and a higher F_c, they can be expected to produce 10–100 times more colonies than the chemical methods from the same amount of DNA under nonsaturating conditions.

C. Saturation and Optimization of Colony-Forming Potential

For large-scale transformations, it is often valuable to titer the experimental DNA to maximize the colony-forming potential for each discrete transformation while avoiding saturation of the cells with excessive DNA, which might produce double transformants and waste precious DNA (see also the discussion on titering cDNA ligations in Ref. 3). Although F_c is secondary in importance to the *XFE*, it does influence the overall colony-forming potential of a competent cell aliquot. To determine the transforming potential of a DNA preparation with a particular preparation of competent cells, add increasing amounts of DNA in a series of transformations. The point at which the total number of transformed colonies produced begins to level off is close to the optimum for both the number of transformants produced per individual transformation and the most conservative use of the DNA. In practice the optimum often turns out to be in the period when the number of DNA molecules has just exceeded the number of cells. For many cDNA transformations using chemical induction of competence, this optimal range has proved to be between 5 and 20 ng of ligated plasmid DNA.

D. Competition from Nontransforming DNA

In many cloning situations only a fraction of the DNA will comprise plasmids which are capable of effecting transformed cells. The nontransforming DNA competes with the plasmids for DNA uptake.[5] However, with chemical transformation there appear to be on the order of 100 separate channels into a cell, and with electroshock induction of competence it seems likely that a similar or even greater number of plasmid uptake channels (pores) are formed. Thus, nontransforming DNA competes as it is able to occupy a majority of such channels on every competent cell. One can approximate the competition with the molar excess of DNA molecules over the total number of cells (since all cells appear to compete for DNA[5]). Typically this competition becomes evident at about 100 ng of DNA and becomes pronounced at 500 ng of DNA per discrete chemical transformation of approximately $2–5 \times 10^8$ cells. With electroshock transformation, competition is not likely to be a factor, since the concentration of total DNA used for the transformation (of which 1 μl is combined with 20 μl of cells) would need to be in excess of 5 μg/μl before competition becomes significant for 2×10^9 cells. The effects of different types of nontransforming DNA cannot be predicted unambiguously, but, in general, restricting the total DNA mass per unit transformation to less than 500 ng when nontransforming DNA is in excess proves to be a good approximation.[5] Of course, when a majority of the DNA is capable of transforming, then the issue of competition changes to one of saturation (and potential transformation by multiple plasmids), as discussed in Section V,C. It is always recommended that a saturation curve of the transformantion efficiency (*XFE*) obtained with increasing amounts of DNA be determined prior to conducting large-scale transformations intended to generate high-complexity plasmid libraries. Then an amount of DNA which cooptimizes *XFE* and total colony number for each discrete transformation can be selected. This value is necessarily below levels where competition or saturation are significant.

E. Effects of Plasmid Size and Form on Transformation

The extensive characterization of the properties of chemical transformation of *E. coli*[5] has revealed that very large plasmids, up to 66 kilobases (kb), can be efficiently transformed into strains such as DH1 and DH5 with the high-efficiency chemical transformation method (Protocols 6 and 7). If one considers the molecular transformation probability, namely, the probability that a plasmid molecule will effect a transformed cell, then transformation frequency declines linearly with increasing size. If one instead considers the *XFE* value, which is scaled as mass and not molecules, then the transformation efficiency declines with the inverse of the mass squared. A similar relation exists for electroshock transformation.[14] In addition, the discovery that the *deoR* mutation selectively improves transformation of large DNA molecules (see below) will facilitate the transformation of plasmids used for genomic cloning of large DNA fragments. Strains such as DH5α, DH5αMCR, and DH10B carry the *deoR* mutation, and the latter two also lack the methylation-dependent restriction activities (see below). Thus, it seems probable that plasmid-based vectors can be developed for cloning larger DNAs than can be accommodated by bacteriophage packaging. Preliminary studies in which an F' plasmid of about 200 kb was readily transformed into DH10B by electro-shock transformation encourage the pursuit of this possibility.

A second relevant property of transformation is that there is no significant requirement for DNA supercoiling in plasmid transformation.[5] The fact that relaxed DNAs transform at frequencies similar to supercoiled forms is significant for cloning experiments involving the ligation of large DNAs. The existence of high-efficiency transformation procedures and of improved strains for transformation of heterologous DNA together suggest that plasmid cloning in the 100-kb range could be a valuable interface between λ/cosmids and yeast artificial chromosomes in large-scale genomic cloning efforts.

Although closed circular double-stranded plasmids transform *E. coli* readily, neither linear plasmids nor single-stranded DNA are effective at transformation. It is clear from work on integrative transformation that *recBC*[18,19] or *recD*[20] mutants are dramatically better than *rec*[+] or recA strains for linear DNAs. However, even in a *recBC* strain background, transformation by linear plasmids is significantly reduced relative to closed circular forms.[18] This result may relate to the other exonucleases present as part of the recombination and repair system (see below). Similarly, single-stranded phagemid DNA transforms *E. coli* at frequencies about 10^4 lower than double-stranded forms using either electroshock or chemical transformation methods. However, the possibility clearly exists that a combined genetic and chemical study seeking to improve transformation by linear and single-stranded plasmids will produce improved conditions for transformation by these forms of DNA.

F. Application of Plasmid Transformation to Large-Scale cDNA and Genomic DNA Libraries

The improvements in plasmid transformation efficiencies and in the *E. coli* strains used as hosts have together dramatically enhanced the applicability of this methodology. Specifically, it is now evident that plasmid-based cDNA libraries can be constructed that are at least 10 times larger per input amount of RNA than can be achieved with bacteriophage λ-based systems. For example, we have recently produced several primary, size-selected cDNA libraries at an efficiency of over 10^8 clones per microgram of input mRNA, using electroshock transformation of DH10B at an *XFE* of $10^{10}/\mu$g pBR322 (G. Christofori and D. Hanahan, unpublished results; J. Jessee and C. Gruber, unpublished results). As mentioned

[18] E. Conley and J. Saunders, *Mol. Gen. Genet.* **194**, 211 (1984).
[19] S. Winans, S. Elledge, J. Krueger, and G. Walker, *J. Bacteriol.* **161**, 1219 (1985).
[20] D. Shevell, A. Abou-Zamzam, B. Demple, and G. Walker, *J. Bacteriol.* **70**, 3294 (1988).

above, the cloning in plasmids of large genomic fragments now seems feasible as well, given the improvements in efficiency and host strains. Thus, it is pertinent to mention methodology for manipulating large numbers of plasmid clones.

A high-density colony-screening procedure has been developed.[21-23] This method maintains colonies (plasmid clones) on nitrocellulose filters at densities up to 20,000 per 10 cm plate. Colonies at this density can be accurately replicated by making transient sandwiches of the colony-bearing filter and a clean one, which when separated and inoculated on an agar plate will produce a replica of the colony distribution. The ability to make numerous accurate replicas means that some copies can be stored as a frozen bank and others used for various screening procedures. The principal critique of this method is that it is quite laborious, which renders distribution and extensive duplication of libraries a time-consuming business. However, it should be noted that a more convenient method for amplifying plasmid libraries exists. This method involves growing colonies as distributions on agar plates, after which the colonies are dispersed and combined to give a suspension of viable cells that comprise the entire library in thousands of copies, relative to the primary transformation. Portions of such a randomized library can be extracted to give a DNA copy of the library, while other portions can be frozen as dozens of aliquots that represent all of the transformants. This method is straightforward and is described in Protocol 14 below. Its principal limitation is that the library is randomized, and, as such, when it is replated for screening purposes, more colonies (150–200%) than the complexity of the original primary library should be screened to assure that every clone is reviewed. This method avoids any liquid growth of the cells comprising the library, which is important because liquid growth can produce substantial distortions in the representation of individual clones because of overgrowth by fast clones. The growth of colonies on plates restricts clone size and thereby maintains relatively uniform clone distributions.

Protocol 14. Large-Scale Plate Amplification, Storage, and Retrieval of Plasmid Libraries

A. Amplification and storage of plasmid libraries
   1. Transform the recombinant plasmid plus cDNA or genomic DNA

[21] D. Hanahan and M. Meselson, *Gene* **10,** 63 (1980).
[22] D. Hanahan and M. Meselson, this series, Vol. 100, p. 333.
[23] D. Hanahan and M. Meselson, *in* "Recombinant DNA Methodology; Selected Methods in Enzymology" (R. Wu, L. Grossman, and K. Moldave, eds.), p. 267. Academic Press, New York, 1989.

preparation to establish the library in *E. coli*; distribute the transformed cells on SOB agar (+drug) plates to give a colony density
of 5000–20,000 per 10-cm plate (or 20,000–50,000 for a 15-cm plate).
If the library is to be (or is being) maintained as a set of colonies on
nitrocellulose filters, then prepare a fresh set of filter replicas as
described by Hanahan and Meselson.[21–23]

2. Incubate the primary platings or the fresh nitrocellulose replicas at
   37° for 1–2 hr and then at 20°–30° until the colonies are 0.5–1 mm
   in diameter, assuming a density of 5000–20,000 colonies per filter.
   (The lower temperature growth allows better control of colony size.)
   It is recommended that the plating, replication, and subsequent
   steps be performed in a sterile hood, so as to minimize contamination with other organisms.

3. If colonies are on agar plates, then add 10 ml of SOB plus 20%
   glycerol. In the case of colonies on filters, first transfer each filter
   with the colonies uppermost to an empty petri dish. In either case
   disperse the colonies, initially by placing the plates on a slow gyra-
   tory shaker and then by gently rubbing them with a sterile rubber
   policeman.

4. Collect the cell suspension by pipetting up and down and then
   transfer it to a sterile tube and place on ice. Vortex to complete the
   dispersal. The cell density should be about 10^9/ml, and at this point
   each plate or filter of colonies has been dispersed in a separate tube.
   These sets of cells can either be combined at this stage to produce
   a complete mixture of the library or be maintained separately as *N*
   sets derived from the *N* master filters of colonies that carried the
   library.

5. Distribute 100- to 500-μl volumes to polypropylene cryotubes. Flash
   freeze in liquid nitrogen or in a dry ice/ethanol bath, then place in
   liquid nitrogen or at −70°.

6. For a DNA copy of the library, divide the set or sets of dispersed
   colonies from Step 4 in half. Take one half and freeze down the
   transformed cells as specified in Step 5. Pellet the other half in a
   clinical centrifuge and wash in an equal volume of SOB. Then pellet
   the cells a second time and proceed with a standard plasmid DNA
   isolation procedure. It is recommended that the DNA copy of the
   library be purified by banding on a CsCl density gradient and then
   stored as an ethanol precipitate.

7. Alternatively, a separate set of replica filters can be prepared for
   DNA isolation, and the colonies can be subjected to plasmid amplification in the presence of chloramphenicol.[21] Each replica is transformed onto an SOB agar plate containing 200 μg/μl chlorampheni-

col and incubated at 37° for 24–48 hr, after which the colonies are collected for DNA isolation as described in Steps 3–6.

B. *Retrieval of frozen aliquots of an amplified library*
1. Remove an aliquot (or a set of aliquots) that represent the library.
2. Dilute the aliquot(s) $10\times$ into SOC medium, producing a set of primary aliquots.
3. Prepare a set of serial dilutions of the primary aliquot(s) so as to determine the density of viable cells capable of forming colonies (e.g., combine 10 μl into 90 μl SOB for a 10^{-1} dilution, or 10 μl into 1 ml for a 10^{-2} dilution). Since the original cell density was about 10^9/ml, the primary aliquots should now have a density of about 10^8 cells/ml, assuming no cell death. However, the titer of viable cells will decline slowly on storage, and therefore the titer will be less than that when collected.
4. Plate 10-μl aliquots of the 10^{-2}, 10^{-3}, 10^{-4}, and 10^{-5} serial dilutions from the primary aliquot onto plates carrying the appropriate selective drug. Incubate overnight to establish colonies. Store the primary aliquot(s) on ice and discard the serial dilutions.
5. The titer of the primary aliquots can be determined from the platings of the dilutions. The titer of the primary aliquot(s) will exceed 90% of its original value during 24 hr of storage on ice. Prepare a fresh dilution from the primary aliquot such that there are about 10^5 colony-forming units/ml. Thus, 50–250 μl can be applied to each plate to produce a colony density of 5000–25,000. Plate on nitrocellulose filters, incubate to establish colonies, and replicate the distributions as described.[21-23] Since the original distribution of colonies has been dispersed and thus randomized, more filters should be prepared than the original library required, so as to assure representation of all clones. Each of the 100 separate aliquots derived from a filter carrying 10^4 colonies will initially have 10^7 cells in it, or about 1000 copies of the distribution. Thus, even very long-term storage should allow one aliquot to restore the clones carried in the original distribution.

VI. Genetic Factors Influencing Choice of Host Strain

A considerable number of *E. coli* genes have been found to influence transformation experiments. These genes are briefly described below, and the knowledge of their effects is then used to evaluate various strains for different types of cloning experiments. The other bacterial organisms that either are used as hosts for DNA cloning or are being studied via DNA transformation are in general less well characterized in this regard, but their genes known to influence transformation are mentioned as well.

A. *Escherichia coli*

The *E. coli* genes influencing transformation fall into two major classes: genes whose products mediate DNA repair and recombination and those constituting genetic restriction systems for discriminating what is self DNA from that which is foreign. In addition, a diverse set of genetic loci affect specific applications of this technique or a particular method of transformation.

1. Repair and Recombination Genes

recA. Studies from many laboratories (reviewed in Refs. 24–26) indicate that *E. coli* has three recombination pathways: *recBCD, recE,* and *recF*. All three pathways are dependent on the product of the *recA* gene, which serves as a master regulator of recombination. The *recA* gene maps at minute 58 on the *E. coli* K12 chromosome,[27] and *recA* alleles can be introduced into *recA*[+] strains via cotransduction with the *srl*::Tn*10* transposon.[28] *recA* mutants are deficient in general recombination[24] and are extremely sensitive to agents which damage DNA, such as ultraviolet irradiation and alkylating chemicals. Nitrofurantoin sensitivity[29] serves as a convenient test to determine whether the *recA* marker has been introduced into a strain.

Escherichia coli recA strains are valuable for the propagation of vectors containing cloned inserts. The *recA* mutation stabilizes most but not all DNA sequences carried in cloning vectors. However, certain DNAs, especially ones containing inverted repeats (palindromes), are subject to deletion in *recA* strain backgrounds. Several different *recA* alleles are found in the commonly used *E. coli* host strains, and these strains differ in the

[24] A. Clark and K. Low, *in* "The Recombination of Genetic Material" (K. Low, ed.), p. 155. Academic Press, New York and London, 1988.

[25] G. Smith, *Annu. Rev. Genet.* **21,** 179 (1987).

[26] G. Smith, *Microbial. Rev.* **52,** 1 (1988).

[27] B. J. Bachmann, *in* "*Escherichia coli* and *Salmonella typhimurium*: Cellular and Molecular Biology" (F. C. Neidhardt, J. L. Ingraham, B. Magasanik, K. B. Low, M. Schaechter, and H. E. Umbarger, eds.), p. 807. American Society for Microbiology, Washington, D.C., 1987.

[28] C. M. Berg, and D. E. Berg, *in* "*Escherichia coli* and *Salmonella typhimurium*: Cellular and Molecular Biology" (F. C. Neidhardt, J. L. Ingraham, B. Magasanik, K. B. Low, M. Schaechter and H. E. Umbarger, eds.), p. 1071. American Society for Microbiology, Washington, D.C., 1987.

[29] S. Jenkins and P. Bennett, *J. Bacteriol.* **125,** 1214 (1976).

amount of residual recombination activity.[30] For example, strain HB101[31] carries the *recA13* allele, which has more residual recombination activity than strains with the *recA1* allele, such as DH5α or DH10B,[7] or strains containing the *srl-recA306* deletion.[32]

recBCD/sbcBC. The *recBCD* genes map at minute 61 on the *E. coli* chromosome[27] and encode the three subunits of exonuclease V.[33–37] *recBC* mutants of *E. coli* K12 are recombination-deficient. However, the recombination-deficient phenotype in *recBC* strains can be suppressed by mutations in *sbcA*, which activate the synthesis of the *recE* protein, exonuclease VIII,[24] or by mutations in both *sbcB* and *sbcC*,[38] which activate the *recF* pathway. Thus, *recBC sbcA* strains catalyze recombination via the *recE* pathway, whereas *recBC sbcBC* strains achieve recombination via the *recF* pathway. In fact, *recBCD* deletion strains are hyperrecombinogenic (Refs. 33–35; F. Bloom & J. Jessee, unpublished results).

Strains containing *sbcB* and *sbcC* mutations alone or in certain combinations have proved useful for stabilizing recombinant plasmids or λ phage that contain inserts with palindromic regions.[39–42] Inverted repeats (palindromes) produce cruciform structures when supercoiled that retard DNA replication, and consequently palindromes tend to be deleted in most strain backgrounds. Together the data indicate that inactivation of *recBC* and either the *sbcB* or *sbcC* gene products allows replication (and hence stable transformation) of many DNA sequences containing palindromes. However, strains containing *recBC sbcB* mutations often also carry cryptic mutations in *sbcC*,[38] and the *sbcB* and *sbcC* mutations together restore recombination proficiency through activation of the *recF* pathway. Thus, both *recBC sbcBC* and *recBC sbcA* strains are recombination-proficient, a possibility which must be considered when using *recBC* strains as recombination-deficient hosts for molecular cloning purposes.

[30] B. Xu, C. Paszty, and P. Lurguin, *BioTechniques* **6**, 752 (1988).
[31] H. Boyer and D. Roulland-Dussoix, *J. Mol. Biol.* **41**, 459 (1969).
[32] L. Csonka and J. Clark, *Genetics* **93**, 321 (1979).
[33] A. Chaudhury and G. Smith, *Proc. Natl. Acad. Sci. U.S.A.* **81**, 7850 (1984).
[34] D. Biek and S. Cohen, *J. Bacteriol.* **167**, 594 (1986).
[35] S. Amundsen, A. Taylor, A. Chaudhury, and G. Smith, *Proc. Natl. Acad. Sci. U.S.A.* **83**, 5558 (1986).
[36] P. Goldmark and S. Linn, *J. Biol. Chem.* **247**, 1849 (1972).
[37] I. Hickson, C. Robson, K. Atkinson, L. Hutton, and P. Emerson, *J. Biol. Chem.* **260**, 1224 (1985).
[38] R. Lloyd and C. Buckman, *J. Bacteriol.* **164**, 836 (1985).
[39] D. Leach and F. Stahl, *Nature (London)* **305**, 448 (1983).
[40] J. Collins, G. Volckaert, and P. Nevers, *Gene* **19**, 139 (1982).
[41] A. Wyman, L. Wolfe, and D. Botstein, *Proc. Natl. Acad. Sci. U.S.A.* **82**, 2880 (1985).
[42] A. Chalker, D. Leach, and R. Lloyd, *Gene* **71**, 201 (1988).

recF, recN, recJ. Several other *E. coli* genes (*recF, recJ, recN*) have been found to inhibit certain types of recombination in the recombination-proficient backgrounds of *recBC sbcA* or *recBC sbcBC*.[43–46] Mutations in these genes have proved useful in stabilizing certain DNA sequences during molecular cloning.[47,48] The *recJ* gene encodes a single-stranded exonuclease which may mediate deletion events between direct repeat sequences.[49] Although the experiences with these genes are as yet insufficient to determine their general importance for DNA transformation and cloning, they remain candidates in attempts to clone or stabilize otherwise unstable sequences.

2. *Classic Restriction–Modification System.* *Escherichia coli* K12 contains two general restriction systems that function to exclude foreign DNA: the classic K type restriction–modification system (*hsdRMS*) and the methylation-dependent restriction system (*MDRS*). The *hsdRMS* locus maps at minute 98.5 on the *E. coli* K12 chromosome.[27] The *hsdR* gene encodes the *Eco*K type I endonuclease which will cleave DNA containing an *Eco*K cleavage site[50] (5' A^{Me}ACNNNNNNGTGC 3' and its complement 5' GC^{Me}ACNNNNNNGTT 3') unless the noted adenine residues are methylated by the *Eco*K methylase. This methylase, encoded by the *hsdM*$^+$ gene, protects *E. coli* DNA from degradation; however, any unmethylated foreign DNA containing *Eco*K cleavage sites will be degraded. Therefore, *E. coli* K12 host strains which lack *hsdR* should be used when constructing primary libraries of non-*E. coli* DNA. Most common cloning strains (Table II) already contain this mutation. Additionally, *E. coli* C was recently shown to lack the entire *hsd* locus.[51] *hsdR* strains can be constructed by P1-mediated transduction by taking advantage of the linkage between the *hsdR* locus and the *zjj202*::Tn*10* transposon.[28]

A plasmid established in a host strain which is *hsdR*$^-$ *M*$^+$ will be methylated at the *Eco*K sites and can be transformed into strains which are either *hsdR*$^+$ or *hsdR*$^-$. However, some *E. coli* strains, such as DH5αMCR and DH10B,[7] contain deletions eliminating the entire *hsdRMS*

[43] Z. Horii and A. Clark, *J. Mol Biol.* **80,** 327 (1973).

[44] S. Lovett and A. Clark, *J. Bacteriol.* **157,** 190 (1984).

[45] R. Lloyd, S. Picksley, and C. Prescott, *Mol. Gen. Genet.* **190,** 162 (1983).

[46] S. Picksley, P. Attfield, and R. Lloyd, *Mol. Gen. Genet.* **195,** 267 (1984).

[47] R. Boissy and C. Astell, *Gene* **35,** 179 (1985).

[48] M. Ishiura, N. Hazumi, T. Koide, T. Uchida, and Y. Okada, *J. Bacteriol.* **171,** 1068 (1989).

[49] S. Lovett and R. Kolodner, *Proc. Natl. Acad. Sci. U.S.A.* **86,** 2627 (1989).

[50] E. A. Raleigh, this series, Vol. 152, p. 130.

[51] A. S. Daniel, F. V. Fuller-Pace, D. M. Legge, and N. E. Murray, *J. Bacteriol.* **170,** 1775 (1988).

region. Strains which are *hsdM* and lack the *Eco*K methylase will not methylate cloned sequences at the *Eco*K cleavage site. Nonmethylated plasmid DNA prepared from these *hsdM* strains should only be transformed into *hsdR*⁻ strains since the DNA will be restricted when transformed into an *hsdR*⁺ strain.

   3. Methylation-Dependent Restriction Systems. *Escherichia coli* K12 also contains several restriction loci which genetically restrict incoming DNA, but *only* if the DNA is methylated.[52] If the DNA contains methylcytosine, degradation can occur owing to restriction endonucleases specified by the *mcrA* or *mcrBC* genes. Alternatively, if the DNA contains methyladenine, degradation can occur owing to the *mrr* gene product. The *mcrA* gene maps at 25 min[27] and the *mcrBC* and *mrr* genes map at 98.5 min near the *hsdRMS* genes.[27]

   The genotypes of *E. coli* strains for the *mcrA, mcrBC,* and *mrr* loci can be determined by methylating appropriate plasmids *in vitro* or *in vivo* and transforming the test strain with both methylated and unmethylated plasmids. For example, methylation with either the *Alu*I or *Hae*III methylases is diagnostic for the *mcrB* gene.[52] Transformation of an *mcrB*⁺ strain with an *Alu*I or *Hae*III methylated plasmid results in a 100-fold decrease in transformation efficiency relative to the efficiency with the unmethylated plasmid. Both methylated and unmethylated plasmids transform an *mcrB*⁻ strain with similar efficiencies.

   An analysis of the *mcrB* region reveals that this locus is actually composed of both an *mcrB* gene (51 kDa protein) and an additional gene (*mcrC*), encoding a 39-kDa protein.[53] Current evidence suggests that *mcrB* encodes a restriction endonuclease and that *mcrC* acts to broaden the range of methylated DNA sites which are recognized and restricted by *mcrB*. For example, restriction of an *Hae*III methylated plasmid requires only the *mcrB*⁺ gene product, whereas restriction of an *Hae*II or *Alu*I methylated plasmid requires both the *mcrB*⁺ and *mcrC*⁺ gene products (E. Raleigh, personal communication; D. Chatterjee, personal communication). Thus, the *mcrC* gene product can be envisioned as a protein which alters the DNA sequence specificity of the *mcrB* restriction endonuclease. Recently, a new methylation-dependent restriction locus, *mcrD*, has been localized in the region between *mcrB* and *mrr*; its range of specificity and genetic characteristics are under investigation (E. Raleigh, personal communication; E. Achberger, personal communication). The *Hpa*II methylase can be used as a diagnostic test for *mcrA*,[52] and the *Hha*II or

[52] E. Raleigh and G. Wilson, *Proc. Natl. Acad. Sci. U.S.A.* **83,** 9070 (1986).
[53] T. K. Ross, E. C. Achberger, and H. D. Braymer, *J. Bacteriol.* **171,** 1974 (1989).

*Pst*I methylases can be used as diagnostic tests for *mrr*[54]. The *mcr* and *mrr* genotypes of a number of commonly used host strains can be found in several recent reviews.[50,55,56]

The methylation-dependent restriction system (MDRS)[57] can affect the ability to clone methylated genomic sequences from a wide variety of higher eukaryotic organisms as well as from many bacteria which methylate their genomes. Plant DNA may contain up to 40–50% of all cytosine as methylcytosine. In mammalian DNA the values are 4–10% (see references cited by Blumenthal[57]). The facts of the existence of an MDRS in *E. coli* and of the methylation of other genomes argue that genomic DNA from any source, prokaryotic or eukaryotic, which contains methylated sequences should be cloned into a host strain that is deficient in all of the MDRS genes. Numerous examples of cloning experiments which resulted in recovery of clones only in MDRS-deficient strains are given by Raleigh *et al.*[56] Plasmid and bacteriophage rescue experiments, in which integrated bacterial vector sequences are recovered from eukaryotic cells, are also best performed in MDRS-deficient strains.[7,58] Woodcock *et al.*[55] have evaluated a number of *E. coli* strains for their ability to tolerate methylated cytosine sequences. The strains were analyzed using primary genomic libraries in a bacteriophage λ vector prepared from either petunia or a mouse cell line DNA. The highest methylation tolerance was obtained using *E. coli* K12 strains which were *mcrA*⁻ and contained a deletion eliminating the *mrr, hsdRMS,* and *mcrBC* genes. This conclusion is consistent with the analysis of a series of MDRS mutants for their ability to rescue plasmids from transgenic mouse DNA.[7]

Escherichia coli C, which lacks most if not all of the MDRS,[51] also has a high methylation tolerance. At least a 10-fold improvement in the titer of a λ phage library containing methylated sequences can be expected by using MDRS-deficient host strains.[55] Woodcock *et al.*[55] recommend the following strains for cloning in λ vectors: NW2, ER1647, and ER1648. Each is *mcrA*⁻ Δ(*mrr, hsdRMS, mcrBC*). Another useful strain is DL538 [*C600 hsdR mcrB recD sbcC*] which has a lower methylation tolerance but is the only known *mcrB sbcC* host strain (see discussion above on the use of *sbcC* to stabilize recombinants containing inverted repeats[42]).

[54] J. Heitman and P. Model, *J. Bacteriol.* **169**, 3243 (1987).
[55] D. Woodcock, P. J. Crowther, J. Doherty, S. Jefferson, E. DeCruz, M. Noyer-Weidner, S. S. Smith, M. Z. Michael, and M. W. Graham, *Nucleic Acids Res.* **17**, 3469 (1989).
[56] E. Raleigh, N. E. Murray, H. Revel, R. M. Blumenthal, D. Westaway, A. D. Reith, P. W. J. Rigby, J. Elhai, and D. Hanahan, *Nucleic Acids Res.* **16**, 1563 (1988).
[57] R. Blumenthal, *BRL Focus* **11**, 41 (1989).
[58] J. Gossen, W. J. F. DeLeeuw, C. H. T. Tan, E. L. Zwarthoff, F. Berends, P. H. M. Lohman, D. L. Knook, and J. Vijg, *Proc. Natl. Acad. Sci. U.S.A.* **86**, 7971 (1989).

4. Other Relevant Genetic Markers

deoR. A mutation designated *deoR*, originally described in the early 1970s,[59,60] has recently been found to improve plasmid transformation of *E. coli* K12 strains (D. Hanahan, unpublished observations, 1984). The original phenotype of *deoR* mutant strains involved the ability of these strains to grow on a minimal medium containing inosine as the sole carbon source. This property was used by Hanahan to select a *deoR* derivative of *E. coli* strain DH1. The *deoR* strain, designated DH5, was found to be approximately 3- to 4-fold more competent than DH1 using test plasmids pBR322, pUC8, and pUC9. Of more practical significance, DH5 was 30-fold more transformable than DH1 with a 66-kb plasmid. The *deoR* mutation therefore enhances the uptake of larger plasmids and allows the construction of gene libraries containing large inserts. The *deoR* mutation maps at minute 19 of the *E. coli* K12 chromosome[27] and is found in strain DH5 and its derivatives DH5α and DH5αMCR[7] as well as strains DH10B[7] and DH11S.

dam, dcm. *Escherichia coli* K12 strains contain two methylases that methylate adenine (DNA adenine methylase, *dam*) or cytosine (DNA cytosine methylase, *dcm*).[61] Methylation at the *dam* recognition site (GATC) or *dcm* recognition sites (CCAGG and CCTGG) can inhibit digestion of the DNA by certain restriction endonucleases because the sites overlap. For example, *dam*-sensitive enzymes include *Bcl*I, *Cla*I, *Hph*I, *Mbo*I, *Mbo*II, *Nru*I, *Taq*I, and *Xba*I. *dcm*-sensitive enzymes include *Ava*II, *Eco*RII, *Sau*96I, and *Stu*I. Therefore, to ensure digestion by these restriction endonucleases one should passage the DNA through a *dam⁻ dcm⁻ E. coli* strain. However, these modifications exist to assist in correct DNA repair of daughter (unmethylated) strands during replication, since the parent strand is methylated. As a result, strains with *dam⁻* and *dcm⁻* are inherently mutagenic and therefore not advised for original cloning or general plasmid or phage amplification.

endA. The *endA* gene specifies a DNA endonuclease and maps at minute 64 on the *E. coli* chromosome.[27] The absence of this endonuclease improves the quality of plasmid DNA prepared by the quick-boiling method of Holmes and Quigley,[62] and the *endA* product may have other subtle effects on the stability of transforming DNAs. The *endA* mutation

[59] A. Munch-Peterson, P. Nygaard, K. Hammer-Jespersen, and N. Fill, *Eur. J. Biochem.* **27**, 208 (1972).
[60] S. Ahmad and Pritchard, *Mol. Gen. Genet.* **111**, 77 (1971).
[61] M. Marinus, *in* "DNA Methylation: Biochemistry and Biological Significance" (A. Razin, H. Cedar, and A. Riggs, eds.), p. 81. Springer-Verlag, Berlin and New York, 1984.
[62] D. Holmes and M. Quigley, *Anal. Biochem.* **114**, 193 (1981).

can be introduced via linkage to the *nupG*::Tn*10* transposon.[27] Common *E. coli* strains which are *endA*⁻ include MM294[8] and its derivatives DH1,[5] DH5, and DH5α, as well as DH10B.[7]

gyrA. The DNA gyrase of *E. coli* is composed of two subunits encoded by the *gyrA* and *gyrB* genes. The *gyrA* gene maps at minute 48 and the *gyrB* gene maps at minute 83 on the *E. coli* chromosome.[27] Strains containing mutations in the DNA gyrase subunits can be isolated by selecting for mutants resistant to nalidixic acid (*gyrA*) or to coumermycin (*gyrB*). DNA gyrase has been implicated in the generation of deletions between sequences containing direct repeats in a *recA*-independent pathway.[63,64] These results suggest that plasmids with cloned inserts containing direct repeats may be stabilized in *E. coli* strains which are *gyrA*⁻.[64] Common cloning hosts which are *gyrA*⁻ include DH1 and its derivatives DH5, DH5α, DH5αF′, and DH5αMCR.

lacZ ΔM15. A partial deletion of the *lacZ* gene was originally described in 1967[65] as a mutation which could be complemented by the amino-terminal α peptide of β-galactosidase. With the advent of recombinant DNA technology, cloning vectors such as M13mp18 and pUC18 were constructed to take advantage of this complementation. A multiple cloning site (MCS) in the vector was incorporated in-frame into the amino-terminal DNA fragment of *lacZ* carried by the vector. When this vector is transformed into a host strain containing the *lacZ ΔM15* mutation, active β-galactosidase is produced, and the colonies turn blue on indicator plates containing X-Gal (5-bromo-4-chloro-3-indolyl-β-D-galactoside). If DNA fragments are cloned into the MCS of the vector, the α subunit is disrupted; therefore, it cannot combine with the β subunit to restore β-galactosidase, and consequently white colonies develop. Blue–white colony screening can be performed to identify vectors containing cloned inserts. *Escherichia coli* strains can carry the *lacZ ΔM15* locus on an F′ (JM103), on the chromosome alone (JM83), or as part of the φ80*dlacZ ΔM15* transducing phage (DH5α, DH5αMCR, DH10B).

EcoPI. Some *E. coli* host strains, such as JM103, are lysogenic for bacteriophage P1.[50] This phage specifies the *EcoPI* restriction system, which will reduce the plating efficiency of λ phage and may affect plasmid transformation as well. Therefore, cloning experiments should not utilize strains which are P1 lysogens. For this reason, JM103 cannot be recommended.

[63] K. Saing, H. Orii, Y. Tanaka, K. Yanagisawa, A. Miura, and H. Ikeda, *Mol. Gen. Genet.* **214,** 1 (1988).

[64] A. Miura-Masuda and H. Ikeda, *Mol. Gen. Genet.* **220,** 345 (1990).

[65] A. Ullman, F. Jacob, and J. Monod, *J. Mol. Biol.* **24,** 339 (1967).

p3 complementation by supF cloning vehicles. p3 is a 60-kb plasmid[66] derived from the plasmid RP1. Plasmid p3 carries a *tra⁻* mutation to prevent conjugation, amber mutations in genes coding for resistance to ampicillin and tetracycline, and an intact kanamycin resistance gene. Cells containing p3 are therefore Km^r Ap^s Tet^s. The amber mutations are suppressible by *supF*, and thus cells that contain p3 and are transformed with cloning vectors carrying *supF* are resistant to both ampicillin and tetracycline.

This system was designed to facilitate screening λ libraries for sequences which are homologous to a probe sequence cloned in *supF* vectors such as πVX.[66] More recently πVX-derived plasmids such as pAN7 and pAN13[67] or pHM3[68] have been used as plasmid vectors for other purposes where minimal plasmid size (830 base pairs) is desirable, which again requires host strains carrying p3 (e.g., MC1061/p3 and DH10B/p3).

B. Strain Recommendations for Transformation of Escherichia coli

1. General Considerations. Three factors are important in choosing a strain for generting cDNA or genomic libraries in plasmid vectors. The first is the transformation efficiency of the particular strain, since the competence of the cells can affect the ability to isolate the clone of interest. In general, fewer transformation reactions with a ligation mix are required to generate a representative library if the cells are highly competent. The second consideration is possible restriction of the clone by the bacterial cell (see Sections V1,A,2 and VI,A,3). The final consideration is the stability of the vector containing the insert (see Section VI,A,1).

2. cDNA Cloning in Plasmid Vectors. The preparation of cDNA libraries does not usually include procedures which would result in restriction by the *hsd* system or the methylation-dependent *mcrA, mcrBC,* or *mrr* systems. In addition, most cDNA clones should be stable in *recA* host strains. Therefore, for most cDNA applications, standard *recA* strains such as DH1, DH5, DH5α, and HB101 will suffice. However, some cDNA cloning procedures outlined by Sambrook *et al.*[69] involve methylation of the DNA. Procedures of this type would motivate the use of a MDRS-deficient strain in order to prevent possible restriction of the clone of interest. For cDNA procedures involving methylation of the DNA, MDRS-

[66] B. Seed, *Nucleic Acids Res.* **11,** 2427 (1983).

[67] C. Lutz, W. C. Hollifield, B. Seed, J. M. Davie, and H. V. Huang, *Proc. Natl. Acad. Sci. U.S.A.* **84,** 4379 (1987).

[68] A. Aruffo and B. Seed, *Proc. Natl. Acad. Sci. U.S.A.* **84,** 8573 (1987).

[69] J. Sambrook, E. Fritsch, and T. Maniatis, *in* "Molecular Cloning: A Laboratory Manual, 2nd Edition" (C. Nolan, ed.), p. 4. Cold Spring Harbor Laboratory, Cold Spring Harbor, New York, 1989.

deficient *recA*⁻ strains such as DH5αMCR and DH10B are recommended. The genotypes and transformation efficiencies of the more frequently used strains are listed in Table II. Strains which may be useful for the generation of cDNA subtraction libraries in phagemid vectors are described in Table IV. A number of the strains listed in these tables are commercially available as frozen competent cells.

3. Genomic Cloning/Plasmid Rescue in Plasmid Vectors. Many of the same considerations used to select a host for constructing cDNA libraries also apply to the construction of genomic libraries. As with cDNA libraries, most but not all genomic sequences will be stable in a *recA*⁻ host strain. However, in contrast to cDNA cloning, most eukaryotic genomic DNA as well as many prokaryotic genomic DNA sequences are methylated. Therefore, the preferred strain for generating representative genomic libraries should have the *recA*⁻, *mcrA*⁻ mutations and contain a deletion eliminating *hsdR, mcrBC,* and *mrr*.[55] The strains which can be recommended for genomic cloning include DH5αMCR and DH10B (Table II).

4. Genetic and Physiological Stabilization of Unstable Inserts. Strains carrying various mutations in the repair and recombination genes have been successfully used to stabilize cloned sequences containing direct repeats or palindromes, as is summarized in Table V. However, no data presently exist on the transformation efficiency of these strains, and therefore they cannot be unambiguously recommended for constructing cDNA or genomic libraries at this time. However, one or another of these could be considered as a supplement for a cloning experiment where one of the well-studied strains is employed as the primary host. Moreover, the knowledge of the effects of the various recombination/suppressor genes described above and summarized in Table V is presently being used to derive a series of strains in genetic backgrounds showing high competence for transformation and general utility for molecular cloning (e.g., DH5αMCR and DH10B). It is expected that such strains will prove useful as a set to be used for genomic DNA cloning, such that if the genome of an organism is transformed into the set, among the collection of strains all sequences will be clonable and stable. It is our current experience that no single strain can stabilize every DNA sequence in a cloning vector, and thus a set which collectively will stabilize any sequence is the best genetic alternative.

In addition to genetic stabilization of DNA sequences during molecular cloning, we have observed that many unfavorable sequences can be preferentially stabilized by growing transformed cells at lower temperatures (25°–30°) in very rich medium, and by not allowing such cultures to reach saturation. Thus, if difficult sequences are being cloned, or if a representative genomic library is being generated, incubate the cells following trans-

TABLE IV

Escherichia coli Strains for DNA Cloning Using M13/Phagemid Vectors

Strain	hsdR	recA	Genes carried by F' plasmid	Other markers	Typical yields (µg/ml culture) for SS phagemid DNA	Maximal chemical competence (XFE; colonies/µg pBR322)	Ref.
JM109	−	−	traD36 proAB+ lacIq lacZ ΔM15		5.2	$1–5 \times 10^8$	a
DH5α/F'	−	−	F'	lacZ ΔM15	2.4	$1–5 \times 10^8$	b
DH5α/F'IQ	−	−	proAB+ lacIq lacZ ΔM15 Tn5 [Kmr]		2.1	$1–5 \times 10^8$	c
71/18	+	+	proAB+ lacIq lacZ ΔM15		(n.d.)	$1–5 \times 10^8$	d
MV1184	+	Δ	traD36 proAB+ lacIq lacZ ΔM15				e
CJ236	+	+	pCJ105 (Cmr)	dut1 ung1	(n.d.)	$0.1–1 \times 10^8$	f
NM522	Δ	+	proAB+ lacIq lacZ ΔM15		3.9	$1–3 \times 10^8$	g
DH11S	Δ	Δ	proAB+ lacq lacZ ΔM15	mcrA Δ(mrr hsdRMSZ mcrBC) deoR	3.0	$1–3 \times 10^8$	h

[a] C. Yanisch-Perron, J. Vieira, and J. Messing, *Gene* **33**, 103 (1985).

[b] L. R. Liss, *BRL Focus* **9**(3), 13 (1987).

[c] J. Jessee and K. Blodgett, *BRL Focus* **10**(4), 69 (1988).

[d] L. Dente, G. Cesareni, and R. Cortese, *Nucleic Acids Res.* **11**, 1645 (1983).

[e] J. Vieira and J. Messing, this series, Vol. 153, p. 3.

[f] T. A. Kunkel, J. D. Roberts, and R. A. Zakour, this series, Vol. 154, p. 367.

[g] J. A. Gough and N. E. Murray, *J. Mol. Biol.* **166**, 1 (1983).

[h] DH11S was derived from strain NM522 by M. Smith, J. Jessee, and F. Bloom, and is available from BRL, Gaithersburg, MD.

TABLE V
GENETIC LOCI THAT HAVE STABILIZED OTHERWISE UNSTABLE DNA SEQUENCES
DURING CLONING IN *Escherichia coli*

Marker	Enzymatic activity	DNA structures stabilized	Ref.
recA	(DNA binding)	Most sequences	a
recBC	Exonuclease V	Palindromes in λ vectors[h]	b,c
recD	Exonuclease V	Palindromes in λ vectors[i]	d
recE	Exonuclease VIII	None identified	
recF	None identified	Palindrome in plasmid vector[j]	e
sbcB	Exonuclease I	Inverted repeat sequences in pBR322[k]	f
sbcC	None identified	Palindromes in λ or plasmid vectors	d
recN	None identified	Cosmid clones derived from mouse, Chinese hamster, or human DNA[j]	g
recJ	Single-stranded exonuclease	Cosmid clones derived from mouse, Chinese hamster, or human DNA[j]	g

[a] A. J. Clark and K. B. Low, *in* "The Recombination of Genetic Material" (K. B. Low, ed.), p. 155. Academic Press, New York and London, 1988. References given are for stabilization. The primary references for the various genetic loci can be found in the text.

[b] D. R. F. Leach and F. W. Stahl, *Nature (London)* **305**, 448 (1983).

[c] A. R. Wyman, L. B. Wolfe, and D. Botstein, *Proc. Natl. Acad. Sci. U.S.A.* **82**, 2880 (1985).

[d] A. F. Chalker, D. R. F. Leach, and R. G. Lloyd, *Gene* **71**, 201 (1988).

[e] R. Boissy and C. R. Astell, *Gene* **35**, 179 (1985).

[f] J. Collins, G. Volckaert, and P. Nevers, *Gene* **19**, 139, (1982).

[g] M. Ishiura, N. Hazumi, T. Koide, T. Uchida, and Y. Okada, *J. Bacteriol.* **171**, 1068 (1989).

[h] Stability of clone tested also requires *sbcBC* mutations.

[i] Stability of clone also requires *sbcC* mutation.

[j] Stability of clone also requires *recBC sbcBC* mutations.

[k] Stability of clone also requires *recB* mutation.

formation in SOC medium at 30°; then establish colonies at 30° on agar plates composed of SOB medium, and expand individual colonies in TB medium at 30°, collecting the cells for DNA isolation mid to late in the period of logarithmic growth. The media are described in Section II,C,2 above.

5. Strain Recommendations: M13/Phagemid Cloning. Several factors are important in selecting a strain for transformation with filamentous bacteriophages. The first is the transformation efficiency of the strain, and the second is the yield of single-stranded M13 or phagemid DNA. The third is the stability of the insert. Most of the strains listed in Table IV can be transformed with efficiencies exceeding 10^8 transformants/μg M13 RF

TABLE VI
Escherichia coli Strains Lacking dam and dcm Methylases

Strain	hsdR	hsdM	dam[a]	dcm	Other markers	Maximal chemical competence (*XFE*, colonies/μg pBR322)
GM37 .	−	+	−	−		$<10^6$
GM2159	+	+	dam13::Tn9	+	recF	$<10^6$
GM2163	−	+	dam13::Tn9	−	mcrA,mcrB	$<10^6$
DM1	−	+	dam13::Tn9	−	mcrB	$>10^7$

[a] Some *dam⁻* alleles (including the one in GM37) are subjects to reversion to *dam⁺*. *dam13*::Tn9 (Cm^r) is more stable in this regard and thus is recommended. The GM strains are from M. G. Marinus (University of Massachusetts Medical School, Worcester, MA 01655). The *dam13*:Tn9(Cm^R) marker is described in M. Marinus *et al.*, *Mol. Gen. Genet.* **192**, 288 (1983). DM1 is from Fred Bloom (BRL, Gathersburg, MD).

DNA. The recommended strains for M13 cloning are the *recA* strains JM109, DH5αF' and DH5α/F'IQ. In addition, most of these strains yield comparable quantities of single-stranded M13 DNA. However, the yields of single-stranded phagemid DNA can vary. The *recA⁻* strains that produce the highest yields of phagemid DNA are JM109 and DH11S.

6. *Strain Recommendations: dam⁻ dcm⁻ Strains*. Several *dam⁻ dcm⁻* strains are listed in Table VI. Strains GM2163, GM37, and DM1 are recommended for preparing DNA for digestion with methylation-sensitive restriction enzymes, since each is *hsdR, dam⁻*, and *dcm⁻*. None transform with high efficiencies, however, and, because of their inherent disposition to mutation, none is recommended for routine cloning or plasmid DNA preparation.

7. *Sources of Escherichia coli Strains*. There are two primary repositories of *E. coli* strains for distribution to the scientific community. The most extensive collection is that of the *E. coli* Genetic Stock Center (ECGS), which is located at Yale University (Department of Biology, 255 OML, P.O. Box 6666, New Haven, CT 06510-7444). A more limited collection is distributed by the American Type Culture Collection (12301 Parklawn Drive, Rockville, MD 20852-1776). Several of the MDRS mutants can be obtained from New England Biolabs (32 Tozer Road, Beverly, MA 01915-5510) or Bethesda Research Laboratories (Dr. Fred Bloom, Life Technologies, Inc., 8717 Grovemont Circle, P.O. Box 6009, Gaithersburg, MD 20877). In addition, it is notable that an increasing number of *E. coli* strains are commercially available as preparations of frozen chemically competent cells (from BRL, Stratagene, La Jolla, CA, and Clontech, Palo Alto, CA) or as frozen cells prepared for electroshock transformation (BRL).

C. Other Gram-Negative Bacteria

Of the other gram-negative bacteria only *Salmonella typhimurium* has been characterized with regard to genes that can influence transformation.[70,71] Many of the genetic markers which are relevant to the choice of an *E. coli* strain for transformation/cloning are also found in *Salmonella typhimurium*.[72] The correlation of the linkage maps of *S. typhimurium* and *E. coli* indicates that *recA*, *recBC*, *endA*, and *hsd* mutants have been isolated in *Salmonella*.[72] *Salmonella typhimurium* LT2 carries three chromosomally encoded restriction systems, *LT*, *SA*, and *SB*[73] which can influence cloning of heterologous DNA. An improved chemical transformation protocol for *S. typhimurium* has been developed.[74] Recently, it has been shown that transformation of *galE⁻* strains occurs at a 100-fold higher efficiency than in *galE⁺* strains using this calcium chloride-based chemical transformation method.[75] Knowledge of the *hsd* and *galE* effects has led to the development of an effective *Salmonella* cloning host, JR501, which carries *hsdSA29*, *hsdSB121*, *hsdL6*, and *galE719*.[75]

D. Gram-Positive Bacteria

Many gram-positive bacteria have been described which are capable of taking up exogenous DNA during certain portions of their growth phase. These naturally competent species include *Streptococcus pneumoniae*, *Streptococcus sanguis*, *Bacillus subtilis*, and *Haemophilus influenzae*. Other genera include *Streptomyces*, *Pseudomonas*, *Acinetobacter*, *Moraxella*, *Neisseria*, *Achromobacter*, *Azotobacter*, and *Deinococcus*. A number of genera including *Bacillus*, *Streptomyces*, and *Pseudomonas* have proved useful for recombinant DNA experiments. A recent review summarizing transformation in these systems can be consulted for references on individual genera/species.[76]

Plasmid Stability in Bacillus subtilis. One mutation in *B. subtilis*, *recE4*,[77] reduces homologous recombination and is useful for stabilizing plasmids employed for cloning purposes. *recE⁻* strains transform as well

[70] E. Lederberg and S. Cohen, *J. Bacteriol.* **119,** 1072 (1974).

[71] P. MacLachlan and K. Sanderson, *J. Bacteriol.* **161,** 442 (1985).

[72] K. Sanderson and J. Hurley, *in* "*Escherichia coli* and *Salmonella typhimurium*: Cellular and Molecular Biology" (F. C. Neidardt, ed.), J. L. Ingraham, B. Magasanik, K. B. Low, M. Schaechter, and H. E. Umbarger, eds.), p. 877. American Society for Microbiology, Washington, D.C., 1987.

[73] L. Bullas, C. Colson, and B. Neufield, *J. Bacteriol.* **141,** 275 (1980).

[74] S. Tsai, R. Hartin, and J. Ryu, *J. Gen. Microbiol.* **135,** 2561 (1989).

[75] J. Ryu and R. Hartin, *BioTechniques* **8,** 43 (1990).

[76] G. Stewart and C. Carlson, *Annu. Rev. Microbiol.* **40,** 211 (1986).

[77] D. Dubnau and C. Cirigliano, *J. Bacteriol.* **117,** 488 (1974).

as $recE^+$ strains,[78] however the plasmid DNA must be concatenated or multimeric.[79] *Bacillus subtilis* can be transformed by monomeric plasmid DNA provided the cell is $recE^+$ and contains a resident plasmid with sequences homologous to the incoming DNA.[80] Monomeric plasmid DNA can also be introduced into protoplasts of *B. subtilis* at high efficiency.[81] It is likely that other genes will influence the transformation and stability of recombinant plasmids in gram-positive bacteria, given the precedents from *E. coli*. This possibility should motivate the screening of different strains of gram-positive bacteria with unstable plasmids to identify phenotypically those strains most suitable for cloning.

For example, Haima *et al.*[82] studied the factors which influence the shotgun cloning of heterologous DNA in *B. subtilis* in an attempt to improve the efficiency of the process. The authors constructed a restriction-deficient strain, 6GM, in a highly competent strain background and then compared stain 6GM (r_m^-, m_m^+) to an isogenic restriction-proficient strain. The strain 6GM was found to improve the cloning of heterologous DNA. Haima *et al.*[83] also extended the use of strain 6GM by incorporating into this strain a plasmid, pGHS1, which expresses both *lacI*q and *lacZ* ΔM15. A second compatible plasmid, pHPS9, which expresses the *lacZα* fragment was also constructed. The transformation of this second plasmid into 6GM/p6HS1 resulted in α complementation, thus providing a cloning system that discriminates similarly to that used extensively in *E. coli*.

Acknowledgments

We wish to thank Leslie Spector for editorial work and preparation of the manuscript. D.H. acknowledges research support from the National Cancer Institute (RO1 CA47632, RO1 CA45234) and the Monsanto Company.

[78] T. Gryczan and D. Dubnau, *Proc. Natl. Acad. Sci. U.S.A.* **75,** 1428 (1978).
[79] U. Canosi, A. Iglesias, and T. Trautner, *Mol. Gen. Genet.* **181,** 434 (1981).
[80] S. Contente and D. Dubnau, *Plasmid* **2,** 555 (1979).
[81] S. Chang and S. Cohen, *Mol. Gen. Genet.* **168,** 111 (1979).
[82] P. Haima, S. Bron, and G. Venema, *Mol. Gen. Genet.* **209,** 335 (1987).
[83] P. Haima, D. Van Sinderen, H. Schotting, S. Bron, and G. Venema, *Gene* **86,** 63 (1990).

[5] *In Vivo* Mutagenesis

By PATRICIA L. FOSTER

Introduction

Mutation of bacteria *in vivo* by chemical and physical means is a powerful method by which to generate genetic variants. When searching for mutants in an unknown gene or in uncharacterized regions of a cloned gene, chemical and/or physical mutagenesis can be combined with insertional mutagenesis to obtain the maximum number of alleles for phenotypic analysis. In general, base substitutions are the most useful mutations with which to investigate both the *in vivo* activities and the structure–function relationship of the gene product of interest. By using the appropriate mutagen and bacterial strain, it is even possible to generate specific classes of mutations. The methods outlined below are designed to aid the geneticist as well as those who wish to investigate mutagenic and DNA repair processes in *Escherichia coli* and *Salmonella typhimurium*.

Bacterial Strains

Escherichia coli and *S. typhimurium* have mutagenic and accurate repair processes that act on DNA lesions. After DNA damage both types of repair are induced as part of the SOS response.[1] In general, mutagens can be divided into those which do not require the SOS response to be mutagenic, such as S_N1 alkylating agents (e.g., *N*-methyl-*N'*-nitro-*N*-nitrosoguanidine) and base analogs (e.g., 2-aminopurine), and those which do, such as S_N2 alkylating agents (e.g., methyl methane sulfonate) and mutagens that produce bulky lesions in the DNA (e.g., UV light). *Escherichia coli* is proficient for SOS mutagenesis unless mutated in *recA* or *umuDC*, but SOS mutagenesis in *S. typhimurium* is considerably weaker. This deficiency can be overcome by using strains with plasmid pKM101,[2] which carries *mucAB*[+], a more active version of the *umuDC* operon.[3] Derivatives of this plasmid that are deleted for conjugal and slow-growth functions[4] also can be used to enhance SOS mutagenesis in *E. coli*. Indeed, with some mutagens, such as aflatoxin B_1 and benzo[*a*]pyrene, the yield

[1] G. C. Walker, *Microbiol. Rev.* **48**, 60 (1984).
[2] K. E. Mortelmans and B. A. D. Stocker, *J. Bacteriol.* **128**, 271 (1976).
[3] G. C. Walker and P. P. Dobson, *Mol. Gen. Genet.* **172**, 17 (1979).
[4] P. L. Langer, W. G. Shanabruch, and G. C. Walker, *J. Bacteriol.* **145**, 1310 (1981).

TABLE I
DNA REPAIR PATHWAYS

Repair pathway	Substrates	Mutations
UvrABC excision repair	Bulky lesions	*uvrA,B,C*
Adaptive response	O^6-Alkylguanine, O^4-alkylthymine	*ada*
Adaptive response	N^3-Alkylpurines, O^2-alkylpyrimidines	*alkA*
Photoreactivation	Pyrimidine dimers	*phr*
Abasic site repair	Abasic sites	*xth, nfo*
Uracil glycosylase	Uracil residues	*ung*
Methyl-directed mismatch repair	Mismatches	*mutH,L,S, uvrD*
Proofreading	Replication errors	*dnaQ/mutD*

of mutations even in *E. coli* is low unless *mucAB*[+] is present.[5] It is likely that the error-prone replication process elicited by the SOS response inserts adenines opposite noninformational DNA lesions[6]; thus, increasing the activity of the mutational process should amplify certain classes of mutations. From the limited data available, enhancing SOS mutagenesis by prior induction, use of SOS-constitutive mutants, or by supplying *mucAB*[+] activity increases the proportions of GC to TA transversions and all mutations at AT sites, although the specific effect obtained varies among mutagens and treatment.[7–12]

Strains mutant in various accurate DNA repair pathways are widely available and can be used to increase the yield of mutations.[13] Table I lists several such repair pathways and mutations that inactivate them. Strains defective in UvrABC excision repair are particularly useful to increase mutagenesis by a variety of mutagens. Some agents that make bulky DNA adducts, such as aflatoxin B_1 and benzo[*a*]pyrene, are barely mutagenic at all unless this repair pathway is inactivated. In contrast, the mutagenic-

[5] P. L. Foster, J. D. Groopman, and E. Eisenstadt, *J. Bacteriol.* **170,** 3415 (1988).

[6] L. Loeb, *Cell (Cambridge, Mass.)* **40,** 483 (1985).

[7] R. G. Fowler, L. McGinty, and K. E. Mortelmans, *J. Bacteriol,* **140,** 929 (1979).

[8] R. G. Fowler, L. McGinty, and K. E. Mortelmans, *Genetics* **99,** 25 (1981).

[9] P. A. Todd, C. Monti-Bragadin, and B. W. Glickman, *Mutat. Res.* **62,** 226 (1979).

[10] J. H. Miller and K. B. Low, *Cell (Cambridge, Mass.)* **37,** 675 (1984).

[11] L. B. Couto, I. Chadhuri, B. A. Donahue, B. Demple, and J. M. Essigmann, *J. Bacteriol.* **171,** 4170 (1989).

[12] E. Eisenstadt, J. K. Miller, L.-S. Kahng, and W. M. Barnes, *Mutat. Res.* **210,** 113 (1989).

[13] S. R. Kushner, *in "Escherichia coli and Salmonella typhimurium:* Cellular and Molecular Biology" (F. C. Neidhardt, J. L. Ingraham, B. Magasanik, K. B. Low, M. Schaechter, and H. E. Umbarger, eds.), p. 1044. American Society for Microbiology, Washington, D.C., 1987.

ity of DNA cross-linking agents such as mitomycin C and *cis*-diamminedichloroplatinum(II) (cisplatin) is greatly enhanced by UvrABC excision repair.[14,15] The photoreactivation pathway is only active against pyrimidine dimers, but, because these lesions induce the SOS response, photoreactivation can result in a virtual elimination of UV-induced mutation.[16] Since photoreactivation requires visible light, it can be circumvented by keeping UV-irradiated cells in the dark. Repair pathways can also be manipulated to yield specific classes of mutations. For example, after exposure to an alkylating agent, strains defective in O^6-alkylguanine DNA alkyltransferase (*ada*) have high frequencies of GC to AT transitions.[17] However, when strains defective in 3-methyladenine glycosylase (*alkA*) are induced for O^6-alkylguanine DNA alkyltransferase by exposure to low levels of an alkylating agent, a high proportion of mutations are AT transversions.[18]

Assays for Mutagenesis

Drug resistances are the most versatile assays for mutagenesis in most genetic backgrounds. Resistances to rifampicin (100 mg/liter), streptomycin (200 mg/liter), and nalidixic acid (40 mg/liter) are due to base substitutions.[19] Reversions of the amino acid auxotrophies common to many *E. coli* strains are particularly convenient for monitoring base substitutions (*hisG4, argE3*) and, in some cases, frameshift mutations (*trpE9777*).[20] In *S. typhimurium* a variety of revertible mutations in the *his* operon are available.[21] Specific classes of mutations can also be monitored. A set of revertible mutations in the *lacZ* gene of *E. coli* can detect each of the six possible base changes.[22] By using simple screens, the base changes that revert specific *his* alleles in *S. typhimurium* can also be identified.[23] A

[14] S. Kondo, H. Ichikawa, K. Iwo, and T. Kato, *Genetics* **66**, 187 (1970).

[15] J. Brouwer, P. van de Putte, A. M. J. Fichtinger-Schepman, and J. Reedijk, *Proc. Natl. Acad. Sci. U.S.A.* **78**, 7010 (1981).

[16] D. E. Brash and W. Haseltine, *J. Bacteriol.* **163**, 460 (1985).

[17] P. F. Schendel and P. E. Robins, *Proc. Natl. Acad. Sci. U.S.A.* **75**, 6017 (1978).

[18] P. L. Foster and E. Eisenstadt, *J. Bacteriol.* **163**, 213 (1985).

[19] J. H. Miller, "Experiments in Molecular Genetics." Cold Spring Harbor Laboratory, Cold Spring Harbor, New York.

[20] E. Eisenstadt, *in* "*Escherichia coli* and *Salmonella typhimurium:* Cellular and Molecular Biology" (F. C. Neidhardt, J. L. Ingraham, B. Magasanik, K. B. Low, M. Schaechter, and H. E. Umbarger, eds.), p. 1016. American Society for Microbiology, Washington, D.C., 1987.

[21] P. E. Hartman and S. L. Aukerman, *in* "Mechanisms of DNA Damage and Repair" (M. G. Simic, L. Grossman, and A. C. Upton, eds.), p. 407. Plenum, New York, 1985.

[22] C. G. Cupples and J. H. Miller, *Proc. Natl. Acad. Sci. U.S.A.* **86**, 5345 (1989).

[23] D. E. Levin and B. N. Ames, *Environ. Mutagen.* **8**, 9 (1986).

plasmid is available which will allow GC to TA transversions to be monitored by the induction of ampicillin resistance.[24]

Although only a few mutagenic events can give rise to drug resistances or revert amino acid auxotrophies, a wide variety of mutagenic events lead to loss of gene function. However, assays for gene knockout are generally less convenient. In *E. coli* LacI⁻ mutants can be selected for by growth on the noninducing substrate phenyl-β-D-galactoside. More commonly, loss of gene function, for example, loss of the ability to metabolize a carbohydrate, must be screened for on indicator plates or identified by replica plating. A variety of additional mutational assays are given in Eisenstadt[20] and Miller.[19]

General Methods

The sensitivity and reproducibility of mutagenesis experiments are greatly enhanced by using exponentially growing (log-phase) cells. This is easily accomplished by diluting a saturated culture 1 : 50 into fresh Luria broth (LB)[19] and growing the cells at 37° for 2–3 hr. The cell density should be about 6×10^7 cells/ml, which corresponds to Klett 65 or OD_{600} 0.5. The cells are chilled, centrifuged, washed once, resuspended in the appropriate buffer (see below), and kept on ice until treatment. Most of the methods given below are designed so that disposable microcentrifuge tubes can be used for mutagen exposure. To generate dose–response curves, either the mutagen concentration or the time of exposure can be varied. After exposure to the mutagen, cells are diluted with cold buffer, centrifuged, washed twice, resuspended in cold buffer, and kept on ice. Appropriate dilutions (in 0.85% NaCl plus 0.001% gelatin to prevent clumping) are plated on LB plates to obtain the level of survival relative to untreated control cells.

Mutations are "fixed" into the DNA by replication; thus, a period of outgrowth under nonselective conditions is necessary after mutagen exposure to obtain the maximum number and an unbiased yield of mutations. As soon as possible after treatment, the washed cells are diluted into LB and grown to saturation at 37° with good aeration. These cultures are then titered and plated for mutants. The mutation frequency is the number of mutants divided by the total number of viable cells plated (assuming that all the cells, including the mutants, had equal growth rates in the nonselective medium).

If too few viable cells are inoculated for outgrowth, the progeny of a

[24] P. L. Foster, G. Dalbadie-McFarland, E. F. Davis, S. C. Schultz, and J. H. Richards, *J. Bacteriol.* **169**, 2476 (1987).

small number of mutants can dominant the resultant population, giving a large proportion of siblings. To avoid this problem, conditions are adjusted so that at least 10^6 viable cells are inoculated. If high mutation rates are desired for a general mutant search, or if a particularly sensitive strain is being used, the number of cells treated and the volume of culture for outgrowth must be increased to compensate for the low cell survival.

Reversions can be assayed after outgrowth, as above, or immediately after mutagen exposure by plating the cells on selective plates with a small amount of the required nutrient.[25] For example, to assay for reversion of an amino acid auxotrophy, approximately 10^7 viable cells are plated on a minimal plate containing 100 nmol of the required amino acid. The cells will grow on the plate until the amino acid is exhausted; in 48 hr revertant colonies will appear on the background lawn. Since with this procedure every mutant arises independently, the mutation frequency is the number of mutants divided by the number of viable cells originally plated (although a "cell density artifact" can be introduced[26]).

"Plate mutagenesis" can be used to quickly test the mutagenic response of a strain or to check the potency of a mutagen. This procedure works well when the mutagenic assay is reversion of an amino acid auxotrophy. Both the cells and the mutagen are added to top agar and plated on minimum medium with a limiting amount of the required amino acid (see above). Alternatively, a small volume of the mutagen is added to a filter-paper disk placed on the surface of the plate.[25]

Some agents are only mutagenic if activated to short-lived, reactive intermediates. Bacteria do not have cytochrome P-450 enzymes; thus, when using compounds that require these enzymes, activation must be done externally. Aroclor-induced S9-fraction liver microsomes work well for most mutagens and are commercially available. The procedure is described by Ames *et al.*[25] The reaction mixture is 0.1 M sodium phosphate buffer, pH 7.4, with 8 mM MgCl$_2$, 33 mM KCl, 5 mM glucose 6-phosphate, 4 mM NADP, and 20–80 μl/ml microsomes. Small aliquots of glucose 6-phosphate (1 M) and microsomes are kept at $-20°$ and $-80°$, respectively, thawed on ice, and discarded after use. NADP (0.1 M) is also kept at $-20°$ and thawed on ice, but it can be refrozen and used several times. Cells and mutagen are added to the reaction mixture and incubated for 1 hr at $37°$. A range of concentrations of microsomes should be tested at each mutagen concentration to obtain the best combination. The S9 microsome

[25] B. N. Ames, J. McCann, and E. Yamasaki, *Mutat. Res.* **31,** 347 (1975).
[26] E. M. Witkin, *Bacteriol. Rev.* **40,** 869 (1976).

mixture can also be incorporated into the top agar when doing plate mutagenesis (see above).

Specific Methods

N-Methyl-N'-nitro-N-nitrosoguanidine

A stock solution of N-methyl-N'-nitro-N-nitrosoguanidine (MNNG) is prepared at 1 mg/ml (7 mM) in 100 mM citrate buffer, pH 5.5 (23 mM citric acid, 77 mM sodium citrate). MNNG is inactivated at higher pH and is unstable in phosphate buffer. The solution can be warmed briefly to 37° to dissolve the MNNG. The stock is then dispensed in small aliquots, stored at $-20°$, used once, and discarded.

Mid log-phase cells are centrifuged, washed once with 1/2 volume cold citrate buffer, and resuspended at $10\times$ in the citrate buffer. MNNG is added to 100 μl of cells at 5–100 μg/ml for 5 min at room temperature. One-half milliliter of cold 100 mM sodium phosphate buffer, pH 7 (39 mM NaH$_2$PO$_4$, 61 mM Na$_2$HPO$_4$), is added, and the cells are centrifuged, washed twice, and resuspended in 0.5 ml of the phosphate buffer. After plating for survival, the entire 0.5 ml is added to 5 ml of LB and grown overnight. With reversion assays, mutation frequencies of 10^{-5} per viable cell with little killing are typical.

Because the major mutagenic lesion induced by MNNG, O^6-methyl-guanine, is not very lethal, high mutation frequencies can be produced with this mutagen. For example, exposure to 1 mg/ml for 15 min gave a survival of 10% and a mutation frequency to Rif^r of about 10^{-4} per viable cell. For a mutant search, this level of mutagenesis will give a good probability of recovering mutations in the desired gene. However, MNNG-induced mutations tend to cluster at the replication point, giving a high frequency of double mutants.[27] In addition, the mutations that result will be almost exclusively GC to AT transitions.

The adaptive response to alkylating agents is induced by low, nonmutagenic concentrations of MNNG.[28] For example, we achieved good induction by diluting overnight cultures grown in Vogel–Bonner minimal medium[29] 1 : 5 into fresh minimal medium containing 0.5 to 1.0 μg/ml MNNG and incubating the cells with gentle aeration for 4 hr at 32°. Wild-type cells were not mutated by this treatment, but cells mutant in *alkA* were, with a high proportion of the mutations occurring at AT sites.[18]

[27] N. Guerola, J. L. Ingraham, and E. Cerdá-Olmedo, *Nature (London)* **230,** 122 (1971).
[28] L. Samson and J. Cairns, *Nature (London)* **267,** 281 (1977).
[29] H. J. Vogel and D. M. Bonner, *J. Biol. Chem.* **218,** 97 (1956).

Methyl Methane Sulfonate

Methyl methanesulfonate (MMS) is a volatile liquid (11.8 M when pure). It is stored at 4° in the dark and used only in a chemical hood.

Mid log-phase cells are centrifuged, washed once with 1/2 volume cold E salts[29] (57 mM K$_2$HPO$_4$, 9.5 mM citric acid, 17 mM NaNH$_4$HPO$_4$, 0.8 mM MgSO$_4$, pH 7; this buffer is used for convenience—an equivalent 0.1 M phosphate buffer can be used), and resuspended at 10× in the E salts. To 100 μl of cells, 1–10 μl of MMS is added, mixed well, and incubated for 5 min at room temperature. Alternatively, 5–25 μl of a 1 : 100 dilution of MMS in distilled water is added to 100 μl of cells for 30 min at 37°. After exposure, 0.5 ml of cold E salts is added, and the cells are centrifuged, washed twice, and resuspended in 0.5 ml of E salts. The MMS is not inactivated, only diluted, so the entire procedure should be as rapid as possible. Plating for survival and outgrowth are as above. With reversion assays, mutation frequencies of 10^{-6} per viable cell at 30% survival are typical. MMS induces all classes of base substitutions with a high proportion of GC to TA transversions.[18,29] Unlike MNNG, mutagenesis by MMS is SOS-dependent.[9]

Ethyl Methane Sulfonate

Ethyl methane sulfonate (EMS) is also a volatile liquid (9.2 M when pure). It should be handled as is MMS (see above).

The following method is from Cupples and Miller.[22] Mid-log cells are centrifuged, washed twice in cold A buffer[19] [60 mM K$_2$HPO$_4$ 33 mM KH$_2$PO$_4$, 7.6 mM (NH$_4$)$_2$SO$_4$, 1.7 mM sodium citrate, pH 7], and resuspended at 2× in cold A buffer. EMS is added at 1.4% to aliquots of the cell suspension in culture tubes, the tubes are sealed with tape, and the cultures are incubated at 37° with gentle aeration. After various times up to 60 min, the cells are centrifuged, washed twice with A buffer, and resuspended in the same volume of A buffer. After plating for survival, 0.5 ml is inoculated into 10 ml of LB and grown overnight. With a 30-min exposure, this procedure gave a mutation frequency of 4 × 10^{-4} Rif^r per viable cell with 56% survival.[22] Like MNNG, EMS is mutagenic in the absence of SOS activity[14] and induces predominately GC to AT transitions.[30]

UV Light

Germicidal (shortwave) UV lamps vary greatly in their intensity, but a new 8-W bulb gives a fluence of about 1 J/m^2/sec at 20 inches. The UV

[30] C. Coulondre and J. H. Miller, *J. Mol. Biol.* **117,** 577 (1977).

fluence can be measured with a UV light meter, but it is often easier to generate a dose–response curve with the strain to be used. To stabilize the light fluence, the lamp must be turned on at least 20 min before use.

Mid log-phase cells are centrifuged, washed once with E salts, and resuspended at $1 \times$ in the E salts. One-milliliter aliquots are distributed to sterile 50-mm glass petri dishes; it is important that the bottom of the dish be completely covered with a thin layer of the cell suspension. Working in the dark, the top of each dish is removed and the dish placed under the lamp for an appropriate time (a sheet of foil can be used as a shield before and after exposure). To prevent shading the cells are gently swirled or stirred during exposure. For uvr^- strains, $2-5$ J/m^2 corresponds to approximately 50% killing, whereas 10 times this fluence is required for uvr^+ strains. Immediately after exposure and for all subsequent manipulations, the cells must be protected from visible light. No washing is necessary. Plating for survival and outgrowth are as above except plates and cultures are incubated in the dark. With reversion assays, mutation rates of 10^{-5} at 50% survival are typical. UV light induces predominantly GC to AT transitions but also induces all other base changes, frameshifts, and deletions.[31] Mutagenesis by UV light is SOS-dependent.[26]

Depending on the mutational target, *S. typhimurium* may be poorly mutated by this method.[32] For reversion of the *hisG46* allele, good mutagenesis was achieved by plating 10^8 mid log-phase cells on minimal plates containing 100 nmol histidine, incubating the plates for 2 hr at 37°, and then exposing the plates to UV light.[12]

Aflatoxin B₁

Aflatoxin B$_1$ (AFB) is inactivated by water and light. The stock is prepared by adding the appropriate volume of dichloromethane (CH_2Cl_2) to the vial of AFB as received to give a 10 mM (3.12 mg/ml) solution (this is safer than weighing out an aliquot). Twenty-microliter aliquots are dispensed, evaporated under inert gas (N_2 or Ar) or vacuum, and stored at $-20°$ under dry N_2 or Ar in tightly capped tubes. Immediately before use 40 μl of dimethyl sulfoxide (DMSO) is added to an AFB aliquot, resulting in a 5 mM solution. This working stock is further diluted in DMSO to the appropriate concentrations. If immediately stored at $-20°$ under dry gas, the 5 mM stock can be used several times, but more dilute stocks are discarded after one use. Gas is dried by passing it through a column of $CaSO_4$; dichloromethane and DMSO are purged with and stored under the dry gas. Although extremely light-sensitive, AFB can be handled

[31] J. H. Miller, *J. Mol. Biol.* **182**, 45 (1985).
[32] C. M. Smith and E. Eisenstadt, *J. Bacteriol.* **171**, 3860 (1989).

under yellow light. It fluoresces blue under UV light, which property can be used to check for contamination.

Mid log-phase cells are centrifuged, washed with E salts, and resuspended at $10\times$ in cold E salts. Cells are diluted $1:5$ into S9 mixture (see above) and a 0.5-ml aliquot dispensed for each dose to be used. Five microliters of an appropriate AFB dilution in DMSO is added to give 5–100 μM final concentration. For each treatment, including the control, the same amount of DMSO (1% of the total volume) is added to the cells. The cells are then incubated 60 min at 37°, diluted with 0.5 ml of cold E salts, washed twice, and resuspended in 0.5 ml of cold E salts. Plating for survival and outgrowth are as above. Mutation rates vary greatly among different targets, but a frequency of 10^{-5} for Rif^r at 50% killing is typical for a uvr^- strain carrying a $mucAB^+$ plasmid (see above). AFB induces primarily GC to TA transversions.[33]

ICR-191

ICR-191, a chlorinated alkylacridine, can be obtained from Raylo Chemicals (8045 Argyll Road, Edmonton, Alberta, Canada T6C 4A9). It is dissolved in distilled water at 1 mg/ml (2.2 mM) and stored at $-20°$ in the dark. It can be thawed and refrozen several times.

The following method is from Miller.[19] Approximately 10^4 cells from a saturated culture are inoculated into 3-ml aliquots of minimal medium supplemented with 2% LB. Then 5–20 μg/ml ICR-191 is added (protect from light), and the cultures are grown with aeration overnight at 37° in the dark. The cells are then titered and plated for mutants.

Initially, a dose–response curve must be generated. The maximum yield of mutations occurs at a concentration that gives a half-saturated culture after 12 hr. Since many of the mutants will be siblings, several cultures at the optimum dose are grown and one mutant from each selected to obtain independent mutants. True mutation frequencies cannot be obtained with this method, but a ratio of mutants to nonmutants of $1:100$ is possible using a forward mutation assay. ICR-191 induces frameshift mutations.[19]

2-Aminopurine

2-Aminopurine (2-AP) is an analog of adenine. To make a 0.6 mg/ml stock solution from the free base, 1.2 mg/ml is dissolved in 0.1 N HCl and titrated to pH 7 with NaOH. An equal volume of $2\times$ LB (but $1\times$ NaCl) is added, and the stock is filter sterilized and stored at 4°.

[33] P. L. Foster, E. Eisenstadt, and J. H. Miller, *Proc. Natl. Acad. Sci. U.S.A.* **80,** 2695 (1983).

Cells are mutated when growing in the presence of 2-AP, and the method is essentially the same as for ICR-191.[19] A dose–response curve is generated by inoculating 10^2–10^3 cells from a saturated culture into LB with 0–600 μg/ml 2-AP. After the cultures reach saturation (which may take 48 hr), the cells are titered and plated for mutants. As with ICR-191, several cultures at the optimum dose are then grown to generate independent mutants. Mutation frequencies 100-fold above spontaneous are typical. 2-AP induces GC and AT transitions[30]; since 2-AP is a DNA base analog, 2-AP mutagenesis is SOS-independent.

1,2-Dibromoethane

1,2-Dibromoethane (EDB) is an extremely volatile, light-sensitive liquid. It is stored at room temperature in the dark and used only in a chemical hood.

The highest levels of mutation are obtained by mutating the cells with gaseous EDB. The following method is adapted from Rosenkranz.[34] Using a reversion assay, 10^8 cells are plated in top agar on minimum medium containing a limiting amount of the required nutrient (see above). A filter paper disk is fixed to the top of the inverted petri dish with a small amount of top agar. Working in a chemical hood, 1–20 μl of EDB is added to the disk, and the plate is closed and immediately sealed with Parafilm. The time elapsed between adding the EDB and sealing the plate is a critical variable and should be as short as possible. Mutants are scored after 2 days of incubation at 37°.

With this method we obtained up to 6000 EDB-induced mutants per plate, which was 10- to 100-fold higher than levels achieved with more conventional treatment methods.[35] EDB is poorly soluble in aqueous solutions unless dispersed in a solvent such as DMSO. Because of its lipid solubility, it disrupts cell membranes.[36] Thus, its acute toxicity to cells may be unrelated to DNA damage, and the gas-phase treatment may be successful because it allows long exposures to subtoxic doses. Although this method is not widely applicable, it may be useful for other lipid-soluble volatile mutagens. EDB induces predominantly transitions at GC and AT sites, most of which are SOS-independent.[35]

Mutator Strains

Mutator strains can be used to generate mutations in plasmid-borne genes. The most powerful mutators are those with defects in methyl-

[34] H. S. Rosenkranz, *Environ. Health Perspect.* **21,** 79 (1977).
[35] P. L. Foster, W. G. Wilkinson, J. K. Miller, A. D. Sullivan, and W. M. Barnes, *Mutat. Res.* **194,** 171 (1988).
[36] H. Brem, J. E. Coward, and H. S. Rosenkranz, *Biochem. Pharmacol.* **23,** 2345 (1974).

TABLE II

MUTATOR ALLELES

Allele	Increase in mutation frequency	Major type of mutation	Ref.
mutH,S,L	10^2-10^3	Transitions, frameshifts	a
mutT	10^3-10^4	AT to CG transversions	a
mutY	$>10^2$	GC to TA transversions	b
mutD5	10^3-10^4	Transitions, transversions frameshifts	a
dnaQ49	10^3-10^4	Transversions	c,d

[a] E. C. Cox, *Annu. Rev. Genet.* **10**, 135 (1976).
[b] Y. Nghiem, M. Cabrera, C. G. Cupples, and J. H. Miller, *Proc. Natl. Acad. Sci. U.S.A.* **85**, 2709 (1988).
[c] T. Horiuchi, H. Maki, and M. Sekiguchi, *Mol. Gen. Genet.* **163**, 277 (1978).
[d] R. Piechocki, D. Kupper, A. Quinones, and R. Langhammer, *Mol. Gen. Genet.* **202**, 162 (1986).

directed mismatch repair (*mutH, mutL,* and *mutS*), in G–A mismatch repair (*mutT* and *mutY*), or in the proofreading activity of DNA polymerase (*dnaQ/mutD*). Table II lists the increases in mutation frequency and the classes of mutations induced by these defects. To generate mutants, plasmid-containing strains of the mutators are grown to saturation from a small inoculum (10^3-10^4 cells/ml), and the DNA is isolated and transformed into another strain.

The mutation rates conferred by *mutD5* and *dnaQ49*, both alleles of *dnaQ*, which encodes the ε subunit of DNA polymerase III, are dependent on growth conditions. When grown in rich medium, *mutD5* strains have mutation rates 10- to 100-fold greater than when grown in minimal medium.[37] The mutation rates of *dnaQ49* strains are similarly increased by temperature and by growth in salt-free rich medium.[38] It has been shown for *mutD5*, and is likely to also be true for *dnaQ49*, that the increase in mutation rate is in part due to saturation of the methyl-directed mismatch repair pathway.[39] Thus, the specificity of these mutators will tend to change from transversions to transitions as the mutation rates increase.

Safety

All mutagens are potential carcinogens, and many are also acutely toxic. The safest procedure is to confine all work with mutagens to a chemical hood with designated pipettors, centrifuges, water baths, etc.

[37] E. C. Cox and D. L. Horner, *Genetics* **100**, 7 (1982).
[38] T. Horiuchi, H. Maki, and M. Sekiguchi, *Mol. Gen. Genet.* **163**, 277 (1978).
[39] R. M. Schaaper, *Proc. Natl. Acad. Sci. U.S.A.* **85**, 2432 (1988).

Gloves, laboratory coats, and safety glasses (UV opaque for work with UV light) should be worn. After cells are washed free of mutagen, work can be continued at the laboratory bench. Disposal of contaminated material should be as designated in the safety sheet supplied by the manufacturer.

Acknowledgments

To assemble the methods described here, I drew upon the experience and knowledge of E. Eisenstadt, J. H. Miller, C. Mark Smith, and J. Cairns, all of whom I thank. The work was supported by U.S. Public Health Service Grant CA37880 awarded to the author by the National Cancer Institute.

[6] Efficient Site-Directed Mutagenesis Using Uracil-Containing DNA

By THOMAS A. KUNKEL, KATARZYNA BEBENEK, and JOHN McCLARY

Introduction

Oligonucleotide-directed mutagenesis is a widely used procedure for studying the structure and function of DNA and the macromolecules for which it codes.[1] The most commonly used strategy for site-directed mutagenesis is to clone the segment of DNA to be mutated into a vector whose DNA can be obtained in single-stranded form. An oligonucleotide partially complementary to the region to be altered, but containing the mutation to be introduced, is hybridized to the single-stranded DNA. A complementary strand is synthesized by DNA polymerase using the oligonucleotide as a primer. Upon completion of synthesis, ligase is used to seal the nick in the new strand, and the double-stranded DNA, homologous except for the intended mutation, is then used to transform an *Escherichia coli* host strain. This yields progeny carrying the oligonucleotide-directed mutation as well as progeny that have the original genotype.

The efficiency of site-directed mutagenesis, that is, the proportion of progeny containing the desired sequence alteration, depends on the quality of each of the steps in the procedure. The number of progeny clones that must be monitored to obtain the desired mutant increases as the efficiency of mutagenesis decreases. This has led to the development of improved methods to screen mutants at the phenotypic and genotypic level,[1,2] per-

[1] M. J. Zoller and M. Smith, *in* "Recombinant DNA Methodology" (R. Wu, L. Grossman, and K. Moldave, eds.), p. 537. Academic Press, San Diego, 1991.

[2] M. Smith, *Annu. Rev. Genet.* **19**, 423 (1985).

mitting mutant identification from even a low-efficiency experiment. Alternatively, rather than improving the screening techniques themselves, several approaches have been taken to improve efficiency, thus decreasing the number of candidates to be screened.[3] A common feature of these approaches is that they all exploit an intentionally produced asymmetry between the two strands of the heteroduplex to permit selection against the original unaltered DNA sequence, thus permitting the efficient recovery of the desired mutant genotype. Here we describe one of these methods, based on the use of uracil-containing DNA.

Principle

The method[4] uses a DNA template containing a small number of uracil residues in place of thymine (reviewed in Ref. 3). This DNA is produced within an *E. coli dut⁻ ung⁻* strain. *Escherichia coli dut⁻* mutants lack the enzyme dUTPase (dUTP pyrophosphatase) and therefore contain an elevated concentration of dUTP, which competes with TTP for incorporation into DNA. *Escherichia coli ung⁻* mutants lack the enzyme uracil *N*-glycosylase, which normally removes uracil from DNA. In the *dut⁻ ung⁻* double mutant, uracil is incorporated into DNA in place of thymine and is not removed. Thus, standard vectors can be grown in a *dut⁻ ung⁻* host to prepare uracil-containing single-stranded DNA templates for site-directed metagenesis.

For the *in vitro* reactions typical of site-directed mutagenesis protocols, uracil-containing DNA templates are indistinguishable from normal templates. Since dUMP has the same coding potential as TMP, the uracil in the template is not mutagenic, either *in vivo* or *in vitro*. Furthermore, the presence of uracil in the template is not inhibitory to DNA synthesis or ligation. The uracil-containing DNA can therefore be used as a template for the synthesis of a complementary strand containing the oligonucleotide-directed mutant DNA sequence and TMP but no dUMP. The products of the *in vitro* reaction are then used to transfect competent cells prepared from any desirable *E. coli ung⁺* strain. The action of uracil *N*-glycosylase *in vivo* releases uracil, producing apyrimidinic (AP) sites in the template strand. AP sites inhibit DNA synthesis and provide positions for strand incision by AP endonucleases, thus greatly reducing the biological activity of the template strand. The majority of the progeny thus arise from the complementary strand containing the desired mutation.

[3] T. A. Kunkel, *Nucleic Acids Mol. Biol.* **12**, 124 (1988).
[4] T. A. Kunkel, *Proc. Natl. Acad. Sci. U.S.A.* **82**, 488 (1985).

Materials and Reagents

Bacterial Strains, Bacteriophage, and Phagemids. No special vectors are required. Any single-stranded bacteriophage or phagemid vector that can be passaged through an *E. coli* dut^-ung^- strain can be employed. Similarly, any *E. coli* strain that has a functional gene for the uracil *N*-glycosylase can be used for transformation with the products of the *in vitro* reaction. Listed here are examples of strains and vectors we routinely use.

Escherichia coli CJ236: *dut1, ung1, thi1, relA1*/pCJ105 (Cm^r). This strain is used to grow phage or phagemid stocks for preparation of uracil-containing DNA. In order to maintain the F′, CJ236 should be grown in the presence of chloramphenicol (30 μg/ml). It should be streaked on an LB (Luria broth) (or H) plate containing chloramphenicol and colonies from this culture used to inoculate liquid medium containing chloramphenicol. A working culture is obtained by streaking onto an LB (or H) plate containing chloramphenicol. Well-isolated single colonies from this plate are used to start an overnight culture. These plates can be used for 2–3 months, then a new plate is made by streaking from the original stock [stored frozen in glycerol or dimethyl sulfoxide (DMSO)].

Escherichia coli MV1190: Δ(*lac-proAB*), *thi, supE,* Δ(*srl-recA*)*306*: Tn*10*(*tet*^{r})[*F′ traD36, proAB, lacI*^{Q}*Z* Δ*M15*]. This strain is used for transfection (or transformation), using the DNA products of the *in vitro* mutagenesis reactions. Grow on a glucose–minimal medium plate to maintain the F′. For a description of general techniques for handling and storing bacterial strains, see [10] in this volume.

M13mp vectors: These derivatives of the single-stranded DNA-containing bacteriophage M13 are described in Refs. 5 and 6.

pTZ18/19U/R: This phagemid vector contains the genes required for replication as a plasmid, the gene coding for ampicillin resistance, and the f1 origin of replication. When a bacterial strain harboring this vector as a plasmid is superinfected with a helper phage, phagelike particles containing single-stranded pTZ DNA are produced. The plus (or coding) strand is packaged. For a review of phagemid vectors, see Ref. 6.

M13K07: This is the helper phage for production of infectious phagemid particles. Normally, when wild-type M13 superinfects a

[5] J. Messing, this series, Vol. 101, p. 20.

[6] T. Maniatis, E. F. Fritsch, and J. Sambrook, "Molecular Cloning: A Laboratory Manual," 2nd Ed. Cold Spring Harbor Laboratory, Cold Spring Harbor, New York, 1989.

host carrying a phagemid, neither is replicated to any extent, a phenomenon called interference. M13K07 has been constructed from an M13 mutant which partially bypasses this interference.[7] Its own DNA replication has been partially disabled by insertion of DNA carrying the origin of replication of p15A and the kanamycin resistance gene from Tn*903*. (The presence of kanamycin selects for cells infected with helper.) This, plus the high copy number of the phagemid, leads to packaging of the phagemid at the expense of the helper phage. Upon entry of phagemid DNA into a bacterial cell, either by infection or transformation, the phagemid resumes its plasmidlike replication.

Growth Media

YT medium: Bacto-tryptone (Difco, Detroit, MI), 8 g; Bacto yeast extract (Difco), 5 g; NaCl, 5g. Add water to 1 liter and sterilize in an autoclave.

2× YT medium: Bacto-tryptone, 16 g; Bacto yeast extract, 10 g; NaCl, 5 g. Add water to 1 liter and sterilize in an autoclave.

LB medium: Bacto-tryptone, 10 g; Bacto yeast extract, 5 g; NaCl, 5 g. Add water to 1 liter and sterilize in an autoclave.

H medium: Bacto-tryptone, 16 g; NaCl, 5 g. Add water to 1 liter and sterilize in an autoclave.

Glucose–minimal medium: Na_2HPO_4, 6 g; KH_2PO_4, 3 g; NaCl, 0.5 g; NH_4Cl, 1 g. After autoclaving, add the following filter-sterilized solutions: 1 M $MgSO_4 \cdot 7H_2O$, 1 ml; 2% thiamin hydrochloride in water, 0.5 ml; 20% glucose, 10 ml. Then add water to 1 liter.

VB salts (50×): $MgSO_4 \cdot 7H_2O$, 10 g; citric acid (anhydrate), 100 g; K_2HPO_4, 500 g; $Na_2HPO_4 \cdot 2H_2O$, 75 g. Dissolve in 1 liter of water and sterilize in an autoclave.

Soft agar: NaCl, 9 g; Difco (Detroit, MI) agar, 8 g. Add water to 1 liter and sterilize in an autoclave. (Alternatively, soft agar can be made with LB or H medium.)

For more detailed instructions on preparation of media, see Refs. 8 and 9.

Enzymes and Reagents. T4 DNA polymerase, T4 gene 32 protein, and T4 DNA ligase were from Bio-Rad Laboratories (Richmond, CA) T4

[7] J. Vieira and J. Messing, *in* "Recombinant DNA Methodology" (R. Wu, L. Grossman, and K. Moldave, eds.), p. 225. Academic Press, San Diego, 1989.

[8] J. H. Miller, "Experiments in Molecular Genetics." Cold Spring Harbor Laboratory, Cold Spring Harbor, New York, 1982.

[9] R .W. Davis, D. Botstein, and J. R. Roth, "Advanced Bacterial Genetics." Cold Spring Harbor Laboratory, Cold Spring Harbor, New York, 1980.

polynucleotide kinase was from Pharmacia, Molecular Biology Division (Piscataway, NJ). Wild-type T7 DNA polymerase was from United States Biochemical Corporation (Cleveland, OH). Deoxynucleoside triphosphates (HPLC grade, 100 mM solutions) were purchased from Pharmacia, Molecular Biology Division, and used without further purification. Other chemicals were obtained from standard suppliers of molecular biological reagents.

Stock Solutions

X-Gal: 50 mg/ml 5-bromo-4-chloro-3-indolyl-β-D-galactoside in N,N-dimethylformamide, stored at $-20°$. Avoid exposure to light.

IPTG: 100 mM isopropyl-β-D-thiogalactopyranoside. Sterilize by filtration and store at $-20°$.

PEG/NaCl (5×): Polyethylene glycol 8000, 150 g; NaCl, 146 g. Add water to 1 liter and filter sterilize.

High salt extraction buffer; 100 mM Tris-HCl (pH 8.0); 300 mM NaCl; 1 mM EDTA.

Phenol: Equilibrated versus multiple volumes of high salt extraction buffer until the pH of the aqueous phase is approximately 8.0.

TE buffer: 10 mM Tris-HCl (pH 8.0); 0.1 mM EDTA.

Kinase buffer (10×): 500 mM Tris-HCl (pH 7.5); 100 mM MgCl$_2$; 50 mM dithiothreitol (DTT).

SSC (20×): 3 M NaCl; 300 mM sodium citrate.

SDS dye mix (10×): 10% sodium dodecyl sulfate (w/v); 1% bromphenol blue (w/v); 50% glycerol (v/v).

TAE buffer (50×): Tris base, 242 g; glacial acetic acid, 57.1 ml; EDTA, 100 ml of a 500 mM solution (pH 8.0). Add water to 1 liter.

Ampicillin (sodium salt): 25 mg/ml in water. Sterilize by filtration and store in aliquots at $-20°$. Working concentration, 50 μg/ml.

Chloramphenicol: 30 mg/ml in 100% ethanol. Store in aliquots in $-20°$. Working concentration, 30 μg/ml.

Kanamycin: 50 mg/ml in water. Sterilize by filtration and store in aliquots at $-20°$.

Preparation of Uracil-Containing Single-Stranded Template DNA

Bacteriophage M13 Vectors

Using a sterile pipette tip, remove one plaque [for M13 vectors this is approximately 10^9–10^{10} plaque-forming units (pfu)] from a plate and place it in 1 ml of sterile YT medium in a 1.5-ml microcentrifuge tube. Incubate the tube for 5 min at 60° to kill bacterial host cells, vortex vigorously to release the phage from the agar, then pellet cells and agar with a 2-min spin in a microcentrifuge. Place 100 μl of the resulting supernatant into a 1-liter flask containing 100 ml of 2× YT medium containing 30 μg/ml chlorampenicol. (In the initial publication of this technique,[4] the medium was supplemented with thymidine, deoxyadenosine, and uridine. Subsequent studies have demonstrated that this is not necessary.) Add 5 ml of a mid-log culture of *E. coli* CJ236 (grown in YT medium containing 30 μg/ml chloramphenicol). The resulting multiplicity of infection (moi, the ratio of phage particles to cells) does not exceed 1. Since most or all phage are "passaged" through the *dut⁻ ung⁻* strain, a single cycle of growth results in a sufficient survival difference (as measured by titers on *ung⁺* and *ung⁻* hosts) to make the DNA suitable for *in vitro* mutagenesis.

The flask is incubated with vigorous shaking at 37°. We have prepared phage from cultures incubated for as short as 6 hr or as long as 24 hr. Shorter times are recommended for vectors that contain unstable inserts, since this will help to avoid the growth advantage of phages which have deleted the insert. After incubation at 37°, the culture is centrifuged at 5000 g for 30 min. The clear supernatant contains about 10^{11} pfu/ml. Before preparing viral template DNA, phage titers can be compared on *ung⁻* (e.g., CJ236) and *ung⁺* (e.g., MV1190) host strains. Phage that contain uracil in the DNA have approximately 100,000-fold lower survival in the *ung⁺* host than in the *ung⁻* host. High efficiency can be obtained even if the survival difference is much less. For example, we have observed a 41% efficiency of mutagenesis with only a 100-fold survival difference.[4]

Phagemid Vectors

To obtain single-stranded uracil-containing phagemid DNA, first the phagemid vector containing the desired insert must be introduced into an *E. coli dut⁻ ung⁻* strain (e.g., CJ236). This may be done by infection with phagemid particles. To obtain the particles, inoculate 5 ml of LB medium containing ampicillin with MV1190 containing the phagemid and grow with shaking to early log phase. Add 1 × 10^9 pfu of M13K07 to the culture and continue incubation for 1 hr. Remove 1 ml of this culture and pellet the cells by centrifugation for 5 min. The supernatant contains the phagemid particles. To infect CJ236, grow a 20-ml culture of this strain in LB supplemented with chloramphenicol to an OD_{600} of 0.30–0.35 and then add 10 μl of the supernatant containing the phagemid particles. Continue incubation for 2 hr. Dilute the culture 10-, 100-, and 1000-fold and then spread 50 μl of each dilution onto ampicillin-containing H or LB plates. After incubation overnight, pick a colony and restreak it onto another ampicillin-containing plate.

As an alternative to infection, the phagemid vector may be introduced by transformation with phagemid DNA, using electroporation[6] or cells made competent using either $CaCl_2$[6] or another of the procedures described in [4] in this volume.

The *dut⁻ ung⁻* strain, now containing the phagemid, is streaked on an LB or H plate supplemented with chloramphenicol and ampicillin. The plate is incubated in 37° until distinct colonies appear. An isolated colony is picked and placed in 20 ml of LB containing chloramphenicol and ampicillin, then incubated with shaking at 37° overnight. A volume of 50 ml of 2× YT medium with chloramphenicol and ampicillin is inoculated with 1 ml of the overnight culture of CJ236 containing phagemid and incubated with shaking at 37°. At a cell density of approximately 10^8 colony-forming units (cfu)/ml (OD_{600} 0.3), the helper phage M13K07 is added to obtain a M.O.I. of around 20. After 1 hr of further incubation with shaking at 37°, 70 μl of the 50 mg/ml kanamycin stock solution is added (the presence of kanamycin selects for cells infected with helper phage), and the incubation is continued for 6–24 hr. After this time, 30 ml of the culture is transferred to a 50-ml centrifuge tube, and centrifuged at 5000 g at 4° for 30 min. The clear supernatant contains the phagemid particles.

As for bacteriophage (above), before extracting single-stranded template DNA, phagemid titers can be compared on *ung⁻* (e.g., CJ236) and *ung⁺* (e.g., MV1190) strains. Overnight cultures of MV1190 and CJ236 are used to start fresh cultures in 2× YT medium. Add 0.5 ml of an overnight culture of MV1190 to 50 ml of 2× YT medium; to another 50 ml of 2× YT add 0.5 ml of an ovenight culture of CJ236. Incubate the cultures at 37° with shaking, until the OD_{600} is 0.3 to 0.35. If one culture reaches this optical density before the other, it is placed on ice. To each culture 10 ml of uracil-containing phagemid supernatant is added. Incubation at 37° is then continued for 2 hr. The MV1190 culture is then diluted 10- and 100-fold; 50 μl of each dilution and the undiluted culture is spread onto ampicillin-containing H or LB plates. The CJ236 culture is diluted 10^3-, 10^4-, and 10^5-fold, and 50 μl of these dilutions is also spread onto ampicillin-containing plates. Phagemids containing uracil in the DNA typically produce approximately 10,000-fold more ampicillin-resistant colonies in CJ236 than in MV1190.

Extraction of Single-Stranded DNA

Phage or phagemid particles are precipitated from the clear supernatant by adding 1 volume of 5× PEG/NaCl to 4 volumes of supernatant, mixing, and incubating the phage at 0° for 1 hr. The precipitate is collected by centrifugation at 5000 g at 4° for 15 min, and the well-drained phage pellet is resuspended in high-salt extraction buffer. After vigorous vortexing, the resuspended phage solution is placed on ice for 60 min and then centrifuged as above. Any debris is discarded. This incubation on ice has proved useful in reducing the level of endogenous low molecular weight DNA, which can nonspecifically prime DNA synthesis in subsequent DNA polymerase reactions. If desired, one can treat the resuspended particles with RNase A prior to extracting the single-stranded DNA.

The supernatant still contains the intact phage. The single-stranded DNA can be extracted from the phage using any standard extraction procedure. We have routinely extracted twice with phenol and twice with chloroform/isoamyl alcohol (24 : 1). More recently, to avoid the use of organic solvents, we have extracted the DNA with cetyltrimethylammonium bromide (CTAB).[10] After precipitating the phage from the supernatant (1 ml) with PEG as described above, and collecting the precipitate by centrifugation, the phage pellet is dissolved in 690 μl distilled water, 100 μl of 1.0 M Tris-HCl (pH 8.0), 100 μl of 0.5 M EDTA, and 10 μl of 1 mg/ml proteinase K. This mixture is incubated at 37° for 30 min. To this is added 100 μl of CTAB solution [5% (w/v) in 0.5 M NaCl]. After at least 3 min of incubation at room temperature, the CTAB–DNA precipitate is collected by centrifugation. The pellet is resuspended in 300 μl of 1.2 M NaCl.

The DNA is then precipitated by standard procedures. We routinely add either 0.1 volume of 3 M sodium acetate (pH 5) or 0.33 volume of 7.8 M ammonium acetate and 2 volumes of ethanol and chill the mixture to −20° for 30 min or longer. The same procedure is used when the DNA is extracted with CTAB except that the addition of sodium acetate or ammonium acetate is not necessary. The DNA is collected by centrifugation and resuspended in TE buffer. The DNA concentration can be determined by standard procedures. If quantitation is by spectrophotometry, 1 OD_{260} unit is equal to 36 μg single-stranded DNA/ml. For phagemid DNA preparations, the relative amounts of phagemid DNA versus helper phage DNA can be determined by agarose gel electrophoresis in the presence of ethidium bromide, comparing band intensities to a known amount of a single-stranded DNA standard. Since it is sometimes difficult to distinguish which band is phagemid DNA and which is helper phage DNA, include a helper phage DNA standard in a separate lane of the gel.

It is advisable to analyze the DNA by electrophoresis in an agarose gel containing ethidium bromide. Not only can the DNA concentration be quantitated by comparison to a known standard, the purity of the DNA can be estimated. We routinely analyze a substantial excess (1–5 μg) of the DNA preparation in one lane of the gel, in order to detect minor species and/or contaminants. Of particular concern are low molecular weight

<hr />

[10] G. Del Sal and C. Schneider, *Nucleic Acids Res.* **15**, 10047 (1987).

contaminants that could serve as nonspecific primers during *in vitro* DNA synthesis. Normally, further purification of the DNA is unnecessary in order to achieve high efficiencies of mutagenesis. If problems related to template purity are encountered, the DNA can be subjected to standard purification procedures (e.g., gel filtration, density gradient centrifugation, chromatography), since the substitution of a small percentage of thymidine by deoxyuridine does not affect the physical properties of the DNA.

In Vitro Polymerase Reaction and Product Analysis

The oligonucleotide containing the DNA sequence of the desired mutation(s) is phosphorylated and then hybridized to the single-stranded template DNA. It is converted to a full-length, circular DNA strand by polymerization and ligation. The resulting products are analyzed by agarose gel electrophoresis.

Phosphorylation of Oligonucleotide

The 5'-OH end of the oligonucleotide is phosphorylated (for subsequent ligation) in a microcentrifuge tube in a 20-μl reaction containing 50 mM Tris-HCl (pH 7.5), 10 mM MgCl$_2$, 5 mM DTT, 0.4 mM ATP, 2 units of T4 polynucleotide kinase, and 10–100 ng oligonucleotide. The amount of oligonucleotide that is phosphorylated can be varied, depending on the desired molar ratio of oligonucleotide to single-stranded DNA template (see below). The reaction is incubated at 37° for 1 hr and terminated by adding 1 μl of 300 mM EDTA and then incubating at 65° for 10 min.

Hybridization of Oligonucleotide to Uracil-Containing Template

To the phosphorylated oligonucleotide is added 0.2–1 μg of the single-stranded circular uracil-containing DNA template and 1/20th volume of 20× SSC. [Alternatively, annealing can be performed in a 10-μl volume with 200 ng of the single-stranded circular uracil-containing DNA template, the oligonucleotide, and 1 μl of 10× annealing buffer (200 mM Tris-HCl, pH 7.4, 20 mM MgCl$_2$, 500 mM NaCl)]. As a control to examine the amount of oligonucleotide-independent synthesis, a second tube should be prepared that contains the same components but lacks the oligonucleotide.

The exact molar ratio of oligonucleotide to template can vary. Ratios of about 5:1 to 10:1 are routine, but in some instances (e.g., for less stable heteroduplexes), ratios greater than this may be useful. However, a large excess of oligonucleotide can interfere with ligation, and high ratios can result in a significant level of hybridization of the oligonucleotide to positions other than the site to be mutated.

The precise hybridization temperature depends on the stability of the heteroduplex formed between the mutagenic oligonucleotide and the template DNA. Typically, after mixing and centrifuging the sample briefly (5 sec) in a microcentrifuge, the tube can be placed in a beaker of water at 70° and allowed to cool to room temperature, at a rate of cooling of about 1°/min. After a 5-sec centrifugation to spin down condensation, the tube is placed on ice and is ready for use in the synthesis reaction.

The ability of the oligonucleotide to prime synthesis at the desired position under the chosen hybridization conditions can be determined by using the template primer for chain-terminator DNA sequencing reactions. For additional comments on the design of the mutagenic oligonucleotide and the stability of heteroduplexes, see Refs. 1 and 2.

In Vitro DNA Synthesis

Reactions (100 μl) are performed in a microcentrifuge tube containing the above hybridization mixture, 20 mM Tris-HCl (pH 7.5), 2 mM DTT, 10 mM MgCl$_2$, 500 μM each of dATP, dTTP, dGTP, and dCTP, 0.4 mM ATP, 0.5–1 unit of native T7 DNA polymerase (the T7 gene 5 protein/thioredoxin complex), and 2–3 units of T4 DNA ligase. All components are mixed at 0°. For convenience, the Tris buffer, DTT, MgCl$_2$, dNTPs, and ATP can be premixed as a 10× stock and then added as 1/10th the volume of the reaction. Enzymes are added last.

The reactions are incubated at 37° for 1 hr and then terminated by addition of EDTA to 15 mM. In instances when the stability of the heteroduplex formed between the mutagenic oligonucleotide and the template is of concern (e.g., multiple mismatches, large loops), it may be useful to incubate the reaction for a short time (e.g., 5–15min) at a lower temperature (e.g., room temperature or even 0°), to permit some synthesis to stabilize the heteroduplex before shifting the reaction to a 37° incubation. Since starting the reactions at the lower temperatures increases the probability that priming may occur with endogenous contaminants in the template DNA preparation or may arise from the mutagenic oligonucleotide hybridizing to partially homologous sites, this should only be done if really necessary.

The products of the reactions are examined by subjecting 20 μl (to which is added 2.5 μl of SDS dye mix) to electrophoresis in a 0.8% (w/v) agarose gel in 1× TAE buffer containing 0.5 μg/ml ethidium bromide (Fig. 1). For comparison, adjacent lanes should contain the appropriate standards: double-stranded, covalently closed circular DNA (cccDNA) and double-stranded, open circular DNA (ocDNA) (see lane 1 of Fig. 1), and single-stranded circular DNA (ssDNA, lane 2). The product of the *in*

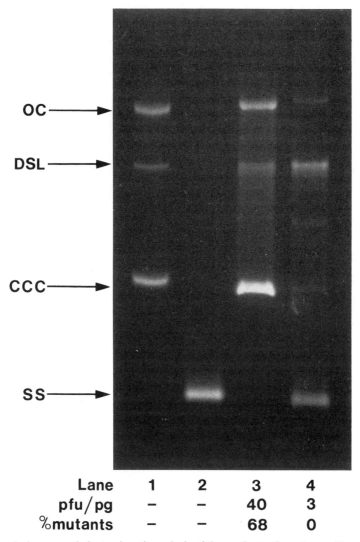

Lane	1	2	3	4
pfu/pg	–	–	40	3
%mutants	–	–	68	0

Fig. 1. Agarose gel electrophoretic analysis of the products of reactions with native T7 DNA polymerase. The template was M13mp2 uracil-containing DNA that has an amber mutation in the *lacZ* α-complementation sequence, yielding a colorless plaque phenotype on plates containing X-Gal and IPTG. The oligonucleotide primer was a 20-mer containing a single nucleotide change that converts the amber codon to a sense codon that yields a blue color plaque phenotype. Hybridization and synthesis reactions were carried out as described in the text, and reactions were incubated for 1 hr at 37°. The gel contained 0.8% agarose and 0.5 μg/ml ethidium bromide in TAE buffer. Lane 1, double-stranded covalently closed circular (CCC), double-stranded open circular (OC), and double-stranded linear (DSL) DNA; lane 2, uracil-containing single-stranded (SS) DNA; lane 3, products of reaction with 1.0 unit of native T7 DNA polymerase and oligonucleotide-primed DNA; lane 4, products of reaction with 1.0 unit of native T7 DNA polymerase and template DNA that had not been primed with the mutagenic oligonucleotide. Plaques were recovered by electroporation of *ung*[+] *E. coli* cells, and mutants were scored as blue plaques. The pfu values are expressed per picogram of input single-stranded DNA.

vitro DNA synthesis reaction with the oligonucleotide-primed template (lane 3) should migrate at the same rate as the double-stranded DNA standard. Migration as cccDNA indicates that the product is covalently closed (but relaxed) circular DNA, owing to the combined action of DNA polymerase and ligase. Those products that are double-stranded but not ligated migrate at the position of ocDNA. Typically, a small amount of a third band, migrating coincidently with double-stranded linear DNA standard (dslDNA), is also seen. The reaction containing template that has not been primed with the oligonucleotide (lane 4) should yield substantially less double-stranded DNA products than does the primed reaction (lane 3). For additional comments on the quality of the *in vitro* reaction, see the section on troubleshooting below.

Transfection and DNA Sequence Analysis

Based on an estimate from the gel analysis, enough double-stranded DNA product is used to obtain the desired number of candidate plaques or colonies by transfection (phage DNA) or transformation (phagemid DNA) of any desired *ung*[+] strain of *E. coli*. This can be done using the products of the *in vitro* reaction without purification to remove reaction components. Bacterial cells can be made competent by any of the procedures described in [4] in this volume or by treatment with $CaCl_2$.[6] The DNA can be introduced into cells with even higher efficiency by electroporation.[6] Resulting clones (as phage plaques or bacterial colonies) can be selected or, if no phenotype is known, chosen randomly for isolation of pure genetic stocks and preparation of DNA for sequencing. Individual mutants can be identified by DNA sequence analysis of a limited number of candidate clones, since the efficiency of mutagenesis is typically above 50% (lane 3, Fig. 1).

Troubleshooting *in Vitro* Reactions

Unsuccessful experiments are usually characterized by too few plaques (or colonies) recovered upon transfection, or by a low efficiency of mutagenesis. By far the most likely possibilities relate to the quality of the *in vitro* reaction. A careful product analysis (as in Fig. 1) should routinely be performed. Production of double-stranded DNA that migrates coincidently with the cccDNA and/or ocDNA standards usually yields a successful result.

Two synthesis problems could explain low plaque numbers. One is contamination of the dNTP substrates with dUTP (e.g., by deamination of dCTP). When incorporated into the newly synthesized strand *in vitro*, dUTP residues provide a target for the production of lethal AP sites. For this reason, high-quality dNTP substrates should be used. The second and more likely explanation for low biological activity is incomplete synthesis of the complementary strand. Remaining regions of uracil-containing single-stranded DNA are sites for lethal attack once the DNA enters the bacterial cell.

Incomplete synthesis can result from inefficient hybridization of the oligonucleotide primer to the template. This can be examined by systematically varying the hybridization conditions and the oligonucleotide to template ratio (see Ref. 1). Other explanations for incomplete synthesis include inactive (or excess) DNA polymerase, inhibitory contaminants in the DNA, the polymerase, or the reagents, and a DNA template which contains structures (e.g., hairpins) that impede polymerization. There are simple solutions to the latter problem (see below).

A low percentage of mutants can also result from impurities in the template DNA which provide endogenous primers for complementary strand synthesis. The amount of endogenous priming can be determined by performing a DNA synthesis reaction without added oligonucleotide and examining the products by gel electrophoresis (e.g., lane 4 in Fig. 1). The amount of cccDNA and ocDNA produced *in vitro* in the absence of oligonucleotide should be small relative to that observed with oligonucleotide-primed synthesis (e.g., compare lanes 3 and 4 in Fig. 1). Since DNA that is converted to double-stranded product without added primer yields nonmutant progeny (lane 4), high efficiency requires a clean template DNA preparation (see procedure described above).

Another source of low mutant yield is displacement of the oligonucleotide during *in vitro* DNA synthesis of the strand which carries the mutation. The choice of DNA polymerase and the efficiency of ligation determine the extent of this problem (see below).

Choice of DNA Polymerase

Site-directed mutagenesis procedures were developed at a time when the highest quality commercially available DNA polymerase for converting an oligonucleotide to a complete complementary DNA strand was the large (Klenow) fragment of *E. coli* DNA polymerase I. Although this enzyme is still in widespread use, we do not recommend it. This is because, once the Klenow polymerase completes synthesis with a circular DNA template, it can displace the mutagenic oligonucleotide. In an oligonucleotide-directed mutagenesis experiment, when the DNA ligase operates inefficiently relative to the DNA polymerase, strand displacement of the oligonucleotide reduces the mutant frequency.

Consistently higher efficiencies can be obtained using T4 DNA polymerase, which does not perform strand displacement synthesis.[11,12] It usually synthesizes a complete complementary strand quite efficiently when reactions are performed at 37° (a result similar to that shown in lane 3 of Fig. 1). However, at lower temperatures, or even at 37° with some template primers, this enzyme may have difficulty in completing synthesis of the complementary strand. In these situations, complete synthesis can usually be achieved by including T4 single-stranded DNA-binding protein (the product of T4 gene 32) in the reaction (for examples, see Refs. 13 and 14).

The commercially available DNA polymerase that has the most favorable properties for complementary strand DNA synthesis in a site-directed mutagenesis reaction[15] is the native T7 DNA polymerase, that is, the T7 gene 5 protein/*E. coli* thioredoxin complex. Like the T4 DNA polymerase, native T7 DNA polymerase does not perform strand displacement synthesis.[16] Polymerization by this enzyme is fast and highly processive,[17] even on templates that contain secondary structures which inhibit polymerization by the T4 DNA polymerase. We find that, in reactions incubated at 37°, native T7 DNA polymerase completely copies single-stranded circular DNA templates (e.g., lane 3 in Fig. 1). Although we routinely incubate the reaction for 1 hr, we have confirmed the original observation[17] that essentially complete synthesis occurs in the first 15 min.

Derivative forms of T7 DNA polymerase that have reduced exonuclease activity[18,19] can also be used for site-directed mutagenesis reactions. The reduction in exonuclease activity actually improves the rate and processivity of polymerization. Whereas these parameters are important for other uses of T7 DNA polymerase (e.g., DNA sequencing), we do not routinely use these modified forms of T7 DNA polymerase for site-directed mutagenesis reactions, because at least one of these performs strand displacement synthesis.[20]

T4 DNA ligase is used to ligate the newly synthesized strand to the

[11] Y. Masamune and C. C. Richardson, *J. Biol. Chem.* **246**, 2692 (1971).
[12] N. G. Nossal, *J. Biol. Chem.* **249**, 5668 (1974).
[13] C. S. Craik, C. Largman, T. Fletcher, S. Roczniak, P. J. Barr, R. Fletterick, and W. J. Rutter, *Science* **228**, 291 (1985).
[14] P. J. Yukenberg, F. Witney, J. Geisselsoder, and J. McClary, *in* Book, in press. IRL Press, London, 1990.
[15] K. Bebenek and T. A. Kunkel, *Nucleic Acids Res.* **17**, 5408 (1989).
[16] M. J. Engler, R. L. Lechner, and C. C. Richardson, *J. Biol. Chem.* **258**, 11165 (1983).
[17] S. Tabor, H. E. Huber, and C. C. Richardson, *J. Biol. Chem.* **262**, 16212 (1987).
[18] S. Tabor and C. C. Richardson, *J. Biol. Chem.* **262**, 15330 (1987).
[19] S. Tabor and C. C. Richardson, *J. Biol. Chem.* **264**, 6447 (1989).
[20] R. L. Lechner, M. J. Engler, and C. C. Richardson, *J. Biol. Chem.* **258**, 11174 (1983).

5′ end of the oligonucleotide primer. Consistent with a lack of strand displacement synthesis by T4 DNA polymerase or native T7 DNA polymerase, we have obtained mutant frequencies in excess of 50% even when DNA ligase is intentionally omitted from the reaction. Nevertheless, we routinely do include ligase, because at least under some circumstances it does improve the efficiency of mutagenesis.

Conclusion

Uracil-containing DNA can be prepared for any vector that can be passaged through an *E. coli dut⁻ ung⁻* strain. We have presented here a simple oligonucleotide-directed mutagenesis protocol to demonstrate the utility of the uracil selection technique for efficiently generating mutants. This DNA can be used in conjunction with a variety of established methodologies for site directed mutagenesis (e.g., gapped duplexes, double priming, degenerate oligonucleotides). If more details are sought on any aspect of site-directed mutagenesis, or on alternatives to the use of uracil-containing DNA for improving efficiency, the comprehensive reviews by Smith are recommended.[1,2]

[7] Uses of Transposons with Emphasis on Tn*10*

By Nancy Kleckner, Judith Bender, and Susan Gottesman

I. General Considerations

A. Introduction

As transposable elements have become indispensable tools for bacterial genetics, many different types of specialized transposon derivatives have been constructed. The most widely used constructs are derived from insertion sequence (IS)-based elements (Tn*10* and Tn*5*) or from bacteriophage Mu; constructs based on cointegrate-forming elements (Tn*3* and gamma-delta) are also available. Details regarding the transposition mechanisms of these elements can be found in a recent collection of review articles.[1]

One goal of this chapter is to summarize the major types of transposon

[1] D. E. Berg and M. M. Howe (eds.), "Mobile DNA." American Society for Microbiology, Washington, D.C., 1989.

METHODS IN ENZYMOLOGY, VOL. 204

constructs available and to provide general guidance as to how best to choose the construct which is most appropriate to the desired application. Classic applications of transposable elements to bacterial genetics were originally outlined by Kleckner *et al.*[2] General considerations for transposon mutagenesis have also been reviewed by Berg and Berg.[3] A second goal is to describe in some detail both the methods used for Tn*10*-derived transposon vehicles and the most recent set of useful Tn*10* vehicles themselves. With respect to the latter goal, this chapter supplements and updates a previous article.[4] Vehicles derived from Tn*5*, Mu, and Tn*3*/gamma-delta are described in detail elsewhere.[3,5–10] Also, a Tn*3* derivative specially adapted for making short in-frame insertions has recently been described.[11] The reader is also referred to the chapter on construction and analysis of fusions by Slauch and Silhavy[12] in this volume.

The use of transposon insertions or transposon-promoted deletions to provide mobile priming sites for DNA sequence analysis of cloned genes is not considered here. Specific vector systems for this purpose are described by Liu *et al.*,[13] Nag *et al.*,[8] Phadnis *et al.*,[6] and Ahmed.[14] In general, the most important parameter limiting the use of transposons for this purpose will probably be the extent to which insertions or deletions occur preferentially at particular sites or into particular regions (see below).

B. Types of Insertions

1. General. Transpositions of an element from one DNA molecule to another are usually isolated by selecting for stable maintenance of a genetic marker present on the transposon under conditions where the DNA mole-

---

[2] N. Kleckner, J. Roth, and D. Botstein, *J. Mol. Biol.* **116**, 125 (1977).

[3] C. M. Berg and D. E. Berg, *in* "*Escherichia coli* and *Salmonella typhimurium*: Cellular and Molecular Biology" (F. C. Neidhardt, J. L. Ingraham, B. Magasanik, K. B. Low, M. Schaechter, and H. E. Umbarger, eds.), Vol. 2, p. 1071. American Society for Microbiology, Washington, D.C., 1987.

[4] J. C. Way, M. A. Davis, D. Morisato, D. E. Roberts, and N. Kleckner, *Gene* **32**, 369 (1984).

[5] C. Sasakawa and M. Yoshikawa, *Gene* **56**, 283 (1987).

[6] S. H. Phadnis, H. V. Huang, and D. E. Berg, *Proc. Natl. Acad. Sci. U.S.A.* **86**, 5908 (1989).

[7] W. Y. Chow and D. E. Berg, *Proc. Natl. Acad. Sci. U.S.A.* **85**, 6468 (1988).

[8] D. K. Nag, H. V. Huang, and D. E. Berg, *Gene* **64**, 135 (1988).

[9] H. S. Seifert, E. Y. Chen, M. So, and F. Heffron, *Proc. Natl. Acad. Sci. U.S.A.* **83**, 735 (1986).

[10] E. A. Groisman, [8], this volume.

[11] M. Hoekstra, D. G. Burbee, J. D. Singer, E. E. Mull, E. Chiao, and F. Heffron, *Proc. Natl. Acad. Sci. U.S.A.* in press (1991).

[12] J. M. Slauch and T. J. Silhavy, this volume [9].

[13] L. Liu, W. Whalen, A. Das, and C. M. Berg, *Nucleic Acids Res.* **15**, 9461 (1987).

[14] A. Ahmed, *Gene* **75**, 315 (1989).

cule which donates the transposable element (the delivery vehicle) is lost. Transposition of an element from a nonreplicating phage or plasmid genome into a stable replicon are the two most popular approaches.

2. *Fusion Transpositions.* A number of specialized transposon derivatives are available in which transposition of the element to an appropriate target site results in activation of an otherwise silent gene. Fusion vehicles can be used in three different ways. (1) They are most commonly used as genetic tools for the isolation of particular desired fusion constructs. (2) Such vehicles are also very useful for obtaining insertions without the necessity of a specialized delivery vehicle. If the transposon originates from a context in which the marker gene is not expressed, transpositions into new sites can be specifically selected without destruction or elimination of the donor molecule (see below). This approach is particularly useful for isolating transposition events in organisms or situations where no general delivery system exists, although it does yield only a subset of all possible insertions. (3) Such vehicles also facilitate analysis of the transposition process per se. They make it possible to compare endogenous transposition rates in different strains or under different conditions without having to resort to a "mating-out"[15] or "λ-hop"[16] assay.

Four different types of fusions can be isolated with existing tools (see also [9], this volume): promoter fusions in which transposition has placed the target gene under control of a transposon-borne *lac* operon promoter, transcriptional fusions between the target gene and a *lacZ* gene on the transposon, translational fusions between the target gene and either the *lacZ* gene or a *kan* gene on the transposon, and translational fusions to the *phoA* gene on the transposon. Expression of the *phoA* gene requires that the target gene product cross the inner membrane to the periplasm, and the Tn5–*phoA* construct can thus be used to identify specifically genes whose products are secreted or localized to the membrane.[17]

C. Choosing Delivery Vehicle: Choice of Target Molecule Determines Choice of Donor Molecule

1. Insertions into Bacterial Chromosome. For isolation of transposition events into the bacterial chromosome, bacteriophages are the most convenient type of delivery vehicle. A phage carrying the transposon can be introduced into the host cell under conditions where the phage genome neither replicates, kills, nor (in many cases) stably integrates into the host cell. λ vehicles are used for isolation of Tn*10*, Tn5, and Mu insertions; Mu

[15] T. Foster, V. Lundblad, S. Hanley-Way, S. Halling, and N. Klecker, *Cell* (*Cambridge, Mass.*) **23**, 215 (1981).

[16] N. Kleckner, D. F. Barker, D. G. Ross, and D. Botstein, *Genetics* **90**, 427 (1978).

[17] C. Manoil and J. Beckwith, *Proc. Natl. Acad. Sci. U.S.A.* **82**, 8129 (1985).

is also used directly. For λ vehicles, the donor is crippled by nonsense mutations in phage replication genes (which necessitate the use of a non-suppressing host for isolation of transposition events), by a mutation in the λ repressor gene, and often by deletion of the phage integration system. These λ derivatives are referred to below as hop phages. For Mu, every lysogen is a transposition event, so lysogenization of phage carrying Mu ends is sufficient to produce insertions. Mu vectors generally are defective for phage growth and therefore must be grown with a helper phage. Tn*3*/gamma-delta vectors are not useful for this purpose because the bacterial chromosome is specifically immune to insertion of these elements.

An alternative to phage delivery vehicles are so-called suicide plasmids which are thermosensitive for replication or which replicate in a donor strain but fail to replicate in the recipient strain where transposon insertions are to be isolated. Such plasmids, some of which have a broad host range, are widely used in strains other than *Escherichia coli* that are not sensitive to bacteriophage λ; they are available but used less frequently in *E. coli* or *Salmonella*.[5]

2. Insertions into Multicopy Nonconjugative Plasmids. For isolation of insertions into nonconjugative multicopy plasmids, bacteriophages are also the delivery vehicles of choice. In the most general approach, a strain harboring the target plasmid of interest is infected with the phage vehicle, and a large number of colonies resulting from transposition are selected, exactly as for chromosomal insertions. About 1% of such colonies contain a transposition event into the plasmid unless a specific enrichment for transposon insertions is used (see below); also, it should be remembered that cells in a single colony may contain a mixture of plasmids with and without the transposon insertion. Plasmid insertions are specifically identified in a subsequent step. Many independent pools of about 1000 such transposition colonies are made, and plasmid DNA isolated from each pool is used to transform a new host. Transformants selected for expression of a marker on the transposon contain insertion-bearing plasmids.

Specific plasmid vehicles have also been constructed for the isolation of Tn*3*-based transpositions into target genes on multicopy plasmids.[9] In this case, isolation of insertions into a plasmid involves multiple steps. First cointegrate insertion products are identified, and then cointegrates are resolved to simple insertion products. Standard pBR-based Amp[R] cloning vectors cannot be used as target plasmids in this system; they are all immune to Tn*3* transposition by virtue of the presence of a single Tn*3* terminal inverted repeat sequence.

3. Transpositions into Bacteriophage λ or Conjugative Plasmid. When the target molecule is a phage or conjugative plasmid, the transposon delivery vehicle can be any type of molecule or replicon other than the target itself. The most efficient transposon donor molecule is a multicopy nonconjugative plasmid because such vehicles usually provide the highest levels of transposition. Also, when there are limitations on the size of the phage genome, as for λ, use of a small minitransposon is almost always essential. Transpositions into a bacteriophage are obtained by growing a stock of the phage on a cell harboring the transposon, and phage derivatives carrying the transposon marker are identified in a subsequent step using any of several approaches (see below). Transpositions into a conjugative plasmid are obtained by constructing a strain carrying both the transposon and the target plasmid, mating this strain with a suitable recipient, and selecting for transfer of the transposon marker into a new host, with appropriate counterselection against the donor strain.

D. Choosing Transposon: General Properties of Transposon Derivatives

The specific choice of transposon element will depend on the ultimate goal of insertion mutagenesis and on the specific features of the transposon derivatives available. In this section we discuss the general features of transposon derivatives which are important if the primary goal is the isolation of stable insertions into a target gene or region of interest. Specific elements for fusion analysis and for isolation of transposon-promoted rearrangements are considered subsequently.

1. Stability of Inserts: Use Minitransposons Whenever Possible. Stable transposon insertions that are unable to undergo additional rounds of transposition can be obtained by using a delivery system in which the transposase gene is located outside of the transposon itself. In the ideal case, the transposase gene is located on the transposon donor molecule and is thus lost along with that molecule following transposition. In a less ideal case, the transposase gene is located on a separate replicon, usually a multicopy plasmid, from which the transposon insertion can eventually be separated (by plasmid segregation or by transfer of the insertion to a new strain). Transposons which do not contain a transposase gene within their boundaries are generically referred to as minitransposons; most of these elements are in fact also smaller than the corresponding wild-type transposon.

It is always preferable to use a minitransposon construct whenever other considerations permit. Since the transposase gene can be separated from the insertion, problems resulting from secondary transposition events are eliminated. Four types of problems are particularly troublesome. (1) Attempts to move an insertion to a new strain or to map an insertion will be confounded if the transposon marker is present at more than a single location at any stage. (2) Intact transposons have the capacity to promote the rearrangement (deletion or inversion) of adjacent material; such rearrangements occur at significant frequencies, often as high or higher than the frequency of transposition of the element itself. (3) For composite transposons such as Tn*10* and Tn*5*, transposition of the individual component insertion sequences (IS) also occurs at a much higher frequency than transposition of the entire transposon, about 10^{-3} and 10^{-2} per element per cell per generation for IS10 and IS50, respectively (e.g., Shen *et al.*[18]). Thus, a strain will tend to accumulate multiple copies of the (unmarked) insertion sequence which have the potential to cause complications during subsequent genetic or physical analysis. (4) P1 transduction of transposition-proficient insertions does not always lead to cotransduction of the transposon marker and the donor site mutation, probably because the transducing lysate contains significant numbers of P1 phages carrying the transposon which yield nonfaithful transductants carrying P1::Tn lysogens (N. Kleckner and S. Gottesman, unpublished observations, 1977; Berg *et al.*[19]). The proportion of "unfaithful" transductants can be very large, especially when the desired insertion is transduced at a relatively low frequency. This problem is eliminated by use of a minitransposon; for transposition-proficient insertions, the problem can be reduced by use of P1*vir* instead of P1*kc* or P1*clr*100 to minimize formation of P1 lysogens or by use of P1-HFT[20] to specifically increase the frequency of true transductants.

2. Insertion Specificity. The hallmark of a transposable element is its ability to insert in many different locations. However, no transposon really chooses its target sites completely at random, and some transposons exhibit a significant degree of target site preference. In all cases, the degree of specificity is low enough that insertions in a several kilobase (kb) region of interest or a single insertion in a specific gene of interest can always, or almost always, be identified. For Tn*5*, Mu, Tn*3*, Tn*9*, and the newly isolated Tn*10*-ATS derivatives described below, the specificity of insertion is sufficiently low that isolation of insertions at many different sites within a single gene is straightforward. However, these elements do still exhibit some preference for particular sites. Wild-type Tn*10* exhibits the highest degree of specificity; an occasional gene may be "cold" for insertion altogether or may be dominated by insertions into a single favored hot spot. If an available Tn*10*-ATS derivative cannot be used, these problems

[18] M. Shen, E. A. Raleigh, and N. Kleckner, *Genetics* **116,** 359 (1987).
[19] C. M. Berg, C. A. Grullon, A. Wang, W. A. Whalen, and D. E. Berg, *Genetics* **105,** 259 (1983).
[20] N. L. Sternberg and R. Maurer, this volume [2].

can usually be overcome by screening a larger than usual number of insertions, since the element does insert into lower affinity sites at a lower frequency.[21] Tn*10* insertion specificity is discussed in detail below.

For Mu and Tn*3*/gamma-delta, insertion within any particular region can be considered essentially random. However, these elements do exhibit preferences for certain regions over others, for reasons that are not clear (reviewed by Kleckner[22]).

3. Selectable Markers. Most transposons contain as an expressed selectable marker either an antibiotic resistance determinant, a *supF* gene, or *lacZ*. In choosing a selectable marker, it is important to consider the types of transposon delivery systems available, the need to save particular markers for other aspects of the analysis, and other factors. Transposons used to generate fusions usually contain an expressed selectable marker in addition to the determinant involved in generating the fusion; thus, random insertions can be isolated in the usual way and then screened in a subsequent step for presence of an active fusion.

Transposon insertions carrying drug resistance markers are especially useful for analysis of mutations in the bacterial chromosome that produce phenotypes which are difficult to screen or that involve hosts with multiple modifications (see Kleckner *et al.*[2] for general discussion). Once an appropriate linked insertion is available, the drug resistance marker can be used to map the location of the mutation and to move the mutation from strain to strain. Sets of mutations from a given selection can be shown to be linked to each other by demonstrating their linkage to a common transposon insertion, even if the transposon itself has not been mapped.

4. Other Issues. Several other considerations may be important in selecting a transposition strategy or in analyzing the insertions obtained.

(a) Size. Smaller transposons are generally more tractable than larger ones. Size may be an overriding factor if insertions are isolated in, or eventually will be transferred to, bacteriophage λ, whose genome size is limited to the amount that can be packaged efficiently. Large transposons will also be harder to handle on multicopy plasmids, since larger plasmids will have lower transformation efficiencies and may give rise to deletion variants more frequently than smaller plasmids. Also, for Tn*10*- and Tn*5*-based transposons, the frequency of transposition decreases about 40% for every kilobase of transposon length. Thus, for example, a 9-kb element will transpose at approximately 1% the frequency of a 2-kb element. The smallest available transposons contain only a 265–390 base pair (bp) *supF* gene between the ends of the transposon (see Phadnis *et al.*[6] and below).

[21] A. Wang and J. R. Roth, *Genetics* **120,** 875 (1988).
[22] N. Kleckner, *Annu. Rev. Genet.* **15,** 341 (1981).

(b) Inverted repeats. All of the transposable elements except for Mu have inverted repeats at their ends. The lengths of these repeats vary from 38 bp (Tn*3*/gamma-delta) to 70 bp (mini-Tn*10* elements) to 400–1400 bp for other Tn*10* and Tn*5* elements. The existence of such repeats is limiting in two ways. First, long inverted repeats are very unstable in single-stranded phage vectors and are somewhat unstable even in multicopy plasmid vectors; short inverted repeats (<100 bp) are not a significant problem. Second, even short inverted repeats interfere with DNA sequence analysis of transposon/target insertion junctions if the dideoxy method is used. On single-stranded templates used for dideoxy sequencing, intramolecular pairing between inverted repeats precludes DNA sequencing across the transposon/target DNA junction. The same may be true of double-stranded templates as well. Furthermore, on a double-stranded template, where the same sequence is present at both termini of the element, synthesis must be primed from inside the element and thus must extend across the inverted repeat before the target junction is reached. However, all of these problems are eliminated either by subcloning each transposon/target junction fragment into a separate vector prior to sequence analysis or by using Maxam–Gilbert sequencing of isolated restriction fragments containing the junction of interest.

(c) Rearrangements that occur on insertion. A small but significant fraction of bacteriophage Mu insertions (<10%) are accompanied by deletion or duplication of sequences adjacent to the insertion site. For other transposons, the initial insertion event is virtually always correct. All elements can give rise at low frequencies to rearrangements in secondary events subsequent to transposition (discussed in detail below), but this is not usually a major practical concern.

(d) Polarity and gene turn-on. Large transposons (the wild-type versions of Tn*10*, Tn*5*, Tn*3*, and Mu) are usually polar on expression of distal genes when inserted into an operon. Minitransposons may also be polar. The degree of polarity of an element depends on the nature and strength of transcription termination signals within the transposon, the presence or absence of promoters within the element which direct transcription outward beyond its ends, the precise location of the insertion with respect to internal Rho-dependent termination sites in the target gene, and the possibility that new promoters might be created or revealed by juxtaposition of transposon and target sequences at the new insertion junction.[21,23-26]

[23] M. S. Ciampi, M. B. Schmid, and J. R. Roth, *Proc. Natl. Acad. Sci. U.S.A.* **79**, 5016 (1982).

[24] D. E. Berg, A. Weiss, and L. Crossland, *J. Bacteriol.* **142**, 439 (1980).

[25] P. Prentki, B. Teter, M. Chandler, and D. J. Galas, *J. Mol. Biol.* **191**, 383 (1986).

[26] R. W. Simons, B. Hoopes, W. McClure, and N. Kleckner, *Cell (Cambridge, Mass.)* **34**, 673 (1983).

Transposon insertions have been used to get rough information about operon structure and have successfully identified internal promoters in several cases. However, the general usefulness of this approach is limited by the many possible complications.

(e) Specific features of Tet^R, Kan^R, and supF selections. Each of the three selectable markers that are most commonly used to select insertions have certain peculiarities which influence the selection procedures. Selection of insertions into the *E. coli* genome is discussed here; selection of insertions into phage λ is discussed in Section III below.

Tet^R. The tetracycline resistance determinant of Tn*10* has the peculiar property that the level of tetracycline resistance conferred is dramatically reduced when the determinant is present on a multicopy plasmid.[27-29] Thus, use of this determinant as the sole selectable marker on such a plasmid should be avoided whenever possible. In particular, tetracycline resistance should not be used as a marker for isolation of transposon insertions into a multicopy plasmid. However, Tn*10–tet* insertions present in single copy in the bacterial chromosome can be routinely cloned into a multicopy cloning plasmid by selection for tetracycline resistance if certain precautions are followed. (1) The level of tetracycline should be reduced to the lowest possible level (2–5 μg/ml). (2) A favorable strain background should be used, as the magnitude of the effect varies considerably from strain to strain; *E. coli* MM294 is particularly good. (3) Transformants should be selected first using another marker on the cloning vector. Tetracycline-resistant transformants can be identified among these by replica plating onto (low) tetracycline plates; inclusion of (low) tetracycline as a second drug in the transformation plates is also a possible strategy but does not work very well and often yields junk.

Kan^R. (1) A multicopy plasmid carrying a kanamycin resistance gene confers resistance to a higher level of antibiotic than that conferred by a single copy of the resistance gene in the chromosome. Thus, when insertions of a *kan* element are made into a strain carrying a multicopy plasmid, the proportion of selected colonies containing an insertion of the transposon into the plasmid can be increased by including a high concentration of antibiotic in the selective plates (300 versus 50 μg/ml). (2) Most *E. coli* strains cannot become kanamycin resistant by spontaneous mutation. However, many or all streptomycin-resistant strains can mutate to kanamycin resistance. If transposon insertions are relatively rare, the frequency of spontaneous Kan^R derivatives may be high enough to cause significant background. The frequency of such derivatives is lower if the

[27] L. D. Smith and K. P. Bertrand, *J. Mol. Biol.* **203,** 949 (1988).
[28] H. S. Moyed, T. T. Nguyen, and K. P. Bertrand, *J. Bacteriol.* **155,** 549 (1983).
[29] D. C. Coleman and T. J. Foster, *Mol. Gen. Genet.* **182,** 171 (1981).

selective medium is rich [e.g., Luria Bertani broth (LB)] than if it is minimal. Increasing the kanamycin concentration is not effective in eliminating spontaneous mutants.

supF. Insertions of the *supF* gene into the *E. coli* chromosome can be selected using a host strain which carries two *supF*-suppressible nonsense mutations (see, e.g., Seed[30]).

(f) Nonsense mutations and su⁺ hosts. All λ phage delivery vehicles are disabled by one or more nonsense mutations in the phage replication genes *O* and *P*. Most vehicles contain amber mutations and thus cannot be used in hosts containing an amber suppressor. However, if use of such a host is unavoidable, one λ::Tn*10* vehicle is available which is disabled with a UGA mutation (λNK370, see below); since UGA mutations are not suppressed by amber or ochre nonsense suppressors, this phage can be used to isolate insertions of wild-type Tn*10* in a host containing an amber or ochre suppressor.[4]

E. Using Tn10 to Create Deletions

1. Tn10-Promoted Adjacent Deletions. Most transposons are capable of generating deletions of chromosomal sequences adjacent to their site of insertion. Tn*10* is the transposon of choice for generating such deletions. Formation of Tn*10*-promoted deletions requires that the transposon contain two intact IS10 sequences; this type of rearrangement occurs by an interaction between the two "inside" IS10 ends of the element and a target site located in adjacent sequences outside of the element.[31] Wild-type Tn*10* is almost always used for this purpose. However, the IS10 elements may be present either in inverted orientation or in direct orientation, so other types of Tn*10* derivatives could also be used.[32]

Tn*10*-promoted deletions result in removal of a continuous segment extending from one of the inside ends of the transposon, across the internal region of the transposon including the tetracycline resistant determinant, across the distal IS10 sequence, and into adjacent chromosomal sequences.[31] Such deletions occur at a frequency of about 10^{-4} in an overnight culture grown from a single colony.[33] Since the Tet[R] determinant is lost, a population highly enriched in such deletions can be obtained either by penicillin selection in the presence of bacteriostatic levels of tetracy-

[30] B. Seed, *Nucleic Acids Res.* **11**, 2427 (1983).
[31] N. Kleckner, in "Mobile DNA" (D. E. Berg and M. M. Howe, eds.), p. 225. American Society for Microbiology, Washington, D.C., 1988.
[32] E. A. Raleigh and N. Kleckner, *J. Mol. Biol.* **173**, 437 (1984).
[33] N. Kleckner, K. Reichardt, and D. Botstein, *J. Mol. Biol.* **127**, 89 (1979).

cline[2] or by a selection for fusaric acid resistance under conditions where the *tetA* gene is induced.[34,35]

Tetracycline-sensitive derivatives obtained by either of these methods include not only TnI0-promoted deletions, but also related rearrangements which occur at roughly the same frequency, TnI0-promoted deletion/inversions (see Kleckner *et al.*[33]; reviewed in Kleckner[31]). In the case of the deletion/inversions, one IS10 element and a segment of adjacent chromosomal material has been inverted relative to its original orientation, rather than deleted as is the case in TnI0-promoted adjacent deletions. Deletions must be specifically distinguished from deletion/inversions by secondary tests. All such tests rely on the fact that deletions will have eliminated a contiguous segment of adjacent DNA, whereas inversions will have only a single new break point and will still retain all of the material between the transposon and that breakpoint. Any derivative which inactivates two adjacent independently expressed genes can be assumed to be a deletion. More generally, the presence or absence of sequences originally adjacent to the transposon can be determined either physically, by Southern blotting with a probe specific to those sequences, or genetically, by asking whether the Tet^S derivative can donate markers from the adjacent region in a phage-mediated transductional cross.[33]

Tetracycline-sensitive derivatives will also include a number of other types of variants. Nearly precise excisions,[15] which are specific deletions within the transposon, occur at about 10% the frequency of deletions and deletion/inversions. Spontaneous deletions and precise excisions of the transposon (see below) are much rarer.

2. Selecting Deletions with Predetermined End Points by Recombination between Modified Tn10 Insertions. Transposons can be used as portable regions of homology.[2] Recombination between two TnI0 elements located at different sites in the genome as either direct or inverted repeats can generate either a deletion or an inversion of the intervening material. Special tools which facilitate the isolation of such recombinants have been developed.[36] These tools permit the replacement of the original *tet* region in a chromosomal insertion with either a *tet::str* or *tet::kan* disruption. The *str* and *kan* disruptions are at different locations in the *tet* segment. Thus, in a strain carrying a *tet::str* insertion at one site and a *tet::kan* insertion at another site, the desired deletion or inversion between the two sites can be obtained by selecting for Tet^R recombinants and then analyzing the recombinants to identify those having the desired structure. These

[34] B. R. Bochner, H.-C. Huang, G. L. Schieven, and B. Ames, *J. Bacteriol.* **143,** 926 (1980).
[35] S. R. Maloy and W. D. Nunn, *J. Bacteriol.* **145,** 1110 (1981).
[36] V. François, J. Louarn, J. Patte and J.-M. Louarn, *Gene* **56,** 99 (1987).

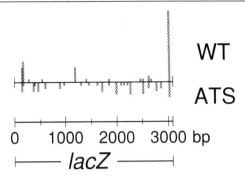

FIG. 1. Target specificity of wild-type and altered target specificity (ATS) mutant transposases in the *lacZ* gene. Fifty independent mini-Tn*10* Kan^R insertions into the 3.1-kb *lacZ* gene carried on a pGEM-3 plasmid vector were isolated from either a "hop phage" carrying wild-type transposase [λNK1105; J. C. Way, M. A. Davis, D. Morisato, D. E. Roberts, and N. Kleckner, *Gene* 32, 369 (1984)] or from a hop phage carrying ATS transposase (λNK1316, derivative 103 in Fig. 2). Insertion sites were determined by sequencing. Approximate positions of insertion are shown; base pair 1 of the scale is at the A of the ATG start codon for *lacZ*. The strongest insertion site for both transposases is at bp 3026–3034.

tools are as applicable to insertions of mini-Tn*10*–*tet* elements as to the full Tn*10* elements on which they were tested. Use of mini-Tn*10* elements should eliminate a residual background of Tet^R recombinants that appear to have undergone Tn*10*-promoted rearrangements.

II. Tn*10* Transposition Vehicles

A. Tn10 Transposase Mutation Which Decreases Target Site Specificity

Wild-type Tn*10* inserts preferentially into so-called hotspots.[37–40] This phenomenon is illustrated by the spectrum of Tn*10* insertions into the 3.1-kb *lacZ* gene which contains a single very strong hotspot plus several less preferred sites (Fig. 1, top). Two factors contribute to Tn*10*'s selection of particular sites. (1) Tn*10* insertion involves recognition, cleavage, and duplication of a specific 9-bp target site sequence. Comparison among many different 9-bp target sites reveals a consensus sequence, 5' NGCTNAGCN 3'. Particular target sites can differ significantly from this sequence, but hotter sites match more closely while colder sites match

[37] N. Kleckner, D. A. Steele, K. Reichardt, and D. Botstein, *Genetics* 92, 1023 (1979).

[38] S. M. Halling and N. Kleckner, *Cell* (*Cambridge, Mass.*) 28, 155 (1982).

[39] O. Huisman, W. Raymond, K. Froehlich, P. Errada, N. Kleckner, D. Botstein, and M. A. Hoyt, *Genetics* 116, 191 (1987).

[40] S. Y. Lee, D. Butler, and N. Kleckner, *Proc. Natl. Acad. Sci. U.S.A.* 84, 7876 (1987).

less well. (2) The efficiency of insertion into a particular consensus sequence varies over several orders of magnitude according to the sequence of the 8–10 bp located immediately adjacent to the consensus sequence on either side. These base pairs may influence DNA structure, since no simple correlation with DNA sequence is obvious by inspection.

The nature and combination of these two factors make it difficult or impossible to predict potential sites of Tn*10* insertion from inspection of the target DNA sequence. However, one general rule follows from the fact that four of the six base pairs in the consensus sequence are GC pairs: very AT-rich DNA will have relatively few hot spots, so Tn*10* insertions will generally be more randomly distributed except in the unfortunate case where a single GC-containing hotspot occurs.[39]

A significant improvement in Tn*10* transposon mutagenesis is provided by the recent isolation of mutant IS*10* transposases that exhibit a much lower degree of insertion specificity than wild type. These altered target specificity (ATS) transposases, *ats1* and *ats2*, result from G to A transition mutations at base pairs 508 and 853 of IS*10* (TRN10IS1R.BACTERIA, GenBank J01829). The mutations change cysteines to tyrosines at amino acids 134 and 249 of the protein (bp 400 and 745 of the transposase gene[41]). The double-mutant ATS transposase exhibits only a slight decrease in transposition activity (3-fold), so its usefulness for insertion mutagenesis is not affected. The spectrum of insertion sites selected by the double-mutant ATS transposase in the *lacZ* gene is compared with that of wild type in Fig. 1. The mutant transposase utilizes a much larger number of sites than wild type; 50 insertions are relatively evenly distributed among 23 different sites in the 3.1-kb region. Tn*10* derivatives for transposon mutagenesis that contain the *ats1 ats2* double mutation in transposase are described in the following section and in Fig. 2 (derivatives 102–108).

B. Current Tn10 Transposon Derivatives

1. Obtaining Tn10 Derivatives. The Tn*10* derivatives described below and *E. coli* strain NK5336 can be obtained by communicating *in writing* with Dr. Nancy Kleckner, Department of Biochemistry and Molecular Biology, Harvard University, 7 Divinity Avenue, Cambridge, MA 02138 (FAX: 617-495-8308). (Requests made by telephone will not be accepted. If you are not sure which elements are appropriate for your needs or if you have general questions, write a letter or a FAX explaining your problem and giving your telephone and/or FAX number.) Bacterial strains N6377 and SG12021 and further information about λD69 procedures can

---

[41] S. M. Halling, R. W. Simons, J. C. Way, R. B. Walsh, and N. Kleckner, *Proc. Natl. Acad. Sci. U.S.A.* **79**, 2608 (1982).

Derivative	Length of Transposon (kb)	Markers in Transposon	Plasmid Vehicle	Phage Vehicle
101	9.3	Tet^R	pNK81	λNK561, λNK370
102	---	---	pNK2881, pNK2882	---
103	1.8	Kan^R	pNK2859	λNK1316
104	2.9	Tet^R	pNK2883	λNK1323
105	1.4	Cam^R	pNK2884	λNK1324
106	3.4	Kan^R, URA3	pNK2885	λNK1325
107	3.0	Cam^R, URA3	pNK2886	λNK1326
108	1.9	Kan^R, Plac	pNK2887	λNK1327
109	---	---	pNK474	---
110	0.4	supF	pNK1759	---
111	4.8	Kan^R, lacZ	pNK2804	---
112	4.9	Kan^R, lacZ	pNK1207	λNK1205
113	6.1	Kan^R, lacZ, URA3	pNK2809	λNK1224
114	2.2	Kan^R, Erm^R	pNK2811	---

FIG. 2. Useful Tn*10* derivatives. The structure of each transposon-containing restriction fragment (or for derivatives 101, 112, and 113, the transposon itself) is drawn to scale. The backbones into which these restriction fragments (or transposons) are inserted to create each transposon vehicle are described in Fig. 3. In *lacI*+ or *lacI*Q strains, *Ptac* or *Plac–UV5* promoters on these constructions can be fully induced by the addition to the medium of IPTG (isopropyl-β-D-thiogalactopyranoside) to a final concentration of 1 m*M*. Derivative 101 (wild-type Tn*10*) constructions have been described previously [J. C. Way, M. A. Davis, D. Morisato, D. E. Roberts, and N. Kleckner, *Gene* **32**, 369 (1984)]. The open reading frame for transposase protein is bp 108–1313 of IS10 Right (TRN10IS1R.BACTERIA, GenBank J01829). Derivative 109 (*Ptac*–wild-type transposase) was constructed by cleaving Tn*10* in IS10 Right with *Bcl*I (bp 66, Fig. 4a) and ligating a *Pvu*II to *Eco*RI fragment containing the *Ptac* promoter [E. Amann, J. Brosius, and M. Ptashne, *Gene* **25**, 167 (1983)] to this filled-in site so that the transposase gene is under *Ptac* control. The derivative extends to the *Eco*RI site at bp 3140 of Tn*10* (Fig. 4a). Derivative 109 provides a backbone for derivatives 110 and 111. Derivative 102 (*Ptac*–ATS transposase) is identical to derivative 109 with two exceptions. First, the transposase gene in derivative 102 carries two ATS mutations (*ats1*, a G to A transition at bp 508 of IS10 Right, and *ats2*, a G to A transition at bp 853 of IS10 Right). Second, the sequence between a *Xho*II site at bp 1319 and a *Bgl*II site at bp 1942 of Tn*10* has been deleted from derivative 109 and an *Xba*I linker inserted at this deletion junction. Derivative 102 provides a backbone for derivatives 103–108. Mini-Tn*10* derivatives 103–108, 110, and 111 are each bounded by identical inverted repeats of the outermost 70 bp of IS10 Right (generated by cleaving IS10 Right with *Bcl*I and converting the *Bcl*I site to a *Bam*HI site). The 70-bp transposon end in these derivatives is embedded in 40 bp of λ *cI* gene sequence terminating in a *Hin*dIII site. Thus each complete transposon is carried on a *Hin*dIII fragment which is inserted into the *Hin*dIII site (bp 2272, Fig. 4a) of derivative 102 (*Ptac*–ATS transposase) or of derivative 109 (*Ptac*–wild-type transposase) to put the mini-Tn*10* in cis to a transposase source. Derivative 103 (mini-Tn*10* kan/*Ptac*–ATS transposase) carries a *Bam*HI Kan^R fragment from Tn*903* (Fig. 4b, TRN903.BACTERIA, GenBank J01839, bp 697 to 2392, *Pvu*II fragment converted to a *Bam*HI fragment with linkers), oriented in the backbone so that the *kan* gene promoter is transcribing in the same direction as the *Ptac* promoter. Derivative 104 (mini-Tn*10* tet/*Ptac*–ATS transposase) carries a *Bam*HI Tet^R fragment from Tn*10* (Fig. 4a, bp 1942–4717, TRN10TETR.BACTERIA, GenBank J01830, bp 3402 to 627, *Bgl*II fragment converted to a *Bam*HI fragment with linkers), oriented in the backbone so that the *tetR* gene promoter is transcribing in the same direction and the *tetA* gene promoter is transcribing in the opposite direction as the *Ptac* promoter. Derivative 105 (mini-Tn*10* cam/*Ptac*–ATS transposase) carries a *Bam*HI Tn9-derived Cam^R fragment from pACYC 184 (P18XCYC18.SYN, GenBank X06403, bp 3500 to 580, *Hae*II fragment converted to a *Bam*HI fragment with linkers), oriented in the backbone so that the *cam* gene promoter is transcribing in the same direction as the *Ptac* promoter. Derivative 106 (mini-Tn*10* kan URA3/*Ptac*–ATS transposase) is identical to derivative 103, and derivative 107 (mini-Tn*10* can URA3/*Ptac*–ATS transposase) is identical to derivative 105 except that in each case the element also carries a *Saccharomyces cerevisiae* *Bgl*II to *Bam*HI URA3 gene fragment from YEP24 (YEP24.VEC, GenBank VB0067, *Eco*RI site at bp 2241 converted to a *Bgl*II to the *Bam*HI site at bp 3784). The URA3 gene is inserted upstream of the *kan* gene (for derivative 107) or the *cam* gene (for derivative 108) and is oriented so that it is transcribed in the opposite direction from the *kan* or *cam* gene. Derivative 108 (mini-Tn*10* kan Plac/*Ptac*–ATS transposase) is identical to derivative 103 except that it also carries a *Plac–UV5* *Bam*HI fragment downstream of the *kan* gene oriented so that the promoter is transcribing in the same direction as the *kan* gene promoter out across the transposon end. The sequence of the transposon end through the promoter fragment is shown in Fig. 5e. Derivative 110 (mini be obtained by communicating in writing with Dr. Susan Gottesman, Bldg. 37 Rm 4B03, Laboratory of Molecular Biology, National Cancer Institute, National Institutes of Health, Bethesda, MD 20892 (FAX: 301-496-0260).

2. *Descriptions of Tn10 Derivatives.* The structures of a number of Tn*10* derivatives useful for various types of transposon mutagenesis are described briefly below and in detail in Fig. 2 and its legend. These derivatives are carried on plasmid and/or phage vehicles, the structures of which are shown in Fig. 3.

Wild-type Tn*10*: Derivative 101. Wild-type Tn*10* has inverted repeats of insertion sequences IS10 Left and IS10 Right at its ends. The intervening material includes genes for tetracycline resistance (Tet^R) and other unknown determinants. This transposon can be used to generate insertions, deletions, or deletion/inversions. A partial restriction map of wild-type Tn*10* is shown in Fig. 4a; further details about the intervening material are summarized by Kleckner.[31] Wild-type Tn*10* is available on a pBR322-derived ampicillin-resistant (Amp^R) plasmid (pNK81), on a λ $O_{am}29\ P_{am}80$ "hop phage" vehicle (λNK561), and on a λ O_{UGA} "hop phage" vehicle

---

Tn*10 supF/Ptac*–wild-type transposase) carries a 248-bp *Xho*II *supF* fragment from bp 845 to 208 of PiAN7 (PIAN7.VEC, GenBank VB0066). Ligation of this *Xho*II fragment between the *Bam*HI sites of the transposon ends restores the *Bam*HI sites. The *supF* gene is oriented in the backbone so that its promoter transcribes in the opposite direction from the *Ptac* promoter. Derivative 111 (mini-Tn*10 lacZ kan/Ptac*–wild-type transposase) carries a promoterless *lacZ Bam*HI fragment and a Kan^R *Bam*HI fragment. The promoterless *lacZ* fragment consists of leader sequences from the *trpA* gene followed by the ribosome binding site and coding sequence of the *lacZ* gene [R. W. Simons, F. Houman, and N. Kleckner, *Gene* **53**, 85 (1987)]. The sequence from the transposon end through the leader region into the *lacZ* gene is shown in Fig. 5d. The fragment extends to the end of the *lacZ* gene (converted to a *Bam*HI site) at bp 4373 of ECOLAC.BACTERIA, GenBank J01636. The *lacZ* fragment is oriented in the backbone so that it would be transcribed in the same direction as read by the *Ptac* promoter. Downstream of this promoterless *lacZ Bam*HI fragment is a 1.5-kb Tn*903*-derived Kan^R *Bam*HI fragment from pUC4K (Pharmacia, Piscataway, NJ) oriented so that the *kan* gene promoter is transcribing in the same direction as the *Ptac* promoter. (A version of derivative 111 that is marked only with the promoterless *lacZ* fragment is also available as pNK2803.) Derivative 112 (Tn*10*–LK) carries the '*lacZ* fragment (ECOLAC.BACTERIA, GenBank J01636, bp 1309 to 4373, missing the first eight codons of the *lacZ* gene) oriented so that the gene is fused to the IS10 Right-derived end of the transposon. The sequence across the transposon end into the '*lacZ* gene is shown in Fig. 5b. The same Kan^R *Bam*HI fragment that marks derivative 103 is carried within the element downstream of the '*lacZ* gene but oriented so that it is transcribed in the opposite direction from the '*lacZ* gene. The second end of this element consists of the outermost 70 bp of IS10 Left. Derivative 113 (Tn*10*–LUK) is identical to derivative 112 except that an *S. cerevisiae URA3 Bgl*II fragment (YSCODCD.PL, GenBank K02206, bp 1 to 1170) has been inserted at the *Bam*HI site (λNK370). In pNK81 the transposon is inserted in the *hisG* gene of *Salmonella typhimurium*; in both λ phages, the transposon is inserted in the *cI* gene.

Unlike the other λ vehicles described in Fig. 2, λNK561 and λNK370 are both suitable for transposon mutagenesis at any reasonable temperature. The transposon insertion in these phages confers an absolute defect in the *cI* gene, and the phage attachment site is deleted; other vehicles carry the *cI857* mutation, so they can form abortive lysogens at temperatures below 37°.

ATS transposase fused to *Ptac*: Derivative 102. Derivative 102 contains the *ats1 ats2* transposase gene fused to the strong isopropyl-β-D-thiogalactopyranoside (IPTG)-inducible *Ptac* promoter (*Ptac*–ATS transposase). This derivative should be used to complement in trans mini-Tn10 constructs carried on vehicles that lack transposase (derivatives 112 and 113 below). It is available on a pBR322-derived Amp^R plasmid (pNK2881) and on a pACYC184-derived Tet^R plasmid (pNK2882). Derivative 102 also

---

between '*lacZ* and the Kan^R markers oriented to be transcribed in the same direction as the '*lacZ* gene. Derivative 114 carries a 'Kan^R fragment and an Erm^R fragment between the outermost 70 bp of IS10 Right. The 'Kan^R fragment consists of a leader sequence and the Tn5 neomycin resistance gene starting at the second codon (TRN5NEO.BACTERIA, GenBank J01834, bp154) and extending through a *Sal*I site 1130 bp downstream [S. J. Rothstein, R. A. Jorgensen, K. Postle, and W. S. Reznikoff, *Cell* (*Cambridge, Mass.*) **19**, 795 (1980)]. The sequence of the transposon end through the start of the '*kan* gene is shown in Fig. 5c. A 1-kb selectable erythromycin resistance (Erm^R) fragment [B. Martin, G. Alloing, V. Mejean, and J. Claverys, *Plasmid* **18**, 250 (1987); K. Josson, T. Scheirlinck, F. Michiels, C. Platteeuw, P. Stanssens, H. Joos, P. Dhaese, M. Zabeau, and J. Mahillon, *Plasmid* **21**, 9 (1989)] has been inserted into the *Sma*I site 970 bp away from the start of the '*kan* gene oriented so that it is transcribed in the same direction as the '*kan* gene. The entire transposon has been cloned on a *Hind*III to *Eco*RI fragment into *Salmonella hisG* and *hisD* sequences (indicated by a heavy line). Cloned upstream of the '*kan* end of the transposon are four tandem repeats of a 180-bp transcriptional termination sequence from the *rrnB* operon [J. Brosius, T. J. Dull, D. D. Sleeter, and H. F. Noller, *J. Mol. Biol.* **148**, 107 (1981)] that prevent expression of the '*kan* gene by nonspecific transcription from the vector (represented by a box containing TT). Transposase is provided to the transposon from a *Ptac*–wild-type transposase fusion (analogous to derivative 109) missing the innermost end of IS10 Right from a *Xho*II site at bp 1319 to a *Bgl*II site at bp 1942 of Tn10 and extending to the *Cla*I site at bp 2591 of Tn10 (Fig. 4a). This *Ptac*–transposase fusion is inserted immediately upstream of the transcriptional terminators and is oriented so that it is transcribed in the opposite direction from the '*kan* gene. Beyond the transposase-proximal end of the construction is another 740 bp of *Salmonella his* DNA terminating in an *Xba*I site. Beyond the mini-Tn10-proximal end of the construction is another 250 bp of *his* DNA terminating in an *Xba*I site. B, *Bam*HI; Bc, *Bcl*I; Bg, *Bgl*II; C, *Cla*I; H, *Hind*III; R, *Eco*RI; X, *Xho*I; Xb, *Xba*I. Open triangle, *Ptac*; filled triangle, *Plac*–*UV5*.

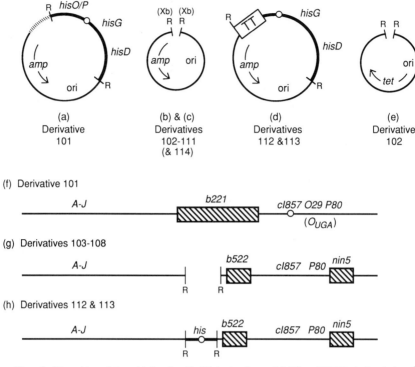

FIG. 3. Plasmid and λ vehicles for Tn*10* derivatives. (a) The pBR322-derived Amp^R plasmid vehicle for derivative 101 (pNK81) has been previously described [T. J. Foster, M. A. Davis, D. E. Roberts, K. Takeshita, and N. Kleckner, *Cell (Cambridge, Mass.)* **23**, 201 (1981)]. It carries 3.1 kb of the *Salmonella his* operon (indicated by a heavy line) containing the promoter/operator (*hisO/P*), *hisG*, and *hisD* genes on an *Eco*RI fragment with the transposon inserted by transposition into a hotspot in the *hisG* gene (STHISOP.EMBL, X13464, bp 379 to 387) oriented so that IS*10* Right is closest to the *his* operon promoter. The site of transposon insertion is marked by an open circle. The *his* fragment is ligated at the *Eco*RI site of a deletion derivative of pBR322 described in (b) below which carries *amp* and the plasmid origin oriented so that the *amp* promoter is transcribing in the opposite direction from the *his* promoter. Between the promoter end of the *amp* gene and the promoter end of the *his* DNA on this plasmid are 1050 bp of uncharacterized "mystery DNA" (indicated as a shaded line). (b) The Amp^R plasmid vehicle for derivatives 102–111 is a deletion of bp 75 to 2353 of pBR322 (PBR322.VEC, GenBank VB0001). The *Hind*III site in this plasmid has been destroyed by filling in. The *Eco*RI fragments shown in Fig. 2 are inserted into the *Eco*RI site of this plasmid (at bp 4360 of the pBR332 sequence) oriented so that the *Ptac* promoter in the insert is transcribing in the opposite direction from the *amp* promoter. (c) Derivative 114 is inserted as an *Xba*I fragment into a derivative of the plasmid described in (b) where a 100-bp deletion has been made between the *Cla*I site (at bp 22 of pBR322) and the *Aat*II site (at bp 4285 of pBR322) and an *Xba*I linker has been inserted at the deletion junction. The insert is oriented so that the *Ptac* promoter is transcribing in the same direction as the *amp* promoter. (d) The Amp^R plasmid vehicle for derivatives 112 and 113 is derived from pNK81 [see (a) above] as follows. The transposon elements are inserted in the same *hisG*

forms the starting point for derivatives 103–108 below. The *Ptac*–transposase fusion is carried on a *Eco*RI fragment which extends from a site just upstream of the promoter to the *Eco*RI site at bp 3140 of Tn*10* (Fig. 4a) except that bp 1329–1942 at the inside end of IS10 Right have been deleted; this deletion removes the transposase binding site but does not affect the transposase structural gene.

Mini-Tn*10* with ATS transposase provided in cis: Derivatives 103–108. Mini-Tn*10* elements are generally short (400–3000 bp), can be engineered to carry a wide assortment of markers, and give rise to stable insertions because they do not carry a transposase gene. Mini-Tn*10* elements have at each terminus short segments carrying Tn*10* ends in inverted orientation; these Tn*10* ends flank one or more selectable markers. For mini-Tn*10* derivatives 103–108, the two ends are perfect inverted repeats of a 70-bp segment carrying the outside end of IS10 Right. These six elements are constructed on delivery vehicles that also carry a *Ptac–ats1 ats2* transposase gene in cis. Each element is carried on a *Hin*dIII fragment which is cloned in appropriate orientation into the *Hin*dIII site of derivative 102

---

site as wild-type Tn*10* in pNK81 and are oriented so that the '*lacZ* gene in each transposon would be transcribed in the same direction as the *his* operon. The site of transposon insertion is marked with an open circle. All of the "mystery DNA" and the *his* promoter sequences upstream of the transposon have been replaced with four tandem repeats of the transcriptional terminator sequences (represented by a box containing TT) from the *rrnB* operon which prevent nonspecific transcription from expressing the '*lacZ* gene on this plasmid. (e) The 4.2-kb Tet^R pACYC 184 plasmid vehicle for derivative 102 contains the *Eco*RI insert cloned in the *Eco*RI site at bp 1 of pACYC 184 (P18XCYC18.SYN, GenBank X06403) oriented so that *Ptac* is transcribing in the same direction as the *tet* gene. (f) The $O_{am}29\ P_{am}80\ \lambda$ hop phage vehicle for derivative 101 (λNK561) consists of the transposon inserted by transposition into the *cI* gene of a λ phage that is *b221 cI857* $O_{am}29\ P_{am}80$; λNK370 is the same phage except that it carries an O_{UGA} mutation rather than the $O_{am}29\ P_{am}80$ mutations and can be used for making insertions into strains carrying amber or ochre nonsense suppressors. The site of transposon insertion in each phage is marked with an open circle. (g) The $P_{am}80\ \lambda$ hop phage vehicle for derivatives 103–108 is λ *b522 cI857* $P_{am}80\ nin5$ with the appropriate *Eco*RI fragment from Fig. 2 substituted for the *Eco*RI fragment of λ DNA between bp 21226 and 26104 [R. W. Hendrix, J. W. Roberts, R. W. Stahl, and R. A. Weisberg (eds.), "Lambda II." Cold Spring Harbor Laboratory Press, Cold Spring Harbor, New York, 1983]. All inserts have been oriented in this phage vehicle so that the mini-transposon-proximal end of the insert is closest to the λ *J* gene. The *b522* deletion removes the phage attachment site. (h) The $P_{am}80$ hop phage vehicle for derivatives 112 and 113 is λgt7–*his b522 cI857* $P_{am}80\ nin5$ [λNK780; T. J. Foster, M. A. Davis, D. E. Roberts, K. Takeshita, and N. Kleckner, *Cell* (*Cambridge, Mass.*) **23**, 201 (1981)]. The λgt7–*his* backbone carries the same *Eco*RI fragment of the *Salmonella his* operon DNA as described for pNK81 [see (a) above] substituted in an unknown orientation for the λ DNA between *Eco*RI sites at bp 21226 and 26104. The transposons are inserted in the same *hisG* site (marked with an open circle) as in the corresponding plasmid constructions. The *b522* deletion removes the phage attachment site.

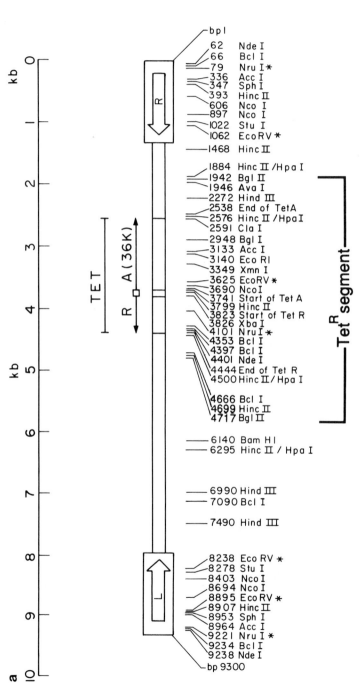

FIG. 4. Partial restriction maps for Tn*10* and the Tn*903* Kan^R segment. (a) Tn*10* includes IS10 Right, IS10 Left, and the tetracycline resistance determinant, which is composed of two divergently transcribed genes: *tetR*, the repressor, and *tetA*, the structural gene for a 36-kDa protein. The positions and orientations of the two genes are indicated by the two correspondingly marked divergent arrows. This map was compiled as described by Way *et al.* (1984), except that sites beyond the HincII/HpaI site at bp 4500 have been repositioned based on recent sequence data. The sequence of IS10 Right from bp 1 of this map to bp 1329 is given in TRN10IS1R.BACTERIA, GenBank J01829. Sequences of the *tet* genes corresponding to bp 5435 to 1406 on this map are given in TRN10TETR.BACTERIA, GenBank J01830. Asterisks indicate that Tn*10* contains EcoRV and/or NruI sites in addition to those shown. The BglII Tet^R segment excised from Tn*10* to mark derivative 104 is indicated. (b) The Tn*903* Kan^R segment that marks derivatives 103, 106, 112, and 113 is shown. The first base pair of this map corresponds to TRN903.BACTERIA, GenBank J01839, bp 697. At the ends of the segment are 360-bp inverted repeats of the inner termini of two IS903 elements. The *kan* gene in this segment starts at bp 465 and extends to bp 1277 (GenBank J01839, bp 1162 to bp 1974). The StuI* sites in the IS903 inverted repeats overlap with sites of *dcm* methylation and therefore can only be cleaved if DNA is prepared from a *dcm*⁻ strain.

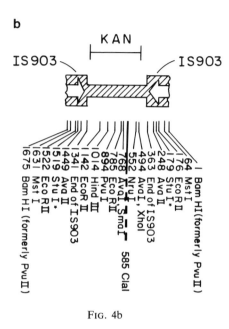

FIG. 4b

(bp 2272 of Tn*10*, Fig. 4a). The six mini-Tn*10* derivatives are thus exactly analogous to each other in structure.

Derivatives 103–105 are mini-Tn*10* constructs useful for general transposon mutagenesis in *E. coli*. Derivative 103 is marked with a kanamycin resistance (Kan^R) fragment from Tn*903* (Fig. 4b), derivative 104 is marked with a Tet^R fragment from Tn*10* (Fig. 4a), and derivative 105 is marked with a chloramphenicol resistance (Cam^R) fragment from Tn*9*.

Special care must be taken to verify the structure of derivative 104 because a portion of the Tet^R marker inside the transposon [from the *Bgl*II site at bp 1942 of Tn*10* (converted to a *Bam*HI site at the right end of the Tet^R marker as drawn in Fig. 2) to the *Eco*RI site at bp 3140 of Tn*10*] is directly repeated in the adjacent sequences outside the transposon. The sequence to the right of the transposon as drawn in Fig. 2 extends from the *Bgl*II site at bp 1942 to the *Hin*dIII site at bp 2272 of Tn*10*. Homologous recombination between this flanking sequence and the Tet^R marker will result in loss of the intervening transposon end. The sequence to the left of the transposon as drawn in Fig. 2 extends from the *Hin*dIII site at bp

2272 to the *Eco*RI site at bp 3140 of Tn*10*. Homologous recombination between this sequence and the Tet^R marker will result in loss of *tet* genes and the other transposon end. The λNK1323 version of derivative 104 is particularly unstable since λ recombination functions are extremely active. Thus, stocks of this phage must each be grown from a single plaque and carefully checked for transposition activity.

Derivatives 106 and 107 are mini-Tn*10* constructs designed specifically for generating insertion mutations into cloned yeast genes. They carry both a marker that is selectable in *E. coli*, for isolation of insertions, and a marker that is selectable in yeast, for subsequent integration of the disrupted gene back into the yeast genome. Derivative 106 is marked with a Kan^R fragment from Tn*903* and the *Saccharomyces cerevisiae URA3* gene. Derivative 107 is marked with a Cam^R fragment from Tn*9* and the *S. cerevisiae URA3* gene.

Derivative 108 is a mini-Tn*10* marked with a Kan^R fragment from Tn*903* and contains the strong IPTG-inducible *Plac–UV5* promoter oriented to read out across one end of the transposon. This derivative can be used to generate promoter fusions of target genes to *Plac–UV5*, which will arise when the transposon inserts in the proper orientation upstream of the gene. The DNA sequence from the end of the transposon through the *Plac–UV5* promoter is shown in Fig. 5e.

Derivatives 103–108 are each available on a pBR322-derived Amp^R plasmid (pNK2859, pNK2883, pNK2884, pNK2885, pNK2886, and pNK2887) and on a $P_{am}80$ λ hop phage vehicle (λNK1316, λNK1323, λNK1324, λNK1325, λNK1326, and λNK1327). The λ vehicle for these elements is deleted for the phage attachment site and carries the temperature-sensitive *c*I857 repressor gene; insertions should be isolated at 37° or above, at which temperature the *c*I857 mutation will prevent formation of abortive lysogens. Plasmid and phage versions of mini-Tn*10* constructions analogous to derivatives 103–105 (mini-Tn*10 kan*, mini-Tn*10 tet*, and mini-Tn*10 cam* complemented to transposase by *Ptac*–ATS transposase) but each carrying a polylinker containing rare-cutting restriction enzyme sites are available to facilitate physical mapping of insertions by pulsed-field gel electrophoresis (J. Mahillon and N. Kleckner, unpublished observations).[41a]

Wild-type transposase fused to *Ptac*: Derivative 109. Derivative 109 consists of a wild-type transposase gene from IS*10* Right fused to the strong IPTG-inducible *Ptac* promoter (*Ptac*–wild-type transposase). It is analogous to derivative 102 except that it lacks the *ats1* and *ats2* mutations and the small deletion at the end of IS*10*. It is carried on a pBR322-derived Amp^R backbone (pNK474). pNK474 is used as the starting point for construction of derivatives 110 and 111. The *Eco*RI fragment carrying

(a) Outside End IS10 Right

bp 1

CTG ATG AAT CCC CTA ATG ATT TTG GTA AAA ATC ATT AAG TTA AGG TGG ATA CAC ATC TTG TCA TAT GAT C . . .

(b) *'lacZ* Translational Fusion

. . . *CC GTC GTT* . . .
            codon 11

(c) *'kan* Translational Fusion

. . . CG GCC AAG CTA GCT TGG *ATT GAA CAA GAT GGA TTG CAC GCA GGT TCT* . . .
      |_____leader sequence_____| codon 2

(d) *lacZ* Transcriptional Fusion

. . . (*) GGATC CGGAC CGATG AAAGC GGCGA CGCGC AGTTA ATCCC ACAGC CGCCA GTTCC

GCTGG CGGCA TTTTA ACTTT CTTTA ATGTT CACAC AGGAA ACAGC T *ATG ACC ATG ATT ACG*

GAT TCA CTG GCC GTC GTT . . .
                    codon 11

(e) *Plac-UV5* Promoter Fusion

                                                    ◄──── +1
. . . (*) GGATC CTGTT TCCTG TGTGA AATTG TTATC CGCTC ACAAT TCCAC AC ATC ATACG
        |_____operator_____| -10'

AGCCG GAAGC ATAAA GTGTA AAGCC TGGGG TGCCT AATGA GTGAG AATTA ATTCC GGATC C . . .
              -35'

FIG. 5. Sequences through the transposon end into fusion constructions. (a) The outermost 70 bp of IS10 Right is shown in bold type. This sequence is fused to the following sequences in the derivatives indicated. (b) Sequence into the *'lacZ* gene of derivatives 112 and 113. The bases in the *lacZ* coding region are italicized. The *'lacZ* fragment is missing the first eight amino acids of the wild-type protein. (c) Sequence into the *'kan* gene of derivative 114. The bases in the *kan* coding region are italicized. The *'kan* fragment is missing the first amino acid of the wild-type protein. (d) Sequence into the promoterless *lacZ* gene of derivative 111. The bases indicated by (*) are three possible sequences C, CG, or CGC of the *Bam*HI linker ligated at the *Bcl*I site of the outer end of IS10 Right. It is not known which of the three sequences is present in the transposon. The bases in the *lacZ* coding region are italicized. Shine–Dalgarno sequences are underlined. (e) Template/antisense sequence of the *Plac–UV5* promoter in derivative 108. The bases indicated by (*) are three possible sequences C, CG, or CGC of the *Bam*HI linker ligated at the *Bcl*I site of the outer end of IS10 Right. It is not known which of the three sequences is present in the transposon. The positions corresponding to the first base of the transcript from this promoter, the operator, the − 10′ region, and the − 35′ region are indicated.

Ptac–wild-type transposase extends from a site just upstream of *Ptac* to an *Eco*RI site at bp 3140 of Tn*10* (Fig. 4a). The mini-Tn*10* elements of derivatives 110 and 111 are cloned on *Hin*dIII fragments into the *Hin*dIII site of derivative 109 (bp 2272 of Tn*10*, Fig. 4a).

New mini-Tn*10* constructs with wild-type transposase provided in cis: Derivatives 110 and 111. Derivative 110 is a mini-Tn*10* marked with the *supF* gene from PiAN7 (C. Jain and N. Kleckner). This derivative can be used to mutagenize a cloned gene in a λ phage carrying one or more amber mutations; phages that have acquired the *supF* transposon can be selected by virtue of their ability to make plaques on a nonsuppressing host strain. This derivative is also useful when a very short transposon is desired. Derivative 110 is only available on a pBR322-derived Amp[R] plasmid (pNK1759).

Derivative 111 is a mini-Tn*10* marked with a promoterless *lacZ* gene and a *kan* gene from Tn*903* (J. Oberto and R. Weisberg). This derivative can be used to generate "operon fusions" in which the *lacZ* gene in the transposon is transcribed from the promoter of the gene into which it inserts. The sequence from the transposon end up through the start of the *lacZ* gene in this derivative is shown in Fig. 5d. Derivative 111 is only available on a pBR322-derived Amp[R] plasmid (pNK2804). A version of this element that lacks the *kan* gene is also available on a pBR322-derived Amp[R] plasmid (pNK2803). In derivatives 110 and 111 the markers are located between inverted repeats of the outermost 70 bp of IS10 Right, as in derivatives 103–108.

Other mini-Tn*10* elements complemented by wild-type transposase have been described previously. Some of these elements are complemented by wild-type transposase in cis.[4] Others require transposase provided in trans and contain various *E. coli* and yeast markers.[4,42] Most of these constructions are superseded by the derivatives above.

Mini-Tn*10* constructs which generate translational fusions of *lacZ* to target genes: Derivatives 112 and 113. Derivatives 112 and 113 are used to make *lacZ* translational fusions to a target gene. These two derivatives have been described previously as Tn*10*–LK[43] and Tn*10*–LUK,[39] respectively. Each derivative contains a *lacZ* gene lacking the appropriate transcription and translation start signals (*'lacZ*). The *'lacZ* gene is not expressed when the transposon is in its original (donor) site on the plasmid or phage vehicle, but LacZ[+] fusions can result from transpositions into an expressed target gene in appropriate orientation and reading frame. Such fusions can be isolated either as pure red colonies on MacConkey

[42] M. Snyder, S. Elledge, and R. W. Davis, *Proc. Natl. Acad. Sci. U.S.A.* **83**, 730 (1986).
[43] O. Huisman and N. Kleckner, *Genetics* **116**, 185 (1987).

lactose medium or, if they arise during growth of a clone, as red (Lac$^+$) papillae within white (Lac$^-$) single colonies.

Derivative 112 (Tn*10*–LK) carries an unexpressed 'lacZ fragment and a Kan^R fragment from Tn*903*; derivative 113 (Tn*10*–LUK) is identical to derivative 112 except that it contains a URA3 fragment from S. cerevisiae between the 'lacZ and Kan^R markers. For these two mini-Tn*10* constructs, the ends of the transposon are 70-bp segments from the left and right ends of Tn*10*. The 'lacZ gene is fused to the 70-bp end derived from IS*10* Right. The sequence from the end of the transposon through the 'lacZ gene for both derivatives is shown in Fig. 5b. Derivatives 112 and 113 are available on pBR322-derived Amp^R plasmids (pNK1207 and pNK2809) and on λ $P_{am}80$ hop phage vehicles (λNK1205 and λNK1224). Neither of these mini-Tn*10* constructions carries its own transposase. Transposase must be supplied in trans to these constructions from a plasmid carrying derivative 102.

Mini-Tn*10* which generates translational fusions to *kan* gene: Derivative 114. Derivative 114 carries a *kan* gene lacking the appropriate transcription and translation start signals ('*kan*).[44] The '*kan* gene is not expressed when the transposon is in its original (donor) site on the plasmid or phage vehicle, but Kan^R fusions can result from transpositions into an expressed target gene in appropriate orientation and reading frame. This derivative thus allows selection of transpositions without requiring destruction or elimination of the transposon donor molecule.

Derivative 114 is a mini-Tn*10* (L. Signon and N. Kleckner) which is marked with both an unexpressed '*kan* gene from Tn*5* and an erythromycin resistance (*erm*) gene from pAMβ1; these markers are located between inverted repeats of the outermost 70 bp of IS*10* Right as in derivatives 103–108, 110, and 111. A Ptac–wild-type transposase fusion is present in cis on the transposon vehicle. The sequence through the transposon end into the '*kan* gene is shown in Fig. 5c. Derivative 114 is available on a pBR322-derived Amp^R plasmid vehicle (pNK2811).

C. Construction of New Mini-Tn10 Derivatives with Ptac–ATS Transposase

Plasmid-borne mini-Tn*10* derivatives analogous to derivatives 103–108 can be constructed in a single step from the plasmid version of derivative 103 (pNK2859) by substitution of any desired fragment of interest for the *Bam*HI fragment carrying the *kan* gene. Substitutions are particularly easy because the *Bam*HI backbone fragment does not give a viable plasmid if it is religated without an insert at the *Bam*HI site; in this case, the pair of inverted IS*10* ends forms an inverted repeat which is lethal to the replicon.

[44] J. K. Sussman, C. Masada-Pepe, E. L. Simons, and R. W. Simons, *Gene* **90**, 135 (1990).

A λ $P_{am}80$ hop phage version of such new plasmids can also be constructed by crossing the appropriate segment of the plasmid onto λNK1316 (or onto λNK1324 if the new plasmid is marked by Kan^R) using homology to each side of the mini-Tn*10*. (Note that mini-Tn*10* elements longer than about 5 kb are too big to fit into this λ genome.) Such crosses involve several steps. First, a stock of λNK1316 must be grown on a *supE* strain transformed with the new plasmid to allow recombination between the phage and plasmid. Second, the resulting λ stock should be titered for single plaques on a *supE* strain lacking the new marker at 30°. Under these conditions the *att⁻ c*I857 phage particles can form abortive lysogens, which are present as cells in the center of each (slightly turbid) plaque. Third, these plaques should be replica plated at 30° onto medium that selects for the new transposon marker. Replica plating of a phage lawn is exactly analogous to replica plating of bacterial colonies except that care should be taken that the layer of top agar remains on the original plate. The desired recombinants should occur at a frequency of about 1% of all plaques; if the frequency is much lower, something is wrong. Phages carrying the new marker can be recovered either by picking the corresponding plaque on the original plate or by purifying lysogenic colonies and recovering phage after heat induction. For heat induction, cultures should be grown in LB medium to early log phase at 30°, heated to 42° for 15 min, grown with aeration at 37° 1 h, and treated with chloroform and centrifuged to pellet debris; phage are recovered in the supernatant. The recombinant phage thus recovered should be tested for transposition by the usual procedure (below). Recombinant phage can also be selected in abortive lysogens by mixing phages with *supE* host cells and plating directly on selective medium at 30°. Phage can then be recovered by heat induction.

III. Procedures for Isolation and Processing of Transposon Insertions

A. Generating Transpositions from λ into
   Escherichia coli chromosome

   1. Preparing Transposon Delivery Phage Lysate. A stock of the λ transposition vehicle should be grown on a suitable host in either a liquid or plate lysate. One procedure is given below; others are provided by Way *et al.*,[4] Silhavy *et al.*,[45] Arber *et al.*,[46] and Maniatis *et al.*[47] For vehicles

[45] T. J. Silhavy, M. L. Berman, and L. W. Enquist, "Experiments with Gene Fusions." Cold Spring Harbor Laboratory, Cold Spring Harbor, New York, 1984.

[46] W. Arber, L. Enquist, B. Hohn, N. E. Murray, and K. Murray, *in* "Lambda II" (R. W. Hendrix, J. W. Roberts, F. W. Stahl, and R. A. Weisberg, eds.), p. 433. Cold Spring Harbor Laboratory, Cold Spring Harbor, New York, 1983.

[47] T. Maniatis, E. F. Fritsch, and J. Sambrook, "Molecular Cloning, a Laboratory Manual." Cold Spring Harbor Laboratory Press, Cold Spring Harbor, New York, 1982.

that contain amber mutations $P_{am}80$ and/or $O_{am}29$, an *sulI* (*supE*) strain is best; *E. coli* C600 and LE392 are both usable. Strain NK5336, containing a UGA suppressor, is available on request for growth of λNK370. For λ stocks to be useful in generating transpositions, the titer of the stock should be at least 5×10^9 per ml, but titers of 1–2×10^{10} should be obtainable routinely. Stocks should always be grown from a single plaque to reduce the frequency of O^+ or P^+ revertants and to reduce the possibility of losing the transposon construct, which occurs at a significant frequency for various vehicles for a variety of different reasons. The presence of the transposon construct in the lysate should also be verified either by measuring the ability of the lysate to give transpositions at a reasonable frequency or by scoring genetically for the presence of an associated transposon marker.

Stocks should be stored in the presence of 10 mM MgSO$_4$ at 5° and should be stable for many months. Protocols for freezing phage lysates are also available.[45] However, because λ phages of abnormal genome size are sometimes unstable, losses in titer of a phage stock can mean enrichment for aberrant types of phages (deletions or duplications). One of the most important reasons that phage lysates should always be made from single plaques is to avoid accumulation of such aberrant derivatives.

Procedure 1: Growing Phage λ Lysates

Solutions

TBMM: Tryptone Bl broth with maltose and magnesium. Composition per liter distilled water, 10 g tryptone, 5 g NaCl; autoclave, add filtered maltose to final concentration of 0.2% (w/v) and MgSO$_4$ from sterile stock to final concentration of 10 mM. Add filter-sterilized thiamin to final concentration of 1 μg/ml.

LB (Luria broth): Per liter distilled water, 10 g tryptone, 5 g yeast extract, 5 g NaCl.

TB1 agar plates: Make up tryptone broth without maltose or MgSO$_4$; add 11 g/liter Difco (Detroit, MI) agar before autoclaving; pour into petri plates when agar is partially cooled. Use the plates while relatively fresh (within 1 week). For growing phage that make very small plaques, substitute BBL trypticase (Fisher Scientific, Pittsburgh, PA) for tryptone.

Top agar: Tryptone broth with 7 g/liter agar.

TMG buffer: Tris, magnesium, gelatin. Composition per liter of distilled water, 1.2 g Tris base, 2.46 g MgSO$_4$ · 7H$_2$O, 0.1 g gelatin. Adjust to pH 7.4; heat to dissolve gelatin and autoclave.

Preparing the Lysate. Grow permissive host to saturation in TBMM. Make serial dilutions of phage lysate in TMG and plate appropriate quantities in 2.5 ml top agar with 0.1 ml of an overnight culture of a permissive bacterial host (C600) on a TB1 agar plate. Incubate overnight at 37°. Pick a single plaque with a micropipette, transfer it to a 50-ml flask containing 10 ml LB plus 10 mM MgSO$_4$ and 0.1 ml of a fresh overnight culture of the permissive host, and shake the flask at 37°–39° for 4–5 hr. The culture should gradually become somewhat cloudy, then clear. Add a few drops of chloroform, shake, let sit 10 min, and then centrifuge at 5000 rpm for 10 min; save the supernatant.

Checking the Lysate. The phage lysate should be checked in the following ways: (1) It should have the appropriate titer on the permissive host. (2) It should have a titer below 10^{-4} of the permissive host titer on a nonsuppressing host. This confirms that the phage still carries amber mutations in essential genes. (3) The simplest way to check for the presence of the transposon is to test the lysate for transposition. Colonies resulting from stable incorporation of the transposon marker should arise at a reasonable frequency; furthermore, approximately 1% of such colonies should have acquired a new auxotrophic marker. Separate lysates made from individual plaques can be made in parallel and checked with a small-scale experiment; once the best lysate and appropriate multiplicity of infection (moi) are defined, a larger scale experiment can be done. A test for the antibiotic resistance marker itself can be done by mixing a sample of the lysate with a host which is lysogenic for λ and incubating for 1 hr to allow time for expression; double lysogens will form by homologous recombination between the infecting phage and the prophage at a frequency of 10^{-3} to 10^{-5} per infecting phage, and they can be detected on the appropriate antibiotic plates.

2. Isolating Insertions. In addition to considerations specific to the experiment of interest, the bacterial host should have the following characteristics. (1) It must be able to adsorb and inject λ. (2) It must not be permissive for lytic phage growth, that is, must not contain an inappropriate nonsense suppressor (see above). (3) Also, it should not contain a λ prophage. If having a prophage is unavoidable, then the strain should be *recA*$^-$ to prevent transfer of transposon markers to the prophage by homologous recombination. (4) The strain should be able to grow at 39°, since transpositions should be carried out at this temperature whenever possible. At lower temperatures, many of the phage vehicles are able to make abortive (unintegrated) lysogens because most of them carry the *c*I857 mutation which renders the repressor protein temperature-sensitive. λNK561, which carries a full Tn*10* element inserted into the λ *c*I gene, can be used at all temperatures. (5) The host must be sensitive to the transposon marker to be selected.

Transpositions from λ delivery vehicles are isolated by infecting sensitive cells with the phage under conditions which preclude lysogen formation and selecting for cells which have acquired the marker from the transposon element. Usually, the marker will confer antibiotic resistance, so the selection can be on rich plates containing the appropriate antibiotic.

One general problem in isolating transpositions from λ vehicles is that a significant number of replication-proficient O^+/P^+ revertant phage may also be present. One can determine the frequency of revertants in the phage stock by titering on a nonsuppressing host; a reversion frequency of 10^{-7} or less is desirable. Even with this low frequency, revertants may be a problem; from the total number of phage present on the plate and the reversion frequency one should be able to calculate the number of revertants per plate. If there are more than a few, these revertants will grow on the plated cells and lyse the cells containing the desired transpositions. A second problem is that the presence of a large number of unadsorbed phage on the selective plates can result in nonspecific killing of cells containing the desired transpositions. Both of these problems can be addressed by including sodium pyrophosphate in the selective plates to destabilize phage particles by chelation of Mg^{2+}; excess unadsorbed phage can also be removed by washing infected cells prior to plating; both procedures are described below.

Procedure 2: Transpositions from λ into Chromosome of *Escherichia coli*

Materials

LB–antibiotic plates: Add to LB broth 15 mg/ml agar and autoclave. Just before pouring, add the appropriate antibiotic from a sterile solution to a final concentration as indicated and mix well: 15 μg/ml tetracycline (filter-sterilized 1% stock in ethanol); 30 μg/ml kanamycin (filter-sterilized 3% stock in water); 25 μg/ml chloramphenicol (filter-sterilized 2.5% stock in ethanol); 100 μg/ml ampicillin (filter-sterilized 5% stock in water).

If sodium pyrophosphate is to be included, add after autoclaving to a final concentration of 2.5 m*M* by dilution from a 125 m*M* sterile stock solution. Plates which do not contain pyrophosphate can be used for 1–3 weeks after pouring if stored at 5°. Plates containing pyrophosphate must be used within about 24 hr of being poured, as continued chelation of divalent cations eliminates growth of all cells regardless of whether they contain transpositions.

Protocol

(See Way *et al.*[4] for a slightly different procedure.)

1. Grow cells overnight in TBMM, concentrate by centrifugation, and resuspend in 1/10 volume LB.

2. Adsorb 0.1 ml of concentrated cells and various quantities of phage for 15 min at room temperature and 15 min at 37°. Assume a concentration of cells of about 10^9–10^{10}/ml in the concentrated culture, and add phage to give an moi of between 0.1 and 1 phage per cell.

It usually pays to do a small-scale experiment first to determine the multiplicity of infection which maximizes transpositions and minimizes killing by the phage. The experiment can then be repeated with multiple tubes at the best multiplicities of infection.

3. As for removing unadsorbed phage (which is optional), it is usually not necessary to wash away free phage or allow time for expression of transposon markers. However, both steps can be accomplished by the following simple procedure. Add 5 ml LB with sodium citrate (50 mM), centrifuge the cells, and resuspend in 5 ml fresh LB plus citrate. Citrate will inactivate free phage that have not adsorbed; washing also removes free phage. Grow the infected culture for about 1 hr at 37°; this allows expression of antibiotic resistance.

4. Plate 0.1 ml of the infected cell mixture on antibiotic selection plates at 39° overnight. If transposition is low (i.e., as with wild-type Tn*10*) the cells can be concentrated before plating. The frequency of insertions obtained per infecting phage should be approximately 10^{-3} to 10^{-4} for most of the minitransposon constructs, somewhat lower for constructs in which transposase is provided from a plasmid, and about 10^{-7} for wild-type Tn*10*.

3. Obtaining Insertions of Interest by Direct Screening or Selection.
In some cases, colonies that arise on selective plates as the result of transposition will be screened for a phenotype introduced by the insertion mutation, either by visual inspection of the original selective plates or by replica plating. On the original selective plates, colony morphology markers can be scored. Expression of *lacZ* or other sugar fermentation genes can also be scored on pH indicator plates such as MacConkey lactose as long as the total number of cells plated is not too high. However, screening for β-galactosidase or alkaline phosphatase activity with 5-bromo-4-chloro-3-indolyl-β-D-galactoside (X-Gal) or 5-bromo-4-chloro-3-indolyl phosphate (XP) is not practical at this stage because of background expression from transiently infected cells which do not give rise to stable transpositions. If colonies are screened by replica plating, the condition (i.e., antibiotic selection) used to select the primary transpositions should be included in the screening plates as well. Cells which have not undergone a transposition event are always present on the original selective plate along with the colonies of interest.

Identification of a transposon insertion in a particular typical gene in

E. coli usually requires screening of about 10,000 transposition events. There are approximately 1000 such nonessential target genes. For Tn*5*, Tn*10*–ATS, Mu, and Tn*3*, a modest level of redundancy is required to ensure that the gene of interest is hit. For insertions of Tn*10* elements with wild-type transposase insertion specificity, it may be necessary to screen a larger number. An example of the screening of initial transposon insertions for colony morphology is found in Trisler and Gottesman[48]; an example of screening fusions for expression under particular circumstances can be found in Kenyon and Walker.[49]

4. Screening and Selecting for Insertions in Secondary Hosts. For many applications, identification of insertions of interest requires transfer of the insertions from the host in which they were selected to a secondary host. One general approach, described below, is to pool a large number of cells carrying independent insertions, grow P1 on the pool, and then transduce the mixture of insertions into a secondary host by selecting for the transposon marker. The resulting transductants are screened or selected for the phenotype of interest. A second approach, described by Silhavy *et al.*,[45] involves direct isolation of transposon insertions in a strain lysogenic for a temperature-inducible P1 followed by pooling and generation of a transducing lysate by induction. In either case, a good P1 lysate grown on a large number of pooled transposition events generated in a wild-type host should be a standard reagent for any laboratory carrying out genetic analysis in *E. coli*.

A frequent application in which transduction of pooled insertions is necessary is for isolation of an insertion which is linked to, rather than within, a gene of interest. In this case, the secondary recipient carries a marker in the region of interest and transductants are screened for loss of that marker. Since P1 carries about 1 min of an *E. coli* chromosome, even a pool of a few hundred independent transpositions is likely to contain at least one insertion that is within P1 transducing distance of any point of interest. Insertions into a particular gene may also be identified in a secondary host if the phenotype of interest requires a property incompatible with isolation of insertions, for example, *su*$^+$ or the presence of a λ prophage. In this case, the number of pooled transposition events must be as great as for direct isolation of insertions (see above).

For identification of insertions in very specific locations, either within or tightly linked to a gene of interest, a two-step approach is often useful. Insertions into the general region of interest are selected by looking for

---

[48] P. Trisler and S. Gottesman, *J. Bacteriol.* **160**, 184 (1984).
[49] C. J. Kenyon and G. C. Walker, *Proc. Natl. Acad. Sci. U.S.A.* **77**, 2819 (1980).

transduction of a general marker such as an auxotrophy, and the resulting transductants are then screened in a second step to identify the specific subset desired.

Procedure 3: Making P1 Lysates on Pools of Chromosomal Insertions

Select at least a few thousand independent chromosomal insertions; more is better. Flood each selective plate with 2–3 ml of LB plus 0.1 M citrate which has been chilled to 4°. Use a glass spreader to resuspend the colonies in the broth. Pipette off the broth into a centrifuge tube and wash the plate once again. Pooled cells will usually give a dense cell suspension. Cells should be washed with LB at 4°, at least twice and preferably 3 or 4 times; 0.1 M citrate should be included in the first washes but omitted from the last one. Cells can then be resuspended at about 10^9 cells/ml and frozen with dimethyl sulfoxide (DMSO) or in 50% glycerol for future use.

For a P1 lysate of the pool, resuspend cells at about 10^{10}/ml in 0.1 ml LB plus 10 mM MgSO$_4$ and 5 mM CaCl$_2$ with P1vir (or P1 HFT[20]) at an moi of about 1; let adsorb 10 min, dilute into 10 ml LB with 10 mM MgSO$_4$ and 5 mM CaCl$_2$, and shake at 37°–39° for 2 hr, or until lysis. Treat with chloroform, centrifuge at 5000 rpm, and save the supernatant. This lysate, which should be stored at 5°, can be kept for many months.

5. Identifying Linked Transposon Insertions. To identify an insertion near a gene of interest, one should select a strain which carries a mutation in that gene and in which loss of the mutation can be selected or screened. This strain is used as a recipient for P1 transduction with the lysate grown on the transposon insertion library from a wild-type host (see above). Transductants receiving the marker on the transposon are selected and screened or selected for loss of the mutation in question. Approximately 1/100 transductants should bring the wild-type allele of the selected marker in along with the transposon insertion. Linkage of the transposon to the marker of interest should be verified by purifying candidate transductants and using them as donors in P1 transductional crosses with the original mutant strain as recipient. Insertions which show a linkage of at least 30% are easiest to use. Insertions that exhibit less than 5% linkage can be used if necessary, but further confirmation that the linkage is real is wise. From crosses that exhibit significant linkage, it is useful to save transductants which both have and have not lost the marker allele; the latter can be used to transfer the mutation to new hosts. Examples of the identification of insertions in or near a gene of interest by transposition followed by P1 transduction from pooled insertions can be found in Trun and Silhavy[50] and Lemotte and Walker.[51]

B. Isolating Transpositions from λ to Multicopy Plasmid

The general procedure for isolating transpositions into a multicopy plasmid is the same as for isolating transpositions into the *E. coli* chromosome (Procedure 2) except that the recipient strain contains the target plasmid of interest. It is best if the plasmid is present only as a monomer in the population of cells used to isolate transpositions. When dimers or higher multimers are present, isolation of insertion mutations by loss of function is impossible, many insertions go into essential plasmid backbone sequences rather than into the region of interest, and restriction mapping of insert plasmids is more difficult. Plasmids growing in *recA*+ hosts frequently exist as head-to-tail dimers, so special procedures are required to ensure that the host cells used for transposition contain primarily monomer plasmids. Essentially pure monomer plasmid populations can be obtained by transforming a *recA*− or *recJ*− host and examining the DNA from a handful of transformants; monomers are easily distinguished from higher forms by gel electrophoresis of undigested plasmid DNA. These purified monomers can then be used to transform a second strain in which transpositions occur. If possible, a *recA*− or *recJ*− strain should be used. However, it is acceptable to transform a *recA*+ strain with pure plasmid monomers and carry out transpositions in the resulting transformants as long as they are not subcultured any more than necessary before the transposition experiment. In a small culture grown directly from a single transformant, the majority of plasmids are still monomers. The general strategy for identification of plasmid inserts involves extraction of DNA from pools of insertion-containing colonies and retransformation into a new host.

Procedure 4: Generating Insertions into Plasmids

Carry out Procedure 2 using a suitable strain carrying the plasmid of interest. It is not necessary to select for a marker on the plasmid, but it is desirable in order to eliminate the possibility of transposon insertions into that marker. Make many independent pools of cells and isolate plasmid DNA; minipreps are sufficient. Approximately 0.1% of the plasmid DNA molecules in these preparations will contain an insert, since 1% of insertions occur into plasmid DNA and since only a subset of the plasmids in a given colony will contain the insert. Use these

[50] N. J. Trun and T. J. Silhavy, *Genetics* **116**, 513 (1987).
[51] P. Lemotte and G. C. Walker, *J. Bacteriol.* **161**, 888 (1985).

DNAs to transform a strain in which the transposon marker can be selected. Also, it is best if this recipient strain is unable to adsorb λ, as there may be a surprisingly significant amount of phage DNA and/ or phage particles in the DNA preparation, and infection of the strain plasmid and bacterial collections with phage is not desirable. Ideally, insertions into the target gene of interest are scored genetically; less ideally, physical analysis is used.

C. Isolating Insertions into Phage λ

To isolate insertions into bacteriophage λ, a lysate of the target phage is made on a host strain carrying the transposition vehicle. Phage which have acquired the plasmid transposon marker are then selected in one of several possible ways, some of which are described below and none of which is foolproof.

1. Integration of Transposon-Carrying Phage into Chromosome by Phage-Mediated Integration. If the phage in question is integration-proficient (*att*$^+$ *int*$^+$), the phage lysate can be used to infect an appropriate λ-sensitive host and lysogens carrying the transposon marker selected. If the phage is defective for making repressor, Int protein, and/or in the phage attachment site, these determinants can be provided in trans by a helper phage. Recovery of the desired phage from resulting lysogens may be complicated. The target phage may recombine with the helper phage prior to integration. Furthermore, the helper phage may still be present in the final lysogen; in fact, if it is providing *att* function, it must be present. [Technical note: Efficient lysogenization of λ requires that each infected cell receives at least 5 phages.] In some cases, use of a defective prophage to provide some functions can minimize the problems of helper phage (see below for such a procedure using the λ D69 vector).

2. Integration of Transposon-Carrying Phage into Chromosome of λ Lysogen by Homologous Recombination with Resident Prophage. In this case, the phage lysate is used to infect a previously established λ lysogen, and cells that acquire the transposon marker are selected. This method has the advantage that it will work for a phage of any genotype, regardless of whether it has the ability to replicate, lysogenize, or establish repression on its own. It has the disadvantage that the efficiency of recovery of marker-containing phage is low (10^{-3} to 10^{-5} per transposon-containing genome), which means that a large volume of phage lysate may have to be processed. Furthermore, the complications inherent in the helper coinfection approach above apply here as well.

3. Selection of Transductants Arising from nin$^+$ *Phage Vectors.* If the phage vector has an intact *nin* region (*nin*$^+$), the phage will grow as a stable plasmid in a host which interferes with N function (*nusA*), and stable transductants can be selected in such a host. Alternatively, if a *nin*[+] phage vector carries nonsense mutations in the *N* gene, transductants can be selected in a standard *su*[-] host.[16] Unfortunately, most standard cloning phage carry a deletion in the *nin* region in order to reduce the size of the phage genome.

4. *Direct Detection of tet Transductants.* A procedure has been described by R. Maurer to detect phage carrying *tet* genes directly in plaques.[52] A tetracycline-sensitive bacterial lawn is grown in LB with maltose (0.2%) and $MgSO_4$ (10 m*M*) to 2–3 × 10[8] cells/ml, then concentrated 10-fold in 10 m*M* $MgSO_4$. The phage lysate is titered on LB plates containing 7 μg/ml tetracycline in top agar without drugs, after adsorption to 0.5 ml of the concentrated cells. A very faint lawn forms, with only the Tet[R] phage forming plaques. The precise concentration of tetracycline should be checked with each batch of drug.

5. *Direct Selection of supF-Carrying Phage.* λ phage which have acquired an insertion of a *supF* element can be selected directly. If the phage carries *supF*-suppressible nonsense mutations, phage carrying the insertion can be isolated as plaque-forming phage on a *su*[-] host.[6,53] If the phage carries no nonsense mutation, the desired phage can be isolated using conditions in which plaques form only if the phage suppresses a nonsense mutation in the host strain used as the plating bacteria. One such selection is described by Phadnis *et al.*[6] Alternatively, if the host strain carries an auxotrophic mutation in a gene such as *his* or *trp*, plaques arise on selective minimal plates only if the phage permits expression of the defective gene; strains bearing nonsense mutations in such genes might be used for selection. The analogous approach should also be applicable for host strains carrying a tetracycline resistance determinant with a nonsense mutation (Maurer *et al.*[52]; D. Botstein, personal communication, 1990).

Procedure 5: Generating Insertions into λ

Transform the plasmid donor into a λ-sensitive host. Because many of the minitransposon donors use a *Ptac*–transposase fusion, it is best to use a *lacI*[Q] host such as *E. coli* JM101,[54] to avoid plasmid rearrangements which might occur with long-term expression of transposase. Grow the transformed cells in TBMM with the appropriate antibiotic to select for the plasmid itself and infect with the target phage by the method of Proce-

[52] R. Maurer, B. C. Osmond, E. Shecktman, A. Wong, and D. Botstein, *Genetics* **108**, 1 (1984).
[53] M. Snyder, S. Elledge, and R. W. Davis, *Proc. Natl. Acad. Sci. U.S.A.* **83**, 730 (1986).
[54] J. Messing, this series, Vol. 101, p. 20.

dure 1. IPTG can be included during preparation of the phage lysate to increase transposase expression. The resulting lysate should be considered a low-frequency transducing lysate for the transposon insertion in question; about $1/10^6$ phage will be likely to contain the desired marker.

Procedure 6: Isolating Transpositions into λ D69 Carrying Cloned Insert

λ D69 is an *att*$^+$ *imm21* phage which has been widely used for cloning purposes.[45,55,56] An *imm* λ *c*I857 derivative of λ D69 is also available.[56] In these phage, foreign DNA is cloned into the phage *int* gene, so the resulting transducing phage are *int*$^-$. Integration of such transducing phage can be specifically targeted to *attB*, the normal site of phage integration, by providing Int function in trans; this approach is used for the initial identification of phage derivatives carrying transposon insertions as described here. Alternatively, integration can be targeted into the bacterial region homologous to the cloned *E. coli* insert by forcing integration in the absence of Int function. In this case, integration usually occurs by homologous recombination at the desired site, although occasional integrations into other sites also occur. Integration via bacterial homology is useful for obtaining transposon insertions in the relevant region of the bacterial genome as described in the following section.

In Procedure 6, a pool of transposon insertions into the phage is generated by growing a lysate of λ D69 on a strain carrying the desired transposon construct as in Procedure 5, using this pool to infect an appropriate host strain, and selecting for a marker on the transposon. Phage can then be recovered from such lysogens, individually or in pools, by an appropriate induction procedure. Phage carrying transposon insertions in the cloned segment can be distinguished from those carrying insertions in nonessential regions of the phage genome by subsequent tests.

The host strain used for isolation of λ D69::Tn insertions is N6377, which carries a defective prophage that provides both Int function and the bacterial attachment site (*attB*). λ D69 phage integrate very efficiently into this host, which makes it easy to collect large numbers of λ D69::Tn derivatives. Subsequent recovery of integrated phages by prophage induction is also easy because the prophage carries the temperature-sensitive *c*I857 repressor allele.

It may be desirable to select for λ D69::Tn phages in a host other than N6377 where, for example, the phenotypic effects of the transposon insertion can be assayed. The defective prophage carried in N6377 can be transferred to another strain by P1 transduction using as a donor strain

[55] S. Mizusawa and D. Ward, *Gene* **20**, 317 (1982).
[56] J. A. Brill, C. Quinlan-Walshe, and S. Gottesman, *J. Bacteriol.* **170**, 2599 (1988).

SG12021, which is identical to N6377 except that it carries a *nadA*::Tn*10* insertion linked to the prophage. If tetracycline-resistant transductions are selected in the new host, 20% will have acquired the defective prophage and can be detected by virtue of a *bio*⁻ marker which accompanies the prophage (see genotype below). The Tn*10* element can be eliminated from the new strain by selecting for a Nad⁺ (Tet^S) revertant.

Bacterial Strains

N6377: *att B.B' bio936 int⁺ xis⁺ Δ(Sal–Xho) c*I857*Δbio r⁻m⁺* (SG12021 is N6377 *nadA*::Tn*10*)

Protocol

1. To generate and recover λ D69::Tn phage, grow a lysate of the transducing phage on the transposon-donating host (Procedure 5). Grow N6377 or another host carrying the defective prophage at 32° in TBMM supplemented with biotin (1 μg/ml) to approximately 2 × 10⁸ cells/ml, heat to 42° for 15 min to induce Int synthesis, and return to 32°. Mix 0.1 ml cells and 0.1 ml of the phage lysate at 32° for up to 1 hr, then select lysogens on LB–antibiotic plates.

2. Recovery of λ D69::Tn phage from lysogens. Induction of the lysogens will yield lysates predominantly containing the desired phage. If a *c*I857 derivative of λ D69 was used, grow the lysogenic cells in LB plus 10 m*M* MgSO₄ to a density of about 2 × 10⁸ cells/ml, heat to 40°, and shake at this temperature for 90 min. Add a few drops of chloroform, centrifuge at 5000 *g* for 10 min, and save the supernatant. If a λ *imm21 c*I⁺ D69 derivative was used to isolate the lysogens, induction of lysogens will require both raising the temperature to 40° (to induce the defective prophage and synthesize Int) and treating the cells with ultraviolet light to induce the *imm21 c*I⁺ phage.

About 5–50% of the λ D69 phage produced on induction of an N6377 (λ D69) lysogen will have undergone recombination with the defective prophage in such a way as to lose the cloned insert; such recombinants will necessarily have become Int⁺. Such Int⁺ phage can be distinguished from the Int⁻ phage by analysis of individual plaques in the red plaque test of Enquist and Weisberg.[57] Comparison of phage containing and lacking the cloned insert provides a simple way of determining whether the transposon insert is within the cloned segment or outside. If the transposon was inserted within the bacterial DNA insert, *int⁺* recombinant phage should also have lost the transposon. All phage which are white on the red plaque test should carry the insertion.

---

[57] L. W. Enquist and R. A. Weisberg, *Virology* **72,** 147 (1976).

D. Moving Insertions After They Are Isolated

1. From Plasmid to Chromosome. An insertion isolated in a cloned bacterial fragment on a plasmid can be transferred directly to the chromosome in several ways. (1) The insert region can be transferred by direct DNA transformation after cleavage of the plasmid on one or both sides of the transposon; a *recBC sbcB, recD* host must be used.[58-60] (2) The insert region can be integrated by recombination between the intact plasmid and the chromosome, with appropriate selection for both integration and excision events[61]; a *polAts* host must be used. (3) The insert can also be moved into the chromosome by first moving it into λ and then following the procedure in the next section. In all of these cases, use of a minitransposon is advantageous because it prevents stable transposon integration into the host chromosome by transposition and thus eliminates an unwanted source of background. In the case of linear transformation, the frequency of desired events is so low that use of a minitransposon is essential.

2. From Phage λ to Chromosome

(a) Moving insertions one at a time. A transposon insertion into a bacterial segment carried on λ can be moved into the chromosome in a two-step process. First the phage is integrated into the bacterial chromosome by recombination between the cloned insert and the homologous region of the chromosome, creating a nontandem duplication of the cloned segment, one copy of which is inactivated by the transposon insertion. Then, in a second step, the prophage is eliminated by recombination between the repeated segments. The resulting recombinants may have retained or lost the transposon insert depending on where recombination occurred; recombinants carrying the transposon can be identified by the presence of the transposon marker and/or their mutant phenotype. Occurrence of recombinants of the latter type at high frequency is taken as reasonable evidence that the disrupted gene is not essential.

This procedure is most easily carried out using a *c*I857 (temperature-inducible) derivative of *int*⁻ phage such as λ D69 (see Maurizi *et al.*[62] for an example). For the first step, lysogens are selected at low temperature in a wild-type, nonlysogenic host, and a handful of such lysogens is isolated

[58] S. Kanaya and R. J. Crouch, *Proc. Natl. Acad. Sci U.S.A.* **81**, 3447 (1984).

[59] S. C. Winans, S. J. Elledge, J. H. Krueger, and G. C. Walker, *J. Bacteriol.* **161**, 1219 (1985).

[60] C. B. Russell, D. S. Thaler, and F. W. Dahlquist, *J. Bacteriol.* **171**, 2609 (1989).

[61] N. I. Gutterson and D. E. Koshland, *Proc. Natl. Acad. Sci. U.S.A.* **80**, 4894 (1983).

[62] M. R. Maurizi, P. Trisler, and S. Gottesman, *J. Bacteriol.* **164**, 1124 (1985).

and purified. Lysogens arise via the desired integration events if normal phage integration is blocked by inactivation of the phage *int* gene. If the phage is *int*⁺, the majority of integrations by the Int pathway must be eliminated by deletion of the bacterial attachment site in the host chromosome; however, since insertions occur at a significant frequency into secondary attachment sites, lysogens isolated in an *att*Δ*int*⁺ situation must be characterized to be sure that the prophage is located in the appropriate place.

For the second step, temperature-resistant derivatives of appropriate lysogens can be isolated by subjecting a culture of cells to a heat treatment and looking for survivors. It is best to identify cured cells first by virtue of their loss of phage immunity and then to score for the transposon marker in a second step. If one selects directly for temperature resistance and antibiotic resistance, abnormal events or secondary suppressing mutations may be obtained, particularly if the disrupted gene is essential.

(b) Isolating multiple transposon insertions in a cloned bacterial segment in a single step. If a bacterial gene has been cloned in λ *c*I857 D69 (or an equivalent *c*I857 integration-deficient phage vector), large numbers of transposon insertions into the cloned bacterial gene can easily be obtained from an unpurified pool of λ D69::Tn phage. In brief, a large number of λ D69::Tn lysogens are made (Procedure 6, Step 1), pooled, and induced (Procedure 6, Step 2). The resulting mixed lysate, which will contain a number of different transposon insertions into the cloned segment as well as insertions into the phage genome, is used to infect a nonlysogenic host; cells which have stably acquired the transposon marker are selected at low temperature. Most of the resulting colonies will contain phage integrated by bacterial homology within the cloned segment. However, some will contain no prophage but will have arisen because a transposon insertion present in the cloned segment of a λ D69::Tn phage has been substituted for the homologous region in the host chromosome by virtue of recombination on both sides of the insert. Cells of the latter type will be viable at temperatures which induce the *c*I857 prophage; they can be identified by pooling colonies carrying the transposon marker, replating on agar plates at 40°–42°, and screening survivors for retention of the transposon marker. (In fact, additional cells of the desired type may also be generated during the temperature induction procedure itself.) Some of the clones that survive heat selection will have become resistant to λ owing to the presence of phage released by induction of nonsurviving cells. It should be noted that insertions into an essential gene cannot be isolated in this way. Applications of this method are described by Maurizi *et al.*[62] and Brill *et al.*[56]

3. Cloning Insertions out of Bacterial Chromosome. In general, cloning insertions out of the bacterial chromosome is straightforward. Special precautions that must be taken when cloning out *tet* insertions are discussed in Section I above. Recently, new tools which facilitate rapid cloning and subsequent sequencing of Tn*10* and mini-Tn*10* insertions have been developed.[63] A recombinant M13mp vector carrying the central portion of the appropriate *tet* or *kan* segment is integrated into the chromosome at the site of the insertion by recombination; recombinants are selected using a chloramphenical resistance determinant present on the vector. Appropriate digestion, ligation, and transformation of chromosomal DNA from such integrants yields M13 phage carrying the segment of interest in a form suitable for sequencing and probing of other libraries for the wild-type gene.

A PCR (polymerase chain reaction) strategy for cloning genes disrupted by Tn*10* or mini-Tn*10* insertions has also recently been described.[64]

E. Eliminating Transposon Insertions

If a linked transposon insertion has been used to bring in a mutation of interest, it is often desirable to then eliminate the transposon (and its associated selectable marker). This can be done in several ways. The easiest and most general method is to plan strain constructions in such a way that markers can be removed by subsequent P1 transduction. For example, one can introduce a *recA* mutation by a single transduction using a linked Tn*10* insertion; however, a better strategy is to first introduce a *srl*::Tn*10* or *srl*::Tn*5* insertion, which is closely linked to *recA*, and then use a non-transposon-containing *recA*⁻ donor strain to bring in that mutation linked to Srl⁺.

However, transposon insertions which are located in appropriate genes can be eliminated by direct selection for restoration of gene function. These precise excision events occur at frequencies of 10^{-6}–10^{-10} depending on the particular insertion site. Revertants can be isolated by growing cells overnight in rich broth, concentrating each culture 25-fold, and plating 0.2 ml onto a selective plate. For very low reversion frequencies, larger volumes and multiple cultures (to take advantage of possible jackpots) may be necessary. For transposon insertions that carry tetracycline resistance, the drug marker (but rarely the transposon) can be eliminated by direct selection for tetracycline sensitivity as described in Section I,E,1 on Tn*10*-promoted adjacent deletions.

[63] M. L. Michaels, *Gene* **93**, 1 (1990).
[64] C. S. J. Hutton, A. Seirafi, J. C. D. Hinton, J. M. Sidebotham, L. Waddell, G. D. Pavitt, T. Owen-Hughes, A. Spassky, H. Buc, and C. F. Higgins, *Cell* **63**, 631 (1990).

F. Mapping of Insertions in Bacterial Chromosomes: Special Approaches

It is easy to use transposon insertions as mapping markers because of the presence of the associated easily selectable marker. In addition to the classic approaches, some additional approaches have been developed which take special advantage of transposon insertions as genetic tools.

1. Mapping Transposon Insertions Located Near or Within Gene of Interest. The general location of the insertion can be determined by transducing it into a number of different Hfr strains having different points of origin. The resulting derivatives can be analyzed in either of two ways. They can be tested for their ability to transfer the transposon marker in a short interrupted mating (30 min); the marker will be transferred efficiently only in those Hfr strains in which it is proximally located and transferred early. Alternatively, the derivatives can be used to transfer a series of prototrophic markers and the selected exconjugants screened for the transposon marker; this is essentially the classic approach.

2. Mapping Nontransposon Mutations Using a Standard Set of Hfr Strains Bearing Transposons. Singer *et al.* have constructed a set of Hfr strains, each of which contains an antibiotic resistance transposon less than 20 min from its origin of transfer.[65] Crosses are performed using these strains as donors and an appropriate mutant strain as the recipient; exconjugants that have received the transposon insertion are selected and then screened for loss of the mutation of interest. Hfr strains which yield wild-type recombinants at a frequency of 50% indicate that the mutation is between the point of origin of the Hfr and the transposon. Both Tet^R and Kan^R Hfr derivatives are available.

3. Mapping Nontransposon Mutations with Hfr Strains and Minitransposon Plasmids. The same approach can be used even without employing a set of stably isolated transposon-containing Hfr derivatives. In this case, an active transposon is introduced on a multicopy plasmid into a series of standard Hfr derivatives. Cultures of such strains contain random transpositions into the Hfr genome. Thus, when they are mated with an appropriate mutant strain for a short period of time, the resulting exconjugants all carry proximally located transposon insertions, and Hfr strains with points of origin near the mutation of interest will frequently transfer the marker of interest along with these transposon insertions as well.[66]

[65] M. Singer, T. A. Baker, G. Schnitzler, S. M. Deischel, M. Goel, W. Dove, K. J. Jaacks, A. D. Grossman, J. W. Erickson, and C. A. Gross, *Microbiol. Rev.* **53**, 1 (1989).
[66] D. Roberts, "Genetic Analysis of Mutants of *Escherichia coli* Affected for Tn*10* Transposition." Ph.D. Thesis, Department of Biochemistry and Molecular Biology, Harvard University, Cambridge, MA (1986).

4. Physical Mapping of Insertions by Pulsed-Field Gel Electrophoresis. If a chromosomal transposon insertion contains a site for a restriction enzyme that cuts only rarely in the bacterial chromosome, it is possible to map the position of the insertion physically by cleaving chromosomal DNA with the enzyme and analyzing the restriction fragments by pulsed-field gel electrophoresis. For example, the *E. coli* chromosome is cleaved into 22 fragments by the enzyme *Not*I, and these fragments have been ordered around the chromosome; a Tn*5* insertion, which contains *Not*I sites in each of its IS*50* inverted repeats (or any other transposon insertion containing a *Not*I site), can be mapped to the appropriate fragment of the chromosome by determining which fragment in the insertion-bearing strain is cleaved internally by *Not*I.[67] The same strategy can be used for a number of other rare-cutting enzymes with sites of cleavage that have been positioned on a chromosomal map.[68] Tn*10* derivatives that carry a polylinker containing rare-cutting restriction enzyme sites are described in Section II,B,2.

Acknowledgments

We are grateful to Nancy Trun, James Kirby, and Douglas Bishop for comments on the manuscript and to Howard Benjamin for helping with computer graphics. Research on transposons by J. Bender and the laboratory of N. Kleckner are supported by grants from the National Institutes of Health (5 RO1 GM25326-12) and the National Science Foundation (DMB-8820303).

[67] C. L. Smith and R. D. Kolodner, *Genetics* **119**, 227 (1988).
[68] Y. Kohara, K. Akiyama, and K. Isono, *Cell (Cambridge, Mass.)* **50**, 495 (1987).

[8] *In Vivo* Genetic Engineering with Bacteriophage Mu

By EDUARDO A. GROISMAN

Introduction

Bacteriophage Mu was discovered as a temperate phage which upon lysogenization generated mutations in the host with a high frequency (Taylor, 1963); its name stands for mutator phage. Mu has become the subject of study of several groups and the starting material for a series of derivatives and strategies for *in vivo* genetic engineering which are discussed in this chapter. A brief review on Mu biology is presented first to facilitate understanding of the particular features of phage Mu which have been crucial for the development of the different tools and techniques.

The reader interested in a more detailed treatment of bacteriophage Mu biology is referred to other comprehensive monographs: Pato's "Bacteriophage Mu" (1989), Harshey's "Phage Mu" (1988), and the book *Phage Mu* edited by Symonds, Toussaint, van de Putte, and Howe (1987).* I use these reviews as the source or reference for most findings on Mu biology and only provide original citations for recent data or when Mu is being used as a genetic tool. A general introduction to the use of transposable elements for genetic engineering (Berg *et al.*, 1989) and a discussion of Mu as a tool (van Gijsegem *et al.*, 1987) have been published.

Overview of Mu Biology

Phage Mu is a temperate phage with a linear, double-stranded DNA genome. Like other temperate phages, when Mu infects a sensitive host it can enter either the lytic cycle or the lysogenic state. In the first case, most phage functions are expressed, the virus replicates and synthesizes its coat proteins, the DNA is packaged into the coat, and the host cell membrane is lysed, releasing 150–200 particles. In the second case, a repressor is synthesized that blocks expression of most viral functions, and the viral DNA forms a stable association with the host. In that state, the phage genome is called a prophage, and the bacterium that propagates the prophage is called a lysogen. The repressor synthesized by the prophage blocks the expression of a superinfecting Mu, so that a lysogen is immune to superinfection by Mu.

A striking difference between Mu and other temperate phages is that the Mu genome integrates in the host chromosome (in a quasi-random fashion) whether it enters the lytic cycle or the lysogenic state. Even Mu DNA isolated from phage particles is found in an integrated state, joined to segments of the host chromosome. When Mu infects a bacterial cell both strands of the phage DNA are inserted into the host genome by a conservative mechanism without replication taking place. Analysis of the products of the integration event indicate that the phage and prophage maps are identical. During subsequent rounds of replication or when a prophage in a lysogen is derepressed, copies of the Mu genome are made and inserted into the different DNA molecules present in the bacterial cell (of chromosomal, plasmid, or phage origin). However, the Mu insertion never leaves its original location in the genome.

Mu can generate different types of chromosomal rearrangements including deletions, inversions, duplications, and transpositions of host DNA segments. It can also fuse independent circular molecules such as

---

* In this chapter references are cited by name and date, and a reference list appears at the end of the text.

two plasmids, or a plasmid and a chromosome, or a circular phage DNA and a plasmid or the chromosome. These properties are shared by IS (insertion sequence) elements and transposable elements (Berg and Howe, 1989; [7], this volume). Mu insertions result in the duplication of 5 base pairs (bp) of the target site, probably owing to repair of the staggered cuts generated during the transposition reaction.

Transposition requires the presence of cis- and trans-acting Mu DNA sequences (Charconas, 1987). The ends of Mu in their proper relative orientation are absolutely required in cis for transposition; an enhancer sequence (also called internal activation sequence), located approximately 1000 bp from the Mu left end, is required for high transposition frequencies (Leung *et al.,* 1989; Mizuuchi and Mizuuchi, 1989; Surette *et al.,* 1989). The Mu A product is absolutely required for both conservative and replicative transposition reactions. In the absence of the Mu B product replicative transposition is reduced to about 1%. Truncated forms of *B* can promote conservative integration but not replicative transposition. Certain deletions at the C-terminal end of Mu *A* result in loss of interaction with Mu *B* (Harshey and Cuneo, 1986). Mu has the highest transposition frequency (>100/cell/generation) and the most random insertion specificity of any known transposon. Mu was the first transposable element for which a defined *in vitro* transposition reaction was described, and the first for which evidence consistent with a replicative transposition mechanism model (Shapiro, 1979) was obtained (Mizuuchi and Higgins, 1987). Among the host factors required for replicative transposition are the products of the *dnaB, dnaC, dnaE, dnaZ, gyrA, gyrB,* and *dnaG* genes (Pato, 1989; Chaconas, 1987). The histonelike protein HU is required to bring the ends of Mu together in the *in vitro* transposition reaction. Strains harboring mutations in both genes for the two subunits of HU are not permissive for Mu growth, lysogenization, or transposition, but those with mutations in only one of the subunit genes behave like wild-type cells (Huisman *et al.,* 1989). The requirement for integration host factor (IHF: encoded by the *himA* and *himD* genes) is due in part to its participation in the transcriptional regulation of the p_{CM} and the p_E promoters (see below).

Mu transposition is controlled by the products of three genes: *c, ner,* and Mu*A*. The Mu repressor (*c*) controls transposition by regulating transcription from the phage early promoter (p_E; an mRNA that includes *ner,* Mu*A,* Mu*B,* as well as other downstream loci) and by binding to the ends of the Mu genome, the sites of action of the transposition proteins. The *ner* gene product negatively regulates both *c* and early transcription (Goosen and van de Putte, 1987). Mu transposase (Mu*A*), which has been shown to be partially homologous to the Mu repressor, can also bind the operator region that controls transcription from p_{CM}. This region coincides in part with the enhancer element for Mu transposition described above. A posttranscriptional regulatory role has been postulated for MuA based on the ability to bind its own message (Parsons and Harshey, 1989) and on its intrinsic instability, being used stoichiometrically in the transposition reaction (Pato, 1989).

Mu DNA isolated from phage particles is found joined to cellular DNA segments owing to the fact that it is packaged from Mu copies randomly integrated in the host genome. Packaging proceeds from the Mu left end toward the right end, terminating when the phage head is full; the packaging signal (*pac* site) is located within sequences 35–58 bp from the Mu left end. Because the head is able to accommodate about 39 kilobases (kb) of DNA and the wild-type Mu genome is only 37 kb long, packaging proceeds approximately 2 kb farther than the actual Mu right end into host DNA, generating a variable end. Mu phage containing insertions have a smaller right variable end equivalent to the size of the insertion. Similarly, large deletions in the Mu genome are compensated for by lengthening the host DNA segment at the right end. The small (50–100 bp) segment of host DNA sequences present at the left end of Mu is probably generated by another mechanism (Howe, 1987a, 1987b).

Bacteriophage Mu has two different host ranges depending on the orientation of an invertible DNA segment region (named G loop) with genes for two different types of tail fibers. Mu with the G region in the (+) orientation can infect *Escherichia coli* K12, *Salmonella typhi,* and *Salmonella arizonae.* Mu G (−) phage can infect *E. coli C, Shigella sonnei, Citrobacter freundii, Erwinia carotovora, Erwinia herbicola,* and *Enterobacter cloacae* (Koch *et al.,* 1987). Mu-sensitive isolates of *Klebsiella pneumoniae* and *Shigella flexneri* have been found (Groisman and Casadaban, 1987a; E. A. Groisman and M. J. Casadaban, unpublished results, 1986). The structural component of the cell wall that is responsible for the adsorption of Mu G (+) phage has been located within the lipopolysaccharide (LPS) of *E. coli* K12. The receptor site is on the N-acetylglucosamine β1-Glu(α1-2)Glu-α part of the outer end of the LPS (Sandulache *et al.,* 1984). The receptor for Mu G (−) phage has been identified as a terminal glucose in $\beta(1 \rightarrow 6)$-linkage (Sandulache *et al.,* 1985). Lysates prepared by induction of a lysogen have 50% of the phage in the G (+) orientation and 50% in the G (−) orientation; lysates propagated lytically consist of virtually 100% of particles in one G orientation (Koch *et al.,* 1987). *Escherichia coli* strains that are resistant to phage P1 (which has a homologous invertible segment named C region) are also resistant to Mu, and 50% of the strains selected for Mu resistance are also resistant to

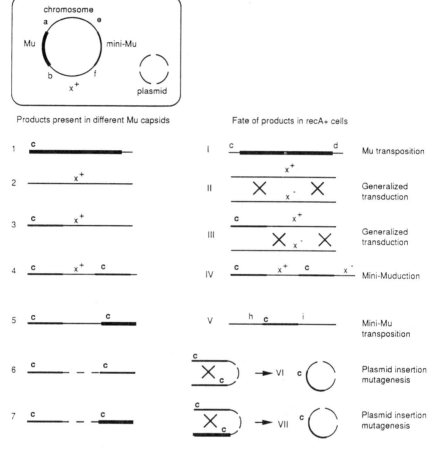

FIG. 1. Gene transfer events mediated by Mu (very thick line) and mini-Mu (intermediate thick line). Induction of a Mu/mini-Mu double lysogen will result in the transposition of both elements and the encapsidation of different products, the fate of which will depend on the genotype of the recipient cells and the genetic selection that is used. The phage lysate will include particles harboring the following: (1) the wild-type Mu genome linked to approximately 2 kb of host DNA [this element will be inherited most certainly by transposition (I) in a nonimmune host, regardless of the *recA* allele]; (2) chromosomal DNA (thin line) that can only be inherited by homologous recombination in *recA*⁺ cells in a generalized transduction event (II); (3) the mini-Mu DNA linked to a host genome segment [the host DNA can be inherited by homologous recombination in a generalized transduction event (III) in *recA*⁺ cells; if the mini-Mu code harbors the MuA and B transposition genes then it can transpose to new sites in both *recA*⁺ and *recA*⁻ cells (V)]; (4) two mini-Mu elements in the same orientation and flanking a host DNA fragment [this whole structure can transpose to new sites in a *recA*-independent process called mini-Muduction (IV) to give rise to cells which will be phenotypically X⁺ but which can easily revert to X⁻ by recombination between the repeated mini-Mu sequences; generalized transduction and mini-Mu transposition can

P1.Mu can acquire P1 host range and vice versa when recombination between the Mu G and the P1 C regions takes place (Toussaint *et al.*, 1978; Csonka *et al.*, 1981).

The Mu *mom* gene product modifies Mu and cellular DNA sequences providing protection against cleavage by various endonucleases. Transcriptional activation of the Mu *mom* gene depends on methylation of its promoter region (Kahmann and Hattman, 1987).

Use of Mu in Genetic Analysis

Several features make bacteriophage Mu the transposable element of choice to carry out *in vivo* genetic engineering. Mu transposition occurs at levels several orders of magnitude higher than in other well-studied transposable elements such as Tn3, Tn5, and Tn10. Moreover, the use of phage with temperature-sensitive alleles of the Mu repressor allows synchronous induction of Mu transposition. Unlike other transposons, Mu shows very little insertion sequence specificity (Pato, 1989). However, analysis of mini-Mu insertions in small plasmid targets showed (i) the presence of preferred sites of insertion ("hot-spots") having a G/C in the center of the 5-bp duplication generated during the transposition reaction, and (ii) the phenomenon of transposition immunity: plasmids with bacteriophage Mu sequences receive additional Mu insertions several times less frequently than those without them. Mu has a broad host range with respect to both infection and transposition, and the preparation and handling of phage lysates is very easy. There are close to 100 different Mu derivatives harboring a variety of drug resistance markers to facilitate genetic selection, and many of them are small enough to make DNA manipulations easy (see next section).

A wide miscellany of genetic manipulations can be performed with the different Mu elements, the fate of which will depend on the recipient cells and the type of genetic selection that is utilized (see Figs. 1 and 2). They include generation of chromosomal mutations and gene fusions;

---

also take place as described above]; (5) a host DNA segment flanked by a mini-Mu element and part of a helper Mu phage can also be inherited by homologous recombination in $recA^+$ cells (III) [a mini-Mu deleted for the Mu*A* and *B* genes can be inherited by transposition (V) in both $recA^+$ and $recA^-$ cells if the segment of the helper Mu that is included in the particle corresponds to the left end of Mu that codes for the transposition functions]; (6) the plasmid DNA flanked by two similarly oriented mini-Mu elements; (7) the plasmid DNA flanked by a mini-Mu element and part of a helper Mu in the same orientation which can circularize in $recA^+$ cells to give rise to plasmids harboring single inserts of the mini-Mu element. The letter **c** in bold type indicates the left end of the Mu or mini-Mu element. The letters a, b, c, d, e, f, h, i, and x indicate different chromosomal loci.

mutagenesis of previously cloned genes; gene localization; determination of transcriptional orientation; reverse genetic experiments; mapping; cloning; strain constructions, involving generation of duplications, deletions, and merodiploids; generation of DNA rearrangements in a lysogen; and DNA sequencing.

Bacteriophage Mu Derivatives for Genetic Analysis

Virtually all Mu-based vectors used for genetic engineering purposes harbor (a) the cis-acting terminal DNA sequences necessary for transposition, (b) the *pac* site, and (c) a selectable marker. Because there are no environmental signals known that can induce Mu lysogens (such as UV or DNA-damaging agents for λ induction), most derivatives contain a thermosensitive (ts) allele of the repressor (usually *c*ts62), and lysates are prepared by thermoinduction of lysogens. Derivatives with only 1001 bp from the left end (up to the *Hin*dIII site) contain the complete repressor coding region that is deleted for its main promoter; they do not confer immunity when present in single copy in the cell (i.e., in the chromosome) but do so when present in multicopy plasmids. Most elements carry the Mu*A* and Mu*B* transposition genes and may contain additional genetic determinants as well.

Mu derivatives which can still grow as phage are designated by the letter p, indicating that they can form plaques (e.g., MupAp1); the letter d indicates that they are defective and unable to grow as phage, but such derivatives can usually mediate the rearrangements typical of transposable elements (e.g., MudI1). Mu derivatives that can form transcriptional or transcriptional–translational fusions are usually designated MudI and MudII, respectively. Table I presents a list of useful derivatives of bacteriophage Mu. The sequence corresponding to approximately 16 kb of Mu DNA is available (Priess *et al.*, 1987; Kahmann and Kamp, 1987). It has recently been reported that the MuR end in some of the most frequently used Mu derivatives, Mud1, is composed of a large inverted repeat which includes sequences of the *trp* operon. This unexpected structure, which probably arose during the *in vivo* genetic construction of Mud1, may also be present in other MudI derivatives such as MudI1681 and MudI1734 (Table I; Metcalf *et al.*, 1990).

Delivery Systems

Mu elements can be introduced into bacteria by transduction, conjugation, or DNA transformation. The choice of route will depend on the particular bacterial species (and even particular strain) used and the type of genetic experiment to be performed.

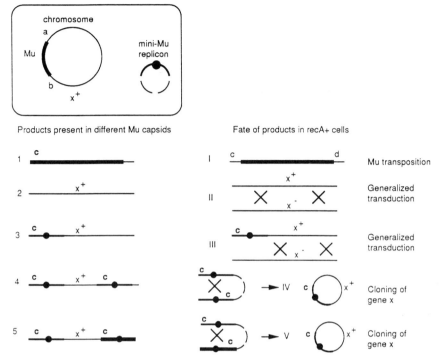

FIG. 2. Gene transfer events mediated by Mu and mini-Mu replicons. Induction of a Mu/mini-Mu replicon double lysogen will result in the transposition of both elements and the encapsidation of different products, the fate of which will depend on the genotype of the recipient cells and the genetic selection that is used. The phage lysate will include particles harboring the following: (1) the wild-type Mu genome linked to approximately 2 kb of host DNA [this element will be inherited most certainly by transposition (I) in a nonimmune host, regardless of the *recA* allele]; (2) chromosomal DNA that can only be inherited by homologous recombination in *recA*⁺ cells in a generalized transduction event (II); (3) the mini-Mu replicon DNA linked to a larger host genome segment [the host DNA can be inherited by homologous recombination in a generalized transduction event (III) in *recA*⁺ cells; transposition events of multicopy replicon-containing mini-Mu elements can be recovered in both *recA*⁺ and *recA*⁻ cells harboring in their cytoplasm a plasmid of the same incompatibility group as the one present in the mini-Mu replicon, or by using *polA*⁻ recipients]; (4) two mini-Mu replicon elements in the same orientation and flanking a host DNA fragment that can circularize efficiently in recombination-proficient recipients to form a plasmid clone of gene x (IV) [clones are obtained in *recA*⁻ recipients at 1000 times lower frequency; selection for both the mini-Mu marker and the gene x would yield plasmid clones, but selection for gene x only would give rise to generalized transductants >90% of the time]; (5) a host DNA segment flanked by a mini-Mu replicon and part of a helper Mu phage (in the same relative orientation), which will also give rise to plasmid clones of gene x by homologous recombination in *recA*⁺ cells (V). The letter **c** in bold type indicates the left end of the Mu or mini-Mu replicon element; the circle in the mini-Mu replicon indicates the plasmid origin of replication. The letters a, b, c, d, and x indicate different chromosomal loci.

TABLE I

Mu Derivatives Useful for Genetic Analysis[a]

Designation	Transposition genes	Selectable marker	Plasmid replicon	Fusion element	Other elements	Size (kb)	Reference
Mucts	A, B	—	—	—	—	37.5	Howe, 1973a
MupAp1	A, B	amp	—	—	—	37.5	Leach and Symonds, 1979
MupAp5	A, B	amp	—	—	—		Leach and Symonds, 1979
Mud1	A, B	amp	—	'lacZYA	—	37	Casadaban and Cohen, 1979
Mud1-8	Aam, B	amp	—	'lacZYA	—	37	Hughes and Roth, 1984
Mud2-8	Aam, B	amp	—	'lac'ZYA	—	37	Hughes and Roth, 1985
MudX	A	amp, cat	—	'lacZYA	—	39.5	Baker et al., 1983
MudI301	A, B	amp	—	'lac'ZYA	—	35.6	Casadaban and Chou, 1984
MudII701-1	A, B	amp, kan	—	'lacZYA	—	35.6	Castilho et al., 1984
MudII701-301	A, B	amp, kan	—	'lac'ZYA	—	35.6	Castilho et al., 1984
Mu3A	A	—	—	—	—	7.5	Résibois et al., 1981
Mu18A	A	—	—	—	—	10	Résibois et al., 1981
Mu18	—	—	—	—	—	8.6	Résibois et al., 1981
Mu18A-1	A	amp	—	—	—	10	Résibois et al., 1981
Mu18-1	—	amp	—	—	—	8.6	Résibois et al., 1981
Mu18A-2	A	cat	—	—	—		Résibois et al., 1981
Mini-Mu-cat	—[b]	cat	—	—	—		Patterson et al., 1986
Mini-Mu-tac	A, B	kan	—	—	p_{tac}	7	Gramajo et al., 1988
Mud5-3	—	kan	—	—	p_{T7}		E. A. Groisman, unpublished
Mud4041	A, B	kan	—	—	—	7.8	Castilho et al., 1984
MudI1678	A, B	amp	—	'lacZYA	—		Castilho et al., 1984
MudII1678	A, B	amp	—	'lac'ZYA	—	22.4	Castilho et al., 1984
MudI1681	A, B	kan	—	'lacZYA	—	15.8	Castilho et al., 1984
MudII1681	A, B	kan	—	'lac'ZYA	—	14.2	Castilho et al., 1984
MudI1734	—	kan	—	'lacZYA	—	11.2	Castilho et al., 1984
MudII1734	—	kan	—	'lac'ZYA	—	9.7	Castilho et al., 1984
Mini-Mu-tet	A, B	tet	—	'lacZYA	—		Belas et al., 1984

Strain	Mu ends	Selection	Replicon	Gene fusion	Other	kb	Reference
Mini-Mu-*lux*	A, B	*kan*	—	'*lux*	—	15.5	Engebrecht *et al.*, 1985
Mini-Mu-*lux*(*tet*)	A, B	*tet*	—	'*lux*	—	18.5	Engebrecht *et al.*, 1985
Mud13-1	—	*kan*	—	'*lacZYA*	$oriT_{RK2}$	10.3	E. A. Groisman, unpublished, 1987
Mud5345	—	*cat*	—	'*lacZYA*	$oriT_{RK2}$	10.5	E. A. Groisman, unpublished, 1987
Mini-Mu121	—	—	—	—	—	9.2	Chaconas *et al.*, 1981
Mini-Mu121*kan*	—	*kan*	—	—	—	15.8	Chaconas *et al.*, 1981
Mini-Mu121*amp*	—	*amp*	—	—	—	9.6	Chaconas *et al.*, 1981
Mini-Mu121*lacL*	—	*lac*⁺	—	—	—	21.5	Chaconas *et al.*, 1981
Mini-Mu222	—	—	—	—	—	3.7	Chaconas *et al.*, 1981
Mini-Mu222*amp*	—	*amp*	—	—	—	5.5	Chaconas *et al.*, 1981
Mini-Mu222*lacL*	—	*lac*⁺	—	—	—	16.0	Chaconas *et al.*, 1981
Mini-Mu222*lacR*	—	*lac*⁺	—	—	—	16.0	Chaconas *et al.*, 1981
MudF	—	*kan, lac*⁺	—	—	—	14.5	Sonti *et al.*, 1991
Mud(*lacZ–nptII*)	—	—	—	*lac'Z'nptII*[c]	—	—	Lång *et al.*, 1987
MudIIPR3	—	*cat*	—	'*nptI*	—	4.2	Ratet and Richaud, 1986
MudIIPR2	—	—	—	'*nptI*	—	2.6	Ratet and Richaud, 1986
MudIIPR13	—	*cat*	—	'*lac'ZYA*	—	9.2	Ratet *et al.*, 1988
MudPR40	—	*cat, gen*	—	'*lac'ZYA*	—	12.1	Ratet *et al.*, 1988
MudIIPR46	—	*cat, gen*	—	'*lac'ZYA*	—	13.7	Ratet *et al.*, 1988
MudIIPR48	—	*cat, gen*	RiHRI	'*lac'ZYA*	$oriT_{pSUP5011}$	21.5	Ratet *et al.*, 1988
MudII4042	A, B	*cat*	p15A	'*lac'ZYA*	—	16.7	Groisman *et al.*, 1984
Mud5005	A, B	*kan*	pMB1	—	—	7.9	Groisman and Casadaban, 1986
MudII5085	—	*cat*	p15A	'*lac'ZYA*	—	13.4	Groisman and Casadaban, 1986
MudII5086	A, B	*kan*	pMB1	'*lacZYA*	—	14.9	Groisman and Casadaban, 1986
MudII5117	A, B	*kan, sm/sp*	pSa	'*lac'ZYA*	—	21.7	Groisman and Casadaban, 1986
MudII5155	A, B	*kan*	pMB1	'*lacZYA*	$oriT_{RK2}$	15.6	Groisman and Casadaban, 1986
MudII5166	A, B	*cat*	pMB1	'*lacZYA*	$oriT_{RK2}$	15.8	Groisman and Casadaban, 1986
Mud5060	A, B	*kan*	pMB1	—	res_{Tn3}	8.3	Groisman, 1986
Mud5260	A, B	*kan*	pMB1	—	$loxP_{P1}$	7.8	Groisman, 1986
Mud5294	A, B	*kan*	pMB1	—	p_{T7}	7.7	Groisman *et al.*, 1991a,b
MudII5119	A, B	*kan*	pMB1	'*lacZYA*	cos_{λ}	16.3	Groisman and Casadaban, 1987a,b

(*continued*)

TABLE I (*continued*)

Designation	Transposition genes	Selectable marker	Plasmid replicon	Fusion element	Other elements	Size (kb)	Reference
MudII79	A, B	amp	pMB1	'lac'ZYA	cos_λ		Gramajo and de Mendoza, 1987
MudP[d]	—	cat		—	P22	36.4	Youderian et al., 1988
MudQ[d]	—	cat		—	P22	36.4	Youderian et al., 1988
λplacMu1	A'			'lac'ZYA	imm_λ		Bremer et al., 1984
λplacMu3	A'			'lac'ZYA	imm_{21}		Bremer et al., 1984
λplacMu5	Aam			'lac'ZYA	imm_λ		Bremer et al., 1988
λplacMu9	A'	kan		'lac'ZYA	imm_λ		Bremer et al., 1985
λplacMu13	Aam			'lac'ZYA	imm_{21}		Bremer et al., 1988
λplacMu15	Aam	kan		'lac'ZYA	imm_λ		Bremer et al., 1988
λplacMu50	A'			'lacZYA	imm_λ		Bremer et al., 1985
λplacMu52	Aam			'lacZYA	imm_λ		Bremer et al., 1988
λplacMu54	Aam			'lacZYA	imm_{21}		Bremer et al., 1988
λplacMu55	Aam	kan		'lacZYA	imm_λ		Bremer et al., 1988
MudE	A, B	kan, erm		—	—		Kuramitsu, 1987
MudαTK	—	kan, tk		'lac'ZYA	—	9.7	Jenkins et al., 1985
MudIIZZ1	—	cat	2 μm ori	'lac'ZYA	LEU2	15	Daignan-Fornier and Bolotin-Fukuhara, 1988

[a] Mu derivatives are listed according to their most frequent use in genetic analysis without attempting a strict classification. For example, an element that was originally designed for *in vivo* cloning (i.e., mini-Mu replicon) can also be used for mutagenesis, isolation of *lac* gene fusions, mapping, and generalized transduction. Mu derivatives that harbor only the first 1001 bp corresponding to the left end contain all the coding region corresponding to the *c* repressor gene but are deleted for the main promoter. These derivatives do not confer immunity when present in a single copy in the chromosome but do confer immunity when present in multicopy plasmids. Genetic and physical maps of some of the Mu elements listed can be found in Faelen (1987).

[b] The MuA and B genes are provided in cis in the same plasmid.

[c] This element can be used to isolate simultaneous translational fusions to *lacZ* and operon fusions to *nptII*.

[d] MudP and MudQ differ in the orientation of the P22 DNA present between the Mu ends.

Transduction has been widely used because Mu has a very broad host range, and Mu-sensitive strains have been found in several bacterial species (see above). Mu can be introduced into species which are resistant to Mu infection but still harbor the host factors necessary for transposition, by transduction with a phage of a different host range (e.g., MuhP1; see below), conjugation, or transformation. Alternatively, mutants that become sensitive to Mu can be isolated (Bachhuber et al., 1976; Muller et al., 1988). A transposon Tn10 insertion linked to a mutation leading to Mu sensitivity in Salmonella typhimurium facilitates the genetic transfer of this allele to resistant strains of the same genus (Faelen et al., 1981; Soberon et al., 1986).

Transduction with phage Mu is usually used because of the presence of the pac site in all Mu derivatives. Mu lysates are often prepared from double lysogens harboring both the defective Mu derivative to be selected for and a helper Mu phage that will complement the Mu derivative for the morphogenic functions. Sometimes the helper Mu also provides transposition functions; however, trans complementation is not very efficient (Reyes et al., 1987). Lysates corresponding to plaque-forming Mu are prepared by heat induction from simple lysogens. The MupAp1 phage is locked in the G (+) orientation and therefore can only infect a subset of the Mu-sensitive species (see above).

Mu can also be delivered by other phage. P1 has been used to transduce Mu elements into S. typhimurium (Rosenfeld and Brenchley, 1980) and Vibrio parahaemolyticus (Engebrecht et al., 1985). Derivatives of phage P1 into which a drug-resistant determinant (such as cam or kan) has been genetically engineered are particularly useful in selecting P1-sensitive mutants of enteric bacteria (Goldberg et al., 1974). The MuhP1 recombinant (which carries the C loop region of phage P1 in place of the G region of Mu) has been used in S. typhimurium galE⁻ mutants (Csonka et al., 1981), Yersinia pestis (Goguen et al., 1984), and Klebsiella aerogenes (E. A. Groisman and M. J. Casadaban, unpublished results, 1986). In S. typhimurium, phage P22 has been used for delivery of different Mu elements (Holley and Foster, 1982; Hughes and Roth, 1988), and in E. coli, cosmids carrying mini-Mu elements were transferred using phage λ (de Mendoza and Rosa, 1985). Though λ host range is fairly narrow, resistant strains can be rendered λ-sensitive by introduction of plasmid clones of the lamB (receptor) gene (De Vries et al., 1984; Ludwig, 1987).

Conjugation is a very efficient way of transferring DNA between strains. The recent examples of conjugation between gram-negative bacteria and yeast (Heinemann and Sprague, 1989), plant cells (Zambryski et al., 1989; Buchanan-Wollaston et al., 1987), and gram-positive bacteria (Trieu-Cuot et al., 1987) exemplify its power at interkingdom and interspecies transfer of DNA. Mu derivatives present in broad host range plasmids or in plasmids that harbor the mobilization site (*oriT*) transpose when mated into nonimmune hosts ("zygotic induction"), and they have been used in the mutagenesis of *Legionella pneumophila* (Mintz and Shuman, 1987).

Transfection with naked Mu DNA is very inefficient, presumably because the isolated phage DNA is missing a Mu protein necessary for circularization and protection of the Mu ends from host exonucleolytic attack. Transformation has been used to introduce mini-Mu plasmids destined for subsequent preparation of genomic libraries, but not as a strategy to deliver mini-Mu elements for mutagenesis of host DNA.

Delivery of transposition-proficient Mu derivatives by means other than the use of a Mu virion may have unanticipated consequences. Transposition of MudI1 (Casadaban and Cohen, 1979; see Table I) delivered by phage P22 is enhanced by the presence of a helper P22 phage and requires host *recA*, *recB*, and *recC* functions, although Mu virion-mediated delivery is independent of host recombination (Hughes *et al.*, 1987).

Mutagenesis: Isolating Mutants and Gene Fusions

Chromosome as Target

Transposon mutagenesis has become the procedure of choice to isolate mutants because it results in a single, unique physical alteration in the gene that has been mutated. Most transposon insertions lead to total inactivation of the gene, and the presence of markers in the transposable element can be used to isolate genes by first identifying an insertion in or very near the gene of interest on the basis of phenotype or linkage and then cloning a fragment of DNA from the mutant genome that harbors the transposon.

Obtaining Stable Insertions. Insertions generated with MuA^+B^+ phage do not normally revert (frequency $<10^{-10}$). Excision of MuA^+B^- phage (owing to polar insertions in *B*) occurs at frequencies of 10^{-6} to 10^{-5} in *recA*$^+$ cells and requires the presence of a functional *A* gene. When isolating Mu transposon insertions, approximately 10% of the lysogens obtained may harbor more than one insertion per haploid genome. This number can be reduced by plating the cells at low temperature ($<28°$) during the initial step. The presence of single insertion can be confirmed by Southern hybridization analysis using probes internal to the Mu derivative and digesting the target DNA with restriction endonucleases that do not cut within the Mu element. The isolated mutation can also be transferred to a "clean" genetic background to demonstrate 100% linkage between the transposon marker and the new phenotype.

Secondary transposition events of MuA^+B^+ phage occur at frequencies high enough that they are not useful for the isolation of second site mutants affecting the expression of a gene fusion, or for the transduction- or conjugation-mediated transfer of Mu insertions between strains. Insertions can be stabilized by recombining a resident transposition-proficient element and a transposition-defective element, or by performing a series of genetic steps leading to the isolation of deletions that remove one end of Mu (Komeda and Iino, 1979). Transposition by a cts mini-Mu element can be greatly reduced by introducing into the strain a wild-type Muc^+ phage (Shapiro and Higgins, 1989) or a clone of the wild-type repressor gene.

Stable insertions can be obtained by performing the mutagenesis with a derivarive that is defective for transposition but that can be complemented in trans. This can be accomplished by using phage that only transpose in suppressor backgrounds (Hughes and Roth, 1984, 1985; Bremer et al., 1988), by using phage that are MuA^+ and coinfecting with a phage that carries the MuB^+ gene in trans (Bremer et al., 1984), or by using MuA^-B^- derivatives (see Table I).

Schemes for cis complementation and high-efficiency transposition of MuA^-B^- derivatives have been developed for E. coli and S. typhimurium, and they should be applicable to other Mu-sensitive and P22-sensitive species, respectively. In E. coli, lysates are prepared from a strain dilysogenic for the mini MuA^-B^- derivative and a helper MuA^+B^+ phage. Some of the phage particles would carry the mini-Mu derivative and also the left end of the helper phage (carrying the A^+B^+ genes) and would yield mini-Mu transpositions in the recipient cell (Castilho et al., 1984). In S. typhimurium, a phage P22 lysate is prepared in a strain that harbors both a mini-MuA^-B^- derivative and a MudI1 phage in the his operon. The mini-MuA^-B^- is located approximately 4 kb from the left (transposition gene-containing) end of the MudI1 phage in an inverted relative orientation. Because P22 packages approximately 44 kb by a headful mechanism and the mini-Mu element and MudI1 phage are 10 and 37.5 kb, respectively, many phage particles that harbor the mini-Mu element will also include DNA corresponding to the left end of MudI1. Selection for the mini-Mu marker yields primarily transposition events, and the background of drug-resistant transductants that are inherited by homologous recombination of the mini-Mu insertion at the his locus (which may be as high as 40%) can be eliminated by plating the cells in medium lacking histidine (Hughes and Roth, 1988). For such mini-MuA^-B^- derivatives, reduced numbers of transductants are recovered in S. typhimurium Rec^- hosts.

The surviving lysogens usually have deletions at the site of insertion of the mini-Mu element (Sonti *et al.*, 1991; see Generating Deletions and Duplications, below).

Isolating Gene Fusions. Genetic fusions have provided an important means of analyzing basic biological problems. Although different reporter genes have been used for fusions, the most common procedure involves the *E. coli lac* operon (Silhavy and Beckwith, 1985; Silhavy *et al.*, 1984; [9], this volume). Two types of *lac* gene fusions can be formed: transcriptional, where the mRNA is initiated at an exogenous promoter and continues into a promoterless *lac* operon to form a hybrid message, and transcriptional–translational (hybrid protein fusions), requiring the presence of both an exogenous promoter as well as translational initiation signals that are fused in frame to a *lac* operon DNA segment that is missing its promoter, ribosome binding site, and an amino-terminal part of *lacZ*. A detailed description of the use of gene fusion is presented elsewhere in this volume [9], this volume). Mu elements with a variety of reporter genes to generate either type of fusion or to allow the simultaneous isolation of both types of fusions are available (Table I).

Isolating Conditional Lethals. Mu derivatives with either the *tac* (*trp–lac*) (Gramajo *et al.*, 1988) or a phage T7 promoter (Groisman *et al.*, 1991a; E. A. Groisman, unpublished data, 1987) can promote transcription in a regulated manner. Genes adjacent to the site of insertion of these Mu elements may be turned on if "sense" RNA is made and off if "antisense" RNA is made. These elements may be useful to define loci which are essential under different physiological conditions.

Targeting Mu Insertions. Mu insertions can be isolated in a particular region of the chromosome by first isolating a large pool of insertion mutants (>20,000) in one strain and then selecting those of interest in a second (transduction or conjugation) transfer step. After the pool of insertions is isolated, a generalized transducing phage is grown in that pool and then used to infect a recipient cell selecting simultaneously for the Mu marker and a gene tightly linked to the region of interest. Alternatively, the first mutagenesis step is performed in an Hfr strain that transfers the region of interest early, and then the strain is mated to a recipient cell with selection for the Mu marker and screening for the phenotype of interest.

Cloning of Adjacent DNA. Mu derivatives bring a selectable marker and a source of restriction sites to the mutated gene which can be used for purification of the adjacent DNA by direct cloning or polymerase chain reaction (PCR) amplification.

Plasmids, Cosmids, and Phage as Targets

Mini-Mu elements have been particularly useful for the mutagenesis, localization of open reading frames, and determination of transcriptional orientation of genes present in small plasmids harboring cloned DNA segments. Althought this procedure is usually performed *in vivo*, mutagenesis of plasmid DNA has also been carried out by mini-Mu *in vitro* transposition using purified components (Waddell and Craig, 1988; Mizuuchi, 1983). The *in vivo* mutagenesis procedure involves the following steps: (1) introduction of the plasmid to be mutagenized into a *recA*⁻ strain harboring a mini-Mu element and a helper phage (a *recA*⁻ strain is used to prevent the formation of multimers of the plasmid); (2) preparation of the lysate from the strain harboring the plasmid; and (3) infection of a recipient cell where selection is carried out simultaneously for both the mini-Mu and the plasmid antibiotic resistance markers (to select against insertions in the vector). The resulting transductants harbor plasmids with mini-Mu insertions that can be used for further studies (e.g., DNA sequencing and overproduction of fusion proteins).

The original procedure requires the presence of a mini-Mu and a Mu helper phage in the strain where the lysate will be prepared, along with the use of *recA*⁺ (to mediate the circularization of the plasmid molecule harboring the mini-Mu insertion) Mu lysogenic (to prevent zygotic induction of transposition-competent mini-Mu elements) recipient cells (Castilho *et al.*, 1984). Cosmids can be mutagenized if a helper λ is used in place of the helper Mu, together with a self-transposable mini-Mu in the donor cell (de Mendoza and Rosa, 1985). The requirement for *recA*⁺ recipient cells can be relieved by using mini-Mu elements containing the P1 *loxP* site-specific recombination site and infecting cells harboring the *cre*⁺ recombinase (Sauer and Henderson, 1988). A couple of warnings regarding the mutagenesis of plasmids are warranted: first, monomers should be used to transform the strain in which the phage lysate will be prepared, to prevent loss of the mutated plasmid; second, care must be taken in the choice of vector because, for example, pUC derivatives contain *lac* DNA sequences that can recombine with *lac* DNA present in many mini-Mu derivatives; and, third, plasmids harboring either end of phage Mu are immune to insertions by a Mu derivative and thus insertions take place at 50–100 times lower frequencies (Reyes *et al.*, 1987; Darzins *et al.*, 1988; Adzuma and Mizuuchi, 1988). Immunity is a cis phenomenon and does not affect insertions into other targets in the same cell.

Plasmid mutagenesis is usually performed in *E. coli* using the enormous variety of mini-Mu derivatives harboring different selectable markers, including the Tn5-derived kanamycin/neomycin resistance gene that works in a number of gram-negative bacteria and an erythromycin resistance gene for selection in gram-positive bacteria (Kuramitsu, 1987). Some derivatives contain gene-fusing elements useful to establish transcriptional orientation of cloned genes, and others harbor regulatable promoters to facilitate gene expression (see Table I). Foreign DNA clones can be efficiently mutagenized in *E. coli* and the mutation then transferred to the appropriate host. During this type of reverse genetics experiment with *Azorhizobium caulinodans,* it was observed that mutagenized plasmids integrated by a single crossover event in the chromosome were very stable (even without selection) only if they contained the Mu right end. The presence of the two ends and of the repressor gene increased recombination in the vicinity of the transposon favoring double crossovers (P. Ratet, personal communication, 1989). I am not aware of similar situations occurring in other bacteria.

Mutagenesis of large, self-transmissible or mobilizable plasmids can be achieved by infecting a cell harboring the plasmid and mating the pool of transductants to another cell with selection for the Mu marker. If a transposition-competent Mu derivative is used as the mutagenic agent, it is recommended that the recipient cells used in the mating be Mu-lysogenic (to avoid inheritance of the Mu element alone, not linked to the plasmid); if a Mup derivative (see above) is used, the recipient cells used should be Mu-lysogenic and Mu-resistant. Nonmobilizable plasmids can be mutagenized in cells harboring a transmissible plasmid with which they can form a cointegrate structure that can be transferred to a recipient cell and resolved by the host recombination machinery.

Mu derivatives, owing to their relatively large size, have not been widely used to isolate simple insertions into DNA of other phage. However, Mu-mediated deletions of phage λ have been isolated using as starting material λ *plac* lysogens with an inserted Mu phage (Bukhari and Allet, 1975), and mini-Mu replicon elements (see below) have been used to obtain clones of phage λ containing deletions of different length (Groisman and Casadaban, 1987b). The size of the deletion can be preselected by choosing a mini-Mu derivative of the appropriate size.

Mu-Mediated Cloning, Mapping, and Transfer of DNA between Strains

Cloning of DNA

The study of a variety of biological problems often relies on the availability of cloned DNA sequences and/or gene products. Taking advantage of the Mu "way of life," several *in vivo* cloning schemes have been developed where Mu transposition replaces the traditional restriction endonuclease digestion and DNA ligation steps, with the buffers and energy sources being provided by the bacterial cell.

Mini-Mu Replicons with Helper Mu. In one system, a mini-Mu replicon (derivative harboring a plasmid replicon and a selectable drug resistance gene) can transpose at a high frequency when derepressed and is complemented by a helper Mu prophage for lytic growth. DNA sequences that become flanked by two copies of a mini-Mu replicon (or by a mini-Mu replicon and a helper Mu) may be packaged in the Mu capsid. After infection (which replaces the DNA transformation step of *in vitro* constructed libraries), homologous recombination can occur between Mu sequences, resulting in the formation of plasmids carrying the transduced DNA (Groisman *et al.*, 1984; Groisman and Casadaban, 1986).

This system requires the presence of both a mini-Mu replicon and a helper Mu prophage in the strain from which the library will be prepared. The strain into which the library will be introduced needs to be Mu-sensitive, $recA^+$, and a Mu lysogen. Recipients which are $recA^-$ can be used; however, the frequencies are 100–1000 times lower unless recipients harbor the phage P1 cre^+ recombinase and the library is prepared using a mini-Mu replicon containing the P1 $loxP$ site-specific recombination site (E. A. Groisman and M. J. Casadaban, unpublished results, 1986) or, alternatively, unless a helper λ phage is used (see below). The Mu lysogenic state of the recipient cells prevents "zygotic induction" of the incoming mini-Mu element and increases the overall frequency about 10-fold. This system has been successfully used to clone genes from several members of the family Enterobacteriaceae by complementation directly into other mutant enterics without having to go through an *E. coli* intermediate (see Table II; Groisman and Casadaban, 1987a). Libraries can be prepared from Mu-resistant strains by introducing into them (by conjugation or transformation) plasmids harboring the helper Mu or the mini-Mu replicon, and then preparing the phage lysate (Groisman and Casadaban, 1987a). Libraries can be introduced into Mu-resistant strains by either isolating Mu-sensitive derivatives (see above), using other helper phages (e.g., MuhP1), or going through a Mu-sensitive recipient (usually *E. coli*) and then mobilizing the library by conjugation or transduction. We have prepared libraries in $galE^-$ *S. typhimurium* strains using a MuhP1 helper phage and then transferred the libraries to other *Salmonella* strains by transduction with a phage P22 lysate grown in the library (E. A. Groisman, unpublished data, 1987).

Cloning frequencies range from below 10^{-4} to above 10^{-2} (selected gene/drug-resistant transductant) for different genes but are constant for a given gene (probably a reflection of Mu insertion specificity). The size of the plasmid clones for a given gene varies, but it is always limited by the 39-kb Mu headful packaging mechanism. When the 16.7-kb mini-Mu replicon was used, the size of the DNA segment cloned for 12 $proC^+$ clones ranged from 2 to 15 kb with an average of 6.2 kb (Groisman *et al.*, 1984). The smallest mini-Mu element is only 5.1 kb long, allowing cloning of DNA fragments of up to 33.9 kb (see Table I).

TABLE II

TRANSMISSIBLE PLASMIDS HARBORING MU DERIVATIVES

Designation	Plasmid::Mu element	Replication origin		Drug resistance	
		Plasmid	Mu element	Plasmid	Mu element
Self-transmissible plasmids					
RP4::Mu	RP4::Mu*cts*	rep_{RP4}		*tet, amp, kan*	
pULB113	RP4::Mu3A	rep_{RP4}		*tet, amp, kan*	
pULB21	RP4Tc::Mu3A	rep_{RP4}		*amp, kan*	
pULB110	RP4Kn::Mu3A	rep_{RP4}		*tet, amp*	
pRK24M4	pRK24::Mu*cts*	rep_{RK2}		*tet, amp*	
pRK241	pRK24::MudI1681	rep_{RK2}		*tet, amp*	*kan*
pREG2	R388::MupAp1	rep_{R388}		*tp*	*amp*
pULB108	pRM5::Mu3AΔTc	rep_{pRM5}		*amp, kan*	
pULB18	pUZ8::Mu18A-1	rep_{pUZ8}		*tet, kan, hg*	*amp*
pULB19	pUZ8::Mu*c*$^+$	rep_{pUZ8}		*tet, kan, hg*	
pULN20	R300B::Mu18A-2	rep_{R300B}		*cat, su, sm*	
pEG5150	pRK24::MudII5117	rep_{RK2}	rep_{pSa}	*tet, amp*	*kan, sp/sm*
pEG5152	pRK24::MudI5086	rep_{RK2}	rep_{pMB1}	*tet, amp*	*kan*
Mobilizable plasmids					
pEG5155	pBC0::MudI5155	rep_{pMB1}	rep_{pMB1}	*amp*	*kan*
pEG5166	pBC0::MudI5166	rep_{pMB1}	rep_{pMB1}	*amp*	*cat*
pEG5092	pLAFR1::Mud5005	rep_{RK2}	rep_{pMB1}	*tet*	*kan*
pEG5117	pSC101::MudII5117	rep_{pSC101}	rep_{pSa}	*tet*	*kan, sp/sm*

The location of the cloned gene in the plasmid can be determined by comparing the minimum overlapping fragment present in all the plasmid clones (replacing the tedious and time-consuming steps of deletion mapping; see Wang *et al.*, 1987b). The presence of additional genetic elements in several mini-Mu replicons can facilitate further analysis: those harboring *lac* gene fusing segments allow cloning of *lac* gene fusions to selected loci (Groisman *et al.*, 1984); elements with an *oriT* determinant facilitate the transfer of plasmid clones by conjugation; and an element that harbors a phage T7 promoter near the Mu right end facilitates expression and determination of transcriptional orientation of cloned genes (Groisman *et al.*, 1990). New open reading frames can be identified by using derivatives with either *lac* gene fusing segments (Wang *et al.*, 1987b) or a T7 promoter (Groisman *et al.*, 1990).

Mini-Mu Replicons with Helper λ. Cosmid libraries can be generated *in vivo* with mini-Mu replicons containing the λ *cos* site and a helper λ instead of a helper Mu. (Groisman and Casadaban, 1987b; Gramajo and de Mendoza, 1987). Libraries can be prepared either by simultaneous thermoinduction of a mini-Mu replicon and a helper λ with a *ts* repressor, by first allowing for mini-Mu replicon transposition and then superinfecting with a helper λ, or by thermoinducing a mini-Mu replicon in a strain also harboring a wild-type λ that is later induced with a DNA-damaging agent such as mitomycin C (Groisman, 1986). Clones obtained with a helper λ have larger and nearly constant size inserts (because λ can package up to 55 kb between two similarly oriented *cos* sites), and they can be isolated efficiently in both *recA*$^+$ and *recA*$^-$ cells (because circularization of the cosmid molecule at the *cos* sites does not require host recombination functions). An additional feature of the λ helper phage cloning system is that λ lysates (unlike Mu lysates) are very stable when stored at 4°. Although λ has a more limited host range than Mu, species of the family Enterobacteriaceae that can be infected by λ (De Vries *et al.*, 1984; Ludwig, 1987) and even support both its lytic and lysogenic growth have been described (Harkki and Palva, 1984; Wehmeier *et al.*, 1989). Another cloning procedure involves the use of mini-Mu elements which can "sandwich" a chromosomal locus and "jump" onto a cosmid, which is then packaged *in vivo* by a helper λ phage (de Mendoza *et al.*, 1986). The resulting clones arise by what has been called Mini-Muduction, and their cointegrate structure is potentially unstable in recombination-proficient cells (see below).

R-Prime Generation and Conjugal Transfer. F-prime (F') and R-prime (R') plasmids can be formed in a strain harboring Mu or mini-Mu derivatives in the chromosome or as part of self-transmissible plasmids (Faelen and Toussaint, 1976). The procedure involves partial induction for Mu transposition in the donor strain and conjugation with a Mu lysogenic recipient cell where selection for the desired trait is carried out (recipients need not be Mu immune when using MuA^+B^- derivatives). The clones obtained vary in size but are usually large (because there is no size constraint imposed by encapsidation into phage particles) and harbor two copies of the Mu derivative in direct relative orientation flanking both the transmissible plasmid and the cloned DNA segment. In *E. coli*, R' were formed with RP4::mini-Mu at frequencies of 10^{-5}, and recombinants were recovered at frequencies of 10^{-4} (van Gijsegem and Toussaint, 1982). A potential drawback of this approach is that in *recA*$^+$ recipient cells the cloned DNA can be lost if recombination between the directly repeated Mu derivatives takes place. This can be prevented by continuous selection for the selected gene or by using Mu derivatives that harbor an origin of replication (Table II; Groisman and Casadaban, 1987a). Plasmid pULB113 (RP4::Mu3A) has been used to generate R' plasmids in a variety of organisms including *S. typhimurium*, *Proteus mirabilis*, *K. pneumoniae*, *Erwinia*, and *Enterobacter cloacae* (van Gijsegem and Toussaint, 1982,

1983). When *E. coli* was used as recipient in heterospecific matings, R' plasmids were recovered at frequencies ranging from 10^{-6} to 10^{-7}. The dosage of the cloned gene is determined by the replicon present in the plasmid harboring the Mu element; derivatives of different copy number are available (Table I).

Mud–P22 Hybrids. Mud–P22 hybrids contain a segment of approximately two-thirds of the *S. typhimurium* phage P22 genome (including the *pac* site) between the ends of Mu (Youderian et al., 1988). They can transpose when complemented for Mu transposition or can recombine with residing Mu derivatives in the host chromosome. Induction of a Mud–P22 prophage gives rise to *in situ* replication under P22 control (without transposition). It is followed by packaging into P22 heads, which occurs by a headful mechanism, starting at the *pac* site to include approximately 2–3 min of adjacent DNA. Mud–P22 hybrids come in two forms with the P22 DNA fragment in both possible orientations with respect to the Mu ends, allowing the isolation of specific chromosomal fragments 5' and 3' to the site of insertion.

placMu Phage. λ p*lac*Mu phage are plaque-forming derivatives of phage λ that are deleted for their attachment sites but can integrate using the Mu transposition machinery (Bremer et al., 1984, 1985). They have been primarily used for the isolation of mutants and gene fusions. These λ–Mu hybrids can integrate in an essentially random fashion and in this way bypass the limitations imposed by previously described cloning systems, where the gene of interest needed to be placed near the attachment site for a lambdoid phage. They contain the Mu attachment sites and a truncated Mu*A* gene that is altered at its carboxy-terminal end but can still catalyze the transposition of the phage (Bremer et al., 1988). To achieve high transposition levels during the isolation of mutants, coinfection is performed with a specialized helper λ phage carrying both Mu*A* and *B* genes. Induction of a λ p*lac*Mu lysogen results in its excision by an illegitimate event and the generation of a specialized transducing phage carrying DNA adjacent to either or both sides of the insertion. The λ p*lac*Mu phage can be used in the cloning of *lac* gene fusions or intact adjacent genes (Bremer et al., 1984; Trun and Silhavy, 1987). Derivatives of some of these phage with amber mutations in Mu*A* have been constructed but were found not to be efficiently suppressed by *supF* (Bremer et al., 1988).

Mapping

Formation of Transient Hfr and Conjugal Transfer of Large DNA Segments. Conjugation is an efficient way of transferring DNA between strains and has been widely used in the mapping and cloning of genes from different gram-negative bacteria. Polarized transfer can be achieved in $recA^+$ donor cells harboring two Mu prophages: one in the chromosome and the other in a self-transmissible plasmid (Zeldis et al., 1973). Homologous recombination between the Mu prophage will give rise to an Hfr strain whose orientation, and hence polarity of transfer, will be determined by the orientation of the Mu prophage. If the chromosomal Mu forms a lac gene fusion with the gene of interest, the transcriptional orientation of the latter can be determined.

Additional mapping strategies involve nonoriented transfer of chromosomal markers and mapping by cotransposition (Faelen and Toussaint, 1976). A genetic map of Erwinia carotovora was constructed by generating transient Hfr strains with strains containing plasmid pULB113 (RP4::Mu3A) and measuring the frequency of cotransfer of two markers, relative to the transfer of each marker (Schoonejans and Toussaint, 1983). One can also estimate genetic linkage by measuring the frequency with which two markers cotranspose onto a self-transmissible plasmid to form a cointegrate. The recovered plasmid can be digested with restriction endonucleases to obtain an accurate figure of the distance between the markers. This approach allows mapping of markers separated by more than the 2 min allowed by phage P1 transductional crosses.

Use of Mud–P22 Hybrids in Salmonella typhimurium. Mu insertions can be mapped by converting the resident Mu into the two possible Mud–P22 hybrids (see above). A lysate is prepared by induction of the prophage in the newly constructed strains and used to infect a collection of approximately 16 strains with evenly spaced markers (Youderian et al., 1988). The map position of the original insertion can be established by comparing the complementation frequencies for the different markers.

Generalized Transduction and Mini-Muduction

A lysate prepared from a Mucts lysogen transduces chromosomal markers at frequencies of 10^{-7} to 10^{-8} per plaque-forming unit (pfu) (Howe, 1973b) and 10 times higher for markers present in the 0–2 min region of the E. coli genome (Bade et al., 1978). The frequency of generalized transduction is increased about 10 times for lysates prepared from strains doubly lysogenic for a Mu and a mini-Mu element (Faelen et al., 1979), and it is slightly higher if mini-Mu replicons are used (Wang et al., 1987a). The higher frequency obtained with double lysogens is probably due to the encapsidation of chromosomal DNA present adjacent to the mini-Mu element (see Fig. 1).

Lysates prepared from Mu/mini-Mu double lysogens can also give rise to transductants in recombination-deficient recipients in a process called mini-Muduction. The transductants obtained harbor the selected marker between two similarly oriented mini-Mu elements inserted in a random location in the genome of the recipient. The transductants obtained are not stable in Rec$^+$ hosts since recombination can take place between the directly repeated mini-Mu elements. Mini-Muduction allows the formation of merodiploids for complementation analysis and the transfer of DNA between different bacterial species because the process does not require homologous recombination. The frequency of this type of transduction event is increased about 100 times if the helper Mu phage is *kil*$^-$ or if the mixed lysate is irradiated with UV light (Cronan, 1984). Mini-Mu replicons do not give rise to mini-Muductants because the insertion of high copy number replicons in the chromosome is lethal unless particular genetic backgrounds (such as *polA*$^-$ recipients or cells harboring a plasmid of the same incompatibility group as the one present in the mini-Mu replicon) are used (Groisman, 1986).

Cloning versus Generalized Transduction

Infection of a recipient cell with a phage lysate prepared from a strain doubly lysogenic for a mini-Mu replicon and a helper Mu yields different products depending on the genotype of the recipient cell and the type of genetic selection applied (see Fig. 2). Infection of a *recA*$^+$ Mu lysogenic recipient with selection for a chromosomal marker will primarily give rise to generalized transductants; selection for both the chromosomal and mini-Mu replicon markers yields plasmid clones (Wang *et al.*, 1987a). If there is low homology between donor and recipient DNA (such as with lysates prepared in a species different from the recipient cell, e.g., *S. typhimurium* into *E. coli*), selection for the chromosomal marker yields almost exclusively plasmid clones. The ability of mini-Mu replicon elements to act for both *in vivo* cloning and generalized transduction can then be used to identify bacterial genes which are lethal when cloned in multicopy plasmids (Berg *et al.*, 1988).

Other Uses

Generating Deletions and Duplications

Deletions and duplications are very useful tools in genetic analysis. Deletions facilitate the genetic mapping of genes, and duplications are useful in complementation and dominance studies. Deletions and duplications can be easily isolated in *S. typhimurium* owing to the observation that P22-mediated transduction of a Mud prophage by homologous recombination requires two simultaneously transduced segments (Hughes and

Roth, 1985). Probably as a consequence of the large size of the Mud phage (37.5 kb) relative to the encapsidation capacity of P22 (44 kb), inheritance of such Mud prophage involves recombination between overlapping Mud sequences carried by two different particles as well as recombination of the flanking chromosomal DNA to enable integration of the Mud phage into the recipient chromosome. The technique to isolate deletions and duplications involves the mixing of lysates grown in two different Mud insertions and selection for the Mud marker. Four types of recombinants will be recovered with equal frequency: each of the two parental types, deletion, and duplication (Hughes and Roth, 1985). This technique, which has allowed the isolation of duplications as large as 3 min, can also be used in the determination of the transcriptional orientation of a gene if a reference Mud insertion near the one of interest is available (Hughes and Roth, 1985). It is useful in gene regulation studies as well because one can have both a wild-týpe and mutant (with a *lac* gene fusion) copy of the gene integrated in the chromosome. In *E. coli,* a related method (albeit more time consuming) involving Mu–Mud double lysogens allows the isolation of deletions and duplications along with determination of transcriptional orientation (MacNeil, 1981).

Transposition of certain mini-Mu elements (such as MudI1734, see Table I) into the chromosome of recombination-defective *S. typhimurium* occurs at a low frequency, and over 50% of the recovered lysogens harbor deletions at the site of insertion of the mini-Mu (Sonti *et al.,* 1991). It has been proposed that this is the result of the inability of Rec$^-$ strains to repair the breaks in the host chromosome generated by the replicative transposition of the mini-Mu element. MudI1734 shows this behavior regardless of the phage that is used for its delivery, Mu or P22; the MudI phage, however, only shows the Rec$^-$ effect when injected by P22 virions. The generated deletions can be easily transferred into different genetic backgrounds by selecting for the mini-Mu marker (Sonti *et al.,* 1991).

DNA Sequencing with Mini-Mu Elements

We and others have used mini-Mu derivatives to facilitate chain-termination DNA sequencing strategies (Sanger *et al.,* 1977). One can effectively saturate a plasmid clone with mini-Mu insertions and then use the ends of the element as portable sequences for the priming of DNA synthesis on a double-stranded DNA template. We have sequenced the *S. typhimurium phoP* (Groisman *et al.,* 1989) and *phoN* (Groisman *et al.,* 1991b) genes using this approach, obtaining over 600 bp of DNA sequence information for each insertion (300 bp from each end). A different strategy involves the isolation of mini-Mu insertions in a M13 hybrid plasmid which are then packaged as single strands. By using a mixture of primers, one for each of the two Mu ends, the need to first determine the orientation of the insert is avoided (Adachi *et al.*, 1987).

Other Uses of Mu Elements

Mini-Mu derivatives harboring *lac* gene fusing segments have also been used in *E. coli* to study colony development and bacterial differentiation (Shapiro, 1984; Shapiro and Higgins, 1989). Mini-Mu elements harboring multicopy (ColE1-type) replicons may also be used for chromosomal mutagenesis by infecting a recipient cell harboring a multicopy plasmid of the same incompatibility group as the one present in the mini-Mu replicon element (Groisman, 1986).

Other Useful Mu Derivatives

Knowledge of the sites and proteins required for transposition and the availability of DNA sequence information for over 16 kb of Mu DNA allow the design of elements tailored to the needs of the investigator. There are a number of mini-Mu derivatives harboring selectable markers for different species that have been used to construct defined mutations in different organisms by reverse genetics or "shuttle mutagenesis." They include derivatives with the erythromycin resistance gene for selection in gram-positive bacteria (Kuramitsu, 1987), the thymidine kinase gene from herpes simplex virus for selection in animal cells (Jenkins *et al.*, 1985), and the *Saccharomyces cerevisiae LEU2* gene for selection in yeast (Daignan-Fornier and Bolotin-Fukuhara, 1988; E. A. Groisman and F. Heffron, unpublished data, 1988).

Mu-Like Phage in Other Species

Several Mu-like phage have been described in gram-negative bacteria (Dubow, 1987). They may be useful genetic tools in species where Mu transposition is inefficient, or because they differ from Mu in their host range, target specificity, and the immunity phenomenon. Mutator phage have been described in *E. coli* (D108 and B278; Grinberg *et al.*, 1988), *Vibrio cholerae* (VcA1; Gerdes and Romig, 1975), and *Pseudomonas aeruginosa* (D3112). Derivatives of the *Pseudomonas* phage have been developed for genetic analysis (Darzins and Casadaban, 1989a) and *in vivo* cloning (Darzins and Casadaban, 1989b). Drug-resistant mini-D108 are also available (Résibois *et al.*, 1981).

Protocols

Preparation of Mu Lysates

Lysates from Mucts Single Lysogens or Mini-Mu/Mucts Double Lysogens

1. A single colony corresponding to a Mucts lysogen is used to inoculate 2 ml of LB (Luria broth) and incubated overnight at 32°.
2. The overnight culture is diluted 1 : 100 in 15 ml of LB in a 250-ml Erlenmeyer flask and incubated at 32° with shaking until early exponential phase (approximately OD_{600} 0.5).
3. The culture is shifted to a 42° shaking water bath for 20 min and then placed in a 37° shaking water bath and incubated until visible lysis is observed (approximately an additional 25 min).
4. The lysate is transferred to a 45-ml Sorvall tube. Chloroform (1% of the volume), $MgSO_4$ to a final concentration of 2 mM (30 μl of a 1 M solution), and $CaCl_2$ to a final concentration of 0.2 mM (3 μl of a 1 M solution) are added to the lysate, and the tube is vortexed for 10 sec.
5. Cell debris is removed by centrifugation at 8000 rpm for 10 min in the SS-34 rotor, and the supernatant is transferred to a clean sterile tube.
6. The centrifugation step is repeated once again, and the supernatant is stored at 4°.

Comments. Typical phage titers are around 5×10^9 pfu/ml. The titer is affected by the way the lysate is prepared: high cell density and poor aeration usually result in lower titers. Mu lysates are unstable (owing in part to the blockage of the tail fibers by cell debris) and should be used within 1–2 weeks of preparation.

Lysates from Mini-Mu/Mucts Double Lysogens Harboring Plasmids: In Vivo Cloning and Plasmid Insertion Mutagenesis

1. A single colony of a mini-Mu/Mucts double lysogen harboring the plasmid is used to inoculate 2 ml of LB containing the antibiotic to which the plasmid confers resistance, and the culture is incubated overnight at 32°.
2. The overnight culture is diluted 1 : 100 in 15 ml of LB (no antibiotic) in a 250-ml Erlenmeyer flask and incubated at 32° with shaking until early exponential phase (approximately OD_{600} 0.5).
3. The culture is shifted to a 44° shaking water bath and incubated until visible lysis is observed, or up to 150 min.

4. The lysate is transferred to a 45-ml Sorvall tube. Chloroform (1% of the volume), $MgSO_4$ to a final concentration of 2 mM (30 μl of a 1 M solution), and $CaCl_2$ to a final concentration of 0.2 mM (3 μl of a 1 M solution) are added to the lysate, and the tube is vortexed for 10 sec.

5. Cell debris is removed by centrifugation at 8000 rpm for 10 min in the SS-34 rotor, and the supernatant is transferred to a clean sterile tube.

6. The centrifugation step is repeated once again, and the supernatant is stored at 4°.

Comments. To mutagenize plasmid-borne genes, a monomer form of the plasmid should be used to transform the mini-Mu/Mu*c*ts double lysogenic strain. The presence of mini-Mu elements in multicopy plasmids (i.e., mini-Mu replicons) seems to inhibit cell lysis (no clearing is observed on thermoinduction of the culture). We have found that for plasmid-containing strains induction and incubation at 44° (rather than 42° and shift back to 37°) yield higher phage titers.

Infection of Mu-Sensitive Cells

Making Mucts Lysogens

1. A single colony of cells to be made Mu lysogenic is used to inoculate 2 ml of LB and incubated at 37° overnight.

2. The next day, a soft agar lawn is prepared by mixing 0.3 ml of the overnight culture with 2.7 ml of LB soft agar (0.6%) at 45° and pouring on top of an LB agar plate containing 1 mM $MgSO_4$. A drop of a fresh Mu*c*ts lysate is spotted onto the soft agar lawn, and the plate is incubated overnight at 32°.

3. Cells are recovered from the clearing zone with a sterile toothpick, streaked to single colonies on a fresh LB agar plate, and incubated overnight at 32°.

4. The next day, individual colonies are patched first onto an LB agar plate (which is incubated overnight at 32°) and then cross-streaked against a Mu-sensitive cell which had been previously streaked onto an LB agar plate containing 1 mM $MgSO_4$. After 6–8 hr of incubation at 42°, results are scored: Mu lysogens are thermosensitive and able to produce phage. The Mu lysogens are then saved from the plate incubated at 32°.

Comments. Another test to confirm the Mu lysogenic state of the newly constructed strains is to prepare a soft agar lawn with an overnight culture of the lysogen (grown at 32°), spot a drop of a Mu*c*ts lysate, and incubate the plate overnight at 32°. A Mu lysogen would show no clearing, indicating its immune state. Different lysogens are likely to harbor the Mu element in a different position in the genome, and it is advisable to save more than one lysogen for any given strain. Lysogens of Mucts phages harboring drug resistance genes are prepared as described above.

Chromosomal Mutagenesis and Isolation of Gene Fusions

1. A single colony of cells to be mutagenized is used to inoculate 2 ml of LB and incubated at 37° overnight.
2. The overnight culture (0.1 ml) is mixed with 0.1 ml of lysate (see Preparation of Mu Lysates) with a multiplicity of infection (moi) of 0.1 to 1 and incubated at 32° for 30 min with no shaking (to allow phage adsorption).
3. Then, 2 ml of LB is added to the tube, and incubation is carried out for 75 min at 32° with shaking (to allow for expression of the drug resistance marker in the Mu derivative).
4. Cells are plated in the appropriate medium containing the corresponding antibiotic and incubated overnight at 32°.

Comments. We routinely introduce the divalent cations Mg^{2+} and Ca^{2+} into the lysates and in this way avoid growing the recipient cells in their presence. For transposition-defective Mu derivatives (which are being complemented for both Mu*A* and *B* genes), incubations may be carried out at 37° (instead of 32°) and for 60 min in Step 3 (rather than 75 min).

Plasmid Insertion Mutagenesis

1. A single colony of cells in which the mutagenized plasmid is to be recovered is used to inoculate 2 ml of LB and incubated at 32° overnight.
2. The overnight culture (0.1 ml) is mixed with 0.1 ml of lysate (moi 0.1 to 1) and incubated at 32° for 30 min with no shaking (to allow phage adsorption).
3. Then, 2 ml of LB is added to the tube, and incubation is carried out for 75 min at 32° with shaking (to allow expression of the drug resistance markers present in both the plasmid and the mini-Mu).
4. Cells are plated in the appropriate medium containing the antibiotics to select for the plasmid and mini-Mu markers and incubated overnight at 32°.

Comments. We routinely use strain M8820Mucts (Castilho *et al.*, 1984) because it is a good Mu recipient and gives high yields of plasmid DNA by the boiling method (Holmes and Quigley, 1981). Other (even non-*E. coli* K12) strains where the phenotype of the mutagenized plasmid can be assessed directly may also be used. The comments regarding transposition-defective mini-Mu derivatives are also applicable here.

In Vivo Cloning with Mini-Mu Replicons

1. A single colony of recipient cells is used to inoculate 2 ml of LB and incubated at 32° overnight.
2. The overnight culture (0.1 ml) is mixed with 0.1 ml of lysate (moi 0.1 to 1) and incubated at 32° for 30 min with no shaking (to allow phage adsorption).
3. Then, 2 ml of LB is added to the tube and incubation is carried out for 75 min at 32° with shaking (to allow for expression of the drug resistance marker in the mini-Mu replicon).
4. Cells are plated in the appropriate medium containing the corresponding antibiotic and incubated overnight at 32°.

Comments. Genes to be cloned can be selected directly in Step 4 or by replica plating onto selective medium from a library of drug-resistant transductants isolated in Step 4. Particular genes are isolated at frequencies of less than 10^{-2} to greater than 10^{-4} per drug-resistant transductant. Although up to 35% of the drug-resistant transductants carry the original mini-Mu replicon plasmid, complete libraries can be prepared with as little as 1 μl of lysate. Nevertheless, scale up may be necessary when preparing libraries from strains where Mu transposition is not very efficient and/or when the recipient cells are heavily restricting the incoming DNA.

Miscellaneous Techniques Involving Mu Derivatives

Conjugation. Transfer of broad host range plasmids harboring Mu elements (see Table II) into bacteria may be performed by patching onto an LB agar plate colonies from fresh overnight streaks corresponding to both recipient and donor cells (donors for triparental matings). Incubate at 32° for 1–2 hr. Pick cells from the patch and streak out to single colonies in selective medium. This procedure works fine for both RK2 (RP4) and R388 types of plasmids, both of which mate better in solid than in liquid medium. This procedure is satisfactory for transferring a plasmid, but it is not suggested for mobilizing whole libraries (mating on filters overnight is recommended).

Transformation. Mu*c*ts lysogens of enteric bacteria can be efficiently transformed by performing the incubations and heat shock at 32°, rather than at 37° and 42°. Transformation of non-Mu lysogens with plasmid DNA harboring transposition-competent mini-Mu derivatives (e.g., mini-Mu replicons) occurs at frequencies over 300 times lower than into the isogenic Mu*c*ts lysogens (Groisman and Casadaban, 1987b).

Cross-Streaking. The Mu lysogenic character of a strain and its ability to produce phage can be easily tested as follows. A streak of a Mu-sensitive strain is carried out with a sterile toothpick along the length of an LB agar plate containing 1 mM MgSO$_4$. Individual candidates for Mu$c$ts lysogens are tested by picking a colony with a toothpick, touching an LB agar plate, and then making a streak against the Mu-sensitive strain (in the LB MgSO$_4$ agar plate). The first plate is incubated overnight at 32° and the one with the cross-streaks at 42° for 6–8 hr. Mu$c$ts lysogens show no growth at the beginning of the streak or at the intersection with the Mu-sensitive cells as a result of cell death owing to induction of the Mu$c$ts phage, which kills the susceptible cells present at the intersection. The growth observed after the cross-streaks is due to carryover of the Mu-sensitive cells. One can usually test over 30 colonies per plate, which should include both a positive and negative control.

Acknowledgments

I thank C. M. Berg and R. Harshey for valuable comments on the manuscript. My research has been aided in part by The Jane Coffin Childs Fund for Medical Research and by grants from the National Institutes of Health to Malcolm Casadaban, Fred Heffron, and Milton Saier.

References

Adachi, T., Mizuuchi, M., Robinson, E. A., Appella, E., O'Dea, M. H., Gellert, M., and Mizuuchi, K. (1987). *Nucleic Acids Res.* **15,** 771.
Adzuma, K., and Mizuuchi, K. (1988). *Cell (Cambridge, Mass.)* **53,** 257.
Bachhuber, M., Brill, W. J., and Howe, M. M. (1976). *J. Bacteriol.* **128,** 749.
Bade, E. G., Howe, M. M., and Rawluk, L. (1978). *Mol. Gen. Genet.* **160,** 89.
Baker, T. A., Howe, M. M., and Gross, C. A. (1983). *J. Bacteriol.* **156,** 970.
Belas, R., Mileham, A., Simon, M., and Silverman, M. (1984). *J. Bacteriol.* **158,** 890.
Berg, C. M., Liu, L., Wang, B., and Wang, M.-D. (1988). *J. Bacteriol.* **170,** 468.
Berg, C. M., Berg, D. E., and Groisman, E. A. (1989). *in* "Mobile DNA" (D. E. Berg and M. M. Howe, eds.), 879. American Society for Micribiology, Washington, D.C.
Bremer, E., Silhavy, T. J., Weisemann, J. M., and Weinstock, G. M. (1984). *J. Bacteriol.* **158,** 1084.
Bremer, E., Silhavy, T. J., and Weinstock, G. M. (1985). *J. Bacteriol.* **162,** 1092.
Bremer, E., Silhavy, T. J., and Weinstock, G. M. (1988). *Gene* **71,** 177.
Buchanan-Wollaston, V., Passiatore, J. E., and Cannon, F. (1987). *Nature (London)* **328,** 172.
Bukhari, A. I., and Allet, B. (1975). *Virology* **63,** 30.
Casadaban, M. J., and Chou, J. (1984). *Proc. Natl. Acad. Sci. U.S.A.* **81,** 535.
Casadaban, M. J., and Cohen, S. N. (1979). *Proc. Natl. Acad. Sci. U.S.A.* **76,** 4530.
Castilho, B. A., Olfson, P., and Casadaban, M. J. (1984). *J. Bacteriol.* **158,** 488.
Chaconas, G. (1987). *in* "Phage Mu" (N. Symonds, A. Toussaint, P. van de Putte, and

M. M. Howe, eds.), 137. Cold Spring Harbor Laboratory, Cold Spring Harbor, New York.

Chaconas, G., de Bruijn, F. J., Casadaban, M., Lupski, J. R., Kwoh, T. J., Harshey, R. M., Dubow, M. S., and Bukhari, A. I. (1981). *Gene* **13**, 37.

Cronan, J. E., Jr. (1984). *J. Bacteriol.* **158**, 357.

Csonka, L. N., Howe, M. M., Ingraham, J. L., Pierson III, L. S., and Turnbough, C. L., Jr. (1981). *J. Bacteriol.* **145**, 299.

Daignan-Fornier, B., and Bolotin-Fukuhara, M. (1988). *Gene* **62**, 45.

Darzins, A., and Casadaban, M. J. (1989a). *J. Bacteriol.* **171**, 3909.

Darzins, A., and Casadaban, M. J. (1989b). *J. Bacteriol.* **171**, 3917.

Darzins, A., Kent, N., Buckwalter, M., and Casadaban, M. (1988). *Proc. Natl. Acad. Sci. U.S.A.* **85**, 6826.

de Mendoza, D., and Rosa, A. L. (1985). *Gene* **39**, 55.

de Mendoza, D., Gramajo, H. C., and Rosa, A. L. (1986). *Mol. Gen. Genet.* **205**, 546.

De Vries, G. E., Raymond, C. K., and Ludwig, R. A. (1984). *Proc. Natl. Acad. Sci. U.S.A.* **81**, 6080.

Dubow, M. S. (1987). *in* "Phage Mu" (N. Symonds, A. Toussaint, P. van de Putte, and M. M. Howe, eds.), p. 201. Cold Spring Harbor Laboratory, Cold Spring Harbor, New York.

Engebrecht, J., Simon, M., and Silverman, M. (1985). *Science* **227**, 1345.

Faelen, M. (1987). *in* "Phage Mu" (N. Symonds, A. Toussaint, P. van de Putte, and M. M. Howe, eds.), p. 309. Cold Spring Harbor Laboratory, Cold Spring Harbor, New York.

Faelen, M., and Toussaint, A. (1976). *J. Mol. Biol.* **104**, 525.

Faelen, M., Toussaint, A., and Résibois, A. (1979). *Mol. Gen. Genet.* **176**, 191.

Faelen, M., Mergeay, M., Geritis, J., Toussaint, A., and LeFebvre, N. (1981). *J. Bacteriol.* **146**, 914.

Gerdes, J. C., and Romig, W. R. (1975). *J. Virol.* **15**, 1231.

Goguen, J. D., Yother, J., and Straley, S. C. (1984). *J. Bacteriol.* **160**, 842.

Goldberg, R. B., Bender, R. A., and Streicher, S. L. (1974). *J. Bacteriol.* **118**, 810.

Goosen, N., and van de Putte, P. (1987). *in* "Phage Mu" (N. Symonds, A. Toussaint, P. van de Putte, and M. M. Howe, eds.), p. 41. Cold Spring Harbor Laboratory, Cold Spring Harbor, New York.

Gramajo, H. C., and de Mendoza, D. (1987). *Gene* **51**, 85.

Gramajo, H. C., Viale, A. M., and de Mendoza, D. (1988). *Gene* **65**, 305.

Grinberg, D. R., Boronat, A., and Guinea, J. (1988). *J. Gen. Microbiol.* **134**, 1333.

Groisman, E. A. (1986). Ph.D. Thesis, University of Chicago, Chicago, Illinois.

Groisman, E. A., and Casadaban, M. J. (1986). *J. Bacteriol.* **168**, 357.

Groisman, E. A., and Casadaban, M. J. (1987a). *J. Bacteriol.* **169**, 687.

Groisman, E. A., and Casadaban, M. J. (1987b). *Gene* **51**, 77.

Groisman, E. A., Castilho, B. A., and Casadaban, M. J. (1984). *Proc. Natl. Acad. Sci. U.S.A.* **81**, 1480.

Groisman, E. A., Chiao, E., Lipps, C. J., and Heffron, F. (1989). *Proc. Natl. Acad. Sci. U.S.A.* **86**, 7077.

Groisman, E. A., Pagratis, N., and Casadaban, M. J. (1991a). *Gene* **99**, 1.

Groisman, E. A., Saier, M. H., Jr., and Ochman, M. (1991b). Submitted.

Harkki, A., and Palva, E. T. (1984). *Mol. Gen. Genet.* **195**, 256.

Harshey, R. M. (1988). *in* "The Bacteriophages" (R. Calendar, ed.), Vol. 1, p. 193. Plenum, New York.

Harshey, R., and Cuneo, S. (1986). *J. Genet.* **65**, 159.

Heineman, J. A., and Sprague, G. F., Jr. (1989). *Nature (London)* **340**, 205.

Holley, E. A., and Foster, J. W. (1982). *J. Bacteriol.* **152**, 959.

Holmes, D. S., and Quigley, M. (1981). *Anal. Biochem.* **114**, 193.

Howe, M. M. (1973a). *Virology* **54**, 93.

Howe, M. M. (1973b). *Virology* **55**, 103.

Howe, M. M. (1987a). *in* "Phage Mu" (N. Symonds, A. Toussaint, P. van de Putte, and M. M. Howe, eds.), p. 63. Cold Spring Harbor Laboratory, Cold Spring Harbor, New York.

Howe, M. M. (1987b). *in* "Phage Mu" (N. Symonds, A. Toussaint, P. van de Putte, and M. M. Howe, eds.), p. 271. Cold Spring Harbor Laboratory, Cold Spring Harbor, New York.

Hughes, K. T., and Roth, J. R. (1984). *J. Bacteriol.* **159**, 130.

Hughes, K. T., and Roth, J. R. (1985). *Genetics* **109**, 263.

Hughes, K. T., and Roth, J. R. (1988). *Genetics* **119**, 9.

Hughes, K. T., Olivera, B. M., and Roth, J. R. (1987). *J. Bacteriol.* **169**, 403.

Huisman, O., Faelen, M., Girard, D., Jaffé, A., Toussaint, A., and Rouviere-Yaniv, J. (1989). *J. Bacteriol.* **171**, 1541.

Jenkins, F. J., Casadaban, M. J., and Roizman, B. (1985). *Proc. Natl. Acad. Sci. U.S.A.* **82**, 4773.

Kahmann, R., and Hattman, S. (1987). *in* "Phage Mu" (N. Symonds, A. Toussaint, P. van de Putte, and M. M. Howe, eds.), p. 93. Cold Spring Harbor Laboratory, Cold Spring Harbor, New York.

Kahmann, R., and Kamp, D. (1987). *in* "Phage Mu" (N. Symonds, A. Toussaint, P. van de Putte, and M. M. Howe, eds.), p. 297. Cold Spring Harbor Laboratory, Cold Spring Harbor, New York.

Koch, C., Mertens, G., Rudt, F., Kahmann, R., Kanaar, R., Plasterk, R., van de Putte, P., Sandulache, R., and Kamp, D. (1987). *in* "Phage Mu" (N. Symonds, A. Toussaint, P. van de Putte, and M. M. Howe, eds.), p. 75. Cold Spring Harbor Laboratory, Cold Spring Harbor, New York.

Komeda, Y., and Iino, T. (1979). *J. Bacteriol.* **139**, 721.

Kuramitsu, H. K. (1987). *Mol. Microbiol.* **1**, 229.

Lång, H., Teeri, T., Kurkela, S., Bremer, E., and Palva, E. T. (1987). *FEMS Microbiol. Lett.* **48**, 305.

Leach, D., and Symonds, N. (1979). *Mol. Gen. Genet.* **172**, 179.

Leung, P. C. Teplow, D. B., and Harshey, R. M. (1989). *Nature (London)* **338**, 656.

Ludwig, R. A. (1987). *Proc. Natl. Acad. Sci. U.S.A.* **84**, 3334.

MacNeil, D. (1981). *J. Bacteriol.* **146**, 260.

Metcalf, W. W., Steed, P. M., and Wanner, B. L. (1990). *J. Bacteriol.* **172**, 3191.

Mintz, C. S., and Shuman, H. A. (1987). *Proc. Natl. Acad. Sci. U.S.A.* **84**, 4645.

Mizuuchi, K. (1983). *Cell (Cambridge, Mass.)* **35**, 785.

Mizuuchi, K., and Higgins, N. P. (1987). *in* "Phage Mu" (N. Symonds, A. Toussaint, P. van de Putte, and M. M. Howe, eds.), p. 159. Cold Spring Harbor Laboratory, Cold Spring Harbor, New York.

Mizuuchi, M., and Mizuuchi, K. (1989). *Cell (Cambridge, Mass.)* **58**, 399.

Muller, K. H., Trust, T. J., and Kay, W. W. (1988). *J. Bacteriol.* **170**, 1076.

Parsons, R. L., and Harshey, R. (1989). *Nucleic Acids. Res.* **16**, 11285.

Pato, M. L. (1989). *in* "Mobile DNA" (D. E. Berg and M. M. Howe, eds.), p. 23. American Society for Microbiology, Washington, D.C.

Priess, H., Schmidt, C., and Kamp, D. (1987). *in* "Phage Mu" (N. Symonds, A. Toussaint, P. van de Putte, and M. M. Howe, eds.), p. 277. Cold Spring Harbor Laboratory, Cold Spring Harbor, New York.

Ratet, P., and Richaud, F. (1986). *Gene* **42**, 185.
Ratet, P., Schell, J., and de Bruijn, F. J. (1988). *Gene* **63**, 41.
Résibois, A., Toussaint, A., van Gijsegem, F., and Faelen, M. (1981). *Gene* **14**, 103.
Reyes, O., Beyou, A., Mignotte-Vieux, C., and Richaud, F. (1987). *Plasmid* **18**, 183.
Rosenfeld, S. A., and Brenchley, J. E. (1980). *J. Bacteriol.* **144**, 848.
Sandulache, R., Prehm, P., and Kamp, D. (1984). *J. Bacteriol.* **160**, 299.
Sandulache, R., Prehm, P., Expert, D., Toussaint, A., and Kamp, D. (1985). *FEMS Microbiol. Lett.* **28**, 307.
Sanger, F., Nicklen, S., and Carlson, A. R. (1977). *Proc. Natl. Acad. Sci. U.S.A.* **74**, 5463.
Sauer, B., and Henderson, N. (1988). *Gene* **70**, 331.
Schoonejans, E., and Toussaint, A. (1983). *J. Bacteriol.* **154**, 1489.
Shapiro, J. A. (1979). *Proc. Natl. Acad. Sci. U.S.A.* **76**, 1933.
Shapiro, J. A. (1984). *J. Gen. Microbiol.* **130**, 1169.
Shapiro, J., and Higgins, N. P. (1989). *J. Bacteriol.* **171**, 5975.
Silhavy, T. J., and Beckwith, J. R. (1985). *Microbiol. Rev.* **49**, 398.
Silhavy, T. J., Berman, M. L., and Enquist, L. W. (1984). "Experiments with Gene Fusions." Cold Spring Harbor Laboratory. Cold Spring Harbor, New York.
Soberon, M., Gama, M. J., Richelle, J., and Martuscelli, J. (1986). *J. Gen. Microbiol.* **132**, 83.
Sonti, R., Keating, D., and Roth, J. R. (1991). Submitted.
Surette, M. G., Lavoie, B. D., and Chaconas, G. (1989). *EMBO J.* **8**, 3483.
Symonds, N., Toussaint, A., van de Putte, P., and Howe, M. M. (eds.) (1987). "Phage Mu." Cold Spring Harbor Laboratory, Cold Spring Harbor, New York.
Taylor, A. L. (1963). *Proc. Natl. Acad. Sci. U.S.A.* **50**, 1043.
Toussaint, A., Lefebvre, N., Scott, J., Cowan, J. A., de Bruijn, F., and Bukhari, A. I. (1978). *Virology* **89**, 146.
Trieu-Cuot, P., Carlier, C., Martin, P., and Courvalin, P. (1987). *FEMS Microbiol. Lett.* **48**, 289.
Trun, N. J., and Silhavy, T. J. (1987). *Genetics* **116**, 513.
van Gijsegem, F., and Toussaint, A. (1982). *Plasmid* **7**, 30.
van Gijsegem, F., and Toussaint, A. (1983). *J. Bacteriol.* **154**, 1227.
van Gijsegem, F., Toussaint, A., and Casadaban, M. (1987). *in* "Phage Mu" (N. Symonds, A. Toussaint, P. van de Putte, and M. M. Howe, eds.), p. 215. Cold Spring Harbor Laboratory, Cold Spring Harbor, New York.
Waddell, C. S., and Craig, N. L. (1988). *Genes Dev.* **2**, 137.
Wang, B., Liu, L., Groisman, E. A., Casadaban, M. J., and Berg, C. M. (1987a). *Genetics* **116**, 201.
Wang, B., Liu, L., and Berg, C. M. (1987b). *J. Bacteriol.* **169**, 4228.
Wehmeier, U., Sprenger, G. A., and Lengeler, J. W. (1989). *Mol. Gen. Genet.* **215**, 529.
Youderian, P., Sugiono, P., Brewer, K. L., Higgins, N. P., and Elliott, T. (1988). *Genetics* **118**, 581.
Zambryski, P., Tempe, J., and Schell, J. (1989). *Cell (Cambridge, Mass.)* **56**, 193.
Zeldis, J. B., Bukhari, A. I., and Zipser, D. (1973). *Virology* **55**, 289.

[9] Genetic Fusions as Experimental Tools

By JAMES M. SLAUCH and THOMAS J. SILHAVY

Introduction

Gene fusion technology has revolutionized bacterial genetics. In the past, investigators were limited by the phenotypes and biochemical assays associated with the system under study, and all too often they looked at advances made with *lac*, for example, and wished for equivalent if not similar methodologies. Fusions satisfy this desire because they permit the investigator to adapt a property of choice to the gene of interest.

With the harnessing of transposable genetic elements and the widespread use of recombinant DNA techniques, the available methods for constructing genetic fusions has increased in exponential fashion. It is no longer possible to list all available methods for fusion construction, even if such a compendium is confined to *Escherichia coli*. We have tried, instead, to compile methods for fusion construction that are of interest historically along with those, which seem to us, to be of particular advantage. In addition, we summarize successful strategies employed using fusion strains that we consider of broad general interest. No doubt, our summary is biased and we apologize in advance to our colleagues whose work we have slighted inadvertently.

Fusion Construction

Fusions are constructed by simply creating a novel DNA joint; sequences which were originally separate from one another are made contiguous, such that translational and/or transcriptional signals which affect one, affect the other. The creation of novel joints is most easily accomplished by one of three methods. A commonly used method for the creation of novel joints is recombinant DNA. Table I[1-29] lists a variety of

[1] M. J. Casadaban, J. Chou, and S. N. Cohen, *J. Bacteriol.* **143**, 971 (1980).
[2] T. J. Silhavy, M. Berman, and L. Enquist, "Experiments with Gene Fusions." Cold Spring Harbor Laboratory, Cold Spring Harbor, New York 1984.
[3] E. Wyckoff, L. Sampson, M. Hayden, R. Parr, W. M. Huang, and S. Casjens, *Gene* **43**, 281 (1986).
[4] N. P. Minton, *Gene* **31**, 269 (1984).
[5] V. DeLorenzo, M. Herrero, F. Giovannini, and J. B. Neilands, *Eur. J. Biochem.* **173**, 537 (1988).
[6] R. W. Simons, F. Houman, and N. Kleckner, *Gene* **53**, 85 (1987).
[7] V. DeLorenzo, M. Herrero, and J. B. Neilands, *FEMS Microbiol. Lett.* **50**, 17 (1988).

METHODS IN ENZYMOLOGY, VOL. 204

plasmid vectors for the creation of fusions. Most of the vectors described are designed for the creation of protein fusions, namely, fusions that result in the formation of a hybrid protein. For the majority of vectors, insertion of the target DNA into the cloning sites, such that the reading frame of the target gene is the same as the reporter gene, results in production of a hybrid protein; the NH_2 terminus is composed of the target protein, and the COOH terminus is the reporter gene. A few of the vectors listed are used for the creation of COOH-terminal fusions, in which the target protein composes the COOH-terminal portion of the hybrid molecule. These constructs are most often used for antibody production. The resulting hybrid is often produced in amounts high enough to form inclusion bodies, which are stable, easily isolated, and contain the hybrid protein in nearly pure form.

A second type of fusion, transcriptional or operon fusions, places an intact reporter gene downstream from the transcriptional start signals of

[8] K. McKenney, H. Shimatake, D. Court, U. Schmeissner, C. Brady, and M. Rosenberg, *in* "Gene Amplification and Analysis" (J. G. Chirikjian and T. S. Papas, eds.), p. 383. Elsevier/North-Holland, Amsterdam, 1981.

[9] C. Gutierrez and J. C. Devedjian, *Nucleic Acids Res.* **17**, 3999 (1989).

[10] J. K. Broome-Smith and B. G. Spratt, *Gene* **49**, 341 (1986).

[11] Y. Zhang and J. K. Broome-Smith, *Mol. Microbiol.* **3**, 1361 (1989).

[12] J. K. Sussman, C. Masada-Pepe, E. L. Simons, and R. W. Simons, *Gene* **90**, 135 (1990).

[13] K. Sieg, J. Kun, I. Pohl, A. Scherf, and B. Muller-Hill, *Gene* **75**, 261 (1989).

[14] J. Germino and D. Bastia, *Proc. Natl. Acad. Sci. U.S.A.* **81**, 4692 (1984).

[15] M. R. Gray, H. V. Colot, L. Guarente, and M. Rosbash, *Proc. Natl. Acad. Sci. U.S.A.* **79**, 6598 (1982).

[16] J. Sambrook, E. F. Fritsch, and T. Maniatis, "Molecular Cloning: A Laboratory Manual," 2nd Ed. Cold Spring Harbor Laboratory, Cold Spring Harbor, New York, 1989.

[17] M. R. Gray, G. P. Mazzara, P. Reddy, and M. Rosbash, this series, Vol. 154, p. 129.

[18] G. M. Weinstock, C. Ap Rhys, M. L. Berman, B. Hampar, D. Jackson, and T. J. Silhavy, *Proc. Natl. Acad. Sci. U.S.A.* **80**, 4432 (1983).

[19] G. M. Weinstock, this series, Vol. 154, p. 156.

[20] M. Koenen, A. Scherf, O. Mercereau, G. Langsley, L. Sibilli, P. Dubois, L. P. daSilva, and B. Muller-Hill, *Nature (London)* **311**, 382 (1984).

[21] U. Ruther and B. Muller-Hill, *EMBO J.* **2**, 1791 (1983).

[22] K. K. Stanley and J. P. Luzio, *EMBO J.* **3**, 1429 (1984).

[23] C. L. Dieckmann and A. Tzagoloff, *J. Biol. Chem.* **260**, 1513 (1985).

[24] E. Harlow and D. Lane, "Antibodies: A Laboratory Manual." Cold Spring Harbor Laboratory, Cold Spring Harbor, New York, 1988.

[25] R. A. Young and R. W. Davis, *Proc. Natl. Acad. Sci. U.S.A.* **80**, 1194 (1983).

[26] M. Snyder, S. Elledge, D. Sweetser, R. A. Young, and R. W. Davis, this series, Vol. 154, p. 107.

[27] K. S. Ostrow, T. J. Silhavy, and S. Garrett, *J. Bacteriol.* **168**, 1165 (1986).

[28] P. Gott, M. Ehrmann, and W. Boos, *Gene* **71**, 187 (1988).

[29] S. Knapp and J. J. Mekalanos, *J. Bacteriol.* **170**, 5059 (1988).

TABLE I

PLASMID VECTORS FOR CREATION OF FUSIONS

Vector	Replicon	Marker[a]	Fusion	Type	Sites[b]	Ref.	Comment
pMC1403	ColE1	Amp	LacZ	Protein	RI, S, B	1	Contains *lacZ* without transcriptional or translational start signals. Sequences are cloned into sites in NH$_2$ terminus of *lacZ* to create fusion. Natural *Eco*RI site near COOH terminus of *lacZ* has been removed. Fusions are LacY$^+$
pMLB1034	ColE1	Amp	LacZ	Protein	RI, S, B	2	Similar to pMC1403 except vector does not produce LacY
pTSV series	ColE1	Tet	LacZ	Protein	RI, (S, Xh), B	3	Derivative of pMC1403 containing *lacUV5* promoter but lacking translational start site for LacZ. Allows cloning of translational start signals without need to have adjacent promoter. Series includes cloning sites in all reading frames. Fusions are LacY$^+$
pNM480 series	ColE1	Amp	LacZ	Protein	RI, S, B, Sl, Ps, H	4	Derivative of pMC1403 containing multiple cloning site from pUC8. Replicon is pUC8 derived and has higher copy number than pMC1403. Series includes cloning sites in all reading frames. Fusions are LacY$^+$
pLC1	ColE1	Amp, Cm	LacZ	Protein	S, B	5	Derivative of pMLB1034 containing gene for chloramphenicol resistance and strong *rpoL* transcriptional terminator upstream of fusion cloning sites. Fusions are LacY$^-$
pRS414 series	ColE1	Amp	LacZ	Protein	RI, S, B	6	Series of vectors containing strong transcriptional terminators 5' to *lac* sequences, preventing
pRS415 series	ColE1	Amp	*lac*	Operon	RI, S, B	6	

(*continued*)

TABLE I (*continued*)

Vector	Replicon	Marker[a]	Fusion	Type	Sites[b]	Ref.	Comment
							transcription from upstream promoters; particularly important for operon fusion vectors. Derivatives are available which also contain Kan marker. Cloning sites in all vectors come in either orientation. Series of λ derivatives permit transfer of fusion, including terminator sequences, onto phage vectors by homologous recombination, allowing analysis of fusion in single copy. Plasmid pRS308 allows recovery of fusion from single-copy λ vectors, by homologous recombination. Lac fusions can be switched to Neo fusions using pRS1292 series described below. Fusions are LacY$^+$
pCON5	ColE1	Amp	LacZ	Protein	RI, S, B	7	Derivatives of pMLB1034 which contain M13 *ori* for direct production of single-stranded DNA for sequencing or site-directed mutagenesis. Fusions are LacY$^-$
pCON4	ColE1	Amp	*lac*	Operon	RI, S, B	7	
pKO	ColE1	Amp	*galK*	Operon	RI, H, S	8	Contains intact *galK* gene preceded by cloning sites and translational stop codons in all three reading frames. Used for cloning of promoter sequences. Derivative pKG1800 has *gal* promoter cloned into *Eco*RI and *Hind*III sites. Used for cloning of transcriptional termination signals. λ derivatives allow transfer of fusion onto phage by homologous recombination for analysis in single copy. Fusions can be recombined into normal *gal* locus from phage derivatives
pPHO7	ColE1	Amp	PhoA	Protein	H, Ps, Sl, X, B, S, K	9	PhoA lacks functional signal sequence. Activity is dependent on contribution from target gene of sequences capable of directing export. *phoA*

	Replicon	Selection	Fusion partner	Type	Site	Ref.	Comments
							sequence is bracketed by polylinkers, each containing the restriction sites listed. Cassette-carrying *phoA* can be cloned from pPHOA7 into target gene. *Hind*III site is unique at 5' end of *phoA*. Target gene can be cloned into this site
pJBS633	ColE1	Tet, Kan	Bla	Protein	P (see comment)	10	Vector carries mature portion of β-lactamase from pBR322 with *Pvu*II site that allows insert of foreign DNA. Other unique sites within or upstream of *tet* gene can be used in conjunction with *Pvu*II site for cloning. Use of some of these sites allows In-frame expression of fusion protein from *tet* promoter. In-frame fusions which direct export of the hybrid protein confer Amp^r to individual cells. In-frame fusions which do not export hybrid confer Amp^r when cells are patched at high density owing to lysis of some cells in population. Plasmid also contains origin of replication from phage f1 for production of single-stranded DNA for directly sequencing fusion joints
pYZ1	ColE1	Tet	Bla	Protein	RI	11	Vector carries mature portion of β-lactamase from pBR322 with *Eco*RI site that allows insertion of foreign DNA. In-frame fusions which direct export of the hybrid protein confer resistance to Amp to individual cells. In-frame fusions which do not export hybrid will confer resistance to Amp when cells are patched at high density owing to lysis of some cells in population. Plasmid also contains origin of replication from phage f1 for production of

(continued)

TABLE I (*continued*)

Vector	Replicon	Marker[a]	Fusion	Type	Sites[b]	Ref.	Comment
pYZ4/5	ColE1	Kan	Bla	Protein	See comment	11	single-stranded DNA for directly sequencing fusion joints. Designed for cloning β-lactamase fusions. pYZ4 has *lacUV5* promoter and complementing fragment of *lacZ* with multiple cloning sites. DNA inserts can be identified by loss of α complementation. Plasmid contains f1 origin for production of single-stranded DNA. pYZ5 contains mature portion of β-lactamase with *Pvu*II site on 5′ end and multiple cloning site on 3′ end. Mature β-lactamase cassette can be inserted into pYZ4 clone to form fusions
pRS1292 series	ColE1	Amp	Neo	Protein	RI, S, B, N	12	Analogous to the pRS415 series.[6] Vectors contain transcriptional termination sequences upstream of multiple cloning sites and *neo* gene lacking transcriptional and translational start signals. Derivatives are available with RI, S, B sites reversed. Also, some derivatives contain *lacUV5* promoter/operator in multiple cloning site; introduction into Lac⁺ cells titrates LacI, causing induction of chromosomal *lac* genes. Recombinant clones which remove *lac* operator from plasmid are identified as no longer inducing *lac.* λ derivatives allow transfer of fusion, including transcription termination signals, to single copy by homologous recombination. In addition, plasmid pRS308 allows recombination of fusion back to high-copy plasmid. Neo fusion can be switched to Lac fusion using pRS415 series described above
pKS11X_a	ColE1	Amp	X_a–LacZ	Protein	H, X, B, Bg, S, Sl, K	13	Vector for construction of NH₂-terminal LacZ fusions linked by short sequence encoding cleavage site for

					Sites	Ref.	Comments
	ColE1	Amp					protease blood coagulation factor X_a. Transcription is directed by *lac* promoter. Fusion protein can be purified and subsequently cleaved with factor X_a to yield NH$_2$-terminal protein fragment. Designed for use in conjunction with λJK2 or pJK2. Can also be used as ORF vector. Fusions are LacY⁻
pJG200	ColE1	Amp	Collagen–LacZ	Protein	B	14	Vector for construction of NH$_2$-terminal LacZ fusions linked by short sequence from chicken pro-α2-collagen. Transcription is directed by λP$_R$ promoter under control of temperature-sensitive λcI857 repressor, also carried on vector. Fusion protein can be purified and subsequently cleaved with collagenase to yield NH$_2$-terminal protein fragment. Fusions are LacY⁻
pMR100	ColE1	Amp	cI–ORF–LacZ	Protein	H, B, S, B	15–17	ORF vector. *lac* promoter directs transcription of NH$_2$ terminus of the λ *cI* gene, followed by polycloning site and then out-of-frame *lacI–lacZ* fusion. Insertion of DNA into polycloning site such that frameshift is corrected allows production of cI–ORF–LacZ fusion protein. Fusions are LacY⁻
pORF1,2	ColE1	Amp	OmpF–ORF–LacZ	Protein	Ps, (Sl, Bg), B, S, B	16, 18, 19	ORF vectors. *ompF* promoter directs transcription of NH$_2$ terminus of *ompF* with polylinker and out-of-frame *lacZ*. Insertion of DNA that restores reading frame produces tribid OmpF–ORF–LacZ protein. pORF1 and pORF2 differ in sites in polylinker and reading frame of *lacZ* versus *ompF*. Fusions are LacY⁻
pUK270	ColE1	Amp	LacZ–ORF–LacZ	Protein	H, X, Bg, Ps, B, RI	20	ORF vector. *lac* promoter directs transcription of *lac* operon. Polycloning site has been inserted in NH$_2$

(continued)

TABLE I (*continued*)

Vector	Replicon	Marker[a]	Fusion	Type	Sites[b]	Ref.	Comment
pUR series	ColE1	Amp	LacZ–ORF	Protein	B, Sl, (X, P), H	21	terminus of *lacZ* out of frame. Insertion of DNA that restores frame gives tribid LacZ–ORF–LacZ protein. Fusions are LacY⁺ Series of vectors for construction of COOH-terminal LacZ fusions. *lac* promoter directs transcription of *lacZ* with multiple cloning sites at 3′ end of *lacZ*. Insertion of open reading frame results in formation of active LacZ–ORF fusion protein. Plasmids differ in reading frame of polycloning sites and in sites themselves. Plasmids with *Pst*I sites have had natural *Pst*I site in *amp* gene destroyed. Fusions are LacY⁻
pEX series	ColE1	Amp	Cro–LacI–LacZ	Protein	(RI), S, B, Sl, P₃	16, 22	Series of vectors for construction of COOH-terminal LacZ fusions. λ P_R promoter directs transcription of *cro–lacZ* fusion with multiple cloning sites at 3′ end of *lacZ*, followed by translation and transcription termination signals. Insertion of open reading frame results in formation of inactive Cro–LacZ–ORF fusion protein. Plasmids differ in reading frame of polycloning sites and in sites themselves. When induced, fusion protein can account for up to 30% of total cellular protein, often in insoluble and easily isolated form. Fusions are LacY⁻
pATH series	ColE1	Amp	TrpE–ORF	Protein	See comment	23, 24	Series of vectors for formation of fusions to COOH terminus of TrpE. Vectors differ in cloning sites. Induction with indoleacrylic acid yields high-level synthesis of fusion, which is usually insoluble and stable

λJK2/4	λ	Amp, $imm^{21}ts$	lacZ	Protein	H, Sp, RI	13	λ vectors for formation of active NH_2-terminal (λJK2) or COOH-terminal (λJK4) fusions to lacZ. Phage carry lac promoter and gene for Amp resistance. System exists where fusion can be conveniently cloned from phage to plasmid. Fusions are $LacY^-$
λgt11	λ	$imm^{\lambda}ts$	lacZ	Protein	RI	25, 26	λ vector for formation of inactive COOH-terminal fusions to lacZ under control of lac promoter. Phage carries Sam mutation and cI857. High-level synthesis of fusion protein can be induced at high temperature without lysis of cells. Fusions are $LacY^-$
λRZ5	λ	—	Lac	Either	—	27	λRZ5 contains the 3' half of lacZ and all of lacY adjacent to 3' half of β-lactamase gene such that recombination with any pBR322-based Amp-resistant fusion vector in which bla and lac are transcribed divergently allows transfer of fusion onto phage by homologous recombination. Recombinant phage is Amp resistant and Lac^+. Phage can then be integrated at attλ for analysis
M13mp181/2	M13	—	Lac	Either	—	28	M13 derivatives contain portion of β-lactamase gene and portion of lacZ such that recombination with pBR322-based Amp-resistant fusion vector allows transfer of fusion joint onto M13 phage by homologous recombination for sequencing of fusion joint. Recombinant is Amp resistant but Lac^- (entire lacZ gene is not on phage) Phage contain bla gene in opposite orientations

(continued)

TABLE I (*continued*)

Vector	Replicon	Marker[a]	Fusion	Type	Sites[b]	Ref.	Comment
λLac Tet	λ	Tet	Lac	Either	—	5	Contains deleted *amp* gene and 3′ end of *lacZ* and *lacY*. Tet resistance gene is in between. Recombination with any pBR322-based Amp-resistant fusion vector where *amp* and *lac* are transcribed in opposite orientations allows transfer of fusion to phage. Recombinant phage is Amp resistant, Lac⁺, and Tet sensitive
pMLB524, pMLB1060, pMLB1094	ColE1	Amp	Lac	Either	See comment	2	pMLB524 is derivative of pMLB1034 containing only 3′ end of *lacZ* beginning at naturally occurring *Eco*RI site. Used to clone previously constructed Lac fusions from any vector in which *Eco*RI site is present in *lacZ*. pMLB1060 is deleted for *lacZ* sequences up to *Ssr*I and has addition of a multiple cloning site. pMLB1094 is deleted up to *Cla*I site with a multiple cloning site added. Insertion of appropriate restriction fragment from previously isolated fusions, e.g., carried on specialized transducing phage, results in reactivation of *lacZ*
pSKCAT	R6K	Amp	*cat*	—	—	29	Plasmid contains internal fragment of *phoA* gene followed by promoterless *cat* gene. R6K origin of replication requires function of *pir* gene. In strains lacking *pir* function, plasmid will integrate at PhoA fusions and convert these to transcriptional *cat* fusions. Plasmid contains cis functions for mobilization by broad host range *incP* conjugative functions

[a] Amp, Ampicillin resistance; Cm, chloramphenicol resistance; *imm*, phage immunity; Kan, kanamycin/neomycin resistance; Spc, spectinomycin resistance; Str, streptomycin resistance; Tet, tetracycline resistance.

[b] B, *Bam*HI; H, *Hind*III; N, *Nhe*I; RI, *Eco*RI; S, *Sma*I; Sl, *Sal*I; Sp, *Spe*I; X, *Xba*I; Xh, *Xho*I; Bg, *Bgl*II; K, *Kpn*I; Ps, *Pst*I.

the target gene. In general, vectors designed for the construction of operon fusions by recombinant methods have been problematic because read-through from plasmid promoters interferes with the monitoring of target promoter activity. As noted in Table I, solutions to this problem involve transcription terminators appropriately placed within the vector to prevent readthrough transcription.

A second method for fusion construction takes advantage of the proper-ties of transposable genetic elements. The minimum requirements for a transposon-based fusion generator are the cis sites required for transposi-tion and the placement, between these sites, of the reporter gene such that there is no interference with translation and/or transcription from outside the transposon. Table II[30–51a] lists many of the transposable elements designed for the creation of fusions to a variety of reporter genes. Unless otherwise stated, the transposons carry their own transposase and are therefore fully competent for transposition. This is sometimes problematic in that fusions created with these vectors are unstable (e.g., Mud). Many of the derivatives listed in Table II are designed to overcome this problem.

[30] H. S. Seifert, E. Y. Chen, M. So, and F. Heffron, *Proc. Natl. Acad. Sci. U.S.A.* **83,** 735 (1986).
[31] L. Kroos and D. Kaiser, *Proc. Natl. Acad. Sci. U.S.A.* **81,** 5816 (1984).
[32] C. Manoil and J. Beckwith, *Proc. Natl. Acad. Sci. U.S.A.* **82,** 8129 (1985).
[33] C. Manoil, *J. Bacteriol.* **172,** 1035 (1990).
[34] V. Bellofatto, L. Shapiro, and D. A. Hodgson, *Proc. Natl. Acad. Sci. U.S.A.* **81,** 1035 (1984).
[35] J. K. Broome-Smith and B. G. Spratt, personal communication (1990).
[36] O. Huisman and N. Kleckner, *Genetics* **116,** 185 (1987).
[37] J. C. Way, M. A. Davis, D. Morisato, D. E. Roberts, and N. Kleckner, *Gene* **32,** 369 (1984).
[38] M. J. Casadaban and S. N. Cohen, *Proc. Natl. Acad. Sci. U.S.A.* **76,** 4530 (1979).
[39] K. T. Hughes and J. R. Roth, *J. Bacteriol.* **159,** 130 (1984).
[40] B. A. Castilho, P. Olfson, and M. J. Casadaban, *J. Bacteriol.* **158,** 488 (1984).
[41] E. A. Groisman and M. J. Casadaban, *J. Bacteriol.* **168,** 357 (1986).
[42] R. Belas, A. Mileham, M. Simon, and M. Silverman, *J. Bacteriol.* **158,** 890 (1984).
[43] M. J. Casadaban and J. Chou, *Proc. Natl. Acad. Sci. U.S.A.* **81,** 535 (1984).
[44] E. A. Groisman, B. A. Castilho, and M. J. Casadaban, *Proc. Natl. Acad. Sci. U.S.A.* **81,** 1480 (1984).
[45] E. T. Palva and T. J. Silhavy, *Mol. Gen. Genet.* **194,** 388 (1984).
[46] J. Engebrecht, M. Simon, and M. Silverman, *Science* **227,** 1345 (1985).
[47] P. Ratet and F. Richaud, *Gene* **42,** 185 (1986).
[48] H. Lang, T. Teeri, S. Kurkela, E. Bremer, and E. T. Palva, *FEMS Microbiol. Lett.* **48,** 305 (1987).
[49] E. Bremer, T. J. Silhavy, J. M. Weisemann, and G. M. Weinstock, *J. Bacteriol.* **158,** 1084 (1984).
[50] E. Bremer, T. J. Silhavy, and G. M. Weinstock, *Gene* **71,** 177 (1988).
[51] E. Bremer, T. J. Silhavy, and G. M. Weinstock, *J. Bacteriol.* **162,** 1092 (1985).
[51a] Y. Komeda and T. Iino, *J. Bacteriol.* **139,** 721 (1979).

TABLE II
TRANSPOSABLE ELEMENTS

Element	Size (kb)	Marker	Fusion	Type	Ref.	Comment
m-Tn3(*lac*)	4.5	Amp	LacZ	Protein	30	System designed for transposition into sequences cloned into pHSS series vectors. Because of Tn3 immunity, transposition occurs solely into cloned DNA. Transposase is provided in trans. m-Tn3(*lac*) transposes from F derivative pOX38::m-Tn3(*lac*) to give cointegrate which cannot resolve owing to lack of *res* site. Conjugation into strain lysogenic for λ(p1*cre*) allows resolution at *lox* site carried on transposon. Fusions are LacY$^+$
Tn5-*lac*	12	Kan	*lac*	Operon	31	Transposes at 6% of frequency of wild-type Tn5. Fusions are LacY$^+$
Tn*phoA*	7.7	Kan	PhoA	Protein	32	*phoA* gene in Tn*phoA* lacks functional signal sequence. Activity is dependent on contribution from target gene of sequences capable of directing export. Can be used to convert LacZ fusions made with Tn*lacZ* to PhoA fusions and vice versa
Tn*lacZ*		Kan	LacZ	Protein	33	Analogous to Tn*phoA*. Can be used to convert PhoA fusions to LacZ fusion and vice versa. Fusions are LacY$^-$
Tn5-VB32	5.7	Tet	*neo*	Operon	34	Promoterless *neo* gene from Tn5 is placed such that insertions next to active promoters gives operon fusions. Transposon also contains Tet resistance gene from Tn*10*
Tn*blaM*	—	Spc	BlaM	Protein	35	Tn5 derivative with mature portion of *blaM* cloned into IS50L. In-frame fusions to cytoplasmic proteins or cytoplasmic domains of exported proteins confer Amp^r when cells are patched owing to lysis of some cells in population. Fusions which cause export of β-lactamase confer Amp^r to individual cells. Transposon is delivered

						Comments
Mini-Tn10-LK	4.9	Kan	LacZ	Protein	36	from either low-copy plasmid with temperature-sensitive replicon or conditionally defective λ phage. Does not work well for insertions into cloned genes
						Contains just outermost 63 or 69 bp from IS10R and L, respectively, with truncated *lacZ* and Kan determinant. Transposase is provided in trans from plasmid pNK629. Transposon is carried on conditionally defective λ1205. Fusions are LacY⁻
TRPLAC fusion hopper	11	Tet	*lac*	Operon	37	Derivative of Tn*10* with *lac* operon replacing all but end of IS10L. Transposon is carried on conditionally defective phage λ1045. Fusions are LacY⁺
MudI1	37.2	Amp	*lac*	Operon	38	Prototype for Mu-specialized transducing phage carrying *lac* genes. Constructs are defective for phage production but are transposition competent, i.e., they carry wild-type *A* and *B* genes. Lysogens are temperature sensitive owing to a *cts* mutation. Fusions are LacY⁺. Mud can be packaged with the use of helper phage
MudI-8	37.2	Amp	*lac*	Operon	39	Derivative of MudI with amber mutations in transposition functions. Transposition occurs in a suppressor plus background or under conditions where Mu*A,B* function is provided in trans. Resulting fusions are stable and temperature resistant in suppressor minus background. Fusions are LacY⁺
MudI1678	24	Amp	*lac*	Operon	40	Derivative of MudI1 with internal deletions. Transposition competent and temperature sensitive
MudI1681	15.8	Kan	*lac*	Operon	40	Derivative of MudI1 with deletion of Amp and insertion of *neo* gene from Tn*5*
MudI1734	11.3	Kan	*lac*	Operon	40	Derivative of MudI1681 where transposition functions have been deleted. Functions must be provided in trans. Resulting fusion is stable and temperature resistant

(continued)

TABLE II (*continued*)

Element	Size (kb)	Marker	Fusion	Type	Ref.	Comment
MudI5086	14.9	Kan	*lac*	Operon	41	Mini-Mu which contains *neo* and ColE1 origin of replication. Allows *in vivo* cloning of genes with concomitant fusion formation. Fusions are LacY[+]
MudI5155	15.6	Kan	*lac*	Operon	41	Derivative of MudI5086 that contains *oriT*, the cis site required for RK2 conjugal transfer
MudI5166	15.8	Cm	*lac*	Operon	41	Derivatives of MudI5155 with Cm resistance replacing Kan
Mini-Mu(Tet[r])	17.1	Tet	*lac*	Operon	42	Derivative of MudI1681 with Tet from Tn*10* replacing Kan marker
MudII301	35.6	Amp	LacZ	Protein	43	Analogous to MudI for construction of protein fusions in single step. Phage is transposition competent and temperature sensitive
MudII-8	35.6	Amp	LacZ	Protein	39	MudII301 derivative with amber mutation from Mud1-8. Resulting fusions are stable and temperature resistant. Fusions are LacY[+]
MudII1678	7.5	Amp	LacZ	Protein	40	Derivative of MudII301 with internal deletions
MudII1681	15.8	Kan	LacZ	Protein	40	Derivative of MudII301 with deletion of Amp and insertion of *neo* gene from Tn*5*
MudII1734	9.7	Kan	LacZ	Protein	40	Derivative of MudII1681 where transpositions functions have been deleted. These functions must be provided in trans. Resulting fusion is stable and temperature resistant
MudII4042	16.7	Cm	LacZ	Protein	44	Derivative of MudII1681 containing Cm and *ori* from pACYC184. Allows *in vivo* cloning with concomitant fusion formation
MudII5085	13.3	Cm	LacZ	Protein	41	Derivative of MudII4042 in which MuA,B transposition genes have been deleted. These functions must be

MudII5117	21.7	Kan, Spc-Str	LacZ	Protein	41	provided in trans. Resulting fusions are stable and temperature resistant Derivative of MudII1678 with low-copy, broad host range IncW pSa-derived origin of replication. Contains genes for spectinomycin and streptomycin resistance
MudII(*lacZU131*,Ap)	35.6	Amp	LacZam	Protein	45	Derivative of MudII1301 with LacZ amber mutation U131 (corresponds to amino acid 41 of wild-type LacZ). Allows creation of protein fusions that would normally be detrimental, e.g., fusions to exported proteins. Production of full-length hybrid is dependent on presence of amber suppressor
Mini-Mu*lux*	15	Kan	*lux*	Operon	46	Derivative of MudII1681 that replaces *lac* with the promoterless *lux* operon from *Vibrio fischeri* encoding for two subunits of luciferase and enzymes for production of tetradecanol substrate. Light is produced when *lux* operon is inserted downstream from active promoter. Derivative that contains Tet marker in addition to Kan marker also exists
MudIIPR3	4.5	Cm	Neo	Protein	47	Derivative of MudII1734 with promoterless *neo* gene replacing *lacZ*. Transposition functions must be provided in trans
Mud(*lacZ npt*-II)	5.2	—	LacZ, *neo*	Protein/operon	48	Transposon containing 117 bp at s end and 1006 bp at c end of Mu. Transposition functions must be provided in trans. Creates protein fusion to LacZ and operon fusion to *neo*. Transposon is carried on ColE1 plasmid such that *neo* is not transcribed. Selection for Kan resistance on complementation of MuA,B functions selects for transposition events

(continued)

TABLE II (*continued*)

Element	Size (kb)	Marker	Fusion	Type	Ref.	Comment
λ*plac*Mu1	~45	*imm*λ	LacZ	Protein	49	Constructed from MudI1301 by replacing *amp* and Mu genes with λ genome. Phage is capable of transposition but at reduced frequency owing to deletion of 3' end of MuA gene.[50] Mu*A*,*B* can be supplied in trans by λpMu507. Resulting fusions are stable. In addition, specialized transducing phage can be isolated that carry fusion and adjacent chromosomal DNA (10 kb). Derivatives exist which are *imm*[21] (λ*plac*Mu3) or that also contain Kan resistance marker (λ*plac*Mu9, Ref. 51). Fusions are LacY[+]
λ*plac*Mu5	~45	*imm*λ	LacZ	Protein	50	λ*plac*Mu1 derivative with Mu Aam1093. Transposition is completely defective. Derivatives exist which are *imm*[21] (λ*plac*Mu13) or that also contain Kan resistance marker (λ*plac*Mu15)
λ*plac*Mu50	~45	*imm*λ	*lac*	Operon	51	Analogous to λ*plac*Mu1. Replacement of Amp and Mu in MudI1 with λ genome. Transposition frequency is enhanced by supplying Mu*A*,*B* in trans from λpMu507. Derivatives exist which are *imm*[21] (λ*plac*Mu51) or that also contain Kan resistance marker (λ*plac*Mu53). Fusions are LacY[+]
λ*plac*Mu52	~45	*imm*λ	*lac*	Operon	50	Derivative of λ*plac*Mu50 containing MuAam1093. Completely transposition defective. Derivatives exist which are *imm*[21] (λ*plac*Mu54) or that also contain Kan resistance marker (λ*plac*Mu55)
λp1209	—	*imm*λ	*lac*	Either	2	λ derivative carrying lac operon and c end of Mu. Used for conversion of Mud fusions to λ*plac*Mu fusions by method of Komeda and Iino[51a]

The third method for the creation of novel joints is deletion formation. This can be accomplished by *in vitro* techniques or by selecting for an illegitimate recombination event *in vivo* (the method used initially for the isolation of the fusions).[52] For example, it is not always possible to clone the target gene into a given plasmid vector such that a fusion is created in a single step. A more general *in vitro* approach is to first clone a large portion of the target gene into the vector such that both the target and reporter genes are in the same orientation but not in the same reading frame. In-frame fusions can be subsequently generated by appropriate nuclease treatment and religation. Alternately, the out-of-frame construct can be introduced into a host cell and fusions selected by demanding function and therefore expression of the reporter *in vivo*. These methods are often useful because a large number of fusions containing varying amounts of target gene sequences can be easily isolated.[53]

When choosing a *lac* fusion vector from Tables I or II, several factors, in addition to the comments provided, should be considered. First, many of these vectors do not contain the *lacY* gene, the gene for lactose permease. Without LacY, many of the genetic manipulations which make *lac* fusions so attractive are not available. Although LacZ activity can be monitored with chromogenic substrates or assayed *in vitro* (see below), the host cell is phenotypically Lac⁻ in the absence of the permease. The presence of *lacY* is indicted for each *lac* vector in the comment sections of Tables I and II.

Many of the early *lac* operon fusion vectors actually express a fusion protein containing the first few amino acids of TrpA fused to LacZ.[54,54a] This is purely for historical reasons; the W209 *trp–lac* fusion, originally constructed *in vivo*, was employed because it provided a convenient means to separate the transcriptional control region of the *lac* operon from the *lacZ* structural gene.[55] Indeed, it was not realized originally that this was actually a protein and not an operon fusion. The Trp sequences, however, do not significantly affect LacZ activity.

Reporter Gene Assays

The chief advantage of gene fusion technology stems from the ability it provides to attach the structural gene for a well-characterized enzyme to any target gene of interest. Thus, with fusion strains, expression of the

[52] S. P. Champe and S. Benzer, *J. Mol. Biol.* **4**, 288 (1962).
[53] M. L. Berman and D. L. Jackson, *J. Bacteriol.* **159**, 750 (1984).
[54] W. W. Metcalf, P. M. Steed, and B. L. Wanner, *J. Bacteriol.* **172**, 3191 (1990).
[54a] J. Zieg and R. Kolter, *Arch. Microbiol.* **153**, 1 (1990).
[55] M. Casadaban, *J. Mol. Biol.* **104**, 541 (1976).

target gene can be simply quantitated by assaying for the reporter. This is a powerful advantage because many important and interesting target gene products are difficult, if not impossible, to assay directly. In a similar vein, gene fusion strains exhibit the characteristic phenotype conferred by the reporter gene. Accordingly, expression of the target gene can be monitored on agar media by scoring activity of the reporter gene product using well-established phenotypic tests. This allows qualitative scoring of large numbers of potentially interesting fusion strains and rapid experimental demonstration of environmental or physiological conditions that alter expression of the target gene. In this section we describe assays and phenotypic tests used for the commonly employed reporter genes.

Spectrophotometric Assays

For several of the commonly used reporter enzymes, commercially available substrates which are colorless in solution yield products that are chromophores with distinctive absorption spectra. These substrates provide the basis for a simple assay that can be accurately monitored with a spectrophotometer.

The compound *o*-nitrophenyl-β-D-galactoside (ONPG) is hydrolyzed by β-galactosidase, yielding *o*-nitrophenol, which is yellow in solution. Miller described an assay for β-galactosidase using this compound that is commonly employed.[56] Indeed, this assay is used so routinely that activity units, which are arbitrary, are referred to in the literature as Miller units. Basically this assay is done with cells permeabilized by chloroform treatment and suspended in an appropriate buffer. After incubation with ONPG for an appropriate time, the reaction is terminated by increasing the pH with Na_2CO_3, and the intensity of yellow color is read in the spectrophotometer. Miller units are then calculated using the formula provided. A fully induced wild-type strain produces about 1000 Miller units of β-galactosidase.

Alkaline phosphatase is assayed using a procedure analogous to that described for β-galactosidase except that *p*-nitrophenyl phosphate (PNPP) is used as substrate. The Beckwith laboratory, which developed this method,[57] uses a formula similar to the Miller formula for LacZ to express activity in arbitrary units.

We[58] have adapted assays for LacZ and PhoA for use with microtiter

[56] J. H. Miller, "Experiments in Molecular Genetics," Cold Spring Harbor Laboratory, Cold Spring Harbor, New York, 1972.
[57] E. Brickman and J. Beckwith, *J. Mol. Biol.* **96**, 307 (1975).
[58] J. M. Slauch and T. J. Silhavy, *J. Mol. Biol.* **210**, 281 (1989). [Erratum **212**, 429 (1990).]

plates and a microplate reader (see also Ref. 59). This procedure requires the purchase of additional equipment, but it offers considerable savings in time if large numbers of assays are to be performed. Software for the IBM PC's is available to perform the necessary calculations using data fed automatically from the microplate reader, and the results are expressed as micromoles of product formed per minute.

Certain cephalosporins such as cephaloridine can be used in a spectrophotometric assay for β-lactamase. In this case, hydrolysis is assayed as a decrease in absorbance at 255 nm.[60] In gram-negative bacteria, however, permeation of this compound across the outer membrane can be problematic. In addition, the behavior of β-lactamase that is internalized in the cytoplasm has not been systematically analyzed. Accordingly, this assay method is seldom used. Instead, bioassays similar to that used for other antibiotic resistance genes are employed (see below).

Assays for Other Reporter Genes

Luciferase activity produced by strains carrying lux fusions can be assayed using a scintillation counter in chemiluminescence mode.[46] This assay is extremely sensitive and accurate. However, enzyme activity requires oxygen, reduced flavin mononucleotide (FMNH$_2$), and tetradecanal, and care must be taken to ensure that these do not become rate-limiting.

Galactokinase, the product of the galK gene, can be assayed using suspensions of whole cells permeabilized by toluene treatment. The amount of galactose phosphate formed by the phosphorylation of [^{14}C]galactose is measured by filtering the assay mixture through DEAE filters and counting the radioactivity retained on the filter.[8] This assay is sensitive but rather cumbersome in comparison to the chromogenic assays described above.

Typically the relative levels of antibiotic resistance gene expression are determined by bioassay. These tests determine the minimum inhibitory concentration (MIC) of the relevant antibiotic.[10] Alternatively, they identify the concentration of antibiotic required to inhibit the efficiency of plating by 50% (in essense, an LD$_{50}$).[12] It is important when doing these tests to use cell suspensions of moderate or low cell density; this is especially true for ampicillin-resistant strains.

[59] R. Menzel, Anal. Biochem. 181, 40 (1989).
[60] J. R. Lupski, A. A. Ruiz, and G. M. Godson, Mol. Gen. Genet. 195, 391 (1984).

Specific Activities

In certain cases, particularly when the fusion results in the production of a hybrid protein, activity of the reporter gene product can be markedly different from the cognate wild-type enzyme. Accordingly, it is necessary to correlate the amount of enzyme present with the activity observed. With *lacZ* fusions this can be simply done. β-Galactosidase is larger than most other cellular proteins of *E. coli,* and it migrates on a typical sodium dodecyl sulfate (SDS) gel in a position that is nearly devoid of other proteins. Thus, with a fusion strain exhibiting over 700 Miller units of LacZ activity, the enzyme can be visualized by simply staining the gel with Coomassie blue. With other reporter enzymes, such as PhoA or Bla, or with LacZ when levels are low, immunoprecipitation or Western blots must be used. Knowledge of the specific activity is crucial for some of the uses of gene fusions described below, and this important step is ignored only at great peril.

Phenotypes

The *lac* operon has been extensively studied, and a variety of media have been described that allow the scoring of Lac phenotypes. Four types of media are commonly employed, and recipes for these media are provided in the laboratory manuals of Miller[56] and Silhavy *et al.*[2]

Minimal lactose agar (M63 Lac) can be used with *lacZ* fusion strains that lack the chromosomal *lacZ* gene but are *lacY*[+] to determine if β-galactosidase activities are high enough to support growth on lactose as the sole carbon source. Surprisingly low levels of enzyme activity will suffice. The growth rate of strains expressing 50 Miller units is nearly indistinguishable from wild-type.

Lactose MacConkey agar is a rich medium containing bile salts, to inhibit the growth of gram-positive bacteria, and a pH indicator, phenol red. Lac[−] *Escherichia coli* grow normally on this medium, but they form white colonies. Lac[+] strains form red colonies because lactose fermentation produces acid. This medium is less sensitive than minimal lactose agar for detecting β-galactosidase activity. Strains with low levels of enzyme may grow normally on minimal medium, but form white colonies on MacConkey agar. LacY[+] strains expressing about 100 Miller units of β-galactosidase form pink colonies; a wild-type strain forms dark red colonies surrounded by a hazy precipitate of bile salts.

Lactose tetrazolium agar is a rich medium that contains 2,3,5-triphenyl-2*H*-tetrazolium chloride (tetrazolium). Cells growing on this medium reduce the tetrazolium, forming an insoluble, red formazan dye. If the cells ferment lactose, the acid production inhibits formation of the dye. Accordingly, Lac⁻ colonies are red, and Lac⁺ colonies are white. The insoluble nature of the red dye is advantageous because it will not diffuse. This medium is less sensitive than lactose MacConkey agar for detecting β-galactosidase activity. LacY⁺ strains producing moderate levels of enzyme may form reddish (Lac⁺) colonies on MacConkey agar yet score as Lac⁻ (red) on tetrazolium. The sensitivity of lactose tetrazolium agar makes it particularly useful for scoring changes in *lacZ* expression in strains that produce relatively large amounts of β-galactosidase.[61]

In general, LacY activity is required to score β-galactosidase activity on the media described above. However, strains that produce large amounts of β-galactosidase (>1000 Miller units) will grow on minimal agar even in the absence of the permease. For example, *lacI lacY* strains appear Lac⁺.[62] Phenotypically, LacY activity can be monitored independent of LacZ by scoring growth on melibiose at 42°. At high temperatures, melibiose transport requires LacY.[63]

A very sensitive and commonly used indicator for β-galactosidase activity is the histochemical stain 5-bromo-4-chloro-3-indolyl-β-D-galactoside (X-Gal). On hydrolysis in the presence of oxygen, this compound yields an insoluble blue dye. Strains producing 1 Miller unit of β-galactosidase can be easily detected. Indeed, many *lacZ* nonsense mutants will form blue colonies after several days owing to low-level misreading. X-Gal works well and can be added to all types of solid media.[2] In addition, it can be used with *lacY* strains.

Each of the media described above (lactose minimal, MacConkey agar, tetrazolium, or media with X-Gal) can be used with *lacZ* fusion strains to score for environmental factors that stimulate *lacZ* expression. For example, fusions of *lacZ* to the *araBAD* operon exhibit higher levels of β-galactosidase in media containing arabinose than in media lacking inducer.[55] However, the medium that exhibits the greatest contrast will depend on the basal (uninduced) level of expression exhibited by the particular fusion. Often X-Gal agar is too sensitive; colonies are always blue, in which case MacConkey or tetrazolium agars should be tried.

Expression of *galK* fusions in strains lacking the chromosomal *galK* gene can be monitored using the minimal, MacConkey, and tetrazolium agars described above by substituting galactose for lactose. In general, the properties of the galactose versions of these media is similar to those employing lactose. In addition, strains lacking galactose epimerase (*galE*) are sensitive to galactose owing to a lethal accumulation of galactose

[61] J. Scaife and J. R. Beckwith, *Cold Spring Harbor Symp. Quant. Biol.* **31,** 403 (1966).
[62] J. Beckwith, personal communication (1990).
[63] J. Beckwith, *Biochim. Biophys. Acta* **76,** 162 (1963).

phosphate.[64] Accordingly, fusions that express galactokinase activity will confer a galactose-sensitive (Gal^s) phenotype in *galE* strains.[8] The Gal^s phenotype can provide a simple means for scoring environmental conditions that increase the expression of a *galK* fusion.

Expression of *phoA* fusions is most commonly monitored using the compound 5-bromo-4-chloro-3-indolyl phosphate (XP). XP is analogous to X-Gal and is used in similar manner. For best results one should employ a strain lacking the chromosomal *phoA* gene. However, since most media contain phosphate, expression of this gene is repressed, and wild-type colonies appear pale green. If the fusion expresses significant levels of phosphatase activity, the resulting blue colonies are simply detected. Minimal media requiring PhoA activity for growth have also been devised. For example, β-glycerol phosphate is not transported by *E. coli,* and cells cannot use this compound as a carbon source unless PhoA is expressed.[65] Alternatively, Tris-based minimal agar with XP as the sole phosphate source can be employed.[65]

The phenotype of strains carrying fusions to antibiotic resistance genes are scored on media containing appropriate concentrations of the relevant antibiotic. These drugs generally work in all solid media. However, we have found that increased sensitivity is observed on MacConkey and tetrazolium agar. Presumably this reflects the damage to envelope structure caused by the various dyes that are present. We routinely add less antibiotic to these media.[2] Also, of course, gene dosage affects the level of resistance directly, and drug concentration should be modified accordingly.

Expression of *lux* fusions can often be monitored simply by examining colonies in the dark.[46]

Genetic and Molecular Manipulations with Fusions

Fusions provide a means to label amino acid sequences with reporter enzyme activity, and they serve to tag target DNA sequences with known sequences from the reporter gene. Accordingly, they can facilitate genetic and molecular analysis of the target gene and its product.

Cloning

Fusions constructed *in vivo* are simple insertion mutations. These insertions provide a marker with which the surrounding DNA can be cloned. Indeed, many of the *in vivo* fusion vectors are particularly suited,

---

[64] M. B. Yarmolinsky, H. Wiesmeyer, H. M. Kalckar, and E. Jordan, *Proc. Natl. Acad. Sci. U.S.A.* **45**, 1786 (1959).

[65] A. Sarthy, S. Michaelis, and J. Beckwith, *J. Bacteriol.* **145**, 288 (1981).

or even specifically designed, for this purpose, for example, the mini-Mu vectors containing plasmid origins of replication. The use of these vectors is described in detail in [8] in this volume.

The λplacMu phage are also useful in cloning experiments. As described,[66,67] once the fusion has been isolated, it is a simple procedure to isolate a specialized transducing phage carrying the fusion. Thus, the 5′ end of the gene or operon is cloned. Vectors (Table I) are available for moving these sequences to a plasmid replicon. Alternatively, these sequences can be used for probes to isolate the entire gene by hybridization.

Using Genes to Identify Proteins

Protein fusions expressing a hybrid protein serve to tag the target protein with a domain of known function and antigenic determinants. This provides a convenient method to isolate the target protein sequences. For example, Shuman et al.[68] isolated the hybrid protein produced from a malF′–′lacZ fusion by virtue of its relatively large size and β-galactosidase activity. The fusion protein was then used to raise antibodies. The antibodies recognize wild-type MalF and allowed characterization of the otherwise unknown protein.

Using Proteins to Identify Genes

It is sometimes the case that a protein has been isolated and characterized but the gene has not been defined. The ORF vectors can be used for cloning any gene for which there are antibodies directed against the corresponding protein. An open reading frame cloned into these vectors allows the production of a functional LacZ fusion. Fusion proteins that react with the particular antibody are identified. The cloned DNA can then be used as probe to isolate the rest of the gene. This approach has been used successfully when only 10 amino acids of the target gene were expressed in the fusion.[69] This type of approach can also be used to delineate specific epitopes of a given protein by cloning various portions of the open reading frames.

Expression in Vivo

Given the current extent to which cloning and DNA sequence analysis has been applied, it is often the case that a putative open reading frame is delineated before the gene is defined genetically. Fusions provide a means

[66] T. J. Silhavy, E. Brickman, P. J. Bassford, M. J. Casadaban, H. A. Shuman, V. Schwartz, L. Guarente, M. Schwartz, and J. Beckwith, *Mol. Gen. Genet.* **174,** 249 (1979).
[67] N. J. Trun and T. J. Silhavy, *Genetics* **116,** 513 (1987).
[68] H. A. Shuman, T. J. Silhavy, and J. R. Beckwith, *J. Biol. Chem.* **255,** 168 (1980).
[69] M. Koenen, U. Ruther, and B. Muller-Hill, *EMBO J.* **1,** 509 (1982).

to determine if the open reading frame is indeed expressed *in vivo*. For example, Kiino *et al.*[70] used this approach to prove that the small open reading frame encoding PrlF was expressed.

Direction of Transcription

Expression of a reporter gene in a fusion requires that the orientation of the reporter be the same as the orientation of the target gene. This provides a means to determine the direction of transcription of the target. For example, fusions constructed with various Mud *lac* phage confer temperature sensitivity to the strain. Isolation of temperature-resistant derivatives often yields deletions. By examining chromosomal deletions that remain Lac⁺ for loss of adjacent chromosomal markers, the direction of transcription of the target gene can be determined because only markers downstream of the fusion can be deleted such that the fusion remains intact. Wanner *et al.* used this approach to determine the orientation of transcription of a *psi* gene.[71]

The orientation of target genes containing fusions constructed with λp*lac*Mu phage can be studied in a similar fashion. By recombining the *c*I857 mutation onto the phage, the strain can be made temperature sensitive. Selection for temperature resistance selects for deletions. Again, only chromosomal markers downstream of the fusion can be deleted such that the fusion remains intact. Alternatively, the use of λp*lac*Mu allows the isolation of λ specialized transducing phage carrying nearby chromosomal markers.[66,67] By determining the frequency that any given phage carries a nearby marker and the *lac* genes, the relative position of the marker and *lac* can be determined. Phage carrying markers upstream of the fusion will always be Lac⁺. For markers downstream of the fusion, however, this is not always the case.

The direction of transcription of fusions carried on the chromosome can also be determined by conjugation experiments. For example, F'*lac* can integrate into the chromosome by homologous recombination with *lac* sequences in the fusion, giving rise to an Hfr strain whose direction of transfer is dependent on the orientation of the fusion. By determining the rate of transfer of nearby chromosomal markers, the direction of transfer and hence the direction of transcription of the target gene can be determined.[72]

When fusions are constructed on plasmids, simple restriction analysis

[70] D. R. Kiino, G. J. Phillips, and T. J. Silhavy, *J. Bacteriol.* **172,** 185 (1990).

[71] B. L. Wanner, S. Wieder, and R. McSharry, *J. Bacteriol.* **146,** 93 (1981).

[72] J. Beckwith, E. R. Signer, and W. Epstein, *Cold Spring Harbor Symp. Quant. Biol.* **31,** 393 (1966).

can reveal the direction of transcription, again, because the orientation of the reporter gene is known.

Transcription Termination

Analysis of transcriptional termination signals at the ends of operons is difficult because of the lack of an apparent phenotype. Fusions can provide the necessary tools for this type of study. Guarente et al. used this approach to study the termination signals at the end of the *trp* operon in *E. coli*. A strain containing the promoterless *lacZ* operon downstream of the *trp* terminator is phenotypically Lac⁻. By selecting for expression of the fusion, mutations which affected termination were isolated.[73]

Gene Regulation

Fusion strains provide many useful tools for the genetic analysis of regulatory systems and the target gene, as evidence by the applications cited below.

Regulons

Genes widely separated on the chromosome, yet controlled by a common regulatory mechanism, are said to be components of a regulon. Because fusions allow rapid screening of physiological conditions that increase or decrease target gene expression, they provide a convenient means to identify regulon components. This principle was first demonstrated by Kenyon and Walker,[74] who screened a collection of random Mud-generated *lacZ* fusion strains for those exhibiting induction with DNA-damaging agents. A similar strategy was employed by Kolodrubetz and Schleif[75] and Wanner and McSharry[76] to identify genes that are induced by arabinose or phosphate starvation. All of these strategies led to the discovery of new genes, and in recent years the method has been used to identify genes whose expression is responsive to a wide variety of environmental conditions.

Autoregulation

Gene fusion technology provides a convenient method for examining target gene expression in the presence of the wild-type target gene product, a condition necessary for studies of autoregulation. Often, specialized

[73] L. Guarente, J. Beckwith, A. M. Wu, and T. Platt, *J. Mol. Biol.* **133**, 189 (1979).
[74] C. J. Kenyon and G. C. Walker, *Proc. Natl. Acad. Sci. U.S.A.* **77**, 2819 (1980).
[75] D. Kolodrubetz and R. Schleif, *J. Bacteriol.* **148**, 472 (1981).
[76] B. L. Wanner and R. McSharry, *J. Mol. Biol.* **158**, 347 (1982).

transducing phage carrying the fusion are used to construct the required merodiploids. However, other methods for introducing the fusion in trans can be employed as well. Casadaban,[77] in early work with *araC–lacZ* fusions, employed λ transducing phage to show that AraC is a repressor of its own transcription. In this case, *araC*⁻ lysogens (the haploid *araC–lacZ* fusion strains) exhibited more β-galactosidase activity than did corresponding *araC*⁺/*araC–lacZ* merodiploids.

Transcriptional versus Posttranscriptional Control

As described above, fusions can be of two different types. In the first, operon fusions, the reporter gene contains its own translation start signals and is dependent on the target gene for transcription only. In the second, protein fusions, expression of the reporter gene requires both the transcription and translation start signals of the target gene. This difference can be exploited to provide evidence for posttranscriptional control as in the examples cited below.

Expression of the transposase gene, *tnp*, of IS10 is controlled at the level of translation initiation by an antisense RNA.[78] The antisense RNA is produced from a promoter, pOUT, that is located downstream of the *tnp* promoter, pIN, and oriented in the opposite direction. Thus, the antisense RNA overlaps *tnp* mRNA for 38 base pairs (bp) in the region containing the translation start signals. The double-stranded RNA formed cannot be translated, and this helps maintain transposase at low levels. Evidence for this control mechanism was provided by examining the levels of β-galactosidase produced by both *tnp* operon and protein fusions. Levels of β-galactosidase activity produced by the operon fusion are much higher than from the protein fusion. Moreover, β-galactosidase production from the operon fusion cannot be significantly inhibited by high levels of antisense RNA provided in trans. In contrast, β-galactosidase production from the protein fusion is reduced substantially by high levels of the antisense RNA.[78]

Expression of the *gnd* gene of *E. coli* is regulated in response to growth rate, that is, more enzyme is present in fast growing cells. Baker and Wolf[79] found that β-galactosidase production from a series of *gnd–lacZ* operon fusions was not regulated by growth rate, suggesting a posttranslational mechanism. Subsequent work with *gnd–lacZ* protein fusions established translational control since protein fusions are growth-rate regulated. In addition, the studies allowed identification of a site within the *gnd* gene

[77] M. Casadaban, *J. Mol. Biol.* **104**, 557 (1976).
[78] R. W. Simons and N. Kleckner, *Cell* (*Cambridge, Mass.*) **34**, 683 (1983).
[79] H. V. Baker and R. E. Wolf, *J. Bacteriol.* **153**, 771 (1983).

that is required for proper control.[80] Apparently, this internal site functions as a cis-acting antisense RNA to control translation initiation in response to growth rate.[81]

Translational Coupling

There are examples of adjacent genes within an operon where the translational stop signal of the first (promoter proximal) gene overlaps the translational start signal of the second (promoter distal). In such cases, translational coupling is observed, namely, translation of the second gene requires translation of the first. Protein fusions provide a means to detect such coupling. In this case, mutations causing translation termination in the first gene exhibit extreme polarity on fusions to the second. This was demonstrated initially with *trpB* and *trpA*.[82] Other examples of this overlap include *ompR* and *envZ*.[83]

Bifunctional Fusions

Protein fusions with *lacZ, phoA,* and *bla* as the reporter gene produce a hybrid protein with amino acid sequences of the target gene product at the NH_2 terminus and a large, functional, enzymatically active reporter fragment at the COOH terminus. Most of the time, the deletion substitution event that produced the hybrid gene destroys target function. However, this is not always the case. Fusions that retain a large fraction of the target gene sequences may retain function, in which case the hybrid protein is bifunctional, exhibiting both target and reporter gene activities. In addition, many potential target proteins contain distinct functional domains, and the gene fusion may leave a subset of these intact. Such constructs are particularly useful because they provide an active or partially active target protein covalently labeled with reporter enzyme.

Bifunctional Hybrid Proteins

A surprisingly large number of gene fusions exhibit normal, or near normal, activities of both the target and the reporter gene products. The list of proteins that retain function despite the presence of a bulky reporter enzyme covalently attached at the COOH terminus includes cytoplasmic

[80] H. V. Baker and R. E. Wolf, *Proc. Natl. Acad. Sci. U.S.A.* **81,** 7669 (1984).
[81] P. Carter-Muenchau and R. E. Wolf, Jr., *Proc. Natl. Acad. Sci. U.S.A.* **86,** 1138 (1989).
[82] S. Askov, C. L. Squires, and C. Squires, *J. Bacteriol.* **157,** 363 (1984).
[83] P. Liljestrom, Ph.D. Dissertation, University of Helsinki (1986).

enzymes (TrpA–LacZ[84] and ThrA–LacZ[85]), DNA-binding proteins that regulate transcription (LacI–LacZ[86] and OmpR–LacZ[53]) or function in replication (the replication initiator protein of plasmid R6K–LacZ[87]), and peripheral (MalK–LacZ[88]) and integral cytoplasmic membrane proteins (SecE–PhoA[89]). These novel constructs can facilitate the identification[90] and purification[87] of the target protein. In addition, they can be used to analyze subunit structure[91,92] or interactions with other cellular proteins.[90] A particularly informative example is provided by the MalK–LacZ hybrid protein. MalK is a soluble protein that functions in maltose transport. Using the bifunctional MalK–LacZ hybrid, it was possible to show that it interacts with MalG, an integral cytoplasmic membrane component of the transport system; in *malG⁻* strains the hybrid protein (LacZ activity) was soluble, and in *malG⁺* strains it was membrane-bound.[90]

Identification of Intragenic Export Signals

Gene fusion technology has been particularly useful for the study of protein export or secretion because many intragenic signals that perform a targeting function are small, discrete, linear sequences of amino acids. Accordingly, these signals behave as independent domains that retain function even in the context of a gene fusion.

PhoA and Bla require periplasmic localization for enzymatic activity, and methods for fusion construction employ 5′ truncations of *phoA* and *bla* that are, in effect, signal sequence deletions (Tables I and II). Thus, fusion strains will exhibit enzymatic activity only if the gene to which these reporters are fused contains a sequence that can function as a signal sequence. This localization-sensitive property has been exploited with Tn*phoA* to screen transposon-induced mutant collections for genes that confer a particular phenotype and specify exported proteins. A good example of the utility of this method is provided by the identification of virulence genes in pathogenic bacteria; many of these gene products are exported (for review, see Ref. 93).

In contrast to PhoA and Bla, LacZ appears to require cytoplasmic

[84] D. Mitchell, W. Reznikoff, and J. Beckwith, *J. Mol. Biol.* **93**, 331 (1975).
[85] I. Saint-Girons, *Mol. Gen. Genet.* **162**, 95 (1978).
[86] B. Muller-Hill and J. Kania, *Nature (London)* **249**, 561 (1974).
[87] J. Germino and D. Bastia, *Cell (Cambridge, Mass.)* **32**, 131 (1983).
[88] S. D. Emr and T. J. Silhavy, *J. Mol. Biol.* **141**, 63 (1980).
[89] P. J. Schatz, P. D. Riggs, A. Jacq, M. J. Fath, and J. Beckwith, *Gen. Dev.* **3**, 1035 (1989).
[90] H. A. Shuman and T. J. Silhavy, *J. Biol. Chem.* **256**, 560 (1981).
[91] G. Heidecker and B. Muller-Hill, *Mol. Gen. Genet.* **155**, 301 (1977).
[92] J. Kania and B. Muller-Hill, *Eur. J. Biochem. (Tokyo)* **79**, 381 (1977).
[93] C. Manoil, J. J. Mekalanos, and J. Beckwith, *J. Bacteriol.* **172**, 515 (1990).

localization for activity. If, by gene fusion, LacZ is directed to the cellular envelop, enzymatic activity is decreased dramatically, presumably because requisite oligomerization is prevented.[94] In addition, it has been well documented that the cellular export machinery of E. coli cannot deal effectively with sequences of LacZ. Thus, if by gene fusion a signal sequence is attached to LacZ and the cell attempts to export the hybrid protein, a lethal jamming of the export machinery can result (for review, see Ref. 95). This phenotype, termed overproduction lethality, has been exploited to obtain signal sequence mutations as outlined below.

Analysis of Membrane Protein Topology

A typical, integral cytoplasmic membrane protein consists of a periplasmic domain separated from a cytoplasmic domain by one or more transmembrane α helices of 20 or so hydrophobic amino acids. The localization-sensitive property of PhoA, Bla, and LacZ fusions can be used to map the disposition of periplasmic and cytoplasmic domains along the primary amino acid sequence of the target gene product. If PhoA or Bla is placed by gene fusion at a position corresponding to a periplasmic domain, then reporter enzyme activity will be high. Conversely, if they are placed in a cytoplasmic domain, activity is low. In contrast, LacZ behaves in an opposite manner: activity is high when the enzyme is cytoplasmic. Thus, gene fusion technology provides a relatively simple means to map membrane protein topology, and while exceptions may be found, these techniques have proved remarkably reliable[10,11] (for PhoA review, see Ref. 93).

Perhaps the most common mistake made with gene fusions when developing a topology map is reliance on activity measurements alone. Low activity can be the result of a variety of uninteresting causes, such as hybrid protein degradation. To be meaningful, activity must be expressed as specific activity, and this, of course, requires an assessment of the amount of hybrid protein present (see above).

Analysis of Protein Folding

In general, reporter gene product activity is insensitive to, and unaffected by, target gene product sequences unless they contain targeting signals that alter or prevent correct cellular localization as mentioned above. However, this is not always true, and some caution is advised. In some cases, the conformation or multimerization of a target domain can

[94] D. Oliver and J. Beckwith, Cell (Cambridge, Mass.) 25, 765 (1981).

[95] K. L. Bieker, G. J. Phillips, and T. J. Silhavy, J. Bioenerg. Biomembr. 22, 291 (1990).

prevent or obscure reporter activity. When this occurs, it can provide a means to analyze amino acid sequences that are crucial for target folding.

A clever example of gene fusion technology applied to the problem of protein folding comes from the work of Luzzago and Cesareni,[96] who analyzed the folding of the H chain of human ferritin, which was produced in *E. coli* using recombinant DNA methodologies. Ferritin is a 24-mer that is, in effect, a molecular cage. In a carefully considered set of experiments, they designed a fusion between ferritin and the α-complementing fragment of LacZ such that the α fragment was sequestered within the ferritin cage. Strains producing the ferritin–LacZα hybrid protein are Lac⁻ even in the presence of the ω-complementing LacZM15 mutant protein because interaction between the α peptide and the ω fragment is prevented by the ferritin molecular cage. Amino acid substitutions in ferritin can be detected that result in α peptide exposure since they allow intramolecular complementation and a Lac⁺ phenotype. Some of these changes can be shown to alter the assembly pathway of ferritin.

Mutant Isolation

Fusion technology allows one to tag a particular target gene with a variety of reporter genes, and, as described above, analysis of such strains can provide insights into the regulation, cellular localization, and function of the target gene product. In addition, of course, gene fusions convey to the target gene phenotypic traits of the reporter. Since many target genes of interest confer phenotypes that are difficult and perhaps impossible to select for or against, this can facilitate subsequent genetic analysis substantially. Because the lore of *lac* is vast, and because *lacZ* fusions have been generally available for a longer period of time, they provide an archetype for genetic analysis using gene fusions. Accordingly, the following discussion focuses primarily on *lacZ* fusions. In principle, similar strategies can be carried out with other reporter genes. In fact, in certain situations, other reporters offer distinct advantages, as noted.

Generally speaking, three types of genetic selections or mutant screens can be envisioned for fusion strains. We can look for mutants in which reporter activity is abolished (off), decreased (down), or increased (up). Each of these is discussed in turn. As the examples show, these selections and screens can provide a means to identify genes whose products affect the target in a variety of different ways, from transcription and translation to export or folding.

---

[96] A. Luzzago and G. Cesareni, *EMBO J.* **8,** 569 (1989).

Selections for Mutations That Abolish LacZ Activity

Strains carrying a *galE* null mutation are sensitive to galactose owing to the cytoplasmic accumulation of galactose phosphate.[64] Such strains are also sensitive to lactose if the *lac* genes are expressed at reasonable levels, since lactose will be transported and hydrolyzed to yield glucose and galactose.[97] Accordingly, in a *galE–lacZ* fusion strain, selection for resistance to lactose can yield mutants in which *lacZ* expression is abolished. Although this selection works, it has not been extensively employed. Introducing and working with the *galE* mutation in fusion strains is somewhat cumbersome, and we suspect that a high proportion of mutants obtained with this selection will be uninteresting *lacZ*, *lacY*, or *galK* null mutations. A better strategy may be to employ one of the selections or screens cited in the following section for mutants in which LacZ activity is decreased.

Selections for Mutations That Decrease LacZ Activity

The compound *o*-nitrophenyl-β-D-thiogalactoside (TONPG) is a metabolic poison that is accumulated by *lacY*⁺ cells to toxic levels. The toxicity of TONPG is best observed in media where cell growth is slow, and relatively high levels of *lacY* expression are required. Provided these conditions are met, it can be used to select mutants in which expression of *lacY* is decreased.

Berman and Beckwith[98] used a *tyrT–lacZ* fusion strain and TONPG to identify mutations that decreased *tyrT* promoter function. In this case, TONPG resistance was selected on minimal succinate agar. To avoid uninteresting mutations such as polar *lacZ* or *lacY* nulls, X-Gal was included in the selective medium. This allowed direct identification of mutants in which expression of both *lacZ* and *lacY* was decreased (TONPG-resistant light blue, not white or dark blue, mutant colonies). A similar strategy was employed by Gutierrez and Raibaud[99] to obtain promoter down mutations affecting the *malPQ* operon. In this case, the selection also yields mutations in *malT*, which specifies the transcriptional activator of the maltose regulon. This demonstrates the utility of the method for identifying positive regulatory genes.

Perhaps the most commonly employed method for identifying mutants that exhibit decreased *lacZ* expression involves the use of lactose indicator media such as MacConkey and tetrazolium agar. Following mutagenesis

---

[97] M. Malamy, *Cold Spring Harbor Symp. Quant. Biol.* **31**, 189 (1966).
[98] M. Berman and J. Beckwith, *J. Mol. Biol.* **130**, 305 (1979).
[99] C. Gutierrez and O. Raibaud, *J. Mol. Biol.* **177**, 69 (1984).

and plating, mutant colonies that exhibit decreased *lac* expression can be scored. In both types of media, mutants with decreased activity can usually be distinguished from the parent and from mutants in which *lac* expression is abolished. The medium of choice depends on the levels of activity exhibited by the parent strain. For example, when *lac* expression is high in the parent, tetrazolium agar is preferred. Indeed, the first promoter mutations in the *lac* operon were recognized using lactose tetrazolium agar.[61] If *lac* expression in the parent is moderate, MacConkey agar provides a better choice since colonies with moderate expression on tetrazolium agar may appear Lac⁻ to begin with.

Knowledge of the chromosomal location of regulatory genes permits the use of localized mutagenesis.[2] With this technique, the mutagenic treatment can be focused to a particular chromosomal region that is often distinct from the *lacZ* fusion. Accordingly, every mutant colony scored represents a regulatory mutation. We have used this method extensively to obtain mutational alterations in the positive transcriptional regulatory protein OmpR, for example.[58]

Selections for Mutations That Increase LacZ Activity

It should be clear from the preceding sections that selections for mutants in which *lac* activity is abolished or decreased are fraught with difficulty because of the wide variety of uninteresting mutations that can confer these phenotypes. In general, this is true in all genetic selections for decreased activity. In contrast, selections for increased activity (up) are much more promising since the number of uninteresting mutations that would confer such a phenotype is small. However, in the absence of fusions, this option is often ill-advised or not available. It is not obvious that an up mutation could be obtained in a given operon or regulon by selection for increased activity, because that activity may already be optimum. For example, it is hard to imagine how one would isolate a mutant strain that is more Lac⁺ or Mal⁺ than wild type. Fusions alter this situation since environmental conditions that select for increased operon or regulon activity are different from those that select for increased reporter gene activity. Most of the useful genetic methods provided by gene fusions stem from this opportunity to select or screen for mutants in which reporter gene activity is increased.

Strains which contain *malPQ–lacZ* fusions are basically Lac⁻ in the absence of maltose since operon expression is not induced. Indeed, most of these strains grow poorly on lactose minimal agar. Selection for faster growth yields (in nearly every case) mutants that express the operon constitutively, at least to some degree. In this manner, dominant mutations in *malT*, which enable this positive transcriptional activator to function in the absence of maltose (*malT^c*), were obtained.[100] Using a similar approach with a strain harboring a *lacZ* fusion to the gene for Tn*3* transposase, Chou *et al.*[101] were able to identify recessive null mutations in a linked gene, *tnpR*, that specifies a transcriptional repressor. These successes demonstrate the utility of gene fusions for identifying regulatory genes and characterizing their mode of action. In both cases, analogous mutations would have been much more difficult to select without the aid of *lacZ* fusions.

The basal level expression of many *lacZ* fusion strains is high enough to permit growth on lactose minimal agar. In this situation, the competitive β-galactosidase inhibitor phenylethylthiogalactoside (TPEG) can be added to the lactose minimal agar to effectively reduce LacZ activity to the point where growth cannot occur. The amount of TPEG required is determined empirically and depends on the particular fusion strain. However, once appropriate conditions are found, mutants that exhibit increased LacZ activity can be selected. Riggs *et al.*,[102] for example, used this method to identify mutations that increase expression of a *secA–lacZ* fusion strain. The report is particularly informative because it also describes two additional methods for identifying mutations that confer LacZ(up) and compares results obtained with the three different methods.

A second method used by Riggs *et al.*[102] utilizes the trisaccharide raffinose. Growth on raffinose requires *lacY*, and, accordingly, the compound can be used to select mutants that exhibit increased *lac* expression. The third method involved EMS (ethyl methanesulfonate) mutagenesis followed by a screen on rich agar containing X-Gal and TPEG. In this case, TPEG addition helped to maximize the color difference between colonies of the parent *secA–lacZ* fusion strain and the LacZ(up) mutants. Riggs *et al.*[102] report 13 LacZ(up) mutants among a population of 3600 mutagenized colonies screened.

It is apparent from the summary provided by Riggs *et al.*[102] that different results are obtained with the different methods. Lactose minimal agar with TPEG yielded more than 1000 mutants with a LacZ(up) phenotype, but a vast majority (96%) were found to be linked to the *secA–lacZ* fusion. Presumably these mutations will help define cis-acting sequences that are involved in *secA* regulation. In contrast, most of the 13 mutants obtained by mutagenesis and colony screening were unlinked to the fusion. Many of these mutations decrease the functional activity of known *sec* gene

---

[100] M. Debarbouille, H. A. Shuman, T. J. Silhavy, and M. Schwartz, *J. Mol. Biol.* **124,** 359 (1978).

[101] J. Chou, M. J. Casadaban, P. G. Lemaux, and S. N. Cohen, *Proc. Natl. Acad. Sci. U.S.A.* **76,** 4020 (1979).

[102] P. D. Riggs, A. I. Derman, and J. Beckwith, *Genetics* **118,** 571 (1988).

products. Such mutants answer this screen because *secA* expression, and thus *secA–lacZ* expression, is derepressed in response to the secretion needs of the cell. In other words, mutations that compromise the export machinery cause *secA* derepression. Among this collection, the authors found a conditional lethal mutation that defined a new *sec* gene, *secE*. The raffinose selection was somewhat disappointing because 288 of the 300 mutant colonies that originally came through the screen did not actually exhibit increased *lacZ* expression. Apparently, these mutations uncovered another permease for raffinose. Among the 12 that exhibited a LacZ(up) phenotype, the distribution of linked to unlinked was intermediate between the other two methods. It is not obvious why the different methods yield different distributions of mutations, but the results underscore the need for investigators to consider alternative approaches.

A final method for obtaining mutants that exhibit a LacZ(up) phenotype utilizes the lactose MacConkey and tetrazolium indicator agars. It is not generally appreciated, but these indicator agars can be used for mutant selection. The lactose concentration in these media is 1%, and, accordingly, mutant colonies with increased *lac* activities have a distinct growth advantage. In addition, since the parent grows, the total number of cells that can be screened on a single plate is enormous. For example, recall that if β-galactosidase is directed to a membrane location by gene fusion, enzyme activity is inhibited (see above). Thus, certain *malE–lacZ* protein fusion strains are basically Lac⁻. Oliver and Beckwith[94] spread a lawn of such a fusion strain on lactose tetrazolium agar, and after 5 days Lac⁺ colonies growing out of the parental law could be detected. These colonies are red; for reasons we do not completely understand the color reaction is reversed for colonies on a lawn. Under normal conditions with isolated colonies, a red color indicates Lac⁻. Mutations identified in this manner include *malE* signal sequence mutations and mutations that decrease the functional activity of components of the cellular protein export machinery. In both cases, a fraction of the LacZ hybrid protein is retained in the cytoplasm, where specific activity is increased substantially. Indeed, the *secA*[94] and *secB*[103] genes were first discovered in this manner.

This section on up mutants would not be complete without reference to the antibiotic resistance genes. When these are used as reporters, up selections are quite simple. One simply increases the antibiotic concentration to the point where the parent fusion strain cannot survive, and under these conditions up mutants can be selected directly.

[103] C. A. Kumamoto and J. Beckwith, *J. Bacteriol.* **154**, 253 (1983).

Novel Phenotypes

In certain cases, fusion strains can exhibit novel phenotypes that are not conferred by either target or reporter gene mutations. When, and if, this occurs, important new strategies for genetic analysis are provided. Perhaps the most profitable example of this exposure is the overproduction lethality observed with *lacZ* fusions to genes that specify exported proteins.

The cellular export machinery of *E. coli* cannot deal effectively with sequences of β-galactosidase. If, by gene fusion, this machinery is presented with β-galactosidase in large amounts, a lethal jamming occurs, and the cell dies. With *lamB*– or *malE–lacZ* protein fusions, this lethality is evidenced as maltose sensitivity (Mal^s), since maltose addition causes induction of hybrid gene expression (for review, see Ref. 95). The most common mutations that relieve Mal^s are those that prevent hybrid gene expression either by inactivating a regulatory protein or by causing premature translation termination. By devising strategies to avoid these more common lesions, mutations affecting protein export or translation initiation can be uncovered.

Starting with a Mal^s, *lamB–lacZ* fusion strain, Emr and Silhavy[88] selected mutants resistant to maltose. To avoid those that cause expression defects, a Lac$^+$ phenotype was demanded as well. Nearly all of the mutants that answer these criteria contain mutations that alter the LamB signal sequence and prevent export of the hybrid protein. These mutations provided direct evidence that the signal sequence was an important intragenic export signal. Schwartz and colleagues[104,105] employed a similar scheme except that mutations that decreased hybrid protein synthesis were sought; these mutants yielded light blue colonies on agar containing X-Gal. To avoid mutations that cause transcriptional defects, they demanded growth on melibiose (Mel$^+$) at high temperature; under these conditions, melibiose transport is dependent on LacY. Consequently, the phenotype sought was LamB–LacZ(down) LacY$^+$. Mutants which answered this scheme contained mutations that altered the signals required for translation initiation at *lamB*. These mutations provided evidence that mRNA secondary structure in the region near the Shine–Dalgarno[106] sequence can prevent ribosome binding.[104,105]

[104] M. Schwartz, M. Roa, and M. Debarbouille, *Proc. Natl. Acad. Sci. U.S.A.* **78**, 2937 (1981).
[105] M. N. Hall, M. Gabay, M. Debarbouille, and M. Schwartz, *Nature (London)* **295**, 616 (1982).
[106] J. Shine and L. Dalgarno, *Nature (London)* **254**, 34 (1975).

Conclusions

It is not possible to predict with certainty how fusions will be useful in all situations. Based on past experiences fusions will continue to provide opportunities for the geneticist that were heretofore not available. Given that a variety of different fusions can now be simply constructed, their application should always be considered.

[10] Storing, Shipping, and Maintaining Records on Bacterial Strains

By Kenneth E. Sanderson and Daniel R. Zeigler

Storage Methods

It is extremely important that wild-type and mutant strains be stored by methods which not only assure survival, but which also make certain that the genotype and hence the phenotype of the strains do not change. It is particularly important that the strains be rechecked after they have been stored in permanent culture and before being reported to assure that they still have the properties which are to be described in publication. Storage methods which do not require metabolic activity and cell growth are preferable; methods which allow growth greatly increase the chances of mutation and variation in the cultures.

Freezing

Freezing is the simplest and most common method of storage. For most bacteria, storage in ultracold mechanical freezers ($-70°$ to $-90°$) is very effective. Storage in $-20°$ commercial freezers is adequate for many bacteria for periods up to 1–2 years but is not recommended for long-term storage. Storage in liquid nitrogen at temperatures from $-156°$ to $-196°$ is superior for cells of some microbial species and for cell lines, but to our knowledge all species of bacteria used extensively for genetic investigation can be maintained at the temperatures achieved by mechanical freezers. Snell,[1] however, reports that the reduced storage temperatures achieved by liquid nitrogen are worth the extra expense.

In order to protect cells from the effects of freezing, cryoprotective

---

[1] J. J. S. Snell, *in* "Maintenance of Microorganisms" (B. E. Kirsop and J. J. S. Snell, eds.), p. 11. Academic Press, London, 1984.

agents must be added. The most commonly used are glycerol (at 10–20%, v/v) or dimethyl sulfoxide (DMSO) (at 5%, v/v), both of which seem to be equally effective in preserving bacteria[2]; both agents pass through the cell membrane and provide both intracellular and extracellular protection against freezing. Other agents which appear to exert their protective effect external to the cell membrane, such as sucrose, lactose, glucose, polyvinylpyrrolidone, and polyglycol, are not recommended.

Ultracold Mechanical Freezers. The following method is used at the Salmonella Genetic Stock Centre (SGSC) (University of Calgary, Calgary AB, Canada) for storing enteric bacteria (usually cultures of *Salmonella* or *Escherichia coli*). The strain to be stored is grown to saturation overnight in L (Luria) broth. Then, 0.8 ml of 50% glycerol, previously sterilized by filtration, is placed in a half-dram vial (~2 ml) with a rubber-lined screw-top (Wheaton, Millville, NJ, Cat. 224881, or Kimble, Vineland, NJ, Cat. 60910L12) to which approximately 1.4 ml of broth culture is added (leaving a small air gap at the top of the vial). After the tube is shaken well, it is labeled on the bottle with water-insoluble ink on label tape and on the cap with pencil on tape or paper label, and the label on the cap is covered with clear nail polish. The vial is placed into a freezer box made by Revco Co. (Asheville, NC, Cat. 5954, box divider 5958), which holds 100 of the above tubes in a 10 × 10 grid. The boxes holding the vials are placed directly into the freezer without precooling. The cooling rate during freezing is important in assuring survival for the larger cells of eukaryotes, and even for bacteria it has been reported that a slow cooling rate gives the best survival (e.g., 1° per minute).[2] Programmed freezers with adjustable cooling rates are available, but it is our experience that the freezing rate of the vial inside a box achieves high survival of the bacterial species tested, so no special measures are used to control freezing rate. Duplicate tubes are made with strains intended for long-term storage, and these are put into two separate mechanical freezers (−70° to −90°) which are in separate buildings on different electrical circuits.

To recover viable cells from frozen cultures, the vial is removed from the freezer and opened, and a sterilized loop is used to scrape out frozen cells, which are streaked directly to a petri plate containing a nutrient agar (L agar with selective antibiotic where appropriate) and incubated at 37°. Thus thawing is very rapid, but no special precautions are used to assure a specific rate. The frozen vial is kept out of the freezer for as short a time as possible. If a box must be kept out of the freezer for a longer period, it is placed on dry ice in an ice bucket. In this way a frozen vial can serve

---

[2] R. T. Gerna, *in* "Manual of Methods for General Microbiology" (P. Gerhardt, ed.), p. 208. American Society for Microbiology, Washington, D.C., 1981.

as a "working stock" and can be used for an indefinite number of separate inoculations over a period of years; we have inoculated from the same tube scores of times with no evidence of loss of viability or changes in the culture. However, it is important to have other backup stocks as well.

It is desirable to have the freezer connected to both regular and emergency power supplies (if available) through a "power failure switch" that will automatically switch to the emergency supply if the regular supply fails. It is also important to have battery-powered alarm systems in operation and to maintain charged batteries in the system. If possible, it is desirable to have carbon dioxide or liquid nitrogen backup systems attached to the freezer in case of power or mechanical failures. Ultracold mechanical freezers cost from $5000 to $8000 to buy and only electrical costs to operate. They normally operate trouble-free for 10 to 20 years and hold over 10,000 strains stored in vials as described above. It is important to defrost the freezer completely every year, by allowing it to remain at room temperature for 1 or 2 days to permit ice to melt off the cooling coils, and to allow the oil in the system to equilibrate.

Storage in Liquid Nitrogen. A good description of liquid nitrogen storage methods can be found in Chang and Elander.[3] The methods of preparing cells, the types of cryoprotective agents, and types of cryogenic containers (vials and ampoules) are not necessarily different from those used in ultracold mechanical freezers. However, it has been reported that controlled-rate freezing is superior to rapid rate freezing for cells in liquid nitrogen. Cells from an agar slant are suspended in a broth–10% glycerol suspension dispensed into 2-ml plastic screw-cap vials, then placed in a refrigerator at 5° for 30 min to cool. The ampoules are subsequently placed in aluminum cans in the freezing chamber of a controlled-rate freezer, and a cooling rate of 1° to 2° per minute is maintained to just above the freezing point (about −30°). The cells are cooled rapidly through the freezing point, then cooled slowly to −50°, at which time they are transferred quickly to their final storage space in a liquid nitrogen refrigerator at −156° to −196°.

Liquid nitrogen storage is used at the Bacillus Genetic Stock Center (BGSC, Ohio State University, Columbus, OH). A midsize liquid nitrogen chest, which can store up to 12,000 vials in vapor phase, costs about $8000, and a 160-liter tank for delivering liquid nitrogen costs $1000. With evaporation of 10 liters per day, and costs of $0.50 per liter, total expenses per year for liquid nitrogen approach $2000. Since the chest holds 30–60

[3] L. T. Cheng and R. P. Elander, *in* "Manual of Industrial Microbiology and Biotechnology" (A. L. Demain and L. A. Solomon, eds.), p. 49. American Society for Microbiology, Washington, D.C., 1986.

liters of nitrogen, it remains cold for several days even during a power loss and is less affected by "incidents" than mechanical freezers.

Commercial Freezers. Commercial freezers maintain a temperature of about $-20°$. The cells can be prepared as for an ultracold mechanical freezer (discussed above). A 15% glycerol solution will remain liquid at this temperature, so cells are readily aliquoted. Moreover, because cells are reported to maintain viability and stability for periods of months or years, this is a convenient way to keep working stocks (B. Bachmann, personal communication, 1989). However, this method is not recommended for long-term storage.

Lyophilization

Lyophilization or freeze-drying involves the removal of water from frozen cell suspensions by sublimation under reduced pressure, and it is an effective method for long-term preservation of many microorganisms. However, there are reports that the freeze-drying process may induce mutations in bacteria. Ashwood-Smith and Grant[4] found a highly significant increase of Trp$^-$ to Trp$^+$ mutation in *E. coli* strains on freeze-drying, apparently associated with the drying process, not the freezing or the thawing. Tanaka *et al.*[5] reported that freeze-drying significantly increased the reversion of amino acid auxotrophy to prototrophy in some strains of *E. coli,* postulated to be due to mutations induced during rehydration by the product of the *exr* gene. The reversion of His$^-$ to His$^+$ in some of the strains of the *Salmonella typhimurium* Ames test set was increased by freeze-drying.[6] However, it appears that freezing and thawing cycles, without drying, are not mutagenic, since this treatment did not induce an auxotroph of *E. coli* to revert, and the sensitivity of *Pseudomonas* to a wide range of antibiotics remained unchanged.[7] The data suggest that some caution should be used in relying only on lyophilized cultures for long-term storage and indicate that frozen stocks are the method of choice. However, lyophilized cultures have advantages. They can be maintained for many years at ambient temperature or 4° without loss of viability; thus, strains are not lost due to equipment accidents as in frozen cultures. Lyophilized cultures are readily transported with reduced need for special precautions, and some jurisdictions permit their transport when cultures of metabolically active cells are prohibited.

A typical freeze-drying apparatus includes the following components:

[4] M. J. Ashwood-Smith and E. Grant, *Cryobiology* **13**, 206 (1976).
[5] Y. Tanaka, I. Yoh, T. Tanaka, and T. MiWatani, *Appl. Environ. Microbiol.* **37**, 369 (1979).
[6] R. J. Heckly and J. Quay, *Cryobiology* **18**, 952 (1981).
[7] M. J. Ashwood-Smith, *Cryobiology* **2**, 39 (1965).

a condenser, a vacuum pump able to generate pressures of approximately 10 μm of mercury, a vacuum gauge, and a manifold or freeze-drying chamber. Serum bottles or vials with stoppers are used for batch or shelf drying in commercial operations. Most smaller operations use ampoules which are sealed under vacuum. The following procedure is used at the SGSC for storage of *Salmonella* and other enteric bacteria. The cells are grown in a slant culture containing L agar, with antibiotic selection where appropriate. Whatman #1 filter paper is cut into small pieces, labeled in pencil with the appropriate numbers, and stuffed into lyophil tubes (Virtis, Gardiner, NY, 1-ml centrally constricted freeze-drying ampoules, Cat. 10-196), and the tubes are sterilized. Just before use, sterilized 10% glucose and 10% peptone are mixed in equal amounts and approximately 1 ml is used to wash cells from the surface of the slant. Enough cell suspension is transferred to the ampoule to moisten the filter paper. The ampoules are connected to the manifold of a Virtis 10-010 lyophilizer with a 12-part manifold when it has been reduced to 10 μm of pressure. After all samples are in place and the vacuum is reduced to 10 μm, the ampoules are sealed off with an air–gas double-flame torch. As soon as an ampoule is sealed, it is "annealed" by slow cooling in the flame of the torch set on low gas. The tubes are tested for vacuum with a Tesla coil a few days after they have been sealed; any which do not show a vacuum are discarded.

Storage of lyophils at 4° in the dark rather than at ambient temperature is generally recommended but may not be necessary. A set of lyophils of *S. typhimurium* made by J. Lederberg in the 1950s and stored at room temperature all retained viability when opened at the SGSC in the 1980s. However, asporogonous *Bacillus subtilis* strains at the BGSC lost viability between the first and the second year of storage at ambient temperature, though not at 4°, but spore suspensions in lyophils showed no loss of viability after a decade of storage at room temperature.

Other Methods of Storage

Methods other than freezing and freeze-drying have been used extensively, but none are as suitable for long-term storage, though some are useful in special situations or for working cultures. Many of these methods allow or indeed require cell growth, with resultant high probability of genetic changes.

Stab Cultures. Maintaining stab cultures is not an appropriate method for many species and strains, although it works well for many *Salmonella* strains, providing they are basically stable, that is, have no plasmids or unstable chromosomal mutations. Many stab cultures of *S. typhimurium*, made as early as 1955 by M. Demerec, are held by the SGSC where they have retained viability and genotype for 35 years. Difco (Detroit, MI) nutrient broth or L broth with 0.7% agar was placed in 1-ml vials, inoculated, plugged with a cork, and dipped in hot wax to seal. The tubes are stored at room temperature. If opened for use the tubes are discarded, so several tubes of each strain must be made. The wax gradually sublimes, and the tubes must be redipped in hot wax every few years to prevent drying of the agar. Most workers who use the method at present use screw-cap vials rather than tubes with corks. Although most auxotrophic mutants stocked by this method have retained the appropriate phenotype, plasmids such as the F-factor are usually lost.

Subculturing. Cultures can be preserved by periodic transfers to fresh agar media. Broth cultures are the least desirable storage method; storage on slant cultures or on plates, at 4°, is more appropriate. This method usually succeeds in preserving viability, but the chance of genetic change is magnified. It is often used, even in genetic laboratories, for short-term maintenance of working cultures taken from permanent stock, or for cultures just isolated while the decision is made on whether to make permanent stocks. However, frozen stocks are easy to make and practical to use as a source of working stocks.

Miscellaneous Methods. Several other methods are described by Chang and Elander.[3] These include immersion in mineral oil and drying on soil, silica gel, or porcelain beads, and they may be satisfactory for fungi, *Streptomyces,* or for spore-forming bacteria. They are not recommended, however, for most bacteria.

Exchanging Bacterial Strains between Laboratories

The rapid advances in bacterial genetics since the 1940s have been dependent not only on the ability to exchange information between laboratories, but also on the ability to exchange living microbial cultures. The capacity to exchange information improves every day: postal systems transfer scientific journals and letters, and added to this are telephones, FAX machines, computer discs, and on-line database searches by computer. It is crucial for researchers to obtain not only information but also the living strains used to generate the data on new materials. The publication of reports on new mutants of bacteria is followed immediately by requests from other laboratories for these strains; the exchange of these strains leads not only to rapid confirmation (and occasionally disproof!) of the hypotheses stated in the reporting paper, but to other types of research which the original author(s) could not have originated. However, the exchange of living microbial cultures is threatened by two problems.

First, authors sometimes do not act to make their mutant strains available to other workers because they may want to do more research work before sharing the material, they may have lost the material because of inappropriate storage measures, or they may be unable to expend the effort or cost of sending out the strains. In addition, in some cases, economic considerations enter, for the strains may be considered valuable in producing or testing a product. However, it should be recognized by scientists that there is an implicit assumption in the scientific community that after a strain is reported in the scientific literature it will be made available to other workers. This implicit assumption was made explicit by the Editors of the *Journal of Bacteriology* by stating in their "Instructions to Authors" that "when authors describe mutants for which genetic stock repositories have not been established or strains that have not been deposited in publicly accessible collections, the *Journal* expects that the authors will make such strains available to other microbiologists." Thus the responsibility is clearly placed on the author to see that crucial material is maintained in viable and genetically unaltered form, and in addition that this material is provided to other workers either through a public culture collection or genetic stock center, or from the reporting laboratory itself. Thus either deposit in a public collection or maintenance and distribution from the originating laboratory itself is expected by the scientific community.

Second, the exchange of cultures between laboratories requires compliance with a series of regulations on packaging, mailing, national import permits, etc. These regulations have been developed because of concerns for the safety of persons involved in transport and for public safety. Some of these concerns were sparked originally by worries about recombinant DNA but are now more widely targeted. In some countries transport of all "perishable biological substances" or of "infectious substances" through the postal system is prohibited. In these cases private courier systems or systems such as Air Freight must be used, with resultant increased costs and sometimes long delays. The ability of microbiologists to exchange living cultures requires that they know and follow the regulations, and on occasion it requires that they make representation to assure that the rules are workable.

Classification of Bacteria in the United States

The geneticist is usually working with many strains of a few species, and most genetic work with bacteria has been with species of low pathogenicity in disease of humans or other species. The regulations for transport differ by country, and the rules established by the United States are discussed here. The U.S. Department of Health, Education, and Welfare established a Classification of Etiologic Agents on the Basis of Hazard in June 1972,[8] and this has been used, with modifications and extensions, as a basis for classification up to this time. Most of the bacterial species which serve as the basis for genetic investigations are Class 1, "agents of no or minimal hazards under ordinary conditions of handling." *Escherichia coli* (nonenteropathogenic serotypes), *Bacillus subtilis,* and several other species commonly studied fall in this class. Class 2 are "agents of ordinary potential hazard: agents which produce disease of varying degrees of severity—but which are contained by ordinary laboratory techniques." These include *Salmonella, Borrelia, Mycobacterium, Staphylococcus, Streptococcus, Haemophilus* (all strains of these genera), as well as some strains of *Streptomyces* and *Pseudomonas.* There are a few bacterial species in Classes 3 and 4, and genetic studies on some of these are beginning. In genetic work a specific strain or a few strains usually serve as the parents for the isolation of large numbers of mutants, many of which have lower pathogenicity than the parent species. For example, most mutants of *E. coli* are derived from the strain K12, a nonenteropathic strain, and most *S. typhimurium* mutants are derived from strain LT2, which has been in culture for many years and is of very low pathogenicity to humans.

Regulations on Packaging

The U.S. Department of Health and Human Services, Centers for Disease Control (CDC), has published regulations on the Interstate Transport of Etiologic Agents as 42 CFR Part 72.[9] A summary of some of these regulations, and instructions on packaging and shipping of Biological Agents, prepared by Alexander and Brandon[10] for use at the American Type Culture Collection (ATCC, Rockville, MD) for shipment within the United States and from the United States on a worldwide basis, is available from the ATCC. The regulations state that there are no specific requirements for shipping of agents of class 1. However, for etiologic agents of class 2 and above there are certain minimum packaging requirements for interstate transportation; these have tended to become *de facto* requirements for all transport. These rules state, "Material shall be placed in a securely closed, watertight container (primary container (test tube, vial etc.)) which shall be enclosed in a second, durable watertight container (secondary container). Several primary containers may be enclosed in a

---

[8] *Federal Register* **37,** 127 (June 30, 1972), updated on several occasions.

[9] *Federal Register* **45,** 141 (July 21, 1980).

[10] M. T. Alexander and B. A. Brandon, "Packaging and Shipping of Biological Materials at ATCC." American Type Culture Collection, Rockville, Maryland, 1986.

single secondary container, if the total volume of all the primary containers so contained does not exceed 50 ml. Each set of primary and secondary containers so contained shall then be enclosed in an outer shipping container constructed of corrugated fiberboard, cardboard, wood, or other material of equivalent strength.'' (See Fig. 1a,b). The outer shipping container must bear a label as shown in Fig. 1c. Transport of some microorganisms which are classed above level 2 require that registered mail or an equivalent system be used. Many other countries have regulations for the transportation of microorganisms within the country that are similar to those of the United States, although some have no well-defined regulations and others are considerably more restrictive. An extensive discussion on packaging and shipping both within the United States and internationally is given by Alexander and Daggett.[11]

Cultures of hardy class 1 organisms can be mailed inexpensively. A drop of dense cell suspension (in 15% glycerol) or of a spore suspension is placed on a filter paper disk and wrapped in sterile plastic wrap or aluminum foil, placed in a regular business envelope, and sent through the postal service. This system is used extensively by the _E. coli_ Genetic Stock Center and the Bacillus Genetic Stock Center.

It is sometimes useful and necessary for microbiologists to give advice to government on transport regulations, an example being recent negotiations on the use of the postal system in the United States. In 1988–1989 the U.S. Postal Service proposed to ban shipment of etiologic agents of disease in humans; a number of agencies, especially the Committee on Culture Collections of the American Society for Microbiology, the Communicable Disease Centers, and others, made strenuous representations which resulted in preventing the ban and modifying the proposed rules.[12]

Importation into a country normally requires an import permit, at least for bacteria of Class 2 and above. For example, for importation into the United States the CDC provides form CDC 0.728 (F13.40), ''Permit to import or transfer etiological agents or vectors of human disease.'' This form must be obtained from the CDC (Centers for Disease Control, Attn: Biohazards Control Office, 1600 Clifton Road, Atlanta, GA 30033, tel. 404-329-3883) and must be filled out by the consignee in the United States; the CDC will then provide a label to be placed on the outside of the package by the consignor (Fig. 2), together with a PHS (Public Health Service) permit number, which should assure acceptance by Health Officers and the U.S. Quarantine Service. The importation of material which may be

[11] M. T. Alexander and P.-M. Daggett, _ASM News_ **54,** 244 (1988).

[12] _ASM News_ **55,** 301 and 660 (1989); _Federal Register_ **54,** 55. p. 11970, March 23, 1989; P.-M. Daggett, _ASM News_ **54,** 650 (1988).

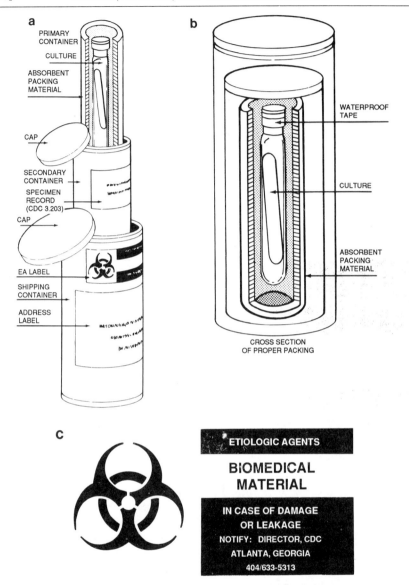

FIG. 1. Packaging and labeling of etiologic agents for interstate transport in the United States. The Interstate Shipment of Etiologic Agents in the United States (42 CFR, Part 72) was revised July 21, 1980, to provide for packaging and labeling requirements for etiologic agents of Class 2 and above. One approved method of packaging and labeling of etiologic agents is shown in (a) and (b). The primary and secondary containers should both be sealed with waterproof tape. (c) shows the label which should be affixed to all shipments of etiologic agents. For further information on the regulations, contact the Centers for Disease Control, Office of Health and Safety, Biosafety Branch, Atlanta, GA 30333.

IMPORTATION OR TRANSFER AUTHORIZED BY

PHS Permit No. _____

Expiration Date _____

TO:

DO NOT OPEN IN TRANSIT
BIOMEDICAL MATERIALS
ETIOLOGICAL AGENTS OR VECTORS

NOTICE TO CARRIER: If inspection on arrival in U.S. reveals evidence of damage or leakage, immediately notify: Director, Centers for Disease Control, Atlanta, Georgia 30333 – Telephone 404–633–5313.   CDC 0.1007 6/85

FIG. 2. Label required for import into the United States. The label is affixed to shipments of etiologic agents of Class 2 and above destined for importation into the United States. The label is obtained by the consignee from the Biosafety Office of CDC and sent to the consignor with the request for strains.

agricultural pests is under the control of other offices. Other countries have other systems, which are more or less restrictive than the U.S. system; it is the responsibility of scientists who want strains sent to their country to assure the sender that appropriate information or forms are provided to that sender.

Computerization of Culture Collection Data

Laboratories engaged in bacterial genetics can generate hundreds, even thousands of strains. It is imperative, then, for the geneticist to develop basic techniques of strain management. We emphasize, however, that strain management is more than just proper storage and retrieval of bacterial cultures; it necessarily includes proper storage and retrieval of strain data, without which the cultures would be virtually useless. Consider the volume of data associated with each strain: its accession number; its origin, date of deposition, and original strain code; its genotypic and phenotypic description; and references to publications dealing with it. It is important that records on strains should be consistent with the proposals for genetic nomenclature stated by Demerec et al.[13] Genetic stock centers and other reference collections must also manage distribution records and personal communications regarding each strain. It is obvious that the number of data statements a laboratory must manage may exceed the number of cultures by an order of magnitude or more.

[13] M. Demerec, E. A. Adelberg, A. J. Clark, and P. E. Hartman, *Genetics* **54,** 61 (1966).

In many laboratories, data about a given strain are scattered throughout notebooks, file cards, even human memories, and efforts to retrieve and summarize these data are inconvenient at best. At a time when almost every scientists has ready access to a personal computer, there is no need for this confusing system of data management to continue. We suggest that any laboratory which stores its own cultures computerize its database at the earliest possible stage. Several software packages designed for strain management now exist (see below). If one does not wish to use these packages, one can at least store strain data, using a tabular format, in a word processing file. The word processing packages allow searches for data strings. These search routines are rather inflexible but are considerably more efficient than a manual search of scattered notes. It is also a simple matter to retrieve these data and insert them into form letters to accompany strain shipments. If one wishes later to use more sophisticated data management software, a well-constructed word processing file can typically be converted to an ASCII file and imported by the new software, so that the information need not be retyped.

A greater degree of automation and searching speed can be achieved through the use of commercially available database management systems. These software packages are typically quite user friendly, requiring only a moderate degree of training and programming skill. Some years ago, Anderson and colleagues published a BASIC program package facilitating data entry and strain searches for collections of up to several hundred strains.[14] Shortly thereafter Bryant described the use of MicroPro International Corporation's Datastar, SuperSort, and Wordstar to accomplish these same tasks as well as to print a formatted catalog of the culture collection.[15] The few statistics available suggest that a large number of laboratories have independently written similar programs for their private use, and this option remains a viable one.[16] More recent relational database management packages are much more powerful and efficient than those of a few years ago, and programming with them is no more difficult than programming with BASIC.

Available Software Packages

There are currently three generally distributed, English language software packages designed specifically for manipulating culture collection data. System requirements, cost, and sources for each package are pre-

[14] L. M. Anderson, S. Borabeck, and K. F. Bott, *ASM News* **48,** 464 (1982).
[15] T. N. Bryant, *J. Appl. Bacteriol.* **54,** 101 (1983).
[16] A nonscientific survey of researchers with interests in plant pathogens indicated that 44% of 308 respondents used computers to assist in documenting their cultures [J. Hanus, *Microbial Germplasm Data Net Newsletter* **1**(3), 1].

TABLE I
SPECIFICATIONS FOR CULTURE COLLECTION DATA SOFTWARE

Parameter	CLONES	STOX	MICRO-IS
Program size	257K	230K	3–4M[a]
RAM required	256K	348K	512K
DOS version	2.1 or higher	2.0 or higher	2.0 or higher
Cost	$10	Free	$153[b]
Source[c]	Hal B. Jenson	Daniel R. Zeigler	Microbial Strain Data Network

[a] Amount of disk space required depends on user selections.
[b] MICRO-IS is "share-ware"; the disks may be freely copied or may be purchased with a user's manual for $118. The fee for registration, which entitles the user to automatically receive updated versions, is an additional $45. The combined fee for disks and registration is $153.
[c] Hal B. Jenson, Department of Pediatrics, Yale University School of Medicine, 333 Cedar Street, New Haven, CT 06510; Daniel R. Zeigler, Department of Biochemistry, The Ohio State University, Columbus, OH 43210; MSDN Secretariat, Institute of Biotechnology, 307 Huntingdon Road, Cambridge CB3 0JX, United Kingdom.

sented in Table I. At present, each package requires the IBM PC/XT/AT/ PS2 or 100% compatible personal computer. These packages arose from different laboratory needs, so each has a unique range of applications and a unique set of advantages and disadvantages.

MICRO-IS (the Microbial Information System) is essentially an applications package designed to manipulate strain data encoded with the RKC (Rogosa–Krichevsky–Colwell) code.[17] The RKC code stemmed from efforts to standardize and computerize microbial strain descriptions.[18] This code consists of a series of statements regarding strains; each statement is assigned a six-digit code number. Some statements call for a character string ("What is your laboratory strain number or designation?"); others require a numeric answer ("The minimum inhibitory concentration [MIC] of Penicillin G is X units/ml."); still others demand a binary (yes/no) response (Cells are gram-positive). Currently some 12,000 codes have been defined[19]; together, they provide a portrait of a strain in terms of determinative microbiology. Each of these codes is a potential field name for a MICRO-IS database. In practice, because no single database requires

[17] M. I. Krichevsky, First Canadian Workshop on Bioinformatics, Ottawa, Canada, 1989.
[18] M. Rogosa, M. I. Krichevsky, and R. R. Colwell, "Coding Microbiological Data for Computers." Springer-Verlag, New York, 1986.
[19] The editor of the RKC code is C. McManus, Microbial Systematics Section, Epidemiology and Oral Disease Prevention Program, National Institute of Dental Research, National Institutes of Health, Bethesda, MD 20892.

all these fields, users select appropriate subsets of items when creating new databases. Users may also select mnemonics to serve as field names in place of the RKC code numbers. MICRO-IS has its greatest practical application in facilitating the identification of new microbial isolates from nature. The user may identify known organisms with a high degree of similarity to the new isolate by means of the MICRO-IS Boolean search capabilities. Genotypes are not coded as such by the RKC system. Users may, however, redefine RKC code numbers to specify genotypes. The "lactose utilization" field, for example, could be used to specify the *lac* allele of an *E. coli* strain.

CLONES[20] is, in contrast, a much simpler software package, It arose out of the need to organize collections of recombinant DNA molecules and host bacterial strains. It enables the user to construct and manipulate databases with four kinds of records: bacterial strains, cloning vectors, recombinant clones, and libraries. The field names and specifications are fixed and cannot be altered by the user. Several fields—laboratory designation, grouping, storage location, journal reference, notes, and date—are common to all four kinds of records. Bacterial strains are assumed to be *E. coli;* the additional field names for strains are as follows: genotype (up to five genes), F' factor (yes or no), color assay, and antibiotic resistance. Vector records have fields for antibiotic resistance genes, single cloning sites, insertional inactivation sites, and size. Recombinant clone and library records have fields for size, name, and origin of insert; left and right cloning site; and vector name. The user may combine the four types of record in a single database or may disperse them among a number of databases. Searches are rapid and simple; any field or combination of fields may be specified as key for searching. The user can only specify a single allele when searching by genotype. CLONES does not support Boolean searches. CLONES is menu-driven; there are no individual help screens, but a user's manual can be viewed on-screen or printed by a selection from the main menu.

STOX[21] is a response to the needs of the Bacillus Genetic Stock Center (BGSC), an organization that maintains some 1200 strains from the genus *Bacillus* and distributes them without prejudice to academic, industrial, and government researchers throughout the world. STOX assists the BGSC in a number of tasks: storage of strain data, maintenance of mailing lists and strain usage records, printing out form letters to accompany strain shipments, and publishing a hard-copy catalog of strains. The STOX

[20] H. B. Jenson, *BioTechniques* **7**, 590 (1989).
[21] D. R. Zeigler and D. H. Dean, "Strains and Data," 4th Ed., p. ix. Bacillus Genetic Stock Center, Columbus, Ohio, 1989.

programs have undergone a continuous evolution since early 1986. Version 2, which the BGSC is currently making available, dates from February 1990. This version was compiled from dBASE III programs as a stand-alone applications package; dBASE III software is not required for its use.

The STOX main menu allows the user to select submenus for editing strain data, editing the mailing list or printing out mailing labels, processing a strain request, printing the catalog of strains, generating a report of strain usage, or conducting an on-line strain search by original code or genotype. The submenus each specify a variety of tasks. The user stays within the most recent submenu after each task is completed until he or she requests a return to the main menu. Database structure is invisible to the STOX user. STOX prompts the user for all information needed for each routine. No specific organism is assumed. Help screens are available at the main menu or at any submenu. Printed output may be directed either to a printer (the HP LaserJet II is assumed) or to an ASCII file for editing outside of STOX.

To find a record, then to edit it, requires fewer key strokes in CLONES than in STOX. CLONES combines all searches and editing on a single screen. STOX searches, in contrast, are divided among several screens. If a STOX user locates a strain by its genotype, he or she must remember the lab strain designation, return to the main menu, choose the editing menu, and finally call for the strain by its designation. STOX searches can be more informative than those of CLONES, however. If a user searches a CLONES database for a strain with a given allele, only the first such strain is located; subsequent strains must be located by sequentially repeating the search. In contrast, the user searching the STOX databases by allele would be given, on screen or in print, a complete listing of all strains possessing that allele; if there are many such strains, the user would first be invited to narrow the search by specifying additional alleles. Thus, STOX genotype searches are partially Boolean; the user may specify that the strain must have a given allele plus any additional allele from a specified list.

In summary, MICRO-IS might be the software of choice for the investigator who requires a large number of descriptor fields for each strain. Researchers who possess very diverse collections or whose primary interest involves determinative microbiology could take full advantage of this system. Indeed, MICRO-IS has been adopted by the American Type Culture Collection.[22] Yet researchers whose interests are primarily genetic, especially those who collect mutants derived from a small number of parent strains, might find the diversity of the MICRO-IS phenotypic

[22] M. I. Krichevsky, personal communication (1990).

descriptors more burdensome than helpful. CLONES or STOX, which are attempts at more restricted objectives and are therefore easier to learn and to use, might be preferable under such circumstances. All three software packages continue to evolve, and new versions of each are in preparation.[23]

Networks and Bulletin Boards

Researchers who convert their culture collection data to machine-readable form may want to share the data with colleagues through computer networks. A full discussion of networking is beyond the scope of this chapter. We instead describe briefly three networks that may be of specific interest to the microbial geneticist.

The Microbial Strain Data Network (MSDN) was founded by member organizations of the International Council of Scientific Unions as a worldwide, electronic strain-locator service for microbial strains and eukaryotic cell lines. The MSDN is a distributive network. As such it stores no strain data directly, but rather maintains a directory of data centers regarding culture collections and provides access to their databases. Catalogs from the American Type Culture Collection, the Deutsche Sammlung für Mikrobiologie (Braunschweig), and several other culture collections are available on-line. MSDN users may search for strain locations by RKC code descriptors (see above). The MSDN also provides electronic mail and bulletin board services, allows for electronic conferences, and distributes the MICRO-IS software package.[24]

The BioSciences Information Service (BIOSIS) is best known for their printed bibliographic databases, including *Biological Abstracts*. In 1985 BIOSIS introduced a pilot system for a computerized information exchange, the Taxonomic Reference File (TRF). The TRF system allows users to access a variety of databases—the actual file of taxonomic nomenclature is only one of them—through an on-line network. Users may comment on the data, share pertinent research findings, and create data subsets for personal use. Currently, the data files from the Bacillus Genetic Stock Center and the *Staphylococcus aureus* collection of P. Pattee are accessible through the TRF. Users may make use of the TRF bulletin board to request strains or information from these collections. BIOSIS

---

[23] M. I. Krichevsky and H. B. Jenson, personal communications (1990).

[24] M. I. Krichevsky, personal communication (1990). Direct inquiries regarding the MSDN should be addressed to MSDN Secretariat, Institute of Biotechnology, 307 Huntingdon Road, Cambridge CB3 0JX, United Kingdom.

would consider adding other culture collection data files provided that the files are available in a well-organized, machine-readable format.[25]

The two networks described above are designed to give researchers access to databases from rather large reference collections. The Microbial Germplasm Data Net (MGDN), in contrast, is intended to provide access to data regarding private research collections, most of which are relatively small. The scope of the MGDN is limited to microbes, antibodies, and cloned molecules relating to plant pathogenesis. The MGDN is in its first phase of development, resulting in a printed culture collections directory that is now in preparation. The second phase of development will culminate with on-line access to culture collection data. The MGDN is administered by Oregon State University.[26]

[25] R. T. Howey, personal communication (1990). Direct inquiries regarding the TRF to Robert T. Howey, Section Chief, BIOSIS, Special Projects Section, 2100 Arch Street, Philadelphia, PA 19103-1399.

[26] J. Hanus, personal communication (1990). Direct inquiries regarding the MGDN to Dr. Larry Moore, Department of Botany and Plant Pathology, Oregon State University, Corvallis, OR 97331.

[11] Bacteriophage P2 and P4

By Michael L. Kahn, Rainer Ziermann, Gianni Dehò, David W. Ow, Melvin G. Sunshine, and Richard Calendar

Introduction

Bacteriophage P4 is a small temperate bacteriophage that grows lytically only in the presence of a helper bacteriophage such as P2. The life histories of P2 and P4 and the relationship between them have been reviewed recently,[1] and the complete nucleotide sequence of P4 is known.[2] Each phage carries the genes necessary to assure its own DNA replication and integration into the host chromosome, but only P2 has genes that encode proteins needed to produce phage particles and lyse the cell. P4 activates the expression of the P2 structural proteins and is encapsidated within a phage particle that is very similar to the P2 particle with the

[1] L. E. Bertani and E. W. Six, *in* "The Bacteriophages" (R. Calendar, ed.), Vol. 2, p. 73. Plenum, New York, 1988.

[2] C. Halling, R. Calendar, G. E. Christie, E. C. Dale, G. Dehò, S. Finkel, J. Flensburg, D. Ghisotti, M. L. Kahn, K. B. Lane, C.-S. Lin, B. H. Lindqvist, L. S. Pierson III, E. W. Six, M. G. Sunshine, and R. Ziermann, *Nucleic Acids Res.* **18**, 1649 (1990).

exception that the capsid is smaller. P4 can also exist as an integrated prophage or as a plasmid, and exploiting this dexterity leads to many strategies for manipulating DNA *in vitro* and *in vivo*.

The major advantage of P4-derived vehicles is that they permit plasmids to be transferred very efficiently from cell to cell without DNA isolation and with few constraints on the genotype of bacterial strains. P2 forms plaques on many enteric bacteria, including isolates of *Klebsiella*,[3] *Salmonella*,[4-6] *Serratia*,[4] *Shigella*,[4] and *Yersinia*,[1] and it can be used to introduce encapsidated plasmid DNA into other gram-negative bacteria such as *Rhizobium*[7] and *Pseudomonas*.[8] P2 and P4 constructs can therefore also be used to extend the host range of plasmids manipulated in *Escherichia coli*. There are a number of components in the P4 genome that can be used in the construction of cloning vehicles. The first part of this chapter focuses on P4-derived cloning vehicles. At the end of the chapter we discuss the use of P2 in the selection of λ transducing phages and in deleting the *his* region of *E. coli* K12. Bacterial strains used in the various manipulations are listed in Table I.

P4-Derived Phasmids

In one early construction with P4,[9] the phage was combined with a kanamycin-resistant derivative of plasmid ColE1 to yield a phasmid, a recombinant that can be propagated both as a phage and as a plasmid. Subsequent work has shown that P4 can replicate as a plasmid even without the assistance of the ColE1 replicon.[3,10-12] The P4 *vir1* mutant is partially immunity insensitive[1] and highly biased toward the plasmid mode of replication.[10-12] The copy number of P4 *vir1* varies with temperature and growth conditions, from about 1–2 at 30° to about 50 at 37°.[13] The largest P4 *Eco*RI fragment contains all of the genes and *cis*-acting sites needed for replication[9] and the complete *cos* site (the DNA region needed

[3] D. W. Ow and F. M. Ausubel, *Mol. Gen. Genet.* **180,** 165 (1980).

[4] L. E. Bertani and G. Bertani, *Adv. Genet.* **16,** 199 (1971).

[5] E. P. Ornellas and B. A. D. Stocker, *Virology* **60,** 491 (1974).

[6] L. Gutmann, M. Agarwal, M. Arthur, C. Campanelli, and R. Goldstein, *Plasmid* **23,** 42 (1990).

[7] M. L. Kahn and C. R. Timblin, *J. Bacteriol.* **158,** 1070 (1984).

[8] A. Polissi, G. Dehò, and G. Bertoni, unpublished observations (1990).

[9] M. L. Kahn and D. R. Helinski, *Proc. Natl. Acad. Sci. U.S.A.* **75,** 2200 (1978).

[10] P. Alano, G. Dehò, G. Sironi, and S. Zangrossi, *Mol. Gen. Genet.* **203,** 445 (1986).

[11] G. Dehò, D. Ghisotti, P. Alano, S. Zangrossi, M. G. Borello, and G. Sironi, *J. Mol. Biol.* **178,** 191 (1984).

[12] R. Goldstein, J. Sedivy, and E. Ljungquist, *Proc. Natl. Acad. Sci. U.S.A.* **79,** 515 (1982).

[13] R. Lagos and R. Goldstein, *J. Bacteriol.* **158,** 208 (1984).

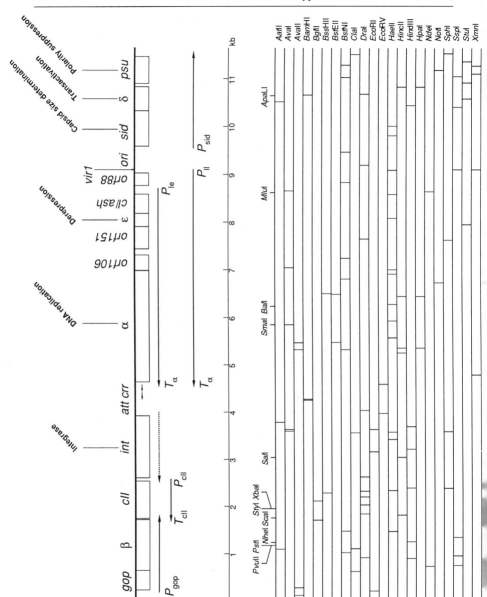

for packaging) near one end[14] (Fig. 1), but, at 8.2 kilobases (kb), it is too small to grow lytically. Insertion of a foreign DNA fragment in the range of 1–5 kb will restore the ability of the phage to grow.[9,15] However, the capacity of P4 *vir1* as a phage cloning vehicle can be increased considerably (to about 17–25 kb) by increasing the size of the phage head to that of the helper, which has a 33-kb chromosome. This can be accomplished by using P4 *sid* mutants, which are defective in size determination[16] and are unable to cause the formation of small heads. Alternatively, P2 *sir* mutants can be used as the helper phage. These mutants do not respond to the P4 size determination factor and produce P2-sized heads.[1]

P4 cloning vehicles have been constructed that contain various antibiotic resistance determinants and restriction sites (Table II). Some phasmids contain an amber mutation in the major replication gene, α. These derivatives can replicate only in an appropriate suppressor strain, such as C-1766, C-1895, or C-2324 (Table I), which carry *supD*. If a *sup°* strain is infected with a P4 αam derivative containing a mutationally altered host gene, it is possible to select for recombinants in which the cloned gene has been integrated into the chromosome.[15] It is also possible to use constructs such as P4::Tn5 AP-2 (Table II) as delivery vehicles for transposons[8] if they are introduced into a nonsuppressing strain.

Cosmids

For P4 DNA to be packaged, it must carry an identifying sequence that allows the phage to recognize it. Early work indicated that the cohesive ends of P2 and P4 were similar[17] and suggested that the *cos* sequence was located in this region. Plasmids, but not linear DNA, containing the "optimal" *cos* site [about 125 base pairs (bp) including the 19 bp long cohesive ends][18] of either P2 or P4 are efficiently packaged into capsids, probably because covalently closed circles are the preferred substrate for

[14] C. Halling and R. Calendar, *in* "Genetic Maps 1990" (S. J. O'Brien, ed.), Book 1, p. 70. Cold Spring Harbor Laboratory, Cold Spring Harbor, New York, 1990.

[15] D. W. Ow and F. M. Ausubel, *J. Bacteriol.* **155,** 704 (1983).

[16] D. Shore, G. Dehò, J. Tsipis, and R. N. Goldstein, *Proc. Natl. Acad. Sci. U.S.A.* **75,** 400 (1978).

[17] J. C. Wang, K. V. Martin, and R. Calendar, *Biochemistry* **12,** 2119 (1973).

[18] R. Ziermann and R. Calendar, *Gene* **96,** 9 (1990).

[19] G. J. Pruss, J. C. Wang, and R. Calendar, *J. Mol. Biol.* **98,** 465 (1975).

FIG. 1. Genetic and restriction map of bacteriophage P4. Arrows below the genetic map indicate transcription units. Positions of cleavages by restriction enzymes that cut P4 DNA once are given along the top of the restriction map.

TABLE I
Escherichia coli STRAINS USEFUL IN MANIPULATION WITH P2 AND P4

Strain	Genotype	Use	Ref.
C-1a	Wild type	Nonsuppressing, nonlysogenic strain for growing P2	a
C-436	Polyauxotrophic, *str*	Good indicator strain for P2	a
C-520	*supD*	Indicator strain for P2, including amber mutants; good host for transductions using P4 lysates	a
C-1706	(P2 *c5*) P2^r	Temperature-sensitive P2 lysogen, induction of derivatives that contain cosmids will generate P2 transducing lysates	b
C-1766	*supD* (P2 *cox4*)	Suppressing, P2-lysogenic strain for growing P4 amber mutants	c
C-1792	*supF*	Indicator strain for P2, including some amber mutants	a
C-1895	*supD* (P2 *lg*)	Suppressing, P2-lysogenic strain for growing P4	d
C-2323	(P2 *lg*)	P2 *lg* lysogen of C-436	e
C-2324	*supD* (P2 *lg*)	P2 *lg* lysogen of C-520; indicator strain for P4, including amber mutants; good host for transductions using P2 *virl* or P2 *c5* lysates	R. Calendar
C-5427	*supD* (P2 *sirl old81*)	Host for P4 phasmids, including amber mutants	f
C-5498	*supD* (P2 *Aam127 sirl ogrl*)	Host for P4 phasmids, produces P2 helper that can be eliminated using nonsuppressing host	f
NM646	(P2 *cox3*)	Galactose nonfermenting, lysogenic for an excision-defective P2 mutant	g

[a] M. G. Sunshine, M. Thorn, W. Gibbs, and R. Calendar, *Virology* **46,** 691 (1971).
[b] R. Calendar, B. Lindahl, M. Marsh, and M. Sunshine, *Virology* **47,** 68 (1972).
[c] W. Gibbs, R. N. Goldstein, R. W. Wiener, B. Lindqvist, and R. Calendar, *Virology* **53,** 24 (1973).
[d] K. J. Barrett, M. L. Marsh, and R. Calendar, *J. Mol. Biol.* **106,** 683 (1976).
[e] D. Ghisotti, S. Finkel, C. Halling, G. Dehò, G. Sironi, and R. Calendar, *J. Virol.* **64,** 24 (1990).
[f] A. Polissi, G. Dehò, and G. Bertoni, unpublished observations (1990).
[g] P. A. Whittaker, A. J. B. Campbell, E. M. Southern, and N. E. Murray, *Nucleic Acids Res.* **16,** 6725 (1988).

DNA packaging.[19] Packaging requires only one *cos* site in the plasmid,[19] rather than at least two *cos* sites on concatemeric DNA required by bacteriophage λ.[20] Although circular multimers contain more than one *cos* site, they can be packaged by P2 and P4.

The number of transducing particles (phage particles containing cosmid

[20] S. M. Rosenberg, M. M. Stahl, I. Kobayashi, and F. W. Stahl, *Gene* **38,** 165 (1985).

TABLE II
P4-DERIVED PHASMIDS

Phasmid	Size (kb)	Special properties	Ref.
P420	12.5	P4 *virl*–ColE1 hybrid, Kan^r,^{e} source of large P4 EcoRI fragment, single PstI site	a
P422	11.0	P4 *virl*–ColE1 hybrid, Kan^r, single EcoRI site	b
P4DO100	10.0	P4 *virl sid1*, Tet^r, single EcoRI, SalI, XhoI, BalI sites	c
P4DO105	11.6	P4 *virl* αam52, Kan^r, Tet^r, single EcoRI, SalI, XhoI, BalI sites	c
P4::Tn5 Ap-2	15.8	P4 *virl* αam1 *del2*::Tn5, Kan^r, integration defective, conditionally replicating transposon delivery vehicle	d
pKGB2	25	P4 *virl rurl*–pKT231 hybrid, Kan^r, Str^r, single SacI, SacII, PvuI, XhoI sites; P4 *virl rurl* is able to grow on strains that produce EcoRI because it has lost EcoRI site 2	d

^{a} M. L. Kahn and D. R. Helinski, *Proc. Natl. Acad. Sci. U.S.A.* **75**, 2200 (1978).

^{b} M. Kahn, D. Ow, B. Sauer, A. Rabinowitz, and R. Calendar, *Mol. Gen. Genet.* **177**, 399 (1980).

^{c} D. W. Ow and F. M. Ausubel, *J. Bacteriol.* **155**, 704 (1983).

^{d} A. Polissi, G. Dehò, and G. Bertoni, unpublished observations (1990).

^{e} Kan^r, Kanamycin resistance; Tet^r, tetracycline resistance; Str^r, streptomycin resistance.

DNA) produced by P2 or P4 infection of a cell that contains a high copy number cosmid can be as much as 30% of the total burst of phage (Fig. 2).[18,21] With the cosmid pKGB4 transduction frequencies (ratio of transducing particles to total burst of phage) of up to 0.75 could be observed.[8] Control experiments have shown that there is no transduction of similar plasmids that do not contain a P4 or P2 *cos* site.[18,21] For a cosmid to be converted to a transducing particle during P2 infection, its length must be within certain size ranges (Fig. 2). Cosmids between 23 and 33 kb can be transduced with high frequencies (>0.1), and cosmids from 17 to 21 kb are probably too small to be packaged as monomers. The increase in transduction frequencies at lengths below 17 kb is likely due to packaging of plasmid dimers,[16] and smaller cosmids must be packaged as higher multimers. The same general pattern can be observed in the production of infectious P4 *virl sid1*[16] phage particles by superinfection of phasmid-containing strains with P2 *virl*.[22] Because P2 and P4 cosmids are packaged at significant frequency ($>1 \times 10^{-3}$) even when smaller than 5 kb, they

[21] M. L. Kahn, unpublished observations (1990).

[22] D. Ow, unpublished observations (1982).

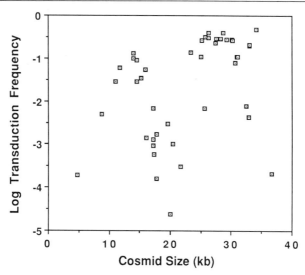

FIG. 2. Various cosmids, each of which had a ColE1 origin of replication and carried an antibiotic resistance gene, were transformed into *E. coli* strain C-1a. Each of the resulting strains was grown in Luria broth, infected with P2 *vir1*, and grown until the cells lysed. Debris was removed by centrifugation, and the lysates were sterilized with chloroform. The titer of transducing particles was then determined by infecting *E. coli* C-2324 (P2 *vir1* is immunity sensitive and cannot propagate in a P2-lysogenic strain) and selecting for antibiotic resistance. The P2 *vir1* titer was determined by infecting C-1a. Values on the abscissa indicate the length of each cosmid as determined by agarose gel electrophoresis of restriction digests; values plotted on the ordinate are $\log_{10}$ and show the resulting transduction frequencies of the different cosmids.

can continue to be used as cosmids during subcloning manipulations. Cosmids such as pcos45[18] (Fig. 3), a 2.8-kb pUC19 derivative that contains a very short (125 bp) P4 *cos* sequence, have a transduction frequency of about 0.1. This high frequency could be due to the high copy number of the pUC19 plasmid or to a difference in the packaging of the small *cos* sequence. Not all small cosmids are so efficiently packaged (Fig. 2), and this discrepancy is not understood.

The size constraints described above can be used to isolate plasmids that carry deletions. Many of the plasmids in Fig. 2 were derived from pMK207,[21] a 37-kb broad host range (*incP-1*) plasmid that carries ampicillin, tetracycline, and kanamycin resistance genes. When a P2 lysate of a strain that contained pMK207 was used to transduce one of these antibiotic resistances, the resulting transductant often was missing another resistance. A number of those that carried all three resistances had plasmid deletions that brought the length of the new plasmid within the P2 size

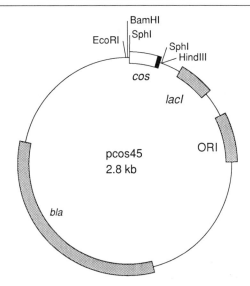

FIG. 3. Cosmid pcos45 contains a 125-bp *cos*-containing fragment of P4 DNA. The plasmid was constructed by first cloning a 325-bp *SphI/EcoRI* fragment of P4 into pUC19 digested with *SphI* and *EcoRI*. This plasmid (pMG1) was digested with *BssHII* and *HindIII* and treated with Klenow polymerase. The *cos*-containing fragment was cloned into pUC19 digested with *HincII*. The *cos* fragment can be resected by digestion with *SphI*. P4 DNA is indicated as an open box. The 19 base long cohesive ends are darkened.

range. The most useful of these was pMK207-9, a plasmid that has been used to screen various gram-negative bacteria to see whether they are within the host range for P2 infection by determining whether they can acquire multiple antibiotic resistance after exposure to a transducing lysate.

ColE1- and P15A-derived cosmids can be used in *E. coli* and other enteric bacteria, and several RK2 (*incP-1*)-derived P4 cosmid vehicles have been developed for genetic analysis in gram-negative organisms such as *Rhizobium* and *Pseudomonas*. Several of these are listed in Table III. They include general purpose cloning vehicles such as P422 HpaI,[23] pcos45,[18] pMK317,[24] and pRZ11,[18] vehicles for analyzing transcriptional regulation such as pMK330 and pMK341,[7] and vehicles for more specialized purposes such as random Tn5 mutagenesis or replacing alleles of particular genes (marker exchange mutagenesis).

Two cosmids, pHX1 and pMK409 (Table III), contain the *Bacillus*

[23] M. Kahn, D. Ow, B. Sauer, A. Rabinowitz, and R. Calendar, *Mol. Gen. Genet.* **177**, 399 (1980).
[24] J. E. Somerville and M. L. Kahn, *J. Bacteriol.* **156**, 168 (1983).

TABLE III
P4-DERIVED COSMIDS

Cosmid	Size (kb)	Special properties	Ref.
P422 HpaI	5.0	ColE1 replicon, Kan^r,^{g} single *Eco*RI, *Bam*HI, *Hpa*I, *Hin*dIII, *Xho*I, *Sma*I, *Cla*I, *Sph*I sites	*a*
pcos45	2.8	pMB1 replicon, Amp^r, contains 125 bp "optimal" *cos*	R. Ziermann
pHX1	28.7	RK2 replicon, *E. coli lac* promoter, *B. subtilis* levansucrase genes, Tn*5* delivery vehicle, Tet^r	M. Kahn
pKGB4	15.5	pKT231 replicon (RSF1010), Kan^r, Str^r, single *Sac*I, *Sac*II, *Hin*dIII, *Pvu*I, *Sma*I, *Xho*I sites	*b*
pMK317	14.3	ColE1 and RK2 replicons, Kan^r, Tet^r, single *Bam*HI, *Bgl*II, *Xho*I, *Sal*I sites	*c*
pMK330	23.0	ColE1 and RK2 replicons, Kan^r, Tet^r, single *Bam*HI, *Bgl*II, *Xho*I, *Sal*I, *Hin*dIII sites, β-galactosidase translational fusion vehicle	*d*
pMK341	15.4	ColE1 and RK2 replicons, Chl^r, Tet^r, single *Bgl*I, *Xho*I, *Sal*I sites, kanamycin operon fusions	*d*
pMK409	23.0	RK2 replicon, Tet^r, *E. coli lac* promoter, *B. subtilis* levansucrase genes, single *Bam*HI site, for expression and marker exchange mutagenesis	M. Kahn
pRZ11	4.5	pMB1 replicon, Amp^r, contains 125 bp "optimal" *cos*; Similar to pMA1,^{e} a pBR322 derivative, with a shorter *cos* and lacking the adenovirus DNA insert	R. Ziermann
pSK1	11.5	RK2 replicon (*incP-1*), Tet^r, single *Bam*HI, *Bgl*II, *Hin*dIII sites	M. Kahn
pSP3	5.8	P15A replicon, Chl^r, single *Hin*dIII, *Bam*HI sites	*f*

^{a} M. Kahn, D. Ow, B. Sauer, A. Rabinowitz, and R. Calendar, *Mol. Gen. Genet.* **177,** 399 (1980).

^{b} A. Polissi, G. Dehò, and G. Bertoni, unpublished observations (1990).

^{c} J. E. Somerville and M. L. Kahn, *J. Bacteriol.* **156,** 168 (1983).

^{d} M. L. Kahn and C. R. Timblin, *J. Bacteriol.* **158,** 1070 (1984).

^{e} M. Agarwal, M. Arthur, R. D. Arbeit, and R. Goldstein, *Proc. Natl. Acad. Sci. U.S.A.* **87,** 2428 (1990).

^{f} L. S. Pierson III and M. L. Kahn, *Mol. Gen. Genet.* **195,** 44 (1984).

^{g} Amp^r, Ampicillin resistance; Chl^r, chloramphenicol resistance; other abbreviations as in Table II.

subtilis levansucrase genes.[25] The presence of these genes make a cell sensitive to 5% (w/v) sucrose. pHX1 can be used to isolate Tn*5* chromosomal insertion mutants. The plasmid is introduced into the recipient strain (*E. coli, Rhizobium*), and, by growing in the presence of sucrose and kanamycin, one selects for cells in which Tn*5* has jumped from the plasmid to the chromosome and the plasmid has been lost. In an analogous way, genes cloned onto pMK409 can be mutated by inserting antibiotic resistance determinants into them. Recombinants that have lost the plasmid but which have transferred the mutant allele to the chromosome can be obtained by selecting for both the drug resistance and sucrose resistance.

P4 Attachment

Some genetic analyses of gene regulation require the use of a single gene copy. Integration of a P4 derivative into the bacterial chromosome causes introduced genes to be present at one copy per chromosome. Reducing the temperature to 30° or less will reduce the P4 plasmid copy number and select for integrated P4.[13] All of the plasmids listed in Table II contain the bacteriophage P4 attachment site (P4 *att*) except P422 and P4::Tn*5* AP-2. Plasmid pSP5[26] expresses P4 integrase but cannot integrate. It can therefore promote the integration of plasmids that contain P4 *att*; pSP5 can be cured from the cell by screening for white colonies on crystal violet plates.[26] These manipulations produce stable, single copy insertions at the chromosomal (P4) *att* site located at the 3′ end of a tRNA$_{Leu}$ gene.[27] In some strains of *Klebsiella*[15] and *Salmonella*,[6] site-specific integration can occur in the absence of the P4 integrase gene, perhaps through the action of integrase carried on cryptic prophage.[27]

Preparation of Bacteriophage P2

High-titer stocks of P2 phage are always grown from phage isolated from a single plaque. Grow P2 *am*$^+$ on C-1a, and grow P2 *am* on C-520 (*supD*) or C-1792 (*supF*), depending on the phage mutant and suppressor specificity (Table I). After growth, mutant stocks should always be tested for revertants. In the method described below, two rounds of growth occur before the culture lyses.

Host Culture. Cells are grown in modified Luria broth (1% (w/v) yeast

[25] P. Gay, D. Le Coq, M. Steinmetz, T. Berkelman, and C. Kado, *J. Bacteriol.* **164,** 918 (1985).
[26] L. S. Pierson III and M. L. Kahn, *Mol. Gen. Genet.* **195,** 44 (1984).
[27] L. S. Pierson III and M. L. Kahn, *J. Mol. Biol.* **196,** 487 (1987).

extract, 0.5% (w/v) tryptone, 0.5% (w/v) NaCl, 0.1% (w/v) glucose, 2 mM MgCl$_2$) at 37° overnight without shaking. For producing phage stocks and titering transductants, it is best to use fresh overnight cultures.

Two-Step Growth. Add 5 ml of a fresh overnight culture to 0.4–0.5 liter of prewarmed modified Luria broth with 2 mM CaCl$_2$ in a 2.8-liter Fernbach flask. Grow at 37° with gentle shaking to 1 × 10^8 cells/ml. This will take just under 2 hr. Infect at a multiplicity of infection (moi) of 0.1 with P2, incubate without shaking for 7 min, then continue shaking. Monitor the density of the culture until it stops increasing (at about 2 hr after infection). When lysis begins, add 8 ml of 4% EGTA (pH 8.8) to chelate calcium and prevent adsorption of the phage to cellular debris and MgCl$_2$ to 10 mM to stabilize the phage.

Extraction of Phage. After lysis add 8 g NaCl. Shake to dissolve, then pellet debris by centrifugation (7000 rpm for 10 min at 4°). While the lysate is spinning add 50 ml of water and 100 g of polyethylene glycol 6000 (or PEG 8000) to the flask and make a slurry. Pour the supernatant into the slurry and mix until the PEG is dissolved. Let stand at least 15 min at 0°. The phage can be left overnight at this stage.

Pellet the phage at 5000 rpm for 10 min. Pour off supernatant and immediately put the bottle back in the rotor with the same side facing out. Spin at 1000 rpm for 1 min. Remove any additional PEG supernatant with a pipette. Resuspend the phage in P2 buffer (10 mM Tris, pH 7.5, 10 mM MgCl$_2$, 1% (w/v) ammonium acetate), about 4 ml per bottle. When the pellet is completely dispersed, transfer to 50-ml tubes. If this is to be the final stage in purification, add 0.5 ml chloroform and mix gently. Pellet debris (10,000 rpm, 10 min). Remove the phage-containing supernatant, then add more chloroform and refrigerate.

If the phage need to be concentrated more, do not add chloroform to the supernatant from the bottles. Pellet debris (10,000 rpm, 10 min). Take the supernatant and add NaCl to 0.5 M and PEG to 10%, let stand for 10 min, then spin in SA-600 rotor at 10,000 rpm for 10 min. Resuspend the pellet in P2 buffer. Add 0.5 ml chloroform and mix gently. Spin to separate phases, then take the supernatant and add a few drops of chloroform to store.

To purify phage using CsCl gradients, add 0.55 g CsCl per milliliter of phage suspension and centrifuge at 40,000 rpm in a Beckman SW50.1 rotor for 36 hr.

Preparation of Bacteriophage P4

The preparation of high-titer P4 stocks, which minimizes the number of revertants, involves infection with phage from a single plaque. *Esche-*

richia coli C-1895 (Table I) can be used for growth of P4 *vir*1 and P4 amber mutants.

Cells (10 ml) are grown overnight at 37° in Luria broth (LB: 1% yeast extract, 0.5% tryptone, 0.5% NaCl) without aeration. To this culture add one plaque of P4 plus $CaCl_2$ to 2 mM to promote adsorption. Incubate at 37° without aeration for 10 min and then add the infected culture to a 2.8-liter Fernbach flask that contains 400 ml modified Luria broth supplemented with 2 mM $CaCl_2$. Incubate at 37°, with gentle shaking. There should be single waves rolling across the flask in order to assure good aeration. Start reading the A_{600} after 90 min. Lysis should begin at about 3 hr, when the culture reaches an absorbance between 0.7 and 1.5. When lysis begins add 8 ml of 4% EGTA (pH 8.8) and $MgCl_2$ to 80 mM to stablize the phage. When lysis is complete, pellet debris by centrifugation in a GSA rotor at 5000 rpm for 10 min. Typical titers are about 10^{10}/ml. Concentration of the phage can be performed by the procedure given for P2, except that the extraction buffer should contain 80 mM $MgCl_2$. To purify using a CsCl gradient, add 0.5 g CsCl per milliliter of phage suspension and centrifuge as for P2 phage.

Transfection with P4 DNA

The calcium chloride procedure[28] is used for making competent cells. Grow P2-lysogenic cells (C-2323 for P4 *am*$^+$ and C-2324 for P4 *am*; Table I) in LB to a density of 5×10^7 colony-forming units (cfu)/ml, pellet by centrifugation (6000 rpm, 4°, 5 min), resuspend in one-half of the original volume of ice-cold 50 mM $CaCl_2$, 10 mM Tris, pH 7.6, and incubate on ice for 15–30 min. Pellet cells again and resuspend in 1/20 of the original volume of ice-cold 50 mM $CaCl_2$, 10 mM Tris, pH 7.6. Dispense in 0.2-ml aliquots to small culture tubes on ice.

Add about 200 ng of P4 DNA to 0.2 ml of competent cells and incubate on ice for 30–45 min. Heat-shock the cells for 2 min at 42° in a water bath, and add 1 ml of LB. Immediately, add 0.2-ml aliquots of the transfected cells and 0.1 ml of a standing overnight culture of the same cell strain to 2.5 ml top agar (LB, 2.5 mM $CaCl_2$, 0.7% agar) and pour onto LC plates (LB, 2.5 mM $CaCl_2$, 1% agar). Incubate overnight at 37°.

Transduction of Cosmids

Transform a suitable strain of bacteria with the cosmid to be packaged into P2 (or P4). Most *E. coli* C or K12 strains are good hosts for both phage, if 2 mM $CaCl_2$ is included in the medium. Grow the cosmid-con-

[28] T. Maniatis, E. F. Fritsch, and J. Sambrook, "Molecular Cloning: A Laboratory Manual," p. 250. Cold Spring Harbor Laboratory, Cold Spring Harbor, New York, 1982.

taining strain under antibiotic selection at 37° overnight without shaking in LB. Add 0.5 ml of the culture to 20 ml of LB containing 2 mM CaCl$_2$. Shake the cells vigorously at 37° for 2–3 hr, which should bring the cell density to 1 × 10^8 cells/ml. Infect with P2 *vir1*[29] at a moi of 0.1. Grow the infected cells at 37° for 2–4 hr, or until lysis occurs. Pellet the cell debris (10,000 rpm, 10 min) and transfer the supernatant to a sterile tube. Add 0.2 ml of chloroform and agitate for 15–30 min to kill surviving cells.

Lysates that contain cosmids can also be prepared using a P2 *c5* lysogenic strain such as C-1706 (Table I). The *c5* mutation causes the P2 repressor to be temperature sensitive, and at 37° prophage gene expression is induced. A standing overnight culture of C-1706 containing the cosmid is grown at 30°, diluted 1 : 100 in LB, and grown at 30° to a density of 3 × 10^8 cells/ml. The culture is shifted to 37° and shaken until the cells lyse. Because of the P2-resistance mutation in C-1706, transducing phage do not readsorb to cell debris.

Small-Scale Preparations. Add 0.1 ml of a standing overnight culture to 2.5 ml of LB top agar and pour the mixture on a LC plate. Spot 10 μl of lysate containing about 10^5 bacteriophage (P2 or P4) onto the surface and incubate overnight at 37°. A plug that contains phage and transducing particles is picked from the center of the clear zone using a Pasteur pipette and is incubated on ice for 3 hr in a small culture tube with 2 ml of P2 or P4 phage buffer and 0.1 ml chloroform.

Determination of Phage Titer. Mix 0.1 ml of a standing overnight culture of a strain that will permit growth of the phage (i.e., C-1a for P2 *vir1*, C-2324 for P4 αam phage, and C-1792 for P2 Aam phage) with 0.1 ml of a serial dilution of the phage lysate (10^{-1}–10^{-3}) and 2.5 ml top agar. Plate out on LC plates and incubate at 37°.

Determination of Transduction Frequency. Mix 0.1 ml of dilutions of phage lysate with 0.1 ml of an overnight culture of C-1a (Table I). C-1a is a wild-type strain and will not permit the growth of any amber phage. When employing a nonamber P2 phage such as P2 *vir1* for transduction, use C-2324 (Table I) as the recipient. C-2324 is lysogenic for P2 *lg* (large burst size), and, because of P2 immunity, it will not support P2 *vir1* lytic growth. Plate with 2.5 ml top agar on an antibiotic plate. For the cosmids listed in Table III, the resistance levels used are 200 mg/liter penicillin, 25 mg/liter tetracycline hydrochloride, and 50–100 mg/liter kanamycin monosulfate or streptomycin sulfate.

P2 Infection of *Rhizobium meliloti* 104A14[24]

Grow *Rhizobium* overnight in 10 ml of YMB (yeast mannitol broth: 1 g yeast extract, 10 g mannitol, 0.5 g K$_2$HPO$_4$, 0.1 g NaCl, 0.2 g MgSO$_4$ ·

[29] L. E. Bertani, *Virology* **12**, 553 (1960).

$7H_2O$ per liter). Dilute the culture 1 : 100 in 20 ml of YMB medium. Shake cells at 30° and let the culture grow up to log phage (OD_{600} 0.5–0.7). Spin the cells down at 5000 rpm for 10 min at 4°. Resuspend the pellet in 10 ml YMB and vortex vigorously for 30 sec. Spin down the cells at 5000 rpm for 10 min at 4°. Resuspend again in 10 ml of YMB and vortex the cells vigorously for 30 sec. Spin down the cells at 5000 rpm for 10 min at 4°. Resuspend the pellet in 1 ml YMB. Mix 100 μl of cells with 100 μl mixed phage lysates in sterile test tubes. Place cells and phage at room temperature for 20 min. Add 2 ml YMB to each tube and shake at 30° for 12 hr. Add 3 ml of YMB top agar to each tube and pour onto YMB plates that contain antibiotic.

Selection of λ Transducing Phages Using the P2 *old* Gene

λ phage cannot make plaques on *E. coli* strains that carry a P2 prophage. This interference phenomenon is due to the action of a nonessential P2 gene called *old* and to λ genes and sites that lie between the prophage attachment site and the *gam* gene.[30,31] Plating of a λ stock on a P2-lysogenic strain selects for λ phage which have lost the *att–gam* region and gained the *E. coli bio* genes that are linked to λ prophage. A large variety of λ transducing phages can be selected using restriction enzyme technology in conjunction with P2 interference.[32] The transducing phage selected by this method are called Spi⁻, because they have lost their sensitivity to P2 interference. Stocks of λ transducing phages selected on a P2-lysogenic strain will contain contaminating P2 phage, owing to spontaneous release of P2 prophage. This contamination can be reduced substantially by the use of an *E. coli* strain such as NM646 (Table I) that is lysogenic for an excision-defective P2 prophage. Alternatively one can prepare λ transducing phages on a strain that expresses the P2 *old* gene on a plasmid such as pSF2, which is derived from pSC101, a low copy number plasmid conferring tetracycline resistance.[33] When the *old* gene and its promoter were cloned into the high copy number plasmid pBR322, interference extended even to λ Spi⁻ phage, presumably owing to overexpression of *old*.[33] In order to overcome this problem, we have separated the *old* gene from its promoter and placed it under the control of a synthetic *lac* operator (Fig.

[30] G. Lindahl, G. Sironi, H. Bialy, and R. Calendar, *Proc. Natl. Acad. Sci. U.S.A.* **66,** 587 (1970).

[31] J. Zissler, E. Signer, and F. Schaefer, *in* "The Bacteriophage Lambda" (A. D. Harshey, ed.), p. 455. Cold Spring Harbor Laboratory, Cold Spring Harbor, New York, 1970.

[32] N. Murray, *in* "Lambda II" (R. Hendrix, J. Roberts, F. Stahl, and R. Weisberg, eds.), p. 395. Cold Spring Harbor Laboratory, Cold Spring Harbor, New York, 1983.

[33] S. Finkel, C. Halling, and R. Calendar, *Gene* **46,** 65 (1986).

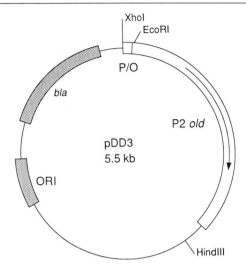

FIG. 4. Plasmid pDD3 has the P2 *old* gene under control of the *lac* promoter (P/O). A 3.0-kb *Bam*HI/*Cla*I pSF1 fragment [S. Finkel, C. Halling, and R. Calendar, *Gene* **46**, 65 (1986)] containing the *old* gene was ligated into pUC119 (J. Vieira and J. Messing, this series, Vol. 153, p. 3.) that had been cleaved with *Bam*HI and *Acc*I. The resulting plasmid was called pDD1. Using site-directed mutagenesis [T. A. Kunkel, *Proc. Natl. Acad. Sci. U.S.A.* **82**, 488 (1985)], we made a C to T transition (TGAACTC to TGAATTC) at −36 from the *old* gene translation start site [the complete sequence of the *old* region is given by E. Häggård-Ljungquist, V. Barreiro, R. Calendar, D. M. Kurnit, and H. Cheng, *Gene* **46**, 25 (1989)]. This mutation, in pDD2, creates a new *Eco*RI site between the promoter and the translation start site. The *Eco*RI/*Nco*I pDD2 fragment containing the *old* gene was placed under control of the *lac* operator in the high copy number (ColE1), ampicillin-resistant, chloramphenicol-resistant plasmid pUHE21-2 of Lanzer and Bujard [M. Lanzer and H. Bujard, *Proc. Natl. Acad. Sci. U.S.A.* **85**, 8973 (1988); M. Lanzer, Dr. rer. nat. thesis, University of Heidelberg, Germany, 1988]. pUHE21-2 was opened with *Bam*HI and filled in using the Klenow reaction. pDD2 was cleaved with *Nco*I and filled in. Both linearized, blunt-ended plasmids were cleaved with *Eco*RI and the appropriate fragments ligated. pDD3 does not express chloramphenicol resistance.

4). In this construction expression of the *old* gene is inducible, and the level of expression can be controlled by the concentration of isopropyl-β-D-thiogalactopyranoside (IPTG). When 200 μM IPTG was included in 2.5 ml of soft agar spread on a 25 ml plate, λ phage did not make plaques, but the λ *bio* transducing phage λ Spi-2 made small plaques. IPTG at a 5-fold higher concentration interfered with λ Spi-2 (Spi⁻) plaque formation.

Host Deletions Associated with P2-Mediated Eduction

Escherichia coli K12 strains lysogenic for P2 at the histidine-linked location H spontaneously segregate histidine-requiring cells that also contain deletions of several genes flanking the histidine operon.[34] This phenomenon, termed eduction (the deletion strains are called eductants), is dependent on the P2 *int* gene product[35] and causes deletions of approximately 1.5% of the *E. coli* genome.[34,36] Deleted genes include *mglBAC*, *araF*, *dcd*, *udk*, *rfbABD*, *gnd*, *his*, and *sbcB*.[34,37–39] Other genes or functions that may also be deleted, by virtue of their propinquity to the known deleted region, are *non*, *nuvC*, and *att*HK 139.[36]

Eductants can easily be obtained from the central growth of P2 plaques on sensitive *E. coli* K[12] strains[35] and from *Klebsiella pneumoniae*.[40] It seems reasonable that eductants might also be obtainable from other P2-sensitive bacterial strains including *E. coli* B, *Serratia,* and *Salmonella*.

Production and Selection of Eductants

Using only c^+ int^+ derivatives of P2, plate phage P2 with lawns of the host strain to be educed so as to obtain single plaques. Pick bacterial growth from several plaque centers to tubes with 1 ml of LB and incubate overnight at 37°. Dilute the cultures approximately 10^5-fold in 1% NaCl and spread on several glucose minimal plates containing 1–2 μg/ml of histidine plus any other required amino acids or growth factors at normal concentrations. After 48 hr of incubation, such plates usually produce 300 to 500 large, opaque colonies and several small, translucent colonies. Pick the small, translucent colonies to LB, grow overnight, dilute into saline solution, and test for histidine auxotrophy by spotting onto properly supplemented glucose minimal plates with and without histidine (40 μg/ml). Histidine-requiring mutants should then be purified by streaking onto the same selective plates used for their isolation, namely, plates with 1–2 μg/ml histidine (this reisolation step can be performed prior to the initial screening for histidine dependence).

[34] M. G. Sunshine and B. L. Kelly, *J. Bacteriol.* **108**, 695 (1971).

[35] M. G. Sunshine, *Virology* **47**, 61 (1972).

[36] B. J. Bachman, *in* "*Escherichia coli* and *Salmonella typhimurium*" (F. C. Neidhardt, J. L. Ingraham, B. Magasanik, K. B. Low, M. Schaechter, and H. E. Umbarger, eds.), Vol. 2, p. 807. American Society of Microbiology, Washington, D.C., 1987.

[37] A. Templin, S. R. Kushner, and A. J. Clark, *Genetics* **72**, 205 (1972).

[38] J. Neuhard and E. Thomassen, *J. Bacteriol.* **126**, 999 (1976).

[39] A. F. Clark and R. W. Hogg, *J. Bacteriol.* **147**, 920 (1981).

[40] S. L. Streichert, E. G. Gurney, and R. C. Valentine, *Nature* (*London*) **239**, 495 (1972).

Acknowledgments

We thank Erich Six for advice on performing transduction experiments and for providing P2 *sir* mutants prior to publication; Dago-Arne Dimster-Denk for the construction of pDD1, pDD2, and pDD3; Giovanni Bertoni for the construction of pKGB2 and pKGB4; Alessandra Polissi for constructing P4::Tn*5* AP-2; and Martin Gonzalez for cloning pMG1. Nora Linderoth and Richard Goldstein are acknowledged for careful reading of the manuscript.

[12] Special Uses of λ Phage for Molecular Cloning

By Noreen E. Murray

Introduction

In the early 1970s bacteriophage λ was an obvious choice as a vector for cloning foreign DNA; it was, and may still be, the best understood DNA phage. Its chromosome was known to be a linear DNA molecule of approximately 50 kilobases (kb), with nearly 40% of the genome being unnecessary for the propagation of the phage. The combined use of known deletions and substitutions together with the ready selection of mutants lacking particular restriction targets led to the generation of many vectors, while the technology of packaging its DNA *in vitro* promoted the efficient use of λ vectors for cloning the smallest DNA fragments to those in excess of 20 kb. The screening of large libraries of recombinants quickly became effective. However, despite efforts to provide essential λ functions in trans, all λ vectors retain at least 30 kb of the λ genome in order to maintain plaque-forming ability. Inevitably, this high proportion of vector sequence complicates restriction mapping of the cloned DNA.

The increased efficiency of recovery of plasmids following electroporation of bacteria[1] and the arrival of vectors able to incorporate much larger fragments of DNA—45 kb for cosmids,[2] 100 kb for phage P1[3] (see also [2] in this volume), and a few hundred kilobases for yeast artificial chromosomes (YACs)[4]—challenge the importance of λ vectors. Their current relevance probably resides in three features: (1) efficient recovery of relatively representative libraries, (2) efficient screening of plaques, and (3) ease of genetic analysis. These features suffice to maintain the usefulness of λ vectors and roles in which they are used to advantage will be illustrated.

[1] W. J. Dower, J. F. Miller, and C. W. Ragsdale, *Nucleic Acids Res.* **16**, 6127 (1988).
[2] J. Collins and B. Hohn, *Proc. Natl. Acad. Sci. U.S.A.* **75**, 4242 (1978).
[3] N. Sternberg, *Proc. Natl. Acad. Sci. U.S.A.* **87**, 103 (1990).
[4] D. T. Burke, G. F. Carle, and M. V. Olson, *Science* **236**, 806 (1987).

Many articles review the technicalities of DNA manipulations,[5-7] but the judicious choice and use of λ vectors require an understanding of the phage and its interaction with *Escherichia coli*. I have chosen, therefore, to emphasize genetic aspects and preface this chapter with an overview of the relevant biology (see Ref. 8 for detailed reviews).

Overview of Bacteriophage λ

The λ genome is encapsidated in an icosahedral head from which projects a tail ending in a single tail fiber that adsorbs to a receptor site in the outer membrane of the bacterium. The receptor sites, which are encoded by the *E. coli lamB* gene, are stimulated by growth in medium containing maltose as the carbon source, and adsorption is facilitated by magnesium ions. λ Phage normally remain infective for decades if kept at 4° in L (Luria) broth or phage buffer in the presence of 1-10 mM Mg^{2+} and the absence of detergent. A wealth of information on the sensitive handling of phage λ is given in Ref. 9.

On entering the cell the λ genome circularizes via its complementary single-stranded 5′ ends, the cohesive ends (*cos*), and is converted to a covalently closed circular molecule, the substrate for both replication and transcription. Being a temperate phage, λ may follow either the productive (lytic) or temperate (lysogenic) pathway. In either case, transcription is initiated from two "early" promoters, p_L and p_R (see Fig. 1), to provide functions essential for DNA replication, genetic recombination, establishment of lysogeny, and the transcriptional activation of the late genes. In a productive (lytic) infection transition from early to late transcription is achieved, virion proteins are made, the replicated genomes are packaged, and lysis of the cell releases about 100 infective particles. In the temperate response, on the other hand, most phage functions become repressed, either directly or indirectly, by λ repressor molecules bound to the operator regions associated with p_L and p_R. If repression occurs in time to prevent activation of the late genes, lysis is avoided and lysogeny may

[5] A. M. Frischauf, N. Murray, and H. Lehrach, this series, Vol. 153, p. 103.
[6] K. Kaiser and N. E. Murray, *in* "DNA Cloning" (D. Glover, ed.), Vol. 1, p. 1. IRL Press, Oxford and Washington, D.C., 1985.
[7] J. Sambrook, E. F. Fritsch, and T. Maniatis, *in* "Molecular Cloning" (C. Nolan, ed.), 2nd Ed., Vol. 1. Cold Spring Harbor Laboratory, Cold Spring Harbor, New York, 1989.
[8] R. Hendrix, J. Roberts, F. Stahl, and R. Weisberg (eds.), "Lambda II." Cold Spring Harbor Laboratory, Cold Spring Harbor, New York, 1983.
[9] W. Arber, L. Enquist, B. Hohn, N. E. Murray, and K. Murray, *in* "Lambda II" (R. Hendrix, J. Roberts, F. Stahl, and R. Weisberg, eds.), p. 433. Cold Spring Harbor Laboratory, Cold Spring Harbor, New York, 1983.

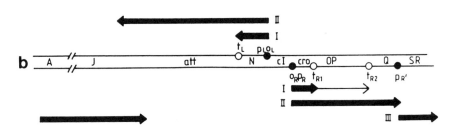

FIG. 1. The λ genome and its transcripts. (a) Genes which encode related functions are clustered in the λ genome. Black regions are inessential and have been deleted during construction of many vectors. The *cI* gene encodes the λ repressor. *imm*434, the immunity region of phage 434, includes the operator/promoter regions and the cognate repressor gene *cI* and *cro*. (b) The genome circularizes before transcription is initiated. Transcription of the genome proceeds initially from the promoters, p_L and p_R. Early during infection transcripts initiated at p_L and p_R (I) terminate at t_L and t_{R1} respectively (some rightward transcripts escape termination at t_{R1} and reach t_{R2}). This transcription allows synthesis of the *N* gene product, which alters RNA polymerase so that transcription may proceed through t_L and t_{R1} into adjacent genes (II). This in turn allows synthesis of the *Q* gene product, in the presence of which "late" transcription (III), initiated from $p_{R'}$, continues through the lysis genes and, since the genome is circular, through the head and tail genes. In many λ vectors extra space has been made by a deletion (*nin5*) that removes t_{R2} and consequently also the gene *rap* whose product is necessary for recombination between λ and plasmids. (●) Major promoters; (○) major termination signals.

ensue. Stable, as opposed to abortive, lysogeny also requires integration of the λ genome into the host chromosome by site-specific recombination in order that the phage genome (prophage) is replicated as part of the *E. coli* chromosome. The temperate nature of λ, namely, its ability to repress the phage genome, is manifest by surviving cells in the center of a plaque and the consequent turbid plaque morphology. Most λ vectors, however, lack a functional repressor gene, are therefore obliged to embark on lytic growth, and produce clear plaques. In some vectors the cloning site is within the gene, *cI*, encoding the repressor, and the successful insertion of a DNA fragment inactivates this gene, resulting in a clear plaque morphology.

In lytic growth, activation of the late genes of λ is normally associated

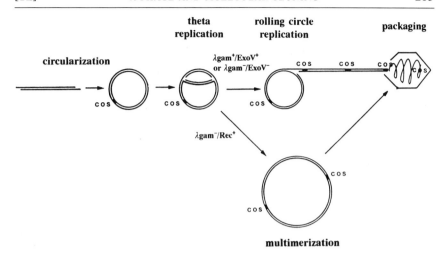

FIG. 2. The lytic pathway of phage λ. Phage DNA enters the cell as linear molecules. Inside the cell the linear DNA circularizes via its cohesive ends, and a covalently closed circular DNA molecule is formed by the action of *E. coli* DNA ligase at the now double-stranded *cos* sites. For the first 15 min or so after infection, replication proceeds bidirectionally (θ replication) to generate a number of monomeric circular copies of the phage DNA. The transition to rolling circle replication from the circular templates is blocked in *recBCD⁺* *E. coli* by exonuclease V, the product of the *recBCD* genes, unless the phage retains a functional *gam* gene. Gam is not necessary for the transition if the RecBCD nuclease is defective. The product of rolling circle replication is a linear concatenated DNA molecule, the usual substrate for packaging *in vivo*. Monomeric circles cannot be packaged. During processing of the concatenated DNA, the packaging machinery recognizes two suitably spaced *cos* sites (37.7–51.5 kb apart) at which staggered breaks are made, thereby creating a mature linear phage genome that is assembled into an infective phage particle. Phage that lack a *gam* gene can be propagated in *recBCD⁺* cells, albeit less efficiently, if a functional generalized recombination pathway is available. If λ is to be a good substrate for the major recombination pathway of *E. coli* it must include a special site called χ; recombination between monomeric circular forms generates multimeric circular molecules, an alternative substrate for packaging.

with a change in the predominant mode of DNA replication (Fig. 2). During the first 15 min after infection, bidirectional replication of DNA yields monomeric circular daughter molecules (θ replication). At later times the predominant products of replication are linear concatenated (multimeric) molecules presumed to originate by a rolling circle mechanism. Concatenated genomes provide the substrate for *in vivo* packaging, a process that involves the cutting of two appropriately spaced *cos* sequences and encapsidation of the resultant linear (mature) λ genome to give an infective virus. Transition to the rolling circle mechanism is impeded by the action of exonuclease V (ExoV), the product of the *E. coli recB, C,* and *D* genes.

In *recBCD*[+] cells concatenated DNA can still be produced providing the phage possess a functional *gam* gene. The product of this gene binds to ExoV and inactivates its nucleolytic activity. Many λ vectors lack the *gam* gene, but *gam*[-] phage can propagate in ExoV[+] hosts as long as they can undergo recombination to produce di- or multimeric circular DNA molecules, an alternative substrate for packaging (Fig. 2). Most, if not all, λ vectors that are *gam*[-] are also defective in the phage-encoded recombination enzymes (Red), as they lack or have mutant *red* genes, and consequently they cannot rely on phage-mediated recombination to produce multimeric DNA. Rather, *red*[-] *gam*[-] λ phage in *recBCD*[+] cells are forced to depend on host recombination (Rec) functions to make concatenated DNA (but see next section for the use of ExoV-deficient strains). It is paradoxical that ExoV prevents a *red*[-] *gam*[-] phage from replicating its DNA by the rolling circle mechanism and yet is necessary in a wild-type host to generate concatenated DNA by recombination (Fig. 2).

Cutting and packaging of λ DNA from concatenated molecules depend on the recognition of *cos* sequences,[10] and the process is relatively tolerant of the distance between the *cos* sites. DNA molecules ranging in size from 37,700 to 51,500 base pairs (bp) are packaged. Packaging *in vivo* may be less sensitive to deviations from normal size than packaging *in vitro*; indeed, current commercial extracts that package DNA with very high efficiency may package molecules of normal length, 48 kb, 10 times more effectively than those of 40 kb. *In vitro* packaging extracts that are much less size selective can be made,[11] but the size selectivity can be used advantageously to enrich for recombinant phage if the vector genome is of an appropriate size.

Since multimeric DNA may be the obligatory substrate for packaging *in vivo*, it seems reasonable to extrapolate that concatenated DNA is the preferred substrate for packaging *in vitro*. Covalent linkage of λ genomes guarantees neighboring *cos* sites and will protect cohesive ends from exonucleolytic attack. Although linear monomers can be packaged *in vitro*, the experimental evidence suggests that monomeric DNA is preferentially packaged after interaction with other λ genomes.[12] The relative merits of hydrogen-bonded linear polymeric substrates versus covalently linked concatemers warrant reexamination using current *in vitro* packaging ex-

[10] M. Feiss, *Trends Genet.* **2**, 100 (1986).
[11] T. Maniatis, E. F. Fritsch, and J. Sambrook, "Molecular Cloning: A Laboratory Manual." Cold Spring Harbor Laboratory, Cold Spring Harbor, New York, 1982.
[12] S. M. Rosenberg, M. M. Stahl, I. Kobayashi, and F. W. Stahl, *Gene* **38**, 165 (1985).

tracts. Nevertheless, λ genomes packaged *in vitro* are infective even in the absence of their 5'-phosphate groups.[13]

In vitro packaging extracts are prepared following the induction of recombination-deficient lysogens in which the prophage is defective in one gene whose product is essential for packaging. A mixture prepared from two differently defective strains makes a competent extract (see Refs. 9 and 11). Packaging of the resident prophage DNA is prevented by blocking its excision from the chromosome. An alternative system using only one lysogenic strain[14] is not defective in any protein component, but the strain lacks the *cos* sequence and hence is unable to package any endogenous sequences. This lysogen is a derivative of *E. coli* C and has the added advantage that it is devoid of other λ-like sequences and is deficient in all known restriction systems. While many laboratories choose commercial packaging extracts for particularly critical experiments, most make their own extracts for routine use. Some workers continue to use two lysogens, others either an extract of the single *cos*-defective lysogen or two extracts of this strain, one prepared by freeze–thaw lysis and the other by sonication. It is worth noting that many suboptimal packaging extracts are defective in Ter, the complex that recognizes the *cos* sequence and cuts concatenated λ DNA to generate the cohesive ends during packaging. Concentrated extracts of Ter can be made very readily and used to boost the efficiency of these extracts or to complement the extract of a Ter-defective lysogen.[15]

Escherichia coli Genes Relevant to Recovery of Representative λ Libraries

Restriction Systems of Escherichia coli

Most laboratory strains of *E. coli* are derivatives of *E. coli* K12. This strain encodes *Eco*K, a restriction enzyme that cleaves DNA containing unmodified targets of the interrupted heptanucleotide sequence $AAC(N_6)GTGC$. A mutation in the *hsdR* gene results in a strain defective in restriction (r_K^-) but proficient in modification (m_K^+), whereas a mutation in *hsdS* that inactivates the specificity polypeptide common to both the restriction and modification enzymes confers an r_K^- m_K^- phenotype. A λ vector grown on an *hsdS*⁻ strain will form plaques on a K-restricting

[13] E. Vincze and G. B. Kiss, *Gene* **96,** 17 (1990).
[14] S. M. Rosenberg, *Gene* **39,** 313 (1985).
[15] S. Chow, E. Daub, and H. Murialdo, *Gene* **60,** 277 (1987).

strain with a relative efficiency (e.o.p.) of 10^{-3}–10^{-4}, a figure reflecting the presence of at least four unmodified targets. Recombinant phage made using modified vector DNA (i.e., modified against K restriction) will be vulnerable to attack only if the fragment of foreign DNA includes an *Eco*K target, and a phage that includes one unmodified target will have an e.o.p. of 10^{-1}–10^{-2} on an r_K^+ strain. Nonrepresentative libraries resulting from the action of *Eco*K have been avoided by the routine use of r_K^- bacteria. However, the effects of other less well-known restriction systems were not considered until quite recently,[16] and some of the strains recommended in the new edition of *Molecular Cloning*[7] restrict DNA which includes methylated cytosines. Such restriction systems are particularly relevant to the recovery of mammalian and plant DNA sequences, and even some bacterial species. There are at least two systems, McrA and McrB, that attack DNA containing methylated cytosines[16,16a,16b] and one, Mrr, that recognizes methylated adenines.[17] Raleigh *et al.*[18] list the *mcr* genotypes of many relevant strains, and data quantifying the effect of *mcrB* on the recovery of plant libraries[19] and of both *mcrA* and *mcrB* on mammalian libraries[20] have been reported. The plant libraries were more susceptible to McrB than the mammalian one. For the mammalian DNA, McrA and McrB each depressed the total yield approximately 10-fold, but some foreign DNA is restricted by *mcrA mcrB hsd* strains.[19,21] Recent evidence implicates a third Mcr system (M. Noyer-Weidner, personal communication). This may be Mrr since this is now known to direct restriction of DNA modified at cytosine residues (J. E. Kelleher and E. A. Raleigh, personal communication; J. Benner, personal communication).

Restriction by *in vitro* packaging extracts has been detected.[14,21] Only a 2- to 7-fold reduction was observed for *Eco*K despite the presence of five unmodified targets in the λ DNA.[14] D. M. Woodcock, M. W. Graham, and J. P. Doherty (personal communication) have checked for restriction of plant libraries by the McrA and detected modest effects. In contrast, λ DNA isolated from mammalian cells is very sensitive to restriction by Mcr$^+$ packaging extracts.[21]

[16] E. A. Raleigh and G. Wilson, *Proc. Natl. Acad. Sci. U.S.A.* **83**, 9070 (1986).

[16a] R. M. Blumenthal, S. A. Gregory, and J. S. Cooperider, *J. Bacteriol.* **164**, 501 (1985).

[16b] M. Noyer-Weidner, R. Diaz, and L. Reiners, *Mol. Gen. Genet.* **205**, 469 (1986).

[17] J. Heitman and P. Model, *J. Bacteriol.* **169**, 3243 (1987).

[18] E. A. Raleigh, N. E. Murray, H. Revel, R. M. Blumenthal, D. A. Westaway, A.D. Reith, P. W. J. Rigby, J. Elhai, and D. Hanahan, *Nucleic Acids Res.* **16**, 1563 (1988).

[19] D. M. Woodcock, P. J. Crowther, W. P. Diver, M. Graham, C. Bateman, D. J. Baker, and S. S. Smith, *Nucleic Acids Res.* **16**, 4465 (1988).

[20] P. A. Whittaker, A. J. B. Campbell, E. M. Southern, and N. E. Murray, *Nucleic Acids Res.* **16**, 6725 (1988).

[21] J. A. Gossen and J. Vijg, *Nucleic Acids Res.* **16**, 9343 (1988).

Escherichia coli C, and some deletion mutants of *E. coli* K12, lack *Eco*K, Mrr, and both the Mcr restriction systems. Such strains are strongly recommended for the recovery of libraries of genomic DNA and are the preferred source of packaging extracts; their use gave a dramatic increase in the recovery of λ genomes from animal cells.[21]

Recombination Systems of Escherichia coli

Although we may know enough about the restriction systems of *E. coli* K12 to avoid loss of sequences susceptible to degradation by host restriction systems, we do not fully understand why some DNA sequences remain difficult to clone, and currently we are only able to ameliorate these problems (see next section). Rec⁻ conditions do not necessarily prevent genetic instabilities,[22] whereas recombination systems make a major contribution to the health of bacterial strains and the vigor of infective phage. Recombination as well as replication are critical to the production of concatenated phage DNA and consequently influence burst size. The *rec* genotype of any host must be considered in the light of its effect on the yield of phage and its influence on the recovery and maintenance of representative libraries.

Given the better viability of Rec⁺ bacteria, and their successful use in the recovery and maintenance of many recombinant phage, most of the commonly used hosts are Rec⁺, although it may be wise to recover part of a library under recombination-deficient conditions. This, of course, is only possible if the phage vector is *red*⁻ rather than *red*⁺. A few reports document the advantage of growing *red*⁻ phage in *recA*⁻ hosts to stabilize cloned sequences.[23] The disadvantage of a *red*⁻ *gam*⁻ λ is its inability to grow in a *recA*⁻ *recBCD*⁺ host, but this can be overcome by providing Gam in trans from a resident plasmid.[24] The expression of Gam from the plasmid, and the consequent inactivation of the RecBCD nuclease (ExoV), occurs only after infection, since transcription of *gam* has been made dependent on the *Q* gene product of λ. A *recA*⁻ host in combination with the *gam* plasmid gives tolerable yields of phage.

More usually, *red*⁻ *gam*⁻ phage are grown on *recA*⁺ hosts; if the host is *recBCD*⁺, their normal route of producing concatenated DNA is impeded by ExoV (Fig. 2), and they must rely on recombination via the RecBCD pathway. Only phage DNA molecules that possess one or more copies of an octanucleotide sequence known as χ (chi) are an efficient

---

[22] M. Ishiura, N. Hazumi, T. Koide, T. Uchida, and Y. Okada, *J. Bacteriol.* **171,** 1068 (1989).

[23] J. Lauer, C.-K. J. Shen, and T. Maniatis, *Cell (Cambridge, Mass.)* **20,** 119 (1980).

[24] G. F. Crouse, *Gene* **40,** 151 (1985).

substrate for the RecBCD pathway (see Refs. 25 and 26 for reviews). Wild-type λ has no χ sequence; these sequences are not a substrate for Red-mediated recombination. χ sequences have been isolated by single base changes at any of four sites within the λ genome, and the inclusion of one such sequence is recommended for *red⁻ gam⁻* λ vectors.[27] A χ sequence is essential wherever ExoV is active, otherwise poor recombination severely limits burst size and phage titer. This is particularly relevant to the selection of *red⁻ gam⁻* phage by their ability to grow on strains lysogenic for P2 (see [11], this volume). In all RecBCD⁺ strains, recombination is the essential pathway for *gam⁻* phage to make concatenated DNA.

An array of mutant hosts defective in ExoV now exists, thereby re-opening the normal pathway of rolling circle replication to *gam⁻* phage. These include Rec⁻ strains defective in either *recB* or *recC* and derivatives of a *recBC⁻* strain in which suppressor mutations, *sbcA* (suppressor of RecBC) or *sbcB*, open up alternative recombination pathways leaving a recombination-proficient host defective in ExoV. *sbcA* mutations activate the RecE pathway, *sbcB* the RecF pathway (Fig. 3). More recently *recD* mutations have been shown to leave ExoV deficient in nuclease activity but superproficient in recombination.[28] *recD⁻* strains are more vigorous than either *recB⁻* or *recC⁻* strains, although they probably produce more inviable cells than *rec⁺* cultures. In *recD⁻* hosts both routes for making concatenated DNA are open, and λ is a good substrate for RecBCD in this hyperrecombinogenic strain, even in the absence of χ sequences. The *recD⁻* strains are now the most commonly used for growth of *red⁻ gam⁻* λ.[29] For some phage that replicate their DNA poorly, the recombinational route to concatenated DNA may be the preferred one. The χ sequence remains important in *recBCD⁺* strains.

sbcC and Conservation of Palindromic Sequences

Long, perfect palindromic sequences are difficult to clone, and their instability has been studied in model systems. Leach and Stahl[30] observed that a derivative of phage λ containing a 571-bp nearly perfect palindrome could be propagated in certain *E. coli* strains known to be mutant for the *recB, recC,* and *sbcB* genes. More recently it has been shown that

[25] G. Smith, *in* "Lambda II" (R. Hendrix, J. Roberts, F. Stahl, and R. Weisberg, eds.), p. 175. Cold Spring Harbor Laboratory, Cold Spring Harbor, New York, 1983.

[26] F. W. Stahl, *Sci. Am.* **256,** 53 (1987).

[27] K. Murray and N. E. Murray, *J. Mol. Biol.* **98,** 551 (1975).

[28] S. K. Amundsen, A. F. Taylor, A. M. Chaudhury, and G. R. Smith, *Proc. Natl. Acad. Sci. U.S.A.* **83,** 5558 (1986).

[29] A. R. Wyman and K. F. Wertman, this series, Vol. 152, p. 173.

[30] D. R. F. Leach and F. W. Stahl, *Nature (London)* **305,** 448 (1983).

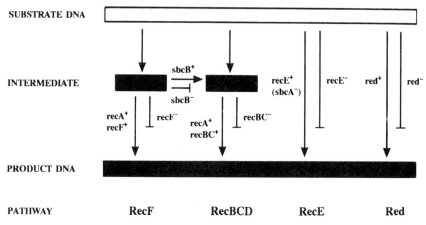

PATHWAY   **RecF**   **RecBCD**   **RecE**   **Red**

FIG. 3. Recombination pathways of *E. coli*. The RecBCD pathway is the predominant (99%) recombination pathway in wild-type *E. coli* K12. *recB⁻* or *recB⁻ recC⁻* (*recBC⁻*) strains are defective in this pathway; some *recD⁻* strains remain active. Generation of an *sbcB⁻* mutation in *recBC⁻* strains appears to prevent shunting from the RecF to the RecBC pathway, enabling efficient recombination (~50% of the wild-type level) to be catalyzed by the *recF* product. *recA⁻* strains, which are defective in both pathways, are the least able to catalyze recombination (10^{-3}–10^{-6} of wild type). The RecE pathway is independent of *recA*. Access to the RecE pathway is gained by an *sbcA* mutation. In terms of biochemistry and genetics, the RecE pathway resembles, in many respects, the Red recombination pathway encoded by wild-type λ; the *recE* gene is carried on a defective lambdoid prophage present in the chromosome of most *E. coli* K12 strains. Normally the *recE* gene is repressed (as are the *red* genes of a λ prophage). *sbcA* mutations, which map close to *recE*, in some way awaken the *recE* gene from repression. The *recBCD*, *sbcB*, and *recE* products are respectively, exonucleases V, I, and VIII.

suppression of *recBC⁻* results from two mutations, *sbcB* and *sbcC*,[31] and furthermore Chalker *et al.*[32] found that it is the SbcC phenotype which is relevant to the stable propagation of DNA replicons containing a long palindrome. While the role of the *sbcC* gene product remains unknown, an *sbcC⁻* strain is recommended in *E. coli* hosts for λ, plasmid, and cosmid derivatives. The efficacy of the *sbcC* mutation is influenced by the genetic background of the strain, but even in the best host it will not guarantee the maintenance of all palindromic sequences.

In summary, hosts for the recovery of genomic DNA should be *hsdR⁻*, *mcrB⁻*, *mcrA⁻*, and possibly *mrr⁻* to prevent restriction of DNA, *recA⁻* if one wishes to propagate *red⁻* phage under recombination-deficient con-

[31] R. G. Lloyd and C. Buckman, *J. Bacteriol.* **164,** 836 (1985).
[32] A. F. Chalker, D. R. F. Leach, and R. G. Lloyd, *Gene* **71,** 201 (1988).

ditions, $sbcC^-$ to ameliorate the problems associated with the replication of palindromic sequences, and probably $recD^-$ if the phage is red^- gam^-.

Special Uses of λ Vectors

Cloning Limited Amounts of DNA

Vectors based on bacteriophage λ have played a leading role when the source of DNA is very limited. This is best illustrated by the extreme example of so-called microcloning in which DNA is dissected from the bands of polytene chromosomes of *Drosophila*.[33] The benefits of λ are at least 2-fold: the very high efficiency of recovery by *in vitro* packaging and the ease of screening for, or selecting, recombinants.[34] Selection is particularly useful since the efficiency of cloning under such disadvantageous conditions is low. Insertion vectors (see Table I[33–40]) were chosen in these experiments since they have no lower limit to the size of fragment that can be cloned and their upper size limit can be as high as 11 kb. Pirrotta and colleagues[33,34] used λNM641, a red^- version of the prototype vectors referred to as "immunity vectors",[35] where successful cloning inactivates the *cI* gene and consequently there is no repressor or "immunity" substance (see Fig. 1). These vectors take advantage of single targets within the *cI* gene of the immunity region of the lambdoid phage 434. Recombinant phage can be selected on a mutant host (hfl^-) in which vector phage are said to have such a high frequency of lysogenization that they fail to make plaques. More accurately, they fail to form plaques because they repress their genomes too successfully, but since they are integration deficient they do not form stable lysogens. Some workers[34] have used an hfl^- strain with a mutant allele designated *lyc7*, others[36] an $hflA^-$ strain; both are effective. Hfl is apparently a protease, a target of which is the

[33] F. E. Scalenghe, E. Turco, J. E. Edstrom, V. Pirrotta, and M. Melli, *Chromosoma* **82**, 205 (1981).

[34] G. Scherer, J. Telford, C. Baldari, and V. Pirrotta, *Dev. Biol.* **86**, 438 (1981).

[35] N. E. Murray, W. J. Brammar, and K. Murray, *Mol. Gen. Genet.* **150**, 53 (1977).

[36] T. V. Huynh, R. A. Young, and R. W. Davis, *in* "DNA Cloning" (D. Glover, ed.), Vol. 1, p. 49. IRL Press, Oxford and Washington, D.C., 1985.

[37] N. E. Murray, *in* "Lambda II" (R. Hendrix, J. Roberts, F. Stahl, and R. Weisberg, eds.), p. 395. Cold Spring Harbor Laboratory, Cold Spring Harbor, New York, 1983.

[38] C. Coleclough, *in* "Immunological Methods" (I. Lefkovits and B. Pernis, eds.), Vol. 4, p. 13. Academic Press, Orlando, Florida, 1990.

[39] A. Swaroop and S. M. Weissman, *Nucleic Acids Res.* **17**, 8739 (1988).

[40] A. Poustka and H. Lehrach, *in* "Genetic Engineering" (J. K. Setlow, ed.), Vol. 10, p. 169. Plenum, New York, 1988.

product of the λ gene *cII*,[41] the obligate activator for transcription of the *cI* gene in the lysogenic pathway. In the absence of Hfl, excess CII inevitably elicits transcription of the *cI* gene and repression of the phage genome. Neither *cI⁻* nor *cII⁻* phage can repress their genomes therefore they form plaques on an *hfl⁻* host.

Rather than the original *red⁻* immunity vector used in the early experiments, some laboratories now use a related *red⁺* vector (λNM1149), which has a single target for both *Eco*RI and *Hin*dIII,[37] or more commonly λgt10.[36] λgt10 is a *red⁺* *Eco*RI vector, but it differs from all its predecessors in having a smaller deletion. Although this reduces the upper limit of the size of the DNA fragment that can be cloned to 7 kb, it has the advantage that, given size-selective packaging, it can greatly increase the efficiency of recovery of small inserts. λgt10 is therefore a preferred vector for cloning cDNA.

Two newer vectors for cDNA, λ*jac* and λ*ecc*,[38] appear to offer an even stronger genetic selection than immunity vectors. Both are *red⁻* but *gam⁺*, and consequently they are sensitive to P2 interference (Spi⁺) and fail to form plaques on a strain of *E. coli* lysogenic for phage P2 (see [11], this volume). The cutting of these vectors inactivates the *gam* gene, and the resulting *red⁻ gam⁻* phage have a weak Spi⁻ phenotype and would form minute plaques on a P2 lysogen. If they are to make respectable plaques they require a χ site to stimulate RecBCD-dependent recombination and the formation of concatenated DNA for packaging. Part of the genetic selection devised by Coleclough is the inclusion of a χ sequence within the synthetic adapter used to initiate cDNA synthesis. An advantage of λ*jac* and λ*ecc* when compared with λgt10 is the modification of flanking sequences to include SP6 and T7 promoters for the specific transcription of cloned sequences. The disadvantages are the need for special adapters that are not commercially available and the fact that these recombinant phages are less vigorous than the remarkably healthy derivatives of λNM1149 and λgt10.

The polymerase chain reaction (PCR)[42] enables limited amounts of DNA to be readily amplified, and, indeed, this technique has already been applied successfully to the products of microdissection (reviewed by Saunders[43]). Nevertheless, the efficient recovery of DNA, the selection against vector (nonrecombinant) phage, and the very efficient screening opportunities[44] leave λ useful in more pedestrian roles.

[41] H. H. Cheng, P. J. Muhlrad, M. W. Hoyt, and E. Echols, *Proc. Natl. Acad. Sci. U.S.A.* **85,** 7882 (1988).

[42] T. J. White, N. Arnheim, and H. A. Erlich, *Trends Genet.* **5,** 185 (1989).

[43] R. D. C. Saunders, *BioEssays* **12,** 245 (1990).

[44] W. D. Benton and R. W. Davis, *Science* **196,** 180 (1977).

TABLE I
INSERTION VECTORS[a]

Vector	Space[b]	Cloning sites	Genotype of recombinants[c]	Biological features[d]	Refs.
λNM641	10.7	*Eco*RI	*att⁻ int⁻ red⁻ imm⁴³⁴ cI⁻ nin5 χ°*	Selection on Hfl⁻; *red⁻ gam⁺*	33, 34, 35
λgt10	6.7	*Eco*RI	*att⁻ imm⁴³⁴ cI⁻ χ°*	Selection on Hfl⁻	7, 36
λNM1149	10.3	*Eco*RI, *Hin*dIII	*att⁻ int⁻ imm⁴³⁴ cI⁻ χ°*	Selection on Hfl⁻	37
λNM1150	11.3	*Eco*RI, *Hin*dIII	*att⁻ int⁻ red⁻ imm⁴³⁴ cI⁻ χ°*	Selection on Hfl⁻; *red⁻ gam⁺*	37
λecc and λjac	~10	Use adapters	*att⁻ int⁻ red⁻ gam⁻ cI857 nin5 χ⁺*	Spi selection; T3 and T7 promoters	38
Charon 21A	8.1	*Eco*RI, *Hin*dIII, *Xho*I	*Wam43 Eam1100 att⁻ int⁻ imm⁸⁰ nin5*	—	7
λgt11	6.3	*Eco*RI	*cI857 nin5 Sam100*	LacZ screen; fusion polypeptides	7, 36
λgt22–23	7.3	*Not*I, *Xba*I, *Sac*I, *Sal*I, *Eco*RI	*red⁻ gam⁻ cI857 nin5 Sam100 χ⁺*	LacZ screen; fusion polypeptides, multiple cloning sites	7
λorf8	7.2	*Eco*RI, *Bam*HI, *Hin*dIII	*intam imm²¹ nin5*	LacZ screen; fusion polypeptides	7
λZAP	10.1	*Sac*I, *Not*I, *Xba*I, *Spe*I, *Eco*RI, *Xho*I	*cI857 nin5 Sam100*	LacZ screen; fusion polypeptides; pBluescript SK (−); T3 and T7 promoters	7
Charon BS	7.9	*Eco*RI, *Hin*dIII	*att⁻ int⁻ imm⁸⁰ nin5*	LacZ screen; fusion polypeptides; pBluescript SK (+ and −); T3 and T7 promoters	39
λNM1151	10.8	*Hin*dIII, *Eco*RI, *Bam*HI	*(int⁻ for Bam*HI) *imm²¹ts nin5 χ°*	Integration proficient	37
λNM1151ABS	10.8	*Hin*dIII, *Eco*RI, *Bam*HI (derivative with *Not*I in place of *Bam*HI is also available)	*Aam32 Bam1 Sam7* derivative of λNM1151	Used in jumping libraries	40

[a] Vectors listed include those mentioned in the text and a few with special properties.

Many insertion vectors are available (Table I), including those commonly used to express fusion polypeptides (see next section), but the most versatile is λZAP,[45] a combination of M13 and plasmid (phagemid) in a phage λ vector. The phagemid, Bluescript SK (−), is a pUC derivative and can be excised *in vivo* from its λ carrier phage following superinfection with phage f1 (or M13) because the plasmid sequence is suitably flanked by signals necessary for excision, namely, a signal for the initiation of (+) strand DNA synthesis and one for termination, both derived from bacteriophage f1. Newly synthesized DNA forms a circular molecule that is converted to a replicative form and maintained in this plasmid form. As with pUC vectors, there are multiple cloning sites within the coding sequence for an incomplete LacZ polypeptide; this polypeptide is detected by its ability to complement a defective polypeptide encoded by the *lacZ* M15Δ gene of the host.[46] The cloning sites are flanked by bacteriophage T3 and T7 promoters and appropriate terminators to facilitate efficient *in vitro* transcription of either strand of the cloned DNA. Because the excised pBluescript phagemid incorporates an f1, M13-like, origin of replication, it can be recovered in single-stranded form on infection with appropriate helper phage. λZAP, like any λ vector, offers the high efficiency of packaging *in vitro* and of screening plaques. It then allows the experimenter to separate the plasmid *in vivo* from its superfluous λ DNA, thereby providing the virtues of both plasmid and M13. The use of some of the cloning sites of the pUC vector is precluded by additional targets in the λ vector, and some workers have found its use in sequencing less reliable than simple M13 vectors. Charon BS (+) and (−) are analogous to λZAP.[39]

Expression of Fusion Polypeptides

Undoubtedly, the expression system that has had the most impact is λgt11, which takes advantage of the *lacZ* gene within a λ vector. Its use is based on the in-frame inclusion of short DNA sequences within *lacZ* to

[45] J. M. Short, J. M. Fernandez, J. A. Sorge, and W. D. Huse, *Nucleic Acids Res.* **16,** 7583 (1988).

[46] A. Ullmann, F. Jacob, and J. Monod, *J. Mol. Biol.* **24,** 339 (1967).

---

[b] The space (in kb) is estimated for the conservative upper size of a λ genome of 50 kb, although longer genomes have been reported.

[c] The relevant mutant phage genes are documented. Charon BS and Charon 21A are presumed to have the *red gam* region of φ80 and consequently are Spi− and Fec+. λgt11, λ*orf8*, Charon BS, and Charon 21A probably have a χ site within their non-λ DNA, but χ sites are unlikely to be relevant to the use of these phage.

[d] The *red− gan+* vectors identify phage that may be grown under recombination-deficient conditions in any *recA−* host.

provide fusion polypeptides that can be detected by antibody probes[36] or even sequence-specific DNA-binding activity.[47] The DNA, frequently cDNA, is inserted at an *Eco*RI site 53 bp from the 3′ terminus of *lacZ*. The large N-terminal sequence of β-galactosidase may stabilize many of the foreign polypeptide sequences and provides a general means of detection by antibody, useful for some analyses and for purification of fusion polypeptides by antibody-dependent affinity systems. The *lacZ* gene has its own promoter, and expression of fusion polypeptides is primarily dependent on the transcription and translation signals of this gene. Consequently, the recombinant phage are recovered and propagated on *lacI*$^+$, or even *lacI*^{q}, hosts to control expression of the fusion polypeptide; derepression of the *lac* promoter is achieved by the addition of inducer. λgt11 is an integration-proficient vector and it encodes a temperature-sensitive repressor. The initial design intended the recovery of recombinants as prophage using an *hfl*$^-$ host to increase the efficiency of lysogeny, and the subsequent screening of induced colonies rather than plaques. This approach is no longer recommended[36] since some recombinant λgt11 phage repeatedly fail to achieve stable lysogeny, even in an *hfl*$^-$ host. It is possible that some of the fusion polypeptides are toxic to the cell. Indeed, the vigor of the recombinant phage is noticeably variable, as if affected by the fusion product. The detrimental effect of novel gene products may be aggravated by some side effect of the initial *lacZ* substitution which extends into the *lom* gene (immediately to the right of gene *J* in Fig. 1). In our experience (A. J. B. Campbell and N. E. Murray, unpublished observation) phage like λgt11, or some Charon derivatives, that are defective in the *lom* gene grow less well than wild type.

Although many vectors (Table I) have been used effectively for cloning cDNA, those resulting in β-galactosidase fusion polypeptides (λgt11, λZAP, and Charon BS) are recommended for probing with antibodies. λZAP was recently exploited for the generation of large combinatorial libraries of the immunoglobulin repertoire.[48] A good account of the use of both λgt10 and λgt11 is provided by Huynh *et al.*[36] A variety of descendants of λgt11 has been described (see Ref. 7) of which λgt18–23 offer extra cloning sites. λgt22 and λgt23 differ only in the orientation of their cloning sites and were designed for cloning between *Not*I and *Sal*I targets, which occur infrequently in mammalian DNA, and should help the recovery of full-length cDNA even in the absence of methylation by the cognate

[47] C. R. Vinson, K. L. Lamarco, P. J. Johnson, W. H. Landschulz, and S. L. McKnight, *Genes Dev.* **2,** 801 (1988).

[48] W. D. Huse, L. Sastry, S. A. Iverson, A. S. Kang, M. Alting-Mees, D. R. Burton, S. J. Benkuvic, and R. A. Lerner, *Science* **246,** 1275 (1989).

modification enzyme. λ*orf8* (see Ref. 7) is an analogous vector derived from a different λ*lacZ* phage. It includes extra cloning sites, though only *Bam*HI and *Hin*dIII, and a *lacI*^{q} gene which should provide tighter control of transcription of the *lacZ* gene.

For expression of any foreign protein the use of protease-deficient hosts is recommended; *lon*$^-$, *htpR*$^-$, or doubly-defective strains are available.[36]

Amplification of Prokaryotic Gene Products

I confine discussion of gene amplification to the use of phage λ itself rather than part of its control system placed within plasmid vectors. Phage λ offers a very simple route to the appreciable amplification of prokaryotic gene products. It has been particularly helpful for enzymes whose products are detrimental to the cell. A gene that cannot be cloned in a multicopy plasmid, unless its expression is very tightly controlled, is often easily and faithfully maintained in λ. In the case of the phage T4 gene encoding polynucleotide kinase, we had no difficulty cloning this gene in an immunity vector in the orientation opposed, rather than reinforced, by the promoters of *c*I.[49] The phage was much less easy to propagate when the gene was located so that it would be transcribed from the late λ promoter, $p_{R'}$, and no phage was recovered with the gene transcribed from the early promoter p_L.

Foreign genes cloned in the central region of λ (between genes *J* and *N*, see Fig. 1) can be transcribed from either their own promoter or, depending on their orientation, the λ promoters p_L or $p_{R'}$. They can be propagated as a prophage with the λ promoters repressed and then induced to activate transcription and amplification of the phage genome. If the cloned gene has a functional promoter, relatively good amplification can be achieved simply by using a mutation in the *S* gene to delay lysis, although better yields have been obtained using phage defective in *Q* and hence in the transcription of all late genes.[50]

In the absence of an efficient promoter within the cloned sequence it is possible to take advantage of a λ promoter, either p_L[50] or the late promoter $p_{R'}$. The latter is easier to harness; using a temperature-inducible prophage (λ*c*I857) defective in genes *W, E,* and *S*, the levels of bacteriophage T4 polynucleotide kinase were boosted to around 7% of soluble cell protein.[49]

[49] C. Midgley and N. E. Murray, *EMBO J.* **4**, 2695 (1985).
[50] A. Moir and W. J. Brammar, *Mol. Gen. Genet.* **149**, 87 (1976).
[51] R. Lathe, J. L. Vilotte, and A. J. Clark, *Gene* **57**, 193 (1987).

Making and Screening Representative Genomic Libraries

The perfect genomic library would contain DNA sequences representative of an entire genome, in a stable form, as a manageable number of overlapping clones. Because the capacity of λ vectors is smaller than that of cosmids, phage P1, or YACs, a representative λ library would need to be correspondingly larger. However, given good *in vitro* packaging extracts, λ vectors provide the easiest and most efficient route to the construction of genomic libraries. In part, the lower efficiency with cosmids may reflect the fact that insert DNA preparations of higher molecular weight are necessary.

The discussion of bacterial hosts forewarns that none guarantees the recovery and maintenance of all DNA sequences. The most likely way of conserving all cloned eukaryotic sequences may prove to be as YACs. Nevertheless, with restriction-deficient hosts ($hsdR^-$ $mcrA^-$ $mcrB^-$ mrr^-) the rate of successful cloning is high and might be even higher if samples of libraries are recovered on a variety of strains, for example, $sbcC^-$; $redD^-$ $sbcC^-$; $recBC^-$ $sbcB^-$ $sbcC^-$; and $recA^-$ $sbcC^-$ (with Gam provided in trans for red^- gam^- phage[24]). If P2 lysogens are to be used for selecting red^- gam^- recombinants by their Spi$^-$ phenotype, an $sbcC^-$, but not Rec$^-$, derivative can be used. However, the use of phosphatase can eliminate the need for genetic selection of recombinant phage,[20] although genetic selection provides an easy monitor for ligation reactions. Reference 20 and various manuals[5-7] offer advice on the preparation of DNAs for making genomic libraries.

Most λ vectors for making genomic libraries are replacement vectors that accommodate approximately 20 kb of foreign DNA (Table II), and many have flanking targets to permit excision of sequences inserted by ligating *Sau*3A partial digestion products to vector arms generated with *Bam*HI. Some recent vectors have T3 and T7 promoters flanking the cloning sites to facilitate the preparation of probes. In EMBL3*cos* and EMBL3*cos–Not* all the essential coding sequences are in the right arm of the vector, and only 200 bp separates the insert from the left cohesive end so that labeling this *cos* sequence with a complementary radioactive oligonucleotide permits the charting of partial digestion products, thereby expediting the construction of restriction maps.[20] Recently, flanking SP6 and T7 promoters have been included in a derivative of EMBL3*cos* (see Table II). One alternative to a replacement vector is a phasmid, a λ genome propagated as a plasmid but recovered as a phage following the insertion of foreign DNA.[52]

[52] N. K. Yankovsky, M. Y. Fonstein, S. Y. Lashina, N. O. Bukanov, K. V. Yakubovich, L. M. Ermakova, B. A. Rebentish, A. A. Janulaitis, and V. G. Debabov, *Gene* **81**, 203 (1989).

TABLE II
REPLACEMENT VECTORS

Name	Space[a] (kb)	Cloning sites in polylinkers	Biological features[b]	Ref.
EMBL3	9–21	SalI, BamHI, EcoRI (flanking polylinkers inverted)[d]	Recombinants red⁻ gam⁻ ChiD, needs Gam in trans in recA⁻ host	7
EMBL4	9–21	EcoRI, BamHI, SalI (inverted)[d]	As for EMBL3	7
λ2001	9–21	XbaI, SacI, XhoI, BamHI, HindIII, EcoRI (inverted)[d]	As for EMBL3	7
EMBL301	9–21	SalI, BglII (NotI, XmaIII), NaeI (SfiI, BglII/XhoI), BamHI, EcoRI (inverted)[d]	As for EMBL3	51
λDASH	9–21	XbaI, SacI, XhoI, BamHI, HindIII, EcoRI (inverted)[d]	λ2001 with T3 and T7 promoters	7
Charon 40	9–21	EcoRI, SacI, KpnI, SmaI, XbaI, SalI, HindIII, NotI, XmaIII, AvrII, SpeI, XhoI, ApaI, BamHI, SfiI, NaeI (inverted)[d]	Central fragment made up of NaeI repeats, recombinants red⁻ gam⁺ and will grow in any recA⁻ host	7
EMBL3cos–Not	9–21	SalI, BamHI, EcoRI—central fragment—EcoRI, BamHI, NotI, SalI	λ cos adjacent to left-hand polylinker	20
EMBL3cosW	9–21	SalI, SfiI, XhoI, BamHI, XmaIII, EcoRI—central fragment—EcoRI, BamHI, XhoI (NotI, XmaIII), SfiI, SalI	As for EMBL3 cos Not, but with SP6 and T7 promoters and transcription terminators	c

[a] There is no significant difference in the cloning capacity of any of these vectors; the figure of 21 kb is a cautious one based on an upper limit of 50 kb for a λ genome. Larger genomes (~52 kb) have been recovered, but they may be less stable than those of normal size.

[b] With the exception of Charon 40, the above vectors can be traced back to λ1059 where the χ sequence was identified as Chi3 (i.e., ChiD rather than ChiC as indicated in many references).

[c] Paul Whittaker, personal communication, 1990.

[d] Inverted indicates that the orientation of the right-hand polylinker is inverted relative to the left.

Libraries in λ are not only easy to make, but the lysed cells within plaques release large amounts of phage DNA, making screening by hybridization to replica filters very effective.[44] An elegant alternative to hybridization depends on recombination between homologous DNA sequences, namely, recombination between an insert in a λ vector and a probe sequence in a plasmid endowed with a genetic tag, for example, a supF tRNA gene.[11] Rare recombinant phage identified by the suppressor are then selected on a suitable bacterial host. Genetic selection is a quick, inexpensive, and sensitive alternative to screens based on hybridization, but the original procedures sometimes failed because of low recombination frequencies and high backgrounds.

Improvements to both the probe plasmid and the selection schemes, together with some appreciation of the systems involved in phage–plasmid recombination, now assure sensitive selection of sequences following *in vivo* recombination with the probe.[53] The modified selection reduces background by depending on two amber mutations in the host—in *dnaB* and *lacZ*—instead of amber mutations within the vector; amber mutations in the vector were sometimes lost by recombination with defective phage produced by the packaging strains. The use of *dnaB* is not simple since its product is essential for replication of both host and phage DNAs. Nevertheless, while a substitute replication product supplied from a resident prophage enables the *dnaB⁻* host to grow, the formation of plaques by phage λ under these conditions of limiting DnaB requires suppression of the *dnaB* mutation. Concomitant suppression of the amber mutation in *lacZ* is easily monitored by hydrolysis of the usual chromogenic substrate; blue plaques confirm the presence of the suppressor, and, subsequently, white plaques reveal phage from which the suppressor plasmid has excised. With hindsight, the low recombination frequencies obtained in screens using most vectors reflected the absence of Rap, a phage-encoded function necessary for phage–plasmid recombination.[54] Most vectors are *rap⁻* since *rap* maps within the region covered by the *nin* deletion (see Fig. 1); Rap provided in trans from a plasmid ensures good recombination frequencies.[53]

Phage λ in Analysis of Escherichia coli

Bacterial genes cloned in λ have often been detected by their expression in the lytic mode irrespective of the presence of their own promoters; this has frequently been the basis for the selection of a particular λ transducing phage from a library of recombinants.[55] A stable lysogen, in which the cloned gene compensates for a genetic lesion in the host, offers an alternative selection and better scope for subsequent analyses. However, only a few λ vectors are integration proficient and *cI⁺*, and thereby permit stable lysogeny without dependence on a "helper." Although λZAP is integration proficient, its plasmid origin of replication precludes maintenance of the prophage in *polA⁺ E. coli*. Some λ functions are necessary only for the establishment of lysogeny; others are also necessary for the maintenance of lysogeny. Integrase, the enzyme necessary for the site-

[53] D. M. Kurnit and B. Seed, *Proc. Natl. Acad. Sci. U.S.A.* **87**, 3166 (1990).
[54] C. T. Lutz, W. C. Hollifield, B. Seed, J. M. Davie, and H. V. Huang, *Proc. Natl. Acad. Sci. U.S.A.* **84**, 4379 (1987).
[55] K. Borck, J. D. Beggs, W. J. Brammar, A. S. Hopkins, and N. E. Murray, *Mol. Gen. Genet.* **146**, 199 (1976).

specific insertion of the λ genome into the *E. coli* chromosome, is required transiently for the establishment of lysogeny, and it is easily obtained by coinfection with a helper phage. In contrast, a cI^- phage can only be maintained as a prophage if the cI^+ helper phage also integrates to give a dilysogen, or if repressor is provided in trans from a plasmid. A structural defect in integration cannot be supplied in trans, and, again, the helper phage is retained as a dilysogen. Alternatively, the phage can take advantage of its cloned genes, rather than a helper phage, and integrate by homology-dependent recombination. Weisberg[56] provides an excellent, concise account of the genetics of λ transducing phages.

A cloned gene that is stably maintained in single copy as a prophage, in some cases even in its normal chromosomal position, more closely approximates the natural situation than one present in a high copy number plasmid; both copy number and supercoiling of DNA can be relevant to the levels of gene expression.[57] Stable lysogens are therefore of paramount importance in some complementation tests, as is the ease with which phage lysates may be tested on many genetically different laboratory strains. Both of these properties, and the ability to reclone in the classic sense by plaque purification, greatly expedite the use of λ derivatives in genetic analyses. The latter is particularly relevant to mutational studies.

Homology-dependent recombination is also fundamental since mutations are readily transferred from chromosome to phage or vice versa. The loss of mutations in cloned sequences by recombination with a wild-type chromosome can be prevented by using a mutant host, preferably one with a suitable deletion. Homology-dependent recombination permits the transfer of mutations from the phage to the bacterial chromosome and may be used as an alternative to transduction by phage P1; it avoids transposons and confines genetic differences to a very small region of the *E. coli* chromosome.[37,56] Reciprocal recombination between homologous sequences can direct the integration of a λ transducing phage into the *E. coli* chromosome or the formation of a cointegrate with a plasmid. Recombination between cloned sequences in λ and plasmid vectors can be used to generate new chimeric sequences.[58]

Stable single-copy lysogens play a valuable role in the analysis of transcription following fusion of the promoter of interest to a reporter

[56] R. Weisberg, *in "Escherichia coli* and *Salmonella typhimurium"* (F. C. Neidhardt, J. L. Ingraham, B. Magasanik, K. B. Low, M. Schaechter, and H. E. Umbarger, eds.), p. 1169. American Society for Microbiology, Washington, D.C., 1987.

[57] C. F. Higgins, C. J. Dorman, D. A. Stirling, L. Waddell, I. R. Booth, G. May, and E. Bremer, *Cells (Cambridge, Mass.)* **52,** 569 (1988).

[58] A. A. F. Gann, A. J. B. Campbell, J. F. Collins, A. F. W. Coulson, and N. E. Murray, *Mol. Microbiol.* **1,** 13 (1987).

sequence. One integration-proficient vector with a polylinker cloning site upstream of a promoterless *lacZ* gene has recently been refined by the inclusion of a RNase III processing site between the polylinker and the reporter gene.[59]

Kohara and colleagues[60] have compiled a physical map of the *E. coli* chromosome by making and assembling the restriction maps of the overlapping fragments from a library of *E. coli* DNA cloned in λ replacement vectors. It is important to identify the genes carried by these phage; they are a reservoir of useful genetic material for those studying *E. coli*,and their analysis will permit the correlation of restriction and genetic maps. The recombinant phage in the Kohara library are *int⁻ red⁻ gam⁻ cI⁻* and defective in their site for Int-mediated integration, but this does not prevent the determination of their genetic content.[55,61] Spotting phage lysates onto lawns of mutant bacteria under conditions where the mutation limits bacterial growth permits detection of a λ phage with inserts overlapping the chromosomal lesion. Surprisingly, recombinant colonies can be detected despite lysis; alternatively, lysogeny can be achieved by coinfection with a homoimmune wild-type phage or by the expression of repressor from a plasmid encoding the *cI* gene. In the latter case integration would depend on homology, but all recombinants would not necessarily be lysogenic. The use of repair-deficient bacteria (*mutS* or *mutL*) may be helpful in detecting recombination between the DNA of *E. coli* and its close relatives.[62] The available restriction map of *E. coli* is already aiding the alignment of cloned sequences for which restriction maps are available and those DNA sequences long enough to identify a sufficient number of restriction targets.[63,64]

A minitransposon has been used for insertional mutagenesis of *E. coli* sequences cloned in λ. Since the mobile sequence is tagged with *sup*FtRNA, phage derivatives containing it can be selected by suppression of amber mutations in either the phage or host. The short (264 bp) inserts are found widely distributed throughout the target DNA sequences. They are readily located in restriction maps and facilitate DNA sequencing.[65] Transposon-induced mutations open the route to "reverse genetics" using the Kohara library of λ phage. The genetic tag permits the identification

[59] T. Linn and R. St. Pierre, *J. Bacteriol.* **172,** 1077 (1990).

[60] Y. Kohara, K. Akiyama, and K. Isono, *Cell (Cambridge, Mass.)* **50,** 495 (1987).

[61] M. Masters, *Am. Soc. Microbiol.* **56,** 6 (1990).

[62] C. Rayssignier, D. S. Thaler, and M. Radman, *Nature (London)* **342,** 396 (1989).

[63] K. E. Rudd, W. Miller, J. Ostell, and D. A. Benson, *Nucleic Acids Res.* **18,** 313 (1990).

[64] C.Médigue, J. P. Bouché, A. Hénaut, and A. Danchin, *Mol. Microbiol.* **4,** 169 (1990).

[65] S. H. Phadnis, H. V. Huang, and D. E. Berg, *Proc. Natl. Acad. Sci. U.S.A.* **86,** 5908 (1989).

of transductants of *E. coli*; Sup$^+$ haploid transductants occur with high frequency only if the transposon disrupts a nonessential gene.[66]

Other Special Uses of Phage λ

Even in the absence of the PCR, DNA sequences cloned in λ vectors have been sequenced without subcloning. Primers flanking the cloning sites of λgt10 and λgt11 have been used, but a generally applicable alternative for sequencing longer inserts relies on minitransposons.[65] The special uses of phage λ for making jumping and linking libraries are summarized elsewhere.[40]

Acknowledgments

I appreciate the constructive criticisms of Annette Campbell, Anne Daniel, Michael Feiss, David Leach, Kenneth Murray, and Damion Willcock, the artwork of Annie Wilson, and the patient expertise of Fiona Govan and Janet Panther in the preparation of the manuscript.

[66] S. H. Phadnis, S. Kulakauskas, B. R. Krishnan, J. Hiemstra, and D. E. Berg, *J. Bacteriol.* **173,** 896 (1991).

Section II

Other Bacterial Systems

[13] Genetic Analysis in *Bacillus subtilis*

By JAMES A. HOCH

Introduction

Of the bacteria *Bacillus subtilis* is the easiest organism to manipulate genetically. The armamentarium of genetic tools includes PBS1 transduction which transfers very large [>150–200 kilobase (kb)] fragments of the chromosome and DNA-mediated transformation which utilizes smaller DNA fragments (40 kb or less). The chromosomal location of new markers can be rapidly determined by transducing a set of auxotrophic strains covering all regions of the chromosome.[1] This transduction system takes the place of conjugation, which has never been developed for this organism. DNA-mediated transformation occurs at high frequency and is most useful for fine structure analysis of chromosomal regions up to 10 kb. In addition, a protoplast transformation system has been developed to allow easy transfer of plasmids between strains.

Strategic Considerations

In a common scenario, an investigator has isolated a series of mutants known to affect a particular pathway or phenomenon and wishes to determine the number of loci involved and their chromosomal locations. The first step in determining the chromosomal locations of the loci is to transduce a set of strains with auxotrophic markers conveniently located around the chromosome such that the entire chromosome will be covered.[1] Figure 1 shows the genetic markers that may be conveniently used for such an analysis. A PBS1 transducing lysate is prepared on the strain of interest and used to transduce each recipient to test for linkage. In practice all of the crosses can be carried out in a single day and a general chromosomal location determined in a week. Once the general location of the locus is obtained, further fine structure analysis of its location should be carried out by transformation. A large variety of genetic markers exist in virtually all regions of the chromosome.[2]

Before embarking on genetic studies in *B. subtilis*, it is strongly suggested that the classic paper by C. Anagnostopoulos and co-workers[3] be

---

[1] M. A. Dedonder, J. A. Lepesant, J. Lepesant-Kejzlarova, A. Billault, M. Steinmetz, and F. Kunst, *Appl. Environ. Microbiol.* **33,** 989 (1977).
[2] P. J. Piggot and J. A. Hoch, *Microbiol. Rev.* **49,** 158 (1985).
[3] M. Barat, C. Anagnostopoulos, and A.-M. Schneider, *J. Bacteriol.* **90,** 357 (1965).

METHODS IN ENZYMOLOGY, VOL. 204

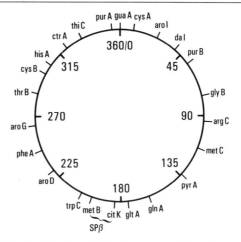

FIG. 1. Landmark loci of the *B. subtilis* chromosome. The chromosome is divided by degrees. Strains bearing these markers are used to locate unknown markers by transduction. Strains are available from the Bacillus Genetic Stock Center, Ohio State University, Columbus, Ohio.

read. This publication is a virtual treasure trove of genetic information and techniques and explores in detail techniques such as the recombination index that are not explained here.

Transformation and Transduction Procedures

Solutions

PAB: 17.5 g antibiotic medium 3 (Difco, Detroit, MI) dissolved in 1 liter deionized water. Autoclave.

TBAB: 17.5 g tryptose blood agar base (Difco) and 8.5 g Bacto Agate (Difco) suspended in 1 liter deionized water. Autoclave and dispense 25 ml in petri dishes.

MG: 2 g ammonium sulfate [$(NH_4)_2SO_4$]; 14 g potassium phosphate, dibasic (K_2HPO_4); 6 g potassium phosphate, monobasic (KH_2PO_4); 1 g sodium citrate ($Na_3C_6H_5O_7 \cdot 2H_2O$); 0.2 g magnesium sulfate ($MgSO_4 \cdot 7H_2O$). Dissolve each of the ingredients successively in 1 liter of deionized water. Autoclave. Add 10 ml of sterile 50% glucose. For MG plates make a 2× solution of salts and a 2× solution of agar (Bacto-agar, Difco) to give 1.5% final agar concentration. Autoclave separately. Mix at 50° and add 10 ml of 50% glucose per liter. Fifty percent of glucose is easily made by dissolving 500 g glucose in 750 ml distilled water.

CH: 5.0 g casein hydrolyzate, acid (Difco), dissolved in 100 ml deionized water. Autoclave. Note that CH does not contain tryptophan, glutamine, purines, pyrimidines, or vitamins.

Amino acid stock solutions: 5 mg/ml in deionized water. Autoclave.

First Growth Period Supplemented MG. The final concentration of additives is 0.02% CH and 50 μg/ml of any amino acid purine or pyrimidine required by the recipient. For one transformation:

1. Dilute 0.1 ml of CH in 2.4 ml of MG. Add 0.25 ml to a 20 × 150 mm culture tube.

2. Dilute 0.1 ml amino acid stock solution in 0.9 ml of MG. Add 0.25 ml of each amino acid required by the recipient to the above tube. Add sufficient MG to bring the volume to 2.5 ml.

Second Growth Period Supplemented MG. The final concentration of additives is 0.01% CH and 5 μg/ml of required amino acid.

1. Make a 1/10 dilution of the required amino acid dilution and a 1/2 dilution of CH dilution from First Growth Period Supplemented MG.

2. To a 20 × 150 mm culture tube add 0.1 ml of CH, 0.1 ml of each amino acid dilution, 0.1 ml 50 mM MgSO$_4$, and MG to 0.9 ml.

Chromosomal DNA Preparation

Inoculate the donor strain lightly in 20 ml of MG supplemented with 20 μg/ml amino acid requirements and 0.05% CH in a 250-ml culture flask. Grow overnight with shaking at 37°. Centrifuge the culture 10 min at 4° at 7000 rpm (SS-34 head in a Sorvall centrifuge). Resuspend the pellet in 2 ml of TE (10 mM Tris, 1 mM Na$_2$EDTA, pH 8.0) in a 16 × 125 mm culture tube and add 5 mg/ml fresh lysozyme and 8 μl of a solution of RNase at 10 mg/ml. Incubate for 30 min at 37°. Add proteinase K to 1 mg/ml and incubate for 15 min at 37°. Next add 200 μl of 10% sodium dodecyl sulfate followed by 2 ml of phenol saturated with TE. Shake gently for 1 min. Centrifuge to separate the layers and extract the top layer with 2 ml phenol–chloroform (1 : 1) followed by 2 ml chloroform. Four milliliters of 95% (v/v) ethanol is layered on the final supernatant, and the DNA is spooled out on a fine glass rod or Pasteur pipette. Dissolve the DNA in 200 μl of TE.

Transformation

Transformation requires the preparation of a cell culture competent to take up DNA. This mystical state occurs at the beginning of stationary phase and is associated with the onset of sporulation. Some mutants are difficult to bring to competence and therefore transform poorly. Many times this is a result of poor growth in the First Growth Period, and the simple addition of a little yeast extract (e.g., 0.01%) will solve the problem. Presented below is the original recipe for transformation devised by C. Anagnostopoulos and Spizizen[4]; which works well on a wide variety of strains.

First Growth Period

1. Streak the recipient strain on a TBAB plate the evening before the experiment. Incubate at 37°.
2. From the fresh overnight TBAB culture sufficient cells are taken to give a lightly turbid suspension in 2.5 ml of First Growth Period Supplemented MG in a 20 × 150 mm sterile culture tube.
3. Incubate with shaking at 37° for 4 hr. The culture should be quite turbid; if not, incubate an additional 1 hr.

Second Growth Period

1. Prepare two culture tubes for the second period by diluting 0.1 ml of culture in 0.9 ml Second Period growth medium in each of two 20 × 150 mm tubes.
2. To one tube add 0.1 ml of DNA solution (1 to 0.0001 μg DNA) diluted in MG. To the second tube (the reversion control) only add MG.
3. Incubate with shaking at 37° for 90 min.
4. Prepare serial dilution tubes containing 0.9 ml MG and dilute the transformation tube 1/10, 1/100, and 1/1000. Spread on separate MG plates 0.1 ml of the undiluted culture or 0.1 ml of each dilution along with 0.1 ml of appropriate amino acid stock solution required by the recipient but not selected for in this experiment.
5. Spread 0.1 ml of undiluted reversion control culture on appropriately supplemented plates.
6. Incubate plates at 37° for 48 hr.

This standard transformation procedure will yield around 100 colonies on the 1/1000 dilution plate for highly competent cells at DNA saturation. This corresponds to a frequency of 1% transformation. DNA saturation under these conditions is around 0.1–1.0 mg DNA/ml.

Protoplast Transformation

The introduction of replicating plasmids into *B. subtilis* is most efficient with plasmid multimers. Furthermore, if the plasmid contains an insert of DNA homologous to the *B. subtilis* chromosome, it is likely to be subjected

---

[4] C. Anagnostopoulos and J. Spizizen, *J. Bacteriol.* **81,** 741 (1961).

to a recombination event that can cause loss of the insert or gene conversion. Therefore, under these conditions it is best to introduce plasmids into recipients using the protoplast transformation procedure. Presented below is a simplified version (M. Zukowski, personal communication, 1990) of the original Chang and Cohen procedure.[5] The procedure works well with minipreparation plasmid DNA.

Solutions

2× SMM buffer: 1.0 M sucrose, 40 mM sodium maleate, 40 mM MgCl$_2$. Adjust to pH 6.5 and autoclave.

SMMP medium: Mix together equal volumes of 4× PAB (autoclaved) and 2× SMM (autoclaved).

PEG solution: 40 g polyethylene glycol (PEG) 6000, 50 ml of 2× SMM buffer. Bring the volume to 100 ml with deionized water and autoclave.

DM3 Regeneration Medium

1. Mix, in 700 ml deionized water, 135 g sodium succinate, 5 g casamino acids (Difco), 5 g yeast extract, 1.5 g KH$_2$PO$_4$, and 3.5 g K$_2$HPO$_4$.

2. Adjust to pH 7.2.

3. Bring the volume to 950 ml with deionized water.

4. Add 8 g agar.

5. Autoclave.

6. Cool to approximately 55°.

7. Add 25 ml of 20% sterile glucose (preheated to 55°).

8. Add 20 ml of 1 M MgCl$_2$ (preheated to 55°).

9. Add 5 ml of 2% bovine serum albumin (BSA, freshly prepared, filter-sterilized).

10. Add antibiotics as required: chloramphenicol, 10 μg/ml final concentration; kanamycin, 100 μg/ml final concentration.

11. Pour plates, leave on the benchtop overnight, then store at 4° for no longer than 1 month.

Procedure

1. A starter culture is grown overnight in 10 ml of PAB at 37° with shaking.

2. The following morning, 0.5 ml of starter culture is inoculated into 50 ml of PAB. Shake the culture at 37° until the cell mass reaches an OD$_{660}$ of 0.4–0.5.

3. Transfer the culture into a 50-ml plastic centrifuge tube. Pellet the

[5] S. Chang and S. N. Cohen, *Mol. Gen. Genet.* **168**, 111 (1979).

cells by centrifugation (e.g., 10 min at 6000 rpm in Beckman TJ-6 centrifuge with a fixed-angle rotor).

4. Resuspend the pellet in 5 ml of SMMP with 2 mg/ml lysozyme freshly added. The solution may be filter-sterilized before adding to cells.

5. Incubate the cells in a plastic centrifuge tube at 37° with gentle rocking (about 100 rpm) for 90 min (60 min may be satisfactory in some cases). Microscopic evaluation of protoplasting efficiency may be called for, but loss of schlieren in the suspension should be indicative.

6. Pellet protoplasts by centrifugation (e.g., 10 min at 5000 rpm at 4° in Beckman TJ-6 centrifuge with a fixed-angle rotor).

7. Pour off the supernatant and add 2.5 ml of SMMP. Resuspend the protoplasts by vigorous agitation of the tube with fingers. Do not use a vortex.

8. a. Mix the plasmid DNA with an equal volume of $2\times$ SMM in a 50-ml plastic centrifuge tube.
   b. Add 0.5 ml of the protoplast suspension. Agitate the tube lightly with fingers.
   c. Add 1.5 ml of the PEG solution. Gently mix by rolling the tube manually.
   d. After 2 min of exposure to the PEG, add 5 ml SMMP.

9. Repeat Step 6.

10. Resuspend the pellet in 1 ml of SMMP; incubate at 37° with gentle shaking for 90 min.

11. Plate 0.1-ml samples onto fresh DM3 regeneration medium plates to which antibiotics have been added as required. Incubate plates at 37° for 48–72 hr. Expect over 10^5 transformants/μg for uncut DNA and 10^4–10^5 transformants/μg for cut and ligated DNA. Expect no transformants if MgCl$_2$ was omitted from solutions.

PBS1 Transduction

PBS1 is a large transducing phage with the capability of encapsulating 150–200 kb of chromosomal DNA in the transducing particle. It depends on motile flagella for attachment and therefore is not useful for Fla⁻, Mot⁻, or any other strain with mutations that impair motility. The vast majority of researchers that use it have never seen a PBS1 plaque since it is an exceedingly poor plaque former. Despite these problems this transducing system works well for locating markers on the chromosome.

PBS1 Stock Lysate. PBS1 stock lysates are prepared on the highly motile *Bacillus licheniformis* 8480 strain or any highly motile *B. subtilis* strain. *Bacillus licheniformis* 8480 (strain 5A1 in the Bacillus Genetic Stock Center Collection) cells from an overnight TBAB plate are inoculated at light turbidity into 2.5 ml of PAB in a 20 × 150 mm culture tube. Shake the tube at 37° for 2.5 hr. Add 0.1 ml of PBS1 stock lysate; allow adsorption for 5 min at room temperature. Dilute the tube into 25 ml of PAB in a 250-ml culture flask. Shake at 37° for 2 hr. Place the flask in 37° incubator overnight without shaking. Centrifuge the lysate at 10,000 g for 20 min at 4° to pellet debris. Add 50 μg/ml DNase I (dissolved in 0.01 M MgSO$_4$) and incubate 30 min at room temperature. Filter-sterilize the supernatant through a 0.45-μm filter. Store at 4°. The lysate will be stable for at least 1 year.

Preparation of Donor Lysates. The donor strain is streaked on a TBAB plate and grown overnight at 37°. Donor cells are inoculated at light turbidity (OD$_{540}$ ~0.1) into 2.5 ml of PAB in a 20 × 150 mm culture tube. Shake at 37° for 2.5 hr. The cells at this time are just beginning to become motile. Add 0.1 ml of the PBS1 stock lysate. Incubate at room temperature for 5 min. Dilute into 25 ml of PAB in a 250-ml culture flask. Incubate with shaking at 37° for 1 hr. Add chloramphenicol to 5 μg/ml (spectinomycin or no antibiotic at all may be used if the strain is Cm^R). Shake for an additional 2 hr at 37°. Let the culture stand without shaking at 37° overnight. Centrifuge the lysate at 10,000 g for 20 min at 4° to remove debris. Add 50 μg/ml DNase I and incubate at room temperature for 30 min. Filter-sterilize through a 0.45 μm filter. Store the lysate at 4°.

Although this procedure seems empirical, and is, it works almost every time, and after several times it seems natural. It is important to treat the lysate with DNase I because it will not filter well without it, and this step avoids transformation of the recipients.

Transduction Procedure. The recipient strain is grown overnight on a TBAB plate at 37° and inoculated at light turbidity into 2.5 ml of PAB in a 20 × 120 mm culture tube. Shake at 37° for 4 hr. The cells at this stage should be motile and uniform in size when examined microscopically. Add 0.5 ml of cells and 0.5 ml of donor lysate to a 16 × 125 mm culture tube. To a second tube add 0.5 ml cells and 0.5 ml of PAB for a reversion control. Shake both tubes for 20 min at 37°. Plate 0.1 ml of transduction and reversion cultures on appropriately supplemented MG plates for selection of recipient markers.

This procedure normally gives between 50 and 500 transductants on the selection plate. Since 0.1 ml of rich medium is plated on the MG plates a slight lawn appears on the plate. This can be a problem with certain markers such as vitamin deficiencies, and, if so, the transduced cells can be centrifuged and washed with MG before plating.

Analysis of Genetic Data

The basic measure relating genetic and physical distance between two markers in transduction and transformation is the recombination unit. This is defined as percent recombination = 100 (1 − cotransfer), where

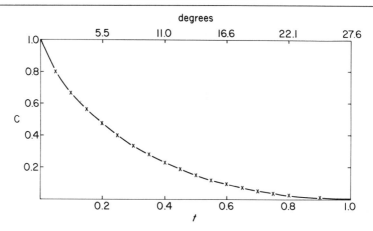

FIG. 2. Relationship between cotransfer values in transduction and physical distance between markers on the chromosome. A graph of the Kemper formula $C = (1 - t) + t(\ln t)$ relates cotransfer values to the degrees separating two markers on the chromosome.[5,6] C is the cotransfer value between markers, and t is the length of the PBS1 transducing DNA fragment.

cotransfer of linked markers is the number of double transformants divided by the single transformants. Markers further apart on the chromosome will have a smaller cotransfer value and a larger recombination value than closely linked markers. To be meaningful in terms of the physical distance separating two markers, the recombination values can be converted to physical distance using the Kemper formula, $C = (1 - t) + t(\ln t)$, where C is the cotransfer value and t is the fractional length of the donor DNA separating the two markers.[6] If the average size of the donor DNA is known (T), the physical distance between the two markers (D) can be calculated from $D = tT$.

Figure 2[7] shows the Kemper formula in graphical form where the t value in PBS1 transduction has been related (approximately) to the degrees separating the markers on the chromosome. The graph makes it clear that cotransfer or recombination values are not additive. For example, a marker, X, located equidistant between two outside markers, A and B, and showing 50% recombination to each would be about $0.2t$ or $5.5°$ from each. The recombination values expected between A and B markers would not be 100% (50% + 50%) but rather 75% ($0.2t + 0.2t$). The assumptions and calculations behind this figure have been detailed elsewhere.[7] The Kemper analysis is most useful in placing markers on the chromosome

[6] J. Kemper, *J. Bacteriol.* **117,** 94 (1974).
[7] D. J. Henner and J. A. Hoch, *Microbiol. Rev.* **44,** 57 (1980).

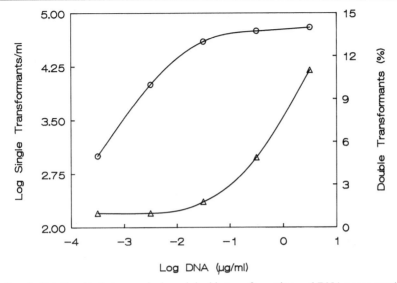

Log DNA (μg/ml)

Fig. 3. Relationship between single and double transformation and DNA concentration in transformation of weakly linked markers. The data (from Ref. 2) show the Leu$^+$ (○) and Phe$^+$ Leu$^+$ (△) transformants for the weakly linked markers *phe-1* and *leu-1*.

map by PBS1 transduction where the size of the transducing fragment is thought to be uniform.

In transformation, the size of the transforming DNA can vary greatly from preparation to preparation, and the values for C are size dependent, particularly the lower values. In general 7–10% recombination between two markers in transformation means that about 1 kb of DNA separates the two markers. The interested reader is referred to Ref. 3 for detailed examples of linkage analysis using both systems and to Henner and Hoch[6] for an in-depth study of the application of the Kemper formula.

Congression and Strain Construction

Congression is a phenomenon unique to transformation and refers to the uptake of more than one transforming DNA fragment by a competent cell, particularly at high DNA concentrations. Congression is a two-edged sword in that it disturbs linkage analysis of distant markers but is extremely helpful in strain construction. Figure 3 shows the effect of congression on the weakly linked *phe* and *leu* markers.[3] As the DNA concentration reaches saturating levels for the competent cell population, the number of double transformants Phe$^+$ Leu$^+$ increases greatly in the cross wild-type

DNA × *phe leu* recipient. The recombination values between closely linked markers are not subject to this effect. The example shown in Fig. 3 is of weakly linked markers where the recombination value stabilizes at nonsaturating DNA concentrations. Completely unlinked markers show no cotransfer at nonsaturating DNA concentrations.

Congression is the major technique used for strain construction if unlinked markers are involved. Any selectable marker in the recipient can be exchanged for one in the donor by transformation using DNA concentrations greater than 1 μg/ml. For example, a *trpC2 leu-1* recipient could be changed to a *phe-1 leu-1* strain by transforming with DNA containing the *phe-1* allele with selection for Trp⁺ and searching among the Trp⁺ transformants for Phe⁻ recombinants.

Piggot and de Lencastre have described a method to generate multiple marked strains where the recipient does not lose the selected marker.[8] This method is based on the observation that the products of certain early sporulation genes, for instance, *spoIIA*, are required only in the mother cell and not in the developing forespore. Thus, if one transforms a *spoIIA* recipient for Spo⁺ with wild-type DNA and selects for Spo⁺, some of the transformants will be genotypically *spoIIA* because the transforming Spo⁺ DNA integrated into the chromosome that remained in the mother cell and, therefore, allowed sporulation to occur but did not convert the forespore chromosome to Spo⁺. In practice, one can transform a *spoIIA* recipient with saturating *phe-1* DNA, for example, selecting Spo⁺ and searching for *phe-1 spoIIA* recombinants among the transformants. Successive markers can then be added to this strain by repeating the procedure with new donor DNA. Thus, congression is a powerful tool to facilitate strain construction with characterized markers. It is also a simple and important means to backcross primary mutant strains to determine if all the phenotypes of a mutant can be ascribed to a single mutation.

Studies of Complementation and Dominance Using Integrative Vectors

Since both complementation and dominance may reflect the subtle interplay of regulatory elements, the vectors to study these phenomena were designed to balance the concentration of the elements involved. The majority of these studies utilize integrative vectors which are basically *Escherichia coli* plasmids carrying an antibiotic resistance gene that is expressed in *B. subtilis*. One of the simplest and most utilized of these vectors is pJH101, which has a CAT (chloramphenicol acetyltransferase)

---

[8] P. J. Piggot and H. de Lencastre, *J. Gen. Microbiol.* **106,** 191 (1978).

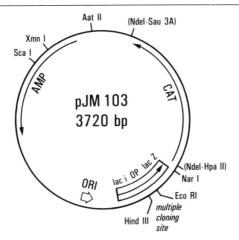

FIG. 4. Map of the integrative vector pJM103. The vector was constructed by ligating a filled-in *Hpa*II–*Sau*3A fragment of pC194 [S. Horinouchi and B. Weisblum, *J. Bacteriol.* **150**, 815 (1982)] into the *Nde*I site of pUC19. A sister plasmid (pJM102) was also constructed from pUC18 (M. Perego and J. A. Hoch, unpublished, 1988).

gene in the *Pvu*II site of pBR322.[9] Integrative vectors are incapable of replication in *B. subtilis* and therefore do not transform *B. subtilis* to chloramphenicol resistance. If a region of homology to the chromosome is cloned on the vector, transformation results in chloramphenicol-resistant colonies which are so-called Campbell recombinants where the inserted vector is flanked by a duplication of the original chromosome fragment carried on the vector. As long as this fragment is big enough to prevent interruption of the transcription unit being studied, meaningful complementation and dominance results between alleles can be obtained since there are only two copies of the region being studied on the chromosome. The vector is also quite stable when integrated.

Figure 4 shows a newer integrative vector, pJM103, which has a CAT gene in pUC19. Kanamycin-resistant derivatives of pUC18 and pUC19 are also available (M. Perego and J. A. Hoch, unpublished, 1988). pJM103 has all the properties of pUC19 in *E. coli,* therefore cloning of *B. subtilis* fragments may be carried out in *E. coli* utilizing the blue/white Lac selection system. After cloning and characterization of the DNA fragment of interest, the plasmid preparation is used to transform *B. subtilis* using the competence regimen outlined above. Minipreparations of the plasmid work very well.[10] Chloramphenicol resistance (Cm^R) may be selected on

[9] F. A. Ferrari, A. Nguyen, D. Lang, and J. A. Hoch, *J. Bacteriol.* **154**, 1513 (1983).
[10] H. C. Birnboim and J. Doly, *Nucleic Acids Res.* **7**, 1513 (1979).

any enriched medium such as TBAB containing 5 μg/ml chloramphenicol. Cm^R transformants are readily obtained with DNA inserts as small as 400–500 base pairs (bp), but the yield falls rapidly below this size range. The smallest DNA insert that we have found to give a Campbell recombinant is 149 bp.

Integrated vectors may be treated as a locus in transformation and transduction studies. PBS1 transduction may be used to locate the vector on the chromosome. In transformation studies the vector plus duplication is a significant fraction of the transforming DNA segment, and therefore larger DNA fragments (i.e., gently prepared DNA) are required to show linkage to closely linked markers. Recombination values are depressed because of the longer DNA fragments, and severe recombinational polarity is observed. Polarity results from the fact that any shearing within the vector in the DNA preparation results in loss of the Cm^R recombinant. Because of these complications, recombination values in transformation are not a reliable indicator of the location of integrated plasmids.

A plasmid system for complementation and dominance studies has been constructed to allow integration in the SPβ prophage found in all *B. subtilis* strains.[11] This plasmid, pFH7, contains a region of homology to SPβ and integrates into it on transformation in *B. subtilis*. The advantages of this system are that the gene of interest is inserted in another location in the chromosome, without disturbing its normal location, and induction of the prophage results in a specialized transducing phage carrying the gene.

Mutagenesis and Cloning with Integrative Vectors

Integrative vectors are highly mutagenic if the insert chromosomal DNA fragments are wholly contained within a gene or transcription unit. Both copies of the gene or transcription unit generated by the integration event are disrupted in this case. Insert fragments that stop within a gene or transcription unit and extend beyond either of its ends are nonmutagenic since one good copy of the gene is generated in the integration event. In a random mutagenesis experiment using *Taq*I chromosomal fragments cloned in the *Acc*I site of pJM102, 1% auxotrophs and 10% sporulation mutants were obtained among the Cm^R transformants (M. Perego, unpublished data, 1989). Mutations generated in this fashion allow the affected locus to be easily cloned by excision of the vector and adjacent sequences with restriction enzymes, ligation *in vitro,* and transformation in *E. coli*.

The procedure for recovery of integrated plasmids is straightforward.

---

[11] E. Ferrari and J. A. Hoch, *Mol. Gen. Genet.* **189,** 321 (1983).

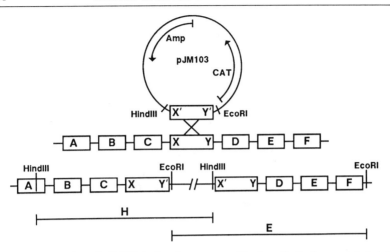

FIG. 5. Integration of pJM103 in the chromosome. The hypothetical genes labeled *A–F* are adjacent to the cloned fragment *X–Y*. The fragments H and E are the expected plasmids recovered by *Hind*III and *Eco*RI digestions, respectively.

An overnight culture of the strain containing the integrated vector is used to prepare chromosomal DNA as described above. Two micrograms of this DNA is digested with the appropriate restriction endonuclease in 30 μl of the appropriate buffer for the enzyme used, and 1 μg is removed to check the digestion on an agarose gel. The remaining 15 μl is diluted to 100 μl with deionized water and extracted with phenol followed by a chloroform extraction. The aqueous layer (~80 μl) is diluted with ligation buffer to 100 μl, ligase (1 unit) is added, and the mixture is incubated at 15° overnight. The entire ligation mixture is used to transform *E. coli* DH5α with selection for ampicillin resistance. Usually 20–50 transformants are obtained.

Figure 5 shows an example of the insertion of pJM103 into a cluster of genes and the expected plasmids that would be obtained by excision with *Eco*RI (E) or *Hind*III (H). Any restriction enzyme with a site within the multilinker of pJM103 can be used, but the direction of walking will depend on the orientation of the site to the inserted fragment. Gene disruption experiments using integrative vectors are standard procedure for the analysis of cloned DNA fragments where the phenotype of a disrupted open reading frame of unknown function is in question.

Mutagenesis by integrative vectors compares favorably in frequency with the best classic mutagenesis procedures. It can be used in a random procedure using total chromosomal DNA or directed specifically to a cloned region or gene. The locus of the mutagenic plasmid can be readily cloned from the affected strain.

Mutation with Chemical Agents

Bacillus subtilis is susceptible to all the chemical mutagenic agents used for induction of mutations in other bacteria. However, a very effective method of mutagenesis unique to spore formers was described by Balassa.[12] If spores are germinated in the presence of chloramphenicol to prevent protein synthesis, they are extremely susceptible to N-methyl-N'-nitro-N-nitrosoguanidine mutagenesis. Up to 25% sporulation mutants and 5% auxotrophs were found among the survivors of such treatment. This procedure is probably applicable to any sporulating bacterium and to a variety of mutagenic agents.

Procedure. A suspension of spores of the strain of interest is prepared in a sporulation medium such as AK agar (BBL, Cockeysville, MD) by streaking the strain on the agar plate, incubating at 37° for 48 hr, and harvesting the spores by scraping the plate with 5 ml of PAB. Vegetative cells are killed by the addition of a drop of chloroform, and the suspension is stored at 4°.

Fifty milliliters of PAB containing 100 μg/ml chloramphenicol in a 250 to 500-ml culture flask is inoculated with the spore suspension to give 10^7–10^8 spores/ml. The suspension is incubated with shaking at 37° for 1 hr to give 90–99% germination. Germinated spores turn phase-dark in a phase microscope in contrast to the highly refractile ungerminated spores. N-Methyl-N'-nitro-N-nitrosoguanidine is added to a final concentration of 50 μg/ml, and incubation is continued for 30 min. The suspension is then centrifuged 6000 g for 15 min at 4° to pellet the spores, and the treated spores are resuspended in sterile MG and stored at 4°. Dilutions of the treated spores are plated on the desired medium for mutant selection or screening.

This procedure is extremely effective when used correctly. The amount of mutagen and time of exposure may need to be empirically determined to give 30–40% survival. Survivals of 10% or less are to be avoided since the ungerminated spore population may be as high as 5%. The treated spores may be stored at 4° with little loss of viability for several days.

Mutagenesis and Cloning with Transposon Tn917

An excellent set of vectors for transposon mutagenesis using Tn917 has been constructed by Youngman and colleagues[13–15] This system allows

---

[12] G. Balassa, *Mol. Gen. Genet.* **104,** 73 (1969).
[13] P. J. Youngman, J. B. Perkins, and R. Losick, *Proc. Natl. Acad. Sci. U.S.A.* **80,** 2305 (1983).

isolation of transposition mutations with selection for Tn917 erythromycin resistance (Em^R) by placing a strain with a temperature sensitive plasmid containing Tn917 at the nonpermissive temperature. The transposon-induced mutations are very stable, and the Em^R marker can be used easily in transduction and transformation experiments. Cloning of the chromosomal regions adjacent to transposition-induced mutations is readily accomplished using a two-step procedure.[16] A pBR322 replicon containing a *cat* gene is first integrated into the transposon, and then the chromosomal DNA is digested with a restriction endonuclease, ligated, and transformed into *E. coli* as described above for recovery of integrative vectors. This system has proved to be very useful for cloning genes of interest from several chromosomal regions even though Tn917 insertions tend to cluster at the terminus of the chromosome, making the mutation search a little more arduous. The details of using Tn917 have been described.[17]

Integrative Vectors for Studying Gene Expression

One of the key techniques for studying gene expression *in vivo* is the coupling of the regulatory regions of a gene to the synthesis of β-galactosidase. This technique is particularly important in sporulation of *B. subtilis,* where determination of the timing of a gene's expression is one indication of its role in the temporal program of developmental transcription. For the identification of new genes that express their information under some controlled conditions, both transposons and integrative vectors may be used. Youngman and colleagues have developed Tn917 derivatives that express β-galactosidase when integrated into a transcribed region of the chromosome.[18] If the transposon happens to be in a gene controlled by the conditions of interest, the mutant strain may be recovered by blue/white screening. This technique has been used to isolate several new sporulation-associated mutations.[19]

Integrative vectors for this purpose have been developed to generate either translation or transcription fusions to β-galactosidase[20] (also M. Perego and J. A. Hoch, unpublished data, 1988). These vectors are used by generating a random library of cloned fragments in the vector *in vitro*

[14] J. B. Perkins and P. J. Youngman, *Plasmid* **12**, 119 (1984).
[15] P. J. Youngman, P. Zuber, J. B. Perkins, K. Sandman, M. Igo, and R. Losick, *Science* **228**, 285 (1985).
[16] P. Youngman, J. B. Perkins, and R. Losick, *Mol. Gen. Genet.* **195**, 424 (1984).
[17] P. Youngman, *in* "Plasmids: A Practical Approach" (K. Hardy, ed), p. 79. IRL Press, Oxford, 1986.
[18] J. B. Perkins and P. J. Youngman, *Proc. Natl. Acad. Sci. U.S.A.* **83**, 140 (1986).
[19] K. Sandman, R. Losick, and P. Youngman, *Genetics* **117**, 603 (1987).
[20] F. A. Ferrari, K. Trach, and J. A. Hoch, *J. Bacteriol.* **161**, 556 (1985).

followed by transformation in *E. coli*. The pooled transformants are then transformed into *B. subtilis* and plated for Cm^R with blue/white screening using the competence regimen. Integrative vectors have the advantage that the gene fused to β-galactosidase may be unaltered in one-half of the duplications generated, and therefore genes whose inactivation would be lethal can be recovered as fusions. Transposons are easier to use, as no *E. coli* step is required, and they are more stable than integrative vectors, but they suffer from a lack of randomness and the vectors available cannot be used in temperature-sensitive strains.

Fusion to β-galactosidase *in vitro* is a mainstay in the study of cloned genes. Several integrative vectors with the capacity for generating translation or transcription fusions to β-galactosidase have been described. These vectors integrate at the region of homology provided by the cloned fragment, which may not be desirable in some instances. In order to avoid this problem a β-galactosidase fusion vector has been described that can be integrated into the amylase gene of *B. subtilis* with loss of amylase activity.[21] This vector, pDH32, places the β-galactosidase fusion opposite to the direction of transcription of the amylase gene, and, because homology to amylase exists on both sides of the fusion and the CAT gene used for selection, the fusion integrates by a double crossover event in a stable configuration with loss of the vector sequences. This system has been used extensively for the study of expression of several genes. β-Galactosidase fusions may also be placed in the ϕ105 prophage to isolate them from the chromosomal region being studied.[22] A clever technique has been devised to transfer Tn*917* β-galactosidase fusions to the SPβ prophage.[23] The SPβ prophage may be induced to form a specialized transducing lysate which can be used to infect strains of interest.

[21] H. Shimotsu and D. J. Henner, *Gene* **43**, 85 (1986).
[22] J. Errington, *J. Gen. Microbiol.* **132**, 2953 (1986).
[23] N. Ramakrishna, E. Dubnau, and I. Smith, *Nucleic Acids Res.* **12**, 1779 (1984).

[14] Genetic Systems in *Haemophilus influenzae*

By GERARD J. BARCAK, MARK S. CHANDLER,
ROSEMARY J. REDFIELD, and JEAN-FRANCOIS TOMB

Introduction

Haemophilus influenzae is a small gram-negative bacterium, frequently commensal in the upper respiratory tract. It has a natural transformation system which, in addition to being very interesting in its own right, is an important tool both for genetic manipulation and for studies of DNA recombination and repair. Six different encapsulated forms (serotypes a through f) of the organism have been identified, as well as nonencapsulated, nontypable (NT) strains.[1] Most studies using transformation are performed in the nonpathogenic strain Rd (a nonencapsulated derivative of a serotype d strain[2]), but serotype b strains are also being intensively studied because of their role in human disease. Recently, many laboratories have adopted molecular genetic techniques for the analysis of gene structure and regulation in *H. influenzae*. It is our goal here to present several of these approaches as well as other techniques useful for the genetic manipulation of this organism.

Growth and Storage of *Haemophilus influenzae*

Haemophilus influenzae is routinely subcultured in either heart infusion (HI) or brain–heart infusion (BHI) broth supplemented with the required cofactors, nicotinamide adenine dinucleotide (NAD, 2–10 μg/ml) and hemin (10 μg/ml). NAD is prepared as a filter-sterilized 10 mg/ml stock and stored at $-20°$. Hemin suspension is stock solution 6 (Table I). Agar medium is prepared by adding 15 g Difco (Detroit, MI) Bacto-agar per liter of broth. To prepare stock cultures, cells are inoculated into supplemented brain heart infusion broth (sBHI) or sHI broth and grown to an OD_{650} of 0.3–0.4. Two milliliters of the culture is added to a vial containing 0.5 ml of sterile 80% glycerol and mixed well. The culture is stored at $-80°$.

Haemophilus influenzae may also be cultured in minimal medium MIc

---

[1] M. Pitman, *J. Exp. Med.* **53**, 471 (1931).
[2] H. E. Alexander and G. Leidy, *J. Exp. Med.* **93**, 345 (1951).

TABLE I
PREPARATION OF MIc MINIMAL MEDIUM

Component	Amount
Solution 1	
Distilled water	800 ml
L-Aspartic acid	5.0 g
L-Glutamic acid	13.0 g
NaCl	58.0 g
K_2SO_4	10.0 g
$MgCl_2$, hexahydrate	4.0 g
$CaCl_2$, dihydrate	0.294 g
Na_2EDTA, dihydrate	0.037 g
NH_4Cl	2.2 g
Adjust pH to 7.2 with 10 N NaOH. Add distilled water to 1 liter. Autoclave	
Solution 2	
(A) L-Arginine	1.5 g
L-Glycine	0.15 g
L-Lysine	0.25 g
L-Methionine	0.5 g
L-Serine	0.5 g
Dissolve in total volume of 100 ml of 0.1 N HCl	
(B) Dissolve 1.5 g of L-leucine in 100 ml of 0.1 N HCl	
(C) Dissolve 1.0 g L-tyrosine in 20 ml of 1 N HCl. Add distilled water to 100 ml. Filter sterilize solutions A, B, and C, and combine equal volumes to make up Solution 2	
Solution 3	
Tween 80	0.1% (v/v)
Poly(vinyl alcohol)	0.1% (w/v)
Glycerol	15% (v/v)
Sodium lactate	4.0 g
Add distilled water to 100 ml. Filter sterilize	
Solution 4	
Uracil	0.2 g
Hypoxanthine	0.04 g
Dissolve in 100 ml 0.1 N HCl. Filter sterilize	
Solution 5A	
Inosine	1.0 g
K_2HPO_4	1.74 g
KH_2PO_4	1.36 g
Add distilled water to 100 ml. Filter sterilize	
Solution 5B	
Inosine	1.0 g
Tris-HCl	1.57 g
KH_2PO_4	0.075 g
Add 90 ml distilled water, adjust pH to 7.4 with NaOH, and bring volume to 100 ml with distilled water. Filter sterilize	

TABLE I (*continued*)

Component	Amount
Solution 6	
Hemin	0.1 g
L-Histidine	0.1 g
Add to 100 ml of 4% triethanolamine (2,2′,2″-nitrilotriethanol, free base, v/v). Sterilize by heating to 65° for 30 min	
Solution 7	
NAD	0.1 g
Thiamin	0.1 g
Pantothenate, Ca salt	0.1 g
Add distilled water to 10 ml, filter sterilize, and store in 1-ml aliquots at −20°	

as described by Herriott *et al.*[3] Stock solutions are prepared as described in Table I.

To Prepare 100 ml of MIc Liquid Medium. To 20 mg of L-cystine dissolved in 1 ml of 1 *N* HCl, add 49 ml of distilled water and the following amounts of the stock solutions: stock 1, 10 ml; stock 2, 6 ml; stock 3, 2 ml; stock 4, 5 ml; stock 5A or 5B, 20 ml; stock 6, 1 ml; stock 7, 40 µl. Adjust the pH to 7.2 using 1 *N* NaOH and bring the volume to 100 ml with distilled water. Filter sterilize.

To Prepare 100 ml of MIc Agar. Mix 50 ml of sterile molten 3.2% agar with the same amount of stock solutions mentioned above. To adjust the pH and bring the volume up to 100 ml, add 2.6 ml of sterile 1 *N* NaOH and 3.4 ml of sterile distilled water. Stock solutions are stored at 4° unless specified.

Genetic Organization of *Haemophilus influenzae* Rd

The *H. influenzae* genome is a 1900-kilobase (kb) circle,[4] with a G + C content of about 37%.[5] Several genetic maps have been published. However, even in a recent rendition,[6] linkage is incomplete, and the map covers only part of the genome.

[3] R. M. Herriott, E. Y. Meyer, M. Vogt, and M. Modan, *J. Bacteriol.* **101,** 513 (1970).
[4] J. J. Lee, H. O. Smith, and R. J. Redfield, *J. Bacteriol.* **171,** 3016 (1989).
[5] P. H. Roy and H. O. Smith, *J. Mol. Biol.* **81,** 427 (1973).
[6] R. B. Walter and J. H. Stuy, *J. Bacteriol.* **170,** 2537 (1988).

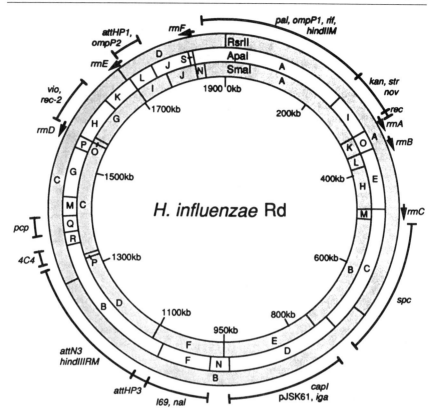

FIG. 1. Physical and genetic map of *H. influenzae* Rd. *Apa*I, *Rsr*II, and *Sma*I restriction fragments of the Rd genome are labeled alphabetically from largest to smallest. *attHP1*, *attHP3*, and *attN3* are attachment sites for bacteriophage Hp1c1, HP3, and N3, respectively; *cap1*, Rd capsule locus; *hindIIM*, HindII methylase; *hindIIIRM*, HindIII restriction endonuclease and methylase; *iga*, IgA protease; *kan, nal, nov*, kanamycin, naladixic acid, and novobiocin resistances; *ompP1, ompP2*, outer membrane proteins P1 and P2; *pal*, peptidoglycan-associated lipoprotein; *pcp*, PAL-comigrating protein; *rec-1*, Rec-1; *rec-2*, Rec-2; *rif*, *spc, str, vio*, rifampin, spectinomycin, streptomycin, and viomycin resistances; *rrnA–F*, ribosomal RNA operons; I69, 4C4, virulence loci.

A restriction map of the genome is shown in Fig. 1. The map was determined by Southern blot hybridization analysis of fragments separated by orthogonal-field alternation gel electrophoresis (OFAGE) and contour clamped homogeneous electric-field gel electrophoresis (CHEF). All of the *Apa*I, *Rsr*II, and *Sma*I sites have been mapped (45 sites in all). The sizes of these fragments are given by Lee *et al.*[4] The map shows the locations of a number of genetic and cloned markers. The identities of these markers are given in the legend. The map also gives the locations and orientations of the six ribosomal RNA operons (*rrnA–rrnF*), identified by Southern hybridizations. The single *Apa*I site shown in each *rrn* operon is actually two sites separated by 1.75 or 1.5 kb (*Apa*I fragments T and U), depending on the presence or absence of a 250-base pair (bp) segment in the spacer between the 16 S and 23 S genes.[4] Note that the small *Apa*I K/*Sma*I O fragment has been repositioned from the location originally reported by Lee *et al.*,[4] based on information obtained when the DNA sequences flanking *rrnB* were cloned.

Using a similar approach, Kauc and co-workers have published a physical map of the *H. influenzae* Rd genome[7] which is significantly different from that shown in Fig. 1, although the strains used by the two groups give identical *Apa*I and *Sma*I patterns.[8] The discrepancies are mainly in the relative positions of genome segments bounded by the *rrn* operons.

Maps of Other Strains and Serotypes

Comparison of restriction patterns in pulsed-field gels can be helpful in identifying or characterizing strains. The *Apa*I and *Sma*I restriction patterns of a clinical type d strain were identical to those of the laboratory Rd strain. However, the widely used Rd strain BC200 has a duplication of about 60 kb,[4] and another Rd strain, Parker, appears to contain a duplication or insertion of about 20 kb.[8] Other serotypes have different patterns.[4] We have examined the *Apa*I and *Sma*I patterns of a number of type b strains; nine of the ten strains had differences in at least one of the two digests. This is consistent with the finding that type b strains are evolutionarily diverse.[9]

Haemophilus influenzae Bacteriophage and Plasmids

Phage

Several temperate phage have been identified in *H. influenzae* isolates, but no virulent phage are known. No transducing phage are known, nor have any phage been developed as cloning vectors. The best studied *H. influenzae* phage are members of the HP1/S2 complex, especially HP1c1,

[7] L. Kauc, M. Mitchell, and S. H. Goodgal, *J. Bacteriol.* **171**, 2474 (1989).
[8] R. J. Redfield, unpublished results (1989).
[9] J. M. Musser, J. S. Kroll, E. R. Moxon, and R. K. Selander, *Infect. Immun.* **56**, 1837 (1988).

an HP1 derivative that gives larger, clearer plaques (unlike λ cI mutants, HP1c1 can still lysogenize host cells). Although HP1 and S2 phage were isolated independently, they have similar morphology and 37-kb genomes with high DNA homology and very similar restriction maps, and lysogens show at least partial heteroimmunity.[10] Two attachment sites are known for S2,[11] but only one of these has been identified for HP1c1.[4,12]

Two other phage are HP3[13] and N3.[14] Their genome sizes are 37 and 42 kb, respectively, and the map locations of their attachment sites are shown in Fig. 1.[8] Their host ranges are more restrictive than that of HP1. HP3 will form plaques on BC200-like strains but not on the standard Rd.[15] The only known host for phage N3 is the Rd9 (or Rd Parker) strain; lysogenization with N3 causes this strain to lose the ability to be transformed.[16]

The plaques produced by all of these phage are small, variable, and often indistinct. For HP1c1, plate lysates of moderate titer [10^9 plaque-forming units (pfu)/ml] can be obtained by overlaying confluently lysed lawns with 10 ml of sBHI. The highest titer lysates are obtained by growing cells to mid log in sBHI, adding mitomycin C at 35 ng/ml,[17] and continuing shaking for 3–5 hr or until the culture appears lysed (high titers can sometimes be obtained even if there is no visible lysis). Lysates should be clarified by centrifugation (10 min at 5000 g), and stored over chloroform.

Several temperature-sensitive mutants of HP1c1 are available, and they can be used for studies of host–cell recombination systems.[18] These mutants make plaques at 30° but not at 40° (note that 40° is close to the upper limit for growth of *H. influenzae,* so precise temperature control is necessary if lawns are to survive.) HP1c1 recombination is much more frequent in competent cells than in log-phase cells.[18]

A defective prophage has been identified in the genome of the standard Rd laboratory strain, and its induction is thought to be responsible for much of the cell death associated with uptake of transforming DNA.[10] Strain BC200 contains a genome duplication which prevents expression of this phage.[19]

[10] M. E. Boling, D. P. Allison, and J. K. Setlow, *J. Virol.* **11,** 585 (1973).
[11] L. Kauc and S. H. Goodgal, *J. Bacteriol.* **171,** 1898 (1989).
[12] A. S. Waldman, S. D. Goodman, and J. J. Scocca, *J. Bacteriol.* **169,** 238 (1987).
[13] J. H. Stuy, *Antonie Van Leeuwenhoek* **44,** 367 (1978).
[14] J. Samuels and J. K. Clarke, *J. Virol.* **4,** 797 (1969).
[15] J. H. Stuy, personal communication (1989).
[16] A. Piekarowicz and M. Siwińska, *J. Bacteriol.* **129,** 22 (1977).
[17] W. P. Fitzmaurice and J. J. Scocca, *Gene* **24,** 29 (1983).
[18] M. E. Boling and J. K. Setlow, *J. Virol.* **4,** 240 (1969).
[19] J. K. Setlow, M. E. Boling, D. P. Allison, and K. L. Beattie, *J. Bacteriol.* **115,** 153 (1973).

Plasmids

Two plasmids, pHVT1[20] and pDM2,[21] have been the workhorse cloning vectors for *H. influenzae* and can also replicate in *Escherichia coli*. Several smaller derivatives of these plasmids that contain new selectable markers and multiple cloning site specificities have been prepared[22,23]; they are shown in Table II[24–27] along with other new cloning vehicles and the recommended antibiotic concentrations for their selection in *H. influenzae* (Table III).

Preparation of Chromosomal and Plasmid DNA

Standard Chromosomal DNA Preparation

This procedure can be scaled to meet the need.

1. Suspend the pellet from a 1-liter overnight culture of cells, grown in sBHI, in 50 ml of 0.15 M NaCl, 0.1 M EDTA (pH 8.0).

2. Add sodium dodecyl sulfate (SDS) to 1%, gently mix, and incubate at 65° for 10 min.

3. Cool to 37° and add proteinase K to a final concentration of 50 to 100 μg/ml. Incubate 1 to 16 hr at 37°.

4. Extract once with an equal volume of phenol buffered with Tris-HCl (pH 8) and once with chloroform–isoamyl alcohol (24 : 1, v/v), centrifuging at 10,000 g for 20 min after each extraction to recover the aqueous phase.

5. Precipitate with 2 volumes of ethanol. Spool out the DNA using a glass rod and rinse 3 times with 70% ethanol.

6. Dissolve the pellet in 25 ml of TE buffer (10 mM Tris-HCl, pH 8.0, 1 mM EDTA).

7. Add RNase A to 50 μg/ml and T1 RNase to 25 units/ml, and incubate at 37° for 30 min to 1 hr.

8. Repeat Steps 4, 5, and 6.

[20] D. B. Danner and M. L. Pifer, *Gene* **18**, 101 (1982).
[21] D. McCarthy, N.-L. Clayton, and J. K. Setlow, *J. Bacteriol.* **151**, 1605 (1982).
[22] J.-F. Tomb, G. J. Barcak, M. S. Chandler, R. J. Redfield, and H. O. Smith, *J. Bacteriol.* **171**, 3796 (1989).
[23] V. N. Trieu and D. McCarthy, *Gene* **86**, 99 (1990).
[24] D. McCarthy, N.-L. Clayton, and J. K. Setlow, *J. Bacteriol.* **151**, 1605 (1982).
[25] J. H. Stuy and R. B. Walter, *Mol. Gen. Genet.* **203**, 288 (1986).
[26] P. J. Willson, W. L. Albritton, L. Slaney, and J. K. Setlow, *Antimicrob. Agents Chemother.* **33**, 1627 (1989).
[27] J. Vieira and J. Messing, *Gene* **19**, 259 (1982).

TABLE II
Haemophilus influenzae Rd Cloning Vectors

Vector	*Haemophilus* plasmid origin[a]	*E. coli* plasmid origin	Size (kb)	Selectable markers[b]	Unique restriction sites[c]	Refs.
pHVT1	pRSF0885	pMB1	10.9	Ap, Tc	*Pst*I, *Cla*I, *Eco*RI, *Mlu*I, *Eco*RV, *Nco*I	20
pGJB103[d]	pRSF0885	pMB1	8.7	Ap, Tc	*Pst*I, *Cla*I, *Eco*RI, *Mlu*I, *Eco*RV, *Nco*I	22
pHJ1[d]	pRSF0885	pMB1	7.0	Ap, Tc	*Pst*I, *Cla*I, *Eco*RI, *Mlu*I, *Eco*RV, *Nco*I	This work
pHK1[e]	pRSF0885	pMB1	9.8	Km, Tc	*Eco*RI, *Eco*RV, *Nco*I, *Nru*I, *Sma*I, *Xho*I	This work
pVT63	pRSF0885	pMB1	6.4	Ap, Km	*Hinc*II, *Pst*I, *Sca*I, *Cla*I, *Hind*III, *Nru*I, *Sma*I, *Xho*I	23
pVT64	pRSF0885	pMB1	6.9	Ap, Sp	*Hinc*II, *Pst*I, *Sca*I, *Ssp*I	23
pVT65	pRSF0885	pMB1	6.7	Ap, Cm	*Hinc*II, *Pst*I, *Pvu*I, *Sca*I, *Ssp*I	23
pVT66	pRSF0885	pMB1	7.2	Ap, Cm	*Hinc*II, *Pst*I, *Pvu*I, *Sca*I, *Sma*I	23
pDM2	pRSF0885		9.8	Ap, Cm	*Bgl*I, *Pst*I, *Pvu*I, *Sma*I	24
pAT4	pRI234 and/or pJS1867	none	8.0	Ap, Tc	*Pst*I, *Eco*RI	25
pLS88	pLS88	none	4.8	Su, Sm, Km	See Ref. 26	26

[a] Plasmids containing the replication origin from pRSF0885 may replicate in *E. coli*.[23]

[b] Ap, Ampicillin; Cm, chloramphenicol; Km, kanamycin; Sm, streptomycin; Sp, spectinomycin; Su, sulfonamides; Tc, tetracycline.

[c] Sites of insertional inactivation.

[d] Deletion derivatives of pHVT1. pHJ1 resulted from a spontaneous deletion of 1.7 kb from pGJB103.

[e] pHK1 was constructed by inserting a 1.1-kb fragment from pUC71K[27] containing the *kan* gene from Tn*903* into the *Bgl*II site of pGJB103.

Quick Microscale Preparation of Chromosomal DNA

1. Pellet 1 ml of cells, grown in sBHI, and resuspend in 0.5 ml of 150 m*M* NaCl, 50 m*M* Tris-HCl, 10 m*M* EDTA, pH 8.0.

2. Add SDS to 1% and RNase A to 50 μg/ml, and incubate at 50°–65° for 10 min or until clear.

3. Extract twice with phenol, once with chloroform.

4. Add 2 volumes of 95% ethanol, invert a few times until the DNA

TABLE III
CONCENTRATIONS OF ANTIBIOTICS USEFUL FOR SELECTION OF
Haemophilus influenzae RD

Antibiotic	Working concentration (μg/ml)	
	Chromosomal marker[a]	Plasmid marker[b]
Ampicillin	—[c]	2–10 (pRSF0885)
Chloramphenicol	2 (Tn9)	2 (p2265, pACYC184)
Erythromycin	5	—
Kanamycin	7	30 (Tn903)
	20 (Tn903)	
Nalidixic acid	3	—
Novobiocin	2.5	2.5[d]
Spectinomycin	20	15 (pRU876)
Streptomycin	250–1000	250[d]
Streptovaricin	5	—
Tetracycline	5	5–10 (Tn10)
Viomycin	150	—

[a] Source of markers are listed in parentheses.
[b] Source of markers are listed in parentheses. Concentrations may vary depending on the plasmid.
[c] Not tested (—).
[d] Cloned chromosomal marker.

precipitates, and remove the clumped DNA with a flame-sealed Pasteur pipette tip. Rinse the spooled DNA with 70% ethanol and allow the precipitate to air dry on the tip. Resuspend in 200 μl of TE. This procedure gives 20–50 μg of DNA.

High Molecular Weight Chromosomal DNA for Pulsed-Field Gel Electrophoresis

We have adapted the method of Sambrook *et al.*[28] with some modifications.

1. Grow 100 ml of *H. influenzae* cells to mid-log phase (OD_{650} 0.4) in sBHI broth.

2. Add chloramphenicol to 20 μg/ml to prevent the initiation of new rounds of chromosomal DNA replication and incubate for an additional 1 hr at 37°.

3. Chill the cells on ice and pellet at 5000 g for 10 min at 4°.

[28] J. Sambrook, E. F. Fritsch, and T. Maniatis, "Molecular Cloning: A Laboratory Manual," 2nd Ed. Cold Spring Harbor Laboratory, Cold Spring Harbor, New York, 1989.

4. Wash the pellet once with cold solution 21 (see Table IV), centrifuge as above, and resuspend in 1/20 original culture volume of solution 21.

5. Embed the cells in low-melt agarose and purify the DNA as described by Sambrook et al.,[28] replacing zymolase with lysozyme at 1 mg/ml and RNase A at 20 μg/ml in the agarose–cell mixture before it sets, and omitting 2-mercaptoethanol from the first incubations.

The resulting agarose plugs contain approximately 50 ng DNA per microliter; at this concentration, gels loaded with 10 μl of plug should reveal fragments as small as 10 kb when examined by ethidium bromide staining. Pulsed-field gels are very sensitive to overloading, so if the lanes are smeared try loading a smaller volume of gel.

Plasmid DNA Preparation

We have found that plasmid pHVT1 and its derivatives can be easily and efficiently prepared from H. influenzae by a modification of the alkaline lysis procedure.[29] The yield of plasmid is dependent on copy number, but with the pHVT1 derivative, pGJB103, sufficient plasmid DNA is obtained from a 10-ml culture to perform 10–20 restriction digests. This protocol can be scaled to accommodate from 1 ml to 1 liter of cells.

1. Grow a 10-ml culture to saturation in sBHI with appropriate antibiotics.

2. Pellet the cells by centrifugation at 5000 g for 10 min at 4°.

3. Resuspend the pellet in 200 μl plasmid buffer (50 mM glucose, 10 mM EDTA, 25 mM Tris-HCl, pH 8) and immediately transfer to a 1.5-ml microcentrifuge tube.

4. Add 400 μl SDS/NaOH solution [0.2 N NaOH, 1% SDS (w/v)]. Mix well by inversion. Place on ice 10 min.

5. Add 300 μl of 3 M sodium acetate (pH 4.8). Mix thoroughly by inversion. Place on ice 15 min.

6. Centrifuge 15 min at room temperature.

7. Remove 850 μl of supernatant and add to 510 μl of 2-propanol. Mix by inversion and place at 4° for 15 min.

8. Centrifuge as in Step 6.

9. Decant the supernatant; rinse the pellet with 70% cold ethanol and dry briefly.

10. Resuspend the pellet in 100 μl of TE (10 mM Tris-HCl, pH 8, 1 mM EDTA).

At this step, the DNA may be digested with restriction enzymes, or it

[29] H. C. Birnboim and J. Doly, Nucleic Acids Res. 7, 1513 (1979).

may be purified further either by centrifugation through a CsCl–ethidium bromide gradient or as described below.

11. Extract the plasmid DNA once with an equal volume of phenol equilibrated with the same buffer in which the DNA is suspended.

12. Remove the aqueous phase to a new microcentrifuge tube and extract with an equal volume of chloroform–isoamyl alcohol (24 : 1, v/v).

13. Precipitate the DNA by adding 2.5 volumes of cold ethanol and placing the sample at −70° for 30 min.

14. Centrifuge to collect the precipitate. Repeat Step 9.

15. Resuspend the pellet in 300 μl of TE. Add 2 μl (20 μg) of RNase A (10 mg/ml in 10 mM Tris-HCl, pH 7.5, 15 mM NaCl, preincubated at 95° for 15 min and slowly cooled) and incubate at 37° for 30 min.

16. Extract once each with an equal volume of Tris-buffered phenol, then chloroform–isoamyl alcohol (24 : 1, v/v).

17. Remove the aqueous phase to a new tube and add 150 μl of 7.5 M ammonium acetate and 2 volumes of ethanol. Repeat Steps 13 and 14.

18. Resuspend the final pellet in 100 μl of TE. Use 5–10 μl per restriction digest.

Transformation Protocols

Several methods have been developed to make *H. influenzae* cells competent for transformation.[30–33] These methods rely on a nutrional shift-down and/or a transient shift to an anaerobic or semianaerobic state. Although many of these protocols employ complex media, the most efficient and reliable method for preparing competent *H. influenzae* employs M-IV synthetic medium.[32] Below we present three competence-inducing protocols and describe their applications.

Method I: Transformation during Overnight Growth in Complex Medium

Method I is simple, yields approximately 10^5 transformants/ml/μg of chromosomal DNA, and is suitable for moving chromosomal markers between strains.

1. Mix 1 μg of DNA with 1 or 2 ml of early log-phase cells in sBHI in a 13 × 100 mm test tube.

[30] S. H. Goodgal and R. M. Herriott, *J. Gen. Physiol.* **44**, 1201 (1961).

[31] J. H. Stuy, *J. Gen. Microbiol.* **29**, 537 (1962).

[32] R. M. Herriott, E. M. Meyer, and M. Vogt, *J. Bacteriol.* **101**, 517 (1970).

[33] R. C. Gromkova, P. B. Rowji, and H. J. Koornhof, *Curr. Microbiol.* **19**, 241 (1989).

2. Grow overnight on a roller drum at 37°.

3. Select or score for transformants by plating different dilutions on the appropriate media. For markers that require extended time for expression, growth in fresh medium for 1 or 2 hr before plating may be helpful.

Method II: Transformation Using Chemically Defined M-IV Medium

Method II gives reproducibly high levels of transformation (0.5–2%; $\sim 10^7$ transformants/ml/μg chromosomal DNA). It is the method of choice for making or screening a plasmid library or for studying the effect of mutations, chemicals, or various physiological states on competence development, DNA uptake, and transformation. Cells prepared by this procedure remain competent for at least 1 year when kept frozen at $-80°$ in 15% glycerol. M-IV competence-inducing medium is prepared fresh before each use as follows, from stock solutions described in Table IV. To 100 ml of solution 21, add 1 ml each of solutions 22, 23, 24, and 40.

Competence Development

1. Inoculate 35 ml of sHI or sBHI with 1.5 ml of cells from a $-80°$ glycerol stock, a loopful of cells from a fresh plate, or 20 to 50 μl from a fresh overnight culture.

2. Shake the culture in a 500-ml flask at 180 rpm at 37° (side-arm flasks are convenient here).

3. When the culture reaches an OD_{650} of 0.2–0.3, pellet the cells at room temperature for 4 min at 6000 g, wash them once with 20 ml of room temperature M-IV medium, pellet again, and resuspend the cells in 35 ml of room temperature M-IV.

4. Transfer the cell suspension to a 500-ml sterile flask and shake for 100 min at 37° and 100 rpm. By this time almost all cells have become competent and will maintain a high level of competence for at least 1 hr. Competent cells may be stored at $-80°$ by the addition of glycerol to a final concentration of 15% (v/v).

DNA Uptake and Transformation

1. To 1 ml of freshly prepared competent cells add 1 μg of DNA and mix by gentle vortexing. If the competent cells are from a frozen glycerol stock, first pellet the cells and resuspend in freshly prepared M-IV medium. Depending on the purpose of the experiment, the donor DNA can be either radiolabeled or genetically marked.

2. Incubate the cells and DNA at 30°–37° for 30 min.

3. To measure the uptake of radioactively labeled DNA, treat the

TABLE IV
PREPARATION OF M-IV
COMPETENCE-INDUCING MEDIUM

Component	Concentration
Solution 21	
Distilled water	850 ml
L-Aspartic acid	4.0 g
L-Glutamic acid	0.2 g
Furmaric acid	1.0 g
NaCl	4.7 g
Tween 80	0.2 ml
K_2HPO_4	0.87 g
KH_2PO_4	0.67 g
Adjust pH to 7.4 with $4N$ NaOH. Add distilled water to 1 liter. Dispense 100 ml per bottle; autoclave. Solution may become cloudy after autoclaving but clears on cooling	
Solution 22	
L-Cystine	0.04 g
L-Tyrosine	0.1 g
Dissolve in 10 ml of 1 N HCl at 37°. Bring to 100 ml with distilled water and add:	
L-Citrulline	0.06 g
L-Phenylalanine	0.2 g
L-Serine	0.3 g
L-Alanine	0.2 g
Filter sterilize	
Solution 23	
$CaCl_2$	0.1 M solution, autoclave
Solution 24	
$MgSO_4$	0.1 M solution, autoclave
Solution 40	
5% (w/v) Solution of vitamin-free casamino acids (Difco) in distilled water. Filter sterilize	

suspension with 100 μg of DNase I (5 mg/ml in 10 mM KPO_4, pH 7, 150 mM NaCl) for 10 min at room temperature, pellet the cells, resuspend in 1 ml of 0.5 M NaCl in M-IV, and pellet again. The radioactivity in the supernatant or pellet can be determined using standard procedures.

4. To select or score for independent transformants, mix 0.1 ml of serial dilutions with 15 ml of sBHI agar (1% agar medium kept molten at 42°) and plate. Allow 2 hr for expression, then overlay with 15 ml of the appropriate medium containing a 2-fold concentrate of the selective agent. Alternatively, allow for phenotypic expression in broth by adding 5 ml of

sBHI to 1 ml of the transformation mix and incubating for 1 to 2 hr at 37° before plating appropriate dilutions.

   5. Incubate at 37° for 16 to 24 hr.

Comments

   When transforming *H. influenzae* with plasmid DNA, it is important to keep several points in mind. First, plasmid vectors containing cloned *Haemophilus* DNA segments give rise to 100–1000 times more transformants than comparable amounts of vector DNA alone.[34] Balganesh and Setlow[34] showed that this effect was probably due to the presence of uptake sites on insert DNA and recombination between homologous segments on plasmid and chromosome. We have taken advantage of this observation, in our own experiments, to exchange cloned minitransposon mutagenized DNA fragments of the *H. influenzae* genome for the corresponding wild-type locus.[35] To evade the high-frequency (30–90%) recombination of cloned segments with homologous regions of the chromosome we use one of the following approaches: (1) perform the transformation in a recombination-deficient (*rec-1*) strain; (2) prepare artificially competent Rd cells using CaCl$_2$ by a method very similar to that used for *E. coli* (see Method III below); (3) transform wild-type, M-IV competent cells by the glycerol-stimulated method of Stuy.[36]

Method III: Calcium Chloride-Induced Artificial Competence for Plasmid Transformation

   The number of plasmid transformants obtained will vary depending on the particular plasmid used and its source. Using this procedure, pGJB103 isolated from *H. influenzae* gives more than 10^6 tetracycline-resistant transformants per microgram DNA, whereas the same plasmid isolated from *E. coli* gives 10^2–10^3 transformants. The following protocol yields enough competent cells for one transformation reaction and can be scaled up.

   1. Grow 10 ml of cells to OD$_{650}$ 0.3; cool on ice. Harvest the cells by centrifugation for 5 min (5000 *g*) at 4°.
   2. Resuspend the cells in an equal volume of ice-cold 25 m*M* CaCl$_2$.

---

[34] M. Balganesh and J. K. Setlow, *J. Bacteriol.* **161**, 141 (1985).
[35] J.-F. Tomb, G. J. Barcak, M. S. Chandler, R. J. Redfield, and H. O. Smith, *in* "Genetic Transformation and Expression" (L. O. Butler, C. Harwood, and B. E. B. Mosely, eds.), p. 113. Intercept Limited, Andover, England, 1989.
[36] J. H. Stuy and R. B. Walter, *Mol. Gen. Genet.* **203**, 296 (1986).

Collect the cells again by centrifugation at 4° and resuspend in 1 ml of 75 mM CaCl$_2$. Let stand on ice for 1 hr.

3. Harvest by centrifugation at 4°. Resuspend in 0.2 ml of 75 mM CaCl$_2$. Add 1 μg plasmid DNA and mix. Let the mixture stand on ice for 30 min.

4. Heat-shock the cells at 37° for 3 min. Return the cells to ice for 10 min.

5. Dilute the cells 1 : 10 in sBHI and incubate at 37° for 1–2 hr with shaking, to allow phenotypic expression.

6. Make the appropriate dilutions and plate the cells on selective medium.

Mapping Genes on Chromosome

Genetic Methods

As mentioned above, no transducing phage are known for *H. influenzae,* and mapping with *H. influenzae* conjugative plasmids has not been reported. Deich and Green have used the *E. coli* F plasmid for conjugative mapping in *H. influenzae,*[37] but this method has not been widely used.

Cotransformation frequencies are very useful for mapping on a small to moderate scale. This approach is aided by the availability multiply marked strains, for example, the 8-fold auxotroph A8[38] and strain MAP7,[39] which carries genes for resistance to seven different antibiotics. Antibiotics concentrations appropriate for selecting for these genes are given in Table III.

However, a severe limitation to this type of genetic analysis in *H. influenzae* is imposed by the paucity of selectable markers distributed about the chromosome. To ameliorate the situation, we have prepared a random set of mini-Tn*10kan* and mini-Tn*10tet* insertion mutation strains and mapped the location of the transposon by Southern hybridization to pulsed-field gel-separated restriction fragments. The locations of the different insertions are shown in Fig 2.

Southern Hybridization of Pulsed-Field Gel-Purified Fragments and Cloned Probes

We usually find it necessary to hybridize each probe to a panel of several filters, because with most enzymes and gel types not all fragments can be resolved in a single gel. The exception is *Sma*I digests in OFAGE-

---

[37] R. A. Deich and B. A. Green, *J. Bacteriol.* **169,** 1905 (1987).

[38] J. Michalka and S. H. Goodgal, *J. Mol. Biol.* **45,** 407 (1969).

[39] B. W. Catlin, J. W. Bendler III, and S. H. Goodgal, *J. Gen. Microbiol.* **70,** 411 (1972).

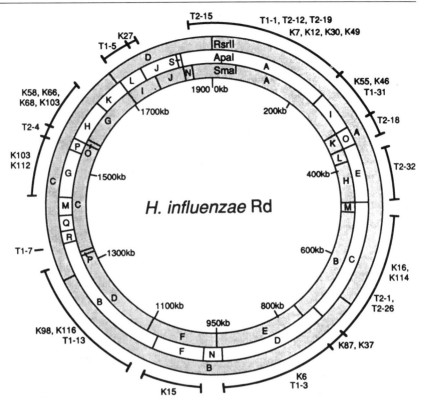

FIG. 2. Physical map locations of the random mini-Tn*10* chromosomal insertions. Mini-Tn*10kan* and mini-Tn*10tet* insertions are denoted as K or T, respectively, followed by the clone isolation number. Insertions were mapped as described in the text.

type pulsed-field gels, where an 8-sec pulse resolves all fragments in a 12-hr run.

Transformation with Gel-Purified Fragments Carrying Selectable Markers

Transformation with gel-purified fragments is a useful method for mapping selectable markers that have not been cloned, and it can also be used for strain construction.

1. Separate the DNA fragments in a low melting point agarose gel, by pulsed-field or conventional electrophoresis (gels containing less than 0.8% agarose often run badly).

2. Stain lightly with ethidium bromide (0.25 μg/ml) and destain for 1 hr or more in water.

3. To prevent DNA damage, illuminate the bands only with long-wave UV light (small bands in genomic digests will be difficult to see unless the gel is overloaded). Cut slices containing the desired fragments with a fresh razor blade. Stored tightly sealed at 4°, fragments in gel slices will keep for at least 1 year.

4. For transformation, melt the gel slice by heating to 65°, cool to 37°, and add 2–100 μl to 0.1 ml or more of M-IV-competent *H. influenzae* (for larger volumes of gel, use larger volumes of cells so the final agarose concentration is no more than 0.2%).

5. Incubate at 37° for 30 min, then plate as usual, allowing time for phenotypic expression if required. Be careful to keep the incubation at 37° before any dilutions, so the agarose does not set. Expect up to 10^4 transformants/μl of preparative gel slice.

Because gel-purified genomic DNA fragments may easily be contaminated with randomly broken pieces of larger fragments, when mapping genes to genomic fragments by transformation it is necessary to compare transformation frequencies with other fragments purified from the same gel.

Mapping onto Cloned Library Phage

We have prepared a genomic library of *H. influenzae* Rd strain KW20, by cloning a 15- to 23-kb fraction of *Sau*3A partially digested DNA fragments into λ GEM-12.[40,41] This vector was selected because the cloned insert DNA is flanked by bacteriophage T7 and SP6 RNA polymerase promoters and both *Sfi*I and *Not*I restriction endonuclease cleavage sites. *Not*I cleaves the *H. influenzae* genome only once.[42] *Sfi*I cleavage has not been detected. Thus, any insert DNA may be subcloned as an *Sfi*I fragment (and all but one segment as a *Not*I fragment) or selectively transcribed *in vitro,* and the DNA mapped using the *Sfi*I linker mapping system.[40,43] The library consisted of approximately 10^7 primary pfu. We shall provide aliquots of the amplified library to interested investigators.

[40] A. M. Frischauf, H. Lehrach, A. Poustka, and N. Murray, *J. Mol. Biol.* **170,** 827 (1983).
[41] G. J. Barcak, unpublished results (1989).
[42] L. Kauc and S. Goodgal, personal communication (1990).
[43] H. R. Rackwitz, G. Zehetner, A. M. Frischauf, and H. Lehrach, *Gene* **30,** 195 (1984).

Genetic Manipulation of *Haemophilus influenzae*

Transposon Mutagenesis

Mutagenesis in Escherichia coli. We have successfully used Tn*5* and the miniature transposons, mini-Tn*10kan* and mini-Tn*10tet*, to mutagenize cloned fragments of the *H. influenzae* genome in *E. coli*,[22,35,41] followed by additive transformation[44] of the mutagenized DNA into a wild-type strain of *H. influenzae*. McCarthy has used a similar approach successfully to identify and clone the *rec-2* locus of Rd.[45] Trieu and McCarthy have reported the preparation of new derivatives of mini-Tn*10*-encoding spectinomycin (mini-Tn*10sp*) and chloramphenicol (mini-Tn*10cm*) resistance.[23] All these minitransposons work well because they transpose at very high frequencies, are small, and lack endogenous transposase, conferring great mutational stability. We selected this approach because, at the time, an *in vivo* method for transposon mutagenesis in *H. influenzae* was unknown. The following procedure resulted in the isolation of 24 mini-Tn*10kan*-induced transformation mutants of *H. influenzae*.[22,35]

1. Prepare a representative library of cloned *Haemophilus* DNA fragments in a plasmid vector incapable of replication in *H. influenzae* (e.g., ColE1 *ori* plasmids). The restriction enzyme used to prepare the library should not cleave the transposon intended for use (e.g., *Pst*I does not cleave mini-Tn*10kan*).

2. Transform the library into a suitable *E. coli* strain. MC1060 has worked well when a λ phage is used to deliver the minitransposon. Alternatively, McCarthy[45] has transformed the library into an *E. coli* strain carrying the minitransposon on a compatible plasmid vector.

3. If using a λ transposon donor, pool the library transformants and subject them to transposon mutagenesis precisely as described by Way *et al.*[46] If using a plasmid transposon donor, add IPTG (isopropyl-β-D-thiogalactopyranoside) to 0.5 mM to induce transposase and incubate at 37° for 90 min.

4. Since only 1–2% of the library clones are likely to contain useful insertions after transposition, we pool approximately 300,000 transductants of a representative library that has an average insert size of 5–6 kilobase pairs.

[44] J. H. Stuy and R. B. Walter, *J. Bacteriol.* **148**, 565 (1981).
[45] D. McCarthy, *Gene* **75**, 135 (1989).
[46] J. C. Way, M. A. Davis, D. Morisato, D. E. Roberts, and N. Kleckner, *Gene* **32**, 369 (1984).

5. Prepare plasmid DNA from the pooled transductants (transformants) and cleave with restriction endonuclease to release the cloned (and mutagenized DNA) from vector sequences.

6. Transform 1 ml of M-IV-competent *H. influenzae* cells (see above) with 1–3 μg of mutagenized plasmid DNA. Select for the antibiotic resistance phenotype conferred by the transposon.

7. Screen for the desired mutant phenotype.

8. Once putative mutants are identified, prepare chromsomal DNA from these strains (see above) to check that acquisition of the drug resistance marker cotransforms with the mutant phenotype. We routinely observe that 5–10% of the time it does not. Additionally, we find that on transformation with certain mutagenized fragments, transformants contain both the wild-type and mutant loci, forming a tendem duplication in the genome.

9. Check that the insertion mutations are unique and of the predicted structure by Southern hybridization analysis.

Mutagenesis in Haemophilus influenzae. Recently, Kauc and Goodgal[47] reported the first *in vivo* transposon mutagenesis of *H. influenzae* and *H. parainfluenzae* using Tn*916*, an element conferring tetracycline resistance. The transposon is carried on two different replicons, pAM120 (Amp^r, Tet^r; carried by *E. coli* PY327) and pAM180 (Ery^r, Tet^r; carried by *Bacillus subtilis* PY200). Transposition is accomplished by transforming cells made competent by the M-IV procedure to tetracycline resistance. The authors observed a transposition frequency of $1-5 \times 10^3$ Tet^r clones per micorgram of transforming plasmid DNA. Apparently neither pAM plasmid is capable of replication in *H. influenzae* since the authors failed to detect either Amp^r or Ery^r phenotypes, nor were vector DNA sequences detected by Southern hybridization in Tet^r clones.

Cassette Mutagenesis

The preparation of polar and nonpolar insertion mutations by *in vitro* methods can be useful in the analysis of gene structure and regulation. We have found that the *kan* gene derived from Tn*903* is an excellent selectable marker in *H. influenzae*. It is available as a versatile cloning cartridge on a series of vectors described by Barany.[48] The antibiotic resistance cartridges encoding spectinomycin and chloramphenicol resistance used by Trieu and McCarthy[23] in the preparation of new derivatives of mini-

[47] L. Kauc and S. H. Godgal, *J. Bacteriol.* **171**, 6625 (1989).
[48] F. Barany, *DNA Protein Eng. Tech.* **1**, 29 (1988).

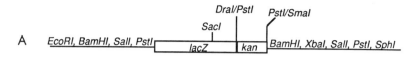

FIG. 3. *lacZ–kan* protein (A) and operon (B) fusion cassettes. The restriction endonuclease sites listed can be used for subcloning the cassettes. The *Dra*I/*Pst*I and *Pst*I/*Sma*I junctions are no longer cleavable by *Dra*I, *Pst*I, or *Sma*I. There are three protein fusion cassettes, one for each potential open reading frame. These cassettes are cloned in plasmids pLZK80, pLZK81, and pLZK82; the operon fusion cassette is cloned in plasmid pLZK83 [G. Barcak and J.-F. Tomb, unpublished results (1989)].

Tn[10] may also prove useful for this purpose. Stull and colleagues[49] have prepared a new derivative of Tn*5*, "TSTE," for cassette mutagenesis of *Haemophilus*. By deleting internal segments, the Tn*5* element has been reduced to 2.2 kb, while still conferring Kan[r]. The new element also contains a synthetic *Haemophilus* transformation uptake sequence[50] and is available as a cartridge in pUC19.

Gene cartridges are also useful for the preparation of genetic fusions. Fusions can be used to assess the contribution of transcriptional and posttranscriptional regulation to the patterns of gene expression during cellular growth under various parameters. The preparation of fusion proteins may also be key to the identification of the cellular compartment in which the gene product is located by providing an assayable enzymatic activity for a protein of unknown function and by providing a target moiety for commercially available antibody (e.g., anti-β-galactosidase antibody) for Western analysis.

Accordingly, we have adapted the *lacZ* protein fusion cassette vectors of Tiedeman and Smith[51] for use in *H. influenzae* as shown in Fig. 3A. The cassettes were deleted for *lacY* to reduce their size and fused to the *kan* gene derived from Tn*903*.[52] Each cassette permits the preparation of protein fusions between *lacZ* and the C terminus of the gene under

[49] C.Sharetzsky, T. D. Edlind, J. J. LiPuma, and T. L. Stull, *J. Bacteriol.* **173**, 1561 (1991).
[50] D. B. Danner, R. A. Deich, K. L. Sisco, and H. O. Smith, *Gene* **11**, 311 (1980).
[51] A. A. Tiedeman and J. M. Smith, *Nucleic Acids Res.* **16**, 3587 (1988).
[52] G. J. Barcak and J.-F. Tomb, unpublished results (1990).

investigation. The fusions can then be transplaced onto the chromosome by additive transformation with selection for kanamycin resistance. In a similar fashion, we prepared a versatile *lacZ–kan* operon fusion cassette[52] by adapting the improved *lacZ* operon fusion fragment from pRS415K[53] as shown in Fig. 3B. The utility of these cassettes is enhanced by the fact that *H. influenzae* does not possess β-galactosidase activity.[54]

Transposon TnphoA[55] has been shown to be an important new tool for the study of membrane protein topology in bacteria.[56] Although we have prepared active alkaline phosphatase fusions to a cloned *H. influenzae* gene in *E. coli*, we have been unsuccessful thus far in transferring these fusions by additive transformation to the chromosome. We suspect that expression of the *kan* gene of TnphoA is inadequate (at least in this particular fusion) to allow for phenotypic (Kan^r) selection. This problem might be circumvented by preparing such fusions *in vitro* as described previously for *E. coli*.[57,58]

Chemical Mutagenesis

Haemophilus influenzae is not mutable to any significant degree by UV or X-irradiation, by methyl methane sulfonate, or by nitrogen mustard, presumably owing to the absence of an error-prone repair system.[59] However, *N*-methyl-*N'*-nitro-*N*-nitrosoguanidine (MNNG), ethyl methane sulfonate, nitrosocarbaryl, and hydrazine are effective mutagens. A protocol for MNNG mutagenesis is described by Setlow *et al.*,[60] and the use of a number of other mutagens is described by Kimball and Hirsch.[61]

Addendum

While this manuscript was in press, one of us (M.S.C)[62] demonstrated that plasmids containing the P15A origin of replication are capable of replication in *H. influenzae* Rd. Plasmids pACYC177, pACYC184, pSU2718, and pSU2719 were all able to establish in *H. influenzae*. The

[53] R. W. Simons, F. Houman, and N. Kleckner, *Gene* **53**, 85 (1987).
[54] M. Kilian, *J. Gen. Microbiol.* **93**, 9 (1976).
[55] C. Manoil and J. Beckwith, *Proc. Natl. Acad. Sci. U.S.A.* **82**, 8129 (1985).
[56] C. Manoil and J. Beckwith, *Science* **233**, 1403 (1986).
[57] S. Michaelis, L. Guarente, and J. Beckwith, *J. Bacteriol.* **154**, 356 (1983).
[58] C. S. Hoffman and A. Wright, *Proc. Natl. Acad. Sci. U.S.A.* **82**, 5107 (1985).
[59] R. F. Kimball, M. E. Boling, and S. W. Perdue, *Mutat. Res.* **44**, 183 (1977).
[60] J. K. Setlow, D. C. Brown, M. E. Boling, A. Mattingly, and M. P. Gordon, *J. Bacteriol.* **95**, 546 (1968).
[61] R. F. Kimball and B. F. Hirsch, *Mutat. Res.* **30**, 9 (1975).
[62] M. S. Chandler, *Plasmid* **25**, in press (1991).

plasmids were stable, could be purified by standard protocols, and were compatible with a second plasmid carrying the RSF0885 origin of replication.

Acknowledgments

We wish to thank all of our colleagues who generously provided preprints, reprints, and personal communications of their unpublished results. This work was supported by NIH grants 5-PO1-CA16519 and 1-RO1-AI27783 to H. O. Smith, in whose laboratory some of the experimental approaches described herein were initially tested. G. Barcak was supported by a Monsanto Postdoctoral Fellowship. M. Chandler was supported by NIH Training Grant 5-T32-CA09139. R. Redfield was supported by a Medical Research Council of Canada Postdoctoral Fellowship.

[15] Genetic Systems in Pathogenic Neisseriae

By H. STEVEN SEIFERT and MAGDALENE SO

Introduction

Members of the genus *Neisseria* are gram-negative diplococci that inhabit the human body. Two members of this genus are important human pathogens: *Neisseria gonorrhoeae* (the gonococcus, GC) and *Neisseria meningitidis* (the meningicoccus, MN). Owing to their serious disease-causing capabilities, the pathogenic properties of both organisms have been intensively studied. The use of genetic methods has gained popularity in investigations concerning the biology of the Neisseriae, because the tools to conduct molecular genetic studies have begun to be developed. This chapter discusses what is known about the naturally occurring gene transfer systems in the pathogenic Neisseriae and how these systems are used to ask biological questions. More information is available on the manipulation of the gonococcus than the meningicoccus and so the gonococcus is the focus of this discussion.

Most of the methods applicable to one organism can be adapted to the other, and many will also work with commensal species. The only broad reaching exception to this transfer of methods between organisms results from ethical considerations concerning the use of antibiotic resistance markers. Only genes encoding resistance to antibiotics not normally used to treat natural infections, or those naturally found in the organism, should be used in genetic experiments. This is exemplified by the chloramphenicol resistance genes, which can be used in the gonococcus since chloramphenicol-resistant organisms are found naturally and chloramphenicol is not normally recommended for treatment of gonorrhea. In contrast, chloramphenicol is a first-line treatment for meningicoccal infections, so chloramphenicol resistance genes would not normally be used as a marker in the meningicoccus. In addition, when using any antibiotic resistance as a selection, organisms should be maintained under strict Biosafety Level 2 containment conditions[1] to prevent the release of antibiotic-resistant organisms into the environment. It is important that all experiments which will result in antibiotic-resistant Neisseriae be approved by the local biohazard safety and/or recombinant DNA safety committee.

Although Neisseriae are gram-negative bacteria, they differ from other commonly studied gram-negative bacteria in several important aspects. Ribosomal RNA comparisons place the gonococcus closer to the group that includes members of the genera *Nitrosomonas, Nitrosolobus, Nitrosococcus,* and *Spirillum.*[2] Therefore, the Neisseriae are on a different evolutionary branch than other well-studied gram-negative bacteria. The Neisseriae contain several well-characterized genetic anomalies that help to establish the unique properties of these organisms.

One characteristic property of the Neisseriae is the production of outer membrane vesicles called blebs.[3] In a growing culture, blebs are extruded from the cell surface and contain an assortment of outer membrane proteins representative of the whole cell. Blebs are thought to act as antigen-containing "decoys" to subvert the host immune response. A role for blebs in genetic transfer has also been described. Blebs carry both chromosomal and episomal DNA, and they can transfer plasmids into recipient cells in the absence of donor cells.[4] It is not clear if bleb-mediated plasmid transfer occurs by way of transformation or conjugation. The Neisseriae are highly autolytic (see below), an event that is catalyzed by specific enzymes.[5] It is thought that bleb formation and autolysis may share mechanistic properties; however, this hypothesis is unsupported by data, and its proof will require the examination of nonautolytic strains.

Another interesting facet of neisserial genetics is the high degree of cytosine methylation found in the chromosomal DNA of these bacteria. It has been postulated that GC expresses from 8 to 11 type II methylase activities because of the high degree of N-methylcytosine found in palin-

[1] *Federal Register* **51,** 16973 (1986).
[2] C. R. Woess, *Microbiol. Revs.* **51,** 221 (1987).
[3] I. W. De Voe and J. E. Gilchrist, *J. Exp. Med.* **138,** 1156 (1973).
[4] D. W. Dorward, C. F. Garon, and R. C. Judd, *J. Bacteriol.* **171,** 2499 (1989).
[5] B. H. Hebler and F. E. Young, *J. Bacteriol.* **122,** 385 (1975).

dromic sequences.[6,7] This high level of cytosine methylation necessitates the use of *Escherichia coli* strains that do not restrict *N*-methylcytosine DNA when cloning from Neisseriae.[8] All of the palindromic cytosine-containing sites are modified even in the cases where no cognate restriction activities can be detected. The specific biological function of DNA methylation is unknown. A role in DNA transformation has been proposed, but unmethylated gonococcal DNA fragments amplified in *E. coli* transform as efficiently as the same methylated sequences derived from GC[8a] (K. Hoikka and H. Seifert, unpublished). Several genes encoding restriction/modification activities have been cloned.[9,10] One modification methylase gene cloned into *E. coli* was found to express its corresponding restriction activity even though that activity cannot be detected in GC lysates (R. Chien, D. Stein, M. So, and H. Seifert, unpublished). This observation supports the hypothesis that the gonococcus may encode multiple restriction activities that are not expressed under normal *in vitro* culture conditions. Only when cloned into *E. coli* is the regulation relieved and the activity expressed. Every modification activity could presumably be paired with a restriction activity whose expression is regulated. The isolation of more paired methylase and restriction genes will support or refute this hypothesis.

Transformation

Biology of Transformation

The Neisseriae are naturally competent for DNA transformation, with frequencies of 1% easily attainable. When GC is compared to the other well-studied naturally competent species, the transformation system most closely resembles that of *Haemophilus influenzae* in that both bacteria take up double-stranded DNA for transformation[11,12] and only take up DNA containing species-specific transformation uptake sequences.[13,14] The characteristics of natural transformation systems indicate that this

[6] C. Korch, P. Hagblom, and S. Normark, *J. Bacteriol.* **155**, 1324 (1983).
[7] C. Korch, P. Hagblom, and S. Normark, *J. Bacteriol.* **161**, 1236 (1985).
[8] E. A. Raleigh and G. Wilson, *Proc. Natl. Acad. Sci. U.S.A.* **83**, 9070 (1986).
[8a] D. Stein, *Can. J. Microbiol.*, in press (1991).
[9] A. Piekarowicz, R. Yaun, and D. Stein, *Nucleic Acids Res.* **16**, 5957 (1988).
[10] K. M. Sullivan and J. R. Saunders, *Nucleic Acids Res.* **16**, 4369 (1988).
[11] G. D. Biswas and P. F. Sparling, *J. Bacteriol.* **145**, 638 (1981).
[12] J. H. Stuy, *J. Mol. Biol.* **13**, 554 (1965).
[13] S. D. Goodman and J. J. Scocca, *Proc. Natl. Acad. Sci. U.S.A.* **85**, 6982 (1988).
[14] D. B. Danner, R. A. Diech, K. L. Sisco, and H. O. Smith, *Gene* **11**, 311 (1980).

process has evolved mainly for recombination of the incoming DNA with the bacterial chromosome. Plasmid transformation occurs at a lower frequency than chromosomal transformation, and it occurs at much greater frequencies if there are resident, homologous DNA sequences in the recipient. Restriction endonucleases act on plasmid DNA during transformation if the DNA is isolated from a heterologous strain.[15] DNA rearrangements often occur during plasmid transformation. This results either from the requisite recombination of the incoming DNA with homologous resident DNA prior to establishing an independent plasmid, or from cutting of the incoming DNA by restriction endonucleases. In contrast, incoming linear chromosomal DNA replaces its resident homolog via a simple double crossover event.

For these reasons, most experiments utilizing transformation to introduce new DNA sequences into the Neisseriae rely on placing genetic constructions into loci in the chromosome rather than on plasmids. This is in contrast to the situation in the Enterobacteriaceae where plasmid transformation, being much more efficient than chromosomal transformation, is the method of choice for introducing genes into these noncompetent organisms. For many studies chromosomal transformation is preferable to plasmid transformation since stable, single-copy insertions are produced in the former case. This is particularly important when studying gene expression where the relative copy number of genes encoding cis- and trans-acting factors may influence the results. In addition, since the majority of the virulence genes in the Neisseriae are chromosomally encoded, studies on pathogenesis necessarily utilize chromosomal transformation to construct mutations.

The evolution of the transformation process requires that free DNA be present in the environment. Neisseriae, like many other naturally competent bacteria, are autolytic. The control of autolysis remains a mystery, but poor growth conditions and/or dense bacterial cultures appear to increase the frequency of autolysis.[5] Although an enzymatic activity has been implicated in the process of autolysis, the mechanistic details remain obscure.

DNA transformation can be separated into two steps: DNA uptake and recombination. The molecular mechanisms used by the Neisseriae to carry out these two steps are poorly understood. Clearly the process must include the recognition and binding of specific sequences to the bacterium, transport of the bound DNA into the cytoplasm, and recombination of the incoming DNA with homologous sequences. The rest of this section discusses what is known about the biology of transformation.

[15] D. C. Stein, S. Gregoire, and A. Piekarowicz, *Infect. Immun.* **56,** 112 (1988).

The first step in genetic transformation is binding of DNA by the bacterium. When the fate of radiolabeled DNA is followed during transformation, two distinct binding phases have been observed: an initial reversible binding of any DNA molecule and a second irreversible binding of species-specific DNA that renders the DNA resistant to degradation by DNase I.[16] The second phase represents uptake of the DNA into the cell. Uptake of DNA during transformation in the Neisseriae requires that a 10-base pair (bp) recognition sequence (5'-GCCGTCTGAA) be present on the DNA.[13] This surface recognition of self is similar to that seen in *Haemophilus* transformation, where an 11-bp sequence (5'-AAGTGCGGTCA) is required.[14] GC and MN can transform each other,[17] suggesting that their DNA share the same uptake sequence. Many of the Neisseriae transformation uptake sequences are present in palindromic pairs that resemble transcriptional stop signals: inverted repeats downstream of open reading frames.[13] One uptake sequence is sufficient to introduce DNA into an organism (D. Stein, personal communication; C. Elkins, C. E. Thomas, H. S. Seifert, and P. F. Sparling, unpublished), and some uptake sequences are present as one copy in the chromosome. If these sequences play dual roles in transformation and transcription termination, this would suggest that these bacteria have evolved a means to provide a transformation uptake sequence near most genes in the chromosome.

Competence for DNA transformation in GC and MN is dependent on the piliation status of the recipient bacterial cell. The original observations concerning the relationship between piliation and transformation were made before the genetic mechanisms of pilin phase variation (the on/off switch controlling pilus expression) were determined.[18] More recent studies have shown that pilin null mutations (deletions, insertions, or nonsense mutations) lower transformation to undetectable levels if amounts of DNA that saturate transformation of piliated bacteria are used. However, transformation of nonpiliated strains can be achieved if higher amounts of DNA are used, but at levels 100- to 1000-fold lower than transformation frequencies detected with piliated cells.[19] Interestingly, some mutations that affect the assembly of pili, but retain pilin expression,

[16] T. F. Dougherty, A. Asmus, and A. Tomasz, *Biochem. Biophys. Res. Commun.* **86,** 97 (1979).

[17] P. F. Sparling, G. Biswas, and T. E. Sox, *in* "The Gonococcus" (R. B. Roberts, ed.), p. 155. Wiley, New York, 1977.

[18] P. F. Sparling, *J. Bacteriol.* **92,** 1346 (1966).

[19] H. S. Seifert, R. S. Ajioka, D. Paruchuri, F. Heffron, and M. So, *J. Bacteriol.* **172,** 40 (1990).

have been reported to retain full competency for DNA transformation.[20] The report that an essential competence factor from *Bacillus subtilis* shows amino acid similarity to GC pilin implies that the latter may also play an active role in transformation.[21] It remains to be shown experimentally whether the pilus plays a direct role in transformation, or if expression of pilin is linked to expression of other cellular components that are required for efficient transformation.

The requirement for a *recA* homolog in transformation is consistent with a mechanistic requirement for heteroduplex formation.[22] The identities of other enzymatic activities involved in transformation are unknown, but several mutations have been isolated that lower transformation competence.[23] One of these (*dud1*, see below) blocks initial DNA uptake, whereas others interfere with subsequent steps in transformation. That some of these mutations lower both plasmid and chromosomal transformation, whereas the others affect only chromosomal transformation, suggests a more stringent requirement for chromosomal recombination. An 11-kDa outer membrane protein has been identified that binds with high affinity to synthetic oligonucleotides encoding the uptake seqence.[24] This protein does not bind to the uptake sequence in cells that carry the *dud1* allele. A protein of similar size was observed to bind preferential gonococcal chromosomal DNA.[25] Presumably, binding of the uptake sequence by the *dud1* gene product is a major part of the transformation process, but *dud1* strains are only partially disabled for transformation. The occurrence of numerous other DNA-binding proteins on the gonococcal cell surface and the observation that all competence mutations, except *recA*, retain measurable transformation competence suggest that there may be redundancy in DNA uptake pathways for transformation.

A nonspecific nuclease activity has been implicated in transformation since plasmid DNA is opened at random sites during uptake into the cell.[26] It is not known if this cutting occurs during DNA uptake or during recombination. The cutting of every plasmid molecule during uptake into the cell may explain why integration of nonreplicating plasmids into the bacterial chromosome is not observed.

[20] C. P. Gibbs, B.-Y. Reimann, E. Schultz, A. Kaufman, R. Hass, and T. F. Meyer, *Nature* (*London*) **338**, 651 (1989).
[21] R. Breitling and D. Dubnau, *J. Bacteriol.* **172**, 1499 (1990).
[22] J. Koomey and S. Falkow, *J. Bacteriol.* **169**, 790 (1987).
[23] G. D. Biswas, S. A. Lacks, and P. F. Sparling, *J. Bacteriol.* **171**, 657 (1989).
[24] D. W. Dorward and C. F. Caron, *J. Bacteriol.* **171**, 4196 (1989).
[25] L. S. Mathis and J. J. Scocca, *J. Gen. Microbiol.* **130**, 3165 (1984).
[26] G. D. Biswas, K. L. Bernstein, and P. F. Sparling, *J. Bacteriol.* **168**, 756 (1986).

High-frequency DNA transformation in the Neisseriae *in vitro* is well documented. There is evidence that this process also occurs with similar efficiency *in vivo:* a substantial amount of genetic information appears to be horizontally transmitted. Surveys of Neisserial strains have revealed interallelic patchworks of penicillin-binding proteins[27] and IgA protease genes.[28] The cassette-type organization of the *opa* and pilin genes[29-31] and the ample evidence that GC pilin undergoes antigenic variation during a natural infection further suggest that transformation is an important process among the Neisseriae.

Transformation Methods

Transformation of the Neisseriae occurs efficiently with very little help from us. Below are listed a few general considerations that may influence the success of any transformation experiment.

DNA. The physical state of the DNA is largely unimportant. We often use partially purified DNA with great success. The only exception to this purity condition is the presence of substances that are detrimental to the bacterial cell (e.g. detergents). For piliated strains, 1 μg/ml of chromosomal DNA is saturating for transformation, and significantly higher concentrations of DNA may give lower transformation frequencies. For nonpiliated variants, at least 50 μg/ml of chromosomal DNA is required to obtain measurable transformation frequencies. Concentration of the nonpiliated cells and DNA onto a solid surface may increase the transformation frequency of a nonpiliated strain. When the donor DNA is from an *E. coli* clone, the greater the concentration of DNA the higher the transformation efficiency (K. Hoikka and H. Seifert, unpublished).

Cofactors. Divalent cations are the only cofactors required for transformation. Although both Ca^{2+} and Mg^{2+} work equally well, the relative insolubility of Ca^{2+} in phosphate buffers makes Mg^{2+} the cofactor of choice. Even though transformation only requires 1 mM Mg^{2+}, we routinely use 5 mM to guard against the presence of excess EDTA in DNA preparations.

Growth Conditions. Neisseriae are competent for transformation throughout their life cycle. No special growth conditions are required for

[27] B. G. Spratt, *Nature* (*London*) **332**, 173 (1988).
[28] R. Halter, J. Pohlner, and T. F. Meyer, *EMBO J.* **8**, 2737 (1989).
[29] R. Haas and T. F. Meyer, *Cell* (*Cambridge, Mass.*) **44**, 107 (1986).
[30] A. Stern, M. Brown, P. Nickel, and T. F. Meyer, *Cell* (*Cambridge, Mass.*) **47**, 61 (1986).
[31] T. D. Connell, W. J. Black, T. H. Kawula, D. S. Barritt, J. A. Dempsey, K. Kevereland, Jr., A. Stephenson, B. S. Shepurt, G. L. Murphy, and J. G. Cannon, *Mol. Microbiol.* **2**, 227 (1988).

transformation, although most workers grow cells on clear typing media in order to observe piliated colony morphology. Gonococci are very autolytic when placed under poor growth conditions. Therefore, cultures should be maintained in growth medium at 37° during transformation and not placed into buffer solutions or on ice. The divalent cations used in transformation help to inhibit autolysis. This reduces any potential problems that short incubations under less than ideal conditions may cause. Autolysis also occurs when cells reach the stationary phase of growth. The most common mistake made when transforming Neisseriae is the use of too many cells.

Selection. Any chromosomally located marker will transform another *Neisseria* strain. However, the status of unlinked genes in the recipient strain may drastically alter the level of antibiotic needed to select against nontransformed bacteria and to select for the antibiotic resistance phenotype. For example, when transforming GC with the CAT (chloramphenicol acetyltransferase) marker used in shuttle mutagenesis, 1 μg/ml of chloramphenicol is used with strain F62, whereas 10 μg/ml is used with strain MS11. Thus, it is crucial to test the minimum inhibitory concentration (MIC) of the recipient strain in order to determine the level of antibiotic to be used for selection of transformants.

Measuring Transformation. Three conditions have been used to help standardize frequency measurements. The first is treatment of transforming cultures with DNase I after DNA uptake has occurred.[18] This has been used to limit the time that the cells are exposed to DNA in order to block continued DNA uptake during phenotypic expression. Another method to provide more accurate frequencies is the use of double overlays.[18] In this method transformed bacteria are plated in rich top agar, allowed to express phenotypic markers, and then overlaid with more top agar containing the antibiotic. Using this method, differential growth rates of transformed and nontransformed cells during phenotypic expression is removed as a factor that could alter the transformation frequency. The final method used to standardize transformation frequencies is the application of sonication to GC cultures prior to plating on solid medium.[32] Piliated and opaque GC form tight clumps that are difficult to disrupt by vortexing. A low-power sonication step breaks the clumps into individual bacteria, resulting in more consistent plate counts.

The following is a standard transformation protocol for gonococci.

1. Streak several piliated colonies onto supplemented GC medium

---

[32] M. Koomey, E. C. Gotschlich, K. Robbins, S. Bergstrom, and J. Swanson, *Genetics* **117,** 391 (1987).

base (Difco, Detroit, MI) (GCB +, 1% Kellogg's supplements,[33] or Isovitalex). The plates should be incubated no more than 18 hr prior to use at 37°, 5–10% (v/v) CO_2.

2. Before harvesting the bacteria, check the colonial morphology in a stereomicroscope. If most colonies show a piliated morphology, swab the cells from a crowded part of one or more plates into 37° GCB liquid supplemented with 5 mM $MgCl_2$ (GCB–Mg). Use sterile Dacron swabs, not cotton. A general rule is that a laboratory coat, gloves, and safety glasses should always be used when handling the organism. The ocular route is the most common means of laboratory infection.

3. Vortex to disrupt clumps and adjust the concentration to about 10^8 colony-forming units (cfu)/ml. This provides a Klett reading of 10 with a red filter.

4. Dilute the cell suspension 1 : 10 into GCB–Mg with and without DNA. Incubate for 10 min at 37° to allow uptake. If DNase I treatment is appropriate, it can be carried out in the same buffer. We normally add 20 μl of cells to 200 μl DNA solution, but the amounts can be adjusted to provide more cells. We have used polystyrene tubes, 12- or 24-well tissue culture plates, or glass tubes. If glass is used be careful that no detergent residue remains after washing.

5. Dilute the DNA–cell suspension 1 : 10 into liquid GCB + at 37° and incubate in 5–10% CO_2 for expression of phenotypic markers. We often perform this in 25-cm^2 tissue culture flasks or microtiter plates. We have not determined whether the presence of CO_2 is necessary for phenotypic expression, but the CO_2 dependence for growth suggests that it may be helpful.

6. Grow the recipients in liquid medium long enough for expression of phenotypic markers. For most antibiotic resistance markers this requires a minimum of 4 hr. It is important to limit autolysis when the culture reaches the stationary phase of growth in order to avoid retransformation during the expression period.

7. Plate bacteria onto selective media. A minus DNA control is important to rule out spontaneous mutation being responsible for the new phenotype. In addition, this control helps to distinguish the true transformants from the lawn of pale, nonpiliated background colonies that appears with some antibiotic selections. Also plate 100 μl of 10^{-4}, 10^{-5}, and 10^{-6} dilutions in GCB liquid onto nonselective plates to determine the number of colony-forming units.

---

[33] D. S. Kellogg, Jr., W. L. Peacock, W. E. Deacon, L. Brown, and C. I. Pirkle, *J. Bacteriol.* **85,** 1274 (1963).

8. Transformants will appear in 24 to 36 hr, depending on the marker and on the relative frequency with which the donor DNA transforms. If there is a background, these colonies will always appear prior to the bona fide transformants. Often only true transformants will grow when subsequently passaged onto selective medium. If new DNA sequences are introduced into the chromosome by the transformation event, the transformants should also be checked by Southern blotting the chromosomal DNA with the relevant gene probe.

Conjugation

Biology of Conjugation

The rise in antibiotic-resistant bacteria is of growing importance in the treatment of infectious disease. The prescription of broad spectrum antibiotics for conditions that do not require their use, combined with the self-administration of antibiotics, has provided powerful selection for antibiotic-resistant bacteria. This has facilitated the appearance of antibiotic resistance encoding plasmids (R plasmids) in many bacterial pathogens. R plasmids that inhabit the pathogenic Neisseriae have been postulated to have originated in *Haemophilus ducreyi, Haemophilus parainfluenzae,* and *Haemophilus influenzae.*[34,35] In addition, commensal Neisseriae are a potential reservoir for R plasmids, and transfer of plasmids between them and pathogenic Neisseriae has been demonstrated.[36] The pathogenic Neisseriae seem to be refractory to conjugal transfer in the natural setting since the appearance of antibiotic-resistant strains was not as rapid as originally feared. This may be due to infrequent contact between heterologous donor and recipient organisms or to physiological barriers to conjugal transfer. Restriction of the incoming plasmids by resident restriction endonucleases may comprise the greatest barrier to conjugal transfer since methylation of plasmids in *E. coli* greatly increases conjugal transfer efficiencies into several gonococcal strains (T. Butler, personal communication). However, the growing number of penicillinase-producing *Neisseria gonorrhoeae* (PPNG) strains proves that barriers to conjugal transfer can be breached when sufficient selection is applied. It will be interesting to determine how quickly the new tetracycline resistance conjugal plasmid transfers into the gonococcus and meningicoccus.

[34] N. Dickgeisser, *Plasmid* **11**, 99 (1984).
[35] S.-T. Chen and R. C. Clowes, *J. Bacteriol.* **169**, 3124 (1987).
[36] C. A. Genco, J. S. Knapp, and V. L. Clark, *J. Infect. Dis.* **150**, 397 (1984).

Penicillinase-Producing Neisseria gonorrhoeae

Penicillin remains the treatment of choice for gonococcal infections. The rise of PPNG strains has reduced its usefulness in many geographic areas. Most PPNG strains carry three plasmids: the 2.6-Mda (megadalton) cryptic plasmid, the 3.2- or 4.4-Mda penicillinase-producing plasmids, and the 24.5-Mda conjugal transfer plasmid.[37] The cryptic plasmid is present in most gonococcal isolates and does not contribute to antibiotic resistance. The penicillinase-producing plasmids are not self-transmissible, but they require mobilization for conjugal transfer. The 24.5-Mda plasmid fulfills this role in conjugation. The penicillin resistance gene carried on the R plasmids encodes a TEM-type β-lactamase identical to that encoded by transposon Tn2.[38] The R plasmids do not carry a transposable element since only one end of Tn2 is present on the plasmids. Similar plasmids carrying the entire Tn2 sequence are found in *Haemophilus* species,[35,39] but these plasmids do not transfer into the gonococcus at a measurable frequency.[40]

One useful derivative of a β-lactamase-producing R plasmid has been constructed.[41] This plasmid, pLES2, can replicate in both *E. coli* and GC, retains sufficient sequences from the R plasmid to allow mobilization by IncP helper plasmids, and has an α-*lacZ* cloning region from pUC8 to provide restriction endonuclease cloning sites and blue/white screening for inserts in *E. coli*. The drawbacks of this vector are its size (~6 kb), its few cloning sites (several of the sites in the polylinker are repeated elsewhere in the plasmid), and the inefficient selection for the β-lactamase marker carried by the plasmid. Despite these problems, pLES2 remains the only proven vehicle for conjugal transfer between *E. coli* and Neisseriae.

Tetracycline Resistance

Recently, tetracycline-resistant gonococci and meningococci have been isolated from patients who fail to respond to tetracycline treatment.[42] The plasmid carrying the resistance gene is related to the 24.5-Mda conjugal plasmid discussed above. This new plasmid is less than 1 kilobase (kb)

---

[37] Interestingly, plasmid epidemiologists still use megadaltons to define the size of a plasmid. In kilobase pairs these sizes translate to 3.9, 4.9, 6.7, and 37.2 kb, respectively.

[38] M. Roberts, L. P. Elwell, and S. Falkow, *J. Bacteriol.* **131**, 557 (1977).

[39] J. Brunton, D. Clare, and M. A. Meier, *Rev. Infect. Dis.* **8**, 713 (1986).

[40] P. J. McNicol, W. L. Albritton, and A. R. Ronald, *Sex. Trans. Dis.* **13**, 145 (1986).

[41] D. C. Stein, L. E. Silver, V. L. Clark, and F. E. Young, *Gene* **25**, 241 (1983).

[42] S. A. Morse, S. R. Johnson, J. W. Biddle, and M. C. Roberts, *Antimicrob. Agents Chemother.* **30**, 664 (1986).

larger than the wild-type conjugal plasmid (25.2 Mda). An undetermined amount of the normal plasmid sequences has been replaced by the *tetM* gene,[43] a determinant described as part of the streptococcal conjugative transposons Tn*916* and Tn*1545*.[43,44] It is unclear whether the neisserial *tetM* determinant is still part of a transposable element. The *tetM* determinant has also been found in the chromosome of *Mycoplasma hominis*, *Ureaplasma urealyticum*, and *Gardnerella vaginalis*,[45] all which inhabit the human genital tract. How the *tetM* gene became associated with the gonococcal conjugal plasmid is unknown. One intriguing hypothesis is that GC conjugal plasmid DNA and *tetM*-containing DNA from one of the other genital tract inhabitants recombined outside of their respective cells and was then transformed into a GC strain containing the conjugal plasmid. Sequence analysis of the site of the *tetM* insertion should reveal how this plasmid appeared.

Conjugation Methods

We present a general method for conjugation into or out of the pathogenic Neisseriae. This method is not significantly different from the standard filter mating procedures used for broad host range conjugal transfer between other bacterial species. In fact, IncP plasmids, RK2, RP4, and derivatives, are used to mobilize R Factors between *E. coli* and *Neisseria*.[46] Prior to outlining the method, we discuss some of the factors that may influence the success of a conjugal transfer experiment.

Recipient. Not all Neisseriae strains are able to act as recipients in conjugal transfer.[47] The ability of a certain strain to act as a recipient should be determined in each laboratory since different clones of the same isolate may differ in this respect.

Antibiotic Selection. As was the case in transformation, the minimum inhibitory concentration of an antibiotic should be tested for each strain prior to conjugal transfer. These levels can vary as much as 20-fold from strain to strain.

Filters. The type of nitrocellulose filter used in the mating will also influence the results. Some nitrocellulose filters contain surfactants which inhibit the growth of Neisseriae. We have found that the BioTrace filters (Gelman, Ann Arbor, MI) work well, and others have reported good

[43] V. Burdett, J. Inamine, and S. Rajagopalan, *J. Bacteriol.* **149**, 995 (1982).

[44] A. E. Franke and D. B. Clewell, *J. Bacteriol.* **145**, 494 (1981).

[45] J. S. Knapp, S. R. Johnson, J. M. Zenilman, M. C. Roberts, and S. A. Morse, *Antimicrob. Agents Chemother.* **32**, 765 (1988).

[46] J.-C. Piffaretti, A. Arini, and J. Frey, *Mol. Gen. Genet.* **212**, 215 (1988).

[47] C. A. Genco and V. L. Clark, *J. Gen. Microbiol.* **134**, 3277 (1988).

results with Bio-Rad (Richmond, CA) nitrocellulose filters.[48] Regardless of the type of filter, it is important to wet the sterile filter on solid medium prior to placing it onto the agar surface where mating will take place. This saturates the filter with medium and also dilutes out any potentially harmful materials left over from the manufacturing process.

Counterselection. When mating between bacterial strains a means of selecting against the donor is important (counterselection). Antibiotic resistance is the best means of selection. Streptomycin or spectinomycin resistance that results from mutations in ribosomal proteins have been shown to be effective counterselection markers. Nalidixic acid resistance arising from mutations in DNA gyrase constitutes another possible marker. Resistant derivatives of any bacterial strain can be isolated by screening for spontaneous mutations. Antibiotic concentration gradient plates and growth in a series of media containing small increases of antibiotic concentration are two methods that have successfully been used to derive antibiotic-resistant derivatives. It is important to check these mutants for increased MIC of other antibiotics to guard against mutations that result in multiple resistance. The natural resistance of gonococcal, meningicoccal, and *Neisseria lactamica* strains to vancomycin and colistin makes these antibiotics the counterselection agents of choice when mating from *E. coli* or a commensal *Neisseria*. Trimethoprim may also be a good counterselection agent, although not all strains are resistant to this antibiotic.

Cocultivation with E. coli. When mating between *E. coli* and *Neisseria* strains, care must be taken to keep the *E. coli* inoculum low during conjugation. Cocultivation of *E. coli* and Neisseriae results in poor growth and killing of the Neisseriae cells. The molecular basis for this killing has not been determined, but lipids from *E. coli* may be the offending substance. Therefore, since the *E. coli* will outgrow the *Neisseria* during cocultivation, one must start with a much lower inoculum of *E. coli*. Normally, an *E. coli* to *Neisseria* ratio of 1 : 10 works well, but this parameter should be determined empirically for different growth conditions.

The following is a specific method for conjugally transferring plasmids.

1. Grow donor and recipient bacteria in preparation for mating. If antibiotic selection is necessary to maintain an unstable plasmid in the donor strain, contaminating antibiotics should be removed before mixing cells. For *E. coli* this is easily done by pelleting the bacteria and resuspending them in antibiotic-free medium. For many *Neisseria* species, centrifugation leads to significant autolysis.

[48] M. C. Roberts and J. S. Knapp, *Sex. Trans. Dis.* **16**, 91 (1989).

Therefore, growth on solid medium is the best method to reduce contamination by antibiotics.

2. Mix the donor and recipient bacteria in a small amount of liquid medium. GCB broth works well for both *Neisseria* and *E. coli.*

3. Deposit the mixed bacteria onto a solid surface. This can be done by placing less than 0.5 ml of the bacteria onto the surface of a dry GCB + plate. Preferably, the mixed culture should be deposited on a sterile 0.2-μm nitrocellulose filter in a holder using a sterile syringe. The filter is subsequently placed onto a GCB plate to allow conjugation to take place. With this method, the volume used to mix the bacteria can be any amount.

4. Allow the mixed culture to grow and conjugal transfer to take place. The growth conditions for the more fastidious strain should be met to allow for good growth during this period. The time of transfer can be from 2 to 24 hr. In an *E. coli* to *Neisseria* mating, the time should be limited to prevent the *E. coli* from killing the *Neisseria.*

5. Harvest the bacteria into liquid medium. This can be done using a sterile Dacron swab to collect bacteria from the solid surface, or the nitrocellulose filter can be placed in liquid medium and the bacteria resuspended by gentle vortexing.

6. Plate onto selective medium. Dilutions should be plated to determine the frequency. This should be compared to colony-forming units of both the input donor and recipient cells. Frequencies of 10^{-2} to less than 10^{-9} have been found for *E. coli* to *Neisseria* conjugation, depending on the strains used. When transferring pLES2 derivatives between *E. coli* and GC the number of transconjugants can be increased by plating the exconjugants first on medium containing a subinhibitory level of penicillin followed by replica plating the bacteria onto medium containing an inhibitory concentration of the antibiotic. In both steps counterselection against the donor must be maintained. It has been found that plating on a subinhibitory concentration of the antibiotic carried by the plasmid with counterselection against the donor, followed by replica plating onto an inhibitory concentration of antibiotic (and counterselection), raises the conjugation frequency of pLES2 derivatives between *E. coli* and the gonococcus (J. Cannon, personal communication). Replica plating is done using sterile Whatman (Clifton, NJ) 3MM filter paper to transfer colonies between the different media. This procedure may be necessary in order to kill the contaminating *E. coli* associated with the *Neisseria* prior to selecting for the transconjugants.

7. Confirm the physical presence of the conjugal plasmid. Streaking

the transconjugants on selective media and isolating single colonies is highly recommended. β-Lactamase production can be determined using a chromogenic substrate,[49] but isolation of plasmid DNA from the transconjugants is necessary to be sure that the plasmid has transferred and that no rearrangements have occurred.

Other Genetic Systems

Bacteriophage

No bacteriophage for *Neisseria* have been isolated. This does not rule out their existence, but numerous screening attempts have so far been unsuccessful. Reports of autoplaquing that were suspected to result from bacteriophage induction were probably due to induced local autolysis since no bacteriophage particles were found.[50]

Transposons

Attempts to introduce transposons normally found in gram-negative bacteria into Neisseriae have met with limited success. This prompted us to develop the shuttle mutagenesis system which relies on transposition into cloned gonococcal genes in *E. coli* followed by transformation to recombine the mutated DNA into the gonococcal chromosome.[19] This technique has been expanded to the meningococcus by the introduction into the minitransposon of an erythromycin resistance gene from gram-positive bacteria (D. Trees and J. G. Cannon, unpublished).

Recently, transposon Tn916 gram positive has been found to transpose in the meningococcus after being introduced on a suicide plasmid by transformation.[51] Tetracycline-resistant meningococci were isolated at a frequency of about 10^{-7}, and the transposon had inserted at random sites in the chromosome. This exciting new finding should enable workers to obtain transposon-induced mutations in the meningicoccus to study virulence. Unfortunately, attempts to introduce this transposon into the gonococcus have been unsuccessful to date.

[49] C. H. O'Callaghan, C. H. Morris, S. M. Kirby, and A. H. Singer, *Antimicrob. Agents Chemother.* **1**, 283 (1972).
[50] L. E. Campbell, H. B. Short, and V. L. Clark, *in* "The Pathogenic Neisseriae" (G. K. Schoolnik, ed.), p. 228. American Society for Microbiology, Washington, D.C., 1985.
[51] S. Katharious, D. S. Stephans, P. Spellman, and S. A. Morse, *Mol. Microbiol.* **4**, 729 (1990).

Final Thoughts

There is much interest in studying the pathogenic Neisseriae because of their importance as bacterial pathogens and their interesting biological properties. These two areas necessarily overlap since these organisms are obligate human pathogens; they are not found in other natural environments. The existence of a high-frequency DNA transformation system and the availability of a variety of antibiotic resistance markers have allowed the creation of defined mutations in isolated genes in these bacteria. In order to increase the sophistication of the genetic experiments available, new vectors and methods for performing genetics must be developed. For instance, the identification of transposons that can function in the Neisseriae and the development of a system for their efficient delivery are important areas of study. That Tn916 functions in the meningococcus is a hopeful sign that the powerful tool of transposon mutagenesis may be available in the future to investigators studying the biology of the Neisseriae.

The ability to overcome restriction barriers by *in vivo* methylation (T. Butler, personal communication) should improve the efficiencies of transposon mutagenesis to develop this into a useful genetic method. Moreover, the construction of a physical and genetic map of a gonococcal strain (J. Cannon, personal communication) means that neisserial genetics will not be limited to individual genes, but that studies encompassing the entire genome of these unique bacteria will soon be forthcoming.

Acknowledgments

We would like to thank those who have been involved in developing genetic methods for the Neisseriae. We would particularly like to thank Tony Butler, Janne Cannon, Chris Elkins, Steven Morse, Fred Sparling, and Dan Stein for communication of results prior to publication.

[16] Genetic Systems in Myxobacteria

By Dale Kaiser

Utility of Genetics in Myxobacteria

Myxobacteria lie on the boundary between uni- and multicellular organisms. They grow and divide as separate cells, yet they constitute a primitive multicellular organism whose cells feed in multicellular units, and which, in times of starvation, assemble compact and regular structures of about 100,000 cells. These structures, called fruiting bodies, have a variety of shapes. Evidently the shapes are under genetic control because each species produces characteristic fruiting bodies.

Myxobacteria live by secreting hydrolytic enzymes with which they degrade particulate organic matter in the soil.[1] When their only nutrient in culture is a polymeric substance, casein, their rate of growth is found to increase with cell density.[2] Thus, cells feed cooperatively, and the association of cells in groups allows them to feed more efficiently. Perhaps the advantage of cooperative feeding has selected in the course of their evolution for aggregate formation, thus fruiting bodies. When food is again available, the multicellular nature of a fruiting body ensures that a new cycle of growth will be started by a community of cells. The macroscopic size of a fruiting body enhances dispersion. Myxobacteria move by gliding on surfaces; gliding facilitates their feeding and permits cells to move over each other when they construct a fruiting body.[3]

In the history of life on earth, multicellular organisms have evolved from unicells on more than 15 independent occasions.[4] Myxobacteria may represent one of the earliest attempts to build a multicellular organism, and their age is estimated at more than 1 billion years.[5,6] Myxobacteria lend themselves to experimental investigation of the multicellular state and the cell interactions that support it. Such interactions are essential for their multicellular development. However, their development is gratuitous; it is not necessary for the growth of a cell or a population of cells. Mutants that fail to form multicellular aggregates can be propagated indefinitely by maintaining cells in the growth phase. Genetics can be used effectively to investigate multicellular development. Since myxobacteria are gram-negative, many of the *Escherichia coli* and *Salmonella* techniques apply, including transposon-based methods.

Gene Transfer and Mapping

Several families of transducing myxophage and one coliphage that transduces myxobacteria have been found.

[1] E. Rosenberg, "Myxobacteria: Development and Cell Interactions." Springer-Verlag, New York, 1984.
[2] E. Rosenberg, K. H. Keller, and M. Dworkin, *J. Bacteriol.* **129,** 770 (1977).
[3] H. Reichenbach and M. Dworkin, *in* "The Prokaryotes" (M. P. Starr, H. Stolp, H. Trüper, A. Balows, and M. Schlegel, eds.), p. 328. Springer-Verlag, Berlin and Heidelberg, 1981.
[4] R. H. Whittaker, *Science* **163,** 150 (1969).
[5] D. Kaiser, *Annu. Rev. Genet.* **20,** 539 (1986).
[6] H. Ochman and A. C. Wilson, *J. Mol. Evol.* **26,** 74 (1987).

Phage Mx4 Family

Mx4, Mx41, and Mx43 are structurally and serologically closely related. All are general transducing phage.[7,8] Mx4 appears to be a virulent phage even though it forms cloudy plaques because no lysogenic bacteria have yet been isolated from infected cells.[7,9] The natural host range of Mx4, Mx41, and Mx43 appears to be limited to a set of nonfruiting mutants of *Myxococcus xanthus*, but the limitation is at a stage beyond adsorption[7] and can be overcome by a phage mutation, *hrm-1*, isolated by M. Wolfner. A mutation to high-frequency transduction has been combined with *hrm-1* and another mutation to temperature sensitivity that reduces the killing of transductants to give a transducing strain of high efficiency.[9]

Phage Mx8 Family

Mx8, Mx81, and Mx82 were isolated from carrier strains of *Myxococcus*,[8] and particles of similar morphology have been seen occasionally in cultures of *M. xanthus* FB.[10] The host range of these phage includes both fruiting-competent and nonfruiting mutants of *M. xanthus*.[8] Transduction frequencies range from 10^{-5} to 10^{-7} per particles.[11,12] Plaque-forming and transducing particles of Mx8 have the same buoyant densities, and since Mx8 DNA and host DNA have the same mole percent GC, the two types of particles are likely to contain equal amounts of DNA.[8] The DNA molecules isolated from purified phage particles have a length corresponding to 56 kilobases (kb).[8] By digestion of Mx8 DNA with restriction endonucleases, Orndorff *et al.*[13] have shown that it is circularly permuted. Mx8 is a temperate phage, and comparison of the restriction map of the prophage with that of DNA from phage particles has demonstrated the absence of unique end fragments, the circular permutation of phage DNA, and unique prophage attachment sites in Mx8 and *M. xanthus*.[13] Stellwag *et al.*[14] have identified the *Eco*RI restriction fragment of Mx8 DNA that contains the prophage attachment site as well as the genes that specify the catalyst for site-specific integrative recombination.

[7] J. M. Campos, J. Geisselsoder, and D. R. Zusman, *J. Mol. Biol.* **119**, 167 (1978).

[8] S. Martin, E. Sodergren, T. Masuda, and D. Kaiser, *Virology* **88**, 44 (1978).

[9] J. Geisselsoder, J. M. Campos, and D. R. Zusman, *J. Mol. Biol.* **119**, 179 (1978).

[10] H. D. McCurdy and T. H. MacRae, *Can. J. Microbiol.* **20**, 131 (1974).

[11] J. Hodgkin and D. Kaiser, *Mol. Gen. Genet.* **171**, 167 (1979).

[12] J. Hodgkin and D. Kaiser, *Mol. Gen. Genet.* **171**, 177 (1979).

[13] P. E. Orndorff, E. Stellwag, T. Starich, M. Dworkin, and J. Zissler, *J. Bacteriol.* **154**, 772 (1983).

[14] E. Stellwag, J. M. Fink, and J. Zissler, *Mol. Gen. Genet.* **199**, 123 (1985).

Mxα

Mxα was discovered by its capacity to transduce a particular set of Tn*5* insertions in *M. xanthus*.[15] The Mxα transducing particles have a density of 1.495 g/cm^3 and are DNase resistant. The transducible set of insertions are found on two or three 80-kb elements. These elements have related sequences that may be circular or terminally redundant, and they are found in most, but not all, *M. xanthus* strains.[16] No plaques have been observed from Mxα; it may be a defective phage or an ubiquitous prophage.

Intergeneric Transduction by Coliphage P1

Coliphage P1 can adsorb to *M. xanthus* and cause its lysis or inhibit its growth.[17] Lipopolysaccharide is the adsorption receptor for P1 in *E. coli*,[18] and *Myxococcus* possesses lipopolysaccharide resembling that of other gram-negative bacteria.[19] Although P1 does not multiply in *M. xanthus*, it does inject its DNA, as shown by the fact that infection by P1::Tn*5* yields kanamycin-resistant strains of *M. xanthus*.[20] No infectious P1 particles issue from these cells, and no lysogenic bacteria are formed. The kanamycin-resistant transductants contain Tn*5* DNA sequences but no P1 DNA detectable by Southern blot transfer.[20] Transduction thus occurs in the absence of P1 multiplication. P1 will transduce whole plasmids into which a segment of P1 DNA has been incorporated, a specialized transduction as described below.[21,22] P1 can also transfer plasmids from *E. coli* to *M. xanthus* by generalized transduction.[23] It is thus possible to introduce DNA in general and transposons in particular from *E. coli*—and possibly from other P1 hosts[24]—into *M. xanthus* via phage P1. Because P1 is unable to replicate its DNA or to produce phage particles in *M. xanthus* when *M. xanthus* is infected with a stock of specialized transducing P1::Tn*5*, only transposed sequences are reproduced. These conditions facilitate the use of transposon techniques[25] to study *M. xanthus* genetics.

Naked DNA can be introduced into *Myxococcus* aided by electropora-

[15] T. Starich, P. Cordes, and J. Zissler. *Science* **230,** 541 (1985).

[16] T. Starich and J. Zissler, *J. Bacteriol.* **171,** 2323 (1989).

[17] D. Kaiser and M. Dworkin, *Science* **187,** 653 (1975).

[18] A. Lindberg, *Annu. Rev. Microbiol.* **27,** 205 (1973).

[19] G. Rosenfelder, O. Luderitz, and O. Westphal, *Eur. J. Biochem.* **44,** 411 (1974).

[20] J. M. Kuner and D. Kaiser, *Proc. Natl. Acad. Sci. U.S.A.* **78,** 425 (1981).

[21] L. J. Shimkets, R. E. Gill, and D. Kaiser, *Proc. Natl. Acad. Sci. U.S.A.* **80,** 1406 (1983).

[22] R. E. Gill, M. G. Cull, and S. Fly, *J. Bacteriol.* **170,** 5279 (1988).

[23] K. A. O'Connor and D. R. Zusman, *J. Bacteriol.* **155,** 317 (1983).

[24] Y. Murooka and T. Harada, *Appl. Environ. Microbiol.* **38,** 754 (1979).

[25] N. Kleckner, J. Roth, and D. Botstein, *J. Mol. Biol.* **116,** 125 (1977).

tion (electroporation was worked out by J. Rodriguez, and it is described by Kuspa and Kaiser[26]). The following specific procedure is effective. Exponentially growing cells are washed twice at 4° in electroporation buffer [5 mM potassium phosphate, pH 7.6, 0.27 M sucrose, 15% (v/v) glycerol] and suspended to about 5 × 10⁹ cells/ml in the same buffer. Five micrograms of DNA is added to 0.6 ml of cell suspension in 5 μl of TE (10 mM Tris hydrochloride, pH 7.5, 1 mM EDTA), and this mixture is subjected to an electrical field of 6250 V/cm with a capacitance setting of 25 μF. The treated suspension is immediately diluted in 1 ml of CTT broth[26] and then plated on CTT agar containing 40 μg/ml of kanamycin sulfate. Within 18 hr the plates are overlaid to a final concentration of 70 μg/ml of kanamycin sulfate. Colonies usually appear after 4 days of incubation at 32°.

Physical/Linkage Map of Myxococcus xanthus

Many localized regions of the *M. xanthus* genome have been ordered by Mx4 or Mx8 generalized transduction with their DNA-carrying capacity of 56 kb. A physical map of the entire 9.5-Mb (megabase) genome has been constructed by Chen *et al.*[27] It was generated from the 16 *Ase*I and 21 *Spe*I restriction fragments,[28] overlapped by *Eco*RI fragments of *Myxococcus* in yeast artificial chromosomes.[29] Approximately 100 gene loci have been mapped either genetically or physically. The entire genome is circular and at 9.5 Mb is one of the largest known for bacteria.

Transposon Tn5 in *Myxococcus xanthus*

Tn5 is a 5.7-kb transposable element originally isolated from *E. coli*.[30] It carries a gene for aminoglycoside phosphotransferase, whose product catalyzes inactivation of kanamycin to which wild-type *M. xanthus* is normally sensitive. Cells that have Tn5 inserted in their chromosome are consequently highly resistant to kanamycin. Tn5 contains a central segment 2.6 kb long flanked by two end sequences of 1.5 kb each. The first 0.8 kb of the central segment encode for aminoglycoside 3′-phosphotransferase II.[30,31] The terminal sequences of 1534 base pairs each are

[26] A. Kuspa and D. Kaiser, *J. Bacteriol.* **171,** 2762 (1989).

[27] H. Chen, A. Kuspa, I. M. Keseler, and L. J. Shimkets, *J. Bacteriol.* **173,** 2109 (1991).

[28] H. Chen, I. M. Keseler, and L. J. Shimkets, *J. Bacteriol.* **172,** 4206 (1990).

[29] A. Kuspa, D. Vollrath, Y. Cheng, and D. Kaiser, *Proc. Natl. Acad. Sci. U.S.A.* **86,** 8917 (1989).

[30] D. E. Berg, J. Davis, B. Allet, and J. D. Rochaix, *Proc. Natl. Acad. Sci. U.S.A.* **72,** 3628 (1975).

[31] E. Beck, G. Ludwig, E. A. Auerswald, B. Reiss, and H. Schaller, *Gene* **19,** 327 (1982).

arranged in reverse orientation and have almost identical sequences. These termini, identified as IS50L for left end and IS50R for right end, are necessary for transposition of the entire element.[32] The IS50L differs from IS50R by a single base pair at position 1442, and this difference generates a rightward promoter in IS50L at which transcription of the aminoglycoside 3'-phosphotransferase II gene is initiated.[33] When Tn5 transposes in *E. coli*, nine bases of host sequence are duplicated in direct orientation.[34] When *M. xanthus* is infected with P1 carrying the transposon, Tn5 can transpose to the *Myxococcus* chromosome, generating kanamycin-resistant strains.[20] These kanamycin-resistant transductants are stable in the absence of kanamycin and, as described above, contain DNA sequences homologous to Tn5, demonstrated by restriction enzyme cleavage pattern and DNA–DNA hybridization; however, they have no sequences that are homologous to P1.

Tn5 transposes from P1 to many different chromosomal sites in *M. xanthus*.[20] In each of the first nine independent kanamycin-resistant transductants examined, the transposon was located in a different fragment generated by cleavage with restriction endonuclease *Eco*RI. Tests of genetic linkage also gave evidence for an approximately random distribution of insertion sites. A method for isolating Tn5 insertions linked to a mutation of interest has been developed.[20] The frequency with which insertions of Tn5 were found near to any particular locus, 0.5–1%, corresponds to the ratio between the amount of DNA in an Mx8 phage particle (56 kb) and the amount of DNA in the genome of *M. xanthus*, 9.5 Mb.

Once inserted, Tn5 remains at the original site of insertion. Initially, 20 strains were tested by restriction analysis during vegetative growth and 5 during transduction from one *M. xanthus* strain to another.[20] Subsequent examination of many other Tn5 insertions has confirmed their general stability. Tn5 at one site in *M. xanthus* generates a second copy at another site (intrachromosomal transposition) at a frequency of less than 1/1000 cells.[35,36]

The structure of Tn5 lends itself to modification by replacement of its unique central segment with other antibiotic resistance genes. These modifications preserve the overall Tn5 structure but change the encoded

[32] D. E. Berg, C. Egner, B. J. Hirschel, J. Howard, L. Johnsrud, R. A. Jorgensen, and T. D. Tlsty, *Cold Spring Harbor Symp. Quant. Biol.* **45,** 115 (1981).
[33] S. J. Rothstein, R. A. Jorgensen, K. Postle, and W. S. Reznikoff, *Cell (Cambridge, Mass.)* **19,** 795 (1980).
[34] E. A. Auerswald, G. Ludwig, and H. Schaller, *Cold Spring Harbor Symp. Quant. Biol.* **45,** 107 (1981).
[35] E. Sodergren and D. Kaiser, *J. Mol. Biol.* **167,** 295 (1983).
[36] L. Avery and D. Kaiser, *Mol. Gren. Genet.* **191,** 99 (1983).

antibiotic resistance. One modified Tn5, called Tn5-132, encodes resistance to tetracycline and retains almost all of both IS50 units intact.[37] Avery has developed a technique for the replacement of Tn5 by Tn5-132 that keeps the original site of insertion.[15,36] In this technique, Tn5-132 is introduced into *M. xanthus* by P1::Tn5-132. Homologous crossovers occur in each IS50 unit, as these units are the only homologous segments between phage and cell genomes, to generate a replacement.

If Tn5-132 were to transpose to a new site in *M. xanthus* instead of replacing an existing Tn5-wild type (*wt*), the resulting tetracycline-resistant transductant would remain resistant to kanamycin, now having one copy of Tn5-*wt* at the original site and one copy of Tn5-132 at a new site. A replacement strain, by contrast, has only one copy of the transposon, a Tn5-132, and has by virtue of the loss of Tn5-*wt* become sensitive to kanamycin.[36] The Tn5-132 now occupies the same site that Tn5-*wt* did originally, which was shown in three ways. First, Tn5-*wt* in the original strain and Tn5-132 in the replacement strain behave as genetic alleles.[35,36] Second, Tn5-132 is found linked to the same mutations in *M. xanthus* as Tn5-*wt* was originally.[36] Finally, the restriction pattern of *M. xanthus* DNA in the vicinity of the ends of Tn5-132 in the replacement strain is shown to be the same as it was in the vicinity of the ends of Tn5-*wt* in the original strain.[36]

Other modified versions of Tn5 now available[38] can be brought into *M. xanthus* either for transposition or for replacement by homologous recombination. One particular derivative, Tn5*lac*, requires additional description. Tn5*lac* is a "promoter probe" or "enhancer trap." It was constructed by adding a promoterless *lacZ* gene to Tn5. The structure of Tn5*lac* is shown in Fig. 1. The promoterless *lacZ* gene was inserted into a position which does not interfere with Tn5 transposition or the expression of the gene for kanamycin resistance, and it is oriented so that transcription initiated from a cell promoter, outside the Tn5*lac* DNA, proceeds through the gene as shown in Fig. 1. There are also translation stop signals ahead of the *lacZ* gene in all three reading frames. Consequently transcriptional fusions, but not protein fusions, are made.

Many developmentally regulated genes have been located by random insertions of Tn5*lac* into *Myxococcus*.[39] In one set of 2374 Tn5*lac* insertion strains, 548 produced β-galactosidase during growth, indicating that Tn5*lac* had inserted in the appropriate orientations in transcription units

[37] S. J. Rothstein, R. A. Jorgensen, J. C. P. Yin, Z. Young-di, R. C. Johnson, and W. S. Reznikoff, *Cold Spring Harbor Symp. Quant. Biol.* **45**, 99 (1981).
[38] D. Berg, in "Mobile DNA" (D. E. Berg and M. M. Howe, eds.), p. 183. American Society for Microbiology, Washington, D.C., 1989.
[39] L. Kroos, A. Kuspa, and D. Kaiser, *Dev. Biol.* **117**, 252 (1986).

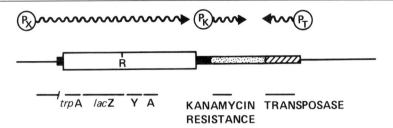

FIG. 1. Structure of Tn5lac. Transcripts (wavy lines) are shown above the DNA, originating from PK, the promoter of the kanamycin resistance gene, from PT, the promoter of the transposase gene, and from PX, the promoter of a transcript into which coding sequence Tn5lac has inserted in the correct orientation to make a transcriptional fusion. Tn5lac consists of sequences from IS50 left (solid), the trp–lac fusion segment (open), the central region of Tn5 (stippled), and IS50 right (hatched). Proteins (solid lines) shown below the DNA include a truncated polypeptide (leftmost) encoded by the gene into which Tn5lac has inserted. [From D. Kaiser, L. Kroos, and A. Kuspa, *Cold Spring Harbor Symp. Quant. Biol.* **50**, 823 (1985); copyright 1985, Cold Spring Harbor Laboratory.]

active during growth. Assuming that Tn5lac inserts with equal frequency in both orientations, one would estimate that twice 548, or about one-half the insertions isolated, were in transcriptionally active regions. This implies that roughly one-half the *Myxococcus* genome is transcribed during growth, under laboratory conditions. Of the 2374 strains, 94 appeared to produce more β-galactosidase during starvation-induced fruiting body development than during growth, and in them *lacZ* is thus developmentally regulated. (The term development refers here to starvation-induced fruiting body development.). In 36 of the 94 strains, *lacZ* expression was found to be strongly regulated in that β-galactosidase activity increased during development by factors of 3–300.

The time courses of strongly developmentally regulated β-galactosidase synthesis in all 36 of these strains have been examined. Three are shown for illustration in Fig. 2. In one of these strains, β-galactosidase increases about 30 min after starvation initiates development. In the second, the increase begins around 12 hr. The third strain accumulates β-galactosidase late, at around 24 hr while spores are forming. As shown in Fig. 2, in the sample of 36 strongly regulated Tn5lac strains, 18 increase β-galactosidase during the first 8 hr, the period of initiation and preparation for aggregation. Another 11 strains show their increased synthesis between 8 and 16 hr, during or after the period of aggregation, and 7 strains show increases during the period (16–24 hr) of sporulation and have their highest activity inside spores.

Most of the 36 Tn5lac insertions indicated in Fig. 2 are in different regions of the *Myxococcus* genome, as demonstrated by their physical

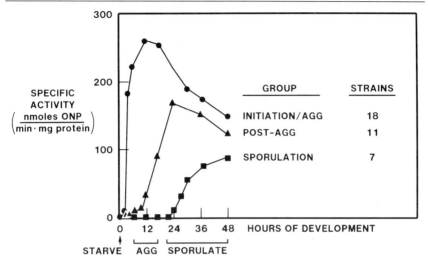

FIG. 2. β-Galactosidase synthesis in developing Tn*5lac* strains. The specific activity of β-galactosidase in sonic extracts of cells harvested at different times during development is shown for three representative strains. [From D. Kaiser, L. Kroos, and A. Kuspa, *Cold Spring Harbor Symp. Quant. Biol.* **50**, 823 (1985); copyright 1985, Cold Spring Harbor Laboratory.] AGG, Aggregate; ONP, *o*-nitrophenol (measures β-galactosidase activity).

mapping. The restriction maps of the DNA adjacent to these insertions do not resemble each other.[39] In five cases, however, two or more insertions do have related restriction maps and also express β-galactosidase at about the same time and level during development, suggesting that these pairs are in the same transcript. Taking these restriction map similarities into account, at least 29 different transcription units are represented by these 36 Tn*5lac* insertions. The timing of gene expression for these 29 units, and for proteins S[40] and H[41] as well, clearly demonstrate a program of differential gene expression during fruiting body development.

Tandem Duplications

Tandem genetic duplications arise spontaneously in *M. xanthus* as they do in other organisms.[42] A tandem duplication contains one new DNA sequence not present in the parental strains, called the "novel joint."

[40] M. Inouye, S. Inouye, and D. R. Zusman, *Proc. Natl. Acad. Sci. U.S.A.* **76**, 209 (1979).
[41] D. R. Nelson, M. Cumsky, and D. R. Zusman, in "Sporulation and Germination" (H. S. Levinson, A. L. Sonenshein, and D. J. Tipper, eds.), p. 276. American Society for Microbiology, Washington, D.C., 1981.
[42] R. P. Anderson and J. R. Roth, *Annu. Rev. Microbiol.* **31**, 473 (1977).

Although duplications arise frequently during growth of a culture, they also decay to the original single-copy state by homologous recombinations between duplicated parts. Therefore, unless duplications are selected in some way they will be lost. Selection for duplications may be natural as well as experimental, and, in fact, wild-type *M. xanthus* has evolved a duplication of the structural gene for protein S.[43]

Tandem duplications can be used for tests of gene dominance and for complementation tests, which define genetic units of function. In *M. xanthus*, where F' strains are not availble, tandem duplications can serve many of the same purposes.[44] Tandem duplications also arise during certain cloning operations described in the next section.

Three properties help to establish the structure of a tandem duplication: segregation, restriction map, and transduction of the novel joint. When homologous recombination occurs between the duplicated parts, it excises a circular DNA molecule, it restores the unduplicated sequence, and it always eliminates the novel joint. When the duplicated copies carry different alleles of a locus, one allele at each duplicate locus is lost during segregation. Segregants may carry parental combinations of alleles or new combinations. The frequency of loss of a mutation is directly proportional to its distance from the novel joint. The map position of the novel joint can thus be determined by measuring the frequency of each segregant class.

Restriction mapping helps to establish the structure of duplication strains. Comparison of the restriction map of the parental strains in the vicinity of the ends of Tn5 with a duplication strain can be made with radiolabeled Tn5 as the DNA-hybridization probe.[20] The ends of Tn5 will be carried in restriction fragments that contain both Tn5 DNA and adjacent chromosomal DNA. These fragments identify the locus of Tn5 insertion. Tn5 and Tn5-*132* at allelic sites, at nonallelic sites, and one or the other at the novel joint of a tandem duplication give predictable sets of Tn5 end fragments.[36,45] The particular set of end fragments observed from the duplication strain and from its segregants gives detailed information about the structure of the duplication.

The novel joint of a tandem duplication uniquely determines its structure, and if a novel joint sequence is transduced into a new strain, a duplication of the same chromosome segment as the donor is recreated in the transductants.[42] Even duplications whose duplicated lengths are greater than the amount of DNA packageable in a single phage particle

[43] S. Inouye, Y. Ike, and M. Inouye, *J. Biol. Chem.* **258,** 38 (1983).
[44] B. Bachmann and K. B. Low, *Microbiol. Rev.* **44,** 1 (1980).
[45] L. Avery and D. Kaiser, *Mol. Gen. Genet.* **191,** 110 (1983).

can be transduced into new strains because the recipient DNA is forced to duplicate. In this case recipient alleles replace donor alleles for DNA sequences more than one phage headful away from the selected marker. Transduction of a novel joint not only helps to characterize a tandem duplication, it also is important in constructing partial diploid strains having the genetic composition necessary for dominance and complementation tests.

Genetic duplications in *M. xanthus* have been isolated in three different ways, two requiring insertions of Tn5. One way, called tandem duplication trapping, is the simultaneous selection in a transduction experiment of two different alleles that occupy the same chromosomal site. Only a duplication provides two copies of the same site. Different forms of Tn5 can be incorporated at the same site by replacement, as described above (Transposon Tn5 in *Myxococcus xanthus*). A transduction is then performed from a donor containing Tn5 (Km^r), for example, into a recipient that contains Tn5-*132* (Tc^r) at the same site, the latter obtained by replacement of the former, and transductants having both Km^r and Tc^r are selected.[36] The duplications are distinguished from phenotypically similar transpositions by segregation and by their restriction maps. One particular duplication spanned at least 45 kb as measured by the frequency of Km^s and Tc^s segregants. A new duplication with the same properties as the old was generated by transduction of the novel joint to a new strain.[36] By trapping, any duplication that covers a predetermined site can be isolated; these duplications may have random end points, provided only that they flank the predetermined site.

A second way, which Avery called construction, permits the formation of tandem duplications with predetermined end points. To construct a tandem duplication, Tn5 is placed at one end of the segment to be duplicated, and Tn5-*132* is placed at the other end in the same strain. The desired duplication forms during growth of this double-insertion strain by (unequal) recombination between the two transposons.[45]

Duplications may also form when a circular *E. coli* plasmid containing a cloned *Myxococcus* DNA segment integrates by a single homologous recombination, or by site-specific recombination, into *M. xanthus*. These processes are described in the next section.

Cloning *Myxococcus xanthus* Genes

No plasmids indigenous to *M. xanthus* have yet been described. Cloning is carried out in *E. coli* with *E. coli* plasmids. To test the function of cloned genes, they are returned to *M. xanthus* either by P1 transduction or by transformation.

Tn*5*, which came to *Myxococcus* from *E. coli*, expresses kanamycin resistance in *E. coli*.[30] Consequently, if we clone random DNA fragments in *E. coli* from a *M. xanthus* strain containing Tn*5* inserted near a gene of interest, the resulting kanamycin-resistant *E. coli* strains are likely to contain the interesting *Myxococcus* gene. Many genes have been cloned with this procedure and have been shown to contain the genes of interest by transferring them from *E. coli* back into *M. xanthus*, where they complement the corresponding defective mutants. Gill *et al.*[22] have developed an *in situ* cloning vector that includes a fragment of Tn*5*. When it is transduced by P1 from *E. coli* into *M. xanthus*, it integrates by sequence homology into the left or right IS50 of a resident Tn*5-132*. The vector design ensures that cleavage of DNA from the kanamycin-resistant transductant releases a single DNA fragment containing both the entire vector and *M. xanthus* DNA from one side of the Tn*5* in *M. xanthus*. This single DNA fragment can form a circle by a monomolecular reaction. The *bsgA* gene[22] and *mgl* gene[46] have been cloned in this way.

Two different transduction methods have been used to return the cloned genes from *E. coli* to *M. xanthus*; both employ P1. Transfer by specialized transduction was developed by Gill and co-workers[21]; they derived a specialized cloning vector by adding a 6.7-kb *Eco*RI fragment encoding P1-specific incompatibility to pBR322.[47] When an infecting P1 establishes lysogeny in an *E. coli* strain carrying Gill's plasmid, P1 cannot lysogenize in the usual way owing to the preestablished incompatibility[48] and instead forms a cointegrate with the plasmid by recombination in the homologous P1-incompatibility sequences. On induction to lytic growth, P1 begins packaging DNA within the P1 segment of the cointegrate, and many of the resulting phage particles contain the entire plasmid.[21,22] This P1 stock transfers the plasmid to *M. xanthus* by specialized transduction, and when the chimeric P1 plasmid DNA is injected into *M. xanthus* homologous recombination between P1 incompatibility sequences regenerates the original circular plasmid. The ColE1-derived plasmid apparently cannot replicate in *M. xanthus*, but a single homologous recombination anywhere within its inserted *M. xanthus* DNA integrates the plasmid into the host chromosome. This is the only event that can give a Km[r] transductant because the *M. xanthus* insert terminates within Tn*5*. The entire plasmid integrates into the chromosome, duplicating in tandem those sequences common to the cloned DNA and to the chromosome. The structure of the

[46] K. Stephens and D. Kaiser, *Mol. Gen. Genet.* **207,** 256 (1987).
[47] N. Sternberg, M. Powers, M. Yarmolinsky, and S. Austin, *Plasmid* **5,** 138 (1981).
[48] R. W. Hedges, A. E. Jacob, P. T. Barth, and N. J. Grinter, *Mol. Gen. Genet.* **141,** 263 (1975).

resulting tandem duplications has been analyzed by restriction enzymes and by segregation and has been found to conform to expectations.[21] *Myxococcus xanthus* DNa segments as large as 50 kb and as small as 1.5 kb have been cloned in *E. coli* and returned to *M. xanthus* in this way.

The route from *E. coli* back to *M. xanthus* via generalized P1 transduction was developed by O'Connor and Zusman.[23] They cloned *M. xanthus* genes in *E. coli* on a pBR322 plasmid by selecting kanamycin resistance of a Tn*5* insert linked to those genes, a selection similar to the method just described. They discovered that P1 grown on *E. coli* that carried the plasmid could transduce the plasmid to *M. xanthus* at a frequency of $10^{-3}-10^{-7}$ kanamycin-resistant transductions per plaque-forming P1.[23] The frequency of plasmid transduction depended on the size of the plasmid and was higher for a 43-kb plasmid than for plasmids of 20–30 kb. One 10.5-kb plasmid was transducible, but another of the same size was not, suggesting that factors other than size may affect transducibility. Some of the kanamycin-resistant transductants were stable, but others were unstable. The unstable ones contained vector–plasmid sequences, as demonstrated by DNA–DNA hybridization. The instability and restriction enzyme cleavage pattern of the unstable transductants suggested that they arose from integration of the plasmid into the chromosome through homology in the cloned *M. xanthus* segment. (This integration event would be similar to that for the specialized P1 transduction route described above.) The stable transductants contained no sequence able to hybridize to the plasmid–vector and behaved like replacements of chromosomal sequences by cloned sequences.[23]

When tandem duplications are generated by either specialized or generalized P1 transduction of a plasmid containing *M. xanthus* sequences, gene conversion is frequently observed.[21,23] For example, about 20% of the kanamycin-resistant transductants from a *csgA*$^+$ donor into a *csgA*$^-$ recipient fail to sporulate. These transductants are *csgA*$^-$/*csgA*$^-$ partial diploids, having restriction fragments characteristic of both the donor and the recipient. They segregate like the sporulation-proficient majority (80%) except that all of their segregants also fail to sporulate and are *csgA*. Shimkets *et al.*[21] suggested that in these cases the *csg*$^+$ allele of the donor was gene-converted to *csg*$^-$ by a mismatch repair that used the *csg*$^-$ strand in a *csg*$^+$/*csg*$^-$ heteroduplex as the repair template. The relatively high frequency of gene conversion in *M. xanthus* is consistent with the high frequency of repair of ultraviolet-induced lesions observed by Grimm and Herdrich.[49]

The two genomic copies of a tandem duplication can be separately

[49] K. Grimm and K. Herdich, *Photochem. Photobiol.* **31**, 229 (1980).

cloned for analysis of their structure[46] or to facilitate further construction. Homozygous partial diploid gene convertants were demonstrated in this way.[46] Clones of known mutant alleles can readily be derived this way as well.[22]

Stellwag *et al.*[14] reported that plasmids which contain the Mx8 attachment site and adjacent genes on an *Eco*RI fragment are able to integrate at the Mx8 prophage attachment site on the *M. xanthus* chromosome. By means of such plasmids, duplications can be formed which are not tandem, seldom undergo gene conversion, and do not reconstruct complete genes from partially deleted ones. This latter property is especially important in carrying out a deletion analysis of regulatory regions that are 5' or 3' to a coding region, since the aim often is to delimit these regions necessary for gene expression by systematic removal of parts.

Furthermore, Li and Shimkets[50] have demonstrated that, when there is a choice between integration at the Mx8 prophage site and integration at the homologous chromosomal site, attachment-site integration almost always occurs. Usually a single copy integrates, but the existence of multiple copies can be distinguished by restriction and Southern analysis.

Extracellular Complementation

Cellular interactions can be explored by a type of complementation test that uses mixtures of whole cells. Such a test is called extracellular complementation to distinguish it from a genotypic complementation test that is carried out inside a cell. A genotypic complementation test performed in a partially diploid bacterium measures the complementary quality of immediate gene products, ones that have not had to pass through any membrane. Extracellular complementation, since it is observed in mixtures of whole cells, by contrast, concerns substances on the cell surface or substances secreted by cells.

To visualize the close connection between cell interaction and extracellular complementation, consider an (idealized) interaction and distinguish a source cell from a target cell. Consider further a mutant defective as a source but still able to function as a target; it would be defective in any process that depended on the interaction. Such a mutant mixed with wild-type cells or with a complementary mutant would be able to interact, and the mixture might have a phenotype approaching that of pure wild type. Extracellular complementation satisfying these expectations has been observed in *M. xanthus* among mutants defective in motility and among mutants defective in fruiting body formation.

[50] S. Li and L. J. Shimkets, *J. Bacteriol.* **170**, 5552 (1988).

Certain nonmotile mutants of *M. xanthus* are conditional mutants of a novel kind: The mutant cells become transiently motile when mixed with nonmutant cells or with cells of a different mutant type.[51] Direct contact between cells appears to be required, because these cells move when contact is permitted but not when contact is prevented by spatial separtion on an agar surface or by the interposition of a membrane filter.[51] This phenomenon, called motility stimulation, has generated six different types of stimulatable mutants. Extracellularly complementing (stimulatable) mutants are found in both motility systems A and S.[11,12,52] Stimulatable mutants, which are initially nonmotile, regain motility when they are mixed with wild-type cells or with mutants. Stimulation is phenotypic and not genetic in that the stimulated cell moves transiently, perhaps 100 μm, and no permanently motile strains are thereby generated.

The genetic significance of extracellular complementation by stimulation is revealed by the fact that mutants of a given stimulation type usually map to the same genetic locus.[11,12,35,51] The mutations found in all 14 mutants that were classified by stimulation tests as type *cglB* map in a cluster that is 90% cotransducible with Tn*5* insertion Ω1931 and define a unique *cglB* locus. Similarly, all *cglC*, *cglD*, *cglE*, and *tgl* mutants map to four discrete loci. The only exception to the one type-one locus rule has been found among the *cglF* mutants by Kalos and Zissler.[52] In general, there is a one-to-one correspondence between stimulation type and genetic locus.

A second type of extracellular complementation has been observed among certain mutants that are defective in fruiting body development. Intercellular coordination of fruiting body development is a major problem for the myxobacteria, because 100,000 cells enter into a fruiting body. Coordination would seem to require the passage of regulatory signals between cells. To look for intercellular signals, mutants were sought that were developmentally defective but could be rescued by wild-type cells. The idea was that wild-type cells would provide a missing signal, allowing mutants that were defective in signal production to develop normally. Fifty such mutants were isolated by Hagen *et al.*[53]

The Hagen mutants fail to form spores in fruiting bodies. However, when they are mixed with wild-type cells they do sporulate. In addition, some pairs of mutant strains can complement each other, and spores are formed. Mutants in mixtures which do not complement belong to the same (extracellular) complementation group. In this way four complementation

[51] J. Hodgkin and D. Kaiser, *Proc. Natl. Acad. Sci. U.S.A.* **74,** 2938 (1977).
[52] M. Kalos and J. F. Zissler, *J. Bacteriol.* **172,** 6476 (1990).
[53] D. C. Hagen, A. P. Bretscher, and D. Kaiser, *Dev. Biol.* **64,** 284 (1978).

groups (A, B, C, and D; now called *asg*, *bsg*, *csg*, and *dsg* for *a* signaling, etc.) are recognized in the initial set of 50 mutants. There appear to be at least four different signals. Each of the four types of Hagen mutants has a different developmental phenotype and corresponds to a different locus or set of loci.[26]

Acknowledgments

Work on motility from the author's laboratory was supported by a grant from the National Science Foundation DCB-8903705 and on cell-to-cell signaling by Public Health Service grant GM23441 from the Institute for General Medical Sciences.

[17] Genetics of *Caulobacter crescentus*

By BERT ELY

Introduction

Caulobacter crescentus is a gram-negative, aquatic bacterium which is characterized by an unusual cell cycle (for a recent review, see Newton[1]). Each cell division results in two different cell types, a sessile stalked cell and a motile swarmer cell. The stalked cell immediately begins to replicate its chromosome and prepare for the next cell division. Midway through chromosome replication, differentiation begins at the pole opposite the stalk, resulting in the synthesis of a flagellum, pill, and membrane phage receptors. This differentiated pole confers motility on the progeny swarmer cell when cell division occurs. The swarmer cell is an immature cell which must go through a maturation process before it can replicate its chromosome and divide. After a period of motility, the polar flagellum is lost, and a stalk is synthesized in its place. This new stalked cell then is capable of DNA replication and cell division. Thus, each cell cycle includes two times when differentiation occurs, the transition from a swarmer to a stalked cell and the differentiation of the flagellar pole of the predivisional cell. Furthermore, since each cell division results in two different cell types, differential segregation of cell components occurs. This differential segregation is most obvious at the poles of the cell but occurs with soluble components as well. Thus, the *Caulobacter* cell cycle has features which are normally found only in more complex organisms.

---

[1] A. Newton, *in* "Bacterial Diversity" (K. F. Chater and D. A. Hopwood, eds.), Vol. 2, p. 199. Academic Press, New York, 1989.

Since a well-developed genetic system has become available for *C. crescentus*, it is a premier organism for the study of the spatial distribution of gene products and the cell cycle control of gene expression.[2]

Growth of *Caulobacter crescentus*

Caulobacter crescentus lives in ponds and streams and is a scavenger in nature. It is adapted to growth in low nutrient conditions and grows poorly, if at all, in media designed for *Escherichia coli*. We grow *C. crescentus* with vigorous aeration at 33° in either a complex medium (PYE) or one of two defined media (M2 or PIG). PYE is a modification of the medium described by Poindexter[3] and consists of 0.2% peptone (Difco, Detroit, MI), 0.1% yeast extract (Difco, Detroit, MI), 0.8 mM MgSO$_4$, and 0.5 mM CaCl$_2$.[4] The addition of calcium is necessary since there is not always enough present in the other components to ensure optimal growth.

The defined medium[4] M2 is prepared from a 10× concentrated base consisting of 17.4 g Na$_2$HPO$_4$, 10.6 g KH$_2$PO$_4$, and 5 g NH$_4$Cl per liter of high-quality distilled water. (We use an Ultrascience Hi-Q glass still. Water prepared in conventional distillation systems contains volatile organics which inhibit the growth of *C. crescentus*, resulting in doubling times of 8–10 hr. Normal doubling times are 2–3 hr in defined medium prepared with high-quality water.) To prepare M2 medium, the 10× M2 stock is diluted and autoclaved. After cooling, 10 ml each of 30% (w/v) glucose, 50 mM MgCl$_2$, 50 mM CaCl$_2$, and 1 mM FeSO$_4$ in 0.8 mM EDTA, pH 6.8, are added aseptically to each liter of medium. For solid medium, equal volumes of 2× M2 and 2% agar are combined after autoclaving, and the CaCl$_2$ is omitted as it is provided by the agar.

Some strains with mutations causing pleiotropic effects do not grow well in M2 medium owing to the high phosphate levels. Therefore, a low phosphate medium, PIG is adopted.[5] The base for PIG medium is 10× I and consists of 5 g NH$_4$Cl and 6.8 g imidazole with the pH adjusted to 7.0 with HCl. PIG solid medium is made in the same fashion as M2 medium except that 10 ml of 0.1 M sodium glutamate, pH 7, and 1 ml of 0.1 M sodium phosphate buffer, pH 7, are added in addition to the other supplements. Liquid medium consists of 1× I plus 20 ml Hutner's mineral

[2] B. Ely and L. Shapiro, *Genetics* **123**, 427 (1989).
[3] J. S. Poindexter, *Bacteriol. Rev.* **28**, 231 (1964).
[4] R. C. Johnson and B. Ely, *Genetics* **86**, 25 (1977).
[5] P. V. Schoenlein, L. M. Gallman, and B. Ely, *J. Bacteriol.* **171**, 1544 (1989).

base,[6] 10 ml of 30% glucose, 10 ml of 0.1 M sodium phosphate buffer, pH 7, and 10 ml of 0.1 M sodium glutamate, pH 7, per liter.

Storage of Genetic Stocks

For long-term studies, it is important to maintain viable genetic stocks in an unchanged state. We have found that *C. crescentus* can be maintained at $-70°$ indefinitely without loss of viability. Cultures of strains to be stored are grown to stationary phase in PYE broth, and 1-ml aliquots are added to sterile vials containing 2 drops of dimethyl sulfoxide (DMSO). Adhesive paper labels are attached to the caps and sealed with clear fingernail polish to provide protection against moisture, and the vials are transferred to a $-70°$ freezer. The dimethyl sulfoxide promotes uniform freezing and prevents the formation of large ice crystals which could disrupt cell membranes. Strains are retrieved by scraping the surface of the frozen stock with a sterile stick and using the stick to inoculate a petri dish or culture tube. In this manner, the same stocks can be used repeatedly without thawing. We have found that stocks prepared in this manner maintain close to 100% viability for at least 15 years. It is important that cultures for frozen stocks be grown to stationary phase since frozen stocks made from log-phase cultures begin losing viability within the first year of storage.

Mutant Isolation

Many kinds of *C. crescentus* mutants can be obtained without mutagenesis. Antibiotic-resistant mutants can be obtained by direct selection, and motility mutants can be obtained by serial enrichment procedures for nonmotile cells on soft afar plates.[7] Auxotrophic mutants can be obtained after enrichment for nongrowing cells. Since wild-type *Caulobacter* is resistant to penicillin, however, fosfomycin or D-cycloserine must be used instead.[4]

The standard procedures described for penicillin enrichment are used for either of the two antibiotics, and prior mutagenesis is unnecessary. Cultures are grown in PYE, centrifuged, and resuspended in defined medium. After one doubling has occurred in defined medium, either 100 μg/ml D-cycloserine or 500 μg/ml fosfomycin is added. Samples are removed 13, 15, and 18 hr after the addition of antibiotic, spun, and resuspended to remove the antibiotic and plated on PYE medium. The resulting colonies are replicated onto defined medium to identify those with nutritional requirements. Auxotrophs are 5 to 10% of the survivors in typical experi-

---

[6] G. Cohen-Bazire, W. R. Sistrom, and R. Y. Stanier, *J. Cell. Comp. Physiol.* **49,** 25 (1957).

[7] R. C. Johnson and B. Ely, *J. Bacteriol.* **137,** 627 (1979).

ments. Auxotrophs and motility mutants can also be obtained by transposon mutagenesis (see below).

Generalized Transduction

Generalized transduction involves the use of a bacteriophage to transfer genetic markers from one strain to another. For *C. crescentus* the bacteriophage ϕCR30 is excellent for this purpose since it can transfer large pieces of DNA and is easy to grow.[8] The basic strategy is to grow ϕCR30 on the donor strain to allow the random mispackaging of fragments of the donor DNA into bacteriophage particles. The resulting lysate is irradiated with ultraviolet light to inactivate the viral genomes and then stored over chloroform to kill any residual bacteria. Transduction experiments are performed by mixing a dilution of the irradiated donor lysate with the recipient strain on the surface of a petri dish containing a selective medium. Colonies which grow should contain the selected marker from the donor.

Preparation of Transducing Lysates

An overnight culture of the strain chosen to be a donor is prepared, and 0.2 ml of this culture is mixed with 0.1 ml of a 10^{-4} dilution of ϕCR30 lysate (unirradiated) in 3 ml of melted PYE soft agar. The mixture is immediately poured onto the surface of a fresh PYE plate and allowed to harden. After overnight incubation at 33°, overlapping plaques should be observed. If the lysis appears satisfactory, 5 ml of PYE broth is pipetted onto the petri dish, which is refrigerated overnight to allow the bacteriophage to diffuse out of the agar into the liquid overlay.

The next day, the broth is decanted into an empty 60-mm-diameter petri dish. The cover is removed and the contents are swirled under a germicidal ultraviolet light at a distance of approximately 45 cm for 5 min. (If the petri dish is swirled by hand, protective gloves should be worn.) After irradiation, the contents of the dish are poured into a glass screw-capped tube, a few drops of chloroform are added, and the tube is shaken vigorously to ensure that all parts of the tube come in contact with the chloroform. A drop of this lysate is placed on a PYE plate and incubated to verify that all of the contaminating bacteria have been destroyed. A good lysate should have a titer of more than 10^{10} plaque-forming units (pfu) per milliliter before irradiation and less than 10^5 pfu/ml after irradiation. Although loss of transducing ability occurs slowly at 4°, phage lysates can be refrigerated for several years and still maintain good donor ability.

[8] B. Ely and R. C. Johnson, *Genetics* **87,** 391 (1977).

Transduction

Transduction experiments are performed by mixing 0.1 ml of a fresh overnight culture of the recipient strain with 0.1 ml of a 10^{-2} or 10^{-3} dilution of the donor lysate. The mixing is done on the surface of a petri dish containing the appropriate selective medium. Colonies appear after incubation for 2 to 5 days at 33°. In the case where tetracycline resistance is used as a selective marker, a 10 ml overlay of melted PYE agar is poured on top of a PYE plate containing 1 mg/ml tetracycline, and the plate is used as soon as the overlay hardens. The overlay allows time for expression of tetracycline resistance before sufficient tetracycline comes in contact with the recipient cells to cause cell death. When transduction experiments are performed on minimal medium, ϕCR30 growth is inhibited and phage-resistant transductants are rarely observed. When transduction experiments are performed on a PYE medium, however, phage replication occurs, and a significant proportion of the transductant colonies will have mutated to phage resistance. In either case, 10^2 to 10^4 transductants are obtained per plate depending on the titer of the original donor phage lysate and the recipient marker being transduced.

Conjugation

Conjugation involves the transfer of DNA from one cell to another through direct cell contact. In the case of *C. crescentus,* this transfer can be mediated by the conjugative plasmid RP4.[9] RP4 is a broad host range plasmid which can be transferred among many gram-negative bacteria and will mediate intraspecies chromosomal transfer. Donor strains containing the RP4 plasmid produce pili which establish contact with the recipient strain and provide a conduit for the actual transfer of DNA from the donor to the recipient.

Plasmid Transfer

Plasmid transfer is an efficient process which can be accomplished by simply allowing contact between the donor and recipient strains. The simplest method is the cross-streak method.[10] A selective medium is chosen which will inhibit growth of both the donor and the recipient strains but will allow growth of those recipient cells which obtain the plasmid from the donor. Overnight cultures of both strains are prepared, and using a 0.1-ml pipette, approximately 20 μl of the donor culture is streaked

[9] B. Ely, *Genetics* **91**, 371 (1979).
[10] B. Ely, *Genetics* **78**, 593 (1974).

across the surface of a petri dish containing a selective medium. The culture is drawn into the pipette by capillary action, and the pipette is held like a pencil to draw a line of culture fluid across the plate. After the fluid in the donor streak has soaked into the medium, streaks of one or more recipient strains are made in a similar manner, perpendicular to the donor streak. As the pipette containing the recipient culture crosses the donor streak, donor cells are dragged along and mixed with the recipient cells. After incubation for 2 days, growth is observed throughout the half of the recipient streak where mixing of the two strains occurred. The other half of the recipient streak and all but the intersection of the donor streak serve as controls to monitor for spontaneous mutation or contamination. Plasmid-containing recipient strains are purified by streaking for single colonies on selective medium.

Chromosome Transfer

Chromosome transfer is less efficient than plasmid transfer and requires a longer time for the actual transfer process to occur. To maximize the efficiency of transfer, the donor and recipient strains are held together on a solid support while the transfer occurs. The filter mating procedure[9] is accomplished by mixing 0.1 ml of an overnight culture of the donor with 0.5 ml of an overnight culture of the recipient, then filtering the mixture through a sterile 2.5-cm-diameter 0.45-μm filter. After the excess liquid is removed, the filter is placed on a fresh PYE plate and incubated at 33° for 2 to 3 hr. The filter is then placed in a tube containing 1 ml of PYE broth and agitated vigorously to resuspend the cells. Aliquots (0.2 ml) of the resuspended cells are spread on the surface of petri dishes containing a medium which is selective for both the donor and the recipient cells. Generally 50 to 200 colonies appear after 3 to 5 days at 33°. Transfer of up to one-sixth of the *C. crescentus* chromosome can be detected with this procedure.

Transposon Mutagenesis

Transposition of both Tn5 and Tn7 occurs in *C. crescentus*.[11] However, Tn7 inserts at a single site on the chromosome and, therefore, is of limited usefulness for genetic studies.[12] On the other hand, Tn5 and its derivatives appear to transpose randomly and provide a means for obtaining a selectable marker in or near any *C. crescentus* gene.

[11] B. Ely and R. H. Croft, *J. Bacteriol.* **149**, 620 (1982).
[12] B. Ely, *J. Bacteriol.* **151**, 1056 (1982).

Isolation of New Transposon Insertions

Transposon mutagenesis occurs after the transposon is introduced into target strain on a vector which is unstable in *C. crescentus*.[11] P-type plasmids containing bacteriophage Mu[13] or based on a ColE1 replicon[14] provide excellent vehicles for the transient insertion of a transposon. Overnight cultures of the recipient strain and an *E. coli* strain containing the donor plasmid are prepared, and the filter mating procedure is carried out as described for conjugation experiments. Approximately 30 to 100 antibiotic-resistant colonies are obtained per plate, and each represents an independent transposition event. If a plasmid containing the Tn*5* derivative Tn*5*-VB32 is used,[15] promoter fusions can be obtained. Tn*5*-VB32 contains a promoterless neomycin phosphotransferase II (NPTase II) gene near one end which can be expressed from an adjacent promoter.[16] If Tn*5*-VB32 is inserted into a *C. crescentus* gene in the proper orientation, the NPTase II gene is expressed from the promoter of that gene and serves as a reporter to monitor the expression of the mutated gene.[17]

Identification of Transposon Insertions Linked to Specific Markers

To obtain transposon insertions linked to a specific marker, the random mutagenesis procedure is used to generate approximately 500 colonies containing independent insertions. The resulting colonies are mixed together and suspended in PYE broth. An aliquot of the resulting suspension is then used as the host for the preparation of a ϕCR30 lysate as described above. Transduction experiments can be performed using the resulting lysate with selection for the antibiotic resistance encoded by the transposon. Subsequent screening for the loss of a recipient marker results in the identification of transposon insertions which are linked to the marker of interest. Phage lysates can then be prepared from individual transposon mutants, and transduction experiments can be performed to determine the proximity (linkage) of the transposon to the gene interest.

[13] J. E. Behringer, J. L. Brynon, A. V. Buchanon-Wallaston, and A. W. B. Johnston, *Nature* (*London*) **276**, 663 (1978).

[14] B. Ely, *Mol. Gen. Genet.* **200**, 302 (1985).

[15] V. Bellafatto, L. Shapiro, and D. A. Hodgson, *Proc. Natl. Acad. Sci. U.S.A.* **81**, 1035 (1984).

[16] R. Champer, R. Bryan, S. L. Gomes, M. Purucker, and L. Shapiro, *Cold Spring Harbor Symp. Quant. Biol.* **50**, 831 (1985).

[17] R. Champer, A. Dingwall, and L. Shapiro, *J. Mol. Biol.* **194**, 71 (1987).

Exchange of Transposon-Encoded Antibiotics Resistance

Sometimes it is useful to change the antibiotic resistance encoded by a specific transposon insertion.[18] The kanamycin resistance encoded by Tn5 can be exchanged for tetracycline resistance using pBEE132 to introduce the Tn5-132 into the Tn5-containing strain. Transfer of pBEE132 is accomplished by the filter mating procedure described above. Colonies which are tetracycline resistant and kanamycin sensitive have the tetracycline resistance of Tn5-132 substituted for the kanamycin resistance of the inserted Tn5 element.

Electroporation

Electroporation involves the use of a brief electric shock to depolarize the cell membrane and allow the uptake of DNA. Electroporation works well with *C. crescentus* for the uptake of plasmid DNA using procedures developed for *E. coli*.[19] However, in contrast to the case for *E. coli*, experiments with cut and religated plasmid DNA have not been successful. To prepare cells for electroporation, a 500-ml culture is grown at 33° to mid-log phase in PYE. The culture is divided into two centrifuge bottles and spun at 6000 g for 5 min. The pellets are resuspended in 100 ml distilled water, spun at 6000 g for 5 min, and then resuspended in 100 ml of distilled water a second time and spun as before. The washed pellets are resuspended in 10 ml of 10% glycerol, combined and transferred to a 50-ml centrifuge tube, spun at 3000 g for 5 min, and resuspended in 2 ml of 10% glycerol. The concentrated cells are divided into 200-μl aliquots and frozen at −70°.

For electroporation, a tube is thawed, and 40 μl of cells is placed in a microfuge tube along with 0.1 to 1 μg of DNA in a small volume of distilled water (or very dilute buffer). After 1 min on ice, the mixture is transferred to an ice-cold Bio-Rad (Richmond, CA) Gene Pulser cuvette with a 0.2-cm electrode gap and subjected to a 2.5-kV shock with the capacitor set at 25 μF (microfarad) and the resistance on the pulse controller set at 200 ohms. Under these conditions, the time constant is 5 msec. After this treatment, the contents of the cuvette are transferred to a culture tube containing 1 ml of PYE, and the cells are incubated at 33° with aeration for 1 hr. After the recovery period, the cells are plated on the appropriate selective medium. We routinely obtain 10^4 to 10^5 transformants per microgram of DNA.

---

[18] B. Ely and T. Ely, *Genetics* **123**, 649 (1989).
[19] A. K. Ellgaard, D. A. Mullin, and S. A. Minnich, *Abstr. Annu. Meeting Am. Soc. Microbiol.*, p. 208 (1989).

Pulsed-Field Gel Electrophoresis

Pulsed-field gel electrophoresis (PFGE) allows one to separate very large DNA fragments. A genomic map of the *C. crescentus* chromosome has been generated by aligning *Dra*I, *Ase*I, and *Spe*I restriction maps with the genetic map.[18] Therefore, the map location of transposon insertions can be determined by PFGE. This procedure is approximately 100 times faster than using a combination of conjugation and transduction to determine map locations.[20] A current version of the genomic map is shown in Fig. 1.

Preparation of DNA Plugs

DNA plugs are prepared by a modification of the method of Smith and Cantor.[21] A 2.5-ml aliquot of an overnight culture of a *C. crescentus* strain is spun at 4000 *g* in a refrigerated centrifuge, and the resulting pellet is resuspended in 2 ml of 10 m*M* Tris, pH 7.6, containing 1 *M* NaCl. The mixture is spun a second time, resuspended in 1 ml of the same buffer, and heated to 37°. One-half of the suspension is then mixed with an equal volume of melted 1% (w/v) agarose (PFG grade) previously equilibrated to 65°. The mixture is immediately transferred to plug molds and allow to cool. After cooling, each set of plugs is removed from the mold and transferred to a disposable plastic test tube (15 × 100 mm) which contains 1.8 ml of 1% (w/v) sarkosyl in 0.5 *M* EDTA (pH 9–9.4) and 0.2 ml of 10 mg/ml proteinase K. After overnight incubation at 50°, the liquid is removed by aspiration and replaced with 2 ml of 10 m*M* Tris, pH 7.5, 1 m*M* EDTA and 30 μl of a phenylmethylsulfonyl fluoride (PMSF) stock (17.4 mg/ml in 100% ethanol). After 1 hr on a rotator at room temperature, the liquid is removed by aspiration and replaced by an identical mixture. After a second hour, the procedure is repeated 3 times without the PMSF. After the final wash, the tube containing the plugs is refrigerated until needed. We have found that DNA-containing plugs can be stored in the refrigerator for several years.

Pulsed-Field Gel Electrophoresis Techniques

PFGE experiments are carried out by digesting DNA-containing agarose plugs with restriction enzymes which cut the genome infrequently. The resulting DNA fragments are then resolved by insertion of the plug into an agarose gel and subjecting the DNA to PFGE. Restriction digests are carried out by placing a small segment of a DNA-containing agarose

---

[20] A. Dingwall, L. Shapiro, and B. Ely, *Methods* **1**, 160 (1990).
[21] C. L. Smith and C. R. Cantor, *Cold Spring Harbor Symp. Quant. Biol.* **51**, 115 (1986).

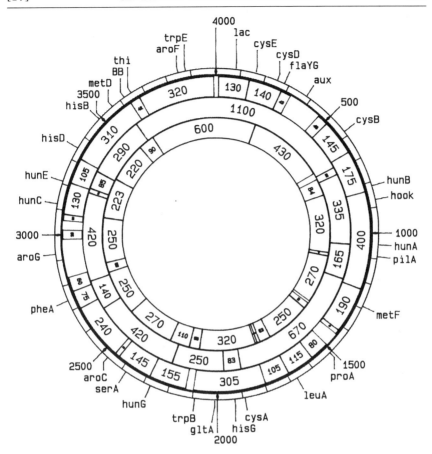

Fig. 1. *Caulobacter crescentus* genomic map. Inner Circles show fragments generated by *Dra*I, *Ase*I, and *Spe*I restriction digests, respectively, with *Spe*I fragments represented on the innermost circle. Fragment sizes are indicated in kilobases.

plug in a microfuge tube along with approximately 10 units of restriction enzyme, 10 μl of 10× restriction buffer, and 50 μl distilled water. After incubation at 37° for 1 hr, the plug fragments are removed with a spatula and inserted directly into the gel wells for PFGE. Table I shows conditions which give good resolution of various size ranges of DNA fragments. After electrophoresis, the gel is stained with ethidium bromide and photographed. DNA fragments containing a Tn5 insertion can usually be identified by the altered migration of the corresponding band on the gel. Differences as small as 2% of the size of the fragment can be resolved under optimal conditions. Alternatively, the DNA can be blotted to nylon membranes and bands identified by hybridization.

TABLE I
CONDITIONS FOR SEPARATING VARIOUS SIZE RANGES OF
DNA FRAGMENTS BY PULSED-FIELD GEL
ELECTROPHORESIS

Fragment size range (kb)	Pulse time (sec)	Voltage (V)	Time (hr)
5–50	1	200	10
20–100	3	200	10
50–250	10	200	14
100–400	20	200	14
200–1000	45	200	16

Complementation

Complementation experiments involve the transfer of cloned DNA into a mutant strain to determine if the transferred DNA can correct the mutant phenotype. Positive results can be obtained as a result of complementation, recombination, or reversion. False positives arising from reversion are detected by the use of appropriate controls, and recombination can be prevented by the use of a recipient strain which is rec⁻.[22]

Construction of rec⁻ Strains

Strains defective in recombination are constructed by use of strain PC7070 for cotransduction of the *recA526* allele with a linked Tn5 insertion.[23] Phage grown on PC7070 is used to transduce a kanamycin-sensitive strain containing the mutation to be complemented. The presence of the *recA526* allele is demonstrated by an increased sensitivity to irradiation with ultraviolet light. Strains containing the mutation are sensitive to a 30-sec exposure to a germicidal lamp at a distance of 45 cm, whereas those containing the wild-type allele are resistant to this exposure. In practice, a master plate of transductant colonies is replicated onto two plates. One of the replicas is exposed to the ultraviolet light, and both are incubated overnight. Colonies which grow only on the unirradiated plate contain the *rec* allele. Alternatively, the *recA* allele can be transferred by cotransduction with the *cysB* gene (2% linkage) in order to avoid the use of a Tn5 insertion (A. Newton, personal communication, 1978).

[22] E. A. O'Neill, R. H. Hynes, and R. A. Bender, *Mol. Gen. Genet.* **198,** 275 (1985).
[23] N. Ohta, M. Masurekar, D. A. Mullin, and A. Newton, *J. Bacteriol.* **172,** 7027 (1990).

Complementation of rec⁻ Strains

Complementation experiments are performed using the cross-streak method described earlier. Complementation of a nutritional mutation can be observed directly by performing the transfer on defined medium. Complementation of motility mutations or other mutations where direct selection is not possible is done by using antibiotic resistance to select for transfer of the plasmid containing the cloned DNA. The plasmid-containing strain is then purified and checked for the loss of the mutant phenotype. Positive complementation results can be verified by growing the strain under nonselective conditions and screening for loss of the plasmid. The resulting strains are then tested for the reappearance of the mutant phenotype. Vectors based on RSF1010 are lost very quickly in *C. crescentus,* so few colonies will contain the plasmid after plating an aliquot of a culture grown overnight on PYE broth. In contrast, when the same procedure is used with vectors based on pRK290, approximately 1% of the resulting colonies have lost the plasmid, so a greater number of colonies need to be screened to obtain the desired segregants.

Inactivation of Chromosomal Genes

Chromosomal genes can be inactivated, or specific mutations can be introduced into the chromosome, through the use of cloned genes. The desired mutation is introduced into a cloned copy of the gene, and the plasmid containing the mutated gene is transferred to the appropriate *C. crescentus* strain. Recombination between the chromosomal and the plasmid copies of the gene usually results in the integration of the plasmid into the chromosome. Thus, the resulting strain will have both the wild-type and the mutated forms of the gene. Subsequent loss of the plasmid antibiotic resistance owing to a second recombination event results in the loss of one copy of the gene. Depending on where recombination occurs these segregants will have either the wild-type or the mutated version of the gene.

Vectors for gene replacement should be incapable of replication in *C. crescentus.* Two types of vectors are appropriate. The first are vectors which contain the sequences necessary for mobilization by the IncP transfer system.[24] These plasmids replicate well in *E. coli* and can be transferred from *E. coli* to *C. crescentus* at a high frequency[25] (also B. Ely, unpublished observations, 1987). However, they cannot be replicated in *C. crescentus,* so they are lost unless recombination occurs to form a cointegrate. Trans-

[24] R. Simon, U. Pfiefer, and A. Puhler, *Bio Technology* **1,** 784 (1983).
[25] S. A. Minnich, N. Ohta, N. Taylor, and A. Newton, *J. Bacteriol.* **170,** 3953 (1988).

duction can then be used to obtain true gene replacements where the resulting strain contains a single copy of the mutated gene.[25] A transducing lysate is prepared on a pool of cointegrate colonies and used to transduce the appropriate *C. crescentus* strain with selection for a marker either within the gene of interest or closely linked to it. The resulting colonies are then screened for the presence of the mutated gene and the absence of the plasmid antibiotic resistance. This procedure can also be used to transfer genes to a new chromosomal location using homology between Tn*5* insertions for recombination (N. Ohta, L. S. Chen, D. A. Mullen, and A. Newton, personal communication). In this case, a plasmid containing a Tn*5* element will integrate at the site of any chromosomal Tn*5* insertion.

A second procedure for the isolation of gene replacements is the use of electroporation to introduce the plasmid containing the mutated gene into the appropriate *C. crescentus* strain (J. Malakooti and B. Ely, unpublished observations, 1990). In this case, any vector can be used as long as it does not replicate in *C. crescentus*. After the plasmid containing the mutated gene is constructed in *E. coli,* plasmid DNA is isolated and used to transform the *C. crescentus* strain with selection for the plasmid antibiotic resistance. If ampicillin is used as the selective marker, a derivative of the ampicillin-sensitive strain SC1107 must be used as the host. The resulting colonies will be cointegrates, and gene replacements can be isolated as described above. Cointegrates are obtained at a frequency of 10^3 per microgram of plasmid DNA. This procedure also can be used to introduce a new restriction site into the genome at a particular location. Homologous recombination between a cloned fragment in the appropriate vector and the chromosome would result in the integration of the plasmid. If the plasmid contains one of the restriction sites used for PFGE mapping, the position of the cloned DNA could be readily determined as described above.

[18] Genetic Analysis of *Agrobacterium*

By Gerard A. Cangelosi, Elaine A. Best, Gladys Martinetti, and Eugene W. Nester

Introduction

Genetic infection by *Agrobacterium* species is the only verified example of natural genetic exchange between the prokaryotic and eukaryotic kingdoms. Infection of plants by *Agrobacterium* strains containing tumor-inducing (Ti) or root-inducing (Ri) plasmids results in transfer of a segment of plasmid DNA, termed the T-DNA, from bacterium to the plant cell, where it becomes stably integrated into the plant nuclear DNA. Genes on the Ti and Ri plasmids code for most of the products required for the infection process, but chromosomal genes also participate in early steps of infection. Ti and Ri plasmids carry over 20 virulence (*vir*) genes, which are expressed in response to plant signals. The *vir* gene products catalyze the transfer of the T-DNA from bacterium to plant cell. T-DNA genes code for the synthesis of phytohormones, which cause the neoplastic growth known as crown gall (*Agrobacterium tumefaciens*) or hairy root (*Agrobacterium rhizogenes*). T-DNA genes also code for the synthesis of unique amino acids called opines, which are metabolized by the tumor-inducing bacteria. The genetics, enzymology, and cell biology of this process have been reviewed recently.[1-4]

Most of the methods used for genetic analysis of crown gall infection are typical of those used for analyzing gram-negative bacteria in general, which are described in detail elsewhere in this volume. The procedures for mutagenesis and gene transfer presented in this chapter are specifically useful for genetic analysis of *Agrobacterium*. We do not discuss the use of *Agrobacterium* as a vector for genetic engineering of higher plants (workers interested in these techniques are referred to recent reviews[5,6]).

The genus *Agrobacterium* includes the species *A. tumefaciens*, which carries a Ti plasmid and causes crown gall tumors, *A. rhizogenes*, which carries an Ri plasmid and causes hairy root, and *Agrobacterium radiobacter*, which is avirulent. *Agrobacterium* strains have been divided into three biotypes based on chromosomally encoded characteristics.[7] *Agrobacterium* strains are also commonly referred to by the type of opines produced by infected plant tissue. This phenotype is determined by the Ti plasmid. Since Ti plasmids are transferable between *Agrobacterium* strains, opine type (as well as species names which depend on the Ti plasmid) may not be an evolutionarily significant classification. Nonetheless, it has been a useful classification in crown gall research. Certain

[1] A. N. Binns and M. T. Thomashow, *Annu. Rev. Microbiol.* **42**, 575 (1988).
[2] Z. Koukilokova-Nicola, L. Albright, and B. Hohn, *in* "Plant Gene Research, Volume IV: DNA Infectious Agents" (T. Hohn and J. Schell, eds.), p. 108. Springer-Verlag, New York, 1987.
[3] S. E. Stachel and P. C. Zambryski, *Cell (Cambridge, Mass.)* **46**, 325 (1986).
[4] P. Zambryski, *Annu. Rev. Genet.* **22**, 1 (1988).
[5] C. Lichenstein and J. Draper, *in* "DNA Cloning, a Practical Approach, Volume 2" (D. M. Glover, ed.), p. 67. IRL Press, Washington, D.C., 1986.
[6] K. Weising, J. Schell, and G. Kahl, *Annu. Rev. Genet.* **22**, 421 (1988).
[7] A. Kerr and C. G. Panagopoulos, *Phytopathology* **90**, 172 (1977).

virulence-related phenotypes, such as host range[8] and the occurrence of the auxilliary *vir* genes *virF*[9] and *virH*,[10] appear to vary between opine classes.

The methods presented here have been developed for analysis of *Agrobacterium tumefaciens* strain A348, a biotype I strain carrying an octopine Ti plasmid.[11] They are probably equally useful for analysis of other *Agrobacterium* strains. *Agrobacterium* is very closely related to the fast-growing *Rhizobium* species, particularly *Rhizobium meliloti*, and few barriers to gene transfer and expression of heterologous genes exist between these genera. Therefore, most genetic techniques that function in one system function equally well in the other. A notable exception to this rule is that bacteriophage transduction systems exist for *Rhizobium* species but not for *Agrobacterium*.

Mutagenesis of *Agrobacterium* Genes Coding for Cell Surface Proteins Using Tn*phoA*

Background

Although *Agrobacterium* is susceptible to a variety of chemical mutagens (one such procedure is described in this section), the many advantages of transposon mutagenesis make it the method of choice for mutational analysis. Because of their low insertional specificity and high stability in most *Agrobacterium* strains, Tn*5* and its derivatives are most useful for shotgun mutagenesis of *Agrobacterium*.

Complex transposons coding for antibiotic resistance and carrying promoterless structural genes for easily assayed enzymes are useful for analysis of gene expression. One such transposon, Tn*3*-HoHo1, was developed for use in genetic analysis of *Agrobacterium*.[12] Tn*3*-HoHo1 is a derivative of Tn*3* (carbenicillin resistance) that carries a promoterless *lacZ* gene near one end of the transposon. Insertion of Tn*3*-HoHo1 into a target gene in the correct orientation causes a transcriptional or translational fusion between *lacZ* and the target gene, placing β-galactosidase activity under the control of the target gene promoter. Tn*3*-HoHo1 has not been

---

[8] L. Otten, G. Piotrowiak, P. Hoykaas, M. Dubois, E. Szegedi, and J. Schell, *Mol. Gen. Genet.* **199**, 191 (1985).

[9] P. J. J. Hoykaas, M. Hofker, H. Den Dulk-Ras, and R. A. Schilperoort, *Plasmid* **11**, 195 (1984).

[10] R. H. Kanemoto, A. T. Powell, D. E. Akiyoshi, D. A. Regier, R. A. Kerstetter, E. W. Nester, M. C. Hawes, and M. P. Gordon, *J. Bacteriol.* **171**, 2506 (1984).

[11] D. J. Garfinkel, R. B. Simpson, L. W. Ream, F. W. White, M. P. Gordon, and E. W. Nester, *Cell (Cambridge, Mass.)* **27**, 143 (1981).

[12] S. E. Stachel, G. An, C. Flores, and E. W. Nester, *EMBO J.* **4**, 895 (1985).

used for shotgun mutagenesis in *Agrobacterium*, but it is useful for mutagenesis of cloned *Agrobacterium* genes in *Escherichia coli*. After mutagenesis, the fusions can be easily reintroduced into *Agrobacterium*. Wild-type *A. tumefaciens* does not express endogenous β-galactosidase activity, which makes it easy to assess β-galactosidase activity expressed from Tn*3*-HoHo1 fusions.

Two hybrid derivatives of Tn*5* (kanamycin resistance) have recently been developed which are useful not only for genetic analysis, but also for analyzing the subcellular localization of bacterial proteins. One, Tn*lacZ*, causes translational fusions between target gene products and β-galactosidase, when inserted in the correct reading frame. The other, Tn*phoA*, causes translational fusions between target gene products and alkaline phosphatase. A fusion protein carrying β-galactosidase will have β-galactosidase activity only if the fusion protein (or the portion of a transmembrane fusion protein containing β-galactosidase) is in the bacterial cytoplasm. Conversely, an alkaline phosphatase fusion will have activity only when the fusion protein (or the portion containing alkaline phosphatase) is exposed to the periplasm or extracellular medium. The construction and use of these elements have been described elsewhere.[13,14] They have been used most extensively to study the transmembrane structure of cell surface proteins.[15] Such applications will be greatly facilitated by recently published methods for interchanging Tn*phoA* and Tn*lacZ* fusions at a single insertion site.[14]

Tn*phoA* has been used for shotgun mutagenesis of genes coding for cell surface proteins involved in the interactions of *Rhizobium meliloti* with its plant host.[16] Colonies expressing active *phoA* fusions are easily detected on indicator plates containing 5-bromo-4-chloro-3-indolyl phosphate (XP). Wild-type *A. tumefaciens* cells express endogenous phosphatase activity under a variety of conditions, resulting in a positive reaction on XP plates. This has limited the usefulness of Tn*phoA* for mutagenesis and analysis of transmembrane protein topology in *A. tumefaciens*. We describe here the isolation of Pho⁻ derivatives of wild-type *A. tumefaciens* by chemical mutagenesis and their use in Tn*phoA* mutagenesis.

Isolation of Phosphatase-Negative Agrobacterium tumefaciens
  Parent Strains

We used the following procedure to isolate derivatives of *A. tumefaciens* which are Pho⁻ on three different media. The procedure is presented

[13] C. Manoil and J. Beckwith, *Proc. Natl. Acad. Sci. U.S.A.* **82**, 8129 (1985).
[14] C. Manoil, *J. Bacteriol.* **172**, 1035 (1990).
[15] C. Manoil, J. J. Mekalanos, and J. Beckwith, *J. Bacteriol.* **172**, 515 (1990).
[16] S. Long, S. McCune, and G. W. Walker, *J. Bacteriol.* **170**, 4257 (1988).

TABLE I
BACTERIAL GROWTH MEDIA AND STOCK SOLUTIONS

Medium or solution	Preparation[a]	Ref.
LB (Luria broth)	10 g Tryptone (Difco Laboratories, Detroit, MI), 5 g yeast extract (Difco), 10 g NaCl, pH 7.5. Sterilize by autoclaving	b
MG/L	500 ml LB, 10 g mannitol, 2.32 g sodium glutamate, 0.5 g KH_2PO_4, 0.2 g NaCl, 0.2 g $MgSO_4 \cdot 7H_2O$, 2 μg biotin, pH 7.0. Sterilize by autoclaving	c
AB	50 ml of 20× AB Salts, 50 ml of 20× AB Buffer, 900 ml of 0.5% glucose (each solution autoclaved separately)	c
Induction broth	50 ml of 20× AB Salts, 1 ml of 20× AB buffer, 8 ml of 0.5 M MES [2-(N-morpholino)ethanesulfonic acid; Research Organics, Inc., Cleveland, OH] pH 5.5, 60 ml of 30% glucose (each solution autoclaved seperately)	d
20× AB Salts	20 g NH_4Cl, 6 g $MgSO_4 \cdot 7H_2O$, 3 g KCl, 0.2 g $CaCl_2$, 50 mg $FeSO_4 \cdot 7H_2O$	c
20× AB Buffer	60 g K_2HPO_4, 23 g NaH_2PO_4 (pH 7.0)	c
2000× XP	80 mg/ml XP (Sigma Chemical Co., St. Louis, MO) in dimethylformamide	—
1000× Acetosyringone	14.6 mg/ml (100 mM) acetosyringone (Aldrich Chemical Co., Milwaukee, WI) in dimethyl sulfoxide	—

[a] Preparation per liter unless noted otherwise. For solid media, add 15 g of agar (Difco) per liter prior to sterilization.
[b] J. H. Miller, "Experiments in Molecular Genetics." Cold Spring Harbor Laboratory, Cold Spring Harbor, New York, 1972.
[c] M. D. Chilton, T. Currier, S. Farrand, A. Bendich, M. P. Gordon, and E. W. Nester, *Proc. Natl. Acad. Sci. U.S.A.* **71,** 3672 (1974).
[d] S. C. Winans, R. A. Kerstetter, and E. W. Nester, *J. Bacteriol.* **170,** 4047 (1988).

in detail for workers interested in isolating additional derivatives.

1. Grow *A. tumefaciens* strain A348 overnight (28° with aeration) in liquid AB minimal medium (Table I), then dilute 10× into 20 ml of fresh AB to an OD_{600} of approximately 0.25. Continue incubation until the OD_{600} reaches approximately 0.35 (2–3 hr).

2. Transfer 4-ml aliquots of the culture to screw-cap tubes. Add ethyl methanesulfonate (EMS) (Sigma Chemical Co., St. Louis, MO) to a final concentration of 4% (v/v) (EMS is a strong mutagen and should be handled and disposed of accordingly). Add an equal volume of water to a control tube to monitor survival. Continue incubation for another 90 min.

3. Wash cells twice with TE buffer (10 mM Tris, 1 mM EDTA, pH

TABLE II
PHENOTYPES OF A348 AND DERIVATIVES ON XP PLATES

		Color reaction[a]			Antibiotic resistance[c]
Strain	Parent	AB	MGL	IB[b]	
A348	—	b	b	b	N
A6001	A348	b	w	b	N
A6003	A348	b	w	b	N
A6004	A348	b	w	b	N
A6005	A348	lb	lb	b	N
A6006	A348	w	b	lb	N
A6007	A6001	b	w	b	N, S
A6009	A6003	b	w	b	N, S
A6010	A6009	w	b	w	N, S

[a] With 40 μg/ml XP: b, blue; lb, light blue; w, white.
[b] IB, Induction broth.
[c] N, Naladixic acid (25 μg/ml); S, streptomycin (500 μg/ml).

7.0), and resuspend in 4 ml AB. Continue incubation for another 2 hr.

4. Plate dilutions of 10^{-5}–10^{-6} (EMS-treated tubes) or 10^{-7}–10^{-8} (water-treated tubes) on AB agar plates and on MG/L plates (Table I) containing 40 μg/ml XP (diluted 1 : 2000 from stock solution; Table I).

5. Count colonies after 2–3 days of incubation at 28° and score color change on XP after 4–5 days. Pho$^+$ colonies are blue on XP, and Pho$^-$ colonies are white to light blue.

In our experiments, survival [number of colony-forming units (cfu) in EMS-treated tubes/number of cfu in water-treated controls) ranged from 10^{-2} to 10^{-1}. Lower concentrations of EMS resulted in higher survival rates but did not yield Pho$^-$ colonies. White or light blue colonies on either AB or MG/L plates were picked and restreaked to score for stable Pho$^-$ phenotypes on the same two media and on induction medium in the presence of XP. Induction medium is a low pH, low phosphate medium favorable for expression of the virulence genes in *Agrobacterium* (Table I). The color reactions of the mutants on these media are summarized in Table II. No single strain was obtained which was Pho$^-$ on all of these media, which suggests that A348 contains several different phosphatases which are differentially regulated. All of these strains were virulent on kalanchoe leaves, assuring that virulence-related genes were not affected by the mutagenic treatment.

To facilitate subsequent genetic manipulations, spontaneous streptomycin-resistant derivatives of some Pho⁻ strains were isolated by plating approximately 10^9 cells on media containing 500 μg/ml streptomycin (Table II). Strain A6010 was a spontaneously occurring derivative of strain A6009 (S. Machlin, unpublished results, 1990).

Isolation of Mutants with TnphoA Insertions in Constitutively Expressed Virulence Genes

Some of the genes involved in early steps of crown gall tumorigenesis, such as attachment and signal recognition, are expressed in bacteria growing in the absence of plant signal molecules[17,18] It is likely that many of the genes involved in such early steps have yet to be identified. The procedure described below was aimed at isolating mutants affected in such constitutively expressed genes (in this case, genes which are expressed during growth of the bacteria on rich medium). Strain A6007 (Table II) was mutagenized with TnphoA using the procedure developed by Long et al.[16] for TnphoA mutagenesis of Rhizobium meliloti. TnphoA is introduced into A6007 on the self-mobilizable suicide vector pRK609,[16] a derivative of RK2 which does not replicate in A. tumefaciens.

1. Grow an auxotrophic E. coli strain such as MM294A (pro-82 thi-1 endA1 hsdR17 supE44)[16] containing pRK609 overnight at 37° in liquid LB (Table I) with kanamycin (50 μg/ml), and grow A. tumefaciens strain A6007 overnight at 28° in liquid MG/L, both with aeration.
2. In the morning, dilute both strains 10-fold into 5 ml of their respective media, and continue incubation for another 4 hr.
3. Pellet 5 ml of each culture by low-speed centrifugation (10 min at 10,000 rpm on a Sorvall SS-34 rotor, or equivalent), then resuspend in 0.5 ml MG/L.
4. Mix 0.1 ml of each suspension in a droplet on the surface of an MG/L plate. To assure the isolation of independent mutants, several matings should be carried out on separate plates. Incubate the plates for 16–24 hr at 28°.
5. Wash the mating mixture off each plate with 0.5 ml MG/L and spread 0.1-ml aliquots of this wash on MG/L plates containing 500 μg/ml streptomycin, 30 μg/ml naladixic acid, 100 μg/ml kanamycin,

---

[17] C. J. Douglas, R. J. Staneloni, R. A. Rubin, and E. W. Nester, *J. Bacteriol.* **161,** 850 (1985).

[18] M.-L. W. Huang, G. A. Cangelosi, W. Halperin, and E. W. Nester, *J. Bacteriol.* **172,** 1814 (1990).

and 40 μg/ml XP. These conditions select for clones of A6007 carrying genomic Tn*phoA* insertions.

6. This procedure normally results in 100–300 colonies per plate, of which 1–4% are blue. Blue colonies are restreaked for purification and analyzed further.

After mutagenesis by this procedure in our laboratory, approximately 2000 clones expressing *phoA* activity on MG/L plates were screened for virulence on kalanchoe leaves as previously described.[19] Over 30 mutants which were avirulent or weakly virulent were isolated. Southern analysis revealed that only three of these had insertions in known Ti plasmid genes (*virB*), and one had an insertion in a known chromosomal virulence gene, *chvB*. The remaining mutants appeared to be affected in as yet unidentified virulence genes, and these mutations are under investigation.

Isolation of Mutants with TnphoA Insertions in Inducible Virulence Genes

Expression *ex planta* of most of the known virulence genes in *Agrobacterium* requires specific signal molecules from plant hosts. Acetosyringone (3′,5′-dimethoxy-4′-hydroxyacetophenone, a phenolic plant metabolite available commercially from Aldrich Chemical Co., Milwaukee, WI), is commonly used to induce the *vir* genes in *A. tumefaciens*. To screen for Tn*phoA* insertions in genes which are expressed only in the presence of acetosyringone, we mutagenized strain A6006 with Tn*phoA* as described above. Unlike strain A6007, strain A6006 forms light blue colonies on induction media. In Step 5, we plated the Tn*phoA*-mutagenized bacteria on induction medium containing 100 μM acetosyringone (diluted 1 : 1000 from stock solution; Table I), 100 μg/ml kanamycin, 200 μg/ml streptomycin, and 40 μg/ml XP. Dark blue colonies were picked to the same medium with and without acetosyringone. Colonies which were blue on acetosyringone and white in its absence were analyzed further. Seventeen such clones were screened for virulence on kalanchoe leaves, and eight were avirulent or weakly virulent. Southern analysis showed that seven of these have insertions in the *virB8, virB9,* and *virB10* genes, suggesting that the products of these genes have extracytoplasmic domains (P. Dion, P. Holland, G. Cangelosi, and E. Nester, unpublished results, 1990). The biology of the other acetosyringone-inducible Tn*phoA* mutations is under investigation.

Plant signals other than acetosyringone and related phenolic compounds may be important inducers of virulence genes. Extracts of plant

---

[19] D. J. Garfinkel and E. W. Nester, *J. Bacteriol.* **144,** 732 (1980).

tissues such as carrot root epidermal shavings [soaked in culture media for several hours, then filter sterilized prior to pouring plates (S. Gelvin, personal communication, 1990)] can be substituted for commercial aceto-syringone in solid media.

Use of TnphoA for Analysis of Transmembrane Protein Topology

TnphoA has been used successfully for the analysis of the transmembrane topology of one of the *vir* proteins, VirA, expressed in a *phoA*⁻ *E. coli* background.[20] The Pho⁻ *Agrobacterium* strains listed in Table II allow such experiments to be conducted in *A. tumefaciens*, circumventing problems associated with expression of *A. tumefaciens* genes in a heterologous host.

Gene Transfer and Marker Exchange by Electroporation

Background

Broad-host range cloning vectors routinely used in our laboratory for genetic analysis of *Agrobacterium* are listed in Table III. All can be mobilized between *E. coli* and *A. tumefaciens* by triparental mating.[21] Table III includes plasmids from three different incompatibility groups, which confers a high degree of versatility in strain construction.

Cloned genes can be introduced into *A. tumefaciens* either by transformation with plasmid DNA or by conjugational transfer of plasmid DNA between *Agrobacterium* strains or between *E. coli* and *Agrobacterium*. *Agrobacterium* can be transformed by a freeze–thaw method which is relatively inefficient, yielding only 10^3 transformants per microgram of DNA under optimal conditions.[22] Introduction of genes into *Agrobacterium* by conjugation is more efficient but requires a means for selecting against the donor strain.

Recently, many gram-negative and gram-positive bacteria have been transformed using high-voltage electrotransformation (electroporation). Using a simple protocol similar to the one developed by Dower *et al.*[23] for electroporation of *E. coli*, we have been able to electroporate plasmid DNA into *A. tumefaciens* at an efficiency of 1×10^6 transformants per

[20] S. C. Winans, R. A. Kerstetter, J. E. Ward, and E. W. Nester, *J. Bacteriol.* **171**, 1616 (1989).

[21] G. Ditta, S. Stanfield, D. Corbin, and D. R. Helinski, *Proc. Natl. Acad. Sci. U.S.A.* **77**, 7347 (1980).

[22] G. An, P. R. Ebert, A. Mitra, and S. Ila, *in* "Plant Molecular Biology Manual" (S. B. Gelvin and R. A. Schilperoort, eds.), p. 1. Kluwer Academic Press, Dordrecht, The Netherlands, 1988.

[23] W. J. Dower, J. F. Miller, and R. Ragsdale, *Nucleic Acids Res.* **16**, 6127 (1988).

TABLE III

BROAD HOST RANGE CLONING VECTORS USEFUL FOR GENETIC ANALYSIS OF *Agrobacterium*

Plasmid	Incompatibility group	Size (kb)	Antibiotic resistance[a] and other features	Ref.
pVK102	IncP	23	RK2 origin, Tc[4], Km[r], *cos*	b
pLAFR1	IncP	21.6	RK2 origin, Tc[r], *cos*	c
pTJS75	IncP	7	RK2 origin, Tc[r]	d
pSP329	IncP	7.5	Derivative of pTJS75 with α-complementation group and multicloning site from pUC18[e]	S. Porter (unpublished data)
pJRD215	IncQ	10.2	RFS1010 origin, Km[r], Sp[r], *cos*, 23 unique cloning sites	f
pUCD2	IncW	13	pSa and ColE1 origins (high copy number in *E. coli*), Km[r], Sp[r], Cb[r], Tc[r]	g
pUCD5	IncW	13	Derivative of pUCD2, Km[r], Cb[r], Tc[r], *cos*	g

[a] Km, Kanamycin (100 μg/ml); Tc, tetracycline (6 μg/ml); Cb, carbenicillin (100 μg/ml), Sp, spectinomycin (50 μg/ml), *cos*, phage λ cohesive ends.
[b] V. C. Knauf and E. W. Nester, *Plasmid* **8**, 45 (1982).
[c] A. M. Friedman, S. R. Long, S. E. Brown, W. J. Buikema, and F. M. Ausubel, *Gene* **18**, 289 (1982).
[d] T. J. Schmidhauser and D. R. Helinski, *J. Bacteriol.* **164**, 446 (1985).
[e] C. Yanisch-Perron, J. Vieria, and J. Messing, *Gene* **33**, 103 (1985).
[f] J. Davison, M. Heusterspeute, N. Chevalier, V. Ha-Thi, and F. Brunel, *Gene* **51**, 275 (1987).
[g] T. J. Close, D. Zaitlin, and C. I. Kado, *Plasmid* **12**, 111 (1984).

microgram of DNA. The method requires only small amounts of DNA, and the cells, once rendered electrocompetent, can be stored for at least 6 months at −70° without loss of transformation competence. Another laboratory recently reported high-efficiency electroporation of both *A. tumefaciens* LBA4404(pRAL4404) and *A. rhizogenes* LBA9402(pRi1855) using a very similar method.[24]

Preparation of Electrocompetent Cells

The procedure described here yields enough cells for eight transformation experiments. It can be scaled as desired.

1. Grow cells overnight in 5 ml of MG/L broth (Table I) at 28° with aeration.
2. Dilute the overnight culture 1 : 100 in 100 ml of MG/L in a 1-liter flask and continue incubation until the cells reach an optical density of 0.4–0.6 A_{600} units (approximately 3 generations).

[24] W. J. Shen and B. G. Forde, *Nucleic Acids. Res.* **17**, 8385 (1989).

3. Divide the culture into four 50-ml Oak Ridge tubes and harvest by centrifugation at 10,000 rpm (Sorvall SS-34 rotor) for 10 min at 4°.
4. Resuspend each of the four cell pellets in 10 ml of ice-cold buffer I (1 mM HEPES, pH 7.0). Recentrifuge as above.
5. Resuspend each of the four cell pellets in 10 ml of ice-cold buffer II [1 mM HEPES, pH 7.0, 10% (v/v) glycerol]. Recentrifuge as above.
6. Resuspend each of the four cell pellets in 0.5 ml of ice-cold buffer II. Combine all of the cells into two 1.5-ml Eppendorf tubes. Centrifuge for 5 min at 4°.
7. Resuspend each cell pellet in 200 μl of ice-cold buffer II to a final cell concentration of approximately 1×10^{11} cfu/ml.
8. Dispense 50-μl samples of the culture into 1.5-ml Eppendorf tubes. The cells may be used immediately or frozen in dry ice/ethanol and stored at $-70°$

Transformation of Electrocompetent Agrobacterium tumefaciens Cells with Plasmid DNA by Electroporation

The following procedure employs a Bio-Rad (Richmond, CA) Gene Pulser electroporation apparatus. Other equipment can probably be substituted with success, provided it can deliver a field strength of at least 12.5 kV/cm.

1. Thaw electrocompetent cells on ice.
2. Incubate 50 μl of cells with 1 μl of plasmid DNA for approximately 2 min on ice. The plasmid DNA may be in either distilled water or a low-salt buffer such as TE (10 mM Tris, 1 mM EDTA, pH 7.0). High-salt buffers should be avoided.
3. Transfer the cell–DNA mixture to a chilled 0.2-cm Bio-Rad electroporation cuvette. Tap the mixture to the bottom of the cuvette. (Note: These cuvettes are disposable, but they can also be reused by rinsing with ethanol and sterile distilled water between uses. There is, however, some risk of contamination or decreased transformation efficiency when cuvettes are reused.)
4. Transfer the cuvette to a chilled Bio-Rad Gene Pulser slide.
5. Set the Bio-Rad Gene Pulser apparatus to the 25-μF capacitor; set the Pulse Controller Unit to 400 ohms. Apply a single 2.5-kV electrical pulse. This should result in a field strength of 12.5 kV/cm with an exponential decay constant of approximately 9 msec.
6. Immediately following the electrical pulse, add 1 ml of MG/L broth (room temperature) to the cuvette and gently resuspend the cells.
7. Transfer the cell suspension to a 150 × 16 mm tube and incubate the culture at 28° with shaking for 2 hr.

8. Plate samples on appropriate selection media. Transformed colonies are visible after 2–3 days of incubation at 28°.

We use this procedure to introduce into *Agrobacterium* plasmid DNA prepared either by CsCl density gradient centrifugation or by mini-alkaline lysis. The efficiency of electroporation with the latter is occasionally lower (about 2.5-fold) than with the former; this may be related to the topological state of the DNA or to the presence of impurities in our plasmid minipreparations. In an experiment to determine the efficiency of transformation, we observed a nearly linear increase in the number of transformants obtained as we increased the amount of plasmid DNA from 1 to 2000 ng, resulting in an efficiency of about 1×10^6 transformants per microgram of DNA throughout this range (data not shown). We have not determined whether the efficiency decreases when plasmids larger than 15 to 20 kilobases (kb) are used.

Use of Electroporation for Introducing and Stably Incorporating Cloned Mutations by Marker Exchange

Analyzing the genetic organization and regulation of gene expression in *A. tumefaciens* frequently involves transposon mutagenesis of cloned *A. tumefaciens* genes in *E. coli*, then reintroducing these genes into their original host. Once the mutagenized genes are reintroduced into *Agrobacterium*, it is often desirable to exchange the mutated gene for the wild-type gene by recombination. Two procedures are normally used to obtain marker exchange. In one procedure, the cloned, transposon-mutagenized gene is introduced into *A. tumefaciens* by conjugation from *E. coli*. An incompatible plasmid is then mated in, selecting for the second plasmid and for the transposon.[11] Use of this incompatibility procedure requires that the mutagenized genes be cloned on broad host range plasmids, which normally have a low copy number in *E. coli* and are therefore inconvenient to isolate and manipulate *in vitro*. It also requires two separate matings and methods for distinguishing donor from recombinant cells. Further, the persistence of the second plasmid in the marker exchange mutants often complicates subsequent genetic manipulations.

In the second procedure, the mutagenized genes are mated into *A. tumefaciens* on a mobilizable suicide vector such as pUC19*mob*.[25] Since such vectors do not replicate in *A. tumefaciens*, a selection for the transposon-mutagenized gene should yield stable recombinants. These vectors are small and have high copy numbers in *E. coli*, making them more convenient to use than the broad host range vectors employed in the

[25] D. Nunn, S. Bergman, and S. Lory, *J. Bacteriol.* **172**, 2911 (1990).

1   2   3   4

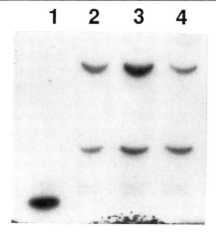

FIG. 1. Southern analysis of marker exchange in the *virH* region (S. Machlin, unpublished, 1990). Genomic DNA from *A. tumefaciens* strains was digested with *Sal*I and *Sst*I, electrophoresed on an agarose gel, and transferred to a Nytran membrane. A 1.45-kb *Sal*I–*Sst*I fragment from pTiA6 containing the *virH* (*pinF*) region was labeled with ^{32}P and used as the probe in Southern hybridization. Lane 1, DNA from wild-type strain A348; lanes 2–4, DNA from three independently isolated marker exchange mutants with a Tn*lacZ* insertion in the *virH* region. Insertion of Tn*lacZ*, which contains one *Sal*I site and one *Sst*I site, converts the 1.45-kb *Sal*I fragment into two larger fragments.

incompatibility procedure. The suicide vector procedure requires only one triparental mating, but its efficiency is frequently very low when the selected mutations are situated close (<1 kb) to one end of the cloned *A. tumefaciens* DNA. A similar limitation was observed using the incompatibility procedure as well.[11]

We have used the electroporation procedure described above to introduce several cloned transposon mutations into the *virH* (*pinF*) region[10] of the *A. tumefaciens* Ti plasmid. *Sal*I fragment 4[26] containing *virH* was cloned into pUC19*mob* and mutagenized in *E. coli* with Tn*lacZ* as described.[14] Insertions were mapped by restriction analysis, and plasmid DNA was introduced into electrocompetent A348 cells as described above, selecting for kanamycin resistance. Resistant colonies were picked to plates containing kanamycin and carbenicillin. Of the kanamycin-resistant clones, 1–4% were sensitive to carbenicillin, indicating that pUC19*mob* DNA was lost. Southern analysis indicated that over one-half of the Km^r Cb^s clones had lost the wild-type *Sal*I–*Sst*I fragment containing the sites of the insertions, and carried instead the Tn*lacZ*-mutagenized region (Fig. 1).

[26] S. E. Stachel and E. W. Nester, *EMBO J.* **5**, 1445 (1986).

Electroporation circumvents triparental conjugation, simplifying the selection and screening of recombinants. We have used this procedure with good success to exchange genes with transposons less than 1 kb from the end of the cloned *A. tumefaciens* DNA fragment. This apparent advantage over the conjugation-based suicide vector procedure may result from a higher efficiency of transformation by electroporation. However, we were not successful in obtaining marker exchanges in a chromosomal region which was refractory to marker exchange mutagenesis by the incompatibility procedure.

Concluding Remarks

We have described mutagenesis and gene transfer techniques which we are currently using to study the mechanism of genetic infection by *Agrobacterium*. As mentioned earlier, the great majority of methods used by researchers in this field are similar or identical to those used in the genetic analysis of enteric bacteria and other gram-negative bacteria. However, not all of the genetic tools available for analyzing other bacteria are effective in *Agrobacterium*. In particular, there is a need for cloning vectors carrying functional promoters that can control the transcription of cloned genes and gene fragments in *A. tumefaciens*.

Recently, two such elements have been developed. One, pED32, carries a *Hin*dIII fragment with the promoter region of *virB* cloned into pTJS75, immediately upstream of a multirestriction site fragment with six unique cloning sites. Transcription of genes and gene fragments cloned into these sites is inducible by acetosyringone in $virA^+$ $virG^+$ *A. tumefaciens* strains (J. Ward, personal communication, 1990). The other, pSW213, is a derivative of pTJS75 carrying $lacI^Q$ and the pUC α-complementation group. The expression of genes under the control of the *lac* promoter of this plasmid can be derepressed by isopropyl-β-D-thiogalactopyranoside (IPTG) (C.-Y. Chen and S. Winans, personal communication, 1990). These plasmids will confer additional versatility for molecular studies of *Agrobacterium* and its interactions with plants.

Acknowledgments

We thank Sarah Machlin, John Ward, Steve Winans, Stan Gelvin, and Steve Farrand for sharing unpublished data with us. Portions of this work were supported by U.S. Department of Agriculture Grant 88-37234-3618, U.S. Public Health Service Grant GM32618-18 from the National Institutes of Health, and National Science Foundation Grant DMB-8704292.

[19] Genetic Techniques in *Rhizobium meliloti*

By JANE GLAZEBROOK and GRAHAM C. WALKER

Introduction

Rhizobia have been studied extensively because of their ability to form nodules on the roots of leguminous plants and fix atmospheric nitrogen.[1] Many genetic techniques have been developed to facilitate analysis of several species and strains of *Rhizobium*. *Rhizobium meliloti* strain SU47, an alfalfa symbiont, has one of the most advanced genetic systems, which is described in this chapter. At the present time, essentially any major genetic strategy that can be used in the study of *Escherichia coli* K12 can also be implemented in the genetic analysis of *R. meliloti*. Many of the techniques described here can be directly applied to other rhizobia or to other gram-negative bacteria in which Tn5 can transpose and IncP plasmids an replicate.

Two basic methods are used to transfer DNA into and out of *R. meliloti* in the course of genetic manipulations, transduction and conjugation. Transduction is carried out using the Ca^{2+}-dependent, virulent bacteriophage ϕM12, in a method very similar to that used for P1 transduction in *E. coli*.[2] A set of conjugation techniques has been developed based on the conjugal transfer (*tra*) functions of the broad host range IncP plasmids. Generally, the IncP origin of conjugal transfer (*oriT*) or mobilization site (*mob*) is inserted in the DNA molecule of interest, and conjugal transfer of that molecule is accomplished by providing the IncP *tra* functions in trans. Conjugation is generally used to transfer recombinant plasmids into and out of *R. meliloti*; a variety of vectors are available that contain IncP *oriT* or *mob* sites and hence can be conveniently mobilized by providing IncP *tra* functions in trans.

Although it is possible to mutagenize *R. meliloti* with chemical mutagens such as N-methyl-N'-nitro-N-nitrosoguanidine (MNNG), transposon mutagenesis has proved to be a very useful tool since mutations induced by transposons are genetically marked by the antibiotic resistance of the transposon and physically marked by the presence of the transposon. Furthermore, a variety of transposon derivatives are now available that permit the facile *in vivo* generation of fusions to genes such as *lacZ, phoA,*

---

[1] S. R. Long, *Annu. Rev. Genet.* **23**, 483 (1989).

[2] T. M. Finan, E. Hartweig, K. Lemieux, K. Bergman, G. C. Walker, and E. R. Signer, *J. Bacteriol.* **159**, 120 (1984).

gusA, and *npt*. There are also Tn*5* derivatives that contain the IncP *oriT* or *mob* sequences, allowing construction of Hfr-like strains which are useful for genetic mapping studies.

Strains and Plasmids

Strains and plasmids are listed in Table I. To simplify descriptions of genetic techniques, it is assumed that, unless otherwise specified, all *R. meliloti* strains are Sm^r, all *E. coli* strains are Sm^s, and all recombinant plasmids carrying cloned *R. meliloti* sequences confer Tc^r. Most of the techniques described can be easily modified to accommodate strains and plasmids with different antibiotic resistances.

Growth Media

Unless otherwise specified, *R. meliloti* is grown at 30° in LB/MC medium which is LB (Luria broth) medium (per liter: 10 g tryptone, 5 g yeast extract, and 5 g NaCl) containing 2.5 mM MgSO$_4$ and 2.5 mM CaCl$_2$, or in M9 medium (per liter: 22 g Na$_2$PO$_4$, 6 g KH$_2$PO$_4$, 1 g NaCl, and 2 g NH$_4$Cl) containing 0.4% sucrose, 1 mM MgSO$_4$, 0.25 mM CaCl$_2$, and 1 μg/ml biotin. *Escherichia coli* is grown at 37° in LB medium. For solid medium 15 g/liter agar is added. Antibiotics are used at the following concentrations: neomycin (Nm) 200 μg/ml, kanamycin (Km) 25 μg/ml, oxytetracycline (Otc) 0.75 μg/ml, tetracycline (Tc) 10 μg/ml, streptomycin (Sm) 500 μg/ml, rifampicin (Rf) 50 μg/ml, gentamicin (Gm) 50 μg/ml for *R. meliloti* and 5 μg/ml for *E. coli*, spectinomycin (Sp) 100 μg/ml for *R. meliloti* and 50 μg/ml for *E. coli*, trimethoprim (Tp) 400 μg/ml, and chloramphenicol (Cm), 25 μg/ml. The concentrations of some antibiotics must be reduced if more than one antibiotic is present in the medium. Some commonly used combinations are: Gm at 20 μg/ml and Sp at 50 μg/ml, Nm at 100 μg/ml and Tc at 5 μg/ml and Km at 25 μg/ml and Tc at 5 μg/ml. 5-Bromo-4-chloro-3-indolyl-β-D-galactoside (X-Gal) and 5-bromo-4-chloro-3-indolyl phosphate (XP) are used at 40 μg/ml. 5-Bromo-4-chloro-3-indolyl-β-D-glucuronic acid (X-Gluc) is used at 20 μg/ml, owing to its expense. Alfalfa is grown in Jensen's medium (per liter: 1 g CaHPO$_4$, 0.2 g K$_2$HPO$_4$, 0.2 g MgSO$_4$ · 7H$_2$O, 0.2 g NaCl, 0.1 g FeCl$_3$, 1 mg H$_3$BO$_3$, 1 mg ZnSO$_4$ · H$_2$O, 0.5 mg CuSO$_4$ · 5H$_2$O, 0.5 mg MnCl$_2$ · 4H$_2$O, and 1 mg NaMoO$_4$ · 2H$_2$O; the pH is adjusted to 7.0 with NaOH).

Isolation of Tn*5* Insertions in *Rhizobium meliloti*

Of the transposons commonly used to mutagenize *E. coli*, Tn*5* is the only one which has been found to both transpose randomly in *R. meliloti* and to encode an antibiotic resistance which can be selected for in *R.*

TABLE I
STRAINS AND PLASMIDS

Strain	Description	Source of ref.[a]
Rhizobium meliloti		
SU47	Wild type	*1*
Rm1021	SU47 *str-21*	F. Ausubel
Rm5000	SU47 *rif-5*	T. Finan
Rm5320	Rm1021 30 ΩTn5-*11* (mobilizable pSymA)	*2*
Rm5300	Rm1021 thi-502::Tn5-*11* (mobilizable pSymB)	*2*
Rm6879	Rm5000 *recA*::Tn5-*Tp*	*3*
Rm8608	Rm5000 601 ΩTn5-*Mob*	J. Glazebrook
Rm8609	Rm5000 602 ΩTn5-*Mob*	J. Glazebrook
Rm8610	Rm5000 611 ΩTn5-*Mob*	J. Glazebrook
Rm8611	Rm5000 612 ΩTn5-*Mob*	J. Glazebrook
Rm8612	Rm5000 614 ΩTn5-*Mob*	J. Glazebrook
Rm8613	Rm5000 615 ΩTn5-*Mob*	J. Glazebrook
Rm8002	Rm1021 *pho-1*	*4*
Rm8501	Rm1021 *lac*	S. Long
Escherichia coli		
MM294A	*pro-82, thi-1, endA, hsdR17, supE44*	G. Walker
MT607	MM294A *recA56*	*2*
MT616	MT607 pRK600	*2*
MT614	MT607 *malE*::Tn5	T. Finan
MT609	P3473 *polA1 thy* Sp^r	T. Finan
Plasmids		
pRK2013	Nm^r, ColE1 replicon with RK2 *tra* genes	*5*
pRK600	Cm^r, Nm^s, pRK2013 Nm^r::Tn9	*2*
pRK602	pRK600 ΩTn5	*6*
pRK604	pRK2013 ΩTn5-*132*	*7*
pRK607	pRK2013 ΩTn5-*233*	*8*
pRK608	pRK600 ΩTn5-*235*	*8*
pRK609	pRK600 ΩTn*phoA*	*3*
pSB387	pRK600 ΩTn5-*gusA1*	*9*
MW03	pRK600 ΩTn5-*Mob*	M. Williams
pTFM1	Cb^r, Km^r, Gm^r, Sp^r, Sm^r, pGS330 ΩTn5-*11*	*2*
pEYDG1	pBR322 derivative containing Tn5-*oriT*	*10*
pSup5011	Ap^r, Cm^r, Km^r, pBR325 ΩTn5-*Mob*	*11*
pPH1JI	Gm^r Sp^r, IncP plasmid	*12*
pSshe	Cm^r, pACYC184 containing *tnpA*	*13*
pHoKm	Ap^r, Km^r, ColE1 derivative carrying Tn*3HoKm*	*13, 14*
pHoGus	Ap^r, Km^r, ColE1 derivative carrying Tn*3Gus*	*15*
pLAFR1	Tc^r, mobilizable RK2 cosmid	*16*
pRK404	Tc^r, mobilizable RK2 vector with *lacZ* α-complementing fragment	*17*

[a] Key to references: (*1*) J. M. Vincent, *Proc. Linn. Soc. NSW* **66**, 145 (1941); (*2*) T. M. Finan, B. Kunkel, G. F. DeVos, and E. R. Signer, *J. Bacteriol.* **167**, 66 (1986); (*3*) S. Klein, Ph.D. Thesis, Massachusetts Institute of Technology Cambridge (1987); (*4*) S. Long, S. McCune, and G. C. Walker, *J. Bacteriol.* **170**, 4257 (1988); (*5*) D. H. Figurski

meliloti.[3–5] In recent years, a large number of Tn5 derivatives have been constructed (Table II), so that Tn5 is now available with a variety of antibiotic resistance markers. There are also Tn5 derivatives that generate fusions to *lacZ, phoA, uidA,* and *npt.* Tn5-*11* and Tn5-*Mob* contain origins of conjugal transfer, allowing Hfr-like mapping techniques to be used in *R. meliloti.*

Tn5 insertion mutations marked with Nm^r, Gm^{r}Sp^r, Otc^r, or Tp^r can be isolated using Tn5, Tn5-*233*, Tn5-*132*, or Tn5-*Tp*, respectively. In cases where it is necessary to construct a double mutant when both mutations of interest were generated by insertions of the same Tn5 derivative, one of the insertions can be replaced by a Tn5 derivative with a different antibiotic resistance marker, as described below.[6] This capability allows great flexibility in the design of genetic experiments.

Fusions can be constructed using the transposons Tn5-*lac*,[7] Tn*phoA*,[8] Tn5*gusA1*,[9] or Tn5-*VB32*.[10] Tn5-*lac* makes transcriptional fusions to the *E. coli lacZ* gene. This transposon has limited usefulness in *R. meliloti,* since nearly all insertions have significant β-galactosidase activity.[11]

[3] J. E. Beringer, J. L. Benyon, A. V. Buchanan-Wollaston, and A. W. B. Johnston, *Nature (London)* **276,** 633 (1978).

[4] E. Bolton, P. Glynn, and F. O'Gara, *Mol. Gen. Genet.* **193,** 153 (1984).

[5] T. M. Finan, personal communication (1990).

[6] G. F. DeVos, G. C. Walker, and E. R. Signer, *Mol. Gen. Genet.* **204,** 485 (1986).

[7] L. Kroos and D. Kaiser, *Proc. Natl. Acad. Sci. U.S.A.* **81,** 5816 (1984).

[8] C. Manoil and J. Beckwith, *Proc. Natl. Acad. Sci. U.S.A.* **82,** 8129 (1985).

[9] S. B. Sharma and E. R. Signer, *Genes Dev.* **4,** 334 (1990).

[10] V. Bellofatto, L. Shapiro, and D. A. Hodgson, *Proc. Natl. Acad. Sci. U.S.A.* **81,** 1035 (1984).

[11] M. Williams, personal communication (1987).

and D. R. Helinski, *Proc. Natl. Acad. Sci. U.S.A.* **76,** 1648 (1979); (6) J. A. Leigh, E. R. Signer, and G. C. Walker, *Proc. Natl. Acad. Sci. U.S.A.* **82,** 6231 (1985); (7) T. M. Finan, A. M. Hirsch, J. A. Leigh, E. Johansen, G. A. Kuldau, S. Deegan, G. C. Walker, and E. R. Signer, *Cell (Cambridge, Mass.)* **40,** 869 (1985); (8) G. F. DeVos, G. C. Walker, and E. R. Signer, *Mol. Gen. Genet.* **204,** 485 (1986); (9) S. B. Sharma and E. R. Signer, *Genes Dev.* **4,** 334 (1990); (10) E. A. Yakobson and D. G. Guiney, Jr., *J. Bacteriol.* **160,** 451 (1984); (11) R. Simon, *Mol. Gen. Genet.* **196,** 413 (1984); (12) J. E. Beringer, J. L. Beynon, A. V. Buchanan-Wollaston, and A. W. B. Johnston, *Nature (London)* **276,** 633 (1978); (13) S. E. Stachel, G. An, C. Flores, and E. W. Nester, *EMBO J.* **4,** 891 (1985); (14) J. Glazebrook and G. C. Walker, *Cell (Cambridge, Mass.)* **56,** 661 (1989); (15) D. Dahlbeck and B. Staskawicz, personal communication (1989); (16) A. M. Friedman, S. R. Long, S. E. Brown, W. J. Buikema, and F. M. Ausubel, *Gene* **18,** 289 (1982); (17) G. Ditta, T. Schmidhauser, E. A. Yakobson, P. Lu, X. Liang, D. R. Finlay, D. G. Guiney, and D. R. Helinski, *Plasmid* **13,** 149 (1985).

TABLE II
Transposon Derivatives

Transposon derivative	Delivery vehicle	Relevant characteristics[a]	Ref.[b]
Tn5's			
Tn5	pRK602	Nm^r/Km^r	1, 2
Tn5-233	pRK607	Gm^r/Km^r, Sp^r/Sm^r	3
Tn5-132	pRK604	Otc^r/Tc^r	4, 5
Tn5-Tp	pGS330 ΩTn5-Tp	Tp^r	6
Tn5-235	pRK608	Nm^r/Km^r, has an expressed lacZ gene	3
Tn5-lac	pMW03	Nm^r/Km^r, makes transcriptional fusions to lacZ	7, 8
TnphoA	pRK609	Nm^r/Km^r, makes translational fusions to phoA	9, 10
Tn5-gusA1	pSB387	Nm^r/Km^r, makes transcriptional fusions to gusA	11
Tn5-VB32	pRK600 ΩVB32	Makes transcriptional fusions to npt	12, 13
Tn5-11	pTFM1	Gm^r/Km^r Sp^r/Sm^r, has oriT of plasmid RK2	14
Tn5-oriT	pEYDG1	Nm^r/Km^r, contains oriT of RK2	15
Tn5-Mob	pSup5011	Nm^r/Km^r, has mob of plasmid RP4	16
Tn3's			
Tn3HoKm	pHoKm	Ap^r, Nm^r/Km^r, makes transcriptional fusions to lacZ	17
Tn3Gus	pHoGus	Nm^r/Km^r, makes transcriptional fusions to gusA	18

[a] When antibiotic resistance is given in the form X/Y, X is used with *R. meliloti*, and Y is used with *E. coli*.
[b] Key to references: (1) D. E. Berg, J. Davies, B. Allet, and J. D. Rochaix, *Proc. Natl. Acad. Sci. U.S.A.* **72**, 3628 (1975); (2) J. A. Leigh, E. R. Signer, and G. C. Walker, *Proc. Natl. Acad. Sci. U.S.A.* **82**, 6231 (1985); (3) G. F. DeVos, G. C. Walker, and E. R. Signer, *Mol. Gen. Genet.* **204**, 485 (1986); (4) D. E. Berg, C. Egner, B. J. Hirschel, J. Howard, L. Johnsrud, R. A. Jorgensen, and T. D. Tlsty, *Cold Spring Harbor Symp. Quant. Biol.* **45**, 115 (1980); (5) T. M. Finan, A. M. Hirsch, J. A. Leigh, E. Johansen, G. A. Kuldau, S. Deegan, G. C. Walker, and E. R. Signer, *Cell (Cambridge, Mass.)* **40**, 869 (1985); (6) S. Klein, Ph.D. Thesis, Massachusetts Institute of Technology, Cambridge (1987); (7) L. Kroos and D. Kaiser, *Proc. Natl. Acad. Sci. U.S.A.* **81**, 5816 (1984); (8) M. Williams, personal communication (1986); (9) C. Manoil and J. Beckwith, *Proc. Natl. Acad. Sci. U.S.A.* **82**, 8129 (1985); (10) S. Long, S. McCune, and G. C. Walker, *J. Bacteriol.* **170**, 4257 (1988); (11) S. B. Sharma and E. R. Signer, *Genes Dev.* **4**, 334 (1990); (12) V. Bellofatto, L. Shapiro, and D. A. Hodgson, *Proc. Natl. Acad. Sci. U.S.A.* **81**, 1035 (1984); (13) R. Waldman, personal communication (1986); (14) T. M. Finan, B. Kunkel, G. DeVos, and E. R. Signer, *J. Bacteriol.* **167**, 66 (1986); (15) E. A. Yakobson and D. G. Guiney, Jr., *J. Bacteriol.* **160**, 451 (1984); (16) R. Simon, *Mol. Gen. Genet.* **196**, 413 (1984); (17) J. Glazebrook and G. C. Walker, *Cell (Cambridge, Mass.)* **56**, 661 (1989); (18) D. Dahlbeck and B. Staskawicz, personal communication (1989).

Tn*phoA* is a probe for genes encoding membrane and periplasmic proteins, owing to the fact that alkaline phosphatase is only active outside the cell cytoplasm. The *phoA* gene in Tn*phoA* lacks a promoter, a ribosome binding site, a translational start site, and a signal sequence. Thus, active fusions are obtained only if the *phoA* gene is fused to a gene encoding a membrane or a periplasmic protein. Tn*5gusA1* generates transcriptional fusions to *gusA*, a gene encoding β-glucuronidase. There is no endogenous β-glucuronidase activity in *R. meliloti* or in plants, making Tn*5gusA1* useful for studying expression of *R. meliloti* genes *in planta*.[12] Derivatives of Tn*5gusA1* which generated translational fusions, or have different antibiotic resistance genes, are also available.[9] These derivatives were constructed very recently, and have not yet been widely used. In Tn*5-VB32*, the *npt* gene of Tn*5* is promoterless, making this transposon useful for identifying promoters.

In addition to those described above, Simon *et al.* have constructed a set of Tn*5* derivatives that contain *mob* sequences and either Gm[r], Sp[r], or Tc[r] selectable markers, generate gene fusions to *lacZ, npt,* or *luc,* or contain *npt* or *tac* promoters reading outward from the transposons.[12a] In very recent work, Herrero *et al.* and deLorenzo *et al.* describe the construction of a set of mini-Tn5s.[12b,12c] These derivatives do not contain the *tnp* gene encoding transposase, and therefore require provision of this function in order to transpose. The collection includes constructs containing selectable markers encoding resistance to the herbicide bialophos, murcuric compounds, and arsenite, as well as the more conventional antibiotic resistances, and derivatives that generate gene fusions. These two extensive collections of Tn*5* derivatives are not included in Table II, rather, the reader is referred to the original publications for descriptions of these transposons.[12a,12b,12c]

Tn*5* insertion mutants of *R. meliloti* are isolated by mating the Sm[r] *R. meliloti* strain to be mutagenized with a Sm[s] *E. coli* strain carrying a derivative of pRK600 containing a Tn*5* insertion. The mating mixture is then plated on LB/MC medium containing Nm and Sm. Since pRK600 cannot replicate in *R. meliloti,* only *R. meliloti* in which Tn*5* has transposed into the genome are able to grow in the presence of Nm and Sm.

Procedure. Overnight cultures of Rm1021 and MM294a/pRK602 are mixed in equal proportions, and 0.1 ml of the mixture is spread on an LB/MC plate and incubated at 30° overnight. Some of the cells are scraped off

---

[12] H. Sittertz-Bhatkar, *BioTechniques* **7**, 922 (1989).
[12a] R. Simon, J. Quandt, and W. Klipp, *Gene* **80**, 161 (1989).
[12b] M. Herrero, V. deLorenzo, and K. N. Timmis, *J. Bacteriol.* **172**, 6557 (1990).
[12c] V. deLorenzo, M. Herrero, U. Jakubzik, and K. Timmis, *J. Bacteriol.* **172**, 6568 (1990).

the plate with a sterile stick and resuspended in 0.85% NaCl. This mixture is diluted 10-, 100-, and 1000-fold into 0.85% NaCl. One-tenth milliliter of each dilution is spread on LB/MC medium containing Nm and Sm. The remainder of the mating mixture is stored at 4°. After 3 days, the number of colonies on each plate is determined. If necessary, the stored mating mixture is used to spread more phates at a colony density which is convenient for screening for the desired phenotype.

Notes. This procedure can also be used to mutagenize *R. meliloti* with Tn5 derivatives, using appropriate delivery plasmids and antibiotics. Tn*phoA* insertions are isolated in strain Rm8002, a *pho* mutant of Rm1021, to allow the alkaline phosphatase activity of fusion proteins to be detected.

Insertion mutants obtained from the same mating mixture may be siblings, so it is advisable to screen multiple independent mating mixtures.

In general, Tn5 will not transpose randomly in a strain which has a resident Tn5.[6] A method for obtaining random Tn5 insertions in a strain with a resident Tn5 derivative is described below.

Introduction of Recombinant Plasmids into *Rhizobium meliloti* by Triparental Mating

The plasmid vectors that we have found to be most useful for working with *R. meliloti* are derivatives of the broad host range IncP plasmid RK2. These vectors contain a gene for Tc[r] and the replication and transfer origins of RK2. They do not contain the *tra* genes of RK2; consequently they are not self-transmissible. Transfer of these plasmids from a donor to a recipient strain is accomplished by supplying the *tra* genes in trans, on the ColE1 plasmid pRK600, which is self-transmissible. The donor, recipient, and an *E. coli* strain carrying pRK600 are mixed, and conjugation is allowed to occur. The recipient strain containing the plasmid is selected by growth on medium containing an appropriate combination of antibiotics. If the recipient strain is *R. meliloti*, the "helper" plasmid, pRK600, will not be present in the final strain since ColE1 replication origins do not function in *R. meliloti*.

Procedure. A procedure to transfer a plasmid from an *E. coli* strain to *R. meliloti* strain Rm1021 is given as an example. Overnight cultures of the *E. coli* donor strain and the helper strain MT616 (an *E. coli* strain containing pRK600) are grown in LB containing appropriate antibiotics. An overnight culture of the recipient strain, Rm1021, is grown in LB/MC. The three cultures are mixed in equal proportions, and 50 μl (or less) of the mixture is spotted onto LB/MC medium and incubated at 30° overnight. Several matings can be spotted onto the same plate. A small amount of the mating mixture is picked from the plate, then streaked out on an LB/

MC plate containing Sm (to select for Rm1021) and Tc (to select for the plasmid).

Notes. We use the cosmid pLAFR1 for construction of genomic libraries,[13] and pRK404 for subcloning.[14] A number of other IncP group derivatives,[15-18] as well as other broad host range cloning vectors,[19,20] are also available.

In cases where the recipient strain is an *R. meliloti* strain which does not have a selectable marker, or the *E. coli* donor strain is Sm[r], the mating mixture is streaked on M9/sucrose medium containing Tc. *Escherichia coli* cannot metabolize sucrose, so only the *R. meliloti* recipient will grow.

Although pRK600 does not replicate in *R. meliloti*, it apparently persists long enough to allow plasmids to be mated from *R. meliloti* into *E. coli* using a procedure similar to the one above. The mating mixture is streaked on an LB plate containing Tc and incubated at 37°. The *E. coli* exconjugants appear after 1 day of growth. The exconjugants may contain pRK600. If this is not desirable, a *polA* recipient is used, since ColE1 plasmids cannot replicate in *polA* mutants.

Isolation of Transposon Insertions in Recombinant Plasmids

Tn5 Insertions

A procedure is available which allows specific regions of the genome to be mutagenized with Tn5. First, a cosmid clone containing the region of the genome to be mutagenized is obtained. Derivatives of this cosmid containing Tn5 insertions are then isolated. Homogenotes in which the Tn5 insertions in the cosmid have recombined into the corresponding position in the genome are obtained by introducing an incompatible plasmid into the *R. meliloti* strains carrying the mutagenized cosmids, while maintaining selection for Tn5.[21] A procedure for isolating such homogen-

[13] A. M. Friedman, S. R. Long, S. E. Brown, W. J. Buikema, and F. M. Ausubel, *Gene* **18**, 289 (1982).

[14] G. Ditta, T. Schmidhauser, E. A. Yakobson, P. Lu, X. Liang, D. R. Finlay, D. G. Guiney, and D. R. Helinski, *Plasmid* **13**, 149 (1985).

[15] R. Simon, U. Prieffer, and A. Puhler, *in* "Molecular Genetics of the Bacteria–Plant Interaction" (A. Puhler, ed.), p. 98. Springer-Verlag, Berlin and Heidelberg, 1983.

[16] F. E. Nano, W. D. Shepherd, M. M. Walkins, S. A. Kuhl, and S. Kaplan, *Gene* **34**, 219 (1984).

[17] G. Selvaraj and V. N. Ayer, *Plasmid* **13**, 70 (1985).

[18] R. A. Rubin, *Plasmid* **18**, 84 (1987).

[19] D. R. Gallie, S. Novak, and C. I. Kado, *Plasmid* **14**, 171 (1985).

[20] U. B. Prieffer, R. Simon, and A. Puhler, *J. Bacteriol.* **163**, 324 (1985).

[21] G. B. Ruvkun and F. M. Ausubel, *Nature (London)* **289**, 85 (1981).

otes is given below. This method can be used with other types of transposons as well, since the mutagenesis of the cosmid is done in *E. coli*, not *R. meliloti*. This technique has been extremely valuable in studies of *R. meliloti*, since many genes with symbiotic functions are clustered. A large number of symbiotic loci have been identified by saturation mutagenesis of cosmids known to contain one symbiotic locus, isolating homogenotes, and screening them for symbiotic phenotyes.[22–25]

The most efficient way to isolate insertions of Tn5 in plasmid sequences is to mate the plasmid to be mutagenized into an *E. coli* strain that has a chromosomal insertion of Tn5.[22] The plasmid is then mated out of this strain, with selection for both Tc^r and Nm^r. Care must be taken to ensure that the resulting Tn5-containing plasmids arise from independent insertion events.

Procedure. To introduce the cosmid to be mutagenized into an *E. coli* strain containing a chromosomal insertion of Tn5, overnight cultures of strain MT614, MT616, and a strain containing the plasmid to be mutagenized are mixed in equal proportions, and 50 µl of this mixture is spotted on an LB/MC plate and incubated overnight. The mating mixture is streaked out on LB medium containing Km and Tc, then incubated at 37° overnight. A subpopulation of the cells in these colonies contain cosmids with Tn5 insertions. In order to isolate mutagenized cosmids arising from independent insertion events, single colonies are picked from the plate and patched onto LB/MC medium. Overnight cultures of Rm1021 and MT616 are mixed in equal proportions, and 20 µl of this mixture is spotted on top of each patched colony. These mating mixtures are incubated at 30° overnight. Each mating mixture is then streaked onto LB/MC medium containing Sm, Nm, and Tc. The colonies which grow on this medium are *R. meliloti* containing cosmids with Tn5 insertions. Only one colony is picked from each streak to be used for further analysis, to ensure that each mutagenized plasmid studied arose from an independent insertion event.

Notes. If it is necessary to determine the positions of the insertions in the mutagenized plasmids by restriction mapping, it is often convenient to isolate the plasmids in an *E. coli* strain, since it is easier to prepare DNA from *E. coli* than from *R. meliloti*. In this case, the *E. coli* polA strain MT609 is used in place of Rm1021 in the second mating, which is performed on LB plates containing 60 µg/ml thymidine. The mating mixtures are

[22] S. Long, J. W. Reed, J. Himawan, and G. C. Walker, *J. Bacteriol.* **170**, 4239 (1988).
[23] J. Glazebrook and G. C. Walker, *Cell (Cambridge, Mass.)* **56**, 661 (1989).
[24] T. W. Jacobs, T. T. Egelhoff, and S. R. Long, *J. Bacteriol.* **162**, 469 (1985).
[25] G. B. Ruvkun, V. Sundaresan, and F. M. Ausubel, *Cell (Cambridge, Mass.)* **29**, 551 (1982).

then streaked out on LB medium containing Sp, Km, Tc, and 60 μg/ml thymidine (MT609 is Sp^r and requires thymidine for growth, even in rich medium). Colonies appear after 2 days of growth at 37°.

Tn3 Derivatives

Tn3 derivatives which create transcriptional fusions to *lacZ* (Tn3*HoKm*)[23,26] or *gusA* (Tn3*Gus*)[27] can also be used to mutagenize cosmids. These transposons require Tn3 transposase activity to be supplied in trans on the plasmid pSshe.[26] To mutagenize a plasmid with Tn3*HoKm* or Tn3*Gus*, the plasmid is introduced by transformation into an *E. coli* strain containing pSshe and pHoKm or pHoGus, then mated out of this strain into *R. meliloti* with selection for Nm^r and Tc^r.

Procedure. An *E. coli* strain carrying pSshe and either pHoKm or pHoGus is transformed with the plasmid to be mutagenized, with selection for Km^r (pHoKm or pHoGus), Cm^r (pSshe), and Tc^r. From this point, plasmids with Tn3*HoKm* or Tn3*Gus* insertions are isolated in the same way that plasmids with Tn5 insertions are isolated.

Notes. *Rhizobium meliloti* has very little endogenous β-galactosidase activity, so *lacZ* fusions can be studied in an Rm1021 background. The endogenous activity can be avoided by using strain Rm8501, a *lac* mutant of Rm1021.

A procedure has recently been described for histochemical staining of plant tissues for expression of *lacZ* fusions, after inactivation of the endogenous plant activities.[28] This may make it possible to study the expression of *lacZ* fusions to *R. meliloti* genes in nodules.

Tn3 and its derivatives do not transpose as randomly as Tn5 does. Occasionally a disproportionately high fraction of insertions will be found to be in one segment of DNA.

Homogenotization of Plasmid-Borne Transposon Insertions

To determine the phenotype of a mutation in a cloned gene, it is convenient to mutagenize the gene with Tn5 on a plasmid and then isolate recombinants in which the Tn5 insertion is in the homologous position in the genome. These recombinants are obtained by introducing another IncP-incompatibility group plasmid, pPH1JI,[3,24] into Rm1021 carrying the Tn5 insertion plasmid. Selection for pPH1JI and Tn5, without selection

[26] S. E. Stachel, G. An, C. Flores, and E. W. Nester, *EMBO J.* **4,** 891 (1985).

[27] D. Dahlbeck and B. Staskawicz, personal communication (1989).

[28] T. H. Teeri, H. Lehvaslaiho, M. Franck, J. Uotila, P. Heino, E. T. Palva, M. VanMontagu, and L. Herrera-Estrella, *EMBO J.* **8,** 343 (1989).

for Tc, results in colonies in which the plasmid-borne Tn5 insertion has recombined into the corresponding position in the genome and the original plasmid has been lost. pPH1JI is present in these colonies.

Procedure. Overnight cultures of Rm1021 carrying the Tn5 insertion plasmid and an *E. coli* strain carrying pPH1JI are mixed in equal proportions, and 50 μl of the mixture is spotted onto LB/MC medium and incubated at 30° overnight. The mating mixture is streaked out on LB/MC medium containing Gm, Nm, and Sm and then incubated at 30° for 3 days. Typically, ten single colonies are screened for Tc^s, and one Tc^s colony is chosen for further work. Occasionally it is necessary to screen a larger number of colonies to obtain one that is Tc^s. This colony should contain an insertion of Tn5 in the genome and the plasmid pPH1JI. The position of the homogenotized Tn5 can be confirmed by Southern hybridization, using the unmutagenized cloned gene as a probe.[29]

Notes. The success of this procedure depends on the frequency of homologous recombination between the DNA sequences flanking the Tn5 insertion being much higher than the transposition frequency of Tn5. This is true when the Tn5 is inserted in a large fragment of cloned DNA, such as a cosmid insert. In these cases, fewer than 1% of the homogenotized Tn5 insertions are in "incorrect" positions. If the fragment is small, or the Tn5 insertion is very close to one end of the fragment, few Nm^r Gm^r Tc^s colonies are obtained. In these cases the fraction of Nm^r Gm^r Tc^s colonies resulting from transposition of Tn5 can be quite high, and the positions of the insertions in such homogenotes should be verified by Southern hybridization. Insertions of Tn3HoKm or Tn3Gus can be easily homogenotized from quite small [3 kilobase (kb)] plasmid inserts.

Exchanging one Tn5 Derivative for Another

It is often useful to be able to exchange a Tn5 insertion for an insertion of a Tn5 derivative carrying different antibiotic resistance genes, so that double mutants can be constructed. When a Tn5 derivative is introduced into a strain with a resident Tn5, a double recombination event can occur between the IS50 sequences that flank the antibiotic resistance genes, so that the central region of the resident Tn5 derivative is replaced by the central region of the incoming Tn5 derivative.[6]

Procedure. A replacement of a Tn5 insertion with a Tn5-233 insertion is given as an example. Overnight cultures of the *R. meliloti* strain containing the Tn5 insertion and the *E. coli* strain MM294A/pRK607 are mixed

[29] J. Sambrook, E. F. Fritsch, and T. Maniatis, "Molecular Cloning: A Laboratory Manual," 2nd Ed. Cold Spring Harbor Laboratory, Cold Spring Harbor, New York, 1989.

in equal proportions. One-tenth milliliter of this mixture is spread on LB/MC medium and incubated at 30° overnight. Some of the cells are then scraped off the mating plate with a sterile stick and resuspended in 1 ml of 0.85% NaCl. Serial 10-fold dilutions of the mixture are prepared and plated on LB/MC plates containing Gm, Sp, and Sm. Gm^r Sp^r colonies are obtained at frequencies between 10^{-3} and 10^{-4}. These colonies are then screened for Nm^r by patching them onto LB/MC plates containing Nm and onto LB/MC plates containing Gm and Sp. Gm^r Sp^r Nm^s colonies are obtained at a frequency of approximately 1%. These colonies have undergone a replacement of Tn*5* by Tn*5-233*.

Notes. Analogous procedures can be used to replace insertions of other Tn*5* derivatives. The Nm^r gene of Tn*phoA* can be replaced by the antibiotic resistance genes of other Tn*5* derivatives, leaving the *phoA* fusion intact. Recombination occurs between an IS50 of the incoming Tn*5* derivative and the portion of IS50L that is downstream of the *phoA* gene in Tn*phoA*, resulting in a Tn*phoA* insertion with altered antibiotic resistance genes. In contrast, the expressed *lacZ* gene in Tn*5-235* is located between the IS50 sequences, so in replacements involving this transposon the *lacZ* gene is exchanged along with the Nm^r gene.

General Transduction Using φM12

General transducing phage for use with *R. meliloti* strains Rm41, GR4, and L5-30 have been reported.[30–32] None of these phage infect SU47. The general transducing phage that we use for SU47 derivatives is φM12.[2] φM12 is a large phage which is morphologically similar to phage T4 of *E. coli*, with a genome size of approximately 160 kb. Chromosomal and megaplasmid markers are transduced at frequencies ranging from 10^{-5} to 10^{-6}. φM12 is a virulent phage, so in order to obtain transductants it is necessary to prevent readsorption of the phage after the first round of infection. Like the *E. coli* transducing phage P1, φM12 requires Ca^{2+} for adsorption, so transduction is accomplished by preincubating the cells and the phage for a short time in the presence of Ca^{2+} and then plating on selective medium with a Ca^{2+} concentration below 0.25 mM.

Procedure. A φM12 lysate is prepared by diluting 0.5 ml of a fresh overnight culture of *R. meliloti* grown in LB/MC into 4.5 ml of LB/MC and incubating it shaking at 30° for 2 hr [optical density (OD) at 600 nm 0.3–0.4]. φM12 is added to a final concentration of approximately 10^8

---

[30] T. Sik, J. Horvath, and S. Chatterjee, *Mol. Gen. Genet.* **178,** 511 (1980).
[31] J. Casadesus and J. Olivares, *Mol. Gen. Genet.* **174,** 203 (1979).
[32] M. Kowalski, *Acta Microbiol. Pol.* **16,** 7 (1967).

plaque forming units (pfu)/ml, and shaking is continued at 30° until lysis occurs (8–16 hr). A few drops of chloroform are added to the culture, and cell debris is pelleted by centrifugation at 10,000g for 10 min. The lysate usually contains between 10^{10} and 10^{11} pfu/ml and can be stored at 4° for several months without any appreciable drop in titer.

Markers are transduced into recipient strains by mixing 0.3 ml of a fresh overnight culture of the recipient strain with 30 μl of an undiluted lysate grown on the donor strain [multiplicity of infection (moi) of about 0.5] and incubating this at room temperature for 15 min. The cells are pelleted, washed with 0.85% NaCl, resuspended in 0.1 ml of 0.85% NaCl, and plated on selective medium lacking $MgSO_4$ and $CaCl_2$. This procedure generally yields 50–300 transductants.

Random Tn5 Mutagenesis of Strains Carrying Resident Tn5

In general, Tn5 does not transpose randomly in a strain carrying a resident Tn5. Nevertheless, Tn5 mutagenesis of a strain with a resident Tn5 can be accomplished by growing ϕM12 on a pool of Tn5 insertion mutants and using this lysate to transduce the strain with a resident Tn5 to the antibiotic resistance of the incoming Tn5. The transductants are then screened for the desired phenotype.

Procedure. The procedure described above (see Isolation of Tn5 Insertions in *Rhizobium meliloti*) is used to obtain Tn5 insertion mutants growing on selective media at a density of about 2000 colonies per plate. One milliliter of 0.85% NaCl is added to each plate, spread with a glass spreader to suspend the bacteria, and removed to a sterile tube. For each pool, colonies washed from 5 plates are combined. We generally work with 5 such pools, derived from independent matings. Each pool of Tn5 insertion mutants is diluted into LB/MC medium to an OD_{600} of 0.1, then incubated at 30° until the cultures reach an OD_{600} of 0.4. ϕM12 is added to the cultures, and lysates are prepared. These lysates are used to transduce the strain with a resident Tn5 insertion to the antibiotic resistance of the incoming Tn5.

Notes. Resident Tn5-233 insertions do not interfere with transposition of incoming Tn5's, so these strains can be randomly mutagenized with Tn5 in the same way that strain Rm1021 is mutagenized.[6]

Isolation of a Tn5 Insertion Linked to a Point Mutation

In general, point mutations cannot be transduced from one strain to another because they have no selectable marker. This makes strain construction and genetic mapping very inconvenient. Isolation of a Tn5 insertion which is cotransducible with a point mutation makes it possible to move the mutation into other strains by transduction, greatly facilitating genetic studies.[33]

Procedure. A ϕM12 lysate grown on a pool of Tn5 insertion mutants is prepared as described above. The lysate is used to transduce the strain containing the point mutation to Nm^r, and the transductants are screened for loss of the mutant phenotype. The transductants occur at a frequency of about 1%. Most of the transductants which have lost the mutant phenotype contain an insertion of Tn5 which is close to the point mutation. ϕM12 is grown on these transductants, and the lysates are used to transduce the point mutant to Nm^r. These transductants are screened for loss of the mutant phenotype, and the linkage of the Tn5 insertions to the point mutation is determined. This linkage is confirmed by choosing a transductant which is Nm^r and retains the mutant phenotype, and demonstrating that the point mutation and the Tn5 insertion can be cotransduced from this strain.

Isolation of Mutants with Deletions between Nearby Tn5 Insertions

When two Tn5 insertions are present in the same DNA molecule in the same strain, recombination can occur between them such that the intervening sequences and one of the Tn5 insertions are lost. This can be used to generate defined deletions between Tn5 insertions whose positions are known.[6] Using this procedure described above one of the Tn5 insertions is replaced by Tn5-233, and the other is replaced by Tn5-235. These insertions are then transduced into the same strain. Deletion events in which the sequences between the IS50 of Tn5-235 which is distal to Tn5-233 and the IS50 of Tn5-233 which is proximal to Tn5-235 are lost result in Lac$^-$, Gm^r Sp^r Nm^s colonies. These colonies are easily identified by their very pale blue color on medium containing X-Gal.

Procedure. The double-insertion strain is plated on LB/MC medium containing 25 μg/ml Gm (if Tn5-233 is used) and 40 μg/ml X-Gal. Most of the colonies appear dark blue owing to the presence of Tn5-235. Colonies that are very pale blue are obtained at a frequency of about 10^{-4}. These are putative deletion mutants. Transduction of the Tn5-233 insertion from the putative deletion mutant into strains containing either of the original Tn5 insertions used as the deletion end points should yield 100% Nm^s transductants. The deletion end points in putative mutants can also be verified by Southern hybridization using Tn5 probes.[29]

---

[33] T. M. Finan, A. M. Hirsch, J. A. Leigh, E. Johansen, G. A. Kuldau, S. Deegan, G. C. Walker, and E. R. Signer, *Cell (Cambridge, Mass.)* **40**, 869 (1985).

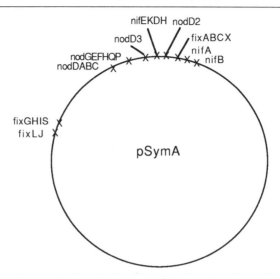

Fig. 1. Genetic map of pSymA. Genes are listed in clockwise order, for example, in "fixGHIS" *fixS* is located clockwise from *fixG*. A large region of pSymA has been physically mapped. The region between *nifE* (sometimes called *fixE*) and *nifB* is described by David et al.,[a] and the region between *nod* and *fixLJ* is described by Batut et al.[b] *fixLJ* is described by David et al.,[c] *fixGHI* is described by Kahn et al.,[d] *nodEFGH* is described by Swanson et al.,[e] *nodPQ* is described by Schwedock and Long,[f] and *nodD2* and *nodD3* are described by Honma and Ausubel.[g] Key to references: (a) M. David, O. Domergue, P. Pogonec, and D. Kahn, *J. Bacteriol.* **169**, 2239 (1987); (b) J. Batut, B. Terzaghi, M. Gherardi, M. Huguet, E. Terzaghi, A. M. Garnerone, P. Boistard, and T. Huguet, *Mol. Gen. Genet.* **199**, 232 (1985); (c) M. David, M. L. Daveran, J. Batut, A. Dedieu, O. Domergue, J. Ghai, C. Hertig, P. Boistard, and D. Kahn, *Cell (Cambridge, Mass.)* **54**, 671 (1988). (d) D. Kahn, M. David, O. Domergue, M. L. Daveran, J. Ghai, P. R. Hirsch, and J. Batut, *J. Bacteriol.* **171**, 929 (1989); (e) J. A. Swanson, J. K. Tu, J. Ogawa, R. Sanga, R. F. Fischer, and S. R. Long, *Genetics* **117**, 181 (1987); (f) J. Schwedock and S. R. Long, *Mol. Plant–Microbe Int.* **2**, 181 (1989); (g) M. A. Honma and F. M. Ausubel, *Proc. Natl. Acad. Sci. U.S.A.* **84**, 8558 (1987).

Genetic Mapping in *Rhizobium meliloti*

The *R. meliloti* genome consists of the chromosome and two megaplasmids each of which contains about 1600 kb of DNA.[34] Many of the genes which are important in symbiosis are located on these megaplasmids; consequently they have been termed pSymA and pSymB. Genes required for nodule induction (*nod*), nitrogenase activity (*nif*), and other functions important in nodulation (*fix*) have been mapped to pSymA (Fig. 1). Loci involved in exopolysaccharide synthesis (*exo, exp*), thiamine synthesis (*thi*), and dicarboxylic acid transport (*dct*), as well as two *fix* loci, have

[34] T. M. Finan, B. Kunkel, G. F. DeVos, and E. R. Signer, *J. Bacteriol.* **167**, 66 (1986).

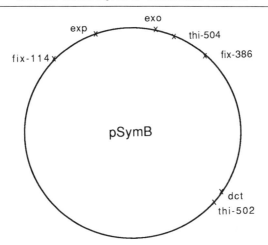

Fig. 2. Genetic map of pSymB, condensed from the complete linkage map of pSymB determined by Charles and Finan.[a] The position of *fix-114* was determined by Finan.[b] Detailed physical maps of the *exo*[c] and *exp*[d] regions are available. Key to references: (*a*) T. C. Charles and T. M. Finan, *J. Bacteriol.* **172**, 2469 (1990); (*b*) T. M. Finan, personal communication (1990). (*c*) S. Long, J. W. Reed, J. W. Himawan, and G. C. Walker, *J. Bacteriol.* **170**, 4239 (1988); (*d*) J. Glazebrook and G. C. Walker, *Cell (Cambridge, Mass.)* **56**, 661 (1989).

been mapped to pSymB (Fig. 2). Some genes required for symbiosis, as well as several amino acid biosynthetic genes, have been mapped to the chromosome (Fig. 3). Regions of the chromosome for which ϕM12 transductional linkage data are available are shown in Fig. 4.

Determination of Whether Mutation Is on Megaplasmid or Chromosome

When mapping a gene in *R. meliloti*, the first question to answer is whether it is on pSymA, pSymB, or the chromosome. This is accomplished using two strains, Rm5320, which contains a Tn5-*11* insertion in pSymA, and Rm5300, which contains a Tn5-*11* insertion in pSymB. Tn5-*11* carries the origin of transfer of plasmid RK2, enabling the megaplasmids to be mobilized from Rm5320 and Rm5300 into other strains when *tra* functions are supplied in trans. In order to determine the location of a mutation of interest, it is first transduced into Rm5320 and Rm5300, and the megaplasmids are then mobilized from each of these strains into a *recA* recipient strain, with selection for Tn5-*11*. The exconjugants are screened for the presence of the mutation of interest. If the mutation can be mobilized from Rm5320, it must be on pSymA. If it can be mobilized from Rm5300, it is

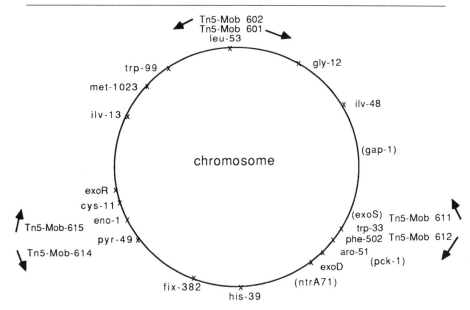

FIG. 3. Genetic map of the SU47 chromosome. 601 ΩTn5-*Mob* and 602 ΩTn5-*Mob* are cotransducible with *leu-53*, 611 ΩTn5-*Mob* and 612 ΩTn5-*Mob* are cotransducible with *trp-33*, and 614 ΩTn5-*Mob* and 615 ΩTn5-*Mob* are cotransducible with *pyr-49*. The exact positions of only two of the Tn5-*Mob* insertions are known (Fig. 4). Parentheses indicate that the position of the marker is not known relative to the other markers in the region. *gly-12* and *phe-502* were mapped by Meade and Signer,[a] *fix-382* and *exoR* were mapped by Meiri and Glazebrook,[b] *exoD* was mapped by Reed,[c] *exoS* was mapped by Doherty et al.[d] *eno-1*, *ntrA71*, *pck-1*, and *gap-1* were mapped by Finan et al.,[e] and all other markers were mapped by Klein.[f] Key to references: (a) H. M. Meade and E. R. Signer, *Proc. Natl. Acad. Sci. U.S.A.* **74**, 2076 (1977); (b) G. Meiri and J. Glazebrook, unpublished results (1989); (c) J. W. Reed, personal communication (1990); (d) D. Doherty, J. A. Leigh, J. Glazebrook, and G. C. Walker, *J. Bacteriol.* **170**, 4249 (1988); (e) T. M. Finan, I. Oresnik, and A. Bottacin, *J. Bacteriol.* **170**, 3396 (1988); (f) S. Klein, Ph.D. Thesis, Massachusetts Institute of Technology, Cambridge (1987).

on pSymB, and if it cannot be mobilized from either strain, it is on the chromosome.

Procedure. The mutation to be mapped is transduced into strains Rm5320 and Rm5300. The Tn5-*11* insertions in these strains confer Gm^r Sp^r, so if the mutation is marked with Tn5-*233*, it must first be replaced with a different Tn5 derivative. Overnight cultures of Rm5320 containing the mutation and Rm5300 containing the mutation are mixed with overnight cultures of MT616 and Rm6879 in equal proportions, and 0.1 ml of the mixtures is spread on LB/MC medium and incubated at 30° overnight. Cells from each mating mixture are scraped off the plates and resuspended

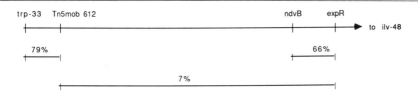

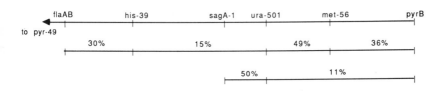

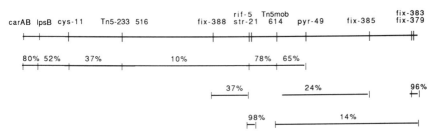

FIG. 4. φM12 transductional linkage maps of three regions of the SU47 chromosome. Distances between markers are given as φM12 cotransduction frequencies. The *trp-33* region was mapped by G. Meiri and J. Glazebrook (unpublished results). In the *his-39* region, *his-39*, *ura-501*, *met-56*, and *pyrB* were mapped by Clover et al.,[a] *sag-1* was mapped by S. Klein,[b] and *flaAB*, along with other tightly linked *fla*, *che*, and *mot* genes not shown in the figure, was mapped by Ziegler et al.[c] In the pyr-49 region, *carAB* and *lpsB* were mapped by Clover,[d] *fix-388*, *fix-385*, *fix-383*, and *fix-379* were mapped by Meiri and Glazebrook, (unpublished results) and the remaining markers were mapped by Williams and Signer.[a] Key to references: (a) In: "Genetic Maps 1987 Volume, 4" (S. J. O'Brien, ed). Cold Spring Harbor Laboratory Press, Cold Spring Harbor, NY, 1987; (b) S. Klein, personal communication (1986); (c) R. J. Ziegler, C. Pierce, and K. Bergman, *J. Bacteriol.* **168**, 785 (1986); (d) R. Clover, Ph.D. Thesis, Massachusetts Institute of Technology, Cambridge, MA (1989).

in 1 ml of 0.85% NaCl. One-tenth milliliter of each mating mixture is spread on LB/MC medium containing Rf (to select for Rm6879), Gm, and Sp, then incubated at 30° for 3 days. Colonies from these plates are screened for the presence of the mutation of interest by patching them onto suitable medium.

Notes. A complete linkage map of pSymB has recently been reported, which makes it possible to determine precisely the position of any mutation which maps to pSymB.[35]

Localization of Mutation to Specific Region of Chromosome

Mutations are mapped on the *R. meliloti* chromosome in a manner very similar to Hfr mapping in *E. coli.* Six strains have been constructed, and each strain contains a Tn5-*Mob* insertion in a different position and/ or orientation in the *R. meliloti* chromosome.[36] The positions of these insertions are shown in Fig. 3. The mutation of interest is transduced into each of the six strains and then mobilized out of each strain in a triparental mating. The frequency with which the mutation is transferred from each Tn5-*Mob* insertion is then determined. The mutation lies between the positions of the two insertions from which it is transferred at the highest frequencies. More precise mapping depends on the availability of mapped mutations in the interval of the chromosome where the mutation of interest is located. If these are available, the position of the mutation can be mapped relative to these markers, in a manner analogous to Hfr mapping in *E. coli.*

Procedure. Tn5-*Mob* confers Nm^r, so if the mutation to be mapped is a Tn5 insertion, it must first be replaced by Tn5-*233*. It is then transduced into each of the six Tn5-*Mob* strains. It is advisable to screen the transductants for Nm^r, to determine if the mutation is cotransducible with any of the Tn5-*Mob* insertions. The six Tn5-*Mob* strains containing the mutation of interest are then mated with MT616 and Rm1021. The mating mixtures are resuspended in 0.85% NaCl, and 0.1 ml of this is spread on LB/MC medium containing Sm, Gm, and Sp (assuming that the mutation of interest is marked by Tn5-*233*). Ten-fold serial dilutions of this mixture are prepared and spread on LB/MC medium containing Sm. Colonies appear after 3 days of incubation at 30°. The number of Gm^r Sp^r Sm^r colonies as a fraction of total Sm^r colonies is calculated for each mating mixture. The mutation lies between the positions of the two Tn5-*Mob* insertions from which it was mobilized at the highest frequencies.

Notes. It is important that the mutation which is being mapped does

[35] T. C. Charles and T. M. Finan, *J. Bacteriol.* **172,** 2469 (1990).
[36] S. Klein, Ph.D. Thesis, Massachusetts Institute of Technology, Cambridge (1987).

not lie very close to the selected (Sm^r) marker. This mutation is 78% linked to 615 ΩTn5-*Mob* by ϕM12 transduction (Fig. 4). If the mutation to be mapped is far enough away from 615 ΩTn5-*Mob* that it is unlinked by ϕM12 transduction, proximity to the Sm^r locus should not be a problem.

Screening pLAFR1 Genomic Libraries for Complementing Cosmids

If a mutant has a phenotype which can be easily detected on petri plates, such as auxotrophy or dye binding, complementing cosmids can be easily isolated by using a triparental mating to transfer a pLAFR1 genomic library *en masse* into the *R. meliloti* mutant and screening for the desired phenotype.[13] If the mutant has a symbiotic phenotype (it is Fix$^-$), it is more difficult to isolate a complementing cosmid. If the mutation was the result of a transposon insertion, the insertion and flanking sequences can be cloned and the flanking sequences used to probe the library.[29] This method is tedious and time-consuming. A much simpler approach is to introduce the library *en masse* into the Fix$^-$ mutant and use the mixture to inoculate a tubful of alfalfa plants growing in medium lacking fixed nitrogen. Mutants that received a complementing cosmid will form Fix$^+$ nodules, and plants with these nodules will grow tall and green. These plants are then uprooted, and the bacteria containing complementing cosmids are recovered from their nodules.

Procedure. A plastic tub (40 × 30 × 12 cm) with small holes in the bottom for drainage is filled three-quarters full with Perlite, moistened with approximately 4 liters of Jensen's medium, and sterilized by autoclaving. Five grams of alfalfa seeds (*Medicago sativa*) are surface sterilized by soaking them in 50% Clorox for 15 min and rinsing 4 times with sterile water. The seeds are then evenly distributed over the surface of the Perlite contained in the tub, the tub is covered with Saran Wrap, and the whole assembly is placed under plant lights.

A triparental mating is performed to transfer the genomic library into the mutant of interest. The mating mixture is scraped off the plate and resuspended in 1 ml of 0.85% NaCl. An aliquot of this mixture is diluted and plated on selective medium to determine the efficiency of the mating. The remainder of the mating mixture is used to inoculate 200 ml of LB/ MC medium containing Sm, Nm (if the mutant is a Tn5 insertion), and Tc. After 1 day of growth at 30°, the cells are pelleted, washed with sterile water, and resuspended in 200 ml of sterile water. This suspension is poured over the alfalfa plants growing in the tub, 5–7 days after the seeds were planted. The plants are grown for 4–6 weeks under lights, and sterile water is added as needed.

If any healthy plants are present, they are uprooted, and their roots are examined for the presence of pink nodules. These are removed, surface sterilized by soaking in 50% Clorox solution for 5 min and rinsed 4 times in sterile water. Each nodule is then crushed in 0.2 ml of LB/MC medium containing 0.4% glucose. This material is plated on LB/MC plates containing Tc. About 10^4 bacteria are recovered from a medium-sized nodule. The cosmids in the recovered bacteria should be retested for complementation by transferring them into the original mutant strain and testing for formation of Fix[+] nodules. Southern hybridization can be used to determine whether the cosmid sequences correspond to the region of DNA that was interrupted by the Tn5 mutation.[29]

[20] Genetic Systems in Cyanobacteria

By ROBERT HASELKORN

Introduction

Bacterial systems for genetic analysis include three essential components: methods for induction and selection of mutants, methods for introduction of DNA into cells, and methods for selection and analysis of complemented mutants and recombinants. We first discuss several features of each of these components that present unusual difficulties for the genetic analysis of cyanobacteria and then describe the protocols, devised in recent years, that appear to surmount most of these difficulties.

All cyanobacteria carry out green plant photosynthesis, using two photosynthetic reaction center assemblies linked by a cytochrome-containing electron transport chain. The cyanobacterial genes and proteins of the photosynthetic apparatus are very similar to their counterparts in plant chloroplasts. The ability to manipulate genes of cyanobacteria makes them preferred organisms for studying structure–function relationships of the proteins of photosynthesis. Many cyanobacteria, in addition, can convert atmospheric N_2 to ammonia. When some filamentous cyanobacterial strains are deprived of a source of combined nitrogen, nitrogen fixation occurs in specialized cells called heterocysts, which differentiate from vegetative cells at regular intervals along the filaments. Heterocyst differentiation provides a unique opportunity to study a simple, experimentally controlled developmental pattern.

METHODS IN ENZYMOLOGY, VOL. 204

Growth and Storage

What are the special difficulties for the genetic analysis of cyanobacteria? Even after decades of study in a large number of laboratories, rapid growth in liquid medium and stability during storage can not be taken for granted for most strains.[1] Long-term storage of both wild-type and mutant strains is a special problem. For example, our wild-type *Anacystis nidulans* R2 (*Synechococcus* 7942), transferred repeatedly at room temperature, lost the ability to be transformed by DNA. We were able to recover the transformable strain from stocks that had been stored at −80°. On the other hand, this organism dies rapidly in the refrigerator. Between 6° and 4°, the cytoplasmic membrane undergoes a phase transition leading to the irreversible loss of K^+, ATP, and other materials essential for life.[2] Other unicellular cyanobacteria, such as *Synechocystis* 6803, survive quite well under the same conditions. Following Wolk, we now store strains and mutants at −80° in 15% glycerol.[3]

Mutagenesis and Mutant Isolation

Cyanobacteria are polyploid. The degree of ploidy depends on growth rate, which in turn is determined by light intensity and CO_2 concentration. Under normal laboratory conditions, there are approximately 10 chromosomes/cell in *Synechococcus* 7942, based on a comparison of the DNA content per cell and the genome size.[4] It is not clear, at present, what this degree of ploidy means in terms of mutant isolation. We have limited experience with both dominant and recessive mutations. Mutation to diuron resistance in the *psbA* gene(s) of *Synechococcus* 7942 appears to be dominant, though it need not be. When DNA prepared from resistant colonies is probed with oligonucleotides that distinguish wild-type from mutant sequences, only mutant sequences are detected at a particular locus,[5] that is, all of the chromosomes contain the same sequence. We do not know whether this result is due to random segregation of chromosomes, nonrandom segregation, or gene conversion. The same result is obtained when cells are transformed by DNA fragments containing antibiotic resistance cassettes. For example, the neomycin phosphotransferase gene from Tn5 can confer resistance if a single copy is present in the cell, yet transformed colonies show, exclusively, complete replacement of the

[1] R. Castenholz, this series, Vol. 167, p. 68.
[2] N. Murata, *J. Bioenerg. Biomembr.* **21**, 61 (1989).
[3] C. P. Wolk, personal communication (1990).
[4] M. Herdman, M. Janvier, R. Rippka, and R. Y. Stanier, *J. Gen. Microbiol.* **111**, 73 (1979).
[5] J. Brusslan and R. Haselkorn, *EMBO J.* **8**, 1237 (1989).

wild-type fragment by the *neo*[r] cassette. Partial merodiploids are instead found when the interrupted gene is essential for growth and the antibiotic concentration is just right.[6]

Aside from the general interest in the mechanism of chromosome partitioning to daughter cells, there is a practical consequence of these considerations for the strategy of mutagenesis. Suppose we wish to isolate a recessive mutant, for example, an auxotroph. Before penicillin selection can be applied, we want to be sure that the "homozygous" mutant exists at a selectable frequency in the population. If the cells contain ten chromosomes, the homozygote cannot appear before at least four cell generations if the chromosomes segregate randomly. In the mutagenesis protocol described below for the isolation of Nif[−] mutants of *Anabaena,* the cells are grown for more than seven generations before selection is applied.

Cyanobacteria must have adapted to growth under DNA-damaging conditions: high light energy flux and oxygen radicals generated during photosynthesis. Consequently, they are expected to have efficient DNA repair systems, and this appears to be the case. The *recA* gene of several unicellular cyanobacteria has been cloned, but it has not been possible to inactivate the gene by insertional mutagenesis, demonstrating that the cyanobacterial *recA* gene is probably essential for normal growth.[6]

Very few nutritional mutants of cyanobacteria have been described. This is probably not due to insufficient time given for partitioning prior to selection, but rather to poor transport of the nutrients. Historically, mutants requiring methionine, uracil, phenylalanine, and tryptophan have been described. It is possible that some amino acids are transported too slowly to sustain growth.

Given the special properties of cyanobacteria, mutants defective in components of the photosynthetic apparatus are much in demand. To propagate such mutants (other than conditional lethals such as temperature-sensitive mutants) they have to be able to utilize external carbon sources. The widely used *Synechococcus* 7942 does not grow on any carbon source other than CO_2. *Synechocystis* 6803 will grow on glucose in the light but not in the dark; mutants defective in photosystem II (PSII) can be obtained but not ones in photosystem I (PSI). *Synechocystis* 6714 has been reported to grow on glucose in the dark, but the strains currently available cannot be transformed with DNA. The marine *Synechococcus* 7002 (*Agmenellum quadruplicatum*) is capable of heterotrophic growth and can be transformed with DNA. Among the filamentous cyanobacteria there are some true heterotrophs but these are not transformable at pres-

---

[6] R. C. Murphy, G. E. Gasparich, D. A. Bryant, and R. D. Porter, *J. Bacteriol.* **172,** 967 (1990).

ent. The conjugation systems currently in use work efficiently only with strains that happen to be obligate phototrophs. To date, electroporation has been used principally to introduce DNA into the unicellular and filamentous strains that can be transformed or can function well as recipients in conjugation. An exception may be *Nostoc* MAC, which has been reported to be transformed by electroporation and is capable of heterotrophic growth.[7] Methods for the selection of photosynthetic mutants are described by Joset.[8]

Mutagenesis

Many studies of mutagenesis of cyanobacteria have been made. Often, the basis of the mutation being sought was not known, so that the efficacy (or lack thereof) of the mutagenesis protocol could not be judged adequately. We present two protocols that have worked well in our laboratory for the isolation of *Anabaena* mutants defective in nitrogen fixation. A more extensive discussion of mutagenesis procedures is given by Golden.[9]

Chemical Mutagenesis

Chemical mutagenesis is the preferred method for creating missense mutations. A protocol for nitrosoguanidine mutagenesis was described by Chapman and Meeks[10] and by Golden.[9] This reagent, however, is thought to induce multiple mutations clustered at the replication fork. We therefore chose diethyl sulfate (DES) to induce mutations in *Anabaena*. The protocol is described for isolation of Nif$^-$ mutants, but variants should work for photosynthetic mutants (of heterotrophs) and for nutritional mutants (of strains capable of transporting the nutrient).

Anabaena PCC 7120 is grown in 1 liter of Kratz and Myers medium containing 3 mM HEPES, pH 7.8, and 2.5 mM $(NH_4)_2SO_4$ (KMN$^+$) gassed with 1–2% CO_2 in air. Cells at mid-log phase (2–6 μg/ml chlorophyll, 0.7–2.0 × 10^7 cells/ml) are collected by centrifugation at room temperature (7000 rpm, 15 min) and resuspended in 30 ml of KMN$^+$. DES is added to a final concentration of 5 μl/ml, and the culture is allowed to stand at room temperature for 5 min. Buffered medium is essential because unbuffered DES at this concentration will kill all the cells. Higher DES concentration or longer time of incubation also leads to excessive killing. The cells are washed several times by centrifugation, then resuspended in 500 ml of

[7] T. Thiel and H. Poo, *J. Bacteriol.* **171**, 5743 (1989).
[8] F. Joset, this series, Vol. 167, p. 728.
[9] S. S. Golden, this series, Vol. 167, p. 714.
[10] J. S. Chapman and J. C. Meeks, *J. Gen. Microbiol.* **133**, 111 (1987).

KMN$^+$. (Plating can be done at this point to determine the extent of killing by DES.) The culture bleaches during the first day, then requires 5 to 7 days to grow up to its original density. During this time, segregation of alleles occurs.

The cells are collected by centrifugation, washed with Kratz and Myers medium without NH$_4^+$ (KMN$^-$), and about one-fifth of the cells are suspended in 1 liter of KMN$^-$ to yield a chlorophyll concentration of 0.12 to 0.25 μg/ml. Two to three days are allowed for consumption of nitrogen storage reserves. (At this point the chlorophyll concentration should be 0.5 μg/ml; ampicillin selection works poorly at higher cell densities.) Then ampicillin is added at 200 μg/ml. After 3 days, most cells have lysed. The survivors are collected by centrifugation, washed to remove ampicillin, then resuspended in 1 liter of KMN$^+$. After 3 days of growth, the cells are collected, washed, suspended in KMN$^-$, grown for 2 days, and then the ampicillin treatment is repeated. The latter procedure is then repeated for a total of three rounds of ampicillin selection. The final pellet of surviving cells is washed with KMN$^-$ and spread on 1% agar plates made with KMN$^-$ containing a small amount of fixed nitrogen. Small green colonies are visible in 1 week; over the next 2 weeks many of these bleach. These are picked onto KMN$^+$ plates and finally tested on KMN$^-$ plates. Many of the mutants isolated by this procedure will be siblings. This problem can be avoided by subdividing the original mutagenized culture prior to growth in KMN$^+$ to allow segregation, but the additional labor involved is significant.

Transposon Mutagenesis

Transposon mutagenesis has numerous advantages, including the introduction of a selectable marker that makes it easy to clone the transposon together with flanking regions. The flanking regions can be used subsequently to isolate the corresponding uninterrupted gene from a libary of wild-type DNA fragments. Transposons such as Tn5-*lac*, Tn5-*lux*, and mini-Mu-*lac* provide the added feature of creating fusions to the genes they interrupt, providing a reporter for further studies on gene expression.[11-13]

Transposons move from one DNA molecule to another as a consequence of the action of genes encoded within the transposon. In the case of Tn5, the gene encodes a transposase. For mini-Mu, the genes *A* and *B* encode Mu DNA replication and integration functions. We have been unable to introduce Tn5 or mini-Mu into unicellular cyanobacteria by

[11] C. Bauer and K. Black, personal communication (1990).
[12] J. Elhai and C. P. Wolk, *EMBO J.* **9**, 3379 (1990).
[13] C. Bauer, personal communication (1990).

transformation. On the other hand, Tn5, Tn5-*lac*, Tn5-*lux*, and mini-Mu-*lac* have all been introduced into *Anabaena* by conjugation.[11–14] The Tn5 derivatives appear able to enter the chromosome at random positions and to remain stably integrated.[14] The mini-Mu-*lac* element is unstable, with such frequent deletions and separation of the *lacZ* gene from the Mu parts that it is not useful as a gene fusion tool in *Anabaena*.[13] We used the conjugation system described by Elhai and Wolk[12] (see below) to transfer pBR322 containing wild-type Tn5 into *Anabaena*.[14] Since pBR322 cannot replicate in *Anabaena*, transposition was essential to create stable neomycin-resistant exconjugants. Wolk and Elhai have improved on wild-type Tn5 by introducing a promoter from the *psbA* gene of *Amaranthus hybridus* (a plant!) to drive transcription of the neomycin phosphotransferase gene, which gives higher and more reproducible resistance to neomycin.[12]

In the absence of active transposition there are still useful applications of transposons in the transformable unicellular cyanobacteria. In those cases for which the target gene has already been cloned, the transposon can be inserted during propagation of the clone in *Escherichia coli,* and then the entire construct can be used to transform the cyanobacterium. If selection is applied only for the marker in the transposon, the vast majority of resistant transformants will result from double recombination events, in which recombination in the flanking sequences yields replacement of the wild-type gene by the gene interrupted by the transposon. The same method can be used to create specific deletions and to replace wild-type genes with site-directed mutations. In the latter case, the transposon is inserted near, not in, the gene in question.[9,15–17] Starting with cloned genes, of course, it is possible to use smaller antibiotic resistance cassettes inserted *in vitro* in place of complete transposons. These are rather more efficient because there is an inverse relationship between transformation frequency and the size of the interrupting element.[16,17]

It has also been possible to exploit antibiotic resistance cassettes as "random" mutagens in the unicellular cyanobacteria by using an extra step.[18] Chauvat and colleagues digested *Synechocystis* 6803 DNA completely with a restriction endonuclease and then ligated the mixture to an antibiotic resistance cassette. Transformation and selection for the antibiotic resistance yielded many deletions (resulting from random joining

[14] D. Borthakur and R. Haselkorn, *J. Bacteriol.* **171,** 5759 (1989).
[15] S. S. Golden, J. Brusslan, and R. Haselkorn, this series, Vol. 153, p. 215.
[16] J. G. K. Williams and A. A. Szalay, *Gene* **24,** 37 (1984).
[17] K. Kolowsky, J. G. K. Williams, and A. A. Szalay, *Gene* **27,** 289 (1984).
[18] F. Chauvat, R. Rouet, H. Botin, and A. Boussac, *Mol. Gen. Genet.* **219,** 51 (1989).

of fragments to the cassette and recombination in the flanking sequences), including genes for PSI and PSII components.[18]

Natural Transformation

Transformation of unicellular cyanobacteria was discovered by Shestakov over 20 years ago.[19] Since then, dozens of articles have appeared on the methodology, but very little insight has been gained into the reasons why some strains can be transformed and others cannot. Restriction may be part of the answer but cannot be all of it. Extensive discussions of the mechanism and procedures for transformation can be found in articles by Porter[20,21] and by Golden.[9] These articles also review methods for preparation of DNA from cyanobacteria, including those strains that have proved to be refractory to conventional lysis steps.

The various transformation protocols are straightforward. They differ mainly in whether cells are incubated with DNA in the dark and how the selective agent is applied. Porter recommends a soft agar overlay containing the selective agent, applied after incubation for a time suitable for phenotypic expression. Golden prefers introducing the selective agent under the agar a day after plating, allowing the agent to diffuse upward to the colonies forming on the surface. Williams and Szalay plate on a filter on nonselective medium, then transfer the filter to selective plates after a period of growth.[16] There has been no problem with direct selection following growth in liquid if the highest efficiency of transformation is not required. Currently, we use one of several variations of the original procedure described by Golden and Sherman[22]:

1. Cells are grown to mid-log phase, collected and washed with 10 mM NaCl, and resuspended in BG11 medium at 5×10^8 cells/ml.
2. DNA is added to 300 μl of cells, and the mixture is incubated in the dark for 4–12 hr.
3. Cells are plated on nonselective 1% agar made with BG11 and allowed to grow for 3–12 hr.
4. Selective agent is distributed underneath the agar, and growth is continued until colonies appear.

The transformation efficiency is reported to be greater than 10^6 cells per microgram DNA and up to one transformant per 10^3 cells at optimal DNA concentration.

[19] S. Shestakov and N. T. Khyen, *Mol. Gen. Genet.* **107,** 372 (1970).
[20] R. D. Porter, *Crit. Rev. Microbiol.* **13,** 111 (1987).
[21] R. D. Porter, this series, Vol. 167, p. 703.
[22] S. S. Golden and L. A. Sherman, *J. Bacteriol.* **158,** 36 (1984).

One variation we use is to incubate the cells and DNA in the dark overnight, followed by 4 hr in the light, and then to spread 100 μl directly on selective plates.[23] The efficiency is lower (10^2 transformants/μg DNA) but sufficient for genetic engineering purposes. Another variation is to grow the cells to 5 $\times$ 10^7/ml, wash with sterile water, and resuspend in BG11 at 10^9 cells/ml. Then 300 μl of this suspension is incubated with 0.5 μg of plasmid DNA overnight in a lighted incubator at 32° with 1% CO_2 in air. The cells are then spread directly on selective plates. Again, this procedure is less efficient than reported by Golden and Sherman[22] or by Williams and Szalay[16] (10^4 transformants/μg DNA), but it is convenient and sufficient for most purposes.[24]

Transformation by Electroporation

Transformation by electroporation has been successfully applied to a large number of animal, plant, and bacterial cells. It is a potentially useful procedure for cyanobacterial transformation as well. The only systematic study of the procedure to date, using cyanobacteria, is that of Thiel and Poo.[7] These authors found that restriction significantly lowers the efficiency of transformation. They used the plasmid pRL6 to transform *Anabaena* strain M-131. The plasmid has a single *Ava*II site. Methylation of that site by passing through an *E. coli* strain carrying the *Eco*47II methylase (which protects *Ava*II sites) increased the frequency of transformation 100-fold, from about 10^{-6} up to 10^{-4} per colony-forming unit (cfu) surviving electroporation.

The procedure of Thiel and Poo uses a Bio-Rad (Richmond, CA) Gene Pulser with the 25-μF capacitor and time constants of 2.5 or 5 msec provided by 100- or 200-ohm resistors, respectively. Cells are electroporated in 40-μl volumes in a chilled, sterile cuvette with a 2-mm gap. Cells are grown to 2 $\times$ 10^7/ml, washed twice in 1 mM HEPES, pH 7.2, and suspended in 1 mM HEPES at 10^9 cells/ml. DNA is dissolved in the same buffer. The number of transformants is roughly proportional to DNA concentration over the range of 0.01 to 10 μg/ml, reaching 10^5 transformants/ml at 10 μg/ml of DNA. The efficiency is also a sharp function of field strength, being optimal at 8 kV/cm with a time constant of 2.5 msec and 6 kV/cm at 5 msec. At higher field strengths, cell killing is significant. Electroporation with linearized pRL6 gives no transformants. No data are yet available on electroporation using chromosomal DNA or plasmid DNA containing chromosomal inserts.

[23] L. DiMagno, personal communication (1990).
[24] G. Ajlani, personal communication (1990).

Preliminary successes in electroporation of *Anabaena* 7120 and *Nostoc* MAC were also reported. *Anabaena* 7120, like *Anabaena* M-131, can receive DNA by conjugation efficiently but *Nostoc* MAC cannot, so in the latter case electroporation has a unique role to play. Among the unicellular cyanobacteria, *Synechococcus* 7942 was successfully transformed by electroporation with replicative plasmid DNA. This strain, however, is efficiently transformed without electroporation. Recently, we have not been able to transform *Synechocystis* 6714 by electroporation.[24]

Conjugation

Bizarre as it may seem, conjugation from *E. coli* to *Anabaena* has become the principal genetic tool for the study of heterocyst differentiation and the rearrangement of nitrogen fixation genes in cyanobacteria. The conjugation system was developed by Wolk and co-workers; the reader should consult Wolk and Elhai for a full discussion of the principles of the method.[24,25]

As currently practiced, conjugation requires the presence, in the ultimate *E. coli* donor strain, of four elements: a broad host range plasmid to provide the transfer functions, a mobilizing gene, a shuttle vector, and a source of modifying enzymes to protect the transferred DNA from restriction. The shuttle vector (cargo vector in the Wolk–Elhai terminology)[26] contains replicons with origins that function in *Anabaena* and in *E. coli*, antibiotic resistance genes for selection in both hosts, an origin (*oriT*) for mobilization, and a multiple cloning site. Many improvements have been made in the original shuttle vectors by J. Elhai, W. J. Buikema, and C. Bauer.[26,27] We now use a vector called pDUCA7 for the construction of cosmid libraries suitable for conjugation *en masse,* in order to isolate wild-type DNA fragments containing genes that complement mutants.[27] pDUCA7 is based on a mini-RK-2 replicon to which was added a λ *cos* site, the *neo*[r] gene from Tn5, the polylinker of pUC19, the *Nostoc* replicon pDU1, and the early transcription terminator of bacteriophage T7. This cosmid vector is low copy in *E. coli* and can be mobilized by RP-1; the T7 terminator is situated so as to prevent transcription of the cloned fragments from the *neo* gene promoter. Partial digests of chromosomal DNA cut with *Sau*3AI can be cloned into a unique *Bam*HI site. A map of pDUCA7 has been published.[27]

[25] C. P. Wolk, A. Vonshak, P. Kehoe, and J. Elhai, *Proc. Natl. Acad. Sci. U.S.A.* **81,** 1561 (1984).
[26] J. Elhai and C. P. Wolk, this series, Vol. 167, p. 747.
[27] W. J. Buikema and R. Haselkorn, *J. Bacteriol.* **173,** 1879 (1991).

The shuttle library is constructed by cutting *Anabaena* chromosomal DNA with *Sau*3A1, then isolating the 30–50 kilobase (kb) size fraction from a preparative agarose gel. This fraction is ligated into the dephosphorylated *Bam*HI site of pDUCA7, and the ligation mixture is packaged into λ *in vitro*. The recombinant phage are used to infect *E. coli* HB101 containing plasmid pRL528, which carries the *Ava*I and *Eco*47II methylase genes.[26] The latter enzyme protects *Ava*II sites. The infected cells are plated, and all the cells from 2×10^4 colonies are pooled, frozen in aliquots, and stored at $-80°$.

For complementation, *Anabaena* cells are grown to mid-log phase in KMN$^+$ (see section on Mutagenesis), washed in KMN$^+$ with 5% LB (Luria broth for *E. coli*), and resuspended at approximately 2×10^8 cells/ml (i.e., concentrated about 10-fold). A portion of the frozen *Anabaena* cosmid library in HB101(pRL528) is thawed and grown in LB containing 50 μg/ml kanamycin plus 10 μg/ml chloramphenicol for 3–4 hr, then washed in KMN$^+$ with 5% LB and resuspended in the same at 5×10^8 cells/ml. A mid-log phase culture of HB101(RP-1) is similarly washed and resuspended. The three suspensions (one of *Anabaena*, two of *E. coli*) are mixed in equal porportions, and then 0.5 ml of the mixture is placed on a detergent-free Millipore (Bedford, MA) HATF 08550 nitrocellulose filter on a KMN$^+$–5% LB agar plate. Following 2 days of incubation in the light, the filters are transferred to fresh KMN$^-$ plates containing 30 μg/ml neomycin. Green colonies of exconjugants appear in about 3 weeks. Single colonies from a complementation are streaked onto KMN$^-$ plates with neomycin. After 2 weeks the streak can be scraped off for the preparation of plasmid DNA, which can be used to transform *E. coli*. Complementing plasmids are stored in *E. coli;* otherwise, they will be lost by recombination with the chromosome in the exconjugant.

Gene Replacement Following Conjugation

Gene replacement in unicellular cyanobacteria can be accomplished by transformation with a cloned DNA fragment in which the target sequence is interrupted with a transposon or other selectable cassette (see Transformation). The necessary double recombination events appear to be extremely rare in *Anabaena*, so that even a nonreplicating shuttle vector (i.e., one lacking the pDU1 replicon) containing a chromosomal fragment interrupted by a selectable cassette does not directly yield the desired gene replacement. Instead, the entire plasmid integrates at the target sequence, resulting in tandemly repeated genes (one of which is interrupted) flanking the vector sequences. The directly repeated genes, with a frequency that depends on the length of the repeated sequence, can undergo subsequent recombination leading to deletion and loss of the vector sequences as well as one copy of the repeated material. If selection is maintained for the cassette, only cells with the interrupted gene will survive.[28] The second recombination occurs with different frequencies at different loci; in some cases it has been possible to recover the single copy interrupted-gene recombinants just by patient waiting for segregation, whereas in other cases the frequency is too low to detect the segregants. In practice, it is impossible to find the double recombinant this way unless there is a convenient screen for the expected phenotype.

Cai and Wolk have introduced a method to counterselect cells that retain the vector sequences, thus improving the probability of finding the desired double recombinants.[29] Their procedure incorporates the *Bacillus subtilis sacB* gene into the vector. This gene encodes levan sucrase, an enzyme that kills many gram-negative bacteria on plates containing 5% sucrose.[30] The enzyme hydrolyzes sucrose and then utilizes the fructose to make levan, in the process yielding a product that is toxic to gram-negative bacteria and to *Anabaena*. Thus, to inactivate gene X in *Anabaena*, the cloned gene X in pBR322 (containing the *sacB* gene) is interrupted by a cassette selectable in *Anabaena*. This could be the neo^r gene from Tn5 driven by a strong promoter (we have used the *woxA* promoter,[31] Elhai has used the *Amaranthus psbA* promoter)[12] or the ω fragment containing spc^r/str^r.[28] Vector protection by methylation and conjugation are carried out as described above. Recombinants are enriched by selection for resistance to the antibiotic conferred by the interrupting cassette. Resistant exconjugants are then streaked on a plate containing the same antibiotic and 5% sucrose. Surviving colonies should have only the interrupted copy of the gene, which should be verified by Southern hybridization.

Reporter Genes

In preceding sections, reference was made to several systems available for the fusion of reporter genes to cyanobacterial promoters for the quantitative analysis of transcription. Reporters provide certain conveniences, but it must be remembered that they are reporters and will differ in some respects from the genes they replace. For example, if the abundance of a message is determined primarily by the rate of degradation at the 3' end,

[28] J. W. Golden and D. R. Wiest, *Science* **242,** 1421 (1988).
[29] Y. Cai and C. P. Wolk, *J. Bacteriol.* **172,** 3138 (1990).
[30] J. L. Ried and A. Collmer, *Gene* **57,** 239 (1987).
[31] D. Borthakur, M. Basche, W. J. Buikema, P. Borthakur, and R. Haselkorn, *Mol. Gen. Genet.* **221,** 227 (1990).

a reporter fusion having a different 3' end will not be degraded in the same way. Moreover, reporters reflect the combined result of transcription and translation, so ideally both types of gene fusion should be used in the analysis of gene regulation.

The use of β-galactosidase (lac), luciferase (lux), and chloramphenicol acetyltransferase (cat) fusions as reporters is described extensively by Friedberg.[32] An additional reporter that is useful for rapid screening of colonies is the catechol 2,3-dioxygenase gene (xylE) from Pseudomonas putida.[33] This enzyme converts catechol to 2-hydroxymuconic semialdehyde, which is assayed by its absorbance at 375 nm.[34] J. D. Lang found that the catechol dioxygenase activity was unstable in Anabaena extracts, so only freshly lysed cell supernatants could be assayed.[35] On the other hand, chloramphenicol acetyltransferase (CAT) activity is stable and can be recovered quantitatively after freezing and thawing extracts.[35]

A promoter assay vector suitable for conjugation into Anabaena was constructed by J. Lang.[35] This vector (pJL3) of 11.7 kb consists of the Nostoc replicon pDU1 and the neo[r] gene, bom site, and ori of pRL25C,[26] a transcription terminator (T1) from an E. coli rRNA gene, a unique BamHI site followed by a promoterless cat gene, and the early bacteriophage T7 transcription terminator. CAT protein was assayed using an ELISA kit obtained from 5 Prime → 3 Prime, Inc. (117 Brandywine Pkwy., Westchester, PA 19380). Alternative assays for CAT activity are given by Friedberg.[32]

For comparative studies of promoter constructions it is essential to know the copy number of the vector. Lang did this by DNA dot blot, extracting total DNA from each of the strains to be assayed and then probing filter-bound total DNA with a unique chromosomal probe, such as a nif gene, and a vector probe, such as the cat gene. Although this method does not yield the actual copy number, it does give a measure of the relative copy number in a series of strains. In a series of deletions around the promoters of the Anabaena 7120 psbB gene, Lang found the copy number of pJL3 derivatives to be the same for inserts of 0, 100, 210, 600, and 650 base pairs (bp); one insert of 840 bp (giving the highest level of CAT activity) reduced the copy number by one-half.[36]

[32] D. Friedberg, this series, Vol. 167, p. 736.
[33] F. C. H. Franklin, M. Bagdasarian, M. M. Bagdasarian, and K. N. Timmis, Proc. Natl. Acad. Sci. U.S.A. 78, 7458 (1981).
[34] M. M. Zukowski, D. F. Gaffney, D. Speck, M. Kaufmann, A. Findeli, A. Wisecup, and J. P. Lecocq, Proc. Natl. Acad. Sci. U.S.A. 80, 1101 (1983).
[35] J. D. Lang, Ph.D. Dissertation, University of Chicago, Chicago, IL (1990).
[36] J. D. Lang and R. Haselkorn, J. Bacteriol. 173, 2729 (1991).

Physical Maps

The ultimate in cyanobacterial genetics would be the complete nucleotide sequence of a genome. Short of that goal is a physical map on which all of the currently identified genes can be located. A map has been obtained by Bancroft *et al.* for *Anabaena* 7120, based on pulsed-field gel electrophoresis of large DNA fragments.[37] The unique chromosomal map contains 6.37 Mb (megabase), and there are three megaplasmids of 410, 190, and 110 kb. A physical map of the *Synechocystis* 6803 chromosome is also being prepared.[38]

Acknowledgments

I am grateful to Ghada Ajlani, Chris Bauer, Kristin Bergsland, Bianca Brahamsha, William Buikema, and Conrad Halling for constructive comments on the preparation of this chapter.

[37] I. Bancroft, C. P. Wolk, and E. V. Oren, *J. Bacteriol.* **171**, 5940 (1989).
[38] S. Shestakov, personal communication (1990).

[21] Genetic Manipulation of *Streptomyces:* Integrating Vectors and Gene Replacement

By Tobias Kieser and David A. Hopwood

Introduction

Host Strains

The *Streptomyces* chromosome is a circle of about 72% G + C DNA at least 1.5 times the size of the *Escherichia coli* chromosome.[1] *Streptomyces coelicolor* A3(2) is genetically by far the most studied streptomycete, with a detailed chromosomal linkage map[2] and a combined physical/genetic map based on pulsed-field gel analysis of large fragments generated by enzymes that recognize permutations of the sequence A_3T_3 such as *Ase*I, *Dra*I, and *Ssp*I.[3] In *S. coelicolor* fundamental studies of differentiation, antibiotic production, and many other aspects of cell and molecular biol-

[1] R. Hütter and T. Eckhardt, *in* "Actinomycetes in Biotechnology" (M. Goodfellow, S. T. Williams, and M. Mordarski, eds.), p. 89. Academic Press, London, 1988.
[2] D. A. Hopwood and T. Kieser, *in* "The Bacterial Chromosome" (K. Drlica and M. Riley, eds.), pp. 147–162. American Society of Microbiology, Washington, D.C., 1990.
[3] H. M. Kieser, T. Kieser, and D. A. Hopwood, unpublished results (1991).

ogy are being made by a combination of *in vivo* and *in vitro* genetics.[4,5] As well as being amenable to a wide range of *in vivo* genetic procedures, the A3(2) strain is an excellent host for the homologous cloned DNA required for many of these studies. However, *S. coelicolor* shows very strong restriction against DNA from *E. coli,* so the closely related *Streptomyces lividans* 66 has been used for most heterologous cloning.[6–8] *Streptomyces lividans* is largely nonrestricting and has the added advantage that it can be used as an intermediate host for cloned DNA manipulated in *E. coli* and destined for *S. coelicolor* A3(2). [*Streptomyces coelicolor* A3(2) has a methylation-dependent restriction system. Bifunctional plasmids isolated from a completely nonmethylating (Dam⁻ Dcm⁻ HsdS⁻) strain of *E. coli* transform *S. coelicolor.*[9] This approach first succeeded for *Streptomyces avermitilis,* which has methylation-dependent restriction systems.[10]] Whether the recently discovered ability of some *E. coli* plasmids to promote conjugation with *Streptomyces,* leading to plasmid transfer,[11] will alleviate restriction barriers remains to be seen, but it clearly has some other interesting practical implications. Differences between *S. coelicolor* and *S. lividans* are listed in Table I.

A Rec⁻ derivative (JT46[12]) of *S. lividans* has particular advantages for cloning with multicopy vectors, which tend to be structurally much more stable in this host than in wild-type *S. lividans.* Unfortunately, structurally unstable derivatives of the low copy number SCP2* plasmid do not seem to profit from being propagated in JT46.[13] (The exact nature of the mutation in JT46 is not known: it greatly reduces intraplasmid homologous and nonhomologous recombination but has no effect on homologous recombination between chromosomal markers in conjugative crosses or protoplast fusions, nor on homologous plasmid–chromosome recombination.[14]) Other hosts that have been recorded as nonrestricting and convenient for heterologous cloning include *Streptomyces ambofaciens*[15] and *Streptomy-*

[4] D. A. Hopwood, *Proc. R. Soc. London, Ser. B* **235,** 121 (1988).

[5] K. F. Chater, *Trends Genet.* **5,** 372 (1989).

[6] P. K. Tomich, *Antimicrob. Agents Chemother.* **32,** 1465 (1988).

[7] D. A. Hopwood, T. Kieser, D. J. Lydiate, and M. J. Bibb, *in* "The Bacteria" (S. W. Queener and L. E. Day, eds.), Vol. 9, p. 159. Academic Press, New York, 1986.

[8] H. Ogawara, *Actinomycetologica* **3,** 9 (1989).

[9] C. P. Smith, personal communication (1989); T. Kieser and R. E. Melton, unpublished (1990).

[10] D. J. MacNeil, *J. Bacteriol.* **170,** 5607 (1988).

[11] P. Mazodier, R. Petter, and C. Thompson, *J. Bacteriol.* **171,** 3583 (1989).

[12] J. F.-Y. Tsai and C. W. Chen, *Mol. Gen. Genet.* **208,** 211 (1987).

[13] T. Kieser and R. E. Melton, unpublished (1989).

[14] H. M. Kieser, D. J. Henderson, C. W. Chen, and D. A. Hopwood, *Mol. Gen. Genet.* **220,** 60 (1989).

[15] P. Matsushima and R. H. Baltz, *J. Bacteriol.* **163,** 180 (1985).

TABLE I
COMPARISON OF *Streptomyces coelicolor* A3(2) AND *S. lividans* 66

Feature[a]	*S. coelicolor* A3(2)	*S. lividans* 66
Linkage map	Very detailed[b]	Rudimentary[c]
Amplifiable DNA elements[d]	Yes	Yes
Fe^{2+}-dependent DNA degradation during electrophoresis[e]	No	Yes[f]
Photoreactivation of UV lesions[g]	No	Yes
Antibiotics known to be produced[h]	Actinorhodin (ACT), undecylprodigiosin (RED),[i] methylenomycin,[j] CDA (calcium-dependent antibiotic)	ACT and RED
Diffusible agarase[k] (Dag)	Dag[+][l]	Dag[-]
Improved sporulation on R2YE medium with added Cu^{2+} (2 μM)[m]	No	Yes
Inducible mls resistance[n]	?	Yes
Unstable argininosuccinate synthase gene (*argG*)[o]	Yes	Yes
Unstable chloramphenicol resistance[p]	Yes	Yes
Unstable tetracycline resistance[q]	?	Yes
Unstable mercury resistance[r]	Yes	Yes
Phage ϕC31 growth limitation (Pgl)[s]	Pgl[+][t]	Pgl[-]
Stability of ϕC31 lysogens	Stable	Unstable[u]
Plasmid content of wild-type strain[v]	SCP1, SCP2,[w] SLP4, SLP1[int x]	SLP2, SLP3
Transformation with bifunctional plasmids from *E. coli*	No[y]	Yes[z]
IS*117* (2527 bp), minicircle (see text)	Two integrated copies together with ccc form at low copy number[aa]	Absent[bb]
IS*110* (1.55 kb)[cc]	At least three copies	Absent
IS*113* (2 kb)[dd]	Absent	One copy
IS*281*-like element (1.4 kb)[ee]	One copy	One copy
IS*466* (1628 bp)[ff]	Two copies in chromosome; one copy in SCP1	Absent
IS*493* (1.6 kb) (see text)	Absent	Three copies

[a] Only leading references, or reviews, are given.

[b] D. A. Hopwood and T. Kieser, *in* "The Bacterial Chromosome" (K. Drlica and M. Riley, eds.), p. 147. American Society for Microbiology, Washington, D.C., 1990.

[c] D. A. Hopwood, T. Kieser, H. M. Wright, and M. J. Bibb, *J. Gen. Microbiol.* **129,** 2257 (1983).

[d] Amplifiable DNA is much further investigated in *S. lividans* than in *S. coelicolor*. There are many different amplifiable sequences. Reviewed by R. Hütter and T. Eckhardt, *in* "Actinomycetes in Biotechnology" (M. Goodfellow, S. T. Williams, and M. Mordarski, eds.), p. 89. Academic Press, London, 1988.

[e] X. Zhou, Z. Deng, J. Firmin, D. A. Hopwood, and T. Kieser, *Nucleic Acids Res.* **16,** 4341 (1988).

ces griseofuscus,[16] but they have not attained widespread use. On the other hand, many other strains have received at least a moderate amount of genetic study because of the interest in their intrinsic properties, usually antibiotic production.[17-20] A list of *Streptomyces* strains that have been transformed or transfected is given by Hütter and Eckhardt.[1]

Streptomyces Plasmids

Covalently closed circular (ccc) DNA plasmids are found in about 30% of wild-type *Streptomyces* strains examined (reviewed in Ref. 7). A few are cryptic, but most carry genes which allow very efficient (up to 100%)

[16] R. N. Rao, M. A. Richardson, and S. Kuhstoss, this series, Vol. 153, p. 166.
[17] E. T. Seno and R. H. Baltz, in "Regulation of Secondary Metabolism" (S. Shapiro, ed.), p. 1. Chemical Rubber Company, Boca Raton, Florida, 1989.
[18] K. F. Chater, *Bio/Technology* **8**, 115 (1990).
[19] J. F. Martín and P. Liras, *Annu. Rev. Microbiol.* **43**, 173 (1989).
[20] I. S. Hunter and S. Baumberg, in "Microbial Products: New Approaches. Society for General Microbiology Symposium 44" (S. Baumberg, I. Hunter, and M. Rhodes, eds.). Cambridge Univ. Press, Cambridge, 1988.

---

f DNA of strain ZX7, a mutant derivative of *S. lividans,* is not susceptible to degradation, making this mutant suitable for pulsed-field gel electrophoresis experiments.
g R. J. Harold and D. A. Hopwood, *Mutat. Res.* **16**, 27 (1972).
h D. A. Hopwood, *Proc. R. Soc. London, Ser. B* **235**, 121 (1988).
i There are at least four different prodigionines produced [S.-W. Tsao, B. A. M. Rudd, X.-G. He, C.-J. Chang, and H. G. Floss, *J. Antibiot.* **38**, 128 (1985)].
j Methylenomycin biosynthesis is encoded by SCP1 (see text).
k D. A. Hodgson and K. F. Chater, *J. Gen. Microbiol.* **124**, 339 (1981).
l NF strains which have SCP1 integrated into the chromosome do not produce diffusible agarase because the *dag* gene is deleted [K. Kendall and J. Cullum, *Mol. Gen. Genet.* **202**, 240 (1986)].
m T. Kieser and R. E. Melton, unpublished results (1989).
n G. Jenkins, M. Zalacain, and E. Cundliffe, *J. Gen. Microbiol.* **135**, 3281 (1989).
o H. Ishihara, M. M. Nakano, and H. Ogawara, *J. Antibiot.* **38**, 787 (1985).
p R. F. Freeman, M. J. Bibb, and D. A. Hopwood, *J. Gen. Microbiol.* **98**, 453 (1977); P. Dyson and H. Schrempf, *J. Bacteriol.* **169**, 4796 (1987).
q H. Schrempf, P. Dyson, W. Dittrich, M. Betzler, C. Habiger, B. Mahro, V. Brönneke, A. Kessler, and H. Düvel, in "Biology of Actinomycetes '88" (Y. Okami, T. Beppu, and H. Ogawara, eds.), p. 145. Japan Scientific Society Press, Tokyo, 1988.
r H. Nakahara, J. L. Schottel, T. Yamada, Y. Miyakawa, M. Asakawa, J. Harville, and S. Silver, *J. Gen. Microbiol.* **131**, 1053 (1985). The mercury resistance genes are encoded by an amplifiable DNA sequence (J. Altenbuchner and M. Brüderlein, in "Book of Abstracts of the 5th International Symposium on the Genetics of Industrial Microorganisms, GIM 86," S6-P10, p. 42. 1986).
s T. A. Chinenova, N. M. Mkrtumyan, and N. D. Lomovskaya, *Genetika* **18**, 1945 (1982).
t øC31 does not form plaques on wild-type *S. coelicolor* A3(2), but lysogenization is possible (J1501 is one of many Pgl⁻ derivatives of *S. coelicolor* which allow normal plaque formation, whereas M130 is temperature-sensitive for Pgl).

plasmid transfer and moderately efficient (10^{-6} to 10^{-3}) exchange of chromosomal genes. (No natural ccc plasmid has been proven to carry genes for other properties such as the antibiotic or metal resistance commonly determined by plasmids of gram-negative or low G + C gram-positive bacteria.) When a small number of plasmid-containing spores are plated with about 10^6–10^9 plasmid-free spores, self-transfer of plasmids results in the formation of "pocks": macroscopically visible, sharply defined circular areas, 1–4 mm in diameter, where the differentiation of the mycelium is retarded.[7,21,22] Most spores arising in a pock contain the plasmid which caused the pock. The pock phenotype can be used instead of antibiotic selection in transformation experiments: one plasmid-carrying trans-

[21] M. J. Bibb and D. A. Hopwood, *J. Gen. Microbiol.* **126**, 427 (1981).
[22] M. J. Bibb, J. M. Ward, and D. A. Hopwood, *Nature (London)* **274**, 398 (1978).

---

[u] Evidence is anecdotal; some results could reflect differences between phage vectors rather than differences between the two species [T. Eckhardt and R. Smith, *in* "Genetics of Industrial Microorganisms, Proceedings of the 5th International Symposium GIM 86, Part B" (M. Alačević, D. Hranueli, and Z. Toman, eds.), p. 17. Pliva, Zagreb, Yugoslavia, 1987; K. F. Chater, personal communication, 1989].

[v] Reviewed by D. A. Hopwood, T. Kieser, D. J. Lydiate, and M. J. Bibb, *in* "The Bacteria" (S. W. Queener and E. Day, eds.), p. 159. Academic Press, New York, 1986.

[w] SCP2* is a spontaneous mutant derivative of SCP2 which gives increased fertility; it is present in many stock strains descended from crosses [M. J. Bibb and D. A. Hopwood, *J. Gen. Microbiol.* **126**, 427 (1981)].

[x] "Plasmid-free" derivatives of *S. coelicolor* A3(2) lack SCP1 and SCP2 and are almost self-sterile, but they contain SLP1 integrated into the chromosome and SLP4 (of unknown structure and properties) [D. A. Hopwood, T. Kieser, H. M. Wright, and M. J. Bibb, *J. Gen. Microbiol.* **129**, 2257 (1983)].

[y] Very strong methylation-dependent restriction. DNA from *dam⁻ dcm⁻ hsdS⁻ E. coli* can be readily introduced into *S. coelicolor* A3(2).

[z] There is a weak methylation-dependent restriction of *E. coli* DNA by *S. lividans*, detected only with certain plasmids [D. J. MacNeil *J. Bacteriol.* **170**, 5607 (1988)].

[aa] J1501 (see footnote *t* has only one copy of IS*117* [D. J. Lydiate, A. M. Ashby, D. J. Henderson, H. M. Kieser, and D. A. Hopwood, *J. Gen. Microbiol.* **135**, 941 (1989)].

[bb] *Streptomyces lividans* contains the preferred chromosomal integration site (deleted in strain DH172) (see text).

[cc] K. F. Chater, C. J. Bruton, S. G. Foster, and I. Tobek, *Mol. Gen. Genet.* **200**, 235 (1985).

[dd] D. J. Henderson, Ph.D. Thesis, University of East Anglia, p. 84. Norwich, England, (1988).

[ee] I. A. Sladkova, *Mol. Biol.* **20**, 1079 (1986).

[ff] R. Di Guglielmo, C. Conzelmann, F. Flett, and J. Cullum, *J. Cell. Biochem.* Suppl. 14A (Abstr. CC 402), 124 (1990); K. Kendall and J. Cullum, *Mol. Gen. Genet.* **202**, 240 (1986). A transposon has been constructed using two copies of IS*466* in inverted orientation (F. Flett, personal communication, 1990).

formant can be seen in a lawn of at least 10^8 plasmid-free individuals.[22] Transfer functions (including pock formation, which is interpreted as a manifestation of the "spreading" of the plasmid within the recipient culture after primary transfer from the donor[23]) require very little DNA [~2.5 kilobases (kb), or three genes, in the most studied example, pIJ101].[23–26] This may reflect the lack of a need for sex pili to ensure prolonged close contact of the nonmotile *Streptomyces* hyphae.

Plasmid-mediated conjugation (sometimes without identification of the plasmid involved) has been used widely for the construction of genetic maps of *Streptomyces* strains.[27] In *S. coelicolor* A3(2), two plasmids have been implicated: SCP2, a 31-kb ccc plasmid (and its SCP2* variants with enhanced transfer ability[7,21]), and SCP1, a linear plasmid of 350 kb.[28] When SCP1 is integrated into the chromosome or forms plasmid primes, very high frequencies of recombinants are produced in matings with SCP1⁻ strains.[2] SCP1 is also the only *Streptomyces* plasmid proven to encode an antibiotic biosynthetic pathway (methylenomycin), together with a specific resistance gene.[29]

Plasmids have also been very widely used as general cloning vectors in *Streptomyces*.[7] Most *Streptomyces* plasmids appear to be confined to the genus, though some will replicate in other actinomycete genera, for example, pIJ101 derivatives in *Micromonospora*,[30] *Amycolatopsis*,[31] *Saccharopolyspora*,[32] and *Thermomonospora*.[33] No wide host range plasmids from gram-negative or low G + C gram-positive bacteria have been found to replicate in streptomycetes.

Streptomyces Phages

All known *Streptomyces* phages have double-stranded DNA; some have cohesive ends, and others presumably use headful packaging. One such phage, SV1, which unfortunately grows only on the chloramphenicol

---

[23] T. Kieser, D. A. Hopwood, H. M. Wright, and C. J. Thompson, *Mol. Gen. Genet.* **185,** 223 (1982).

[24] K. Kendall and S. N. Cohen, *J. Bacteriol.* **169,** 4177 (1987).

[25] D. S. Stein, K. J. Kendall, and S. N. Cohen, *J. Bacteriol.* **171,** 5768 (1989).

[26] K. Kendall and S. N. Cohen, *J. Bacteriol.* **170,** 4634 (1988).

[27] D. A. Hopwood, *in* "Manual of Industrial Microbiology and Biotechnology" (A. L. Demain and N. A. Solomon, eds.), p. 191. American Society for Microbiology, Washington, D.C., 1986.

[28] H. Kinashi and M. Shimaji-Murayama, *J. Bacteriol.* **173,** 1523 (1991).

[29] K. F. Chater and C. J. Bruton, *EMBO J.* **4,** 1893 (1985).

[30] P. Matsushima and R. H. Baltz, *J. Antibiot.* **41,** 583 (1988).

[31] P. Matsushima, M. A. McHenney, and R. H. Baltz, *J. Bacteriol.* **169,** 2298 (1987).

[32] H. Yamamoto, K. H. Maurer, and C. R. Hutchinson, *J. Antibiot.* **39,** 1304 (1986).

[33] K. A. Pidcock, B. S. Montenecourt, and J. A. Sands, *Appl. Environ. Microbiol.* **50,** 693 (1985).

producer, *Streptomyces venezuelae,* is a useful generalized transducer for that strain.[34] There is also an example of a phage (øSF1) which functions as a conjugative plasmid (pUC13) in the prophage form; it is a generalized transducer.[35] *Streptomyces* phages appear to be confined to members of the genus.

Most currently used *Streptomyces* phage vectors come from the 41.4-kb temperate phage øC31 which can form plaques in and lysogenize a wide range of *Streptomyces* species (reviewed in Ref. 36). Up to 9.5 kb of DNA can be cloned into c^- $attP^-$ øC31 vectors, which can form double lysogens by homologous recombination with an integrated wild-type prophage. Vectors that are att^- but c^+ can also lysogenize *Streptomyces* strains by homologous recombination between a cloned piece of DNA and identical sequences in the host. If the cloned fragment is internal to a transcription unit, gene disruption occurs; that is, a mutation is generated at the site of phage integration.[37] Such "mutational cloning" is an elegant way of obtaining mutations and cloned pieces of the disrupted genes in one experiment. (All kinds of øC31 lysogens release phages particles spontaneously at a low frequency, allowing recovery of hybrid phage from lysogens.) *In vivo* site-directed disruption of chromosomal genes by att^- phages has also been used to determine the functions and the transcriptional organization of already cloned DNA fragments.[29,38–40]

Scope of Chapter

Rather than adding a further review on general genetic manipulation procedures with *Streptomyces* to those available in this series[16,41] and elsewhere,[1,27,42–44] we here describe new methods that combine *in vivo* and

[34] S. Vats, C. Stuttard, and L. C. Vining, *J. Bacteriol.* **169**, 3809 (1987).

[35] S.-T. Chung, *Gene* **17**, 239 (1982).

[36] K. F. Chater, *in* "The Bacteria" (S. W. Queener and L. E. Day, eds.), Vol. 9, p. 119. Academic Press, New York, 1986.

[37] K. F. Chater and C. J. Bruton, *Gene* **26**, 67 (1983).

[38] F. M. Malpartida and D. A. Hopwood, *Mol. Gen. Genet.* **205**, 66 (1986).

[39] F. M. Malpartida, S. E. Hallam, H. M. Kieser, H. Motamedi, C. R. Hutchinson, M. J. Butler, D. A. Sugden, M. Warren, C. McKillop, C. R. Bailey, G. O. Humphreys, and D. A. Hopwood, *Nature (London)* **325**, 818 (1987).

[40] E. P. Guthrie and K. F. Chater, *J. Bacteriol.* **172**, 6189 (1990).

[41] D. A. Hopwood, M. J. Bibb, K. F. Chater, and T. Kieser, this series, Vol. 153, p. 116.

[42] D. A. Hopwood, M. J. Bibb, K. F. Chater, T. Kieser, C. J. Bruton, H. M. Kieser, D. J. Lydiate, C. P. Smith, J. M. Ward, and H. Schrempf, "Genetic Manipulation of *Streptomyces:* A Laboratory Manual." John Innes Foundation, Norwich, England, 1985.

[43] M. Okanishi, M. Katagiri, T. Furumai, K. Takeda, K. Kawaguchi, M. Saitoh, and S. Nabeshima, *J. Antibiot.* **36**, 99 (1983).

[44] I. S. Hunter, *in* "DNA Cloning, Volume II: A Practical Approach" (D. M. Glover, ed.), p. 19. IRL Press, Oxford, 1985.

in vitro techniques for introducing cloned genes stably into the *Streptomyces* genome, and techniques for mutating chromosomal genes by transposon insertion or gene replacement. These systems extend the versatile and well-tested possibilities arising from the earlier development of the ϕC31 phage by K. F. Chater and colleagues into a whole range of vectors suitable for integrating DNA into the *Streptomyces* genome. Phage techniques have been fully described.[36,41,42,45]

Cloning with IS*117*, the Minicircle of *Streptomyces coelicolor* A3(2)

IS*117*, the 2527-base pair (bp) minicircle, occurs in *S. coelicolor* A3(2) and most of its derivatives (Fig. 1a) as a ccc molecule of very low copy number (about one per 10–20 chromosomes),[46] together with two linear copies integrated in the chromosome at two distant sites, A and B. [J1501, a frequently used Pgl⁻ strain (i.e., able to support plaque formation by ϕC31[47]) has only copy A; copy B seems to have been lost, together with flanking DNA, by a spontaneous deletion event rather than by specific excision of the element.] *Streptomyces lividans* does not contain any sequence that hybridizes with IS*117*.

When the circular form of IS*117* was linearized at its unique *Bcl*I site and cloned into the *attP⁻ c⁺* ϕC31 derivative KC515 (Fig. 1b),[46] the resulting phage, KC591 (Fig. 1c), efficiently lysogenized *S. lividans* by recombination between the cloned IS*117* and a specific site in the chromosome, called the "preferred integration site" (Fig. 1d). Several other *Streptomyces* strains were also lysogenized in this way by KC591 but not by KC515. KC590, a derivative of KC515 carrying only the 1.8-kb *Bcl*I–*Sst*I fragment of IS*117* and so lacking its attachment site (*att* in Fig. 1), failed to lysogenize. Integration via the IS*117* site was also found when the circular form was linearized at its unique *Hind*III site and cloned into pBR327, together with a 2.4-kb *tsr/neo* resistance cassette for selection of *Streptomyces* transformants, to produce pIJ4210, pIJ4200 (Fig. 1e), and other similar plasmids.[48]

Transformation of *S. lividans* protoplasts with pIJ4210 or pIJ4200 isolated from *E. coli* yielded thiostrepton-resistant colonies at a frequency of 10³ per microgram, about 1000-fold lower than for *Streptomyces* plasmids

[45] K. F. Chater, N. D. Lomovskaya, T.A. Voeykova, I. A. Sladkova, N. M. Mkrtumian, and G. L. Muravnik, *in* "Biological, Biochemical and Biomedical Aspects of Actinomycetes" (G. Szabó, S. Biró, and M. Goodfellow, eds.), p. 45. Akadémiai Kiadó, Budapest, 1986.

[46] D. J. Lydiate, H. Ikeda, and D. A. Hopwood, *Mol. Gen. Genet.* **203**, 79 (1986).

[47] T. A. Chinenova, N. M. Mkrtumyan, and N. D. Lomovskaya, *Genetika* **18**, 1945 (1982).

[48] D. J. Henderson, D. J. Lydiate, and D. A. Hopwood, *Mol. Microbiol.* **3**, 1307 (1989).

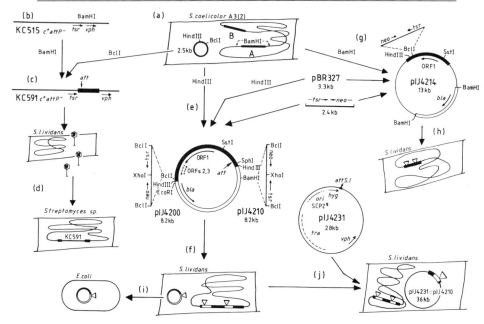

FIG. 1. IS*117* (minicircle), a 2.5-kb transposable element of *S. coelicolor* A3(2). Steps a–j are explained in the text. Black bars are IS*117* sequences, open boxes are pBR327 DNA, and single lines are DNA of the chromosome, øC31, pIJ4231, or the 2.4-kb fragment with the *tsr* and *neo* genes. *att*, IS*117* site which recombines with the chromosome on integration; *att S.l.*, *S. lividans* chromosomal preferred "attachment site" for the minicircle; *attP*, øC31 attachment site; *c*, øC31 repressor gene; *bla*, β-lactamase gene (ampicillin resistance in *E. coli*); *hyg*, hygromycin resistance gene, expressed in *Streptomyces; neo*, promoterless *neo* gene from Tn5; ORF, open reading frame; *ori*, origin of replication; *tra*, transfer genes of SCP2*, also responsible for pock formation; *tsr*, thiostrepton resistance gene, expressed in *Streptomyces; vph*, viomycin resistance gene, expressed in *Streptomyces*.

of similar size, indicating that the minicircle does not replicate autonomously. pIJ4210 was always found integrated as tandem copies in *S. lividans* (Fig. 1f), whereas pIJ4200 gave a mixture of single and tandem inserts, always in the same site. This site has the same sequence over at least 300 bp as the A site of *S. coelicolor* A3(2).[48] pIJ4210 and pIJ4200 integrated in *S. lividans* were stably maintained in the absence of thiostrepton. (The strains grew and sporulated much better in the absence of thiostrepton, which is also observed with øC31 prophages but not for autonomously replicating plasmids containing the *tsr* gene.)

To show that a naturally integrated copy of IS*117*, from which the free ccc forms probably arise, can transpose, integrated copy A, together with flanking *S. coelicolor* chromosomal sequences, was cloned into pBR327 using *Bam*HI, which does not cut within the element, and tagged by inserting into the *Bcl*I site the same *tsr*/*neo* cassette present in pIJ4200 and pIJ4210.[48] The resulting plasmid, pIJ4214 (Fig. 1g), transformed *S. lividans* to thiostrepton resistance, but about 100 times less frequently than these two plasmids. Most transformants contained two tandem copies of the tagged IS*117* transposed (without pBR327 flanking sequences) into the preferred attachment site (Fig. 1h). (About 20% of transformants had the whole of pIJ4214 integrated, presumably by homologous recombination between flanking *S. coelicolor* sequences on the plasmid and the very similar *S. lividans* chromosomal sequences.) The 100-fold lower transformation frequency for pIJ4214 (requiring excision and reintegration) compared with circular forms like pIJ4210 (requiring only integration) suggests that excision of the element from its linear state may be rate limiting. Thus, the circle is deduced to be the normal intermediate in transposition.

The sequence of IS*117*[48] shows one large open reading frame (ORF1), where disruption abolishes integration, whereas insertion of the *tsr*/*neo* cassette into the *Sst*I site 132 bp upstream of the translation start of ORF1 caused reduced transformation frequencies; thus, the ORF1 product probably has transposase/integrase function. [This hypothesis is supported by the finding from nucleotide sequencing that the single large ORFs of each of the insertion sequences IS*110* of *S. coelicolor* A3(2),[49] IS*116* of *Streptomyces clavuligerus*,[50] and IS*900* of *Mycobacterium paratuberculosis*[51] specify proteins similar to the IS*117* ORF1.[50]] There are two additional, smaller, overlapping, potential ORFs in the minicircle; inactivation of these did not depress integration of the circular element,[52] but their possible involvement in excision is under study.

The chromosomal integration site A of *S. coelicolor* and its *S. lividans* homolog share only a short stretch of partial identity with the minicircle attachment site.[48] There is no resemblance to a tRNA gene at the integration point (in contrast to the sites for some other integrating *Streptomyces* elements; see below). Integration of the minicircle does not cause the target site duplication typical of most transposable elements.

DH172, a *S. lividans* strain in which a 200 bp sequence containing the preferred integration site has been replaced by the *ermE* gene (for

[49] C. J. Bruton and K. F. Chater, *Nucleic Acids Res.* **15**, 7053 (1987).
[50] B. K. Leskiw, M. Mevarech, L. S. Barritt, S. E. Jensen, D. J. Henderson, D. A. Hopwood, C. J. Bruton, and K. F. Chater, *J. Gen. Microbiol.* **136**, 1251 (1990).
[51] E. P. Green, M. L. V. Tizard, M. T. Moss, J. Thompson, D. J. Winterbourne, J. J. McFadden, and J. Hermon-Taylor, *Nucleic Acids Res.* **17**, 9063 (1989).
[52] D. J. Henderson and D. A. Hopwood, D.-F. Brolle, R. E. Melton, and T. Kieser, unpublished results (1990).

erythromycin resistance) of *Saccharopolyspora erythraea*,[52a] was transformed by pIJ4210 at a frequency about 10- to 20-fold lower than its parent, yielding plasmid insertions at alternative sites. Insertion is, however, not entirely random since up to five inserts at the same site occurred in a sample of 19 independent events, nine of which were unique.[52a] The promoterless *neo* gene is not expressed when pIJ4210 integrates into the preferred site of *S. lividans* but is expressed at varying levels at secondary sites in DH172, indicating readthrough transcription from chromosomal promoters which might be exploited.

The unique *Xho*I and *Bcl*I sites of pIJ4210 are available for cloning. pIJ4200 has two *Bcl*I sites but a single available *Bam*HI site. The unique *Eco*RI site can presumably also be used in both plasmids. The *Xho*I site is particularly useful for cloning of *Sau*3AI fragments after partial infilling of the ends.[53,54]

Hybrid minicircle constructs that are integrated into the *S. lividans* chromosome, like the wild-type elements in *S. coelicolor* A3(2), produce low abundance ccc molecules which can be recovered by transformation of *E. coli* (selection for ampicillin resistance) (Fig. 1i).[52] Highly competent *E. coli* cells are needed because the free circles are rare. The amounts of free circles isolated from *S. lividans* are not sufficient to allow direct transformation of *S. lividans* protoplasts. pIJ4210 can, however, be mobilized and transferred efficiently between *Streptomyces* strains by pIJ4231, a pock-forming derivative of SCP2* with a 200-bp fragment containing the *S. lividans* preferred integration site.[52] pIJ4210 transposes spontaneously and frequently into this sequence to give pIJ4231::pIJ4210 cointegrates (Fig. 1j), which can transfer by conjugation or transformation to other *Streptomyces* strains or be introduced by transformation into *E. coli*.

IS*117* shows great promise as a cloning vector because of its stable inheritance in the absence of antibiotic selection. Its strong preference for integration into a single site (and in the same orientation) in *S. lividans* ensures comparable results in repeated experiments, for example, when modified forms of a gene are to be compared. Further developments of IS*117* vectors are on the way. An example is pHC2, a promoter–probe vector consisting of an *E. coli* replicon carrying essential minicircle sequences and the *xylE* reporter gene downstream of a multiple cloning site.[55]

[52a] D. J. Henderson, D.-F. Brolle, T. Kieser, R. E. Melton, and D. A. Hopwood, *Mol. Gen. Genet.* **224,** 65 (1990).

[53] E. R. Zabarovsky and R. L. Allikmets, *Gene* **42,** 119 (1986).

[54] C. Korch, *Nucleic Acids Res.* **15,** 3199 (1987).

[55] S. Wong, personal communication (1990).

Cloning with Integrating Plasmids of SLP1 Class

Eight plasmids [SLP1 from *S. coelicolor* A3(2),[56,57] pSAM2 (*S. ambofaciens*),[58,59] pIJ110 (*S. parvulus*),[60] pIJ408 (*S. glaucescens*),[60,61] pSG1 (*S. griseus*),[62] pSE101 and pSE211 (*Saccharopolyspora erythraea*),[63,64] and pMEA100 (*Amycolatopsis mediterranei*),[65,66]] are known that can integrate at specific sites in host chromosomes. Five of them were discovered in their original host as low copy number ccc DNA molecules. The other three (SLP1, pIJ110, pIJ408) were isolated as ccc molecules from *S. lividans* strains that had acquired the ability to form pocks during mixed culture with the parent strains. Probing showed the parents to lack ccc DNA but to have the plasmids integrated into a larger replicon; for SLP1, genetic mapping proved this to be the *S. coelicolor* chromosome.[56]

The host integration sites (*attB*), sequenced for six of the above plasmids, resemble, and probably are, tRNA genes.[63,64,67,68] Integration presumably proceeds via recombination between an *attP* site (*P* stands for plasmid) on a circular form of the plasmid and the *attB* site of the host. The *attP* and *attB* sites are identical over at least 43 bp. Because the *attB* sequence overlaps an end of the tRNA sequence, an intact tRNA gene remains after plasmid integration. Here we concentrate on SLP1 and pSAM2 and their development as integrating cloning vectors, referring only briefly to other members of the plasmid family.

SLP1 Derivatives: Integrating and Autonomously Replicating Vectors

The 17.2-kb SLP1 element is integrated near the *strA* marker in *S. coelicolor* A3(2) (Fig. 2a)[56] in a tRNA$_{Tyr}$ gene.[67] No SLP1 ccc DNA can be detected in *S. coelicolor*. When spores of a SCP1$^-$ SCP2$^-$ *S. coelicolor*

[56] M. J. Bibb, J. M. Ward, T. Kieser, S. N. Cohen, and D. A. Hopwood, *Mol. Gen. Genet.* **184,** 230 (1981).

[57] C. A. Omer and S. N. Cohen, *Mol. Gen. Genet.* **196,** 429 (1984).

[58] J.-L. Pernodet, J.-M. Simonet, and M. Guérineau, *Mol. Gen. Genet.* **198,** 35 (1984).

[59] J.-M. Simonet, F. Boccard, J.-L. Pernodet, J. Gagnat, and M. Guérineau, *Gene* **59,** 137 (1987).

[60] D. A. Hopwood, G. Hintermann, T. Kieser, and H. M. Wright, *Gene* **11,** 1 (1984).

[61] M. Sosio, J. Madon, and R. Hütter, *Mol. Gen. Genet.* **218,** 169 (1989).

[62] A. Cohen, D. Bar-Nir, M. E. Goedeke, and Y. Parag, *Plasmid* **13,** 41 (1985).

[63] D. P. Brown, S.-J. D. Chiang, J. S. Tuan, and L. Katz, *J. Bacteriol.* **170,** 2287 (1988).

[64] D. P. Brown, J. S. Tuan, K. A. Boris, J. P. Dewitt, K. B. Idler, S.-J. D. Chiang, and L. Katz, *Dev. Ind. Microbiol.* **29,** 97 (1988).

[65] P. Moretti, G. Hintermann, and R. Hütter, *Plasmid* **14,** 126 (1985).

[66] J. Madon, P. Moretti, and R. Hütter, *Mol. Gen. Genet.* **209,** 257 (1987).

[67] W.-D. Reiter, P. Palm, and S. Yeats, *Nucleic Acids Res.* **17,** 1907 (1989).

[68] P. Mazodier, C. J. Thompson, and F. Boccard, *Mol. Gen. Genet.* **222,** 431 (1990).

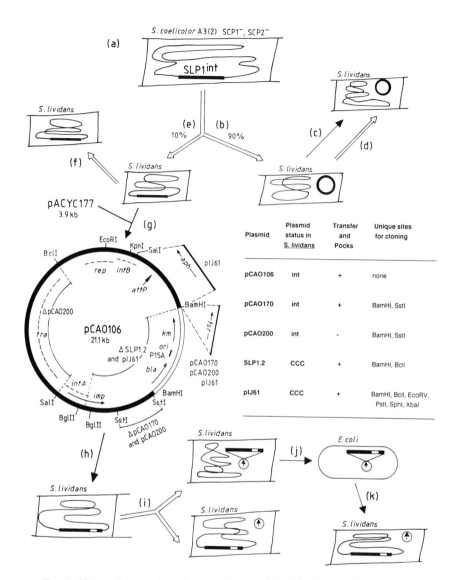

FIG. 2. SLP1 and derivatives. Steps a–k are explained in the text. Open arrows drawn with double lines indicate conjugation; filled arrows with single lines indicate transformation. Dashed lines in the restriction map of pCAO106 indicate the positions of imprecisely located functions. The small circle with an internal arrow (in i–k) is a *Streptomyces* multicopy plasmid containing the *S. lividans* attachment site for SLP1 (tRNA$_{Tyr}$ gene). *aph*, Neomycin phosphotransferase gene from *S. fradiae; attP*, site on SLP1 plasmids (e.g., pCAO106) that recombines with the chromosomal attachment site; *bla*, β-lactamase gene giving ampicillin resistance in *E. coli; imp*, inhibitor of maintenance of SLP1 plasmids; *intA, intB*, SLP1 integration functions; *km*, kanamycin resistance gene, expressed in *E. coli* and in *S. lividans; ori*, origin of replication; *rep*, SLP1 replication functions; *tra*, SLP1 transfer functions, also responsible for pock formation; *tsr*, thiostrepton resistance gene, expressed in *Streptomyces;* Δ, deletion.

A3(2) strain are mixed with a large excess (say, 10^6) of spores of *S. lividans* 66 and plated on R2YE medium, pocks (Ltz$^+$ phenotype) are seen in the developing bacterial lawn. The number of pocks varies with the *S. coelicolor* A3(2) donor but is typically around 1% of the number of *S. coelicolor* spores plated. *Streptomyces lividans* spores isolated from the pock area have acquired the ability to form pocks when plated with plasmid-free *S. lividans* strains. About 90% of the original pocks yield 10- to 14.5-kb ccc plasmids with a copy number of 4–10 per chromosome (Fig. 2b). These deleted forms of SLP1 can be introduced by transformation (Fig. 2c) into *S. lividans* protoplasts with the high efficiency (up to 10^7 per microgram ccc DNA) typical of autonomous *Streptomyces* plasmids; they are self-transmissible by conjugation (Fig. 2d) and can mobilize the *S. lividans* chromosome. SLP1 plasmids with added antibiotic resistance markers can be introduced back into *S. coelicolor* A3(2) at a frequency of only about 50 transformants per microgram ccc DNA, and yielding no pocks; the newly introduced plasmid integrates into the resident SLP1 copy.

The other 10% of pock-forming *S. lividans* exconjugants from matings with *S. coelicolor* A3(2) (Fig. 2e) lack free ccc DNA but contain the 17.2-kb SLP1 sequence integrated into the tRNA$_{Tyr}$ gene of *S. lividans* [identical to that of *S. coelicolor* A3(2)].[57,69] These *S. lividans* strains transfer the integrated SLP1 at 100% frequency to other *S. lividans* strains (Fig. 2f); in contrast to the result of interspecific transfer between *S. coelicolor* A3(2) and *S. lividans*, the 17.2-kb sequence always becomes integrated. No ccc plasmids or deleted forms have been observed to arise from this sequence once integrated in *S. lividans*. The discovery of one *S. lividans* strain with two tandem SLP1 sequences allowed excision of a complete element by digestion of the chromosomal DNA with *Bam*HI, which cuts the 17.2-kb sequence only once. The element was cloned into pACYC177 to give pCAO106,[70] which confers ampicillin and kanamycin resistance on *E. coli* (Fig. 2g). It can be introduced by transformation (Fig. 2h) into *S. lividans* at high frequency ($>10^5$ per microgram of ccc DNA). Transformants can be detected by pock formation or by selection with kanamycin (after replication to minimal medium; kanamycin selection on R2YE protoplast regeneration medium is unsatisfactory).

pCAO106 can be rescued intact from *S. lividans* if the strain containing the integrated plasmid is crossed with another *S. lividans* strain containing a nontransmissible multicopy plasmid with the *S. lividans* SLP1 *attB* site. pCAO106 probably excises from its host chromosome and, when it enters

[69] C. A. Omer and S. N. Cohen, *J. Bacteriol.* **166,** 999 (1986).
[70] C. A. Omer, D. Stein, and S. N. Cohen, *J. Bacteriol.* **170,** 2174 (1988).

the new host, integrates frequently into the *attB* site of the plasmid rather than the *attB* site of the chromosome (Fig. 2i). The resulting cointegrate plasmid can then be rescued by transformation of *E. coli* (Fig. 2j) (selection for ampicillin resistance).[71] It is noteworthy that pCAO106 changes position only when its host is mated with another strain: introduction of an *attB*-carrying multicopy plasmid into *S. lividans* with integrated pCAO106 did not result in plasmid cointegrates detectable after transformation of *E. coli*. When the cointegrate plasmids from *E. coli* are transferred back into *S. lividans,* pCAO106 can excise and regenerate the multicopy *attB* plasmid (Fig. 2k). This indicates that SLP1 moves in a conservative rather than a replicative manner.[71]

Comparison of pCAO106 with autonomous ccc plasmids found in *S. lividans* after mating with *S. coelicolor* A3(2) showed the latter to lack a region encoding a function called *imp* (inhibition of maintenance of SLP1 plasmids) which operates in trans to prevent autonomous replication of SLP1 plasmids.[72] All autonomously replicating SLP1 plasmids, except the largest one, SLP1.2, also lack *attP*. SLP1.2 can be made to integrate into the chromosome of *S. lividans* if *imp*, and possibly another function, *intA*, are supplied in trans.[72] (pIJ61,[73] a popular conventional ccc cloning vector derived from SLP1.2, lacks *attP* and cannot integrate.) Inserts in the *KpnI* site (kb 1.05) of pCAO106 also prevent plasmid integration by destroying the *intB* function. An integrating vector, pCAO170 (see map in Ref. 41), was derived from pCAO106 by insertion of the thiostrepton resistance gene and deletion of the nonessential DNA between the two *SstI* sites (kb 8.3–9.9) of pCAO106.[70] Autonomous replication of SLP1 derivatives, and also their ability to survive integrated into the host chromosome, depend on two noncontiguous functions (*intA* and *intB*) which are destroyed by

[71] S. C. Lee, C. A. Omer, M. A. Brasch, and S. N. Cohen, *J. Bacteriol.* **170,** 5806 (1988).

[72] S. R. Grant, S. C. Lee, K. Kendall, and S. N. Cohen, *Mol. Gen. Genet.* **217,** 324 (1989).

[73] C. J. Thompson, T. Kieser, J. M. Ward, and D. A. Hopwood, *Gene* **20,** 51 (1982).

FIG. 3. pSAM2 and derivatives. Steps a–d and the plasmids are described in the text. The open arrow (step c) indicates conjugation; filled arrows indicate transformation (steps d and f) or changes in the strains (steps b and e). Black bars are pSAM2 sequences. *attB*, Chromosomal attachment site for pSAM2 (tRNA$_{Pro}$ gene); *attP*, site on pSAM2 that recombines with *attB* on integration; *bla*, β-lactamase, gives ampicillin resistance in *E. coli; int*, integrase gene of pSAM2; *ori*, origin of replication; *oriT*, RK2 origin of transfer; P_{tipA}, thiostrepton-inducible promoter; *rep*, two separate regions on pSAM2 required for autonomous plasmid replication; *scm*, small circular molecules; *spc/str*, resistance determinant for spectinomycin and streptomycin, expressed in both *S. lividans* and *E. coli; ter*, transcriptional terminator of phage fd; *tra*, transfer region required for conjugative plasmid transfer; *tsr*, thiostrepton resistance gene, expressed in *Streptomyces; xis*, excision gene of pSAM2.

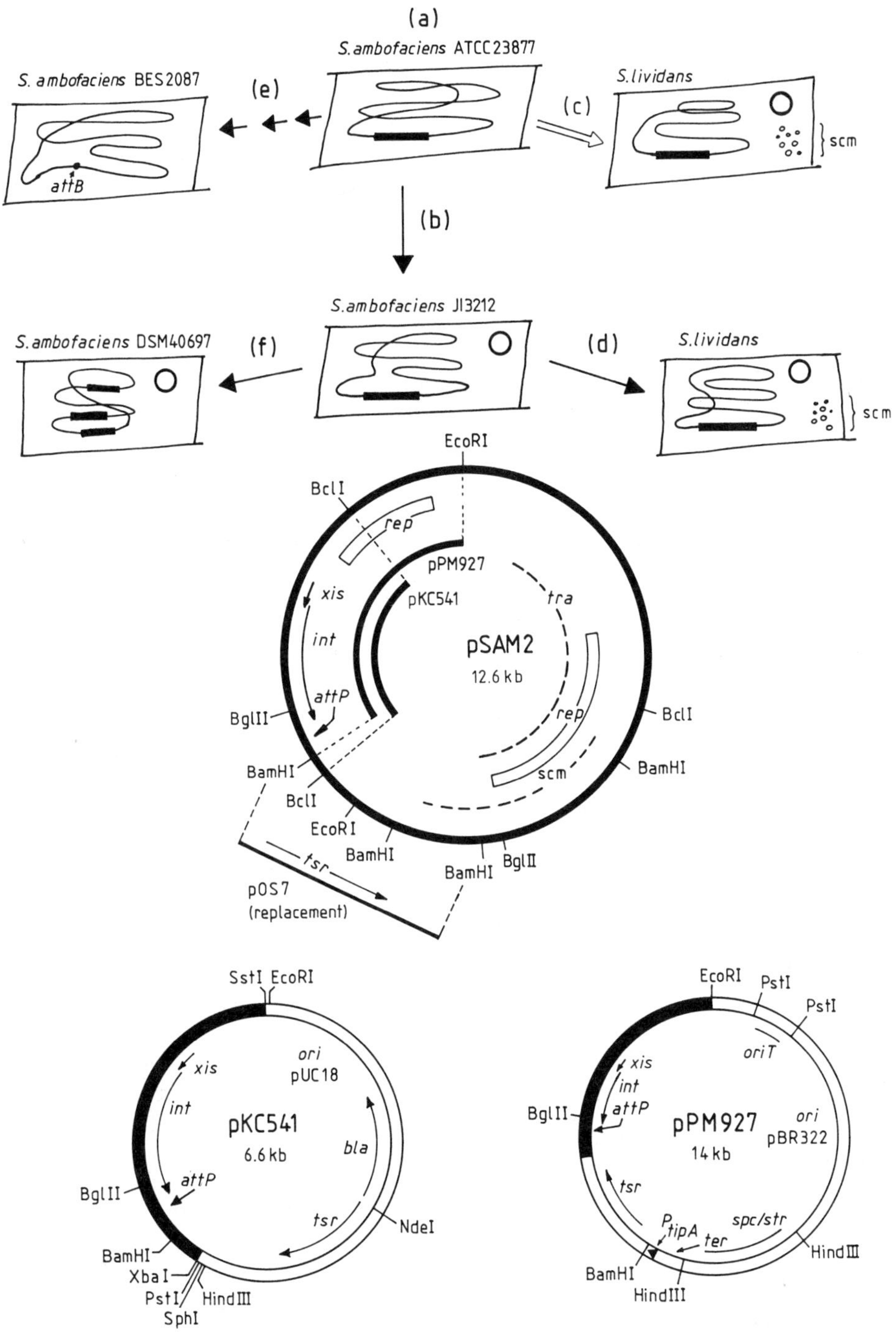

(a)
S. ambofaciens ATCC23877
S. ambofaciens BES2087
(e)
attB
(c)
S. lividans
scm
(b)
S. ambofaciens JI3212
S. ambofaciens DSM40697
(f)
(d)
S. lividans
scm
EcoRI
BclI
rep
pPM927
pKC541
xis
int
attP
tra
pSAM2
12.6 kb
rep
scm
BglII
BclI
BamHI
BamHI
BclI
EcoRI
BamHI
BamHI
BglII
tsr
pOS7
(replacement)
SstI
EcoRI
ori
pUC18
xis
int
pKC541
6.6 kb
bla
attP
BglII
tsr
NdeI
BamHI
XbaI
PstI
SphI
HindIII
EcoRI
PstI
PstI
oriT
xis
int
attP
BglII
pPM927
14 kb
ori
pBR322
tsr
P tipA
spc/str
ter
BamHI
HindIII
HindIII

deleting either the EcoRI–KpnI fragment (kb 0–1.05) or the BglII fragment (kb 11.05–11.55). The transfer ability of pCAO106 was eliminated, either by inserting DNA into the unique EcoRI site of pCAO106 or by deleting the SalI–BclI fragment (kb 12.65–19.0): this might be useful if, for containment purposes, a nontransmissible vector were required.

Integrating Vectors Derived from pSAM2

pSAM2 occurs in S. ambofaciens ATCC 23877 (Fig. 3a) integrated in a tRNA$_{Pro}$ sequence. In another S. ambofaciens strain, JI3212 (a derivative of the 23877 strain; Fig. 3b), pSAM2 occurs as both integrated and low abundance ccc forms.[58,59,68,74,75] In matings with the 23877 strain, plasmid-free S. lividans strains receive with high frequency the ability to form pocks, and they acquire both integrated and ccc forms of pSAM2 (Fig. 3c). Covalently closed circular pSAM2 from S. lividans after either conjugation with S. ambofaciens ATCC 23877 or transformation of S. lividans protoplasts with ccc DNA from S. ambofaciens JI3212 (Fig. 3d) are indistinguishable both in restriction map and in ability to integrate into host chromosomes. The free pSAM2 plasmids in S. lividans have not suffered deletions like those of the SLP1 plasmids in S. lividans. There are, however, small (~2.5 kb) heterogeneous circles (scm, small circular DNA molecules) of high copy number which originate from a specific region of pSAM2.[59] These small molecules occur in S. lividans but not in other strains into which pSAM2 or its derivatives have been introduced.

pSAM2^{int} seems to coexist with ccc pSAM2, because S. ambofaciens ATCC 23877 can be transformed with pOS7 (a thiostrepton-resistant derivative of pSAM2[59]; see map of pSAM2 in Fig. 3) at about the same frequency as strain BES2087 (Fig. 3e), a derivative of this strain lacking all pSAM2 sequences but retaining an attB site.[76]

pSAM2 probably lacks the equivalent of the imp function of SLP1 (see above). Disruption of the pSAM2 int gene (see below) produced ccc plasmids unable to integrate into the host chromosome. This suggests that pSAM2 is a replicon. The ccc form of pSAM2 could also be introduced by transformation into the plasmid-free strain S. ambofaciens DSM40697 (Fig. 3f) where the free ccc form coexists with several integrated copies of unknown location.[74] pSAM2 forms pocks in this strain and can mobilize chromosomal markers.[77]

[74] F. Boccard, J.-L. Pernodet, A. Friedmann, and M. Guérineau, Mol. Gen. Genet. 212, 432 (1988).

[75] F. Boccard, T. Smokvina, J.-L. Pernodet, A. Friedmann, and M. Guérineau, Plasmid 21, 59 (1989).

[76] S. Kuhstoss, M. A. Richardson, and R. N. Rao, J. Bacteriol. 171, 16 (1989).

[77] T. Smokvina, F. Francou, and M. Luzzati, J. Gen. Microbiol. 134, 395 (1988).

The integration and excision functions of pSAM2 have been sequenced and studied in detail.[78] Two genes resembling *int* and *xis* of temperate phages such as λ are within 2.5 kb of the *attP* site. No other functions are required for integration and stable maintenance of the element. (In SLP1, essential functions are located in opposite regions of the circular map; see above.) Small nontransmissible integrating vectors, such as pKC541[76] and pPM927[79] (Fig. 3), have been constructed from pSAM2. pPM927 carries *oriT* from RK2 and can be mobilized from *E. coli* strains carrying the broad host range plasmid RP4 into *Streptomyces*. pPM927 also features the thiostrepton-inducible *tipA* promoter[80] for the expression of cloned genes. Both pKC541 and pPM927 lack the *scm* region and so do not form small circles in *S. lividans*.

The tRNA$_{Pro}$ gene overlapping the pSAM2 *attB* site is conserved in all tested actinomycetes.[68,75] Site-specific integration of a pSAM2 derivative into the genome of *Mycobacterium smegmatis* has been demonstrated,[81] suggesting that these integrating vectors have a wider host range than autonomously replicating *Streptomyces* plasmids.

Other Integrating Plasmids

pSE101 and pSE211. The two unrelated plasmids pSE101 and pSE211 occur as very low abundance ccc molecules in the erythromycin producer *Saccharopolyspora erythraea* NRRL 2338; probing also showed a copy of each plasmid integrated in the host chromosome.[63,64] These plasmids are significant because *Saccharopolyspora erythraea* does not support the stable maintenance of pIJ101, SLP1.2, or SCP2*-derived plasmids (pIJ702 derivatives containing cloned *Saccharopolyspora erythraea* fragments integrate into the genome under selective pressure for the plasmid marker[82]) and pSV1 derivatives replicate in the substrate mycelium but are not inherited at significant frequency through spores.[32]

pSE101 (11.3 kb) does not form pocks. It is integrated in a tRNA$_{Thr}$ gene.[83] The ccc form of the element was cloned into pBR322 with a thiostrepton resistance determinant to give pECT2. When introduced by transformation into *Saccharopolyspora erythraea* NRRL 2359, an inde-

[78] F. Boccard, T. Smokvina, J.-L. Pernodet, A. Friedmann, and M. Guérineau, *EMBO J.* **8,** 973 (1989).

[79] T. Smokvina, P. Mazodiev, F. Boccard, C. J. Thompson, and M. Guérineau, *Gene* **94,** 53 (1990).

[80] T. Murakami, T. G. Holt, and C. J. Thompson, *J. Bacteriol.* **171,** 1459 (1989).

[81] C. Martin, personal communication (1990).

[82] J. M. Weber and R. Losick, *Gene* **68,** 173 (1988).

[83] L. Katz, D. Brown, and K. Idler, *J. Cell. Biochem.* Suppl. **14A** (Abstract CC409), 126 (1990).

pendent soil isolate free of pSE101 sequences, pECT2 again integrated into a single site homologous to the site in strain 2338. In *S. lividans,* pECT2 can replicate autonomously or—interestingly—integrate into several different positions in the genome.[63,64]

pSE211 (18.1 kb) is a conjugative plasmid which forms pocks when introduced by transformation into a cured derivative, strainER720,[84] of *Saccharopolyspora erythraea.* Integration occurs into a $tRNA_{Phe}$ gene. pSE211 also forms pocks in *S. lividans,* where it replicates autonomously but does not integrate into the chromosome. Like pSAM2 (see above), pSE211 has ORFs resembling *int* and *xis* close to the *attP* site.[84a]

pMEA100. The 23.7-kb pMEA100 element, from the rifamycin producer *Amycolatopsis mediterranei* LBGA3136 (formerly *Nocardia mediterranei*), is a pock-forming fertility plasmid in this organism and the only plasmid known for this species. Low abundance ccc DNA occurs in the strain, and all pMEA100 sequences could be eliminated by screening colonies after protoplast regeneration. It is not clear whether pMEA100 has its own replication functions.[65,66] The integration site, in a $tRNA_{Phe}$ gene,[67] is identical to that of pSE211 (see above).

Transposon Mutagenesis in Streptomycetes

The integrating plasmids just described insert into one or possibly a few specific sites. IS117, the minicircle, has a strongly preferred site, as well as recently described secondary sites; whether the latter can be within genes, with insertions being mutagenic, remains to be seen. Site-specific integration is ideal for cloning, where the results of experiments would become irreproducible if the vector inserted into a different sequence every time it was introduced into a host. In contrast, two transposable elements from streptomycetes, Tn4556 and IS493, seem to insert into random sites and have been developed as tools for transposon mutagenesis.

[84] J. P. DeWitt, *J. Bacteriol.* **164,** 969 (1985).
[84a] D. P. Brown, K. B. Idler, and L. Katz, *J. Bacteriol.* **172,** 1877 (1990).

FIG. 4. Transposon Tn4556 and its derivative Tn4560. Black bars are Tn4556 sequences. Steps a–f are explained in the text. MM, Minimal medium; ORF, open reading frame; Thio, thiostrepton; *tsr*, thiostrepton resistance gene, expressed in *Streptomyces;* t[s], temperature sensitive; Vio, viomycin; *vph*, viomycin resistance gene, expressed in *Streptomyces.*

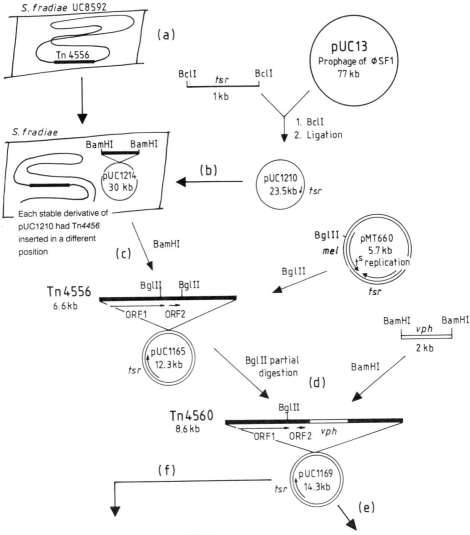

S. fradiae UC8592

Tn 4556

(a)

BclI   tsr   BclI

1 kb

pUC13
Prophage of φSF1
77 kb

1. BclI
2. Ligation

S. fradiae

BamHI   BamHI

pUC1214
30 kb

(b)

pUC1210
23.5 kb tsr

Each stable derivative of
pUC1210 had Tn4456
inserted in a different
position

(c)   BamHI

BglII   pMT660
5.7 kb
mel   t^S replication

tsr

Tn 4556
6.6 kb

BglII   BglII

ORF1   ORF2

BglII

pUC1165
12.3 kb
tsr

BglII partial
digestion

(d)

BamHI   vph   BamHI

2 kb

BamHI

Tn 4560
8.6 kb

BglII

ORF1   ORF2   vph

(f)

pUC1169
14.3 kb
tsr

(e)

TRANSPOSITION INTO POCK-FORMING PLASMIDS

1. Introduce target plasmid into strain
   containing pUC1169

2. Propagate

3. Mix spores containing pUC1169 and the
   target plasmid with a large excess
   plasmid-free spores and plate to obtain
   separate pocks

4. Replica plate to MM + Vio and MM + Thio

5. Purify strains from areas of Vio^R Thio^S
   growth

TRANSPOSITION INTO THE CHROMOSOME

1. Introduce pUC1169 into host by
   transformation (Thio^R or Vio^R selection)

2. Propagate at 29° on Thio

3. Harvest spores and plate for single
   colonies on MM + Vio

4. Incubate 5 days at 39°

5. Replica plate to MM + Vio and MM + Thio

6. Purify Vio^R Thio^S strains

Tn4556 and its Derivative Tn4560

Tn4556, a transposon of the Tn3 family,[85-87] was discovered in a neomycin producer, *Streptomyces fradiae* UC8592 (Fig. 4a), when this strain was transformed with a very unstable plasmid, pUC1210 (Fig. 4b), derived from the 77 kb plasmid prophage of øSF1[35] from the same strain. Occasionally, and reproducibly, stable plasmids with increased copy number appeared, with an insert of 6.6 kb. Over 100 independently generated stable derivatives of pUC1210 were analyzed, and each had a single insert of 6.6 kb in a different position. This insert was designated Tn4556. Insertion caused a 5-bp duplication of the target sequence. One of these plasmids, pUC1214, had Tn4556 inserted into a *Bam*HI site, which was duplicated in the process. This enabled the transposon to be cloned into a smaller vector, pMT660,[88] a temperature-sensitive derivative of the wide host-range plasmid pIJ702, to give pUC1165 (Fig. 4c). Addition of the promoterless *vph* gene (for viomycin resistance) into one of the two *Bgl*II sites of the transposon gave pUC1169 (Fig. 4d) carrying Tn4560. pUC1169 has been used for transposon mutagenesis of chromosomal genes (Fig. 4e) and, with a somewhat different procedure (Fig. 4f), for the mutagenesis of SCP2*-derived plasmids and of genes cloned into these plasmids. The availability of the DNA sequence of the whole element[89] will facilitate the construction of derivatives of Tn4556 with specific properties.

To obtain insertions into the chromosome (Fig. 4e), pUC1169 is introduced into the strain of interest by transformation (the plasmid is nontransmissible and does not form pocks). Selection is for thiostrepton resistance or viomycin resistance. The strains must be grown at low temperature (29°) to ensure maintenance of the plasmid. Spores are harvested and plated on antibiotic-free medium to give single colonies. The plates are incubated at 39° or 38° [the highest temperatures tolerated by *S. lividans* 66 and *S. coelicolor* A3(2), respectively]. After sporulation, the colonies are replica plated to agar containing viomycin or thiostrepton. Viomycin-resistant, thiostrepton-sensitive colonies will have lost the plasmid but retained Tn4560 by its transposition to the chromosome (or another replicon). This procedure has occasionally failed, presumably because the temperature sensitivity of plasmid replication reverted to wild type; it is

---

[85] S.-T. Chung, *J. Bacteriol.* **169**, 4436 (1987).

[86] E. R. Olson and S.-T. Chung, *J. Bacteriol.* **170**, 1955 (1988).

[87] S.-T. Chung and L. L. Crose, *in* "Genetics and Molecular Biology of Industrial Microorganisms" (C. L. Hershberger, S. W. Queener, and G. Hegemann, eds.), p. 168. American Society for Microbiology, Washington, D.C., 1989.

[88] A. W. Birch and J. Cullum, *J. Gen. Microbiol.* **131**, 1299 (1983).

[89] D. R. Siemieniak, J. L. Slightom, and S.-T. Chung, *Gene* **86**, 1 (1990).

therefore advisable to purify a temperature-sensitive variant at the outset of the experiment. It should also be possible to eliminate pUC1169 by exploiting the strong incompatibility between wild-type (Sti$^+$) pIJ101 and its Sti$^-$ derivatives such as pUC1169 (see below).

The selection of Tn*4560* insertions into low copy number pock-forming plasmids has been very successful with SCP2* derivatives. The scheme described here differs slightly from the published example.[90] The plasmid to be mutagenized with Tn*4560* is introduced into a strain containing pUC1169 by either transformation or conjugation. (Strains containing Tn*4560* in the chromosome may be preferable if available.) It does not matter if the plasmid contains the ubiquitous *tsr* gene as the only resistance marker because transformants or exconjugants are recognizable by their ability to form pocks. The resulting strain, with two plasmids, is propagated at normal temperature (29°–30°) to produce numerous spores. These are then diluted with an excess (>10^6) of plasmid-free spores of the same host strain (preferably with a different chromosomal marker to distinguish donor and recipient strains) and plated on R2YE medium. Isolated pocks should develop and are replicated to plates containing viomycin.[91,92] Viomycin-resistant pocks contain Tn*4560* inserted into the target plasmid; multiple isolates from one experiment may be clonal. (Instead of conjugation, transformation might be used to isolate SCP2* derivatives with Tn*4560* inserts, but this has the disadvantage that the smaller multicopy plasmid pUC1169 may reappear, even without selection, in many of the regenerated protoplasts. This problem is minimized in the conjugation experiment because nontransmissible multicopy plasmids like pUC1169 are mobilized only at the low frequency of 0.5% by self-transmissible plasmids,[23] probably by formation of co-integrates.)

Tn*5* has been used to mutagenize bifunctional plasmids in *E. coli*,

---

[90] N. K. Davis and K. F. Chater, *Mol. Microbiol.* **4**, 1679 (1990).

[91] Differentiation in the pock area is often delayed and so the growth may not replicate when the plates are too young. It should be easy to recognize the pocks in which viomycin-resistant growth originates from many spores spread over the whole pock area, and to distinguish them from pocks where the viomycin-resistant growth originates from spores confined to the center of the pocks. The difference between the two types of pocks is greatest in young growth and disappears later when even single viomycin-resistant spores have had time to develop into large colonies.

[92] A potential problem with the use of this system is the dependence on the *vph* gene for selection. Viomycin has not been commercially available, but can now be bought from: Dr. V. Danilenko, Director of BIOAN Association, National Research, Institute of Antibiotics, 3a Nagatinskaya sur., 113105 Moscow, U.S.S.R. Telephone: (Moscow) 111-41-64; Telex: 411718 (answer back Genom SU); Fax: (Moscow) 315-05-01. Viomycin is much more potent on media containing low salt concentrations; we use viomycin exclusively on minimal medium.

followed by transformation into *S. lividans*.[93] However, Tn*4560* has the advantage that *E. coli* is not involved at any stage. Tn*4560* also transposes efficiently from the *S. lividans* chromosome to SCP2* derivatives,[87] making it possible to mutagenize efficiently nontransmissible plasmids, which are purified by transformation.

IS*493* and Derivative Tn*5096*

IS*493* is a 1.6-kb transposable element present in three copies in *S. lividans* 66 but missing from *S. coelicolor* A3(2) and other strains tested. The element was discovered[94] in a directed search using a bifunctional *Streptomyces/E. coli* vector, pCZA126 (Fig. 5), with an apramycin resistance gene [*aac(3)IV*] downstream of the λ promoter P_L. *Escherichia coli* containing pCZA126 is apramycin sensitive because the plasmid also contains the repressor gene, *cI857*, which prevents transcription from P_L and thus expression of the *aac(3)IV* gene. Inactivation of the repressor gene by insertion of a transposable element, for example, leads to apramycin-resistant *E. coli*.

pCZA126 was introduced by transformation into *S. lividans* (Fig. 5a), with selection for thiostrepton resistance. After propagation of the transformant, plasmid DNA was isolated and again introduced into *E. coli* by transformation (Fig. 5b). Ampicillin- and apramycin-resistant colonies were selected (on apramycin alone too many resistant mutants appeared), and the plasmids were analyzed. They contained either inserts that originated from the *E. coli* host or IS*493*. The element was subcloned into pGM160[95] (a derivative of the *Streptomyces ghanaensis* plasmid pSG5 which is naturally temperature sensitive; see below) and tagged by insertion of the *aac(3)IV* gene into the *StyI* site which lies between the two ORFs identified in the DNA sequence.[94] The resulting plasmid (pCZA168, Fig. 5) contains the IS*493* derivative Tn*5096*.

pCZA168 has been used in a simple way to insert Tn*5096* into the

[93] G. Muth, W. Wohlleben, and A. Pühler, *Mol. Gen. Genet.* **211**, 424 (1988).

[94] P. J. Solenberg and S. G. Burgett, *J. Bacteriol.* **171**, 4807 (1989).

[95] G. Muth, B. Nussbaumer, W. Wohlleben, and A. Pühler, *Mol. Gen. Genet.* **219**, 341 (1989).

FIG. 5. IS*493* and derivative Tn*5096*. Black bars are IS*493* sequences. Steps a–c are explained in the text. *aac(3)IV*, Apramycin resistance gene, expressed in both *E. coli* and *Streptomyces;* Am, apramycin; *bla*, β-lactamase gene giving ampicillin resistance in *E. coli;* *cI857*, phage λ repressor; ORF, open reading frame; *ori*, origin of replication; P_L, phage λ promoter, repressed by the *cI857* gene product; Thio, thiostrepton; *tsr*, thiostrepton resistance gene, expressed in *Streptomyces*.

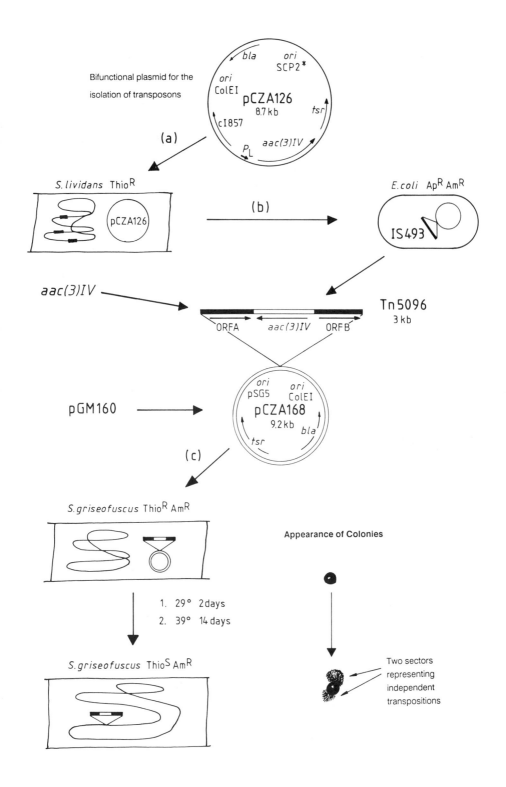

Bifunctional plasmid for the
isolation of transposons

pCZA126
8.7 kb
bla
ori SCP2*
ori ColEI
cI857
PL
aac(3)IV
tsr

(a)

S. lividans ThioR
pCZA126

(b)

E. coli ApR AmR
IS493

aac(3)IV

Tn5096
3 kb
ORFA   aac(3)IV   ORFB

pGM160

pCZA168
9.2 kb
ori pSG5
ori ColEI
bla
tsr

(c)

S. griseofuscus ThioR AmR

1. 29° 2 days
2. 39° 14 days

S. griseofuscus ThioS AmR

Appearance of Colonies

Two sectors
representing
independent
transpositions

genome, using *Streptomyces griseofuscus* C581 as a test system.[96] The plasmid, isolated from *E. coli*, was introduced into C581 by transformation (Fig. 5c). After nonselective growth of the resulting strain, it was plated on agar with apramycin to give about 50 colonies. For the first 2 days the plates were incubated at 29°, after which the temperature was raised to 39° to eliminate the plasmid. Distinct sectors later developed at the edges of the small colonies. Growth in these sectors was thiostrepton sensitive but apramycin resistant. Each sector, even from the same colony, had the transposon inserted into a different sequence.

Tn*5096*, like Tn*4560*, will probably be useful for transposon mutagenesis in many different *Streptomyces* species. The two elements can probably coexist, making it possible to create double mutations without conflict of the resistance determinants.

Marker Exchange between Cloned Sequences and Chromosome

Streptomycetes are normally proficient in generalized homologous recombination. Therefore, in "self-cloning" experiments, recombination between cloned sequences and their chromosomal counterparts occurs, its frequency depending on the length of homology between endo- and exogenote. A single crossover [Fig. 6a (i)] causes integration of a circular plasmid or phage vector and a duplication in the chromosome, by the familiar Campbell mechanism. If the cloned DNA fragment contains a transcription unit truncated at both ends, gene disruption occurs (Fig. 6b), which can be exploited for "mutational cloning."[37] Two crossovers, occurring either simultaneously (Fig. 6c) or in succession [Fig. 6a (ii)], cause reciprocal exchange between the cloned and resident sequences. (The use of linear DNA fragments to force double crossovers immediately after introduction of the DNA, as in fungi and exonuclease-deficient *E. coli*, has not been successful in *Streptomyces*.) "Homogenotization," producing strains with the same allele in chromosome and cloned DNA, can in principle occur in two ways. The first [Fig. 6d (i)] is by segregation (after a reciprocal exchange of markers) of a recombinant chromosome carrying the allele of the cloned DNA into the same cell as a further copy of the original clone [or the converse, with a nonrecombinant chromosome and a recombinant exogenote; Fig. 6d (ii)]. Such events have in principle a high probability in the "multinucleate" cells of *Streptomyces*. The second route to homogenotization is gene conversion (Fig. 6e), in which repair synthesis occurs during recombination. Homogenotization by this route may also occur between duplicated sequences after vector integration (Fig. 6f).

[96] P. J. Solenberg and R. H. Baltz, *J. Bacteriol.* **173**, 1096 (1991).

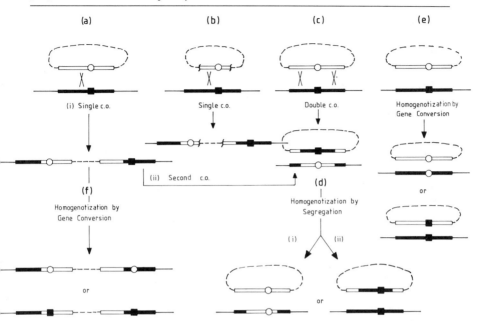

FIG. 6. Vector integration, gene replacement, and homogenotization by gene conversion or segregation. Black and white bars are the same sequences differing in one position indicated by a black square and an open circle, respectively. Continuous and dashed lines represent chromosomal and vector DNA, respectively. In (b) the black bar represents a complete transcription unit and the white bar the same transcription unit truncated at both ends.

Such marker exchanges can be useful for the transfer of all kinds of mutations, including insertions and even large deletions, from cloned sequences into the chromosome, and, in reverse, for the "rescue" of markers from the chromosome. The following examples illustrate how this has been achieved in practice in *Streptomyces*.

Homogenotization during Insert-Directed Lysogenization
of Streptomyces coelicolor A3(2)[97]

A 5.6-kb fragment of *S. coelicolor* DNA containing the complete *bldA*⁺ gene was cloned into an *att*-deleted *c*⁺ øC31 phage vector. Lysogenization of *bldA* strains (the *bldA* mutation is recessive) gave about 90% Bld⁺ and 10% Bld⁻ lysogens. The latter had evidently originated by homogenotization; phages released from them contained the *bldA* mutant allele. They were then used to lysogenize *bldA*⁺ strains, resulting in about 1% Bld⁻ lysogens, again through homogenotization. In this way, a series of isogenic

[97] J. M. Piret and K. F. Chater, *J. Bacteriol.* **163**, 965 (1985).

pairs of strains, with either the bldA⁺ or the bldA mutant allele, were prepared, including S. lividans strains [homology between the DNA sequences of S. coelicolor A3(2) and S. lividans is almost 100% over most genes analyzed to date]. Similar experiments have also been described by Méndez and Chater.[98]

Replacement of Chromosomal Segment of Streptomyces coelicolor A3(2) with Resistance Gene

Buttner et al.[99] replaced a segment of a sigma factor gene (hrdC) in the S. coelicolor A3(2) chromosome with the ermE gene (for erythromycin resistance) from Saccharopolyspora erythraea. This was achieved by using a combination of a glk⁻ host strain (lacking glucose kinase and thus resistant to 2-deoxyglucose) and an attP-deleted phage containing glk⁺,[100] the presence of which makes the host become sensitive to 100 mM 2-deoxyglucose. Loss of the prophage restored 2-deoxyglucose resistance to the host strain and selection for erythromycin resistance ensured retention of the disrupted hrdC gene.

Similar gene replacements have been achieved in S. lividans 66.[52,101] The marked instability of øC31 lysogens in this strain allows easy screening for strains which have lost the phage without the need for specific selection as is required for S. coelicolor.

Replacement of Streptomyces hygroscopicus Chromosomal Gene with Mutant Allele Cloned in Multicopy Plasmid[102]

A 3-kb fragment of S. hygroscopicus DNA containing genes from the bialaphos biosynthetic pathway was mutated by infilling an internal EcoRI site. This fragment, cloned in the multicopy plasmid pIJ680,[42] was introduced by transformation into wild-type S. hygroscopicus. Thiostrepton-resistant transformants were grown in liquid culture, protoplasted and regenerated on nonselective medium. Five (2.5%) of the colonies did not produce bialaphos. Four of them were thiostrepton resistant and still contained the plasmid (presumably homogenotization had occurred); the fifth was thiostrepton sensitive and plasmid free. (The protoplasting step was essential for obtaining single colonies and for the elimination of the vector plasmid, which occurred at ~7% frequency.)

[98] C. Méndez and K. F. Chater, J. Bacteriol. 169, 5715 (1987).
[99] M. J. Buttner, K. F. Chater, and M. J. Bibb, J. Bacteriol. 172, 3367 (1990).
[100] S. H. Fisher, C. J. Bruton, and K.F. Chater, Mol. Gen. Genet. 206, 35 (1986).
[101] T. Eckhardt and R. Smith, in "Genetics of Industrial Microorganisms, Proceedings of the 5th International Symposium GIM 86, Part B" (M. Alačević, D. Hranueli, and Z. Toman, eds.), p. 17. Pliva, Zagreb, Yugoslavia, 1987.
[102] H. Anzai, Y. Kumada, O. Hara, T. Murakami, R. Itoh, E. Takano, S. Imai, A. Satoh, and K. Nagaoka, J. Antibiot. 41, 226 (1988).

Integration of Apramycin Resistance Gene (aac(3)IV) into pSAM2^{int} in Streptomyces ambofaciens ATCC 15154[76]

An *E. coli* plasmid was constructed which contained a 4-kb fragment of pSAM2 (without known integration and replication functions) and, inserted in the middle of this fragment, the *aac(3)IV* gene. This plasmid was introduced into *S. ambofaciens* protoplasts to give apramycin-resistant transformants ($\sim$100/μg DNA). In about 7% of the transformants, the *aac(3)IV* gene was integrated without vector sequences into pSAM2^{int}, presumably by double crossing-over. The other 93% had the complete *E. coli* plasmid integrated by a single crossover. Interestingly, these latter strains, during propagation, did not lose the vector sequences by a second crossover. Thus, enough independent transformants were needed in order to ensure recovery of a product of the primary double crossover event.

In Vivo Cloning of the Entire red Gene Cluster, for Undecylprodigiosin Biosynthesis[103]

Earlier work had led to the isolation of the entire set of *red* genes, but on two only slightly overlapping fragments of *S. coelicolor* A3(2) DNA. First attempts to generate a clone carrying the genes in a contiguous cluster by partial digestion and religation of the two clones failed. Therefore, two segments (10.6 and 4.3 kb) of the cloned DNA, from just outside opposite ends of the cluster, were inserted side by side, and in the same orientation, into a low copy number SCP2* derivative and introduced into wild-type *S. coelicolor*. (The cloning was first done in *E. coli;* the cloned double fragment was then inserted into the *Streptomyces* vector pIJ941 by cloning in *S. lividans,* and finally transferred by transformation to *S. coelicolor*.) On propagation of a transformant, a clone carrying the entire *red* cluster on a 38-kb insert arose, presumably by two crossovers, one in each of the adjacent segments carried by the introduced plasmid. It was recognized by yielding a Red$^+$ colony on transformation of a *red* mutant carrying a mutation in a central gene of the cluster, not included in the regions cloned together in the original *E. coli* plasmid.

General Comments on Gene Replacement

When replacing genes in the *Streptomyces* chromosome, the main problem is elimination of the vector. *Escherichia coli* plasmids with a selectable marker for *Streptomyces* might seem ideal because they cannot replicate in *Streptomyces*. *Escherichia coli* plasmids have been used successfully in *S. ambofaciens*[76] and *S. lividans*.[104] The temperature-sensitive

[103] F. Malpartida, J. Niemi, R. Navarrete, and D. A. Hopwood, *Gene* **93**, 91 (1990).
[104] M. Vögtli, unpublished results (1990).

replicons pMT660[88] and derivatives of the naturally temperature-sensitive *S. ghanaensis* plasmid pSG5[95] are lost at elevated temperature (above 34°). pSG5 (copy number ~15) has the advantage of a very wide host range in *Streptomyces,* replicating in strains that have not been shown to support replication of pIJ101. Bifunctional derivatives of pSG5 equipped with the phage f1 replication origin can be isolated from *E. coli* either in single-stranded or double-stranded form. Both forms gave similar transformation frequencies of several *Streptomyces* species but the single-stranded form integrated 10 to 100 times more efficiently into the *S. viridochromogenes* chromosome by homologous recombination via a cloned sequence than the double-stranded form of the same plasmid.[104a]

Many of the generally used vector plasmids, and in particular most of the bifunctional (*E. coli/Streptomyces*) vectors, will give rise to occasional plasmid-free cells after one or two rounds of nonselective growth. Protoplasting and regeneration often stimulates plasmid loss. Sti⁻ derivatives of pIJ101 (e.g., pIJ486,[105] pIJ680,[42] pIJ699,[106] and pIJ702[107]) can be displaced efficiently with pIJ101 or Sti⁺ derivatives thereof.[108] SCP2*-derived plasmids can be dislodged with unstable SCP2* derivatives that lack the putative partition function carried on the *Pst*I B fragment.[109] An example of such a plasmid is pIJ80[110] carrying the *aph* (neomycin resistance) gene of *S. fradiae* and not the *tsr* gene present on most vectors. On culture in the absence of neomycin, pIJ80 is in turn lost spontaneously at high frequency. SLP1 plasmids can be prevented from replicating by introducing another plasmid with the *imp* function into the same cells.[72]

øC31 derived vectors have proved particularly useful for gene replacements in *S. lividans* where lysogens are unstable (see above). For use in *S. coelicolor glk* mutants a counterselectable phage is available.[100]

Acknowledgments

We thank colleagues who provided unpublished information and F. Boccard, K. F. Chater, C. Khosla, and C. P. Smith for critical comments on the manuscript. Research in the authors' laboratory was supported by grants in aid from the John Innes Foundation and the Agricultural and Food Research Council.

[104a] D. Hillemann, A. Pühler, and W. Wohlleben, *Nucleic Acids Res.* **19,** 777 (1991).
[105] J. M. Ward, G. R. Janssen, T. Kieser, M. J. Bibb, M. J. Buttner, and M. J. Bibb, *Mol. Gen. Genet.* **203,** 468 (1986).
[106] T. Kieser and R. E. Melton, *Gene* **65,** 83 (1988).
[107] A. Katz, C. J. Thompson, and D. A. Hopwood, *J. Gen. Microbiol.* **129,** 2703 (1983).
[108] Z. Deng, T. Kieser, and D. A. Hopwood, *Mol. Gen. Genet.* **214,** 286 (1988).
[109] M. J. Bibb and D. A. Hopwood, *J. Gen. Microbiol.* **126,** 427 (1981).
[110] C. J. Thompson, E. T. Seno, and J. M. Ward, unpublished (1980).

[22] Genetic Techniques in Rhodospirillaceae

By TIMOTHY J. DONOHUE and SAMUEL KAPLAN

I. Introduction

Phototrophic bacteria encompass a diverse group of microorganisms whose rRNA signature places them in at least one-half of the eubacterial phyla.[1] This diversity suggests that bacterial photosynthetic capacity is an ancient trait. True bacterial photosynthesis is an obligately anaerobic process. *Erythrobacter* sp. OCh114 is the only known exception; this species requires oxygen for bacteriochlorophyll synthesis.[2]

Rhodospirillaceae are within the α and β subdivisions of the purple bacteria.[1] They are either obligately or facultatively photosynthetic; the latter have a wide spectrum of growth modes. *Rhodobacter* species (formerly classified as *Rhodopseudomonas*) and *Rhodospirillum rubrum* are most often studied because they grow well under both aerobic and anaerobic conditions. These organisms are capable of virtually all known biological energy transformations under anaerobic conditions.[3] These include photosynthetic growth in the presence of light, anaerobic respiration in the dark if appropriate external electron acceptors are present, and fermentation. They can grow autotrophically if H_2 and CO_2 are present, and many species fix N_2 if they are grown in the absence of combined nitrogen.

Although many species have been characterized by physiological traits[4] and biophysical studies,[5] only a few have been analyzed genetically. Among *Rhodobacter*, *R. capsulatus* and *R. sphaeroides* have been studied from biochemical, biophysical, physiological, and genetic perspectives. More recently, interest in *Rhodopseudomonas viridis* has increased because of its unique photosynthetic membrane structure[6] and because the crystal structure of its reaction center was the first to be solved.[7,8] The fact that *R. viridis* grows poorly under nonphotosynthetic conditions may

---

[1] C. R. Woese, *Microbiol. Rev.* **51,** 221 (1987).
[2] T. Shiba and U. Shimidu, *Int. J. Syst. Bacteriol.* **32,** 211 (1982).
[3] M. T. Madigan, *in* "Biology of Anaerobic Microorganisms" (A. J. B. Zehnder, ed.), p. 39. Wiley, New York, 1988.
[4] N. Pfennig and H. G. Truper, *Ann. Microbiol.* (*Paris*) **134B,** 9 (1983).
[5] J. Amesz and D. B. Knaff, *in* "Biology of Anaerobic Microorganisms" (A. J. B. Zehnder, ed.), p. 113, Wiley, New York, 1988.
[6] G. Drews and P. Giesbrecht, *Arch. Microbiol.* **52,** 242 (1965).
[7] R. Huber, *EMBO J.* **8,** 2125 (1989).
[8] J. Deisenhofer and H. Michel, *EMBO J.* **8,** 2149 (1989).

limit future genetic analysis,[9] since mutations affecting photosynthetic growth may be difficult to analyze.

In this chapter we present genetic techniques used when working with photosynthetic bacteria. This presentation cannot be all-encompassing; rather, we have chosen to focus on those advances made in classic and molecular genetics since this topic was last reviewed.[10-12] Although most techniques have been described for *R. sphaeroides* or *R. capsulatus*, we do not wish to imply that these are the only species that have been studied. We indicate, where possible, if techniques are or should be useful in other species.

II. Relevant Biochemical/Physiological Characteristics

Many photosynthetic bacteria are gram-negative and amenable, in varying degrees, to techniques for genetic analysis used in *Escherichia coli* and *Salmonella typhimurium*. Photosynthetic bacteria typically possess DNA of a guanine plus cytosine (G + C) content between 68 and 70%.[4] The *E. coli* genes that have been tested (β-galactosidase, etc.) are not highly expressed in photosynthetic bacteria from their own promoters,[13] nor are genes from photosynthetic bacteria generally transcribed from their own promoters in *E.coli*.[14] Therefore, expression of genetic information from photosynthetic bacteria in *E. coli* usually requires the use of expression systems. A caveat is that most of the Rhodospirillaceae genes tested for expression in *E. coli* have been highly regulated genes for anaerobically induced components of the photosynthetic apparatus, the Calvin cycle, or nitrogen fixation. Since necessary transcription factors may be limiting in *E. coli*, absence of expression must be interpreted with caution. Indeed, *in vitro* experiments suggest that RNA polymerase from some photosynthetic bacteria may recognize some enteric promoters,[15,16] so the *in vivo* results may reflect the lack of required trans-acting transcriptional activators. To overcome problems associated with poor expression of *R. sphaeroides* DNA in *E. coli*, Chory and Kaplan[17] developed a homol-

[9] F. S. Lang and D. Oesterhelt, *J. Bacteriol.* **171,** 2827 (1989).
[10] B. Marrs, S. Kaplan, and W. Shepherd, this series, Vol. 69, p. 29.
[11] P. A. Scolnick and B. L. Marrs, *Annu. Rev. Microbiol.* **41,** 703 (1987).
[12] J. M. Pemberton, S. Cooke, and A. R. St. G. Bowen, *Ann. Microbiol. (Inst. Pasteur)* **134B,** 185 (1983).
[13] F. E. Nano and S. Kaplan, *J. Bacteriol.* **152,** 924 (1982).
[14] P. J. Kiley and S. Kaplan, *Microbiol. Rev.* **52,** 50 (1988).
[15] J. L. Wiggs, J. W. Bush, and M. J. Chamberlain, *Cell (Cambridge, Mass.)* **16,** 97 (1979).
[16] J. W. Kansy and S. Kaplan, *J. Biol. Chem.* **264,** 13751 (1989).
[17] J. Chory and S. Kaplan, *J. Biol. Chem.* **257,** 15110 (1982).

ogous *in vitro* transcription–translation system that is capable of expressing DNA from other high G + C organisms. This *in vitro* system has been very useful in identifying and analyzing gene products encoded by cloned *R. sphaeroides* DNA.

A. Growth Conditions

Most wild-type species of the Rhodospirillaceae grow in either a mineral salts or a complex medium[3]; optimum growth occurs between approximately 30° and 35°. However, *R. sphaeroides* WS8 will grow at temperatures up to 40°, and some thermotolerant species of the Rhodospirillaceae have been isolated.[18] Highly aerobic cells are obtained either by vigorous shaking of low density cultures ($<5 \times 10^8$ cells/ml) in flasks containing at most 10% of their volume in liquid, or by sparging with a mixture of O_2 (30%), N_2 (69%), and CO_2 (1%). For photoheterotrophic growth tungsten lamps are used to provide wavelengths of light above 800 nm, as required for optimal growth. Photoheterotrophic cells are grown either in completely filled vessels or in cultures sparged with a N_2 (95–99%)–CO_2 (1–5%) mixture. Anaerobic growth on solid media uses BBL (Cockeysville, MD) H_2–CO_2 generators and anaerobic gas pack jars. Nitrogen-free media are used to allow diazotrophic growth in the presence of light. Autotrophic growth is achieved by omitting a fixed carbon source and sparging with an H_2 (95%)–CO_2 (5%) mixture.

Species capable of anaerobic respiration in the dark utilize external electron acceptors such as dimethyl sulfoxide (DMSO; 0.2%) trimethylamine *N*-oxide (TMAO, 0.2%), or N_2O. Growth is slow in minimal medium under these conditions, so complex media are often used. Some strains of *R. sphaeroides* and *R. capsulatus* use nitrate as inorganic electron acceptors under anaerobic conditions,[19] and several *R. sphaeroides* subspecies are denitrifiers.[20,21]

Both wild-type *R. sphaeroides* and mutants unable to grow by DMSO respiration[22] can use tellurite or selenite (30–100 μg/ml) as electron acceptors.[23] These concentrations are significantly higher than the levels tolerated by other bacteria, so such inorganic electron acceptors might be useful to enrich for some *Rhodobacter* sp.

[18] J. Favinger, R. Stadtwald, and H. Gest, *Antonie van Leeuwenhoek* **55**, 291 (1989).

[19] S. J. Ferguson, J. B. Jackson, and A. G. McEwan, *FEMS Microbiol. Rev.* **46**, 117 (1987).

[20] T. Satoh, Y. Hosino, and M. Kitamura, *Arch. Microbiol.* **108**, 265 (1976).

[21] W. P. Michalski and D. J. D. Nicholas, *FEMS Microbiol. Lett.* **52**, 239 (1988).

[22] M. D. Moore and S. Kaplan, *J. Bacteriol.* **171**, 4385 (1989).

[23] M. D. Moore and S. Kaplan, unpublished observations (1990).

B. Resistance Markers

Molecular genetic analyses are based on the ability to select for resistance markers associated with plasmids or transposons. Wild-type strains of Rhodospirillaceae are sensitive to antimicrobial compounds such as kanamycin (Kn), tetracycline (Tc), spectinomycin (Sp), streptomycin (Sm), mercury (Hg), chloramphenicol (Cm), and trimethoprim (Tp).

Resistance to these compounds can be mediated by the genetic elements described in Table I. There are significant differences between species in terms of expression of these markers. For example, *R. sphaeroides*[24,25] and *R. capsulatus*[26,27] do not express resistance to ampicillin, nor does *R. sphaeroides* express resistance to Cm when elements from Tn9 are used.[28] However, Cm resistance is expressed when a promoter which *R. sphaeroides* can recognize is placed upstream of *cat*.[29] In contrast, ampicillin resistance is expressed in *R. viridis*,[30] and *R. rubrum* expresses resistance to Cm.[31]

The source of the drug resistance element can also produce significant differences in expression *in vivo*. For example, Tc resistance determinants from RK2 derivatives work well in most species,[27,31–36] whereas the Tc resistance gene on pBR322[26,37] or Tn*10*[30] is expressed poorly in some species. Similarly, the Tn*501* Hg resistance determinant is expressed well in *R. sphaeroides*,[38] but the Tn*501*-derived *mer* operon cloned in the Hg-Ωcartridge does not allow stable selection for Hg resistance.[39]

[24] W. R. Sistrom, *J. Bacteriol.* **131**, 526 (1977).
[25] W. T. Tucker and J. M. Pemberton, *FEMS Microbiol. Lett.* **5**, 173 (1979).
[26] D. P. Taylor, S. N. Cohen, W. G. Clark, and B. L. Marrs, *J. Bacteriol.* **154**, 580 (1983).
[27] J. C. Willison, G. Ahombo, J. Chabert, J. P. Magnuson, and P. M. Vignais, *J. Gen. Microbiol.* **131**, 3001 (1985).
[28] V. V. Zinchenko, M. M. Babykin, and S. V. Shestakov, *J. Gen. Microbiol.* **130**, 1587 (1984).
[29] H. C. Yen, personal communication (1987).
[30] F. S. Lang and D. Oesterhelt, *J. Bacteriol.* **171**, 4425 (1989).
[31] D. P. Lies, W. P. Fitzmaurice, L. J. Lehman, and G. P. Roberts, personal communication (1989).
[32] J. A. Johnson, W. K. R. Wong, and J. T. Beatty, *J. Bacteriol.* **167**, 604 (1986).
[33] B. S. DeHoff, J. K. Lee, T. J. Donohue, R. I. Gumport, and S. Kaplan, *J. Bacteriol.* **170**, 4681 (1988).
[34] E. J. Bylina, R. V. M. Jovine, and D. C. Youvan, *Bio/Technology* **7**, 69 (1989).
[35] A. Colbeau, A. Godfroy, and P. Vignais, *Biochimie* **68**, 147 (1988).
[36] P. Atvges, R. G. Kranz, and R. Haselkorn, *Mol. Gen. Genet.* **201**, 363 (1985).
[37] T. N. Tai, W. Havelka, and S. Kaplan, *Plasmid* **19**, 175 (1988).
[38] J. M. Pemberton and A. R. St. G. Bowen, *J. Bacteriol.* **147**, 110 (1981).
[39] R. Fellay, J. Frey, and H. Kirsch, *Gene* **52**, 147 (1981).

TABLE I
RESISTANCE PROPERTIES OF RHODOSPIRILLACEAE

Antibiotic	Organism	Sensitivity	Determinant	Expression	Refs.
Ampicillin	R. sphaeroides	S	Tn1, Tn3	−	24,25,28,72
	R. capsulatus	S	Tn1, Tn3	−	26,27,87
	R. rubrum	S	Tn1, Tn3	−	31
	R. viridis	S	Tn1, Tn3	+	30
	R. acidophila	S	Tn1, Tn3	+	77
Chloramphenicol	R. sphaeroides	S	Tn9	−	85
	R. capsulatus	S	Tn9	−	87
	R. rubrum	S	Tn9	+	31
Kanamycin	R. sphaeroides	S	Tn5, Tn903, RP1	+	24,25,28,85,86,129
	R. capsulatus	S	Tn5, Tn903	+	87,89,132
	R. rubrum	S	Tn5, Tn903, RP1	+	31
	R. viridis	S	Tn5	+	30
	R. acidophila	S	Tn5	+	77
Mercuric chloride	R. sphaeroides	S	Tn501	+	38
Spectinomycin	R. sphaeroides	S	R100.1	+	91
	R. capsulatus	S	R100.1	+	68,69
	R. rubrum	S	R100.1, pPH1JI, Tn7	+	31
Streptomycin	R. sphaeroides	S	RSF1010	+	28,76,131
	R. rubrum	S	RSF1010	+	31
Sulfonamide	R. sphaeroides	S	RSF1010	+	76
	R. rubrum	S	RSF1010, R388	+	31
Tetracycline[a]	R. sphaeroides	S	RK2 derivatives	+	85,86,128,139,145
	R. capsulatus	S	RK2 derivatives	+	32,34
	R. rubrum	S	RK2 derivatives	+	31
	R. viridis	S	RK2 derivatives	+	30
	R. acidophila	S	RSF1010 derivative	+	77

[a] We routinely filter light through a <650 nm cutoff filter when using tetracycline with photosynthetic cultures to minimize light-induced degradation of this antibiotic.[85,86]

III. Classic Genetics

A. Mutagenesis

Photosynthetic bacteria are susceptible to chemical {nitrosoguanidine (NTG), ethyl methanesulfonate (EMS), G-chloro-9-[3-(2-chloro-ethyl-amino)propylamino]-2-methoxyacridine (ICR191)} and physical (UV) mutagenesis.[10,31] Ampicillin or D-cycloserine have been used to enrich for auxotrophs, etc.[10] Tetracycline uptake is light-dependent in photosynthetic cells,[10,40] so tetracycline suicide is an effective means to enrich for

[40] J. Weckesser and J. A. Magnuson, J. Bacteriol. 138, 678 (1979).

mutants defective in photosynthesis.[10] Section IV outlines systems for transposon mutagenesis, marker exchange, and gene interconversion which can be used to construct chromosomal mutations.

Little is known about DNA repair in photosynthetic bacteria. Both *R. sphaeroides* and *R. capsulatus* induce a DNA repair system in response to low doses of alkylating agents[41]; one report suggests that photoreactivation and SOS-like repair exists in *R. capsulatus*.[42]

Although there are reports of recombination-deficient mutants of *R. sphaeroides*[43] and *R. capsulatus*,[44] the existence of an *E. coli recA* equivalent has not been proved. An internal probe for *recA* from *Pseudomonas aeruginosa*[45] has been used to clone a homologous DNA fragment from *R. sphaeroides* 2.4.1. When this region of the *R. sphaeroides* genome is replaced with a Sp^R/Sm^R cartridge, there is no increase in sensitivity to UV or EMS.[46] This might suggest that this species lacks a *recA*-dependent SOS-like response. Work is underway to confirm that this is the *R. sphaeroides recA* homolog and to determine the recombination proficiency of this mutant.

B. Phage

Several lytic phage for *R. sphaeroides*[47–55] and *R. capsulatus*[56,57] have been isolated. Virus-mediated genetic exchange has been reported with a temperate *R. sphaeroides* phage.[50,51] Giffhorn and co-workers have demonstrated the existence of several lytic phage in newly isolated *R. sphaeroides* strains.[52–55] *R. sphaeroides* Y contains a temperate phage which is

[41] J. A. Vericat and J. Barbe, *Mutagenesis* **3**, 165 (1988).

[42] J. Barbe, I. Gilbert, M. Llagostera, and R. Ricardo, *J. Gen. Microbiol.* **133**, 961 (1987).

[43] W. R. Sistrom, A. Macaluso, and R. Pledger, *Arch. Microbiol.* **138**, 161 (1984).

[44] F. J. Genthner and J. D. Wall, *J. Bacteriol.* **160**, 971 (1984).

[45] T. A. Kokjohn and R. V. Miller, *J. Bacteriol.* **163**, 568 (1985).

[46] W. D. Shepherd and S. Kaplan, unpublished observations (1990).

[47] A. Abelovich and S. Kaplan, *J. Virol.* **13**, 1392 (1974).

[48] T. J. Donohue, J. Chory, T. E. Goldsand, S. P. Lynn, and S. Kaplan, *J. Virol.* **55**, 147 (1985).

[49] R. J. Mural and D. J. Friedman, *J. Virol.* **14**, 1288 (1974).

[50] W. T. Tucker and J. M. Pemberton, *J. Bacteriol.* **135**, 207 (1978).

[51] W. T. Tucker and J. M. Pemberton, *J. Bacteriol.* **143**, 43 (1980).

[52] M. Duchrow, G. W. Kohring, and F. Giffhorn, *Arch. Microbiol.* **142**, 141 (1985).

[53] M. Duchrow and F. Giffhorn, *J. Bacteriol.* **169**, 4410 (1987).

[54] M. Duchrow, S. Heiteuss, J. Kalkas, M. Hoppert, and F. Giffhorn, *Arch. Microbiol.* **149**, 476 (1988).

[55] S. Heitefuss and F. Giffhorn, *J. Bacteriol.* **135**, 911 (1989).

[56] J. D. Wall, P. F. Weaver, and H. Gest, *Arch. Microbiol.* **105**, 217 (1975).

[57] L. S. Schmidt, H. C. Yen, and H. Gest, *Arch. Biochem. Biophys.* **165**, 229 (1974).

induced after exposure to mitomycin.[52] Transduction of genomic markers with *R. sphaeroides* phage have been unsuccessful. This topic should be reinvestigated now that known mutants with stable genetic defects are available.

C. Restriction/Modification

Restriction endonucleases with unique site specificities are present in different wild-type strains of *R. sphaeroides* (*Rsh*I in 2.4.1, *Rsa*I in 28/5, *Rsr*I and *Rsr*II in RS630).[58–62] The efficiency of plating (eop) of the lytic phage RS1 indicated that phage grown in one strain was restricted in the others.[48] The only verified restriction endonuclease mutant is the RS7001 derivative of *R. sphaeroides* strain RS630. Strain RS630 contains *Rsr*I (an isoschizmer of *Eco*RI),[61,62] an *Eco*RI methylase,[63] and *Rsr*II,[60] and it does not plaque RS1 phage propagated on strain 2.4.1.[48] RS7001 no longer displays *Rsr*I activity *in vitro*,[64] and it plaques phage RS1 grown on strain 2.4.1 with an eop of 1.0.[48] RS1 progeny phage obtained from strain RS7001 plaque with an eop of 1.0 on strain 2.4.1[48]; this suggests that strain RS630 possesses a broad-spectrum methylase.

D. Gene Transfer

In 1974, Marrs reported that *R. capsulatus* spontaneously released a "gene transfer agent" (GTA) toward the end of exponential growth which was capable of generalized transfer of genetic markers.[65] GTA is specific for *R. capsulatus*. Isolation of a GTA-overproducing strain made it possible to isolate the agent and demonstrate its phagelike character (see Ref. 10 for a recent review). GTA was used to map a series of *R. capsulatus* mutations affecting nitrogen fixation[66] and photosynthesis.[67] Recent experiments indicated that GTA packages approximately 4.5–5.0 kb of DNA.[26]

[58] S. P. Lynn, L. K. Cohen, J. F. Gardner, and S. Kaplan, *J. Bacteriol.* **138**, 505 (1979).

[59] S. P. Lynn, L. K. Cohen, S. Kaplan, and J. F. Gardner, *J. Bacteriol.* **142**, 505 (1980).

[60] C. D. O'Connor, E. Metcalf, C. J. Wrighton, T. J. R. Harris, and J. R. Saunders, *Nucleic Acids Res.* **12**, 6701 (1984).

[61] C. Aiken and R. I. Gumport, *Nucleic Acids Res.* **16**, 7901 (1988).

[62] P. J. Greene, B. T. Ballard, F. Stephenson, W. J. Kohr, H. Rodriguez, J. M. Rosenberg, and H. Boyer, *Gene* **68**, 43 (1988).

[63] R. I. Gumport, personal communication (1990).

[64] S. P. Tai and S. Kaplan, unpublished observations (1985).

[65] B. Marrs, *Proc. Natl. Acad. Sci. U.S.A.* **71**, 971 (1974).

[66] J. D. Wall, J. Love, and S. P. Quinn, *J. Bacteriol.* **159**, 652 (1984).

[67] H. C. Yen and B. Marrs, *J. Bacteriol.* **126**, 619 (1976).

GTA-Mediated Gene Transfer

Method for transfer of point mutations from nonoverproducing strain. Recipient cells are grown aerobically to early stationary phase in PYS medium (3 g peptone and 3 g yeast extract per liter containing 10 mM MgSO$_4$ and 10 mM CaCl$_2$[68] and harvested by centrifugation (12,000 g, 10 min at 4°). The cells are suspended in an equal volume of GTA buffer (10 mM Tris-Cl, pH 7.8, 1 mM NaCl, 1 mM MgCl$_2$, 1 mM CaCl$_2$ containing 500 μg/ml bovine serum albumin). These cells are stable as recipients for several hours at 24°.

GTA is obtained by growing donor cells photosynthetically in PYS medium in a screw-capped tube until the culture is in stationary phase. One-half to one milliliter of cell culture is filtered through a Gelman (Ann Arbor, MI) 0.45-μm filter. This phage filtrate can be stored in a plastic vial for up to 12 hr on ice. For long-term storage 9 volumes of the GTA filtrate are mixed with 1 volume of 10× GTA buffer in a plastic vial and stored at −70°.

The transduction is performed by mixing 0.4 ml GTA buffer, 0.1 ml recipient cells, and 0.1 ml GTA filtrate in a plastic vial. The mixture is incubated at 35° for 60 min, then 2.5 ml of molten PY soft agar (0.6%) is added (PY is PYS lacking the MgSO$_4$ and CaCl$_2$). The mixture is poured onto a PY plate and incubated at 35° for 4 hr to allow phenotypic expression of the newly acquired genetic trait. If necessary, the plate is then overlayed with 2.5 ml of molten PY soft agar containing the appropriate drug for selection of the recipient. Colonies are visible after 2 days of incubation at 35°.

Interposon mutagenesis using GTA-overproducing strain. A plasmid containing the interrupted gene of interest is transferred from *E. coli* to an *R. capsulatus* GTA overproducer (e.g., CB1127 *crtD, rif*).[69] GTA overproducers are unstable in complex (PY) medium, so matings are performed on minimal (RCV)[70] plates. The conditions for transfer of plasmid DNA from *E. coli* to *R. capsulatus* are similar to those described below. When exconjugants are visible, the plate is flooded with PYS medium, and the colonies are scraped from the plate with a glass spreader. The liquid is transferred to a screw-capped tube, and the tube is filled with PYS medium and incubated photosynthetically at 35° until the culture reaches stationary phase (~2–3 days). A GTA filtrate is prepared, and the transduction is performed as described above.

---

[68] C. E. Bauer, D. A. Young, and B. L. Marrs, *J. Biol. Chem.* **263**, 4820 (1988).

[69] D. A. Young, C. E. Bauer, J. A. Williams, and B. L. Marrs, *Mol. Gen. Genet.* **218**, 1 (1989).

[70] P. F. Weaver, J. D. Wall, and H. Gest, *Arch. Microbiol.* **105**, 207 (1975).

Discussion. Although GTA is released from aerobic cultures, maximal GTA production occurs when cells are grown photosynthetically to early to mid-stationary phase in complex medium. GTA-mediated gene transfer is very sensitive to temperature fluctuations; therefore, maintaining 35° while performing the transduction is critical. GTA overproducers should be grown and stocked in minimal media. These strains are unstable in complex media, and they revert to normal GTA production levels even if stored in complex media. Finally, since GTA particles stick to glass, phage infection and storage of phage stocks are performed in plasticware.

GTA-mediated gene transfer can be routinely used for transferring genetic markers with a strong selectable phenotype or for determining the linkage of two scorable markers within approximately 4–5 kilobases (kb). However, using GTA for mapping relatively closely linked markers (~7–10 kb apart, or greater) or for placing large constructions (Mu*D* phages, etc.) in the chromosome is impossible because of DNA packaging constraints.

E. R Factors and Plasmids

The first transfer of a promiscuous R factor to the Rhodospirillaceae was reported by Olsen and Shipley, who mobilized R1822 (now called RP1) from *Pseudomonas aeruginosa* to both *R. sphaeroides* and *R. rubrum* by selecting for ampicillin resistance.[71] Subsequently, Miller and Kaplan showed that RP4 could be transferred from *E. coli* to *R. sphaeroides*.[72] Although ampicillin resistance from RP4 was not expressed, other markers could be selected for.

The R factors originally used in photosynthetic bacteria were derivatives of the P1 incompatibility group plasmids RP4, RP1, R68, R751, etc.[73] More recently, other vectors have been used to transfer DNA either from *E. coli* to photosynthetic bacteria or among different photosynthetic bacteria (Table II). Relatively small RK2 derivatives (~10 kb) containing polylinker sites and *lacZα* regions are available to allow efficient construction and screening of recombinants.[74,75] The copy number of the RK2 derivatives pRK404 and pRK415 in *R. sphaeroides* is around 4–6.

Incompatibility group Q plasmids such as RSF1010 are stable in several

---

[71] R. H. Olsen and P. Shipley, *J. Bacteriol.* **113,** 772 (1973).

[72] L. Miller and S. Kaplan, *Arch. Biochem. Biophys.* **187,** 229 (1978).

[73] C. M. Thomas and C. A. Smith, *Annu. Rev. Microbiol.* **41,** 77 (1987).

[74] G. Ditta, T. Schmidhauser, E. Yakobsen, P. Lu, X. W. Liang, D. R. Finlay, D. Guiney, and D. Helinski, *Plasmid* **13,** 149 (1985).

[75] N. T. Keen, S. Tamaki, D. Kobayashi, and D. Trollinger, *Gene* **70,** 191 (1988).

TABLE II
CONJUGATIONAL PLASMIDS USED IN RHODOSPIRILLACEAE

Replicon	Plasmid	Species	Drug resistance	Maintenance	Refs.
IncPα	pRK404/pRK415 (RK2)	R. sphaeroides	Tc	+	86,128,139,145
	pRK404/pRK415 (RK2)	R. capsulatus	Tc	+	32,34
	pRK404/pRK415 (RK2)	R. rubrum	Tc	+	31
	pRK404/pRK415 (RK2)	R. viridis	Tc	+	30
	RP1	R. acidophila	Tc	+	77
IncPβ	pPH1JI	R. capsulatus	Sp/Sm, Gm	+	103
	pPH1JI	R. rubrum	Sp/Sm, Gm	+	31
IncQ	RSF1010/pKT231	R. sphaeroides	Sm, Su/Sm, Kn	+	76,131,135
	RSF1010/pKT231	R. rubrum	Su/Sm	+	31
	pNH2ᵃ	R. acidophila	Tc	+	77
IncW	R388	R. sphaeroides	Tp, Su	+	25
	R388	R. rubrum	Su	+	31
pBR322	pDPT42	R. capsulatus	Kn, Tc	+	26
ColE1	pSup202	R. sphaeroides	Tc	−	85,86,129
	pSup202	R. capsulatus	Tc	−	89
	pSup202	R. viridis	Ap	+	30
ColE1	pRK2013	R. sphaeroides	Kn	−	131
	pRK2013	R. capsulatus	Kn	−	87
	pRK2013	R. rubrum	Kn	−	31
ColE1	pSup2021	R. sphaeroides	Kn/Sm(Tn5)	−	95,115
	pSup2021	R. capsulatus	Kn(Tn5)	−	87,89,132
	pSup2021	R. rubrum	Kn(Tn5)	−	31

ᵃ See Ref. 79 for a description of plasmid pHN2.

species of Rhodospirillaceae[31,76,77]; the copy number in *R. sphaeroides* is approximately 4 to 10. RSF1010 contains sulfonamide and streptomycin resistance genes, and it can be selected for in *R. sphaeroides* if sulfabenzamide is used.[76] However, the relatively high spontaneous frequency of streptomycin resistance negates practical use of this drug. RSF1010 derivatives are now available with different selectable drug markers, additional unique restriction sites, and reporter genes for constructing operon and protein fusions[78,79] (see Section IV,E).

IncW plasmids are stable in several species of photosynthetic bacteria, but they have not been routinely used.[25,28,31]

[76] C. Fornari and S. Kaplan, *J. Bacteriol.* **152**, 89 (1982).
[77] R. Cogdell, personal communication (1990).
[78] M. Bagdasarian and K. N. Timmis, in "Current Topics in Microbiology and Immunology" (P. H. Hofschneider and W. Geobel, eds.), p. 47, Springer-Verlag, Berlin, 1981.
[79] C. N. Hunter, and G. Turner, *J. Gen. Microbiol.* **134**, 1471 (1988).

IncP1[75,80,81] and IncQ[82,83] derivatives with *cos* sites are available to construct mobilizable cosmid banks. *Rhodobacter sphaeroides* cosmid banks are stably maintained in *E. coli recA hsdR* hosts grown on medium containing minimal levels of the appropriate antibiotic. We routinely maintain *R. sphaeroides* banks in *E. coli* strains containing the necessary *tra* functions on the chromosome[84]; this allows direct diparental mobilization into *R. sphaeroides* (see below). Cosmids containing inserts of approximately 30 kb are in single copy in *R. sphaeroides*.

ColE1-based plasmids are not maintained in *R. sphaeroides* unless they become integrated into an *R. sphaeroides* replicon.[76] Mobilizable ColE1 plasmids (pSUP series[84]) have been used as suicide plasmids in *R. sphaeroides*,[85,86] *R. rubrum*,[31] and *R. capsulatus*[87-89] to perform mutagenesis via marker exchange, insertional inactivation, or transposition (see Section IV). Several laboratories use the pDPT vectors derived from pBR322 as stable plasmids in *R. capsulatus*.[26,28] ColE1 derivatives are stable in *R. viridis*.[30]

Plasmid DNA is introduced into Rhodospirillaceae by conjugation[84-89]; transformation[76] and electroporation have also been demonstrated for *R. sphaeroides*. The most efficient and generally applicable system for introducing plasmid DNA is conjugation. One advantage of conjugation is the apparent lack of restriction of incoming DNA. Protocols for conjugation, transformation, and electroporation of *R. sphaeroides* are presented below.

Conjugation

Method. DNA is routinely transferred from *E. coli* to *R. sphaeroides* using the diparental system of Simon *et al.*[84] The *E. coli* donors contain chromosomal copies of the trans-acting elements that mobilize *oriT*-containing plasmids.[84] The advantage of this method over the triparental

[80] A. M. Freidman, S. R. Long, S. E. Brown, W. J. Buijkema, and F. M. Ausubel, *Gene* **18**, 289 (1982).
[81] L. N. Allen and R. S. Hanson, *J. Bacteriol.* **161,** 955 (1985).
[82] J. Frey, M. Bagdasarian, D. Freiss, F. C. H. Franklin, and J. Deshusses, *Gene* **24,** 299 (1983).
[83] J. Davison, personal communication (1987).
[84] R. Simon, U. Priefer, and A. Puhler, *Bio/Technology* **1**, 784 (1983).
[85] T. J. Donohue, A. G. McEwan, S. Van Doren, A. R. Crofts, and S. Kaplan, *Biochemistry* **27**, 1918 (1988).
[86] J. Davis, T. J. Donohue, and S. Kaplan, *J. Bacteriol.* **170**, 320 (1988).
[87] N. Kaufmann, H. Hudig, and G. Drews, *Mol. Gen. Genet.* **198**, 153 (1984).
[88] W. Klipp, B. Masepohl, and A. Puhler, *J. Bacteriol.* **170**, 693 (1988).
[89] B. Masepohl, W. Klipp, and A. Puhler, *Mol. Gen. Genet.* **212**, 27 (1988).

mating systems[90] is that only one *E. coli* donor is necessary. Our protocol differs slightly from those used with *R. capsulatus*,[68,86–89] *R. rubrum*,[31] and *R. viridis*.[30]

Donors are grown overnight at 37° in complex or minimal medium containing the appropriate antibiotic. The donors are subcultured approximately 1 : 10 in fresh medium for 1–2 hr to ensure that exponential phase cells are used. Prior to initiating the mating, donor cells are centrifuged for about 1 min in a sterile Eppendorf tube, then suspended in fresh antibiotic-free medium to minimize carryover of the antibiotic to the mating assay. *Rhodobacter sphaeroides* recipients are grown in the appropriate medium and subcultured to ensure that log phase cells are used.

Miller and Kaplan demonstrated that plasmid mobilization into *R. sphaeroides* was more efficient if the mating pairs were placed on a solid surface.[72] The appropriate volumes of exponential phase donors ($\sim 5 \times 10^6$) and washed recipient cells ($\sim 5 \times 10^8$) are mixed in a sterile Eppendorf tube, centrifuged, and suspended in approximately 100 μl of fresh medium. This mixture is spotted onto a sterile 25-mm 0.45-μm pore size nitrocellulose filter that was previously placed on a petri plate. We generally use LB (Luria broth) plates as supports for matings, however. Use of minimal or a different complex medium (3 g peptone, 5 g yeast extract, and 0.1 g NaCl per liter, supplemented with 50 μM succinate; made and autoclaved separately as an $\sim 40 \times$ stock at pH 7.0) allows cells to mate as well as wild-type cells.[91] The mating pairs are incubated at 32° for about 6 hr, the filter disks are removed from the plates, the cells are washed from the filters into an Eppendorf tube containing approximately 1 ml of growth medium, and the filter pad is discarded. The cells are collected by centrifugation and washed twice before being suspended in 0.1 ml of growth medium. Approximately 10^9 plaque forming units of *E. coli* bacteriophage T_4D is added, and the mixture is incubated at 37° for 30 min to allow infection of the donor cells. The cells are washed twice in 1 ml of ice-cold growth medium prior to plating on the appropriate medium for the selection of exconjugants. Agar plates are incubated at 32°. Drug-resistant colonies appear in about 2–3 days. Conjugation frequency is the number of drug-resistant exconjugants obtained relative to the number of recipients recovered. Another approach is to plate the mating mixture on medium containing 200 μg/ml tellurite; only *R. sphaeroides* will survive.

Discussion. This system produces approximately 1 exconjugant per

[90] G. Ditta, S. Stanfield, D. Corbin, and D. R. Helinski, *Proc. Natl. Acad. Sci. U.S.A.* **77**, 7347 (1980).
[91] B. MacGregor and T. J. Donohue, unpublished observations (1989).

10^4 *R. sphaeroides* recipients (higher frequencies have been obtained in some species). Transfer of stable plasmids is best when dry petri plates are used as a solid substrate for matings. Dry petri plates are obtained by allowing covered plates to sit at room temperature for around 1 week. The frequency of transfer of a self-mobilizable plasmid from a photosynthetic bacterium to *E. coli* or another photosynthetic bacterium is usually significantly lower.[30,31,72] Transfer between different species of photosynthetic bacteria can occur, but the frequency of such events has not been routinely reported.

Transformation. Transformation of a photosynthetic bacterium has been reported only by Fornari and Kaplan for *R. sphaeroides*.[76] This system requires closed circular DNA, $CaCl_2$, polyethylene glycol (PEG), and washing cells with ice-cold, 500 m*M* Tris buffer (pH 7.2). This procedure yields approximately 1–5 × 10^3 transformants per microgram of DNA. This buffer is used to prepare competent cells because it was reported to disturb outer membrane components in gram-negative bacteria. The transformation frequency decreases rapidly with increasing plasmid size. Transformation requires plasmids which lack sites for host restriction enzymes. For example, *R. sphaeroides* 2.4.1 contains *Rsh*I activity and it could be transformed by RSF1010 which lacks a *Rsh*I site. Cloning DNA which contained a *Rsh*I site into RSF1010 prevented transformation.[76]

Method. *Rhodobacter sphaeroides* 2.4.1 is grown aerobically to approximately 8 × 10^8 cells/ml in either complex or minimal medium. The culture is centrifuged at 4° at 6000 *g* for 5 min, and the pellet is suspended in 10 ml of ice-cold 0.5 *M* Tris-Cl (pH 7.2). After harvesting the cells from the Tris wash solution, the pellet is suspended in 1.2 ml of ice-cold 0.1 *M* Tris-Cl (pH 7.2), 0.2 *M* $CaCl_2$. Plasmid DNA is immediately added to 0.2 ml of competent cells, and an equal volume of cold 40% PEG-6000 in 0.1 *M* Tris-Cl (pH 7.2) is added slowly. The suspension is gently shaken to mix the two phases and left on ice for 10 min. A 2-min heat shock at 35° is immediately followed by the addition of 1.0 ml of fresh medium, and the tubes are kept at 35° for 20 min. Three milliliters of fresh medium is added, and the tubes are incubated at 35° for over 6.5 h- to allow phenotypic expression before spreading cells on selective media. Drug-resistant colonies appear after about 2–3 days of incubation at 32°.

Discussion. A more efficient transformation system may be difficult to obtain in photosynthetic bacteria owing to intrinsic properties of their membranes, restriction endonucleases, or other unknown factors. However, higher efficiency transformation systems would facilitate future genetic exploitation of these organisms.

Electroporation. Electroportion has increased the number of bacterial species that can be transformed, and it has increased the transformation of *E. coli* to frequencies as high as 10^{10} transformants per microgram of circular plasmid DNA.[92] *Rhodobacter sphaeroides* can be transformed via electroporation using a protocol based on that developed by Dower *et al.* for *E. coli*.[92] The system we have optimized uses a Bio-Rad (Richmond, CA) Gene Pulser apparatus, a pulse controller, and 0.2-cm electrode gap cuvettes.[93] A resistance of 400 ohms on the pulse controller and settings of 2.5 kV and 25 μF are used to generate the pulses. With 0.2-cm electrode gap cuvettes, these settings generate a field strength of approximately 12.5 kV/cm. Slightly different conditions will probably need to be empirically determined for other species.

Using this system we routinely achieve approximately $1-10 \times 10^4$ drug resistant cells per microgram of circular plasmid DNA (CsCl purified). These frequencies have been obtained by selecting for Kn^R encoded by the IncQ plasmid pKT231.[78] Electroporation with pRK404 or pRK415 has not been successful, probably because they contain restriction sites for *Rsh*I. We are attempting to increase the applicability of electroporation by isolating host mutants deficient in *Rsh*I and by methylating *Rsh*I sites *in vitro* prior to electroporation.

Method. Five hundred milliliters of cells are grown to midexponential phase in minimal salts medium and placed on ice for 10–15 min before harvesting by centrifugation at 10,000 *g* at 4° for 10 min. The cell pellet is washed by resuspending in 500 ml ice-cold sterile distilled water and centrifuging, washed with 250 ml ice-cold sterile water, and then finally washed with 125 ml ice-cold sterile water. This cell pellet is suspended in approximately 15 ml sterile ice-cold 10% (v/v) glycerol and harvested by centrifugation. The final cell pellet is suspended in 10% glycerol to adjust the cell density to around 10^{10} cells/ml. The cells are kept on ice until used or frozen in dry ice and stored at $-80°$ for several months with no significant loss in viability or transformation frequency.

In a chilled electroporation cuvette, 40 μl of the cell suspension is mixed with plasmid DNA. If the sample does not cover the electrodes at the bottom of the cuvette, gently tap the cuvette on the bench top. Optimal electroporation frequencies are obtained using 4 pulses per sample, leaving the cuvette on ice for around 1 min between pulses. The time constant for each pulse should be approximately 8.5 msec. If the sample is too conductive (i.e., time constant too low) dilute the cells, lower the kilovolt setting, or increase the resistance in the pulse controller. If the sample is not conductive enough (i.e., the time constant is too high) increasing the

[92] W. J. Dower, J. F. Miller, and C. W. Ragsdale, *Nucleic Acids Res.* **16**, 6127 (1988).
[93] E. V. Stabb and T. J. Donohue, unpublished observations (1990).

cell concentration or the number of pulses per sample may increase the transformation efficiency.

The sample is added to 1 ml of SOC medium[92] and incubated with shaking for about 6 hr (i.e., ~2 cell doublings) to allow phenotypic expression. Cells are then directly plated on selective and/or nonselective media to monitor transformation frequency and cell survival.

Discussion. We have attempted to increase the electroporation frequency using the washes employed by Fornari and Kaplan[76] prior to the distilled water washes. However, any improvement was low (2–5 fold) and not reproducible. In addition, the samples were often highly conductive; this is not surprising given the high ionic strength of the buffers used to prepare competent cells.

F. Genome Organization

Physical Mapping. Efforts are underway to map genetically and physically the genomes of both *R. capsulatus*[94] and *R. sphaeroides*.[95,96] The many sophisticated genetic techniques and the numerous genetic markers used in genomic mapping in other bacteria are not available in photosynthetic bacteria. Therefore, other approaches are necessary to allow investigators to localize genetic markers efficiently.

The genome size of *R. capsulatus* B10 has been estimated in the 4000-kb range[94]; this includes a cryptic plasmid of approximately 130 kb[97] The *R. sphaeroides* 2.4.1 genome is about 4450 kb[95]; this includes five "small" plasmids of approximately 42, 95, 97, 105, and 110 kb.[95,98] Pulsed-field mapping and Southern hybridization divided the remaining 4000 kb between two circular replicons of approximately 3050 and 914 kb.[96] Suwanto and Kaplan have designated each of these replicons as chromosomes since they both contain *rrn* operons.[96] *Rhodobacter sphaeroides* strains 630, L, Y, 2.4.7, and 8253 also appear to have two chromosomes of approximately the same sizes as those in 2.4.1, even though these strains differ in the number and sizes of their endogenous plasmids.[99]

In physically mapping the *R. sphaeroides* genome, the restriction endonucleases *Ase*I, *Spe*I, *Dra*I, and *Sna*BI are particularly useful. McClelland

[94] J. A. Williams, B. Marrs, and S. Brenner, *in* "Molecular Biology of Membrane Bound Complexes in Phototrophic Bacteria," (G. Drews and E. A. Dawes, eds.), p. 5, Plenum, New York, 1990.

[95] A. Suwanto and S. Kaplan, *J. Bacteriol.* **171,** 5840 (1989).

[96] A. Suwanto and S. Kaplan, *J. Bacteriol.* **171,** 5850 (1989).

[97] J. C. Willison, J. P. Magnin, and P. M. Vignais, *Arch. Microbiol.* **147,** 134 (1987).

[98] C. S. Fornari, M. Watkins, and S. Kaplan, *Plasmid* **11,** 39 (1984).

[99] A. Suwanto and S. Kaplan, unpublished observations (1989).

et al.[100] demonstrated that these enzymes plus *Xba*I should also be useful in mapping of the *R. capsulatus* genome.

Genetic Mapping. Willison, Vignais, and co-workers used the temperature-sensitive RP4 derivative pTH10 to construct a circular linkage map of *R. capsulatus.*[27] This genetic map should soon be aligned with the physical map of Williams *et al.*[94] which was generated by ordering cosmids within a *R. capsulatus* genomic library. Plasmid pTH10 is a temperature-sensitive derivative that is lost in *E. coli* at 42° unless the plasmid integrates into the genome. Once integrated, this plasmid can unidirectionally transfer markers. Willison *et al.* found that pTH10 transferred genomic *R capsulatus* markers at frequencies of around 5×10^{-3} per donor, even though this replicon was not temperature-sensitive in this bacterium.[27] Presumably this occurred because pTH10 spontaneously inserted into the genome by Tn*I*-mediated transposition at a relatively high frequency.

R factors have also been used to isolate R-prime (R′) elements carrying contiguous pieces of DNA from several species.[31,43,101] The formation of either an R′ or a cointegrate is most likely mediated by IS elements on the R factor. Marrs and coworkers isolated pRPS404 as a derivative of RP1.[101] This R′ was subsequently shown to contain about 40 kb of contiguous genomic DNA required for photosynthetic growth. This included the *R. capsulatus puf, puh, bch,* and *car* loci. It is stable in *E. coli* but readily recombines into the *R. capsulatus* genome.

A larger R′ (pWS2) was isolated from *R. sphaeroides* strain WS8.[43] In the same study, Sistrom *et al.* isolated an unrelated R′, pWS1, which complemented several amino acid auxotrophs and a putative recombination-deficient strain. We have determined that pWS2[102] contains approximately 109 kb of contiguous *R. sphaeroides* DNA, and it contains the *puf, puh, cycA,* and *puc* operons as well as the *bch* and *car* loci.[102] Plasmid pWS2 has not been analyzed as extensively as pRPS404, but it is stable in both *E. coli* and *R. sphaeroides.* Although the genetic markers within the photosynthetic gene cluster from *R. capsulatus,*[26] and *R. sphaeroides*[102–104] are similar, there are differences in the chromosomal linkage of *puc, cycA,* and other photosynthesis functions between these two closely related bacteria.[94,96]

Rhodobacter sphaeroides and *R. capsulatus* also differ in the number

[100] M. McClelland, R. Jones, Y. Pagel, and M. Nelson, *Nucleic Acids Res.* **15,** 5985 (1987).
[101] B. Marrs, *J. Bacteriol.* **146,** 1003 (1981).
[102] Y. O. Wu, B. MacGregor, T. Donohue, and S. Kaplan, *Plasmid* (in press).
[103] S. A. Coomber, M. Chaudri, and C. N. Hunter, *in* "Molecular Biology of Membrane-Bound Complexes in Phototrophic Bacteria" (G. Drews and E. A. Dawes, eds.), p. 49, Plenum, New York, 1990.
[104] J. M. Pemberton and C. M. Harding, *Curr. Microbiol.* **14,** 25 (1986).

and organization of other loci. For example, *R. capsulatus* contains only a single copy of *hemA*,[105,106] whereas *R. sphaeroides*[107] contains two distinct regions which complement uncharacterized mutations causing δ-aminolevulinic acid auxotrophy.

An additional difference in genome organization between *R. sphaeroides* and *R. capsulatus* is in those genes which encode proteins involved in the Calvin cycle.[108,109] Both species contain *rbcLS* and *rbcR* loci which code for the Form I and Form II ribulose-1,5-bisphosphate carboxylase enzymes, respectively. The analysis of these regions is not as complete in *R. capsulatus* as it is in *R. sphaeroides*, but it appears that only *R. sphaeroides* has duplicate genes for phosphoribulokinase, fructose-1,6-bisphosphatase, and the *cfx* gene product.[110] We anticipate that many additional differences in genome architecture will be found between these two species given the existence of two chromosomes in *R. sphaeroides* and the apparently single replicon in *R. capsulatus*.

G. Endogenous Plasmids

Virtually all *R. sphaeroides* strains that have been examined possess at least one and as many as five endogenous plasmids,[98] and, for the most part, the function(s) of these plasmids remains unknown. There is considerable sequence homology among several of the 2.4.1 plasmids as well as between the plasmids from 2.4.1 and those from other wild-type *R. sphaeroides* strains which contain different sizes or numbers of plasmids.[99,111]

Rhodobacter capsulatus B10 contains a cryptic plasmid of approximately 130 kb,[97] and Willison has recently reported that this plasmid contains both the assimilatory and dissimilatory nitrate reductases.[112] *Rhodospirillum rubrum* S1 has an approximately 55-kb extrachromosomal element of unknown function, although Kuhl *et al.* have reported that under certain conditions loss of this plasmid is correlated with loss of photosynthetic growth.[113]

S-Factor-Mediated Conjugation. Suwanto and Kaplan have found that

[105] G. Drews, personal communication (1989).
[106] J. W. Biel, M. S. Wright, and A. J. Biel, *J. Bacteriol.* **170**, 4382 (1988).
[107] T. N. Tai, M. D. Moore, and S. Kaplan, *Gene* **70**, 139 (1988).
[108] P. L. Hallenbeck and S. Kaplan, *Photosynth. Res.* **19**, 63 (1988).
[109] F. R. Tabita, *Microbiol. Rev.* **52**, 155 (1988).
[110] J. Shively, P. L. Hallenbeck, and S. Kaplan, unpublished observations (1986).
[111] F. E. Nano and S. Kaplan, *J. Bacteriol.* **158**, 1094 (1984).
[112] J. C. Willison, *FEMS Microbiol. Lett.* **66**, 23 (1990).
[113] S. A. Kuhl, D. W. Nix, and D. C. Yoch, *J. Bacteriol.* **156**, 737 (1983).

the approximately 42-kb plasmid of *R. sphaeroides* 2.4.1 is self-mobilizable (S factor).[99] It is possible that a significant amount of its DNA is dedicated to transfer functions. This process was discovered by monitoring transfer of derivatives of this plasmid which contain selectable markers.[95]

For example, cointegrates with the 42-kb plasmid can be formed if a Tn5-containing pSUP plasmid (pSUP2021)[84] is mobilized into an *R. sphaeroides* strain containing a Tn5 insertion in the S factor. Cointegrate formation between the S factor and pSUP2021 can be monitored by demanding both Kn (Tn5) and Tc (pSUP plasmid) resistance. These cointegrates presumably form by homologous recombination between the Tn5 elements on the 42-kb plasmid and pSUP2021. Once formed, such cointegrates can be transferred to other *R. sphaeroides* strains, and they can also be mobilized from *R. sphaeroides* to *E. coli*. They replicate in *E. coli*, presumably using the pSUP2021 replication origin. However, these cointegrates cannot be mobilized between *E. coli* strains or from *E. coli* to *R. sphaeroides*. A simple explanation is that S-factor transfer function(s) is either not expressed in *E. coli* or, if expressed, those elements contained on the S factor are not sufficient for transfer. The S-factor transfer function(s) can also act in trans to mobilize other plasmids between *R. sphaeroides* strains. In very recent work it has been demonstrated by Suwanto and Kaplan (unpublished) that the S factor is comprised of a contiguous 9-kb region containing all known Tra functions, a 3.3-kb region containing OriV, and a 1.1-kb region encompassing OriT. In the latter instance when OriT is placed within the large chromosome, directional transfer has been observed.

H. Insertion Elements

There are suggestions that endogenous insertion element(s) (IS) exist in *R. sphaeroides* 2.4.1. This would not be surprising given their widespread existence in other cells. Nano and Kaplan observed cointegrate formation between the endogenous *R. sphaeroides* plasmids when defective Mud phage were introduced from *E. coli*.[111] Cointegrate formation was also correlated with the occurrence of pigment mutations. More recently, Suwanto and Kaplan[95] observed that insertion of Tn5 into an endogenous *R. sphaeroides* 2.4.1 plasmid could result in pigment mutants or an auxotrophic requirement (e.g., Phe$^-$). The Phe$^-$ strain reverts to prototrophy without losing Tn5 from the plasmid, and it is complemented by a cosmid containing chromosomal DNA. This suggests that the Phe$^-$ phenotype is not directly related to the presence of Tn5 on the plasmid. One explanation for these observations is that host IS elements can be "activated" by transposable elements under the appropriate conditions.

IV. Molecular Genetic Techniques

A. Transposition

Drug resistance elements and transposition functions found on wild-type Tn*5*, Tn*501*, and Tn*7* are expressed in *R. sphaeroides*. Transposon Tn*5* has been used most often for the apparently random construction of auxotrophs, or other mutants. The mobilizable pSUP plasmids are commonly used to deliver Tn*5* because they can be conjugated from *E. coli* to Rhodospirillaceae using either the di- or triparental mating system described in Section III,E.[84] Tn*5* transposition frequencies have been estimated at around 1–10 in 10^{-5} in *R. sphaeroides*. Tn*5* encodes both Kn and Sm resistance in several nonenteric bacteria.[114,115] Kn is most commonly used because the rate of spontaneous Sm resistance is high.

Tn*7* transposes to specific locations in *R. rubrum*[31]; this has been noted in other bacteria.[116] Finally, Tn*10* transposition in either *R. rubrum*[31] or *R. sphaeroides*[117] has yet to be achieved.

B. Mutant Construction via Recombination

Systems exist for making specific mutations in previously cloned genes. All depend on homologous recombination between an inactive copy of the cloned gene and the genomic copy. The first step is usually the insertion of a selectable drug resistance marker into the gene of interest in *E. coli*. Kn[118,119] or Sp[120] resistance cartridges have been routinely used for this purpose. This provides a direct selection for the desired recombination event with the host genome. Alternatively, genes previously inactivated by Tn*5* in *E. coli* have been used to mutagenize specific regions in *R. sphaeroides*[121,122] and *R. capsulatus*.[123,124]

One system developed for constructing mutants in *R. sphaeroides*,[85,86] which is applicable to other species,[31,88,89] capitalizes on the inability of pSUP202[84] to replicate in this bacterium. Genes inactivated with a se-

---

[114] P. Mazodier, O. Genilloud, E. Giraud, and F. Gasser, *Mol. Gen. Genet.* **204,** 404 (1986).
[115] E. A. O'Neill, G. M. Kiely, and R. A. Bender, *J. Bacteriol.* **159,** 388 (1984).
[116] R. L. McKown, K. A. Orle, T. Chen, and N. L. Craig, *J. Bacteriol.* **170,** 352 (1988).
[117] J. E. Flory and T. J. Donohue, unpublished observations (1989).
[118] J. Viera and J. Messing, *Gene* **19,** 259 (1982).
[119] S. Harayama, R. A. Leppik, M. Rekik, N. Mermod, P. R. Lehrbach, W. Reineke, and K. N. Timmis, *J. Bacteriol.* **167,** 455 (1986).
[120] P. Prentki and H. M. Kirsch, *Gene* **29,** 303 (1984).
[121] D. L. Falcone, R. G. Quivey, and F. R. Tabita, *J. Bacteriol.* **170,** 5 (1988).
[122] C. N. Hunter, *J. Gen. Microbiol.* **134,** 1481 (1988).
[123] K. M. Zsebo and J. E. Hearst, *Cell (Cambridge, Mass.)* **37,** 937 (1984).
[124] R. G. Kranz, *J. Bacteriol.* **171,** 456 (1989).

lectable marker (Kn or Sp resistance) are cloned on pSUP202 and mated from *E. coli* into the appropriate recipient. If the suicide plasmid contains a second scorable drug resistance marker (i.e., Tc resistance), it is easy to distinguish events which occur by integration of the entire plasmid (i.e., cells which become Kn and Tc resistant) from those where reciprocal recombination places the mutated gene in the chromosome (Kn resistant and Tc sensitive). Assuming that the pSUP replicon is mobilized as efficiently as an RK2 plasmid, one can calculate recombination frequencies of approximately 0.5–1.0 in 10^5 per *R. sphaeroides* recipient when there is about 1 kb of homologous DNA flanking the selectable marker.[85,86] As expected, Kn- and Tc-resistant strains represent the major class of mutants obtained; Kn-resistant and Tc-sensitive strains usually represent around 1–10% of the total depending on the amount of homologous DNA flanking the selectable marker.

Lee and Kaplan[125] have used this delivery system to place a *pucB:lacZ* protein or a *puc:kan* operon fusion in the chromosome. In this case, by screening recombinants that remained Tc resistant (the suicide plasmid marker) for a B800-850$^+$ phenotype, they identified cells that were merodiploid for the *puc* operon. Such merodiploids are stable as long as Tc selection is maintained; however, the integrated suicide plasmid is lost at a detectable frequency if antibiotic selection is removed.

Segregation of such merodiploids also allows this technique to be extended to placing nonselectable mutations (point mutations, linker insertions/deletions, etc.) in the chromosome. In this case, the identification of strains containing the desired trait requires either that a simple screen exist for the phenotype associated with the mutation or that genomic Southern blots be performed to screen the mutant allele within the population.

A second system uses two plasmids of the same incompatibility group to force recombination of the inactive copy of the gene with the genomic copy. The inactive copy of the gene of interest (e.g., encoding Kn resistance) is cloned and introduced into *R. capsulatus* on a stable plasmid which encodes a different drug resistance (e.g., Tc resistance). A second plasmid of the same incompatibility group (e.g., pPH1JI; Sp^R) is mated into this strain. Homologous recombination between the inactive gene and the chromosome is forced by demanding both Kn and Sp resistance. As with the first system, scoring for acquisition of the Tc determinant from the first plasmid allows one to identify recombinants which occur by single and reciprocal crossover events. This system has been used in *R. capsulatus*[123]; however, it has not produced reproducible results in *R.*

---

[125] J. K. Lee and S. Kaplan, unpublished observations (1990).

rubrum, and Lies *et al.* have suggested that pPH1JI does not express strong incompatibility in this species.[31]

GTA was the system initially used to select for insertional inactivation of genes in *R. capsulatus*. This process, which was termed "interposon mutagenesis,"[126,127] is limited to *R. capsulatus* because of the host specificity of GTA. In this case, the inactivated gene is first introduced on a stable plasmid into a *R. capsulatus* strain which overproduces GTA. GTA is propagated on this strain and used to infect the appropriate recipient. GTA cannot package the entire plasmid; so selection for the marker used to inactivate the gene of interest demands homologous recombination between the incoming fragment packaged in the phage particle and the genomic copy of the gene.

C. Gene Interconversion

DeHoff *et al.*[33] used homologous recombination to place a single copy of a mutagenized *R. sphaeroides puf* operon in the chromosome. This system should be applicable to placing other nonselectable mutations in the chromosome as long as sufficient information on the operon of interest is available. The parent strain (PUFB1)[86] could grow aerobically, but it was unable to grow photosynthetically (PS$^-$) because it contained a Kn resistance cartridge within the *puf* operon. A stable RK2 derivative, containing a deletion in the *pufAL* intercistronic region flanked by wild-type DNA upstream and downstream of the chromosomal drug resistance gene, was introduced into PUFB1. The RK2 derivative did not contain sufficient cis-acting DNA upstream of the *puf* operon to allow normal transcription from the plasmid-encoded operon; therefore, PUFB1 cells containing this plasmid were PS$^-$ because the plasmid was unable to complement the defective region in trans. Aerobically grown cells were inoculated under photosynthetic conditions, and recombination between the RK2 plasmid and the chromosome was demanded by selecting for PS$^+$ cells. Growth was apparent after about 3 days of incubation under photosynthetic conditions. Cells were plated under aerobic conditions in the absence of both drugs to allow for loss of the RK2 plasmid. Kn- and Tc-sensitive, PS$^+$ cells, which arose by exchange of the mutant *puf* operon, were identified. The resulting strain had the expected *pufAL* intercistronic deletion incorporated into the chromosome; this was confirmed by both genomic Southern blots with the mutagenizing oligonucleotide and DNA sequencing of the mutant *puf* operon. This was the first case where a specific oligonucleotide-induced mutation in a regulatory or structural region has been ana-

[126] P. A. Scolnick and R. Haselkorn, *Nature (London)* **307**, 289 (1984).
[127] D. C. Youvan, S. Ismail, and E. J. Bylina, *Gene* **38**, 19 (1985).

lyzed in single copy in a photosynthetic bacterium. The same strategy has been used to place mutations in the stem loop region around the *pufL* ribosome binding site into the *R. sphaeroides* chromosome.

The success of this gene interconversion depended on three facts. First, the parent strain contained a null mutation that resulted in a PS⁻ phenotype, and it did not revert at a detectable frequency. Second, parallel studies of the *puf* operon indicated that the mutation in the *pufAL* intercistronic region would allow photosynthetic growth if the altered operon were transcribed; this suggested that cells containing this mutation in single-copy would be PS⁺. Finally, there was a convenient restriction fragment to supply the mutated operon in trans with sufficient flanking DNA to allow homologous recombination in the absence of cis-acting sequences required for normal transcription. Such gene interconversions should be feasible in other systems if the phenotype of the parent strain is sufficiently stable and one has a strong selectable phenotype to enrich for cells containing the mutated gene in the chromosome.

D. Plasmids and Mutant Complementation

The most commonly used vector systems for complementation or for maintaining plasmid-encoded genes are those based on the RK2 replicon. Plasmids pRK404[74] and pRK415[75] were initially used for such purposes in *R. sphaeroides*. These and other RK2 derivatives are now being used in *R. capsulatus*. We have noted gene dosage effects on the synthesis of some gene products encoded by these plasmids (e.g., cytochrome c_2[128]). In contrast, synthesis of B875 light-harvesting and reaction center proteins does not appear to be severely affected by the plasmid copy number.[33,129]

The Ω cartridges[39,120] contain translation terminators in all three reading frames and *E. coli* phage T₄ transcription terminators. These can be placed upstream of the gene of interest to avoid transcriptional readthrough from vector sequences.

RK2 plasmids are stable in *R. sphaeroides* as long as selective pressure is maintained, but they are lost from up to 50% of the cells if *R. sphaeroides* is grown for 6–8 doublings in the absence of selection.[33,86,128] This property is useful in scoring gene interconversion events and in differentiating between genomic and plasmid mutations.

In *R. capsulatus*, the ColE1-based pDPT vectors[26] were initially used by some laboratories,[68,69,127] but there appears to be a conflict regarding the stability of the ColE1 or pBR322 replicons in this species. Youvan and

[128] J. P. Brandner, A. G. McEwan, S. Kaplan, and T. J. Donohue, *J. Bacteriol.* **171**, 360 (1989).

[129] R. E. Sockett, T. J. Donohue, A. R. Varga, and S. Kaplan, *J. Bacteriol.* **171**, 436 (1989).

co-workers have engineered plasmids based on either ColE1[130] or RK2[34] replicons with unique restriction sites for the efficient oligonucleotide-directed mutagenesis of the reaction center genes of *R. capsulatus*.

E. Monitoring Gene Expression

β-Galactosidase. Nano *et al.* initially demonstrated that the *E. coli* *lacZ* gene could be used as a reporter group in *R. sphaeroides*[13] to construct either operon[111] or protein fusions.[131] Biel and Marrs used *lacZ* operon fusions based on defective Mu phage in *R. capsulatus*.[132] The activity of *lacZ* can be easily monitored by the use of indicator plates containing 5-bromo-4-chloro-3-indolyl-β-D-galactoside (X-Gal), and many methods developed for use with *E. coli* can also be used in photosynthetic bacteria. Wild-type strains of *R. sphaeroides* and *R. capsulatus* are normally *lac°*. *Rhodobacter sphaeroides* 2.4.1 will not use glucose as a carbon source, but mutants capable of growth on glucose can be isolated at a reasonably high frequency.[13] Glucose-utilizing mutants are able to grow on lactose when *E. coli lacZYA* is expressed.

Kranz and Haselkorn[133] noted that an *R. capsulatus* strain harboring a plasmid-encoded *nifH:lacZ* fusion was able to grow on lactose only when cells are incubated diazotrophic conditions. They used this pheno-type to select for different classes of trans-acting mutants which allow growth on lactose either anaerobically in the presence of ammonia or under aerobic conditions.[133]

The pRS415 plasmid of Simons *et al.*[134] contains restriction sites (*Eco*RI *Bam*HI, *Sma*I) for constructing *lacZYA* operon fusions. Once operon fusions are constructed in *E. coli*, the *Eco*RI and *Stu*I (blunt) sites on this plasmid can be used to move them into the *Eco*RI and *Hpa*I (blunt) sites of pKT231.[78] *lacZ* operon fusions to *R. sphaeroides cycA* promoters are expressed *in vivo*,[135] presumably using the hybrid ribosome binding site upstream of *lacZ*. These plasmids are compatible with those of other incompatibility groups (RK2, etc.).

Several groups have constructed LacZ hybrid proteins to monitor gene expression in both *R. sphaeroides*[37,131] and *R. capsulatus*.[68,124] These fusions have used either Mu*d* derivatives or specific restriction fragments

[130] E. J. Bylina, S. Ismail, and D. C. Youvan, *Plasmid* **16,** 175 (1986).
[131] F. E. Nano, W. D. Shepherd, M. M. Watkins, S. A. Kuhl, and S. Kaplan, *Gene* **34,** 219 (1984).
[132] A. J. Biel and B. L. Marrs, *J. Bacteriol.* **156,** 686 (1983).
[133] R. G. Kranz and R. Haselkorn, *Proc. Natl. Acad. Sci. U.S.A.* **83,** 6805 (1986).
[134] R. W. Simons, F. Houman, and N. Kleckner, *Gene* **53,** 85 (1987).
[135] B. A. Schilke and T. J. Donohue, unpublished observations (1990).

as a source of LacZ. High level synthesis of LacZ fusions to membrane proteins is deleterious in *E. coli*. Unfortunately, experiments have not addressed if potential problems are associated with high-level synthesis of LacZ hybrids involving membrane proteins in photosynthetic bacteria.

RSF1010 plasmids have been modified to allow direct construction of *lacZ* translational fusions in all three reading frames.[37] Derivatives are available with and without a *lac* promoter; those with a *lac* promoter allow one to screen the fusions in *E. coli* prior to mobilization into other bacteria. These plasmids were constructed with Tc resistance genes instead of the less desirable drug resistance elements on RSF1010.

Alkaline Phosphatase. Manoil and Beckwith demonstrated how protein fusions to the *E. coli* alkaline phosphatase gene (*phoA*) could be used for analyzing synthesis of exported proteins[136] and the topology of membrane proteins.[137] Alkaline phosphatase (APase) is an excellent reporter group for constructing translational fusions to exported *R. sphaeroides* proteins. Wild-type *R. sphaeroides* contains a low level of APase in both the periplasm and cytoplasm,[138] but they are phenotypically Pho⁻ on plates containing 5-bromo-4-chloro-3-indolyl phosphate.[138] Thus, it is easy to score the *pho* phenotype of strains carrying hybrid proteins. Plasmids are available to construct hybrid proteins in all three reading frames.[139]

Both purified restriction fragments and Tn*phoA* have been used to construct a series of fusions to the *R. sphaeroides* cytochrome c_2 structural gene in *E. coli*.[139,140] The *R. sphaeroides* cytochrome c_2 signal sequence functions in *E. coli*,[139–141] so export of *phoA* fusions can be directly screened in this host as long as they are downstream of an *E. coli* promoter. Recent experiments have demonstrated that protein fusions between cytochrome c_2 and APase are substrates for heme attachment.[140] Such hybrid proteins should facilitate future studies on export and heme attachment.

Finally, Tn*phoA* has been cloned on a pSUP suicide plasmid and used to create in-frame fusions to exported proteins in *R. sphaeroides*.[22] Thus, Tn*phoA* can be used as a general screen for exported proteins in photosynthetic bacteria.

Other Reporter Groups. Several other genes have been used as reporter groups to monitor gene expression. The *xylE* gene from the *Pseudomonas putida* TOL plasmid[119] has been used to generate operon fusions in *R*.

[136] C. Manoil and J. Beckwith, *Proc. Natl. Acad. Sci. U.S.A.* **82**, 8129 (1985).
[137] C. Manoil and J. Beckwith, *Science* **233**, 1403 (1986).
[138] S. P. Tai, Ph.D. Thesis, University of Illinois,Urbana, (1985).
[139] A. R. Varga and S. Kaplan, *J. Bacteriol.* **171**, 5830 (1989).
[140] J. P. Brandner, E. V. Stabb, R. Temme, and T. J. Donohue, *J. Bacteriol.* (in press).
[141] A. G. McEwan, S. Kaplan, and T. J. Donohue, *FEMS Microbiol. Lett.* **59**, 253 (1989).

sphaeroides with expression being controlled by the *rrn* promoters.[142] A promoterless Kn^R cartridge from pUC4K has also been used to probe for *puc* operon promoter activity.[125] Finally, a truncated *Cellulomonas fimi* cellulase gene has been used to analyze hybrid proteins in *R. capsulatus*.[32]

F. Expression Systems

A limited number of sequences have been used to promote transcription of genes in photosynthetic bacteria. The most well-developed and highly regulated promoter available is the *nifHDK* promoter from *R. capsulatus*.[143] Pollock *et al.* have shown that the *nifHDK* promoter was induced more than 2600-fold under diazotrophic conditions. This promoter was fully derepressed in media lacking fixed nitrogen, and transcription occurred at intermediate levels in the presence of fixed nitrogen sources such as cysteine, arginine, or leucine. Bauer and Marrs subsequently used this promoter to obtain nitrogen-source-regulated synthesis of the *pufQ* gene product under anaerobic conditions in *R. capsulatus*.[144]

The major limitation of the *nifHDK* promoter in wild-type cells is that it is highly expressed only under nitrogen-limiting conditions in the absence of oxygen. The *puc* operon promoter has been identified in both *R. sphaeroides*[145] and *R. capsulatus*.[146] Lee *et al.*[145] have shown that *puc* transcription is induced approximately 100-fold under anaerobic conditions in *R. sphaeroides*; this promoter is induced about 3-fold more by low light intensity. Thus, the *puc* promoter could be used to regulate transcription under anaerobic conditions in the presence of ammonia or other sources of fixed nitrogen.

Promoter sequences are available that allow transcription of a target gene in the presence of oxygen. The *R. sphaeroides* cytochrome c_2 structural gene (*cycA*) contains at least two promoters.[91] When downstream sequences are fused to *lacZ*, LacZ activity is independent of oxygen. In contrast, a *cycA:lacZ* fusion is induced approximately 6-fold under anaerobic conditions when the upstream oxygen-sensitive promoter(s) is used.[135] Finally, Lee and Kaplan have localized the promoter for the pUC4-K Kn resistance gene.[125] Transcription of the Kn resistance gene is high in *R. sphaeroides* under all physiological conditions tested.

It should be emphasized that heterologous promoter systems are still in their infancy in photosynthetic bacteria, and the definition of what

[142] S. Dryden and S. Kaplan, unpublished observations (1991).
[143] D. Pollock, C. E. Bauer, and P. A. Scolnick, *Gene* **65,** 269 (1988).
[144] C. E. Bauer and B. L. Marrs, *Proc. Natl. Acad. Sci. U.S.A.* **85,** 7074 (1988).
[145] J. K. Lee, P. J. Kiley, and S. Kaplan, *J. Bacteriol.* **171,** 3391 (1989).
[146] A. P. Zucconi and J. T. Beatty, *J. Bacteriol.* **170,** 877 (1988).

constitutes a promoter is still subject to speculation. Comparisons between promoters must be viewed with great caution because analysis of promoter activity has not always been performed under identical growth conditions or with fusions present in the same copy number.

V. Future Directions and Needs

This is an exciting time to be studying photosynthetic bacteria. Detailed biochemical, biophysical, physiological, and spectroscopic studies have provided us with unique insights into important basic and applied processes such as photosynthesis, membrane biogenesis, nitrogen fixation, carbon dioxide fixation, and the function of energy-transducing membranes. The two chromosomes in *R. sphaeroides* also presents a unique opportunity to ask questions regarding chromosome evolution and segregation in a prokaryote.[96]

The techniques summarized in this chapter demonstrate that genetic analysis of photosynthetic bacteria has matured significantly in recent years. Tools are now available to dissect the regulatory circuits controlling synthesis or activity of proteins required for the diverse physiological functions of these bacteria. The progress made is especially impressive considering the relatively small number of laboratories studying these bacteria. It is also obvious that more tools are needed to facilitate future genetic analyses.

Mapping of genetic markers would be facilitated if a well developed Hfr-like system and phage capable of transducing relatively large pieces of DNA (i.e., 1 min or more) were available for each species. Such systems would be very useful in mapping mutations conferring new phenotypes, testing the effects of individual mutations in different genetic backgrounds, and aligning new markers with existing physical–genetic maps. A high-frequency transformation system would also greatly facilitate future genetic analysis.

In the absence of lysogenic phage, future challenges include developing a simple way to efficiently construct merodiploids. Genes for widely different physiological functions (i.e., N_2 fixation, Calvin cycle, H_2 metabolism, photosynthetic functions, alternate electron acceptors) are available to serve as targets for construction of merodiploids. Merodiploids could be generated at these target sites if the gene of interest is cloned along with its cis-acting regulatory sequences within the appropriate target sequences on a suicide plasmid. This approach is only feasible if the resulting inactivation of the target operon is neither lethal nor deleterious to the system being studied. Lies *et al.* noted how Tn7 might facilitate construction of merodiploids,[31] especially if Tn7 derivatives with convenient restriction sites can be constructed. The latter approach is particularly powerful since the lesion generated by Tn7 transposition in Rhodospirillaceae and other bacteria does not produce a scorable phenotype other than acquisition of drug resistance from the transposon.

Finally, it is critically important to extend the study of this diverse group of microorganisms beyond traditional areas of focus. A broader physiological analysis of these metabolically versatile bacteria promises to uncover other interesting traits. A parallel genetic analysis of these traits will increase the availability of genetic markers to a point where strain construction, mutant selection, phenotypic analysis, etc., is no longer limited by our inability to analyze and exploit other regions of the genome.

Acknowledgments

The authors would like to thank past and present members of their laboratories for their diligence and insights in developing both genetic and biochemical techniques for analysis of photosynthetic bacteria; their enthusiasm has always made this work interesting. We would also like to thank our many colleagues who sent reprints and preprints of unpublished works. Particular thanks go to Carl Bauer for helping with the section on GTA-mediated gene exchange, and to Patricia Kiley, Gary Roberts, Doug Lies, and current members of our laboratories for critical readings of the manuscript.

Recent research has been supported by grants from the National Institutes of Health (GM15590 and GM31667 to S.K. and GM37509 to T.J.D.), by U.S. Department of Agriculture Grant 89-37262-4939 to S.K. and T.J.D., and by USDA Hatch project WIS3028 to T.J.D.

[23] Genetic Systems in *Pseudomonas*

By R. K. ROTHMEL, A. M. CHAKRABARTY, A. BERRY, and A. DARZINS

Introduction

Pseudomonads are a diverse group of organisms. They range from species pathogenic to humans (*Pseudomonas aeruginosa*) and plants (e.g., *P. syringae* and *P. solanacearum*), to species capable of degrading organic and recalcitrant compounds (*P. cepacia* and *P. putida*), to those that promote agricultural growth (*P. fluorescens*). Many of these species carry the genetic determinants for the utilization of various growth substrates on plasmids. The diversity found among *Pseudomonas* species partially stems from many years of taxonomists categorizing gram-negative organisms, not fitting into other taxonomic groups, into the genus *Pseudomonas*. Recent developments in the use of rRNA homology groups and conserved features of biosynthetic pathways have made phylogeny-based classification possible.[1] It has been suggested that the genus *Pseudomonas* be broken up into five different genera, based on five different rRNA homology groups.[2] In addition to taxonomic variations, pseudomonads also exhibit interesting genetic characteristics quite different from *Escherichia coli* and other gram-negative microorganisms. In this short review, we emphasize the salient genetic features of the pseudomonads that are somewhat unique to this genus.

Genome Organization: Mapping of Genes in *Pseudomonas*

To understand the genetic organization of any bacterial species, effective genetic tools and an accurate chromosomal map are required. For both *P. putida* and *P. aeruginosa*, the most extensively studied *Pseudomonas* species, effective mapping procedures have been developed.[1,3-5] These methods include the classic genetic methods of conjugation, transduction, and transformation. Only since the late 1970s and early 1980s has the circularity of the chromosomal genetic map been demonstrated for *P. aeruginosa* PAO,[6] *P. aeruginosa* PAT,[7] and *P. putida* PPN.[8]

A number of conjugative plasmid systems are available for mapping *P. aeruginosa* PAO. The first available plasmid, FP2, had limited application because there is only one efficient origin of transfer for chromosomal DNA; therefore, only markers less than 30 min from the transfer origin could be accurately mapped.[3] A derivative of the broad host range IncP-1 plasmid R68, R68.45, generates many transfer origins around the chromosome and was used in two- and three-factor crosses to determine a linkage map and establish PAO chromosome circularity.[6] Isolation of temperature-sensitive replication mutants of the IncP-1 plasmids RPI[9] and R68[10] provided the genetic tool for high frequency of recombination mapping. pMO514,[11] a *trfA* mutant of R68 loaded with the transposon

[1] B. W. Holloway and A. F. Mogan, *Annu. Rev. Microbiol.* **40**, 79 (1986).
[2] N. J. Palleroni, *BioScience* **55**, 370 (1983).
[3] B. W. Holloway, V. Krishnapillai, and A. F. Morgan, *Microbiol. Rev.* **43**, 73 (1979).
[4] R. Bray, D. Strom, J. Barton, H. F. Dean, and A. F. Morgan, *Genetics* **133**, 683 (1987).
[5] K. O'Hoy and V. Krishnapillai, *Genetics* **115**, 611 (1987).
[6] P. L. Royle, H. Matsumoto, and B. W. Holloway, *J. Bacteriol.* **145**, 145 (1981).
[7] J. M. Watson and B. W. Holloway, *J. Bacteriol.* **136**, 507 (1978).
[8] H. F. Dean and A. F. Morgan, *J. Bacteriol.* **153**, 485 (1983).
[9] D. Haas, J. Watson, R. Krieg, and T. Leisinger, *Mol. Gen. Genet.* **182**, 240 (1981).
[10] B. W. Holloway, C. Crowther, H. Dean, J. Hagedom, N. Nolmes, and A. F. Morgan, *in* "Drug Resistance in Bacteria" (S. Mitsuhasllhi, ed.), p. 231. Japan Scientific Societies Press, Tokyo, Japan, 1982.
[11] K. O'Hoy and V. Krishnapillai, *FEMS Microbiol. Lett.* **29**, 299 (1985).

Tn*2521*, promotes chromosome transfer from a number of different sites and can mediate polarized chromosome transfers. In addition to conjugation, transductional systems for genetic analysis of *P. aeruginosa* were also developed because of the many natural bacteriophage that can be isolated. Phage F116L[3] and G101[3] have been used to establish the order of closely linked markers. Extracellular DNA produced by *P. aeruginosa* can also be used in transformation experiments for mapping closely linked markers.[12]

Unlike *P. aeruginosa*, the development of conjugative plasmids and transducing phage for *P. putida* has been more difficult. Except for a derivative of the plasmid XYL-K, native chromosome-mobilizing plasmids have not been isolated. XYL-K has been used to map a number of chromosomal genes in *P. putida* strain PpG1.[13] R68.45 can promote chromosome transfer in *P. putida*; however, strains carrying degradative IncP-2 plasmids cause fertility inhibition of the IncP-1 plasmids.[3] In addition, determining isofunctionality of gene products between species, using R-primes carrying *P. putida* PPN chromosomal DNA and *P. aeruginosa* PAO as a Rec⁻ recipient, is far more difficult than the reciprocal use. *Pseudomonas putida* R-prime plasmids, nonetheless, have been isolated.[4] Bacteriophage in *P. putida* are rare, and only Pf16 and its host range mutant Pf16h2 have been used as generalized transducing phage.[14]

Genome Characteristics of *Pseudomonas*

Gene Arrangement

Several features of chromosome arrangement become apparent when the maps of *P. aeruginosa* and *P. putida* are compared.[11] First, there is the noncontiguous arrangement of genes of biosynthetic pathways, as compared to the contiguous arrangement found in Enterobacteriaceae; and, second, genes with related functions are clustered into noncontiguous groups, described as supraoperonic clustering.[15] The lack of continuity of biosynthetic genes can be exemplified by the tryptophan pathway.[1] In *P. aeruginosa trpA* and *trpB* are contiguous at 26 min, whereas *trpC* and *trpD* are found at a region near 34–35 min and *trpF* at 54 min. *trpE* is separated from *trpC* and *trpD* by pyocin genes. In *P. putida* PpG1 *trpA* and *trpB* are

---

[12] Y. Muto and S. Goto, *Microbiol. Immunol.* **30,** 621 (1986).

[13] J. R. Mylroie, D. A. Friello, T. V. Siemens, and A. M. Chakrabarty, *Mol. Gen. Genet.* **157,** 231 (1977).

[14] A. M. Chakrabarty, C. F. Gunsalus, and I. C. Gunsalus, *Proc. Natl. Acad. Sci. U.S.A.* **60,** 168 (1968).

[15] B. J. Leidigh and M. L. Wheelis, *J. Mol. Evol.* **2,** 235 (1973).

part of one transductional group separate from *trpC*, *trpD*, and *trpE*, and *trpF* is separate from either of these groups. The organization of *trp* genes in *P. syringae* and *P. acidovorans* are slightly different.[1] In contrast, the *trp* genes are all contiguous in *E. coli* and *Salmonella typhimurium*.

Supraoperonic clustering has been described in both *P. aeruginosa* and *P. putida*. In *P. aeruginosa* the glucose utilization genes are clustered together in a 6-kilobase (kb) region. A further example is seen in *P. putida* in the 0–14 min region, where almost all known dissimilatory genes are located including genes for catechol, mandelate, histidine, protocatechuate, phenyl acetate, phenylalanine, and quinate dissimilation.[1]

Nucleotide Composition/Codon Usage

The size of a number of *Pseudomonas* genomes has been estimated as well as their G + C content. These include *P. aeruginosa* (5850 kb, 67 mol% GC), *P. putida* (5620 kb, 61–62 mol% GC), and *P. fluorescens* (4107 kb, 59–63 mol% GC). Because of the high G + C content of the *Pseudomonas* species, codon usage is extremely biased. A recent review discusses the codon usage for *P. aeruginosa*.[16] The preferred choice for the third base in the codon sequence is cytosine, which reflects the high G + C content. Exceptions are seen in the preferential codon usage of CTG for leucine and CCG for proline and the equivalent use of GTG and GTC for valine and TCG and TCC for serine. Pilin genes also violate the rule because codons ending in T and G are preferred, and there is an increased use of A. *Pseudomonas aeruginosa* preferentially uses G in the first codon position and nucleotides other than G in the second position for chromosomal genes; *P. putida* shows a similar distribution of codon usage.[17] In contrast, degradative or resistance plasmids show a broader codon usage and are only moderately biased toward G and C.[18] The preferred start codon is ATG, and the preferred stop codon is TGA. Translation initiation in *Pseudomonas* occurs in a similar manner to that of *E. coli*. The consensus sequence of 3' end of the *P. aeruginosa* 16 S RNA, $G(X)_2PyCUCUCCUU(A)$-OH, which encompasses the ribosome binding site is identical to that of *E. coli* for seven nucleotides.[19]

[16] E. H. West and B. H. Iglewski, *Nucleic Acids Res.* **16**, 9323 (1988).
[17] C. Nakai, H. Kagamiyama, M. Nozaki, T. Nakazawa, S. Inouye, Y. Ebina, and A. Nakazawa, *J. Biol. Chem.* **258**, 2923 (1983).
[18] I. P. Crawford and L. Eberly, *Mol. Biol. Evol.* **3**, 436 (1986).
[19] J. Shine and L. Dalgarno, *Nature (London)* **254**, 34 (1975).

Plasmids

Pseudomonads with the ability to utilize various carbon sources are readily isolated by selective growth enrichment procedures. Many of these organisms have been shown to carry plasmids with genes specifying metabolism of compounds such as salicylate, camphor, naphthalene, toluene, xylene, and 3-chlorobenzoate. The genes for these dissimilatory pathways, unlike the biosynthetic genes on the chromosome, are tightly linked on the plasmid. As noted above, the catabolic genes on the chromosome also tend to be more closely clustered. One explanation for the apparent clustering of catabolic genes is that these genes may have originated from plasmids, which have inserted into the chromosome at specific regions.[1]

Transposons and Insertion Sequence Elements in Pseudomonas

Transposable elements including both transposons and insertion sequence (IS) elements are quite common in various *Pseudomonas* species and include those with both antibiotic and heavy metal resistance.[20] A group of elements belonging to class II transposons merits mention here since they incorporate rather large segments of DNA that specify degradation of hydrocarbons such as xylene/toluene or naphthalene/salicylate. Early studies indicated that the genes encoding the degradation of such hydrocarbons were transposable.[21,22] Tsuda and Iino[23] clearly demonstrated that a 56-kb region containing the xylene degradative (*xyl*) genes and a 17-kb derivative of this region (harboring a 39-kb internal deletion) transposed to various sites on target replicons such as pACYC184 and R388 in *E. coli recA* strains. The transposable 56- and 17-kb regions were designated Tn*4651* and Tn*4652*, respectively. Subsequently it was demonstrated[24] that the 56-kb Tn*4651* containing the *xyl* genes was completely contained within a larger 70-kb transposon termed Tn*4653*.

Both Tn*4651* and Tn*4653* encode a transposase gene, *tnpA*, which allows formation of cointegrates as intermediate products of transposition. The resolvase gene products *tnpS* and *tnpT* encoded by Tn*4651* allows resolution of both Tn*4651* and Tn*4653*-mediated cointegrates.[25] Sequence

[20] J. A. Shapiro, "Mobile Genetic Elements." Academic Press, New York, 1983.

[21] A. M. Chakrabarty, D. A. Friello, and L. H. Bopp, *Proc. Natl. Acad. Sci. U.S.A.* **75**, 3109 (1978).

[22] G. A. Jacoby, J. E. Rogers, A. E. Jacob, and R. W. Hedges, *Nature (London)* **274**, 179 (1978).

[23] M. Tsuda and T. Iino, *Mol. Gen. Genet.* **210**, 270 (1987).

[24] M. Tsuda and T. Iino, *Mol. Gen. Genet.* **213**, 72 (1988).

[25] M. Tsuda, K. I. Minegishi, and T. Iino, *J. Bacteriol.* **171**, 1386 (1989).

analysis demonstrated that Tn*4651* and Tn*4653* have 46- and 38-base pair (bp) terminal inverted repeats, respectively, and that both elements generate a 5-bp duplication of the target sequence on transposition. Even though both Tn*4651* and Tn*4653* demonstrated considerable similarity with the class II transposons Tn*3*, Tn*21*, and Tn*1721*, the trans-acting transposition functions of Tn*4651* were not interchangeable with those encoded by any of the class II transposons. In contrast, the Tn*4653* *tnpA* function was interchangeable with the Tn*1721* *tnpA* function. Tsuda *et al.*[25] further demonstrated that Tn*4653* encoded a resolvase (product of *tnpR* gene) that complemented the *tnpR* mutations of Tn*21* and Tn*1721*. In addition, the Tn*4653* *tnpR* gene, located 5′ upstream of the *tnpA* gene, shares extensive sequence homology with the Tn*1721* *tnpR* gene.

It should be noted that the *xyl* genes in Tn*4651* are bounded by a 1.4-kb direct repeat, and these direct repeats have been implicated in the frequent excision of the 39-kb segment harboring the *xyl* genes.[26] An interesting question in this regard is whether the terminal direct repeats represent remnants of an IS element that might have been responsible in the recruitment of the *xyl* gene segment and its integration on the chromosome of the *P. putida* mt-2 strain. In an analogous way, recruitment of the 2,4,5-trichlorophenoxyacetic acid (2,4,5-T) degradative genes in *P. cepacia* AC1100 may have occurred by the IS element IS*931*, which can carry intervening DNA during transposition, as described below.

Another interesting feature of transposable elements found in *Pseudomonas* species are the presence of outwardly directed promoters. *Pseudomonas cepacia* strains, such as *P. cepacia* 249 (ATCC 17616), harbor multiple copies of IS elements which have been implicated in promoting genomic rearrangements to recruit foreign genes by replicon fusion and insertional activation.[27] For example, insertion of IS elements such as IS402 or IS405 upstream of the *bla* gene resulted in 20- to 40-fold higher levels of β-lactamase activity.[28] Similarly, insertion of IS elements such as IS406, IS407, or IS415 upstream of the *lac* operon or specific *lac* genes present in the transposon Tn*951* allowed efficient expression of the *lac* genes in *P. cepacia* 249. So far at least 9 such IS elements have been characterized in strain 249,[27] and they exhibit considerable variations in size as well as specificity.

[26] P. Meulien, R. G. Downing, and P. Broda, *Mol. Gen. Genet.* **184,** 97 (1981).

[27] T. G. Lessie, M. S. Wood, A. Byrne, and A. Ferrante, *in* "*Pseudomonas*: Biotransformations, Pathogenesis, and Evolving Biotechnology" (S. Silver, A. M. Chakrabarty, B. Iglewski, and S. Kaplan, eds.), p. 279. American Society for Microbiology, Washington, D.C., 1990.

[28] G. E. Scordilis, H. Ree, and T. G. Lessie, *J. Bacteriol.* **169,** 8 (1987).

Transposable elements carrying promoters have also been characterized in a 2,4,5-T-degrading strain of *P. cepacia* AC1100. This element, initially designated RS1100, occurs in high copy number (at least 30) on the genome of this strain and is present near the *2,4,5-t* genes.[29] Sequencing of this element revealed it to be 1477 bp in length with 38 to 39-bp terminal inverted repeats immediately flanked by 8-bp direct repeats representing the duplication of the target sites.[30] This highly repeated element from *P. cepacia* AC1100 has not only been shown to be capable of transposing from one replicon to another, but its insertion upstream of a promoterless gene such as *aphC* allows gene expression.[31] RS1100 has therefore been redesignated IS931. Both IS931 and the genes involved in 2,4,5-T degradation do not demonstrate any appreciable homology with the total genomic digests of a large number of pseudomonads, including *P. cepacia* 249. This suggests that such genes might have been recruited from nonpseudomonal ancestors during evolution of the *2,4,5-t* genes in a chemostat.[30] Another interesting feature of IS931 is the fact that, during transposition, IS931 often inserts with multiple copies containing intervening DNA sequences, suggesting that this type of a process might have been involved in the recruitment of the *2,4,5-t* genes from other organisms to *P. cepacia* AC1100.[31]

The methodologies used to study transposition of IS elements take advantage of the fact that many *P. cepacia* IS elements can promote gene expression at the site of insertion in a new replicon. Thus, broad host range plasmids containing a promoterless gene such as *bla* or *lac* can be used to detect a transposition event.[27] Selection on carbenicillin-containing plates or plates with lactose as the sole source of carbon results in colony formation only when the IS element transposes upstream of the promoterless *bla* or *lac* gene. In the case of *P. cepacia* AC1100, the promoter probe plasmid pKT240[32] carrying a promoterless *aphC* gene was used for selection of transposition derivatives of IS931. Isolation of streptomycin-resistant colonies indicated that IS931 transposed upstream of the *aph* gene. This was confirmed by determining the position of the transposons by Southern hybridization.

[29] U. M. X. Sangodkar, P. J. Chapman, and A. M. Chakrabarty, *Gene* **71**, 267 (1988).

[30] P. H. Tomasek, B. Frantz, U. M. X. Sangodkar, R. A. Haugland, and A. M. Chakrabarty, *Gene* **76**, 227 (1989).

[31] R. A. Haugland, U. M. X. Sangodkar, and A. M. Chakrabarty, *Mol. Gen. Genet.* **220**, 222 (1990).

[32] M. M. Bagdasarian, E. Amann, R. Lurr, B. Ruckert, and M. Bagdasarian, *Gene* **26**, 273 (1983).

Mini-D3112 Transposable Bacteriophage System of *Pseudomonas aeruginosa*

The *E. coli* transposable bacteriophage Mu[33] has had an enormous impact on the study of prokaryotic genetics and gene regulation. Derivatives of Mu, known as mini-Mu elements, have been widely used in genetic studies to translocate genes and to transfer genes between bacteria, as well as for studies of gene expression and function.[34,35] However, the Mu genetic system has its limitations. For example, Mu does not work well in those organisms that are significantly different from those on the which basic Mu genetic techniques were developed. As a result, *in vivo* genetic manipulations mediated by Mu have not been possible in many nonenteric hosts.

Recently, a large number of *P. aeruginosa* bacteriophage have been identified which use the homology-independent recombination process of transposition for their replication.[36,37] The *P. aeruginosa* transposable bacteriophage D3112 is similar in many ways to the *E. coli* phage Mu. For example, the recombination of D3112 prophage termini with multiple chromosomal regions is associated with amplification of internal prophage sequences[38] and is consistent with models proposed to explain the duplication and recombination of transposable elements.[39] Furthermore, like Mu, heterogeneous lengths of host DNA are attached to the termini of D3112 phage DNA as a result of a Mu-like headful packaging mechanism.[36] In addition, under certain selection conditions, D3112 is also capable of acting as an insertional mutagen.[40]

The transposable nature of D3112 has been exploited to develop useful mini-D3112 elements capable of carrying out a variety of *in vivo* genetic applications in *P. aeruginosa*. The usefulness of the D3112 *in vivo* cloning replicons has been demonstrated by cloning different genes from the *P. aeruginosa* PAO chromosome involved in both anabolic and catabolic processes.[41] Transductants arise at frequencies ranging from 9.2×10^{-4}

---

[33] A. I. Bukhari, *Annu. Rev. Genet.* **10**, 389 (1976).

[34] M. J. Casadaban and J. Chou, *Proc. Natl. Acad. Sci. U.S.A.* **81**, 535 (1984).

[35] E. A. Groisman and M. J. Casadaban, *J. Bacteriol.* **169**, 687 (1987).

[36] V. N. Krylov, V. G. Bogush, and J. Shapiro, *Genetika* **16**, 824 (1980).

[37] V. Z. Akhverdyan, E. A. Khrenova, V. G. Bogush, T. V. Gerasminova, N. B. Kirsanov, and V. N. Krylov, *Genetika* **20**, 1612 (1984).

[38] S. Rehmat and J. A. Shapiro, *Mol. Gen. Genet.* **192**, 416 (1983).

[39] A. Toussaint and A. Resibois, *in* "Mobile Genetic Elements" (J. A. Shapiro, ed.), p. 105. Academic Press, New York, 1983.

[40] V. N. Krylov, T. G. Plotnikova, L. A. Kulakov, T. V. Fedorova, and E. N. Eremenko, *Genetika* **18**, 5 (1982).

[41] A. Darzins and M. J. Casadaban, *J. Bacteriol.* **171**, 3917 (1989).

to 5×10^{-7} per plaque-forming unit (pfu) of the helper phage. Clones of particular genes can be isolated at frequencies of between 1.7×10^{-3} and 1×10^{-4} per drug-resistant transductant.

Like Mu, the D3112 headful packaging mechanism encapsidates approximately 40 kb of DNA.[36] The size of *in vivo* generated clones is, therefore, limited by the size of the mini-D replicon. The sizes of the mini-D replicons described below are all very similar (10–12 kb). Theoretically, therefore, the upper size limit of cloned DNA one could achieve using these elements is between 27 and 29 kb.

Strains

The D3112-sensitive *P. aeruginosa* strain PAO works well with the D3112 phage system. However, this system should work with any D3112-sensitive strain of *P. aeruginosa*.

Plasmids and Constructions

The termini of transposable elements contain the cis-reactive sequences involved in the transposition reaction. To separate the integrative functions of D3112 from its viral properties, a mini-D3112 element was constructed which contained short sequences from both phage termini. This element, designated mini-D163 (pADD163), was the starting material for all the mini-D constructions and contains the terminal 1.85 kb from the left end and the terminal 1.4 kb from the right end. To select for mini-D elements conveniently, markers in the form of antibiotic resistance cassettes and various drug-resistant, broad host range plasmid replicons were cloned between the two phage ends.[42] Figure 1 shows the various mini-D3112 transposable elements available for *in vivo* cloning in *P. aeruginosa*. Mini-D214 resides on plasmid pADD214, D366 on pADD366, D386 on pADD386, and D948 on pADD948. These cloning elements contain three different but compatible replicons, and, therefore, it is possible to perform up to three cloning experiments using the same recipient. In addition, some of the mini-D replicons also contain the RK2 origin of transfer sequence (*oriT*), which allows *in vivo* generated clones of *P. aeruginosa* to be mobilized by conjugation to many different species of gram-negative bacteria.[41]

These elements lack the genes essential for phage growth but, in the presence of a helper D3112 *c*ts phage, can undergo transposition.[42] By themselves, the mini-D elements are not lethal at elevated temperatures

---

[42] A. Darzins and M. J. Casadaban, *J. Bacteriol.* **171**, 3909 (1989).

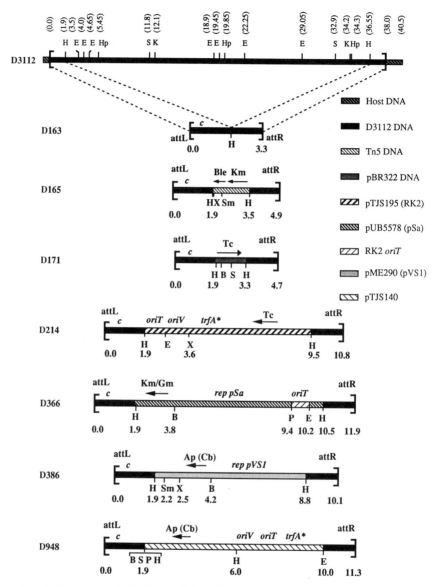

FIG. 1. Genetic and physical maps of the various mini-D3112 elements. The cross-hatched area adjacent to the D3112 left and right ends (top) represents packaged host sequences. The ends of the phage are shown in brackets. Abbreviations: attL, D3112 left terminus; attR, D3112 right terminus; *oriT*, origin of transfer of plasmid RK2; *oriV*, origin of vegetative replication of plasmid RK2; *trfA**, replication protein of plasmid RK2; B, *Bam*HI; Bg, *Bgl*II; E, *Eco*RI; H, *Hin*dIII; Hp, *Hpa*I; K, *Kpn*I; P, *Pst*I, R, *Eco*RI; S, *Sal*I; Sc, *Sac*I; Sm, *Sma*I; X, *Xho*I; Xb, *Xba*I; *c*, D3112 cts repressor; Ap/Cb, ampicillin/carbenicillin resistance; Ble, bleomycin resistance; Gm, gentamicin resistance, Km, kanamycin resistance, Tc, tetracycline resistance. Positions are given in kilobases from the left side of each mini-D element. Plasmids pTJS140, pTJS195, pUB5578, and pME290 have been previously described.[16,17]

owing to the absence of the replication (transposase genes) and are stably maintained in *P. aeruginosa*.

Genetic Procedures

Mini-D elements are introduced into *P. aeruginosa*::D3112 cts lysogens from *E. coli* by triparental matings with pRK2013 as the helper mobilizing plasmid. D3112 cts lysogens are isolated from confluent zones of lysis made by placing a drop of a D3112 cts lysate onto a bacterial lawn spread on LB (Luria broth) agar medium containing 1 m*M* MgSO$_4$. Following incubation at 30° overnight, potential lysogenic survivors of the D3112 infection are purified from the turbid zone by being streaked to single colonies. Lysogens are then confirmed by testing for temperature sensitivity at 42°, ability to release phage at 42°, and immunity to superinfection by D3112.

Preparation of D3112 cts Lysates

1. Grow a culture of a PAO::D3112 cts lysogen harboring a mini-D3112 element overnight at 30° in LB liquid medium with shaking in the presence of an appropriate antibiotic as specified by the mini-D element.
2. Dilute the culture 1 : 100 in 25 ml sterile LB without drugs and grow at 30° with shaking for approximately 3 hr to mid-log phase.
3. Shift the culture to a 42° water bath and shake vigorously for 2 hr or until lysis occurs. After thermal induction of mini-D/D3112 cts replication, a complete clearing is not always observed. However, normal D3112 titers of over 10^9 pfu/ml are obtained routinely.
4. Add chloroform (final concentration 1%), MgSO$_4$ (final concentration 2 m*M*), and CaCl$_2$ (final concentration 0.2 m*M*) to the lysates, vortex briefly, and remove the cell debris by centrifugation at 8000 rpm for 10 min at 4°.
5. Remove the lysate avoiding the pellet containing cellular debris.
6. Repeat the chloroform treatment and centrifuge at 8000 rpm for 10 min at 4°.
7. The resulting lysates are titered and stored at 4° until used. Unlike Mu lysates, which are unstable and generally used within 1 week, D3112 lysates are extremely stable. Virtually no loss in phage titer was detected in a D3112 phage lysate stored at 4° for over a 1-year period.

In Vivo Cloning

The scheme used to clone DNA sequences from *P. aeruginosa in vivo* using the D3112 phage system is shown in Fig. 2. Briefly, a plasmid containing a mini-D phage with a plasmid replicon between the two D3112

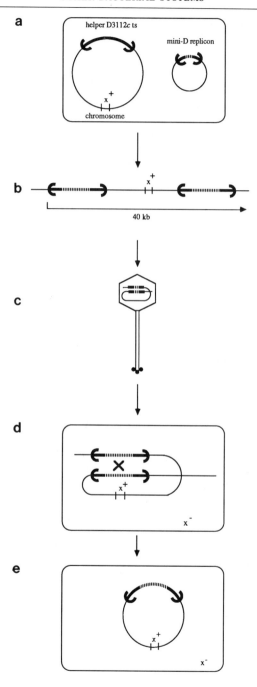

termini is introduced into a D3112 cts lysogen. On thermal induction of phage lytic functions, transposition of the mini-D replicon can occur on both sides of a particular gene to form a structure that can be encapsidated by the headful packaging mechanism (Fig. 2a,b). The packaged DNA is then introduced by phage infection into a D3112-sensitive recipient where homologous recombination can occur between the mini-D replicon sequences to generate a plasmid clone (Fig. 2c,d). Selection for a drug resistance marker carried by the mini-D element results in the isolation of transductants, which have a segment of the donor chromosome incorporated outside the ends of the mini-D replicon. These transductants form a gene library from which particular clones can be selected such as by their ability to complement specific mutations.

Procedure. Mini-D replicon/D3112 cts lysates prepared as described above are used to infect sensitive *P. aeruginosa* auxotrophic recipients. Infections are carried out on solid plates owing to the low infectivity of *P. aeruginosa* cells when grown in liquid media.[42]

1. Equal volumes (0.15 ml) of an overnight culture of recipient cells and lysate are mixed [multiplicity of infection (moi) of approximately 0.1 to 1].
2. Spread the mixture to dryness on a LB agar plate and incubate at 30° for 3 hr.
3. Following incubation, remove the cells from the plates by suspending the growth in 5 ml of phosphate-buffered saline (PBS).
4. Pellet the cells by centrifugation (5000 rpm for 10 min) and repeat wash with PBS.
5. Resuspend the cells in 2 ml PBS and select drug-resistant transductants by plating on *Pseudomonas* isolation agar (PIA) agar supplemented with the appropriate antibiotic. Select recombinant mini-D

FIG. 2. Scheme for *in vivo* cloning with mini-D replicons. (a) A plasmid containing a mini-D replicon is introduced into a D3112 cts lysogenic cell, which contains the gene of interest, x, by transformation, transduction, or conjugation. D3112 c ts (38 kb) and the mini-D replicon (10.1–11.9 kb) are not drawn to scale. (b) The cells are derepressed for D3112 replication and transposition by heat inactivation of the temperature-sensitive repressor. Multiple rounds of transposition of the mini-D replicon frequently result in a copy of the mini-D replicon being integrated near x with another copy of the mini-D element, or a helper D3112, on the other side in the same orientation. Packaging presumably proceeds by a headful mechanism starting from the left side of the mini-D replicon, including x and some part of the other mini-D replicon or helper D3112. A total of about 40 kb is packaged as shown by the arrow. (c) The resulting phage lysate is then used to infect a sensitive *P. aeruginosa* recipient where recombination (d) can occur between the D3112 homologous regions to generate drug-resistant transductants with recombinant plasmids, some of which have gene x (e).

plasmids capable of complementing PAO chromosomal mutants by plating on a suitable minimal medium supplemented with the appropriate antibiotic(s).

Insertion Mutagenesis of the Pseudomonas aeruginosa Chromosome

1. Prepare a mini-D (either D165 or D171)/D3112 cts lysate and infect a sensitive recipient as described above.
2. Remove the cells from the infection plate, wash with PBS, resuspend, and plate on the appropriate selective medium.

Km^r (D165) and Tc^r (D171) transductants arise at frequencies between 10^{-4} and 10^{-5}/pfu of helper phage. Southern hybridization analysis of many mini-D171 transductants has revealed that approximately 80–90% of the Tc^r colonies contain single insertions. These insertions are in many different locations, although auxotrophic mutations are formed much less frequently by D3112 than by Mu. The mini-D insertions, once established, should not excise from the chromosome because they lack the D3112 replication functions and contain a functional repressor gene to repress the transposition of any incoming D3112 helper phage.

The introduction of new genetic material into bacteria is an important step in constructing organisms with novel and useful activities and usually involves introducing recombinant plasmids containing foreign genes which encode desirable gene products. However, plasmids, in the absence of selective pressure, may be unstable with respect to the inheritance of the cloned gene or the continued expression of that gene. An alternative method to introducing genes on extrachromosomal elements is the use of transposable elements as vehicles to introduce new or altered genetic material directly into the bacterial chromosome. The mini-D3112 elements, D165 and D171, described here can be used as excellent vehicles to stably introduce genetic material as a single copy into the chromosome of *P. aeruginosa*.

Insertion of Mini-D Elements into Plasmids

The procedure for localzing mini-D insertions into plasmids is as follows:

1. Transform or mobilize the desired plasmid (≤30 kb) into a strain that harbors both a mini-D element and a D3112 cts prophage.
2. Prepare a D165 or D171 phage lysate by thermoinduction as described above.
3. Infect a sensitive recipient and select transductants which carry the resistance specified by the plasmid.

Using the composite plasmid pTJS140 (8.0 kb; Cb^r), which contains origins of replication from both RK2 and pUC9, transductants arise at a frequency of about 10^{-6}/helper pfu. Of 50 Cb^r transductants tested, all 50 also contained the Tc^r marker of mini-D171. Of the 10 mini-D171::pTJS140 insertions examined more closely, at least 8 contained different insertion sites as determined by restriction and agarose gel analysis. Insertions in both orientations were found at nearly the same frequency.

Mini-D3112-Mediated Generalized Transduction

Bacteriophage D3112 is a generalized transducing phage and can transduce various chromosomal markers at frequencies from 8.0 × 10^{-8} to 2.0 × 10^{-9}.[36] Mini-D171, which is only 4.7 kb (Fig. 1), is capable of packaging up to 35 kb of adjacent host sequence. On injection into a recipient cell, this host DNA can replace resident DNA in a *recA*-mediated generalized transduction.

1. Prepare a mini-D (D171 or D165)/D3112 *c*ts lysate on a prototrophic host and infect a sensitive recipient containing the desired mutation.
2. Plate cells on minimal medium (lacking antibiotic) to select for prototrophic transductants.

A mini-D171/D3112 *c*ts lysate is able to transduce various PAO chromosomal markers, in recA$^+$ host, as high as 700 times more efficiently than was a D3112 *c*ts lysate alone. These transduction values are, at best, 20-fold greater than the frequencies reported for the generalized transducing phage F116L and G101. Mini-D3112-mediated generalized transduction is potentially more useful than generalized transduction by the large DNA phages, such as F116L and G101, for fine-structure mapping because it involves smaller host segments (<35 kb).

Cloning DNA from *Pseudomonas*

Isolation of chromosomal DNA from *Pseudomonas* has been described.[43] Naturally occurring megaplasmids from pseudomonads can also be isolated.[44] The isolated DNA is partially cleaved by one or more restriction enzymes and ligated into a broad host range vector (see below) cleaved with the same or a compatible enzyme(s). A ligation buffer that has worked well in our hands contains a final concentration of 40 mM Tris-HCl, pH 7.5, 8 mM MgCl$_2$, 10 mM dithiothreitol (DTT), 0.5 mM spermidine, 0.1

---

[43] J. B. Goldberg and D. E. Ohman, *J. Bacteriol.* **158**, 1115 (1984).
[44] F. Casse, C. Boucher, J. S. Julliut, M. Michel, and J. Denarie, *J. Gen. Microbiol.* **113**, 229 (1979).

mg/ml bovine serum albumin (BSA). The DNA is incubated overnight with T4 ligase at 15° and then transformed into an appropriate *E. coli* host by the CaCl$_2$ method,[45] creating a library of *Pseudomonas* DNA. The library can then be mobilized into a mutant *Pseudomonas* strain by conjugal transfer. Any host now able to grow on media that restricted growth of the mutant harbors a plasmid containing a fragment of *Pseudomonas* DNA, capable of transcomplementation. The complementing plasmid can then be isolated, for analysis, from the *Pseudomonas* strain by the alkaline lysis procedure described by Maniatis *et al.*[45]

Broad Host Range Cloning Vectors

There are a number of vectors currently available for cloning and expression of *Pseudomonas* genes. These vectors are based on different replicons such as RSF1010, a broad host range IncQ/P-4 group plasmid; RP1 (or RP4/RK2), a broad host range IncP-1 group plasmid; or pSA, a multicopy IncW plasmid. The construction of vectors having different antibiotic resistances and multiple cloning sites have been described.[46–50] pDSK509, pDSK519, and pRK415 are very useful because of the multiple cloning site from pUC19 that was inserted as well as the useful antibiotic resistance genes for kanamycin or tetracycline.[49] In addition, pDSK519 and pRK415 encode the *lacZ* gene, allowing direct screening of recombinant clones in *E. coli*. The *lacZ* gene is also present in the pVD vectors.[48] pMMB66EH and pMMB66HE are expression vectors utilizing the inducible *tac* promoter.[47] Cosmid vectors such as pCP13,[51] pLAF3,[52] and pLAF5[49] have been developed as broad host range cloning vectors. Many of these vectors are self-transmissible, whereas others can be mobilized from *E. coli* to *Pseudomonas* by triparental mating (described below) using the plasmid pRK2013,[53] which carries the RP4 *tra* genes.

[45] T. Maniatis, E. F. Fritsch, and J. Sambrook, "Molecular Cloning: A Laboratory Manual." Cold Spring Harbor Laboratory, Cold Spring Harbor, New York, 1982.

[46] M. Fukada and K. Yano, *Agric. Biol. Chem.* **49,** 2719 (1985).

[47] J. P. Furste, W. Pansegrauu, R. Frank, H. Blocker, P. Scholz, M. Bagdasarian, and E. Lanka, *Gene* **48,** 119 (1986).

[48] V. Deretic, S. Chandrasekharappa, J. F. Gill, D. K. Chatterjee, and A. M. Chakrabarty, *Gene* **57,** 61 (1987).

[49] N. T. Keen, S. Tamaki, D. Kobayashi, and D. Trollinger, *Gene* **70,** 191 (1988).

[50] J. Davison, M. Heusterspreute, N. Chevalier, V. Ha-Thi, and F. Brunel, *Gene* **51,** 275 (1987).

[51] A. Darzins and A. M. Chakrabarty, *J. Bacteriol.* **159,** 9 (1984).

[52] B. Staskawicz, D. Dahlbeck, N. T. Keen, and C. Napoli, *J. Bacteriol.* **169,** 5789 (1987).

[53] D. Figurski and D. R. Helinski, *Proc. Natl. Acad. Sci. U.S.A.* **76,** 1648 (1979).

Conjugal Transfer of Cloning Vectors

Method A: Filter Matings. This method is based on that of Figurski and Helinski.[53]

1. Inoculate 5-ml cultures of donor strain, recipient strain, and "helper" plasmid strain (HB101/pRK2013) in appropriate media with antibiotics and grow overnight.
2. Remove 0.5 ml of each strain and mix briefly in sterile test tube.
3. Pour the mixture into a sterile 10-ml syringe which is attached to a Swinnex (Millipore, Bedford, MA) filter holder containing a 0.5-μm filter.
4. Force the liquid through the filter, depositing the cells onto the filter surface.
5. Open the filter apparatus and remove the filter with sterile forceps.
6. Lay filter on an LB agar plate (cell-side up) and incubate overnight at optimal *Pseudomonas* growth temperature.
7. Remove the filter from the plate with sterile forceps and place into 4 ml of sterile saline [0.9% (w/v) NaCl]. Vortex to suspend the cells.
8. Serially dilute the cell suspension in sterile saline 10^{-1} to 10^{-6}.
9. Plate 0.1 ml each dilution on selective medium.

Method B: Plate (Patch) Matings

1. Grow the donor strain, recipient strain, and "helper" plasmid (HB101/pRK2013) on LB agar plates with antibiotics overnight.
2. With a sterile loop, smear cells from the donor strain onto an LB agar plate in a small circular area, then collect HB101/pRK2013 cells and mix in with the donor stain. Finally, collect a loopful of cells from the recipient cells and mix with the donor and helper strain. Incubate overnight at the appropriate temperature.
3. Using a sterile loop, collect a portion of the mixed cell growth and streak on selective medium.
4. Purify single colonies from the selective medium.

Controlled Expression Vectors for Protein Hyperproduction in *Pseudomonas*

A number of controlled expression vector systems suitable for hyperproduction of foreign proteins in *E. coli* have been developed[54]; these include $P_{lac}UV5$-$lacI^q$, PT5N25/$lacO$-$lacI^q$, P_{tac}-$lacI^q$, λP_R, and λP_R-$cI857$.

---

[54] G. Gross, *Chim. Oggi.* **3**, 21 (1989).

However, the utility of these vectors for overexpression of cloned genes in other bacteria such as *Pseudomonas* is limited because the plasmid replicons are nonfunctional in organisms other than *E. coli*. Analogous controlled expression vector systems have now been constructed using broad host range plasmid replicons that allow overexpression of cloned genes in several gram-negative bacteria.

The most widely used controlled expression vector system for hyperproduction of proteins in *Pseudomonas* has been the hybrid *trp–lac* (*tac*) promoter–*lacI*^{q} system,[55] which has been cloned into the plasmid pKT240.[32] The resulting plasmids, pMMB22 and pMMB24, contain unique *Eco*RI and *Hin*dIII restriction sites, respectively, directly downstream of P_{tac} for introduction of cloned genes. Repression of P_{tac} by the plasmid-encoded *lac* repressor is incomplete in *Pseudomonas* but is nevertheless of the same order of magnitude as the level of repression observed in *E. coli*.[32] In controlled experiments testing the overproduction of catechol 2,3-dioxygenase (C2,3-O) from the *xylE* gene of *P. putida* TOL plasmid PWWO[56] cloned under P_{tac} resulted in protein levels estimated to be about 7% of the total cellular protein both in *E. coli* and P. putida when induced with isopropyl-β-D-thiogalactopyranoside (IPTG).[32] Recently, more versatile derivatives of pMMB22 and pMMB24 have been constructed having several unique cloning sites directly downstream of P_{tac}.[47,48]

The λP_R and λP_R-*c*I857 controlled expression system has also been adapted for use in *Pseudomonas* and other gram-negative bacteria besides *E. coli*. A DNA fragment containing the λ*c*I857 gene, the O_R operator, the λP_R promoter, the *cro* ribosome binding site, and the in-phase *galK* gene fusion was recloned into the broad host range (RSF1010-based) plasmid pJRD215.[50] The resulting plasmid pJRD215K was transferred to a variety of gram-negative bacteria, and most exhibited themosensitive galactokinase activity, although the efficiency of induction at 42° was variable. This is probably due to differences in temperature sensitivity of the host organisms tested.[57] These results showed that the λP_R promoter is recognized by RNA polymerase in gram-negative organisms other than *E. coli* and that the λ*c*I857-encoded repressor is functional in these organisms. This was consistent with a previous report on the regulated expression of the pilin structural gene from *Bacteroides nodosum*, which was achieved in a *P. aeruginosa* host using an RSF1010-based vector containing the λP_R

[55] E. Amann, J. Brosius, and M. Ptashne, *Gene* **25**, 167 (1983).
[56] F. C. H. Franklin, M. Bagdasarian, M. M. Bagdasarian, and K. N. Timmis, *Proc. Natl. Acad. Sci. U. S. A.* **78**, 7458 (1981).
[57] J. Davison, M. Heusterspreute, N. Chevalier, and F. Brunel, *Gene* **60**, 227 (1987).

promoter and the λcI857 gene.[58] In this case, overproduction of pilin could not be achieved in *E. coli*, since the *B. nodosum* prepilin is not processed to mature pilin in *E. coli*. This example obviates the need for controlled expression vectors that function in organisms other than *E. coli*.

One obvious drawback to using P_{tac}-*lacI*^{q}-based controlled-expression vector systems is the high cost of the chemical inducer IPTG. Likewise, thermal induction of the λP_R or λP_L-cI857-based systems becomes less practical as the scale of bacterial fermentations increases. An alternative broad host range (RSF1010-based) controlled expression vector system has been described that is based on the positively activated P_m twin promoters of the *P. putida* TOL plasmid pWWO and the *xylS* gene, the product of which together with the coinducer benzoate positively regulated the P_m promoters.[59] This vector, pNM185, contains unique *Eco*RI, *Sst*I, and *Sst*II cloning sites downstream of the P_m twin promoters. Using *xylE* as an indicator gene, as much as a 600-fold induction of C2,3-O activity was observed on addition of micromolar levels of the inducers benzoate or *m*-toluate, depending on the host bacterium tested. In some cases, C2,3-O comprised up to 5% of the total cellular protein.[59] This controlled-expression vector system functioned in gram-negative bacteria belonging to at least 16 different genera, and the ability to function in any given organism correlated with the phylogenetic distance of the organism from *P. putida*.[59]

Induction of Protein Synthesis Using the P_{tac}–lacI System

1. Grow cells containing pMMB22, pMMB24, or pMMB66 constructs (containing the cloned gene of interest) overnight in selective medium.
2. Inoculate 100 ml fresh medium with a 1% inoculum.
3. Grow until early log phase (1–2 hr).
4. Add IPTG to 1 m*M* to induce *tac* promoter, then continue growing for an additional 7–10 hr.
5. Harvest the cells and wash them with 0.9% saline.
6. Make a cell extract as described for the gel retardation protocol (see Transcriptional Regulation in *Pseudomonas*, below), except add glycerol to 20% in the final step.
7. This extract can then be used for sodium dodecyl sulfate-polyacryl-

---

[58] T. C. Elleman, P. A. Hoyne, D. J. Stewart, N. M. McKern, and J. E. Patterson, *J. Bacteriol.* **168**, 574 (1986).

[59] N. Mermod, J. L. Ramos, P. R. Lehrbach, and K. N. Timmis, *J. Bacteriol.* **167**, 447 (1986).

amide gel electrophoresis (SDS–PAGE)[60] and for protein purification.

Sequencing *Pseudomonas* Genes

Sequencing DNA isolated from pseudomonads can be cumbersome because of the high G + C content, which leads to band compression. Stable intrastrand structures formed by the G and C residues are not fully denatured during electrophoresis, resulting in band compression where fragments differing in size by one or two bases migrate with similar mobilities. In addition, secondary structures can interfere with the processivity of DNA polymerases leading to chain termination. To eliminate band compression we have used the modified T7 DNA polymerase, Sequenase, purchased from United States Biochemicals (Cleveland, OH) in reaction mixes containing 7-deaza-dGTP (c^7dGTP). Premature termination was eliminated by the addition of a chase reaction. A new analog, 7-deaza-dITP (c^7 dITP), has just been made available by Pharmacia LKB (Piscataway, NJ). This analog is reported to eliminate band compression arising from Hoogsteen G–G and G–C base pairing without the limitations of dITP or c^7dGTP.[61] dITP tends to cause chain termination of DNA polymerases, and c^7dGTP does not resolve compressions arising from G–C base pairs. We have not used c^7dITP and cannot comment on its effectiveness. The sequencing protocol described here is a modification of the procedure suppled by United States Biochemicals for use with their Sequenase enzyme.[62]

Solutions and Buffers

Annealing buffer (5 ×): 200 mM Tris-HCl pH 7.5, 100 mM MgCl$_2$, 250 mM NaCl

Dithiothreitol (DTT), 0.1 M

Stop solution: 95% (v/v) Deionized formamide, 20 mM EDTA, 0.05% (w/v)/Bromphenol blue, and 0.05% (w/v) xylene cyanol FF

Labeled nucleotide: [α-^{35}S]dCTP, specific activity 1000–1500 Ci/mmol

Labeling nucleotide mixture 7.5 μM 7-deaza-dGTP, 7.5 μM dATP, and 7.5 μM dTTP

[60] U. K. Laemmli, *Nature (London)* **227**, 680 (1970).
[61] Pharmacia LKB Biotechnology product profiles, Fall 1989.
[62] Protocol supplied with Sequenase kit from United States Biochemical.

TERMINATION NUCLEOTIDE MIXTURES

ddGTP	ddATP	ddTTP	ddCTP
160 μM 7-deaza-dGTP	80 μM 7-deaza-dGTP	80 μM 7-deaza-dGTP	80 μM 7-deaza-dGTP
80 μM dATP	80 μM dATP	80 μM dATP	80 μM dATP
80 μM dTTP	80 μM dTTP	80 μM dTTP	80 μM dTTP
80 μM dCTP	80 μM dCTP	80 μM dCTP	80 μM dCTP
8 μM ddGTP	8 μM ddATP	8 μM ddTTP	8 μM ddCTP
50 mM NaCl	50 mM NaCl	50 mM NaCl	50 mM NaCl

Chase solution: 50 μM each of 7-deaza dGTP, dATP, dTTP, dCTP

Annealing Template and Primer

1. For each template, combine the following in a 0.5-ml Eppendorf tube: primer, containing 0.5 pmol of the standard 17-bp sequencing primer or other suitable oligonucleotide; DNA, approximately 1 μg single-stranded (ss) DNA from an M13 clone isolated as described by LeClerk *et al.*[63] or ss plasmid DNA; annealing buffer (5 ×), 2 μl; and deionized, distilled water to 10 μl.
2. Warm the capped tube to 65° for 2–5 min, then allow the mixture to cool slowly to approximately room temperature.

Labeling Reaction

See Step 1 below before proceeding. To the annealed template–primer add the following: DTT (0.1 M), 1 μl; labeling nucleotide mix (diluted 1/ 20 to read 0–250 bp or 1/5 for greater than 250 bp), 2 μl; [α-^{35}S]dCTP, 1 μl (10 μCi); and Sequenase, 2 units (2 μl). Incubate for 2–5 min at room temperature.

Termination and Chase Reactions

1. Label four 1.5-ml Eppendorf tubes "G," "A," "T," "C," and fill each with 2.5 μl of the appropriate dideoxy termination mixture. DO THIS STEP BEFORE THE LABELING REACTION. Prewarm tubes at 37° while labeling is in progress.
2. When the labeling reaction is over, transfer 3.5 μl of it to each of the four termination tubes. An Eppendorf electronic digital pipette (EDP) works well for this step.
3. After 2–5 min of incubation at 37°C add 1 μl of the chase mixture and place at room temperature. Four microliters of stop solution is added to each tube after an additional 2–5 min.

[63] J. E. LeClerc, N. L. Istock, B. R. Saran, and R. Allen, Jr. *J. Mol. Biol.* **180**, 217 (1984).

4. The samples are boiled for 3 min and 2–3 μl is loaded onto either a 6% nongradient denaturing polyacrylamide sequencing gel (6% (w/v) polyacrylamide, 20 : 1 polyacrylamide–bisacrylamide, 7 M urea) or a 5% gradient gel (5% (w/v) polyacrylamide, 20 : 1 polyacrylamide–bisacrylamide, 7 M urea). The preparation of such gels has been described.[64,65] The TBE buffer used for electrophoresis is the same as described for gradient gels.[65]

Promoter Structure of Pseudomonas Genes

Pseudomonas promoters that have been identified and characterized generally do not function efficiently in *E. coli*.[66] This may be due to the lack of a necessary σ factor or required regulatory proteins necessary for transcription. Comparison of several *Pseudomonas* promoters with the *E. coli* σ^{70} (*rpoD*) consensus sequence, TTGACA-17 ± 2bp-TATAAT,[67,68] revealed a range of similarity.[66] There appears to be three classes of promoters, those having homology to the *E. coli* σ^{70} consensus sequence, those sharing homology with the core σ^{54} (*ntrA, rpoN*) consensus sequence (TGGC-8 bp-TGCT), and those with no apparent homology to any known enteric promoters. For a number of genes, such as *catBC*,[69] *nahG nahR*,[70] major outer membrane proteins,[71] and *argF* mRNA-1,[72] σ^{70}-like promoters have been identified. Promoters with σ^{54}-like sequences include the promoters of the *xyl* operons, pilin gene promoters, the carboxypeptidase G2 gene promoter,[66] isoamylase gene,[73] and the alginate genes *algD* and *algR1*.[66,74] It has been shown that pilin genes[75] and several *xyl* genes[76] require the presence of the RpoN gene product. Promoters for exotoxin

[64] J. Messing, this series, Vol. 101, p. 20.

[65] M. D. Biggin, T. J. Gibson, and G. F. Hong, *Proc. Natl. Acad. Sci. U.S.A.* **80,** 3963 (1983).

[66] V. Deretic, J. F. Gill, and A. M. Chakrabarty, *Bio/Technology* **5,** 469 (1987).

[67] W. R. McClure, *Annu. Rev. Biochem.* **54,** 171 (1985).

[68] W. S. Reznikoff, D. A. Siegle, D. W. Cowing, and C. A. Gross, *Annu. Rev. Genet.* **19,** 355 (1985).

[69] T. L. Aldrich and A. M. Chakrabarty, *J. Bacteriol.* **170,** 1297 (1988).

[70] M. A. Schell and M. Sukordhaman, *J. Bacteriol.* **171,** 1952 (1988).

[71] M. Duchene, C. Barron, A. Schweizer, B.-U. von Specht, and H. Domdey, *J. Bacteriol.* **171,** 4130 (1989).

[72] Y. Itoh, L. Soldati, V. Stalon, P. Falmagne, Y. Terawaki, T. Leisinger, and D. Haas, *J. Bacteriol.* **170,** 2725, (1988).

[73] M. Fujita, A. Anemura, and M. Futai, *J. Bacteriol.* **171,** 4320 (1989).

[74] K. Kimbara and A. M. Chakrabarty, *Biochem. Biophys. Res. Commun.* **164,** 601 (1989).

[75] K. S. Ishimoto and S. Lory, *Proc. Natl. Acad. Sci. U.S.A.* **86,** 1954 (1989).

[76] T. Kohler, S. Harayama, J.-L. Ramos, and K. N. Timmis, *J. Bacteriol.* **171,** 4326 (1989).

A expression, protocatechuate 3,4-dioxygenase,[77] and *argF* mRNA-241[72] have no apparent homology to known promoters. Consensus sequences have been proposed for both inducible[78,79] and constitutive[80] *Pseudomonas* promoters. Both proposed positively regulated consensus sequences in the − 10 region, C--T-$\underline{A}$----T[79] and GCA$\underline{A}$T,[78] include an A at the − 11 position. By site-specific mutagenesis, an A at this position in the *catBC* promoter was found to be critical for function.[81] This base is highly conserved in several *Pseudomonas* promoters[69] and in *E. coli* promoters.[82] This suggests that there are some structural similarities between *Pseudomonas* and *E. coli* σ^{70} promoters. The consensus sequence proposed by Mermod *et al.*[78] included an A at − 12; however, our mutational data indicated that a C or G in this position worked equally well.[81] Inouye *et al.*[80] proposed a consensus sequence for constitutive *P. putida* promoters that included a T located in the first position (− 12) of the Pribnow box. Placing a T in this position of the *catBC* promoter in fact was sufficient to render it constitutive. None of the proposed consensus sequences for *Pseudomonas* have indicated any conserved sequences at the − 35 region. Mutations in the − 35 region of the *catBC* operon have little effect on promoter activity.[81]

Promoter Probe Vectors

The use of promoter probe vectors is very useful for both the identification and characterization of promoters. There are a number of broad host range vectors that have been constructed, which contain promoterless genes. Plasmid pKT240[32] has a promoterless aminoglycoside phosphotransferase (*aph*) gene. Promoter activity is detected as resistance to streptomycin owing to the insertion of a DNA fragment upstream of the *aph* gene. The two limitations of pKT240 are the relatively few restriction sites for cloning and the lack of an assay system to quantitate promoter strength. Insertion of multiple cloning sites and gene cassettes expressing proteins that can be easily assayed have greatly enhanced the use of promoter probe vectors. A number of promoter probe vectors belonging to the incompatibility group IncP1 have been constructed by Farinha and Kropinski[83] using either a promoterless *tet* (tetracycline resistance) or *cat* (chlor-

[77] G. J. Zylstra, R. H. Olsen, and D. P. Ballou, *J. Bacteriol.* **171**, 5915 (1989).
[78] N. Mermod, P. R. Lehrbach, W. Reineke, and K. N. Timmis, *EMBO J.* **3**, 2461 (1984).
[79] B. Frantz and A. M. Chakrabarty, *Proc. Natl. Acad. Sci. U.S.A.* **84**, 4460 (1987).
[80] S. Inouye, Y. Asai, A. Nakazawa, and T. Nakazawa, *J. Bacteriol.* **166**, 739 (1986).
[81] T. L. Aldrich, R. K. Rothmel, and A. M. Chakrabarty, *Mol. Gen. Genet.* **218**, 266 (1989).
[82] O. Raibaud and M. Schwartz, *Annu. Rev. Genet.* **18**, 173 (1984).
[83] M. A. Farinha and A. M. Kropinski, *Gene* **77**, 205 (1989).

amphenicol resistance) gene. A promoterless β-Gal (β-galactosidase) and *lux* (luciferase) gene have also been used.[84] The *lux* operon has also been used in constructing the selection vector pUCD615 based on the pSA replicon.[85] *xylE* encoding catechol 2,3-dioxygenase (C2,3-O), an enzyme that converts catechol to 2-hydroxymuconic semialdehyde, producing a yellow color, has been used to construct the vector, pVDX18[86] to analyze promoter activity. Yellow color develops immediately on spraying colonies with a catechol solution. The activity of C2,3O can be determined as described below.

Catechol 2,3-Dioxygenase Assay[87]

Partial Purification

1. Grow 100-ml cultures to mid-log phase and collect cells by centrifugation.
2. Resuspend the pellet in 5 ml of cold 50 mM potassium phosphate buffer (pH 7.5).
3. Sonicate cells 3 times for 30 sec at 100 mV on ice.
4. Centrifuge at 25,000 g for 10 min at 4°. Discard pellet.
5. Add 1 volume of cold acetone to supernatant.
6. Centrifuge at 10,000 g for 10 min at 4°. Discard pellet.
7. Add 2 volumes of cold acetone to the supernatant.
8. Recentrifuge as in Step 6 and discard supernatant.
9. Resuspend the pellet in 5 ml of 50 mM potassium phosphate buffer (pH 7.5) containing 10% acetone.
10. Centrifuge at 10,000 g for 20 min at 4° to clarify. Discard pellet.
11. Use supernatant for enzyme assays.

Assay. Combine in a spectrophotometer cuvette (final volume 1 ml) catechol (1 mM final concentration), partially purified C2,3-O (varying dilutions depending on protein concentration), and 50 mM potassium phosphate buffer, pH 7.5, containing 10% acetone. Monitor the continuous formation of β-muconic ε-semialdehyde spectrophotometrically (increase in absorbance at 375 nm). One unit of C2,3-O activity is defined as that amount that oxidizes 1 μmol of catechol per minute, based on a molar extinction coefficient of the product of 4.4×10^4.[87] Specific activity is expressed as micromoles of product formed per minute per milligram of

[84] A. M. Kropinski, personal communication (1989).
[85] P. M. Rogowsky, R. J. Close, J. A. Chimera, J. J. Shaw, and C. I. Kado, *J. Bacteriol.* **169**, 5101 (1987).
[86] W. M. Konyecsni and V. Deretic, *Gene* **74**, 375 (1988).
[87] M. Nozaki, this series, Vol. 17A, p. 522.

protein used in the assay. Protein concentration in the partially purified C2,3-O preparations is determined by the method of Bradford.[88]

mRNA Mapping of *Pseudomonas* Genes

The methods for isolation and determination of transcriptional start sites for genes isolated from pseudomonads are similar to those for other bacterial systems. Described here is a slight modification of the guanidinum isothiocyanate–hot phenol method for isolation of mRNA described in Maniatis *et al.*[45] as well as a protocol for reverse transcriptase mapping.

Isolation of mRNA

Reagents

Solution I: 4 M guanidinium isothiocyanate solution, prepared as described in Ref. 45.
Solution A: 0.1 M sodium acetate (pH 5.2), 10 mM Tris-HCl (pH 7.4), 1 mM EDTA disodium salt
Solution B: 0.1 M Tris-HCl (pH 7.4), 50 mM NaCl, 10 mM MgCl$_2$

Protocol

1. Grow a 5.0-ml culture of cells overnight in an appropriate medium.
2. Inoculate 50–100 ml of a fresh culture at a 1/100 dilution.
3. Harvest cells at mid-log phase by a low-speed spin in a Sorvall centrifuge (8000 rpm).
4. Wash the cells with 0.9% (w/v) NaCl and recollect.
5. Resuspend the cells in 2 ml of Solution I per 50 ml of culture used (use a sterile plastic 15-ml test tube). Each 2 ml cell suspension is then treated as follows.
6. Place the tube at 60°. Take an 18-gauge (or smaller) needle–syringe unit and force the lysed cell mixture up and down through the needle to shear the DNA. Repeat until the solution is no longer viscous.
7. Add 2.0 ml phenol equilibrated with Tris-HCl (pH 8) and again force cells through the needle, keeping the solution at 60°.
8. Add 1.0 ml solution A and mix.
9. Add 2.0 ml chloroform–isoamyl alcohol (24 : 1) and mix. At this point the organic and aqueous phases should separate.
10. Spin the tube in a table-top centrifuge and recover the top aqueous phase. Reextract with an equal volume of phenol–chloroform.

---

[88] M. M. Bradford, *Anal. Biochem.* **72**, 248 (1976).

11. To the aqueous phase (~6 ml) add 2 volumns of cold 95% (v/v) ethanol and precipitate nucleic acids at $-70°$.
12. Redissolve nucleic acids in solution B. Use 1 ml per 50 ml original culture.
13. Add 6–10 units of RNasin purchased from Boehringer Mannheim (Indianapolis, IN). Split the 1 ml volume into two 1.5-ml Eppendorf tubes.
14. Add RNase-free DNase, 5–10 units. Incubate for 1 hr at $37°$.
15. Add sodium dodecyl sulfate (SDS) to a final concentration of 0.2%, EDTA to 10 mM, and 1 μg of proteinase K to 1 μg to each tube and incubate an additional 30 min to 1 hr at $37°$.
16. Extract twice with an equal volume of phenol–chloroform and ethanol precipitate at $-70°$. Store RNA in 70% (v/v) ethanol at $-70°$ until needed.
17. The RNA concentration can be determined spectrophotometrically.

Reverse Transcriptase Mapping

Solutions and Materials

0.5 pmol of an oligonucleotide homologous to the 5′ end of the mRNA for each reaction

10× RT buffer: 500 mM Tris-HCl (pH 8.3), 1.0 M KCl, 100 mM MgCl$_2$, 100 mM DTT

[α-^{32}P]dCTP at 3000 Ci/mmol.

25 mM solution of dGTP, dATP, dTTP, and a 25 mM solution of dCTP.

Protocol

1. Anneal 0.5 ng of the labeled primer to 8 μg of RNA in 1.5× RT buffer for 1 hr (total volume 10 μl) at the optimal annealing temperature. Let the solution cool slowly to room temperature. The optimal annealing temperature may vary depending on the GC content of the RNA, but should be between $50°$ and $65°$. The optimal temperature should be determined for each RNA sample. This can be determined by a trial hybridization at various temperatures.
2. To the cooled annealed primer/template mix add 1.0 μl of a 100 mM DTT solution, 1.5 μl of a 25 mM solution of dGTP, dATP, and dTTP, 2.0 μl of [α-^{32}P]dCTP, and 1.0 μl of Moloney murine leukemia virus reverse transcriptase (MMLV RT) (~20 U).
3. Incubate the reaction for 1 hr at $37°$.
4. Add 1 μl MMLV RT and 1.5 μl of 25 mM dCTP.

5. Incubate 1 hr at 37°.
6. Stop the reaction by adding 10 μl stop solution, the same as used in sequencing.
7. The reactions are boiled for 3 min and 5 μl is loaded onto a 6% sequencing gel alongside a sequencing reaction that was done using the same primer.

Transcriptional Regulation in *Pseudomonas*

Gene expression in pseudomonads, like other prokaryotic or eukaryotic systems, can be regulated at the transcriptional, translational, or post-translational level. As in other organisms, this regulation can be either positive or negative. A number of transcriptional control systems exist for *Pseudomonas* genes and are briefly described below.

Environmental Activation

Gene expression can be modulated by the concentration of various solutes in the environment. Response to environmental stimuli can occur through a two-component sensory transduction system that has been described for a variety of prokaryotes.[89,90] One component is comprised of a "sensor" protein that detects the environmental stimuli and relays the information to the second component that acts as a "regulator" controlling expression of particular genes. It is likely that a two-component system exists for the expression of alginate genes in *Pseudomonas aeruginosa*.[91] By using the reporter gene *xylE*, transcription from *algD* responded to several environmental conditions including high osmolarity. This response was dependent on the presence of several regulatory proteins, AlgR1, AlgR2, AlgR3, etc. These regulatory proteins are the likely regulator(s) of the system. AlgR1 has a high degree of homology with the regulator or receiver class of signal transduction proteins. The sensor protein has not as yet been assigned.

Positive Regulatory Proteins: LysR Family

Another general mechanism of gene regulation is through positive regulatory proteins that are required for transcription of structural genes. One such family of proteins are the LysR regulatory family.[92] Many of

[89] C. W. Ronson, B. T. Nixon, and F. M. Ausubel, *Cell (Cambridge, Mass.)* **49**, 579 (1987).
[90] E. C. Kofoid and J. S. Parkinson, *Proc. Natl. Acad. Sci. U.S.A.* **85**, 4981 (1988).
[91] J. D. DeVault, A. Berry, T. K. Misra, A. Darzins, and A. M. Chakrabarty, *Bio/Technology* **7**, 352 (1989).
[92] S. Henikoff, G. W. Haughn, J. M. Calvo, and J. C. Wallace, *Proc. Natl. Acad. Sci. U.S.A.* **85**, 6602 (1988).

these proteins are transcribed divergently from the operons they regulate. They regulate gene expression by binding to the promoter/control region. In the presence of an inducer molecule, the binding conformation presumably alters, priming the DNA for transcription initiation. Binding protein in this region may also autoregulate gene expression for the regulatory protein.[92] Binding studies and footprint analysis using *Pseudomonas aeruginosa* TrpI and its target DNA supports the idea of a conformational change in binding when an inducer is present.[93] In the absence of its inducer, indoleglycerol phosphate, TrpI binds upstream of the *trpBA* operon overlapping its own promoter region. The binding pattern alters in the presence of inducer so that the protected region extends downstream toward the *trpBA* promoter region. The proposed conformational change in binding of the regulatory protein at the target DNA is mediated by an inducer interacting with the regulatory protein. This interaction likely occurs within the C-terminal region of the regulatory protein owing to the divergence in amino acid homology that is found at the C-terminal region between members of the LysR family. Other *Pseudomonas* genes that are members of the LysR family include *P. aeruginosa trpI*,[94] *P. putida nahR*,[67,95] *P. putida catR*,[96] and *P. putida clcR*.[97] Binding of these proteins to target DNA can be assayed by gel retardation[98] as described.

Gel Retardation

Preparation of Probe

1. Anneal at 65° the following: 1 μg ss template DNA containing the promoter fragment, 2 μl sequencing annealing beffer (5×), 5 ng primer, and distilled water to 10 μl.
2. Cool slowly, then add 1 μl of 100 mM DTT, 1 μl Sequenase (2–3 units), 2 μl dNTP (2.5 mM of dATP, dTTP, dGTP), and 1 μl [α-^{32}P] dCTP (800 Ci/mmol) or [α-^{35}S]dCTP (1000 Ci/mmol). Incubate for 10 min 37°. Add 1 μl dCTP (2.5 mM) and incubate for an additional 5 min.
3. Incubate for 10 min at 65° to inactivate Sequenase.
4. Place on ice and add restriction buffer to give correct salt concentration, and distilled water to 20 μl, and restriction enzymes that release the promoter fragment. Incubate 1.5 hr at 37°.

[93] M. Chang and I. P. Crawford, *Absr. Annu. Meet. Am. Soc. Microbiol.*, H-98, p. 186 (1988).
[94] M. Chang, A. Hadero, and I. P. Crawford, *J. Bacteriol.* **171,** 172 (1989).
[95] M. A. Schell and P. Wender, *J. Bacteriol.* **166,** 9 (1986).
[96] R. K. Rothmel, T. L. Aldrich, J. E. Houghton, W. M. Coco, L. N. Ornston, and A. M. Chakrabarty, *J. Bacteriol.* **172,** 922 (1990).
[97] W. M. Coco, R. K. Rothmel, and A. M. Chakrabarty, unpublished results (1990).
[98] M. Fried and D. M. Crothers, *Nucleic Acids Res.* **9,** 6505 (1981).

5. Add loading buffer and electrophorese through a 0.7% agarose gel. Stain with ethidium bromide and cut out the appropriate band. Purify DNA using Gene Clean (purchased from Bio 101, La Jolla, CA).
6. Resuspend in 20 µl distilled water.

Cell Extract

1. Grow 100–200 ml of cells harboring the cloned DNA (under control of an inducible promoter, i.e., *tac*) under induced or noninduced conditions until mid-log phase.
2. Collect the cells and resuspend in 5 ml of 50 mM Tris-HCl, pH 8.0, 5 mM DTT, 1 mM PMSF (phenylmethylsulfonyl flouride).
3. Sonicate cells and remove cell debris by centrifugation at 40,000 g for 30 min at 4°. This can be followed by a high-speed spin at 105,000 g for 1 hr at 4°.
4. Glycerol is added to the supernatant to 50%. The samples are stored at −70°.

Binding and Electrophoresis of Probe

1. Mix 3 µl of 10× binding buffer [1% (w/v) Triton X-100, 40% (v/v) glycerol, 10 mM Na$_2$EDTA, 100 mM 2-mercaptoethanol (2-BME), 100 mM Tris-HCl, pH 7.5], 5 µl of poly(dI–dC) at 1 µg/µl, 2 µl of labeled probe (~5 ng DNA), Cell extract containing 5–10 µg of protein, and distilled water to 30 µl. Incubate for 30 min at 30°.
2. Load the samples onto a 5% (w/v) polyacrylamide gel (50 : 1, acrylamide–*N,N'*-methylenebisacrylamide linkage). Load one lane with restriction loading dye. Electrophorese at 120 V in TBE buffer until the bromphenol blue reaches the bottom of the gel (~2.4–3 hr depending on size of the gel).
3. Remove one glass plate from the gel then soak in 10% (v/v) glycerol for 15 min.
4. Transfer the gel to Whatman (Clifton, NJ) 3M paper, cover with plastic wrap, and dry for 1.5 hr at 80° in a gel dryer.
5. Place gel in X-ray cassette and expose to film overnight.

Other Positive Regulation: xylR

There are other *Pseudomonas* positive regulatory proteins that have been studied which are not part of the LysR family. A very well-characterized system is the control of xylene–toluene degradation, which is encoded on the *P. putida* TOL plasmid. The two operons *xylCAB* and *xylDLEGF* are regulated by two proteins, XylR and XylS. XylR stimulates both operons in the presence of pathway hydrocarbons or alcohols, whereas

XylS is essential for activating *xylDLEGF* in the presence of pathway carboxylic acids.[99]

Negative Regulation

Gene transcription can be turned off in a similar manner to induction, by the interaction of a repressor molecule with an operator region. One example of this control mechanism has recently been described for the *P. putida* histidine utilization genes.[100] A single repressor, *hutC*, binds to the three operons required for histidine utilization, shutting off transcription. The addition of urocanate, the normal inducer, caused dissociation of all operator–repressor complexes.

Another type of negative transcriptional control is catabolite repression. Although catabolite repression has not been intensely studied in *Pseudomonas*, its existence has been observed. The amidase and histidine utilization genes of *P. aeruginosa* are repressed strongly by succinate and weakly by pyruvate. (Amidase synthesis is under positive regulation by the *amiR* gene product.[101]) Succinate strongly represses the *P. cepacia* protocatechuate 3,4-dioxygenase genes; however, catabolite repression by lactate is greater than that by pyruvate.[102]

Transcription of *Pseudomonas* genes is not exclusively regulated by any of the above mechanisms, but may occur by any number of other modes. In addition, gene expression may be under translational or post-translational controls.

Conclusion

Pseudomonads as described are a diverse group of organisms, which have unique features. In this chapter, we have tried to describe some interesting properties of the *Pseudomonas* genome as well as provide detailed genetic methodologies that have been used in studying DNA and proteins from *Pseudomonas*.

Acknowledgments

We would like to thank Lula Johnson for typing the manuscript. This work was supported by a Public Health Service grant (ES04050) from the National Institute of Environmental Health Sciences, a Public Health Service grant (AI 16790-10) from the National Institutes of Health, and in part by a grant (DMB87-21743) from the National Science Foundation.

[99] R. A. Spooner, M. Bagdasarian, and F. C. H. Franklin, *J. Bacteriol.* **169**, 3581 (1987).
[100] L. Hu, S. L. Allison, and A. T. Phillips, *J. Bacteriol.* **171**, 4189 (1989).
[101] D. J. Cousen, P. H. Clarke, and R. Drew, *J. Gen. Microbiol.* **133**, 2041 (1987).
[102] G. J. Zylstra, R. H. Olsen, and D. P. Ballou, *J. Bacteriol.* **171**, 5907 (1989).

[24] Genetic Analysis in *Vibrio*

By MICHAEL SILVERMAN, RICHARD SHOWALTER,
and LINDA MCCARTER

Introduction

Bacteria of the genus *Vibrio* are gram-negative, straight or curved rods, capable of both respiration and fermentation, and most have been isolated from freshwater, estuarine, or marine environments where they exist as free-living forms or in association with mammals, fish, or other organisms.[1] Interest in developing methods for genetic analysis of pathogenic *Vibrio* has been particularly intense, and analysis has focused on elements of pathogenesis such as toxins, hemolysins, and other virulence factors and on antigenic determinants which could be exploited for the production of vaccines. Refined genetic techniques have been developed for the El Tor and classic biotypes of *Vibrio cholerae* which cause diarrheal disease in epidemic proportions in humans. Also, considerable progress on genetic analysis has been made with strains of *Vibrio parahaemolyticus*, a major cause of seafood poisoning, and with *Vibrio anguillarum*, which is a pathogen of fish. Other properties of *Vibrio* have also excited interest, and the genetics of luminescence of bacteria such as the light organ symbiont *Vibrio fischeri* and of chitin degradation by bacteria such as *Vibrio harveyi* and *Vibrio vulnificus* have been investigated. A broad collection of genetic tools including conjugation, transduction, transformation, recombinant DNA manipulation for cloning, sequencing, and expression, transposon mutagenesis, and gene fusion construction have been applied to the *Vibrio* group as a whole. We describe the advances in applying these various categories of genetic analysis to particular species of *Vibrio*.

Genetic Transfer Systems

Conjugation

The P plasmid of *V. cholerae*[2] has been exploited to develop a versatile system for conjugative transfer of DNA. The P fertility factor is 68 kilobases (kb) in length[3] and is capable of high-frequency self-transfer from

[1] P. Baumann, L. Baumann, M. J. Woolkalis, and S. S. Bang, *Annu. Rev. Microbiol.* **47**, 369 (1983).

[2] K. Bhaskaran, *Indian J. Med. Res.* **47**, 253 (1959).

[3] E. J. Bartowsky, G. Morelli, M. Kamke, and P. Manning, *Plasmid* **18**, 1 (1987).

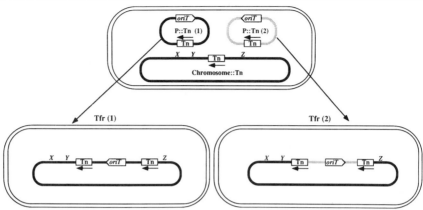

Fɪɢ. 1. Transposon-facilitated transfer strains. High-frequency conjugal transfer of the chromosome of *V. cholerae* was achieved by constructing strains with pairs of transposon (Tn) insertions, one in the conjugative P plasmid and one in the chromosome, which provided homologous regions of DNA required for chromosomal integration of the P plasmid. Integrated P plasmids in such strains mobilized chromosomal genes proximal to the plasmid origin of transfer, *oriT*, and the direction of transfer was reversed by using a P::Tn in which the orientation of the transposon insertion was reversed [e.g., compare P::Tn(*1*) and P::Tn(2)]. Transposons Tn*1*, Tn*5*, Tn*10*, and mutagenic phage VcA-1 have been used to provide the portable homology which directed P plasmid integration.

donor cells, P⁺, to recipient cells, P⁻.[4] However, mobilization of chromosomal DNA occurs very infrequently, apparently because of the absence of insertion sequences in the plasmid which could cause integration into the genome. High-frequency mobilization of the *V. cholerae* genome (10^{-2} to 10^{-3} recombinants per recipient) was achieved by using transposon Tn*1* insertions to provide the homologous sequences required for chromosomal integration of the P plasmid.[5] This high-frequency transfer system, called Tfr for transposon-facilitated recombination, consisted of a derivative of the P plasmid, P::Tn*1*, in a strain of *V. cholerae* with a chromosomal insertion of Tn*1*. Transfer from P⁺ cointegrate strains originated from the chromosomal site of Tn*1* insertion, and the direction of transfer could be reversed by using a different P::Tn*1* with the opposite orientation of Tn*1* insertion (see Fig. 1). Mapping procedures similar to those used for Hfr crosses in *Escherichia coli* were applied to obtain a circular linkage map for *V. cholerae*.[6]

This method for construction of Tfr strains employed a P plasmid

[4] C. Parker and W. R. Romig, *J. Bacteriol.* **112**, 707 (1972).
[5] S. R. Johnson and W. R. Romig, *Mol. Gen. Genet.* **170**, 93 (1979).
[6] R. D. Sublett and W. R. Romig, *Infect. Immun.* **32**, 1132 (1981).

derivative to deliver Tn*1* for mutagenesis of the genome, so subsequent manipulations were required to eliminate the resident plasmid, and transposition of Tn*1* into the genome rarely occurred. The Tfr strain construction method was improved by using insertion mutants generated by *V. cholerae* mutator phage VcA-1[7] or transposons Tn*5* and Tn*10*[8] to provide the homology for chromosomal integration of the P plasmid. In the latter case F' factors temperature sensitive for replication carrying Tn*5* or Tn*10* were conjugated into *V. cholerae* which were then grown at elevated temperature under antibiotic selection to recover transposon insertion mutants. P::Tn*1* derivatives with a VcA-1, Tn*5*, or Tn*10* insertion were also constructed for use with the corresponding insertion mutants of *V. cholerae*. Refined mapping of the genomes of El Tor and classic strains, resulting from use of this improved Tfr methodology, revealed an inversion of the *pro–ura* regions in the respective biotypes which otherwise appeared to be very similar in genome organization. Other applications included the mapping of genes for hemolysis, agglutination, and toxin production (reviewed in Ref. 9).

Conjugal strategies for gene transfer involving broad host range plasmids, particularly vectors derived from RP4, have been applied to diverse *Vibrio* species including *V. anguillarum*,[10] *V. cholerae*,[11] *V. harveyi*,[12] and *V. parahaemolyticus*.[13] Vector designs which incorporate elements of broad host range plasmids are discussed more in the following sections, but general applications include transfer of cloned wild-type or mutated genes into *Vibrio* for complementation analysis, transfer of mutated genes for recombinational replacement of wild-type alleles to analyze mutant phenotypes, transfer of indicator gene fusions into *Vibrio* for examination of protein export and function or control of gene transcription, and delivery of transposons for insertional mutagenesis and other methods based on engineered transposons. The utility of such vectors, based on the promiscuity of transfer into and the ability to replicate in a wide variety of hosts, has already been established, so the arduous process of building a genetic system from a native plasmid for each experimental organism as was done with the P plasmid of *V. cholerae* will probably be unnecessary for new investigations on other *Vibrio* species. For example, a Tfr-like system for

[7] S. R. Johnson, B. C. S. Liu, and W. R. Romig, *FEMS Micribiol. Lett.* **11**, 13 (1981).
[8] J. W. Newland, B. A. Green, and R. K. Holmes, *Infect. Immun.* **45**, 428 (1984).
[9] A. Guidolin and P. Manning, *Microbiol. Rev.* **51**, 285 (1987).
[10] M. E. Tolmasky and J. H. Crosa, *J. Bacteriol.* **160**, 860 (1984).
[11] J. J. Mekalanos, D. J. Swartz, G. D. N. Pearson, N. Hartford, F. Groyne, and M. deWilde, *Nature (London)* **306**, 551 (1983).
[12] M. Martin, R. Showalter, and M. Silverman, *J. Bacteriol.* **171**, 2406 (1989).
[13] L. McCarter and M. Silverman, *J. Bacteriol.* **169**, 3441 (1987).

chromosome mobilization based on broad host range plasmid RP4 and transposon Tn*10* insertion strains was recently developed for use in a marine isolate, *Vibrio* strain 60.[14]

Transduction

Several DNA bacteriophage with transducing activity for marine *Vibrio* have been isolated. Phage hv-1 infected the luminous bacterium *V. harveyi* and could transduce Trp⁻ auxotrophs to prototrophy at a frequency of approximately 1 recombinant per 10^6 plaque forming units (pfu).[15] No transfer of other auxotrophic markers was demonstrated, which suggested a specialized transduction mechanism. However, stable lysogens could not be isolated in this strain of *V. harveyi*, so it was not clear if the mechanism of transduction is related to that of temperate, specialized, transducing phage such as ϕ80 and λ. A generalized transducing phage capable of transducing a wide variety of auxotrophic markers, transposon Tn*10* insertion mutations, and broad host range plasmid RP4 into exohemagglutinin-positive marine isolates such as *Vibrio* strain 60 has been characterized.[14] This phage, AS-3, is similar in morphology to phage T4 of *E. coli,* does not lysogenize, and transduces markers at a frequency of 10^{-5} to 10^{-6}/pfu even without UV irradiation treatment to reduce killing of recipient bacteria by virulent phage. AS-3 is apparently capable of packaging considerable DNA, as more than 30 independent Tn*10* insertion mutations and auxotrophic markers could be mapped by cotransduction analysis to a small number of linkage groups (i.e., four). Phage AS-3 only infects a particular group of marine *Vibrio* which is presently not well-defined, possibly those related to *V. anguillarum.*

CP-T1 is a temperate phage of both the El Tor and classic biotypes of *V. cholerae,* and lysogeny causes serotype conversion, namely, the appearance of a new lipopolysaccharide antigenic determinant. Phage CP-T1 mediates generalized transduction and has been shown to transfer a limited number of markers at frequencies as high as 10^{-5}/pfu.[16] The phage genome of about 43 kb is terminally redundant and circularly permuted,[17] and packaging is thought to occur by a mechanism similar to that of P1 of *E. coli* and P22 of *Salmonella* in which packaging of a headful of DNA is initiated at a specific site on a concatemeric phage genome and proceeds to encapsulate additional genome segments without need for recognition of additional packaging sites. Insertion of the CP-T1 packaging

[14] A. Ichige, S. Matsutani, K. Oishi, and S. Mizushima, *J. Bacteriol.* **171,** 1825 (1989).
[15] A. Keynan, K. Nealson, H. Sideropoulos, and J. W. Hastings, *J. Virol.* **14,** 333 (1974).
[16] J. E. Ogg, T. L. Timme, and M. M. Alemohammad, *Infect. Immun.* **31,** 737 (1981).
[17] A. Guidolin, G. Morelli, M. Kamke, and P. Manning, *J. Virol.* **51,** 163 (1984).

site, *pac*, into a plasmid which was harbored in *V. cholerae* resulted in more than a 100-fold increase in the frequency of CP-T1-mediated transduction of that plasmid. This finding indicated that it should be possible to develop a system for high-frequency transduction with CP-T1 by inserting the *pac* sequence in the chromosome of *V. cholerae*. The *pac* site and a neighboring sequence which could encode a *pac* terminase for endonucleolytic cleavage has been cloned and sequenced,[18] and the *pac* region has been inserted in Tn*5* to produce Tn*pac* with the goal of introducing CP-T1 packaging sites into numerous locations in the chromosome of *V. cholerae*.[19] It is expected that markers could then be transduced at very high frequency and that the gradient of inheritance of markers closely linked to the Tn*pac* insertion could be used to obtain the linear order of the markers.

An alternative gene transfer strategy is to use phage of *E. coli* such as P1 and λ to transduce DNA from *E. coli* into *Vibrio*. Phage P1 was first used to transfer a derivative of transposon Tn*5*, Tn*5-132* encoding tetracycline resistance, into *V. harveyi* strain BB7 for mutagenesis of the luminescence genes, *lux*.[20,21] Phage P1 infects, that is, adsorbs and injects, but does not replicate in this host, so P1, specifically P1*clr100*CM::Tn*5-132*, was used as a suicide vector to deliver Tn*5-132* which must transpose to the chromosome to be stably inherited. The transduction procedure involved heat induction of a temperature-sensitive P1*clr100*CM::Tn*5-132* lysogen of *E. coli*, infection of recipient *V. harveyi* with phage lysate in the presence of 25 m*M* CaCl$_2$ to stabilize phage and promote adsorption, and subsequent plating of infected *V. harveyi* on a medium containing tetracycline for selection of a library of transposon insertion mutants. P1 was also used to transfer transposons mini-Mu*lac* or mini-Mu*lux* for mutagenesis and formation of transcriptional gene fusions in *V. parahaemolyticus*[13,21,22] and *V. harveyi*.[12] Here *E. coli* lysogenic for P1*clr100*CM and containing a temperature-inducible mini-Mu transposon insertion was induced by heat shift to produce a transducing lysate containing mini-Mu packaged in P1 phage particles. The application of this technique was confined to a limited number of *Vibrio* strains which were receptive to P1 infection. Transduction of DNA from *E. coli* can, in principle, be extended to many more *Vibrio* by transplanting the gene for λ phage receptor into the recipient. For example, *V. cholerae* has been

---

[18] A. Guidolin and P. Manning, *Mol. Gen. Genet.* **212**, 514 (1988).
[19] P. A. Manning, *Microbiol. Sci.* **5**, 196 (1988).
[20] R. Belas, A. Mileham, D. Cohn, M. Hilmen, M. Simon, and M. Silverman, *Science* **218**, 791 (1982).
[21] R. Belas, A. Mileham, M. Simon, and M. Silverman, *J. Bacteriol.* **158**, 890 (1984).
[22] R. Belas, M. Simon, and M. Silverman, *J. Bacteriol.* **167**, 210 (1986).

made receptive to λ infection by introducing the cloned λ receptor gene, *lamB*, on a broad host range plasmid.[23] Such receptive strains support infection but not replication of λ, so λ phage can be used as a suicide vector to deliver transposons for mutagenesis of the chromosome. Alternatively, recombinant constructs in λ phage vectors or in vectors with the λ packaging site, *cos*, could be encapsulated *in vivo* by infection with λ phage or *in vitro* with a packaging extract and then delivered by infection to a receptive *Vibrio*.

Transformation

Transformation, reported to have been accomplished only in *V. cholerae* with protoplasts in the presence of polyethylene glycol,[24] was inefficient and limited to small plasmids. Production of extracellular DNases is a significant barrier to transformation in *V. cholerae*. Indirect evidence that production of DNase reduces efficiency of transformation was obtained with *E. coli* containing a gene encoding a 24-kDa DNase cloned from *V. cholerae* El Tor strain 017. Transformation of $CaCl_2$-treated *E. coli* was reduced by approximately 100-fold in the recombinant strain producing the *V. cholerae* DNase.[25] A different DNase gene, *xds*, cloned from *V. cholerae* El Tor strain 26-3, encoded a 100-kDa protein. This gene was mutated in *E. coli* with transposon Tn5 and recombined into *V. cholerae* by a gene replacement method to construct a DNase⁻ mutant JN1001.[26] Transformation of $CaCl_2$-treated mutant JN1001 has recently been evaluated,[27] and approximately 10^4 transformants per microgram of pBR322 DNA was observed with the mutant compared with no transformants for the DNase⁺ *V. cholerae*. Improvement in transformation efficiency was very substantial but was still three orders of magnitude less than that obtained with *E. coli*, so other barriers such as the production of the 24-kDa DNase mentioned above and a restriction–modification system could also be influencing transformation.

"Transformation" by electroporation was an effective method to transfer a plasmid with the iron-regulation gene, *fur*, of *E. coli* into a hemolysin-deficient mutant of *V. cholerae*.[28] A rigorous evaluation of electroporation variables was also performed along with the examination of the DNase⁻

[23] A. Harkki, T. R. Hirst, J. Homgren, and E. T. Palva, *Microb. Pathogen.* **1**, 283 (1986).
[24] A. Hamood, R. Sublett, and C. Parker, *Infect. Immun.* **52**, 476 (1986).
[25] T. Focareta and P. A. Manning, *Gene* **53**, 31 (1987).
[26] J. W. Newland, B. A. Green, J. Foulds, and R. K. Holmes, *Infect. Immun.* **47**, 691 (1985).
[27] H. Marcus, J. M. Ketley, J. B. Kaper, and R. K. Holmes, *FEMS Microbiol. Lett.* **68**, 149 (1990).
[28] J. A. Stoebner and S. M. Payne, *Infect. Immun.* **56**, 2891 (1988).

mutant discussed above,[27] and transfer of plasmids by electroporation into *V. cholerae*, which could be greater than 10^5 transformants per microgram of DNA under optimal conditions, was found to be improved by increasing field strength, by decreasing plasmid size, by adding isotonic sucrose to compensate for low electrolyte concentration, and by using plasmid DNA isolated from the same strain used as a recipient, which latter observation indicated the influence which a host restriction–modification system could have on the efficiency of transformation. Extensive analysis of the biological and electrical variables affecting electroporation and considerable innovation in apparatus development have been made[29] which should facilitate application of this technique to more species of *Vibrio*.

Cloning and Expression

Selective Methods

Procedures for isolating genes encoding selectable phenotypes have been developed for many genes including those from *Vibrio*. For example, the *trp* operon of *V. parahaemolyticus* was isolated by introducing a cosmid library into a Trp⁻ auxotroph of *E. coli* and plating the recombinants onto a minimal medium devoid of tryptophan.[30] A similar strategy was used to isolate cloned genes from *V. alginolyticus* necessary for utilization of sucrose. *Escherichia coli* cannot catabolize sucrose, but recombinants with one cloned DNA fragment encoding both sucrase and sucrose transport functions could use sucrose as a sole carbon source.[31] Isolation of the origin of replication, *oriC*, of *V. harveyi* was based on the property of this locus to compensate for the inability of ColE1 cloning vectors to replicate in a *polA*-defective *E. coli* host.[32]

Such procedures require expression of a function which compensates or complements a defect in the recombinant host, but it is possible to select recombinants with a particular locus without reliance on expression of the function encoded by that locus. This can be achieved by using transposon insertion mutations which physically link the selectable drug resistance marker of the transposon to the target gene. Specifically, this involves generation of a mutant library by transposon mutagenesis of a particular *Vibrio* (see next section); isolation of mutants with insertions in

[29] K. Shigekawa and W. J. Dower, *Bio/Technology* **6**, 742 (1988).
[30] I. P. Crawford, C. Y. Han, and M. Silverman, *Sequence* **1**, 189 (1991).
[31] R. R. Scholle, V. E. Coyne, R. Maharaj, F. T. Robb, and D. R. Woods, *J. Bacteriol.* **169**, 2685 (1987).
[32] J. W. Zyskind, J. M. Cleary, W. S. A. Brusilow, N. E. Harding, and D. W. Smith, *Proc. Natl. Acad. Sci. U.S.A.* **80**, 1164 (1983).

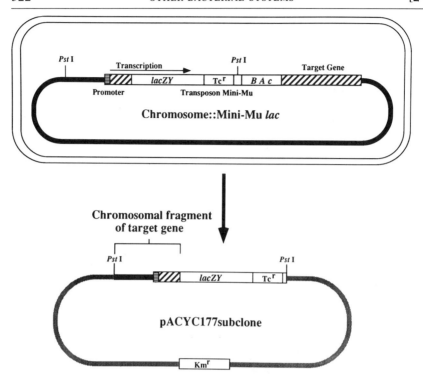

FIG. 2. Transposon-directed cloning. Transposon insertions physically link the drug resistance gene of the transposon to target gene sequences so the target gene DNA can be cloned by selection of recombinants expressing the drug resistance phenotype. Here, *Pst*I cleavage of chromosomal DNA containing a transposon mini-Mu*lac* insertion yields a fragment with both the selectable marker of the transposon, Tc^r, and a portion of the target gene DNA from one side of the insertion site. This fragment can, after ligation into vector pACYC177 and transformation into *E. coli*, be recovered by selection for tetracycline-resistant recombinants.

the locus encoding the desired function, recognition of which is guided by the transposon-induced mutant phenotype; fragmentation of genomic DNA from the insertion mutants with a restriction enzyme which does not cleave the transposon DNA or which generates a fragment containing the transposon drug resistance gene and target gene sequence on one side of the insertion site; and ligation of these genomic fragments into a cloning vector and transfer into *E. coli* for selection of recombinants expressing both vector-encoded and transposon-encoded drug resistance. These steps are shown in Fig. 2.

    The target gene sequence in the cloned DNA is usually inactive because of interposition of the transposon or incomplete because the cloned fragment originated from a restriction enzyme site inside the transposon and one in adjacent chromosomal DNA. However, complete, uninterrupted gene regions can subsequently be obtained by splicing cloned fragments from mutants with insertions in different positions in the same gene region or by using the incomplete gene fragment or "nub" to probe clone libraries of wild-type DNA for a complete gene fragment. This strategy was used to obtain the phosphate regulation locus, *pho*, from *V. parahaemolyticus*,[13] the luminescence regulation gene, *luxR*, from *V. harveyi*,[12] and the luciferase genes, *luxA* and *luxB*, from *V. harveyi*.[20] This method, which requires transposon mutagenesis of the native *Vibrio*, can be applied in principle to the cloning of any nonessential gene which encodes a recognizable phenotype.

Screening Methods

Many genes cloned from *Vibrio* are expressed to some extent in *E. coli*, so assays for gene function can often be employed to detect recombinants. Hydrolysis of the chromogenic substrate *p*-nitrophenyl-2-acetamido-2-deoxy-β-D-glucopyranoside (PNAG) was used to detect recombinants with the chitobiase gene, *chb*, from *V. harveyi*[33] and *V. vulnificus*.[34] Hydrolysis of the fluorogenic substrate 2'-(4-methylumbelliferyl)-α-D-*N*-acetylneuraminic acid was used to detect *E. coli* with the neuraminidase gene, *nanH*, of *V. cholerae*.[35] Recombinants with a gene, *xds*, encoding a 100-kDa extracellular DNase[26] and another gene encoding a 24-kDa extracellular DNase from *V. cholerae*[25] were identified as colonies which produced clear zones on DNA test agar. Lysis of sheep red blood cells in overlays or in blood agar plates was used to detect recombinants with a gene, *tdh*, for thermostable direct hemolysin from *V. parahaemolyticus*,[36] with a gene for cytotoxin-hemolysin from *V. vulnificus*,[37] or with a gene, *hlyA*, for hemolysin from *V. cholerae*.[38,39] Isolation of the *recA* gene of *V. cholerae* was accomplished by screening recombinants in which a *recA* defect in the *E. coli* host was complemented, that is, in which resistance to

---

[33] R. W. Soto-Gil and J. W. Zyskind, *in* "Chitin, Chitosan and Related Enzymes" (J. P. Zikakis, ed.), p. 169. Academic Press, New York, 1984.

[34] A. T. Wortman, C. C. Somerville, and R. R. Colwell, *Appl. Environ. Microbiol.* **52**, 142 (1986).

[35] E. R. Vimr, L. Lawrisuk, J. Galen, and J. B. Kaper, *J. Bacteriol.* **170**, 1495 (1988).

[36] J. B. Kaper, R. K. Campen, R. J. Seidler, M. M. Baldini, and S. Falkow, *Infect. Immun.* **45**, 290 (1984).

[37] A. C. Wright, J. G. Morris, D. R. Maneval, K. Richardson, and J. B. Kaper, *Infect. Immun.* **50**, 922 (1985).

[38] S. L. Goldberg and J. R. Murphy, *J. Bacteriol.* **160**, 239 (1984).

[39] P. A. Manning, M. H. Brown, and M. W. Heuzenroeder, *Gene* **31**, 225 (1984).

the DNA-damaging agent methyl methanesulfonate (MMS) was restored.[40] The presence of iron-uptake genes, namely, those encoding siderophore synthesis, transport, or gene regulation, in fragments cloned from the indigenous plasmid pJM1 from *V. anguillarum* was detected by complementation of defects in the native bacterium.[10] Recombinant cosmids were transferred by conjugation into various iron-uptake mutants, and these exconjugates were then screened for the ability to grow on an iron-depleted (EDDA-treated) medium.

Antibody probes have also been used for screening recombinant libraries for expression of cloned genes. Western immunoblot methods employing rabbit antisera, in concert with ^{125}I-labeled *Staphylococcus aureus* protein A or with goat anti-rabbit IgG coupled to horseradish peroxidase, were used to detect recombinants expressing the following: the gene, *ompP*, for a polyphosphate porin protein,[13] and the flagellar filament genes *flaA,B,C,D*[41] and *lafA*[42] from *V. parahaemolyticus* as well as a hemagglutinin gene,[43] a major outer membrane protein gene, *ompV*,[44] and genes for lipopolysaccharide O-antigen synthesis[45] from *V. cholerae*. Gene expression is not a prerequisite for screening with DNA probes. The *ctxAB* operon encoding *V. cholerae* toxin was cloned by taking advantage of the cross-hybridization which these genes had with the cloned genes for the subunits, LT-A and LT-B, of heat-labile enterotoxin of *E. coli*.[46] Furthermore, once cloned, this or other DNA has been used to search for homologs in the parent strain and other strains, species, and genera for the purpose of analyzing genome organization and rearrangement[47,48] and horizontal gene transfer.[49] Alternatively, synthetic oligonucleotide probes designed from amino acid sequence data can be used to detect cloned genes. For example, an oligonucleotide probe composed of mixed sequences to account for redundancy in codon usage was used to isolate the gene, *luxA*, encoding the α subunit of luciferase from *V. harveyi*.[50]

[40] I. Goldberg and J. J. Mekalanos, *J. Bacteriol.* **165,** 715 (1986).
[41] L. McCarter, M. Hilmen, and M. Silverman, *Cell (Cambridge, Mass.)* **54,** 345 (1988).
[42] L. McCarter and M. Silverman, *J. Bacteriol.* **171,** 731 (1989).
[43] V. I. Franzon and P. A. Manning, *Infect. Immun.* **52,** 279 (1986).
[44] G. Stevenson, D. I. Leavesley, C. A. Lagnado, M. W. Heuzenroeder, and P. A. Manning, *Eur. J. Biochem.* **148,** 385 (1985).
[45] P. A. Manning, M. W. Heuzenroeder, J. Yeadon, D. I. Leavesley, P. R. Reeves, and D. Rowley, *Infect. Immun.* **53,** 272 (1986).
[46] G. D. N. Pearson and J. J. Mekalanos, *Proc. Natl. Acad. Sci. U.S.A.* **79,** 2976 (1982).
[47] J. J. Mekalanos, *Cell (Cambridge, Mass.)* **35,** 253 (1983).
[48] M. Nishibuchi and J. B. Kaper, *Mol. Microbiol.* **4,** 87 (1990).
[49] J. B. Kaper, J. G. Morris, and M. Nishibuchi, *in* "DNA Probes for Infectious Diseases" (F. C. Tenover, ed.), p. 65. CRC Press, Boca Raton, Florida, 1989.
[50] D. H. Cohn, R. C. Ogden, J. N. Abelson, T. O. Baldwin, K. H. Nealson, M. I. Simon, and A. J. Mileham, *Proc. Natl. Acad. Sci. U.S.A.* **80,** 120 (1983).

The isolation of genes involved in regulation of expression is particularly challenging because their gene products are seldom abundant and usually lack enzymatic activity. The approach used to isolate genes regulating toxin production in *V. cholerae* is instructive.[51] The promoter region of the toxin operon, *ctxAB*, was aligned with the *lacZ* indicator gene of *E. coli* so that transcription of the operon could be measured conveniently and fairly accurately *in vivo* by monitoring hydrolysis of a chromogenic substrate of β-galactosidase, 5-bromo-4-chloro-3-indolyl-β-D-galactoside (X-Gal). This construction with the *ctx::lacZ* transcriptional fusion was then inserted in a λ phage cloning vector capable of lysogenization, thereby eliminating the influence of exogenous vector promoters and also lowering the basal level of β-galactosidase activity by reducing copy number. The *ctx::lacZ* fusion in a λ lysogen of *E. coli* was used as a host for screening recombinant plasmids to identify those with genes regulating *ctx* expression. Clones with elevated production of β-galactosidase (more intensely blue with the X-Gal substrate) were found, and these contained a gene, *toxR*, the product of which was shown to be a positive transcriptional regulator of *ctxAB*. Further examination revealed a second gene, *toxS*, in this locus which was also necessary for *ctxAB* expression.[52] Additional uses of gene fusions are discussed later.

Expression

Numerous genes cloned from various *Vibrio* species have been sequenced, and the general impression from examination of sequence data is that elements of promoters and ribosome binding sites are similar to consensus sequences in *E. coli* and that codon usage preferences are also similar. The apparent success in screening for gene function in recombinants also suggests that *E. coli* is a good host for transcription and translation of nucleic acid from *Vibrio*. Interchangeability of regulatory systems has also been demonstrated. For example, a locus cloned from *V. parahaemolyticus* complemented *phoBR* regulation defects of *E. coli*.[13] Also, reports of the usefulness of minicell and maxicell methods for identifying gene products indicate that relatively abundant protein synthesis is often obtained with cloned genes in *E. coli*. However, impressions about *Vibrio* gene expression are biased since cloning failures and difficulties in obtaining gene expression are seldom published. It is also clear that transcription of cloned genes is frequently driven by exogenous plasmid promoters on high copy number cloning vectors such as those derived from plasmid ColE1. For example, transcription of the *luxCDABE* operon for luminescence enzymes from *V. harveyi*, when positioned in the "promoterless"

[51] V. L. Miller and J. J. Mekalanos, *Proc. Natl. Acad. Sci. U.S.A.* **81,** 3471 (1984).
[52] V. L. Miller, V. DiRita, and J. J. Mekalanos, *J. Bacteriol.* **171,** 1288 (1989).

vector pK04, was then dependent on provision of additional regulatory functions in trans.[53] Moreover, transcription of the *ctxAB* operon encoding cholera toxin, cloned in a λ phage vector, required *toxR* and *toxS* as discussed previously. It would appear that, to obtain expression, initial attempts at cloning should be performed with high copy number vectors with strong exogenous promoters. However, some gene products could be detrimental to the host cell and result in poor host viability or cause instability of the cloned DNA. This difficulty appears to have been encountered in the cloning of the lipopolysaccharide O-antigen genes of *V. cholerae*.[45] Remedies for the deleterious effects of foreign gene expression include using low copy number plasmids or λ vectors, using promoter probe type vectors in which insert DNA is shielded from readthrough transcription originating in the vector, and using vectors with regulatable promoters such as p*lac* or pT7. In the latter case expression can be turned off for cloning and then turned on for detection and characterization of gene products.

Extracellular enzymes, hemolysins and toxins are important objects of *Vibrio* research, but secretion of these gene products in recombinant *E. coli* is usually defective, with most of the product localized in the cytoplasm or the periplasm. Extracellular hemolysin encoded by *hlyA* from *V. cholerae* was localized in the periplasm of recombinant *E. coli*, but movement to the extracellular medium was greatly enhanced in mutants of *E. coli* with *tolA* or *tolB* defects.[54] The leakiness of such mutants should facilitate detection of activity encoded by genes for extracellular proteins, but reconstruction of a functionally complete *Vibrio* secretory system in recombinant *E. coli* has not been achieved. Introduction of a mutation in *hlyB*, which is linked to the *hlyA* gene of *V. cholerae*, into the genome by a gene replacement method resulted in defective hemolysin secretion in *V. cholerae*,[55] but the presence of *hlyB* was not sufficient for extracellular export in recombinant *E. coli*. Also, a cloned gene from *V. cholerae* encoded a secretion protein which caused extracellular release of a periplasmic *E. coli* DNase but not of the *hlyA* hemolysin.[56] Mutants of *V. cholerae* specifically defective in toxin secretion have been isolated, and work is proceeding to isolate the genes for secretion and to study their function in the native *V. cholerae*.[57] These and other results indicate that protein export in *V. cholerae* and in other species of *Vibrio* is unique,

[53] R. Showalter, M. Martin, and M. Silverman, *J. Bacteriol.* **172,** 2946 (1990).
[54] A. Mercurio and P. A. Manning, *Mol. Gen. Genet.* **200,** 472 (1985).
[55] R. A. Alm and P. A. Manning, *Mol. Microbiol.* **4,** 413 (1990).
[56] T. Focareta and P. A. Manning, *FEMS Microbiol. Lett.* **29,** 161 (1985).
[57] M. G. Jobling, H. Marcus, and R. K. Holmes, *in* "Advances in Research on Cholera and Related Diarrheal Diseases" (R. B. Sack and Y. Takeda, eds.), Vol. 8. KTK Scientific Publishers, Tokyo, Japan, 1991.

complex, and consists of multiple export pathways composed of both specific and common elements. Significant advances have been made in developing tools for genetic analysis in the native *Vibrio* species, some of which are designed particularly for studying export (see Gene Fusions), so the difficulties encountered in using recombinant *E. coli* can probably be circumvented.

Mutagenesis

Transposon

Transposable elements have proved to be remarkably versatile tools for genetic analysis. Their application in providing portable homology for Tfr strain construction and in linking selectable markers to genes for the purpose of cloning has already been mentioned. Transposons generally produce a null mutant phenotype because insertion disrupts the continuity of the target gene, and insertion can also prevent transcription of genes downstream in an operon. The polar influence of transposon insertion can be used to analyze gene organization and to define the direction of transcription of genes and operons. The insertion of several kilobases of transposon DNA facilitates restriction mapping of the site of a transposon-generated mutation in a cloned gene or in a chromosomal gene, and the transposon-encoded drug resistance also serves as a convenient marker for mapping with genetic crosses. Examples of using transposons such as Tn*3lac*, Tn5, Tn*1725*, and mini-Mu*lac* to mutate, map, and define genes and operons by complementation analysis and to determine the direction of transcription of cloned genes in recombinant *E. coli* include the *hlyABC* locus from *V. cholerae*,[39] the *lux* gene locus from *V. fischeri*,[58] and the iron-uptake genes from *V. anguillarum*.[59] A variety of well-practiced procedures have been used to deliver transposons for mutagenesis of cloned genes from species of *Vibrio* in recombinant *E. coli:* temperature-inducible P1 phage or Mu helper phage have been used to deliver mini-Mu*lac*[58]; an R factor temperature sensitive for replication has been used to mobilize Tn*1725*[39]; a plasmid containing a transposase deficient Tn*3lac*, namely, Tn*3*::HoHo1, was used in conjunction with plasmid pSShe with the transposase gene in trans for Tn*3* mutagenesis[59]; and λ::Tn phage have been used to deliver a large variety of transposons.[60] In the latter case the λ::Tn

[58] J. Engebrecht and M. Silverman, *Proc. Natl. Acad. Sci. U.S.A.* **81,** 4154 (1984).

[59] M. E. Tolmasky, L. A. Actis, and J. H. Crosa, *J. Bacteriol.* **170,** 1913 (1988).

[60] J. C. Way, M. A. Davis, D. Morisato, D. E. Roberts, and N. Kleckner, *Gene* **32,** 369 (1984).

phage, which contain amber mutations, were grown on a Sup$^+$ host and then used to infect a Sup$°$ recombinant in which the phage could not replicate, resulting in suicide of the delivery vehicle. Application of selection for the transposon-encoded drug resistance yielded libraries of transposon insertion mutants, some of which contained insertions in the recombinant DNA. Transposon insertions in the recombinants could subsequently be isolated by purifying the targeted plasmid or cosmid vector pools and transforming another *E. coli* recipient or, alternatively, by conjugating the mutated vectors into another recipient, with selection for the transposon-encoded and vector-encoded drug resistances.

Considerable progress has been made in developing procedures for transposon mutagenesis of native species of *Vibrio*. Some of these methods have been mentioned earlier: F$'_{ts}$ factors or a derivative of the P plasmid of *V. cholerae* have been used to deliver Tn*1*, Tn*5*, or Tn*10* into *V. cholerae*,[5,8] and phage P1 has been used to deliver Tn*5*, mini-Mu*lac*, or mini-Mu*lux* into *V. harveyi*[12,21] or *V. parahaemolyticus*.[22] Also, the mutagenic, Mu-like, temperate phage VcA1, VcA2, and VcA3 have been isolated from *V. cholerae*.[7] VcA1 was used to construct a *ctx*::VcA1 strain of *V. cholerae* for Tfr mapping,[61] and VcA1 and VcA2cts1 insertions flanking the *ctxAB* operon were used to construct a mutant of *V. cholerae* with a toxin operon deletion.[62] *Vibrio cholerae* can be made sensitive to λ infection by provision of the *lamB* receptor gene from *E. coli*,[23] and it should be possible to use the large repertoire of specialized λ::Tn vectors[60] for mutagenesis of *V. cholerae* and other species of *Vibrio* as well.

Vectors based on the broad host range incP1 conjugative plasmid RP4 have been used to deliver transposon Tn*10* into marine *Vibrio* strain 60[14] and a transposon Tn*5* derivative, *TnphoA*, into *V. cholerae*.[63,64] Here, the delivering vector, RP4 with Tn*10* or pRK290 with *TnphoA* which is a derivative of RP4 containing the mobilization site *mob* but requiring provision of *tra* function in trans for conjugal transfer, must be eliminated from the recipient *Vibrio* before selection can be applied for strains with a chromosomal insertion of the desired transposon. This was accomplished by introducing by conjugation a second incompatible replicon, pPH1JI encoding gentamycin resistance, to displace the first replicon while maintaining selection for the drug resistance encoded by the transposon. Mutagenesis of *V. cholerae* with *TnphoA* has also been performed with a

[61] I. Sporecke, D. Castro, and J. J. Mekalanos, *J. Bacteriol.* **157**, 253 (1984).

[62] J. J. Mekalanos, S. L. Moseley, J. R. Murphy, and S. Falkow, *Proc. Natl. Acad. Sci. U.S.A.* **79**, 151 (1982).

[63] R. K. Taylor, V. L. Miller, D. B. Furlong, and J. J. Mekalanos, *Proc. Natl. Acad. Sci. U.S.A.* **84**, 2833 (1987).

[64] R. K. Taylor, C. Manoil, and J. J. Mekalanos, *J. Bacteriol.* **171**, 1870 (1989).

mobilizable vector which replicates in *E. coli* but not in *V. cholerae*.[64] This suicide vector in this delivery system was derived from plasmid pJM703.1, which is discussed later in relation to gene replacement procedures. Conjugal transfer of RP4 has been used for transposon Tn*1* mutagenesis of the plasmid-encoded iron-uptake genes of *V. anguillarum*,[65] although here it was not necessary to employ a second incompatible replicon because exconjugate insertion mutants without the RP4 delivery vector arose spontaneously.

Another transposon delivery system based on the mobilization site of plasmid RP4 should also be generally applicable.[66] The mobilization site has been inserted in common cloning vectors such as pACYC184 and pBR322 so that they can be transferred from *E. coli* to other gram-negative bacteria by providing *tra* function in trans from a derivative of RP4. Transposons such as Tn*5*, Tn*7*, and a derivative of Tn*5* encoding tetracycline resistance, Tc^R Tn*5*, were added to construct a series of "pSUP" plasmids. Because the pSUP plasmids usually do not replicate in hosts other than enteric bacteria, their transfer is abortive or suicidal, and selection for inheritance of the transposon-encoded drug resistance results in isolation of mutants resulting from transposition of the transposon to the chromosome of the recipient. Vectors based on pACYC184 and pBR322 do replicate in *V. cholerae* (see Refs. 26 and 40), but replication is apparently inefficient since plasmidless cells arise in the absence of selection for plasmid-encoded drug resistance. Thus, repeated application of selection for the transposon-encoded drug resistance alone could be expected to result in enrichment for insertion mutants. Other *Vibrio* species could also be good targets for pSUP mutagenesis.

Gene Replacement

Cloned *Vibrio* genes which were mutated in *E. coli* have been recombined into the genome of the parent *Vibrio* to produce specific mutant phenotypes. Mutants of *V. cholerae* with deletions of the *ctxAB* toxin operon[11,67] and of *V. parahaemolyticus* with deletions of the flagellar filament genes, *flaA,B,C,D*,[41] and with Tn*5* insertions in chemotaxis genes, *che*,[68] were constructed by variations of the following gene replacement

[65] M. A. Walter, S. A. Potter, and J. H. Crosa, *J. Bacteriol.* **156**, 880 (1983).
[66] R. Simon, U. Priefer, and A. Puhler, *Bio/Technology* **1**, 784 (1983).
[67] J. B. Kaper, H. Lcokman, M. M. Baldini, and M. M. Levine, *Nature (London)* **308**, 655 (1984).
[68] N. Sar, L. McCarter, M. Simon, and M. Silverman, *J. Bacteriol.* **172**, 334 (1990).

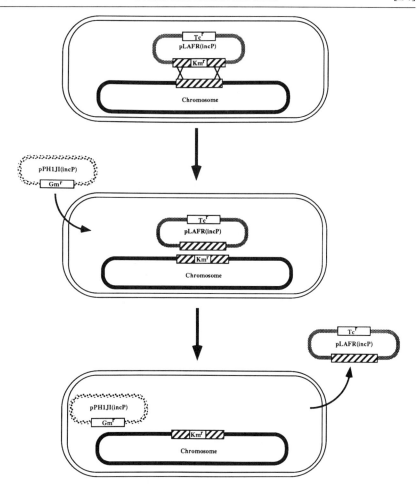

Fig. 3. Gene replacement mutagenesis. Cloned genes which have been mutated by insertion of a selectable marker are used to replace the wild-type allele in the native species of *Vibrio*. Here, a cloned gene (hatched box) in a mobilizable vector pLAFR has been mutated by insertion of a kanamycin resistance gene cassette, Km^r. Homologous recombination between the cloned gene DNA and the chromosomal sequences results in exchange of mutated and wild-type alleles. The exogenote is then eliminated by introducing and selecting for maintenance of a second incompatible replicon, plasmid pPH1JI.

method (see Fig. 3). Cloned *Vibrio* genes were mutated by transposon insertion or by removal of restriction fragments followed by interposition of a drug resistance gene cassette encoding Km^r or Ap^r (deletion–substitution). Regions flanking the cloned genes were retained to promote subsequent homologous recombination into the *Vibrio* genome, and the drug resistance gene was inserted so the mutated target gene could be manipulated as a selectable marker. The mutated loci were then subcloned into a Mob$^+$ derivative of plasmid RP4 such as pRK290 or pLAFRII and transferred by conjugation, with the assistance of an appropriate *tra* donor plasmid such as pRK2013, into a wild-type *Vibrio* recipient. Exconjugates from plate or filter matings were isolated by selection for antibiotic resistances encoded by the vector, the drug resistance marker in the mutated gene, and the recipient *Vibrio*. Counterselection of the *E. coli* donor (or donors if a triparental mating was performed) was usually achieved by adding polymyxin B, streptomycin, or rifamicin to which the donor was sensitive and the recipient was resistant. Isolation of recombinants resulting from the exchange of the mutant for the wild-type allele required elimination of the delivery plasmid, and this was accomplished by conjugal transfer of a second incompatible incP1 replicon, such as pPH1JI encoding gentamycin resistance, into the *Vibrio* exconjugate. Gene replacement mutants, for example, with the *kan* gene inserted in the *ctxAB* operon, could then be selected by plating on a medium containing kanamycin and gentamycin. Loss of the second "kick-out" plasmid, pPH1JI, occurred spontaneously in the absence of gentamycin selection.

DNase$^-$ and RecA$^-$ mutants of *V. cholerae* have also been constructed by gene replacement,[26,40] but with these mutants the defective genes on recombinant pACYC184 or pBR322 vectors were mobilized into the recipient *Vibrio* by F'-mediated transfer in the former case and with *tra* function provided by pRK2013 in the latter case. Gene replacement mutants were then isolated directly by selection or screening procedures since the delivery vectors were lost spontaneously in the absence of selection for the vector-encoded drug resistance. Another approach for introducing mutations into specific genes in the chromosome was used to construct a ToxR$^-$ strain of *V. cholerae*.[69] Inactivation of the *toxR* gene was achieved by selecting for integration of a replication-defective vector into the target gene. Specifically, this involved construction of a broad host range cloning vector which contained *mob*, the mobilization site from RP4, and *oriR6K*, the origin of replication of plasmid R6K, but which was missing the *pir* gene encoding the R6K replication protein π. This vector, pJM703.1, could replicate in *E. coli*, with *pir* function provided in trans on λ *pir*, but could not replicate when transferred into *V. cholerae* which lacks *pir*. Directed mutagenesis of *toxR* was then performed by inserting a restriction fragment from the interior of the *toxR* gene into pJM703.1, mobilizing this construct with the truncated *toxR* gene into *V. cholerae*, and selecting for inheritance of the vector-encoded drug resistance gene, namely, *bla*. Inheritance of

[69] V. L. Miller, and J. J. Mekalanos, *J. Bacteriol.* **170**, 2575 (1988).

drug resistance required integration of the *toxR* plasmid into the chromosomal *toxR* gene by a single crossover event, resulting in the formation of a cointegrate in which the vector DNA interrupted the integrity of the *toxR* gene. Stable maintenance of such a mutation would require continuous antibiotic selection, but this method could in principle be applied to many species of *Vibrio* from which an internal gene fragment had been cloned.

Oligonucleotide

Transposon and gene replacement mutagenesis usually produce defects with null phenotypes, but oligonucleotide-directed mutagenesis has the advantage of generating missense mutations with subtle alterations in phenotype which are extremely useful for analyzing structure–function relationships. Specific amino acids in the B subunit of cholera toxin have been changed by oligonucleotide-directed mutagenesis to identify ganglioside G_{M1} binding domains.[57] This method is also a powerful tool for examining the structural features of cis-acting regulatory sites, and several bases in a putative *lux* operator region of *V. fischeri* have been changed to assess the role of this site in transcription of luminescence genes.[70] Complex alterations in gene organization can also be achieved. The *luxA* and *luxB* genes of *V. harveyi* encode the α and β subunits of luciferase and are cotranscribed in that order in a larger operon. These genes have been fused to form one gene by using mutagenic oligonucleotides to change the termination codon of *luxA* and the initiation codon of *luxB* and to add one base pair to the 26-bp intercistronic region to align translation of *luxA* in frame to *luxB*.[71] Remarkably, the hybrid monomer protein with the N terminus of LuxB fused to the C terminus of LuxA had considerable luciferase activity.

Indicator Gene Fusions

Gene Fusions

Genetic fusions are of two types, gene fusions (also called translational fusions) which encode hybrid proteins and operon fusions (also called transcriptional fusions) which link separate genes into one transcriptional unit. Gene fusions which couple a fragment of the alkaline phosphatase gene *phoA*, missing sequences required for translation and secretion of PhoA into the periplasm, to genes for secreted and membrane proteins

[70] J. H. Devine, G. S. Shadel, and T. O. Baldwin, *Proc. Natl. Acad. Sci. U.S.A.* **86,** 5688 (1989).
[71] A. Escher, D. J. O'Kane, J. Lee, and A. A. Szalay, *Proc. Natl. Acad. Sci. U.S.A.* **86,** 6528 (1989).

can result in formation of hybrid proteins which rescue the secretion defect in the truncated alkaline phosphatase. Alkaline phosphatase is active only if the PhoA portion of the hybrid protein is transported through the cytoplasmic membrane. The exploitation of this property for studying transmembrane protein topology and protein secretion is discussed in another chapter (Silhavy, this volume), but two uses of *phoA* gene fusions are described here. A specialized derivative of Tn*5*, Tn*phoA*, has been constructed for the *in vivo* generation of gene fusions by insertional mutagenesis, and Tn*phoA* was used to isolate a series of *toxR::phoA* mutations with different fusion junctions in the cloned *toxR* gene of *V. cholerae*.[72] Expression of alkaline phosphatase activity by recombinants with different ToxR–PhoA hybrid proteins was governed by the ability of the truncated portion of ToxR to direct secretion of the PhoA portion of the hybrid to the periplasmic face of the cytoplasmic membrane. Alkaline phosphatase activity was related to the position of the fusion point in the *toxR* gene to define the topology of the membrane-spanning domains of the ToxR protein. Further, Tn*phoA* was also used, by a method described earlier,[63] to mutagenize *V. cholerae* and isolate PhoA⁺ insertion mutants, namely, those Km^r strains which hydrolyzed the chromogenic substrate 5-bromo-4-chloro-3-indolyl phosphate (XP). These mutants defined genes encoding secreted and membrane proteins. Those fusion mutants which were dependent on *toxR* for expression, that is, were part of a *toxR* regulon, were subsequently identified by introducing a *toxR* defect into the *phoA* fusion mutants.[73] If inactivation of *toxR* resulted in a PhoA⁻ phenotype the mutation in a particular fusion mutant was in a gene controlled by the ToxR regulatory protein.

Operon Fusions

Operon fusions are useful for studying regulation of gene expression because a phenotype which is convenient to measure is substituted for one which is difficult to measure. The use of a *ctx::lacZ* operon fusion, in which β-galactosidase synthesis was substituted for toxin production, for cloning the *toxR* gene was discussed earlier. The *lacZ* and other indicator genes have also been added to transposons such as Tn*3*, Tn*5*, and mini-Mu so that operon fusion formation and insertional mutagenesis could be accomplished simultaneously *in vivo*. Genes encoding luminescence enzymes from *V. harveyi* and *V. fischeri* are useful indicator genes because light production can be measured simply, sensitively and *in vivo* with minimal disturbance of the bacteria in the environmental context in which they are growing. Five *lux* genes are part of a single operon and encode

---

[72] V. L. Miller, R. K. Taylor, and J. J. Mekalanos, *Cell (Cambridge, Mass.)* **48,** 271 (1987).
[73] K. M. Peterson and J. J. Mekalanos, *Infect. Immun.* **56,** 2822 (1988).

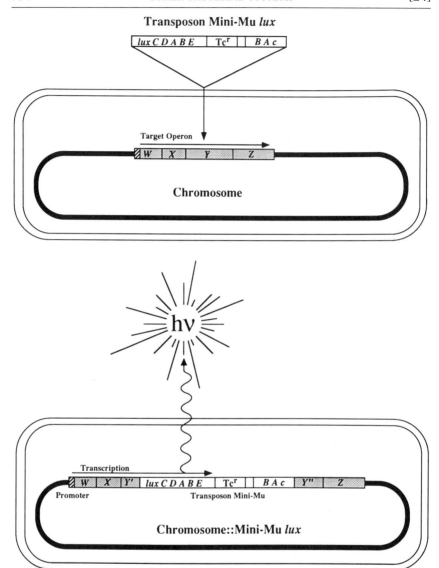

FIG. 4. Constructing *lux* gene fusions *in vivo* with transposon mini-Mu*lux*. Transposition of mini-Mu*lux* results in insertion of *lux* genes into the coding region of target gene Y. Two orientations of insertion are possible, and the orientation of insertion shown aligns transcription of the target gene with that of the *lux* genes. When transcription of gene Y occurs *lux* genes are also transcribed, and light production results.

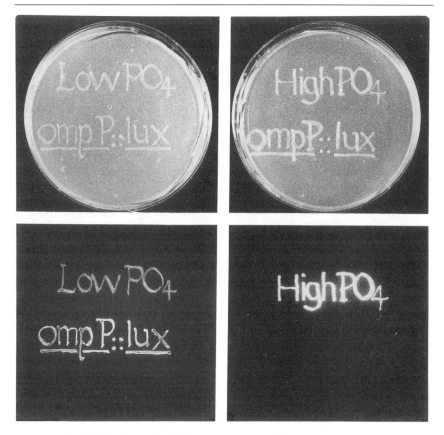

FIG. 5. Physiological control of indicator gene transcription. Two mutants of *V. parahae-molyticus* containing mini-Mu*lux* fusions were grown on high and low phosphate plates. The control mutant, used to write the "High PO₄" and "Low PO₄" designations on each plate, contained a fusion that produced light irrespective of medium composition. Strain LM417, used to write "*ompP::lux*," contained a fusion of the bioluminescence genes to *ompP*, a phosphate-regulated porin gene. The plates at the top were photographed by incident light and those at the bottom by bioluminescence.

the luciferase subunits (*luxA,B*) and the fatty acid reductase subunits (*luxC,D,E*). These are sufficient for light production in bacteria. A fragment containing the five genes, but which was missing the *lux* promoter and regulatory gene region, was positioned at one end of transposon mini-Mu to construct mini-Mu*lux*[74] which in one orientation of insertion aligned *lux* gene transcription with that of the target gene (Fig. 4).

[74] J. Engebrecht, M. Simon, and M. Silverman, *Science* **227**, 1345 (1985).

Mini-Mu*lux* was used to mutagenize *V. parahaemolyticus,* and insertion mutants with a Laf⁻ phenotype, that is, those defective in swarmer cell differentiation, were isolated. Strains with *laf::lux* operon fusions were then analyzed to identify the environmental stimuli which induced differentiation.[22,41] A library of *lux* fusion mutants of *V. parahaemolyticus* was also replica plated onto high and low phosphate media to identify, by visual examination of mutant libraries in a dark room, those strains in which *lux* was fused to phosphate-starvation inducible genes.[13] The effect of phosphate deprivation on expression of luminescence of a mutant of *V. parahaemolyticus* which contains a mini-Mu*lux* insertion in the *ompP* gene is shown in Fig. 5. The *ompP* gene encodes an outer membrane porin protein for polyphosphate diffusion. Fusion strains such as this one define members of regulatory families, in this case the phosphate regulon, and genetic analysis with these fusion strains could be continued to identify the controlling elements of the regulon. For example, a second round of mutagenesis could be applied to search for strains which were Lux⁻ as a result of inactivation of genes required for expression of the target *lux* fusion gene.

Summary

Bacteria of the genus *Vibrio* are remarkably diverse, and until recently the methodology for genetic analysis consisted of a patchwork of different approaches, many of which were narrowly applicable to a single species. The invention of the recombinant DNA technology and the subsequent innovations in transposon mutagenesis and in transductive and conjugative gene transfer techniques have led to the development of very powerful and general strategies for genetic analysis of species of *Vibrio*. The striking synergy of combining recombinant DNA, transposon, and gene transfer methods is particularly evident in the construction of transposons which generate gene fusions and of broad host range plasmids which deliver transposons and mutated genes and which mobilize chromosomes. With such tools it should be possible to perform advanced genetic analysis on the many undomesticated species of *Vibrio* still to be explored.

[25] Genetic Systems for Mycobacteria

By William R. Jacobs, Jr., Ganjam V. Kalpana,
Jeffrey D. Cirillo, Lisa Pascopella, Scott B. Snapper,
Rupa A. Udani, Wilbur Jones, Raúl G. Barletta,
and Barry R. Bloom

Introduction

The ability to perform genetic analyses on bacteria has provided powerful tools and experimental systems to unravel fundamental biological processes. The advances of recombinant DNA technologies have ignited the development of genetic systems for bacteria that are difficult to work with. Clearly, the genus *Mycobacterium* contains a set of the most difficult bacterial species to manipulate experimentally. Despite its discovery over 100 years ago, the leprosy bacillus, *Mycobacterium leprae,* remains unable to be cultivated in the laboratory except in mouse footpads or in the nine-banded armadillo. Although Robert Koch pioneered the work of pure culture with *Mycobacterium tuberculosis,* the ability to extend genetic analyses beyond these initial studies has been greatly hampered by the slow growth of mycobacteria and the stubborn tendency to clump. The tuberculosis vaccine strain, BCG (bacille Calmette Guérin) has been used to vaccinate more individuals than any other live bacterial vaccine, yet little is known about mycobacterial gene structure and expression. The recent development of phage,[1,2] plasmid,[2] and gene replacement[3] systems for the introduction of recombinant DNA into mycobacteria has opened up a new era of research on members of the genus *Mycobacterium.* Indeed, the use of these technologies should pave the way for the elucidation of mycobacterial virulence determinants and a detailed understanding of the mechanisms of drug action and resistance, as well as the development of recombinant multivalent BCG vaccines[1,4] capable of engendering immune responses to a variety of viral, bacterial, or parasitic antigens. The goal of this chapter is to bring together an array of recently developed methods and techniques which enable researchers to genetically manipulate mycobacterial species.

[1] W. R. Jacobs, Jr., M. Tuckman, and B. R. Bloom, *Nature (London)* **327,** 532 (1987).
[2] S. B. Snapper, L. Lugosi, A. Jekkel, R. Melton, T. Kieser, B. R. Bloom, and W. R. Jacobs, Jr., *Proc. Natl. Acad. Sci. U.S.A.* **85,** 6987 (1988).
[3] R. N. Husson, B. E. James, and R. A. Young, *J. Bacteriol.* **172,** 519 (1990).
[4] B. R. Bloom, *J. Immunol.* **10,** i (1986).

Mycobacterial Strains

The mycobacteria fall into two general categories based on their growth rates. The fast-growing mycobacteria, consisting of such strains as *M. smegmatis, M. phlei, M. aurum,* and *M. fortuitum,* have doubling times of 2–3 hr and will yield a colony from a single cell in 3–4 days. In contrast, the slow-growing mycobacteria such as *M. tuberculosis,* BCG, or *M. avium* strains double every 18–26 hr and thus yield colonies from single cells in 14–28 days. Although it is likely that the vectors and methodologies described here are applicable to many mycobacterial species, we have focused our efforts on the fast-growing *M. smegmatis* and the slow-growing BCG. The strains used are (1) *M. smegmatis* strain mc²6, a single colony isolate of the predominant colony type found in the ATCC 607 culture; (2) *M. smegmatis* strain mc²155, an efficient transformation mutant[5] of mc²6; and (3) BCG-Pasteur obtained as a lyophilized pellet from the Statens Serum Institute in Copenhagen, Denmark.

Biohazard Considerations

Biosafety level 3 (BL3) containment facilities and corresponding practices and equipment are appropriate for activities involving the propagation and manipulation of *M. tuberculosis* and *M. bovis;* all other mycobacterial species including BCG and *M. avium* can be safely handled in BL2 containment facilities.[6] Experiments involving the introduction of recombinant mycobacterial DNA into *Escherichia coli* or any mycobacterial host require that the investigator obtain approval from the Institution's Biosafety Committee before the initiation of experiments.[7] The National Institutes of Health Recombinant DNA Advisory Committee (RAC) recommends that experiments involving the introduction of recombinant DNA from a class 3 agent, such as *M. tuberculosis,* into a class 2 agent such as *M. smegmatis* be performed using BL3 containment facilities. The deliberate transfer of a gene conferring drug resistance whose acquisition could compromise the use of a drug to control the pathogen, such as the introduction of isoniazid resistance into *M. tuberculosis,* can only be

[5] S. B. Snapper, R. E. Melton, S. Mustafa, T. Kieser, and W. R. Jacobs, Jr., *Mol. Microbiol.* **4,** 1123 (1990).
[6] U.S. Dept. of Health and Human Services, CDC/NIH. "Biosafety in Microbiological and Biomedical Laboratories." U.S. Government Printing Office, Washington, D.C. HHS Publication No. (CDC) 86-8395 (1986).
[7] U.S. Department of Health and Human Services, *Federal Register* **51,** 16957 (1986).

performed following approval from both the RAC and the Institutional Biosafety Committee.

Growth and Maintenance of Mycobacterial Strains

Media

For Growth of Mycobacteria in Liquid Culture

M–ADC–TW broth: Dissolve 4.7 g of Difco (Detroit, MI) Middlebrook 7H9 broth base and 2 ml glycerol in 900 ml deionized water. Autoclave 20 min. After cooling, add 100 ml albumin–dextrose complex (ADC) enrichment and 2.5 ml of 20% Tween 80 solution. The albumin component of the ADC enrichment is essential for the growth of BCG but is not needed for *M. smegmatis* growth.

ADC enrichment: Dissolve 2 g glucose, 5 g bovine serum albumin fraction V, (Boehringer Mannheim, Indianapolis, IN), and 0.85 g NaCl in 100 ml deionized water. Filter sterilize and store at 4°.

20% Tween 80: Add 20 ml of polyoxyethylene sorbitan monooleate to 80 ml deionized water, heat at 55° for 30 min to dissolve completely, and filter sterilize.

For Colony Titrations, Transformations, and Transductions with Shuttle Phasmids

Middlebrook 7H10 agar: Add 19 g Middlebrook 7H10 agar to 900 ml deionized water, autoclave 20 min, and allow to temper to 55°. Add 100 ml ADC enrichment and pour 40–45 ml per plate. Cycloheximide can be added at a final concentration of 50 μg/ml to inhibit mold contamination. Kanamycin is added to a final concentration of 10 μg/ml for the selection of shuttle phasmid transductants or transformants.

For Propagating Mycobacteriophage and Shuttle Phasmids. It is most important not to have Tween 80 in media used to propagate phage.

DBϕ bottom agar: Add 20 g Dubos oleic acid agar base, 1 g NaCl, and 7.5 g glucose to 1 liter deionized water. Autoclave 20 min, temper to 55°, and add $MgSO_4$ and $CaCl_2$ to final concentrations of 10 and 2 mM, respectively.

Mycobacteriophage top agar: Add 0.5 g Middlebrook 7H9 broth base, 0.1 g NaCl, 0.75 g glucose, and 0.7 g agar to 100 ml deionized water. For BCG, supplement this by adding 1.0 g proteose peptone #3

(Difco), 0.5 g yeast extract (Difco), and 5 ml of a 20% glycerol solution (v/v).

Growth of Mycobacterium smegmatis and BCG

The slow growth of the mycobacteria necessitates that excellent aseptic technique be followed since most contaminants will outgrow mycobacteria, particularly the slow-growing mycobacteria. Since the time required for a single cell to yield a colony might be up to 3 to 6 weeks, which is often essential for genetic analyses, agar plates onto which bacteria cells are plated must be poured thick and wrapped with Parafilm both to prevent desiccation of the media and to reduce possible risks of mold contamination. Tween 80 is an essential component of mycobacterial media that reduces the clumping of mycobacteria and aids in the preparation of single cell suspensions. However, microscopic analyses will usually demonstrate clumps of mycobacterial cells even when the cultures are grown in Tween 80. *Mycobacterium smegmatis* will clump less than most of the other fast-growing mycobacteria and can yield a reasonably satisfactory single cell suspension when grown in M–ADC–TW broth, in baffled flasks, with shaking at 37°. In contrast, BCG will clump considerably if grown like *M. smegmatis*. We have found it best either to grow BCG standing, for phage work, or to grow it in tissue culture roller bottle flasks to yield high levels of reasonably unclumped viable cells.

Mycobacteriophage and Shuttle Phasmids

Historically, mycobacteriophage have been used to type various mycobacterial isolates. Novel phage vectors, termed shuttle phasmids, have been constructed from the mycobacteriophage TM4 and L1.[1,2] Shuttle phasmids[8] can replicate in mycobacteria as phage capable of undergoing lysis or lysogeny. These vectors can also replicate in *E. coli* as cosmids and, thus, be packaged into bacteriophage λ heads to facilitate subsequent cloning of additional genes.[8]

Preparation of High-Titer Lysates of Mycobacteriophage and Shuttle Phasmids

1. Prepare *M. smegmatis* mc^{2}6 cells for a phage infection by inoculating 0.1 ml of the starter culture into 25 ml of M–ADC–TW broth in a 250-ml baffled flask. Shake the flask at 100 rpm on a platform

[8] W. R. Jacobs, Jr., S. B. Snapper, M. Tuckman, and B. R. Bloom, *Rev. Infect. Dis.* **11**, S404 (1989).

shaker at 37° overnight. Although mycobacteriophage infections are inhibited by Tween 80, overnight cultures of *M. smegmatis* contain Tween hydrolase that destroys most of the added Tween. However, the addition of Tween 80 is still necessary to yield an unclumped lawn of mycobacterial cells. We have found it best for BCG phage lawns to use 10-day-old standing cultures of BCG grown at 37° from a 1 : 1000 dilution of an exponential BCG culture.

2. Dilute the phage or shuttle phasmid, 10-fold serially, in MP buffer (10 mM Tris, pH 7.6, 100 mM NaCl, 10 mM MgSO$_4$, 2 mM CaCl$_2$) to obtain approximately 5×10^4 to 10^5 plaque forming units (pfu) per milliliter.

3. Mix 2.0 ml of mc^{2}6 cells ($\sim 3 \times 10^8$ cells/ml) with 1.0 ml of the phage or shuttle phasmid diluted to 5×10^4 pfu/ml in a sterile 13×100 mm culture tube. Incubate at 37° for 30 min to allow adsorption to occur.

4. Pipette 0.3 ml of the phage–cell mixture to a culture tube containing 3.0 ml of DBϕ top agar from *M. smegmatis* or DBϕ supplemented top agar for BCG, mix the phage–cell suspension with the top agar by gently rotating the tube in the palms of your hands, and aseptically pour on top of a DBϕ plate. Allow the agar to solidify. Invert the plate and incubate at 37° for 24–36 hr for *M. smegmatis* or 7–10 days for BCG. For each lysate, we normally prepare five plates.

5. Pipette 5 ml of MP buffer on each plate. Set plates upright at 4° overnight.

6. Pipette off as much liquid as possible (usually ~ 3 ml) from each plate lysate and combine all of the lysate fluids in a 50-ml polypropylene tube. Centrifuge the lysates at 4° at 3000 g for 10 min. Filter the supernatant fluid through a 0.45-μm filter unit. Transfer the filtered phage lysate preparation to a sterile phage bottle. Store at 4°. *Note:* Do not add chloroform to mycobacteriophage lysates as most mycobacteriophage contain lipids and are inactivated by nonpolar solvents.

Phage Purification and Isolation of DNA

Mycobacteriophage can be easily purified and concentrated on CsCl density gradients as described for *E. coli* phage.[9] The procedures described here have given high yields of phage particles from mycobacteriophage D29, L1, TM4, and shuttle phasmid derivatives. DNA can be efficiently

---

[9] R. W. Davis, D. Botstein, and J. R. Roth, "Advanced Bacterial Genetics." Cold Spring Harbor Laboratory, Cold Spring Harbor, New York, 1980.

isolated from the purified phage particles following proteinase K treatment, phenol–chloroform extractions, and ethanol precipitations.

Transfection of Phage DNA by Electroporation

1. Transfect phage DNA by electroporation. A detailed protocol for electroporation of mycobacteria is described below. Basically, the same procedure is carried out for transfection with phage DNA. The following considerations should be taken into account for transfection protocols: to maximize the yield of plaques, it is essential to dilute the cell suspension after electroporation, and incubate further for 1 hr at 37°. The dilution is important to reduce the level of harmful chemicals (radicals) generated during electroporation, and the subsequent incubation allows the mycobacterial cells to recover from the electric shock.

2. Mix 0.1 ml of an appropriate dilution of the electroporation mixture with 0.2 ml of a freshly grown culture (logarithmic to early stationary culture) of mycobacterial cells. Then add 3 ml of soft agar and plate. Usually, transfection frequencies around 10^5 pfu/μg of phage DNA are obtained.

Construction of Shuttle Phasmids

Shuttle phasmids are chimeric constructs between a mycobacteriophage and an *E. coli* plasmid, in which the plasmid is inserted in a nonessential region of the phage genome. Usually, an *E. coli* cosmid is used as a plasmid, since the possibility of performing *in vitro* packaging greatly facilitates the construction of recombinant structures. The steps involved in making such constructs are as follows:

1. Ligate mycobacteriophage DNA at high DNA concentrations ($\geq$100 μg/ml, usually between 250 and 500 μg/ml).

2. Digest the ligated mycobacteriophage DNA partially with a frequent cutter (e.g., *Sau*3AI), so that the majority of the fragments obtained are in the range of 44 kilobases (kb) (if the cosmid used is ~6 kb, like pHC79). At this stage it is possible to fractionate the 44-kb fragments if desired.

3. Linearize the cosmid vector DNA with a compatible restriction enzyme (usually the corresponding hexamer, e.g., *Bam*HI); treat with alkaline phosphatase if desired.

4. Ligate the linearized cosmid vector with the digested mycobacteriophage DNA from Step 2 at high DNA concentrations (250–500 μg/ml) so that concatemeric DNA is obtained, with intermingled fragments of mycobacteriophage and cosmid DNA.

5. *In vitro* package, using commercially available packaging mixes.

The ligated DNA is packaged in λ particles by virtue of the *cos* sites present in the cosmid vector. At this step a transducing phage lysate containing the recombinant DNA is obtained.

6. With the phage lysate transduce an appropriate *E. coli* host containing mutations in the *hsdR*, *mcrA*, and *mcrB* genes, such as ER1451 (New England Biolabs, Beverly, MA). Select transductants containing recombinant cosmid DNA by plating on appropriate antibiotic plates (L agar plus 25 μg/ml of ampicillin for pHC79).

7. Pool transductants and isolate plasmid DNA by standard procedures. The plasmids isolated from this pool represent an insertion library of the cosmid vector in the mycobacteriophage genome. Only those plasmids with insertions within a dispensable region of the mycobacteriophage genome will give rise to shuttle phasmids.

8. Electroporate an appropriate mycobacterial host (mc^{2}6 or mc^{2}155 for L1 and TM4), using the pool of recombinant plasmids, as previously described.

9. Screen the mycobacteriophage plaques obtained using the cosmid vector as probe. Positive hybridizing plaques represent true shuttle phasmids. At this stage, we have observed with the construction of TM4-derived shuttle phasmids that almost all the resulting plaques have lost the insertion of the cosmid DNA (1 positive for every 400 plaques), presumably the result of a recombination event between multiple recombinant cosmid::phage molecules entering a cell and recombining. Shuttle phasmids can be isolated, ligated, *in vitro* packaged into bacteriophage λ heads, and transduced into *E. coli* cells where they replicate as cosmids conferring antibiotic resistance.

Cloning Genes in Shuttle Phasmids

Unique cloning sites within the cosmid vector can be useful for cloning foreign DNA, provided that they are absent from the phage genome. For example, both TM4 and L1 lack *Eco*RI and *Hin*dIII sites within the phage genome, and therefore the unique *Eco*RI site or *Hin*dIII sites in pHC79 can be used to introduce additional foreign DNA. The size of foreign DNA that can be cloned in a shuttle phasmid is constrained by the size requirements of the mycobacteriophage packaging machinery, as well as by the cosmid packaging mechanism. In this way, additional genes, 2 to 4 kb in size, can be efficiently cloned into shuttle phasmids. The procedure can be summarized as follows:

1. Ligate shuttle phasmid DNA so that long concatemers are obtained.

2. Digest shuttle phasmid DNA with an appropriate restriction enzyme (e.g., *Eco*RI for L1 shuttle phasmids) to yield large DNA vector

molecules that can be ligated to a gene of choice with compatible restriction enzyme ends.

3. After ligation, *in vitro* package, and proceed as described above for the construction of shuttle phasmids. By using the temperate shuttle phasmid phAE15,[2] derived from mycobacteriophage L1, one can select or screen for mycobacterial colonies carrying the recombinant DNA with the gene of interest.

Electroporation of *Mycobacterium smegmatis* and BCG

The highest efficiencies for transformation of mycobacteria have been obtained using electroporation. We have found that mycobacteria can withstand very high voltages for extended periods of time. Thus, a high-resistance electroporation buffer such as 10% glycerol is used along with a high parallel resistance. This method results in long time constants, which have been shown to give the highest efficiencies of transformation. The protocol which we use is similar to that used for *E. coli*.[10]

1. Inoculate 1 liter of M–ADC–TW broth with 10 ml of a 10- to 15-day culture of BCG (A_{600} = 0.5–1.0). In the case of *M. smegmatis*, 1 liter of M–ADC–TW broth can be directly inoculated from a frozen stock or colony directly.
2. Incubate at 37° until the A_{600} reaches 0.5–1.0. For BCG this period varies greatly, from 7 to 25 days, depending on the passage of the culture and growth conditions; *M. smegmatis*, however, is almost always at the correct stage of growth by 48 hr.
3. Harvest in 250-ml centrifuge bottles. Centrifuge for 10 min at 10,000 rpm and 4°. Discard supernatant.
4. Resuspend each BCG pellet in 5 ml of ice-cold 10% glycerol, pool them into two 15-ml polypropylene conical tubes, and centrifuge for 10 min at 3000 rpm and 4°. *Mycobacterium smegmatis*, however, should be directly suspended in the centrifuge bottles in 250 ml of 10% glycerol and centrifuged as before.
5. Resuspend the BCG pellets in 10 ml cold 10% glycerol. At this point, *M. smegmatis* pellets may be resuspended in 10 ml cold 10% glycerol, pooled into two 50-ml polypropylene conical tubes, and the volume raised to 50 ml with 10% glycerol. Centrifuge for 10 min at 3000 rpm and 4°.
6. Repeat the wash in Step 5.
7. Resuspend each pellet to a final volume of 1 ml in 10% glycerol.

[10] W. J. Dower, J. F. Miller, and C. W. Ragsdale, *Nucleic Acids Res.* **16**, 6127 (1988).

The protocol can be continued with the freshly prepared cells as shown below, or, at this step, the cells can be aliquoted into Eppendorf tubes and frozen quickly in dry ice–ethanol to be used later by thawing slowly at room temperature (sometimes the thawed cells will contain salts and cause arcing; this can be prevented by washing the cells in the Eppendorf tubes with 10% glycerol once or twice).

8. Place DNA (5 pg–5 μg; 5 μl volume, maximum) and 50 μl of cells in an Eppendorf tube and mix well by pipetting up and down. Place on ice 1 min.

9. Set the voltage of the pulser (Bio-Rad, Richmond, CA) to 2500 V, 25 μF, and the pulse controller (Bio-Rad) to 1000 ohms. Sometimes the cells will arc with a high parallel resistance; washing the cells further with 10% glycerol will usually prevent this. If DNA containing high salt is added, either a lower resistance in the pulse controller must be used or the DNA must be ethanol precipitated and washed with 70% (v/v) ethanol at least once. If a lower resistance is used, lower transformation frequencies will result; therefore, ethanol precipitation is recommended.

10. Transfer the solution containing the cells and DNA into a cuvette with a 0.2-cm electrode gap (Bio-Rad). Tap the cuvette against the bench several times to get the cells to the bottom of the cuvette and to remove as many bubbles as possible.

11. Place the cuvette in the pulser and expose to one pulse (time constants are usually between 15.0 and 25.0 msec).

12. For BCG electroporations add 0.4 ml of M–ADC–TW medium directly to the cuvette and suspend the cells well in it. For *M. smegmatis* add 1 ml of M–ADC–TW broth, suspend the cells in it, and transfer to a round-bottomed 15-ml polypropylene tube.

13. Incubate at 37° approximately 3 hr to allow antibiotic expression. Plate the cells on selective medium.

Both the *M. smegmatis* and BCG electroporation procedures are extremely sensitive to the concentration of salts present in the DNA sample used. The DNA used should be washed with 70% ethanol if there is any possibility of the presence of excess salt. Also, it is important to add as small a volume of DNA as possible to prevent dilution of the bacterial cell concentration in the cuvette. Ligations must be ethanol precipitated and washed with 70% ethanol before they can be used (it is possible to use small amounts of a ligation, but significantly lower transformation efficiencies will result). After the last wash, if the packed cell volume is not very close to or is more than 1 ml, do not raise the volume to 1 ml final, just add enough 10% glycerol to allow the cells to be pipetted easily. Fresh cells in our hands give transformation efficiencies above 10^6 transformants/ μg (using a 10-kb plasmid), and frozen cells give efficiencies which vary from 10^4 to 10^6 transformants/μg.

Chromosomal DNA Isolation and Plasmid Isolation
   from Mycobacterial Cells

*Isolation of High Molecular Weight Mycobacterial Chromosomal DNA
   Using Mini-Beadbeater Cell Disruptor*

1. Grow 100 mi of culture of *M. smegmatis* or BCG cells to approximately 3×10^8 cells/ml. Normally, we use overnight *M. smegmatis* cultures started from a 0.1-ml starter culture or 7- to 9-day-old BCG cultures that have been grown with shaking. Large numbers of freshly grown cells are critical to obtaining good yields of high molecular weight DNA.
2. Pellet the cells by centrifuging in 50-ml polypropylene tubes in a Sorvall RT6000 centrifuge at 3000 rpm for 10 min at 4°.
3. Carefully, pour off the supernatant fluid and resuspend the pellet in 0.5 ml of homogenization buffer (0.3 M Tris, pH 8, 0.1 M NaCl–6 mM EDTA).
4. Transfer the resuspended cells to conical 2-ml screw-cap vial which is one-fourth to one-half filled with 0.5-mm acid-washed sterile glass beads.
5. Place the vial on the Mini-Beadbeater cell disruptor (Biospec Products, Bartlesville, OK) and beat it for 1 min.
6. Transfer the entire contents of the tube to a 15-ml conical polypropylene tube. Wash the 2-ml Beadbeater vial with 1 ml of homogenization buffer and add this to the disrupted cell suspension.
7. Extract with an equal volume of phenol–chloroform (1 : 1) solution. Use Tris-buffered neutralized phenol. Repeat the extraction one more time. After each extraction, spin the tubes in a Sorvall RT6000 (Dupont, Wilmington, DE) centrifuge at 3000 rpm for 10 min.
8. Peform one chloroform–isoamyl (24–1) extraction and then ethanol precipitated with 1/10 volume of 3 M sodium acetate, pH 5.0, and 2 volumes of cold ethanol.
9. Place the tubes at $-70°$ for 30 min or $-20°$ overnight for ethanol precipitation.
10. Spin the tubes for 10–15 min at 3000 rpm in Sorvall RT6000 centrifuge.

11. Pour off the ethanol and wash the pellet with 2 ml of cold 70% ethanol.
12. Spin the precipitate again for 5 min in a Sorvall RT6000 centrifuge at 3000 rpm.
13. Pour off the ethanol and dry the pellet.
14. Resuspend the pellet in 1 ml of TE (10 mM Tris–1 mM EDTA) buffer. Do not shake or vortex vigorously. Let the tube sit for a few minutes and then just gently mix up and down 2–3 times. If the pellet still does not dissolve, add more TE.
15. Digest RNA by adding RNase A at 100 μg/ml using 10 mg/ml boiled stock. Incubate at 37° for 30 min.
16. Extract with an equal volume of phenol–chloroform twice as before.
17. Extract with an equal volume of chloroform–isoamyl alcohol once.
18. Ethanol precipitate with 1/10 volume of 3 M sodium acetate and 2 volumes of ethanol.
19. Pellet the DNA, pour off the supernatant fluid, dry, and resuspend the pellet in 1 ml TE.

Plasmid DNA Isolation from Mycobacterial Cells

We have adapted the alkaline lysis protocol[11] with the following modifications:

1. Grow 5 to 10 ml of the mycobacterial cell culture to approximately saturation (overnight for the fast growers, several days for slow growers).
2. Harvest the cells by centrifugation, resuspend the pellet in 150 μl of GTE (25 mM Tris-HCl, pH 8.0, 10 mM EDTA, 50 mM glucose) containing 10 mg/ml of lysozyme, and incubate at 37° overnight.
3. Add 2 volumes (300 μl) of fresh 0.2 M NaOH–1% sodium dodecyl sulfate (SDS) and mix by inversion. Incubate at 4° for 10 min.
4. Add 1.5 volumes (225 μl) of 5 M potassium acetate and mix by inversion. Incubate at 4° for 10 min.
5. Centrifuge in microcentrifuge for 30 min to 1 hr. This step is critical to obtain a good plasmid preparation.
6. Remove and extract the supernatant with an equal volume of CPI (chloroform–phenol–isoamyl alcohol, 24:25:1).
7. Transfer the aqueous phase to a clean tube and add 2 volumes (1200 μl) of room temperature ethanol, incubate for 5 min at room temperature, spin down the DNA pellet, wash with 70% ethanol, and dry briefly under reduced pressure.

[11] H. Birnboim and J. Doly, *Nucleic Acids Res.* **7,** 1513 (1979).

8. Resuspend in 25 to 50 μl of TE buffer, then incubate at 65° for 15 min.

Insertional Mutagenesis

A method of generating random chromosomal mutations by inserting a piece of DNA containing a selectable marker gene is essential for the mycobacteria since it permits one to select mutants arising from a single mutagenic event directly, even from a pool of clumped cells. Insertional mutagenesis using transposons has been a valuable method for mutational analysis. In the absence of a mycobacterial transposon that transposes at high frequency we have developed an indirect method of random mutagenesis, a random shuttle mutagenesis.[12] This method was developed by extending the strategy of targeted mutagenesis (Fig. 1). Shuttle mutagenesis of a targeted gene involves three steps: cloning of the desired gene from a given organism into a vector that can replicate in *E. coli* but not in the organism of interest; introduction into the cloned gene of a transposon which expresses a marker that can be selected for in the organism of interest; and return of the insertionally mutagenized gene to the chromosome of the organism of interest by homologous recombination. We chose Tn*5 seq1*[13] for performing shuttle mutagenesis because it (1) endoces the *neo* gene that confers kanamycin resistance to both *E. coli* and mycobacteria, (2) permits one to select for insertions into DNA sequences cloned into plasmid vectors using its neomycin hyperresistance phenotype,[14] (3) transposes preferentially into DNA sequences of high guanine plus cytosine content, (4) facilitates sequencing of the gene into which it inserts owing to the presence of T7 and SP6 primer binding sites, and (5) lacks the cryptic gene encoding streptomycin resistance of Tn*5*, an important biohazard consideration for *M. tuberculosis* strains.

Selection for Transpositions into Cloned Genes: Neomycin Hyperresistance Selection

1. Transform any cloned mycobacterial gene into ec²270 (*E. coli* strain χ2338::Tn*5 seq1*) and select for colonies resistant to kanamycin and another antibiotic A (the genetic marker present on the plasmid cloning vector).
2. Inoculate individual colonies into LB (Luria broth) medium adjusted

---

[12] G. V. Kalpana, B. R. Bloom, and W. R. Jacobs, Jr., *Proc. Natl. Acad. Sci. U.S.A.* **88,** 5433 (1991).
[13] D. K. Nag, H. V. Huang, and D. E. Berg, *Gene* **64,** 135 (1988).
[14] D. E. Berg, M. A. Schmandt, and J. B. Lowe, *Genetics* **105,** 813 (1983).

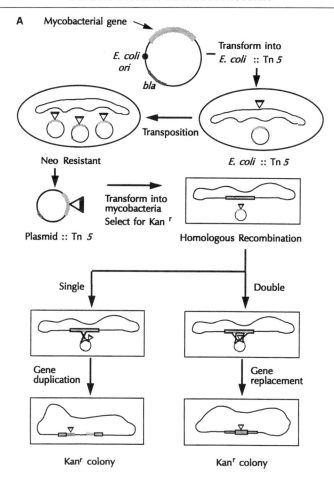

FIG. 1. (A) Targeted shuttle mutagenesis of a mycobacterial gene using Tn5 seq1. (B) Random shuttle mutagenesis of mycobacterial genes.

to pH 7.2 and containing antibiotic A but no kanamycin and incubate overnight.

3. Plate dilutions of overnight cultures onto LB agar plates containing neomycin at 250 μg/ml plates and incubate for 24 hr.

4. Isolate plasmids from colonies that have grown on the neomycin plates. Retransform an *E. coli* cloning strain selecting for resistance to both antibiotic A and kanamycin to enrich for plasmids containing Tn5 seq1.

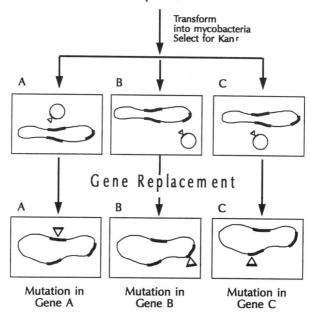

Library of mycobacterial DNA in an
E. coli plasmid vector

FIG. 1. (*continued*).

5. Map the position of Tn*5 seq1* insertion by standard restriction analyses.

Random Shuttle Mutagenesis of Mycobacteria

The method described above can be used to isolate Tn*5 seq1* insertions into any cloned DNA fragment. The description below provides a method to obtain random insertional mutants of *M. smegmatis*. This can be achieved by mutagenizing genomic library cloned mycobacterial DNA fragments in *E. coli* and then reintroducing these mutagenized DNA fragments into the chromosome of *M. smegmatis*.

1. In order to minimize the possibility of obtaining transposon insertions in the vector portion of recombinant mycobacterial clones a

derivative of pBR322,[15] pYUB36,[12] was constructed in which the nonessential 1.9-kb *Eco*Rv to *Pvu*II fragment has been deleted.

2. Chromosomal DNA of *M. smegmatis* strain, mc²6, was prepared as described above and partially digested with *Msp*I.

3. *Msp*I DNA inserts of 4 to 7 kb were isolated by electroelution.

4. The size-selected inserts were ligated to *Cla*I (a unique site)-digested pYUB36.

5. The ligated DNAs were electroporated into ec²270 and plated on L agar containing both ampicillin and kanamycin at 40 and 50 μg/ml, respectively.

6. About 30,000 individual kan^r and Amp^r transformants were pooled and samples diluted 10^{-3} into 20 independent 5-ml L broth containing ampicillin (no kanamycin) and incubated at 37° overnight.

7. A 200 μl sample from each culture yielded approximately 1000 neomycin hyperresistant colonies on L agar containing 250 μg/ml neomycin, which selects for colonies resulting from transposition Tn*5 seq1* into plasmids. Plasmid DNA from the combined pool of neomycin-resistant colonies was retransformed into χ2338.

8. After 1 hr incubation, the transformants were directly inoculated into 1-liter L broth containing kanamycin and ampicillin, grown to saturation at 37°, and plasmid DNA was prepared.

9. This Tn*5 seq1*-mutagenized plasmid library was then electroporated into *M. smegmatis* strain, mc²6, and kanamycin-resistant transformants were selected on K agar [Middlebrook 7H10 agar supplemented with 5 mg/ml casamino acids (Difco), 100 μg/ml diaminopimelic acid, 50 μg/ml thymidine, 40 μg/ml uracil, and 133 μg/ml adenosine].

10. About 800 individual *M. smegmatis* transformants were screened for auxotrophy by streaking onto modified minimal Sauton medium[15a] without asparagine and this procedure yielded three auxotrophs.[12]

Construction of Mycobacterial Genomic Libraries in Shuttle Cosmids

Shuttle cosmids have been constructed that contain an origin of replication that functions in *E. coli*, an origin of replication that functions in mycobacteria, a kanamycin-resistance gene that functions in both *E. coli* and mycobacteria, bacteriophage λ *cos* sequence that permits these mole-

[15] F. Bolivar, *et al.*, *Gene* **2**, 95 (1977).
[15a] L. G. Wayne and G. A. Diaz, *J. Bacteriol.* **93**, 1374 (1967).

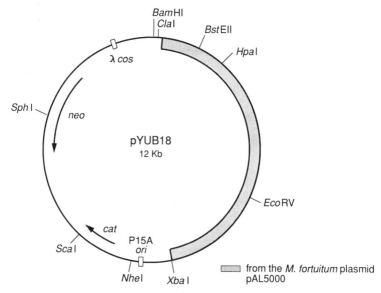

FIG. 2. *Escherichia coli*/mycobacteria shuttle cosmid.

cules to be packaged into bacteriophage λ heads, and unique restriction sites for construction of the genomic libraries. We chose to make mycobacterial genomic libraries in shuttle cosmid vector for the following reasons: the large insert size allows for the genome to be represented in as few as 100 clones, the large insert size provides the possibility of cloning sets of genes responsible for biosynthesis of complex polysaccharides or lipids, and the libraries can be stored stably as phage lysates following *in vivo* cosmid packaging.

The vector pYUB18 was constructed by inserting the *Bcl*I–*Bgl*II fragment of bacteriophage λ into pYUB12[5] (Fig. 2). This vector retains a unique *Bam*HI site to be used for ligating *Sau*3A-digested chromosomal DNAs. In order to prevent the formation of unstable cosmids resulting from the ligation of multiple cosmid vectors within a larger recombinant cosmid, we have found it best to both gel purify large insert fragments for cloning and to treat the vector with alkaline phosphatase prior to ligation with the insert. Following *in vitro* packaging, the constructed libraries are transduced into cosmid *in vivo* packaging strains to permit amplification and efficient repackaging of recombinant cosmids into bacteriophage λ heads thus allowing for storage of the libraries as phage lysates.

1. Digest CsCl-purified pYUB18 DNA with *Bam*H1, phenol–chloro-

form extract, and ethanol precipitate, and resuspend the DNA pellet in a volume of TE which gives a DNA concentration of 2 $\mu g/\mu l$.

2. Treat the vector with alkaline phosphatase as follows: add 45 μg *Bam*H1-cut pYUB18 DNA to each of three tubes containing phosphatase buffer (Boehringer Mannheim) and add three different concentrations of calf intestinal alkaline phosphatase (CIAP) (Boehringer Mannheim) to final concentrations of 0.016, 0.008, and 0.004 units CIAP/μg pYUB18. Incubate for 2 hr at 37°. The efficiencies of the phosphatase reactions should be tested by setting up self-ligations at a concentration of 10 ng/μl of the three different CIAP reactions and of nonphosphatased *Bam*H1-cut pYUB18. Then 10 ng amounts of the ligations should be used to transform *E. coli,* and the number of transformants from each phosphatased ligation should be compared to the number of transformants from 10 ng of the unphosphatased *Bam*H1-cut pYUB18. Use the phosphatased pYUB18 which gives approximately 1% self-ligation compared to the unphosphatased control.

3. Set up atrial *Sau*3A digests of approximately 300 μg of mycobacterial high molecular weight chromosomal DNA in five tubes, the first of which contains 100 μg and the other four tubes which contain 50 μg each. Dilute *Sau*3A from New England Biolabs from its original concentration of 5 to 2.5 units/μl in 1X medium salt buffer. Add 1 μl of the diluted enzyme to the first tube of chromosome such that the final concentration of *Sau*3A is 0.025 units/μg DNA. Transfer 50% of the volume of tube 1 into tube 2 and mix such that the concentration of *Sau*3A in tube 2 is 0.0125 units/μg, and then serially transfer 50% of each tube until the last transfer is made into tube 5. Incubate the reactions at 37° for 30 min. The partial digestion reaction should be stopped by adding EDTA to a final concentration of 20 mM to each tube.

4. Load 2 μl of each reaction tube onto a 0.4% agarose gel next to λ DNA cut with either *Apa*I which gives two bands, 38.4 and 10 kb, or *Xho*I which gives two bands, 33.5 and 15 kb, to determine which partial digest gave the most 35–45 kb pieces of chromosomal DNA. Usually, reaction tubes 2 and 3 give the optimal DNA sizes. Pool the two tubes which give the most 35–45 kb DNA fragments, and load onto a 0.4% preparative agarose gel containing ethidium bromide. The gel should be made up in autoclaved 0.5X TBE buffer to reduce the possibility of nuclease contamination. Run the gel overnight at 20 V. Photograph and then cut out the gel slice that

contains DNA sized at 33.5 kb and larger, and place into a 2.5-cm wide dialysis bag containing 0.5–1.0 ml 0.5X autoclaved TBE buffer.

5. Place the dialysis bag in the gel chamber such that the gel slice is pushed against the side of the bag closest to the cathode, and turn the voltage to 100 V to electroelute the DNA from the gel slice to the opposite side of the dialysis bag. Use a handheld UV lamp to monitor the progress of the ethidium-bromide-stained-DNA out of the gel slice. Once the DNA migrates to the opposite side of the dialysis bag, switch the positive and negative electrodes and run the voltage at 100 V for only a minute or two until the DNA just migrates away from the side of the dialysis tubing and enters the TBE solution in the bag.

6. Pour the TBE and DNA solution out of the tubing and wash the tubing with 0.5X TBE to retrieve any solution that may have lingered. Gently phenol–chloroform extract and ethanol precipitate the DNA. Approximately 30 μg DNA, with an average size of 40 kb, should be recovered.

7. Ligate the 40-kb chromosomal pieces with Sau3A ends to the BamH1-digested, phosphatased pYUB18. Ligations should have a 1 : 1 molar ratio of insert : vector, so add three times as much insert DNA as vector; 250 ng of BamH1-cut phosphatased pYUB18 should be added to 750 ng Sau3A-cut 40-kb pieces of mycobacterial chromosome in a 20 μl ligation reaction whose final concentration is >50 ng/μl DNA. Add 400 units of T4 DNA ligase to the ligation reaction, incubate at 16° overnight, or at room temperature for 3 hr.

8. *In vitro* package 4 μl of the completed ligation reaction using Giga Pack Plus from Stratagene.

9. Transduce the *in vitro* packaged mycobacterial cosmid library lysate into the *in vivo* packaging strains of *E. coli, χ2764*, or χ2819.[16] To five separate tubes, add 0.1 ml of the lysate to 0.2 ml of saturated *E. coli* culture, incubate standing at 30° for 25 min, add 0.7 ml L broth, incubate shaking at 30° for 1 hr, and plate 0.2 ml per plate of L agar containing 25 μg/ml kanamycin. Incubate at 30°. The *in vitro* packaged lysate should yield 5,000–10,000 transductants. Representative recombinant cosmids from a pYUB18: *M. tuberculosis* DNA genomic library are shown in Fig. 3.

10. To amplify the cosmid library, the transductants should be pooled,

[16] W. R. Jacobs, J. F. Barrett, J. E. Clark-Curtiss, and R. Curtiss III, *Infect. Immun.* **52**, 101 (1986).

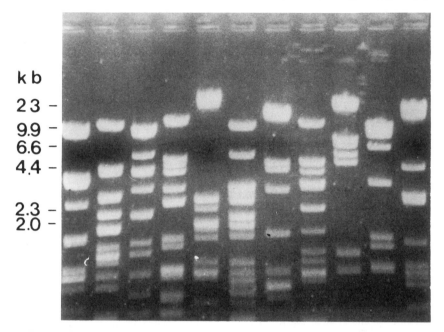

FIG. 3. Plasmid DNA isolated from *E. coli* containing random PYUB18:: *M. tuberculosis* DNA recombinant cosmids was digested with *Pst*I and run on an agarose gel strained with ethidium bromide.

and inoculated into L broth containing 25 μg/ml kanamycin to grow for *in vivo* packaging. *In vivo* packaging is performed as described.[16] Library lysates made in this manner yield titers of approximately 10^{10} transducing particles per milliliter and can be stored at 4° indefinitely.

11. To prepare the plasmid form of mycobacterial cosmid library DNA for electroporation of mycobacterial strains, the *in vivo* packaged lysate should be diluted in phage buffer and transduced into an *E. coli* cloning host to yield a minimum of 1000 kanamycin-resistant transductants. This library should be pooled, grown in liquid media containing kanamycin,[17] and plasmid DNA isolated by standard alkaline lysis methods. This library of plasmid DNA can be efficiently electroporated into *M. smegmatis* or other myobacteria using protocols described above.

[26] Genetic Manipulation of Pathogenic Streptococci

By MICHAEL G. CAPARON and JUNE R. SCOTT

Introduction

The streptococci are a diverse class of gram-positive bacterial species containing many human pathogens. Suppurative diseases, including pharyngitis ("strep throat"), scarlet fever, cellulitis of the skin, impetigo, erysipelas, and the recently recognized "toxic shock-like syndrome,"[1] are associated with infection by *Streptococcus pyogenes* (Lancefield group A) which is also responsible for the serious nonsuppurative diseases of acute glomerulonephritis and rheumatic fever. Members of the viridans group (*Streptococcus mutans, Streptococcus salivarius,* and *Streptococcus sanguis*) inhabit the oral cavity and are a frequent cause of infective endocarditis. A normal inhabitant of the gut, *Enterococcus* (formerly *Streptococcus*) *faecalis* (group D), is also associated with cases of endocarditis and is commonly involved in urinary tract infections. *Streptococcus mutans* plays a central role in the development of dental plaque and dental caries, and *Streptococcus agalactiae* (group B) has become a very important cause of neonatal meningitis. *Streptococcus pneumoniae* remains one of the major causes of morbidity and mortality among debilitated individuals despite the widespread use of antibiotics, and it is an important cause of otitis media in infants.

Current therapies for streptococcal diseases are almost exclusively based on the use of antibiotics. However, resistance to antibiotics is becoming widespread among streptococcal species, and isolates with multiple resistances are not uncommon. Resistances have been described to kanamycin, chloramphenicol, MLS antibiotics (macrolides, lincosamides, and streptogramin), streptomycin, gentamicin, neomycin, tetracycline, and penicillins (for review, see Refs. 2 and 3). The genes encoding these resistances are often present not just on conjugative plasmids, but also on broad host range conjugative transposons (see below), which may

[1] L. A. Cone, D. R. Woodward, P. M. Schievert, and G. S. Tomory, *N. Engl. J. Med.* **317**, 146 (1987).

[2] D. B. Clewell and C. Gawron-Burke, *Annu. Rev. Microbiol.* **40**, 635 (1986).

[3] D. B. Clewell, *Microbiol. Rev.* **45**, 409 (1981).

contribute to the rapid disemination of resistance among different species.[4] To combat the increasing antibiotic resistance of the streptococci, alternative therapies must be investigated.

No effective vaccines exist for streptococcal diseases, with the possible exception of the pneumococcal capsule vaccine. Work on vaccine development and on alternative therapies will require the identification and characterization of streptococcal virulence factors at a molecular level. The techniques described in this chapter provide powerful tools for this analysis.

Because so many serious illnesses are caused by the streptococci, much effort has been directed at understanding the mechanisms by which these organisms cause disease. As for most pathogens, it is likely that streptococcal virulence is multifactorial. No single determinant is sufficient to cause the organism to be pathogenic; many determinants are necessary. For this reason, the genetic approach of ablating individual potential virulence determinants and evaluating the resulting organism for virulence is probably the most powerful method for determining the contribution of each to the ability of the streptococci to cause disease. This method involves subjecting the streptococcal species under investigation to transposon mutagenesis to generate mutants defective in only a single gene. The mutants are then compared to their isogenic parent in an assay for virulence to determine the contribution of the gene to pathogenesis. The gene is cloned and analyzed, often in *Escherichia coli,* and returned to the streptococcus of origin for complementation analyses. If the gene encodes a diffusible factor that contributes to the virulence of the organism, expression of the gene should restore virulence to the mutant. Further complementation analyses can then be employed for studies on the structure and function of the gene and gene product.

Techniques for transposon mutagenesis and complementation analysis requires methods for the introduction of foreign DNA into the streptococcal species under study. Methods for DNA transfer are currently quite limited in these organisms. Like other bacteria, streptococci can undergo transduction, transformation, and conjugation. Transduction is usually of limited value as a means of DNA transfer, primarily because of the narrow host range of most phage. Conjugative plasmids are available for most streptococcal species and in certain instances can be an effective means of DNA transfer. In one approach (see section "Transformation of an Intermediate Host"), a naturally competent streptococcus containing a

[4] T. Horaud, C. Le Bouguenec, and G. De Cespedes, *in* "Streptococcal Genetics" (J. J. Ferretti and R. Curtiss, eds.), p. 74. American Society for Microbiology, Washington, D.C., (1987).

mobilizing plasmid is transformed with the DNA of interest, and conjugation is used to transfer the DNA to the final streptococcal host. There is a report of conjugational transfer from *E. coli* to streptococci,[5] but this technique has not yet been generally adopted.

At this time, the most practical technique for introduction of foreign DNA into streptococci is transformation. Although the discovery of genetic transformation in *S. pneumoniae* played a central role in the development of modern molecular genetics,[6] the use of transformation has been limited. Among the pathogens in this genus, only the pneumococcus, a single strain of *S. sanguis,* and some strains of *S. mutans* are naturally competent to take up foreign DNA. The recent introduction of electroporation for transformation of streptococci (see below) appears to be of general use in this genus and will potentially allow transformation of species that are not naturally competent. However, we have experienced limitations to this technique using some strains and some plasmid DNA.

Instead of attempting to detail all the methods available for genetic manipulation of each organism in this diverse genus, this chapter focuses on the techniques we have used in the analysis of virulence determinants in the group A streptococcus (*S. pyogenes*). Many of these methods should be easily adaptable for use in other streptococcal species.

Transposon Mutagenesis

The isolation of mutants defective in certain traits is central to genetic analysis. For the study of virulence in pathogenic bacteria, this involves generating and detecting mutants that are defective in phenotypes associated with virulence. The use of transposons for insertional mutagenesis is a powerful tool to obtain these mutants. The transposon is introduced into the bacterium by a process that results in random insertions throughout the genome of the recipient, and then mutant candidates are screened for the absence of the phenotype to be studied. The virulence of the mutant isolated this way is then compared with that of its isogenic nonmutagenized parent in an animal model. Meaningful information can then be obtained on the contribution of a specific gene to the pathogenicity of the bacterium.

The transposon Tn917 is similar in structure to Tn3 and is the prototype of a conventional type of transposon that is active in gram-positive bacteria. Tn917, which was originally described in group D streptococci,[7] has been used successfully for mutagenesis in gram-positive species, primarily

[5] P. Trieu-Cuot, C. Carlier, P. Martin, and P. Courvalin, *FEMS Microbiol. Lett.* **48,** 298 (1987).

[6] O. T. Avery, C. M. MacLeod, and M. McCarty, *J. Exp. Med.* **79,** 137 (1944).

[7] P. Tomich and D. Clewell, *J. Bacteriol.* **141,** 1366 (1980).

Bacillus subtilis. Youngman and co-workers have developed several cleverly designed derivatives of Tn917 and delivery systems for them that can be used in *B. subtilis.*[8] Although Tn917 and the vectors derived by Youngman *et al.* have been used to generate insertions in group D organisms,[9,10] Tn917 has not been employed widely for mutagenesis in pathogenic streptococci. However, development of electroporation techniques for the transformation of streptococcal strains that are not naturally competent (see section Introduction of DNA into Streptococci: Methods Based on Electroporation) may make the use of Tn917 practical in this genus in the near future.

The other conventional types of transposons available for gram-negative organisms cannot be used to mutagenize the streptococci because they do not appear to transpose in gram-positive bacteria. Furthermore, as mentioned above, well-defined delivery systems for transposon introduction are not available for most streptococci. Instead, extensive use has been made of a different class of elements known as conjugative transposons.

Conjugative Transposons

Conjugative transposons, represented by Tn916 and Tn1545, were first identified in the streptococci and cause the gram-positive hosts in which they reside to act as conjugative donors (reviewed by Clewell and Gawron-Burke[2]). During the conjugation event, the transposon is transferred from its location in the donor to a new location in the recipient. Donor and recipient do not have to be of the same species or even the same genus, and donors include species not previously thought to be able to conjugate (e.g., *Bacillus subtilis*[11]).

Tn916 and Tn1545 have been used to generate mutations affecting capsule synthesis in *S. agalactiae*[12] and hemolysins in group A,[13] group B,[14] and group D organisms.[15] Mutations have also been generated in the C5a peptidase of group A organisms,[16] the cAMP factor of a group B

[8] P. Youngman, P. Zuber, J. B. Perkins, K. Sandman, M. Igo, and R. Losick, *Science* **228,** 285 (1985).

[9] K. E. Weaver and D. B. Clewell, *J. Bacteriol.* **170,** 4343 (1988).

[10] E. R. Krah and F. L. Macrina, *J. Bacteriol.* **171,** 6005 (1989).

[11] J. R. Scott, P. A. Kirchman, and M. G. Caparon, *Proc. Natl. Acad. Sci. U.S.A.* **85,** 4809 (1988).

[12] C. E. Reubens, M. R. Wessels, L. M. Heggen, and D. L. Kasper, *Proc. Natl. Acad. Sci. U.S.A.* **84,** 7208 (1987).

[13] K. Nida and P. P. Cleary, *J. Bacteriol.* **155,** 1156 (1983).

[14] J. N. Weiser and C. E. Reubens, *Infect. Immun.* **55,** 2314 (1987).

[15] Y. Ike, H. Hashimoto, and D. B. Clewell, *Infect. Immun.* **45,** 528 (1984).

[16] S. P. O'Connor and P. P. Cleary, *J. Infect. Dis.* **156,** 495 (1987).

strain,[17] and a bacteriocin of *S. mutans*.[18] Mutagenesis with Tn*916* has been used to identify a locus (*mry*) required for expression of the M protein in group A streptococci,[19] and "shuttle" transposons, which capitalize on the conjugative abilities of Tn*916*, have been developed for the analysis and generation of specific deletions in cloned genes[20] (see below).

Conjugative Transfer of Tn916 into Streptococci

A Tn*916*-containing donor and an antibiotic-resistant recipient are required for Tn*916* mutagenesis. Any gram-positive host containing Tn*916* potentially can act as a donor of the transposon. The frequency with which any particular strain donates can vary over a 4-fold range, probably depending on the specific location of Tn*916* in the genome of the donor[21] and on the number of copies of Tn*916* present there. Thus, it is advisable to test several donors to obtain the highest possible efficiency of conjugation. The group D streptococcal strain CG110[21] has been used widely as a donor and can donate Tn*916* to *S. pyogenes* at a frequency of approximately 10^{-6} per recipient colony-forming unit (cfu).[19] A *B. subtilis* strain (MEN202) is now used to construct and donate altered derivatives of Tn*916*[20] (see below).

Tetracycline is used to select for transfer of Tn*916*, which carries *tetM*. If the recipient is resistant to tetracycline, as is often encountered with group B strains, it is possible to use derivatives of Tn*916* that have been constructed *in vitro* to encode resistance to erythromycin (Tn*916*ΔE^{22}) or kanamycin (Tn*916*-J2[22a]). To allow selection against the donor, the recipient must be resistant to an appropriate different antibiotic. Spontaneous streptomycin-resistant derivatives of *S. pyogenes* strains to be used as recipients can easily be isolated[23] by plating at least 10^9 cells per plate containing streptomycin (1000 μg/ml).

Transfer of Tn*916* requires that donor and recipient be brought into close contact. This is accomplished by collecting the cells on a membrane

[17] S. K. Hollingshead, P. W. Caufield, and D. G. Prichard, *in* "Abstracts of the Annual Meeting," p. 35. American Society for Microbiology, Washington, D.C., 1989.

[18] C. Caufield, B. Robinson, G. Shaw, and S. K. Hollingshead, *in* "Abstracts of the Annual Meeting," p. 35. American Society for Microbiology, Washington, D.C., 1989.

[19] M. G. Caparon and J. R. Scott, *Proc. Natl. Acad. Sci. U.S.A.* **84**, 8677 (1987).

[20] M. Norgren, M. G. Caparon, and J. R. Scott, *Infect. Immun.* **57**, 3846 (1989).

[21] C. Gawron-Burke and D. B. Clewell, *Nature (London)* **300**, 281 (1982).

[22] C. E. Reubens and L. M. Heggen, *Plasmid* **20**, 137 (1988).

[22a] M. G. Caparon, M. Norgren, and J. R. Scott, manuscript in preparation.

[23] J. R. Scott, P. C. Guenthner, L. M. Malone, and V. A. Fischetti, *J. Exp. Med.* **164**, 1641 (1986).

filter²⁴ or by plating donor and recipient together on the surface of an agar plate as follows.

Plate Mating to Introduce Tn916 into Streptococcus pyogenes

Culture Medium

THY broth: 15 g Todd–Hewett medium (Difco, Detroit, MI), 1 g yeast extract, 500 ml water

1. Overnight cultures of donor and recipient are grown with appropriate antibiotics in Todd–Hewitt medium supplemented with 0.2% yeast extract (THY). When *S. pyogenes* is the recipient in a cross with a *B. subtilis* or *E. faecalis* donor, it is generally necessary to concentrate the overnight culture 100-fold to obtain an adequate number of bacteria. (We grow 100-ml overnight cultures at 37° without aeration.) The *B. subtilis* donor is cultured with tetracycline at 10 µg/ml and requires vigorous aeration.

2. Cells are collected by centrifugation (8000 g, 10 min, 15°) and resuspended in THY broth. The *S. pyogenes* culture is usually resupended to a total volume of 1 ml (to concentrate them 100-fold), whereas the *E. faecalis* or *B. subtilis* cultures are resuspended to their original volumes. Hyaluronidase (Sigma, St. Louis, MO, 50 µl of a 34 mg/ml solution) can be added to the *S. pyogenes* suspension to remove the hyaluronic acid capsule.

3. Equal volumes of donor and recipient are then added to the same sterile test tube. It is advisable, at this point, to assay the number of viable donors and recipients by colony count. Dilutions are made in normal saline containing 18 mM CaCl₂. We try to obtain a ratio of at least 10 recipient cfu per donor cfu. Since streptococci grow in chains, precise quantitation of viable cells requires brief sonication to obtain single cells. For *S. pyogenes*, a 15-sec treatment of ice-cold cells at a power setting of 5 using a Heat Systems W-220 sonicator with a microtip (Farmingdale, NY) is usually sufficient. The absence of chains is confirmed by microscopy.

4. The test tube containing the mixture of donor and recipient is gently mixed and 0.2-ml aliquots spotted in as small an area as possible on the surface of several fresh (<24 hr old) THY plates. For some *S. pyogenes* strains, the plates must be supplemented with 2% defibrinated sheep's blood. As controls, 0.1 ml of a suspension of donors alone and of recipients alone are spotted onto separate THY plates.

5. After incubation overnight at 37°, 2 ml of THY broth is added to each plate, and the bacterial growth is removed as completely as possible

---

²⁴ A. E. Franke and D. B. Clewell, *J. Bacteriol.* **145,** 494 (1981).

by resuspension using a glass rod or pipette. The suspension is drawn off with a pipette and placed in a sterile test tube. Aliquots of the mating mixture are then plated on selective medium which contains tetracycline at 5 μg/ml and streptomycin at 1000 μg/ml for an *S. pyogenes* recipient and an *E. faecalis* or *B. subtilis* donor. Each control plate ("mock mating") is treated in a similar fashion. The number of surviving donor and recipient colony-forming units is determined by plating appropriate dilutions of the mating mixture on media to select only against the recipient (tetracycline only) and only against the donor (streptomycin only). The frequency of transfer is calculated as the number of streptomycin–tetracycline-resistant transconjugants divided by the number of surviving streptomycin-resistant recipients. Transfer frequency is reported per recipient cfu, but the best results are obtained from matings in which the number of surviving donors and recipients is about equal.

Analysis of Tn916 Insertion Mutants Using Southern Blots

Introduction

Organisms with Tn916 insertions are assayed for the absence of the virulence phenotype in question. When a potential insertion mutant is obtained, it is necessary to determine the number and location of the Tn916 element(s). This can be accomplished by Southern blot analysis using a Tn916-specific probe.

Preparation of DNA from Streptococci

In general, the cell walls of the streptococci are resistant to digestion by lysozyme (*E. faecalis* is a notable exception[25]). For some species, efficient digestion may be effected by other enzymes, including mutanolysin for *S. agalactiae*[26] and S. mutants[26,27] and phage lysin for group A streptococci.[28] The susceptibility of most streptococcal species to lysozyme can be enhanced by growth of the cells in glycine (or threonine), which presumably acts by inhibiting the degree of cross-linking in the cell wall.[29] Lysozyme-treated cells are then lysed by treatment with a detergent, usually sodium dodecyl sulfate (SDS). The following method

[25] D. Clewell, Y. Yagi, G. Dunny, and S. Schultz, *J. Bacteriol.* **117,** 283 (1974).
[26] K. Yokogawa, S. Kawata, S. Nishimura, Y. Ikeda, and Y. Yoshimura, *Antimicrob. Agents Chemother.* **6,** 156 (1974).
[27] M. C. Hudson and R. Curtiss III, *Infect. Immun.* **58,** 464 (1990).
[28] V. A. Fischetti, E. C. Gotschlich, and A. W. Bernheimer, *J. Exp. Med.* **133,** 1105 (1971).
[29] S. E. Coleman, I. Van De Rijn, and A. S. Bleiweis, *Infect. Immun.* **2,** 563 (1970).

adapted from Chassy[30] has been successful for the preparation of DNA from *S. pyogenes*.

Preparation of DNA from Streptococcus pyogenes

1. Grow a 100-ml overnight culture at 37° in THY broth supplemented with 20 mM glycine. (The concentration of glycine may need to be adjusted for other group A strains or other streptococcal species).

2. Harvest the cells by centrifugation (10,000–15,000 g, 10 min, 4°), and wash once in 10 ml of Tris buffer (20 mM Tris-HCl, pH 8.2). Resuspend the final pellet in 3.2 ml of the same Tris buffer.

3. Add 7 ml of polyethylene glycol solution (PEG 20,000, Fisher Scientific, Fairlawn, NJ, 24%, w/v, in water, sterilize by filtration). Mix gently and add 3.5 ml of lysozyme solution (Sigma L-6876, 20 mg/ml in water).

4. Incubate for 1 hr at 37° with vigorous shaking, then centrifuge (15,000 g, 10 min, 4°). Carefully pipette off and discard all supernatant fluid. Resuspend the pellet in 5.7 ml TE (10 mM Tris-HCl, pH 8.0, 1 mM EDTA) and add 0.3 ml of a 20% (w/v) aqueous solution of SDS. Incubate for 15 min at 65°.

5. Add 0.2 ml RNase A solution (Sigma R-4875, 5 mg/ml in water) and incubate for 30 min at 37°.

6. Add 0.2 ml of proteinase K solution (Sigma P-0390, 10 mg/ml in water). Continue incubation at 37° for 30 min.

7. Extract the lysate once with TE-saturated phenol, once with phenol–chloroform (1 : 1) and once with chloroform.

8. Allow residual chloroform to evaporate and add 1/10 the aqueous volume of 3 M sodium acetate. Mix and add 2 volumes of 95% ethanol. Mix gently, and a visible precipitate should appear.

9. Collect the DNA by centrifugation (5000 g, 4°, 5 min) and wash it twice with 5 ml of 70% (v/v) ethanol.

10. Dry the pellet under reduced pressure and resuspend it in 1 ml of TE.

Digestion and Transfer of DNA

Chromosomal DNA is digested with the appropriate restriction endonuclease, subjected to agarose gel electrophoresis, and transferred to a membrane by standard methods.[31,32] Because Tn916 is large [16.4 kilobases (kb)], it is advisable to select agarose gel conditions that optimize the

[30] B. M. Chassy, *Biochem. Biophys. Res. Commun.* **68,** 603 (1976).
[31] E. M. Southern, *J. Mol. Biol.* **98,** 503 (1975).
[32] S. Meinkoth and G. Wahl, *Anal. Biochem.* **138,** 267 (1984).

separation of large DNA fragments (i.e., 0.4% agarose). To obtain transposon–chromosome junction fragments of manageable size, it is useful to digest the chromosomal DNA with a restriction enzyme that cuts only once in the transposon, for example, HindIII or SstI.[2]

Preparation of Tn916 Probe

A common source of a Tn916-specific probe is the E. coli plasmid pAM120, which contains a copy of Tn916 cloned from the E. faecalis plasmid pAM211.[33] Because the vector has no homology with streptococcal strains tested so far, it is usually acceptable to use the entire plasmid as a probe without purification of a specific fragment. To confirm that this is appropriate for any particular streptococcal strain, a negative control should be performed with DNA from the unmutagenized recipient prior to transposon insertion. Plasmid pAM120 is easily labeled with ^{32}P by the random priming method of Feinberg and Vogelstein (see below).[34] Owing to the high specific activities obtained by this method, it is usually not necessary to remove the unincorporated nucleotides prior to using the probe in hybridization reactions.

Radiolabeling of Probe

Stock Reagents

TM: 250 mM Tris-HCl, pH 8.0, 25 mM MgCl$_2$, 50 mM 2-mercaptoethanol

DTM: 100 μM dCTP, 100 μM dGTP, 100 μM dTTP

OL: 1 mM Tris-HCl, pH 7.5, 1 mM EDTA, 90 optical density units per milliliter of random oligodeoxynucleotides (Pharmacia, Piscataway, NJ, 27-2166-01, a mixture of random hexadeoxynucleotides produced by pancreatic DNase degradation of calf thymus DNA)

Stop solution: 20 mM NaCl, 20 mM Tris-HCl, pH 7.5, 2 mM EDTA, 0.25% SDS

Labeling mixture: Mix 100 μl DTM (see above), 28 μl OL (see above), and 100 μl of 1 M HEPES, pH 6.6. Prepare aliquots of 20 μl each; store at $-20°$ until needed.

Labeling Reaction

1. If plasmid DNA is used, it should first be treated with a restriction enzyme which digests the DNA into multiple linear fragments. For pAM120, the restriction enzymes Sau3A, HincII, and HpaII can be

---

[33] C. Gawron-Burke and D. B. Clewell, J. Bacteriol. **159**, 214 (1984).
[34] A. P. Feinberg and B. Vogelstein, Anal. Biochem. **132**, 6 (1983).

used.[2,33] If the final concentration of DNA in the reaction mixture is between 70 and 100 ng/μl, it can be labeled directly from the restriction reaction without further purification.

2. Add approximately 70 ng of DNA (i.e., 1 μl of the above digest) to a microcentrifuge tube and denature it by heating for 2 min at 95°–100°. Immediately cool the tube on ice for 5 min.

3. To the denatured DNA add 11.4 μl labeling mixture, 1 μl bovine serum albumin (BSA, 10 mg/ml stock solution, nucleic acid grade), and sterile distilled water to bring the volume to a 25 μl total. Add 5 μl (50 μCi) of [α-^{32}P]dATP (10 mCi/ml, 3000 Ci/mmol). Add 2.5 units of the Klenow fragment of DNA polymerase I, mix gently, and incubate for 4 hr at room temperature.

4. Calculate the incorporation of ^{32}P as described below, then add 200 μl of stop solution. Store at 4°.

Monitoring Incorporation

1. Cut a thin strip of PEI-cellulose thin-layer chromatography (TLC) paper (J. T. Baker, Phillipsburg, NJ, 4-4473) with the dimensions 0.5 by 4 cm.

2. Carefully spot 0.5 μl of the labeling reaction approximately 3 mm from one end of the TLC strip (it helps to mark this end of the strip with a pencil for later identification).

3. Incubate the strip under a heat lamp until the spot is completely dry.

4. Place the strip (with the spot at the bottom) into a 7-ml scintillation vial containing 200 μl of 2 M HCl to begin chromatography.

5. Allow chromatography to continue until the solvent front is about 1–3 mm from the top of the strip.

6. Remove the strip and place it under a heat lamp until completely dry.

7. Cut the dried strip into thirds, place each section into a separate scintillation vial and quantitate the radioactivity of each by liquid scintillation counting.

8. Unincorporated nucleotides will migrate with the solvent front, while nucleotides polymerized into double-stranded DNA will remain at the origin. Usually greater than 90% of the label is incorporated into double-stranded DNA by this method. Calculate the percentage of incorporation using the following formula:

$$\% \text{ Incorporation} = \frac{\text{cpm of bottom segment}}{\text{total cpm (bottom + middle + top)}} \times 100$$

Determining Hybridization Conditions for Streptococcal DNA

Hybridization reactions involving streptococcal DNA must take into account the relatively high A and T content of their genomes (the genome of *S. pyogenes* contains 60% A + T[35]). This high A + T composition means that even stringent hybridization must be carried out under conditions that would not necessarily be considered stringent for an organism that has a lower A + T content in its DNA, like *E. coli*. The following equation can be used to approximate hybridization conditions when pAM120 is used as a probe:

$$T_m = 81.5°C + 16.6 \log M + 0.41(G + C) - 0.61(\% \text{ formamide})$$

where T_m is the melting temperature of the DNA duplex and M, the ionic strength in moles/liter.

Stringency is most easily controlled by allowing the reaction to occur at a temperature several degrees below the calculated T_m at a given ionic strength. The closer the reaction temperature is to the calculated T_m, the more stringent the conditions. If the desired reaction temperature is too high to be practically obtainable, formamide (a helix-destabilizing agent) can be included according to the above equation, to reduce the effective T_m. For a more detailed discussion of factors that affect hybridization, see Meinkoth and Wahl.[32]

Hybridization and Washing of Southern Blots

Reagents

20× SSC (3.0 *M* NaCl, 0.3 *M* sodium citrate, pH 7.0)

50× Denhardt's solution: Prepare 5 g Ficoll type 400 (Sigma F4375), 5 g BSA (fraction V), and 5 g polyvinylpyrrolidone (Sigma PVP360) and dissolve in 500 ml distilled water. Filter sterilize, dispense into 5-ml aliquots, and store at −20°. Use within 1 week of thawing and do not refreeze.

Hybridization solution: 2× SSC (see above), 5× Denhardt's solution (see above), 100 μg/ml yeast tRNA, 0.1% SDS, and 10 m*M* EDTA

Hybridization

1. Place the filter containing DNA in a sealable plastic bag (e.g. Dazey Seal-a-Meal). Add 7 ml of hybridization solution to the bag, carefully remove any bubbles, and seal.

2. Incubate the sealed bag for 2 hr at the temperature at which the hybridization reaction will be carried out (normally 37°–42°).

---

[35] J. Marmur, S. Falkow, and M. Mandel, *Annu. Rev. Microbiol.* **17**, 329 (1963).

3. Calculate the volume of probe solution needed for hybridization. For a 15 by 15 cm filter, approximately 1×10^7 incorporated counts/min (cpm) is required. Be sure to use only the data for incorporated cpm (i.e., the bottom segment of the TLC strip from above) to make this calculation, not the total cpm of the sample.

4. Place the aliquot of the radiolabeled probe in a microcentrifuge tube and denature the DNA by heating at 95°–100° for 2 min. Immediately place on ice for 5 min.

5. Add the denatured probe to 2 ml of hybridization solution.

6. Carefully cut open a corner of the sealed bag containing the filter and, using a disposable plastic pipette, add the solution containing the probe. Gently remove all bubbles and reseal the corner of the bag.

7. Incubate the reaction overnight at the desired temperature.

Washing. The stringency at which the filter is washed is the most important factor in determining the final stringency of the reaction. Thus, the hybridization step is generally carried out under less stringent conditions than the washes.

1. When pAM120 is used as a probe for Tn916, the first set of washes are carried out under conditions of relatively low stringency. The filter is usually washed 5 times in a solution containing $0.2 \times$ SSC and 0.1% SDS. Each wash is for 15 min and is at 42°. This represents a stringency of approximately 30° below the T_m of pAM120 under these conditions.

2. The filter is not dried after the first set of washings. Instead, the wet filter is sealed in a plastic bag and subjected to autoradiography. If the background on the resulting autoradiograph is minimal, no further washing is required. If background is unacceptable or if washing at a higher stringency is required, the blot can be rewashed.

Analysis of Southern Blots

When DNA is restricted with enzymes that recognize a single site in Tn916 (e.g., HindIII or SstI, see Fig. 1), the Tn916-specific probe should hybridize to two fragments for each copy of Tn916 present. These represent the junction of the transposon and the target to the left of the insertion and the junction of the transposon and the target to the right of the insertion. The size of the junction fragments will vary depending on the location of the adjacent restriction sites in the target chromosome. If HindIII is used, one junction should be a minimum of 5.7 kb in size and the other should be at least 10.7 kb owing to the location of the HindIII site in Tn916 (see Figs. 1 and 2).

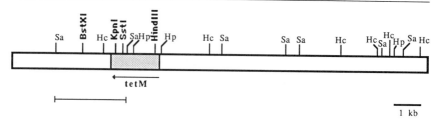

Fig. 1. Restriction map of Tn916. The location of *tetM* is represented by the shaded box and its direction of transcription by the arrow under the box. The thin bar under the transposon delineates a 2.8-kb *Sau*3A fragment used in construction of pVIT vectors (see below). The restriction enzyme sites shown in bold indicate enzymes which cut only once in Tn916. Other restriction enzyme sites: Hc, *Hinc*II; Sa, *Sau*3A; Hp, *Hpa*II.

Use of Transduction to Move Mutation to Unmutagenized Host

Southern blot analysis often reveals multiple Tn916 insertions in the prospective mutant strain.[19] To be sure that the mutant to be studied results from insertion and not spontaneous random mutation, and to determine which of the insertions give rise to the mutant phenotype, it is of critical importance to establish linkage between the mutant phenotype and a particular transposon insertion. In group A streptococci, this can be accomplished by transduction using phage A25.

Transduction of Tn916 Insertions in Streptococcus pyogenes

Media

No. 1 broth[36]: Prepare 30 g proteose peptone #3 (Difco), 1.5 g NaCl, and 0.35 g Na_2HPO_4. Add 500 ml of water and autoclave. Filter sterilize the following components, and add just prior to use (for 500 ml of media): 2.5 ml of 10% dextrose (autoclave to sterilize), 5.0 ml of 0.5 M $CaCl_2$ (autoclave to sterilize), 5.0 ml horse serum (heat inactivate at 56° for 30 min, filter sterilize), and 0.2 ml hyaluronidase (34 mg/ml).

Z6 soft agar[37]: Prepare 20 g proteose peptone #3, 1 g yeast extract, 3 g NaCl, 0.35 g Na_2HPO_4, and 3.5 g agar. Add 500 ml water. Divide into 100-ml aliquots and autoclave.

Wannamaker plates: Prepare 20 g proteose peptone #3, 1 g yeast extract, 4.4 g agar, and 1 g Tris (base). Add 500 ml water and autoclave. Just prior to pouring plates add 10 ml 10% dextrose, 2.5 ml 0.5 M $CaCl_2$, and 1 ml hyaluronidase (34 mg/ml).

[36] L. W. Wannamaker, S. Almquist, and S. Skjold, *J. Exp. Med.* **137**, 1338 (1973).
[37] P. P. Cleary and J. Johnson, *Infect. Immun.* **16**, 280 (1977).

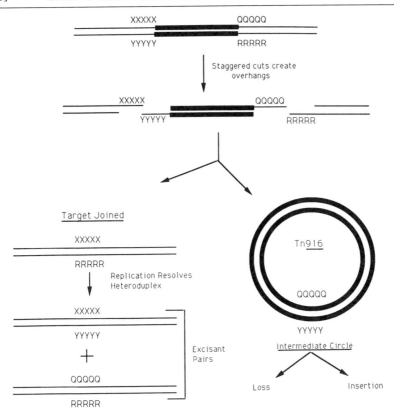

FIG. 2. A model for excision of Tn916. Thick lines represent Tn916, and thin lines represent the DNA adjacent to the transposon. The coupling sequences are indicated by the hypothetical nucleotide pairs X–Y and Q–R. A staggered cleavage of the phosphodiester backbone on the 3' side of the coupling sequence on both strands (first line) generates molecules with 3' single-stranded ends (second line). The target and transposon sequences are joined by their 3' single-stranded regions to generate an excisant molecule and the transposon circle. The transposon circle can serve efficiently as an intermediate in transposition when introduced into an appropriate gram-positive host, but it is usually lost from the cell when present in *E. coli*.[11] Because there is no requirement for homology between the two coupling sequences, both the excisant and the transposon circle contain heteroduplexes when the coupling sequences are joined. Semiconservative replication resolves the heteroduplex present in the excisant. As a result, a pair of molecules (excisant pairs) is generated, one of which has the left coupling sequence at the target site. [Reprinted from "Excision and insertion of Tn916 involves a novel mechanism of recombination," M. G. Caparon and J. R. Scott, *Cell (Cambridge, Mass.)* **59,** 1027 (1989).]

Preparation of Lysates of Phage A25

1. Grow an overnight culture of the potential mutant in No. 1 broth at 30°. The Ca^{2+} in the medium will inhibit tetracycline; however, it is not usually necessary to add antibiotics at this point.

2. Dilute the overnight culture 1/20 in No. 1 broth without Ca^{2+} and grow to mid-log phase (A_{600} 0.35 to 0.45) at 30°.

3. To 10 ml of log-phase cells, add phage A25[38] to a final concentration of 1×10^8 phage/ml. Place the mixture in a 125-ml Erlenmeyer flask and incubate with gentle shaking at 30° for 2–4 hr.

4. Add 100 μl $CHCl_3$, gently swirl for 5 min at room temperature, then centrifuge (5000 g, 10 min, 4°). Save the supernatant in a sterile glass screw-cap tube.

Titration of Lysate

1. Completely melt Z6 soft agar and cool to 47°.

2. Prepare appropriate serial dilutions of the lysate using saline with 18 mM $CaCl_2$.

3. To 18 × 35 mm test tubes add 0.1 ml of lysate dilution and 0.1 ml of indicator streptococci (concentrate an overnight culture grown in No 1. broth 10-fold). Use a group A strain sensitive to phage A25 such as K56.[39] Incubate for 15 min at 37°.

4. Add 2.5 ml of Z6 soft agar and pour onto the surface of a Wanna-maker plate.

5. Incubate overnight at 37° (do not invert plates).

6. Count plaques. The usual yield for a lysate grown for 2 hr is 10^7 to 10^8 pfu/ml, and for one grown 4 hr a titer of 10^8–10^9 pfu/ml can be anticipated.

Transduction. The antibiotic-sensitive parent of the mutagenized strain can serve as the transduction recipient so that transductants can be distinguished from the original Tn916 strain used to generate the transducing lysate.

1. Grow the strain to be transduced to late log phase (A_{600} 0.8) in No. 1 broth.

2. Add 5×10^8 pfu of phage A25 grown on the donor strain as described above to 5 ml of late log-phase recipient cells. Incubate at 30° for 30 min. A negative control consists of 5 ml of recipient culture to which no phage have been added.

---

[38] H. Malke, *in* "Streptococci and Streptococcal Diseases" (L. W. Wannamaker and J. M. Matsen, eds.), p. 119. Academic Press, New York, 1972.

[39] E. Kjems, *Acta Pathol. Scand.* **42**, 456 (1958).

3. Centrifuge (5000 g, 10 min, 15°) to remove free phage and resuspend the pellet in 5 ml of THY broth.

4. Incubate at 30° for 2 hr, then centrifuge as above and resuspend pellet in 1 ml THY broth.

5. Plate 0.1-ml aliquots on each of 10 THY plates containing tetracycline at 5 μg/ml to select for transductants that have acquired the Tn916 insertion. Colonies should appear after 3 to 4 days. Tetracycline-resistant transductants are usually obtained at a frequency of approximately 10^{-7} per plaque-forming unit.

Analysis of Transductants

Tetracycline-resistant transductants are assayed for the appropriate mutant phenotype, and the location of the Tn916 insertion is determined using Southern blots for comparison to the original strain. Transposition of Tn916 to a new locus following transduction has not been observed,[19] indicating that recombination between the chromosomal sequences that flank the inserted transposon and the homologous regions of the genome of the recipient occurs much more frequently than transposition to a new location. Thus, the transductants will have the transposon inserted at the same locus as in the original strain. If the mutation is caused by insertion of Tn916 and did not arise spontaneously, there should be a correlation between the presence of the transposon in a particular locus and the mutant phenotype. Similarly, if the original strain had multiple Tn916 insertions, the insertion responsible for the mutation of interest can be determined by this approach.

Cloning Tn916-Inactivated Allele

The cloning of the Tn916-inactivated allele is simplified by the direct selectability of the tetM gene of Tn916 in E. coli.[33] However, because of the large size of Tn916 (16.4 kb), a vector that can accommodate the insertion of a large DNA fragment must be used. Cosmids, which can accommodate the insertion of fragments up to 40 kb, are the vectors of choice.[40]

To choose an appropriate restriction enzyme for digestion of the mutant chromosomal DNA to be cloned, Southern blot analyses should first be performed. Because streptococcal DNA has such a low G + C content compared to E. coli, enzymes commonly used for genetic engineering in the latter, including BamHI, EcoRI, and PstI, recognize few sites in the

[40] J. Collins and H. J. Bruning, Gene 4, 85 (1978).

former. If one of these three enzymes cuts the cosmid at a unique site, it can be used for digestion of the mutant chromosomal DNA. Neither *Eco*RI, *Bam*HI, not *Pst*I cuts within Tn916,[2] so the transposon insertion and surrounding chromosome can be cloned on one contiguous fragment. If the Southern blot analysis reveals that the transposon has inserted into a fragment equal to or less than 40 kb, a complete digest with the chosen enzyme can be used for cosmid cloning. This method has the advantage of avoiding any rearrangements of the transposon and adjacent target DNA that occurs by ligation of partial digestion fragments to each other in arrangements different from that in the chromosome. If the fragment containing the transposon is much smaller than 40 kb, it is likely that some rearrangement will occur caused by the inclusion of additional fragments from the chromosome. This results from the size selection of λ packaging for molecules of approximately 40 kb. However, the entire transposon and the adjacent target DNA will still be present on one contiguous fragment.

If no enzyme can be identified that can generate a single fragment suitable for cloning, a partial digestion must be used. In this case, the possibility that the cloned DNA contains rearranged fragments must be carefully ruled out. It is important to check that the fragment inserted into the vector contains contiguous DNA fragments in the correct order. This is done by using Southern blot analyses to compare the cloned insert in the chimeric plasmid with the chromosomal DNA from which it is presumed to have been derived. The cloned DNA is used as a probe, and the fragment sizes derived from total chromosomal DNA are compared with those from the cloned piece. Several restriction enzymes that digest the DNA into very small fragments should be used in this analysis. It is not necessary to know the order of the small fragments as long as those in the chromosome that hybridize with the cloned DNA are of the same size as those of the cloned DNA. The more fragments produced in this analysis, the more sensitive the test is and the more likely it is to detect small rearrangements.

Obtaining Wild-Type Allele

Once the transposon-inactivated form of the potential virulence gene has been cloned, further studies on its role in pathogenesis will require analysis of the wild-type allele. As will be discussed, excision of Tn916 from the cloned transposon-inactivated allele can sometimes be used to obtain the wild type. However, because of the unique mechanism of recombination used by Tn916 in excision (see below), only some of the excision events regenerate the wild-type allele.

Excision of Tn916

Tn916 excises at high frequency from most plasmids in *E. coli* by a process that generates a nonreplicating circular form of the transposon[11] and rejoins the ends of the plasmid molecule.[2,11] The rejoined plasmid, which no longer contains Tn916 or confers resistance to tetracycline, is known as an excisant. Because the circular transposon molecule is not a replicon, it is lost from the *E. coli* population and the bacteria become sensitive to tetracycline. In the absence of tetracycline selection, over 85% of the cells in an overnight culture of *E. coli* usually contain excisants.[19,33] However, the circular form can efficiently serve as an intermediate in transposition when introduced into an appropriate gram-positive host.[11]

Excision and Insertion of Tn916 Involve Novel Mechanism of Recombination

Excision of Tn916 involves a novel mechanism of recombination in which noncomplementary bases that flank the inserted element are paired to form a heteroduplex at both the joint of the circular form and at the target site in the excisant molecule.[41] Following the introduction of staggered nicks (as indicated in Fig. 2), the noncomplementary bases of the 5-base pair (bp) "coupling regions" that flank the inserted transposon pair with each other. This generates the heteroduplex structure at the joint of the transposon circle and at the target site in the excisant.[41] Replication of the excisant resolves the heteroduplex in this molecule and produces two daughter excisants that differ in sequence at the site of excision: one contains the sequence from the left coupling region, and the other contains the sequence from the right coupling region.

To understand the derivation of these coupling sequences, the transposon insertion step was investigated more closely. Insertion proceeds by a reversal of the excision process (Fig. 3). The bases present at the joint of the transposon circle pair with noncomplementary bases of the target in a heteroduplex. This means that one of the two coupling sequences that flank the inserted transposon is derived from the new target while the other is introduced with the transposon circle and is derived from the previous target. Thus, of the two types of excisants generated by excision of the transposon, one will have the wild-type sequence at the target site, whereas the other may have a substitution of up to 5 bp depending on the sequences of the coupling regions of the previous insertion.

[41] M. G. Caparon and J. R. Scott, *Cell (Cambridge, Mass.)* **59,** 1027 (1989).

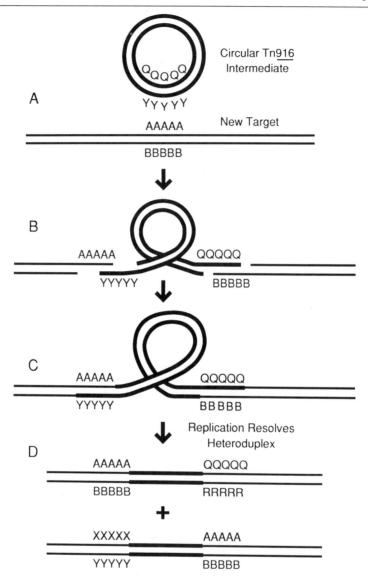

FIG. 3. Model for insertion of Tn916. Symbols are the same as in Fig. 2. A and B represent the complementary bases of the target. Because the transposon circle does not replicate, the heteroduplex consisting of the coupling sequences from the previous insertion are maintained at the joint. A staggered cleavage at the 3' ends of the new target and the heteroduplex-containing joint of the transposon circle (A) generates molecules with 3' single-stranded ends (B). The transposon circle and target are covalently joined by their single-stranded ends to create a new insertion with a heteroduplex at each end (C). Replication resolves the heteroduplexes and generates a pair of molecules in which each member is flanked by the

Excision to Regenerate Wild-Type Phenotype

In most but not all cases (see below), excision of Tn916 either regenerates the wild-type sequence or leads to a substitution of bases. Because insertions and deletions do not usually occur, any potential reading frame in the region containing the transposon will be conserved following excision. The substitution of 5 bp can affect, at most, 3 amino acids. This substitution may or may not alter the amino acid sequence, and, as long as a nonsense mutation is not generated, a functional product may be produced from a mutant sequence in the excisant. Thus, even in cases where it is possible to assay for reversion, an excisant with a wild-type phenotype need not necessarily have the wild-type sequence.

Obtaining Sequence of Wild-Type Allele

A two-step procedure can be used to determine the sequence of the target in the wild-type strain. In the first step, the sequence adjacent to Tn916 in the cloned transposon-inactivated gene is determined. This information is then used to design synthetic oligonucleotides that will hybridize to the wild-type allele in the region adjacent to the target site. The second step involves using the oligonucleotides as primers for the polymerase chain reaction (PCR)[42,43] to determine the target sequence using the wild-type chromosome as a template.

Unlike most transposons, the ends of Tn916 do not contain any long repeated sequences.[44] This means that the sequences adjacent to the insertion can be determined directly from the original cosmid clone of the Tn916-inactivated gene without further subcloning. Synthetic oligonucleotides which hybridize to regions specific for each transposon end (left end, OTL1E = CGTGAAGTATCTTCCTACAG; right end, OTR1E = GAGTGGTTTTGACCTTGATA)[41] can be used in conventional dideoxy

[42] R. Saiki, S. Scharf, F. Faloona, K. B. Mullis, B. T. Horn, H. A. Erlich, and N. Arnheim, *Science* **230,** 1350 (1985).
[43] K. B. Mullis and F. Faloona, this series, Vol. 155, p. 335.
[44] D. B. Clewell, S. E. Flannagan, Y. Ike, J. M. Jones, and C. Gawron-Burke, *J. Bacteriol.* **170,** 3046 (1988).

---

target sequence at one end and a coupling sequence from the previous insertion at its other end (D). The two forms would segregate at the first cell division. This means that of the two coupling sequences, one is derived from the target, and the other is introduced through insertion of the transposon and is derived from a coupling sequence of the previous insertion. [Reprinted from "Excision and insertion of Tn916 involves a novel mechanism of recombination," M. G. Caparon and J. R. Scott, *Cell (Cambridge, Mass.)* **59,** 1027 (1989).]

sequencing reactions with the cosmid as a template to read from each end of Tn916 into the adjacent target DNA.[41]

When the sequences of the regions to both the right and left of the insertion are determined, two new oligonucleotides can be synthesized that will hybridize to the wild-type allele on either side of the target site. The primers should be designed so that both will sequence toward the target site. The two opposing primers are then used to amplify and sequence the target site directly from the wild-type chromosome by the PCR.

As an alternative method, the sequence of the wild-type target can be determined using an inverse PCR.[45,46] Like the conventional PCR method, this technique also requires two oligonucleotide primers that will hybridize in the region adjacent to the target site. However, instead of using two different primers based on two different and opposing sequences, the primers are complements of a single sequence. For the inverse PCR, chromosomal DNA from the wild-type strain is digested with a restriction enzyme that will include on a single fragment of 1–2 kb both the target site and the sequence of the complementary primers. The digest is treated with ligase to circularize all fragments, and the specific fragment is amplified using the complementary primers in a standard PCR.

Cloning Wild-Type Allele

Even if its sequence is known, it may not be possible to use excision to obtain the wild-type allele. Other products can sometimes result from excision of Tn916, including those with deletions of target DNA adjacent to the transposon (M. Norgren, M. G. Caparon and J. R. Scott, unpublished results, 1989). Tn916 can also utilize different coupling sequences for excision than it did for insertion; the right transposon end appears to "slip" in a region of consecutive T residues,[41] which can generate additional types of excisants. For all these reasons, excision cannot be relied on to regenerate the wild-type allele or even to restore the wild-type phenotype.

To obtain the wild-type form of the insertionally mutated allele, it should be cloned from the unmutagenized parental strain. This is done by using an excisant as a hybridization probe to screen a library prepared from the unmutagenized parental streptococcal strain.

Introduction of Cloned DNA into *Streptococcus:* Methods Using Natural
    or Protoplast Transformation

Once a potential virulence gene has been identified by transposon mutagenesis and cloned into *E. coli,* further analysis usually requires

[45] H. Ochman, A. S. Gerber, and D. L. Hartl, *Genetics* **120**, 621 (1988).
[46] J. Silver and V. Keeridatte, *J. Virol.* **63**, 1924 (1989).

techniques for complementation and allelic replacement. Complementation tests are performed to test whether the insertion mutation is in a gene encoding a diffusible protein that is required for virulence. This involves construction of a merodiploid streptococcal strain. To do this, the cloned wild-type allele is introduced into the strain containing the insertion mutation. If the gene encodes a diffusible factor, the virulent phenotype should be restored to the mutant when the wild-type form of the gene is also present.

The relationship of the structure of the virulence protein to its function can be elucidated by altering the coding sequence of the cloned gene in *E. coli*. The altered form is then used to replace the wild-type allele in the streptococcal strain ("allelic replacement"). Alterations of structure can then be correlated with function in an otherwise wild-type background.

Techniques for complementation and allelic replacement involve introduction of foreign DNA or DNA manipulated *in vitro* into the streptococcus. For complementation, the merodiploid strain is constructed by introducing the cloned DNA on a plasmid capable of replicating in streptococci or on a transposon that will integrate randomly instead of at the chromosomal site of homology with the cloned gene. Allelic replacement, on the other hand, requires introduction of the cloned gene on a nonreplicating DNA fragment which has homology at its ends with the sites surrounding the chromosomal region to be replaced (see below, Introduction of Mutation Specific Gene by Targeted Allelic Replacement).

The methods that are available for the introduction of cloned DNA into streptococci include transformation of naturally competent strains (see below), transformation of protoplasts,[47,48] and transformation using electroporation[49-52] (see below). If the strain under study cannot be transformed by any of these methods, it may be possible to first introduce the cloned DNA into an intermediate host that is transformable and then to mobilize it into the final host by conjugation.[23,53] Other methods involve "shuttling" a broad host range conjugative plasmid from *E. coli* to the streptococcal strain[5] or using a "shuttle" transposon that takes advantage of the conjugative properties of Tn916 (see below).

[47] R. Wirth, F. Y. An, and D. B. Clewell, *J. Bacteriol.* **165**, 831 (1986).
[48] M. Smith, *J. Bacteriol.* **162**, 92 (1985).
[49] A. Suvorov, J. Kok, and B. Venema, *FEMS Microbiol. Lett.* **56**, 95 (1988).
[50] S. Fiedler and R. Wirth, *Anal. Biochem.* **170**, 38 (1988).
[51] I. B. Powell, M. G. Achen, A. J. Hillier, and B. E. Davidson, *Appl. Environ. Microbiol.* **54**, 655 (1988).
[52] S. F. Lee, A. Proglulske-Fox, G. W. Erdos, D. A. Piacentini, G. Y. Ayakawa, P. J. Crowley, and A. S. Bleiweis, *Infect. Immun.* **57**, 3306 (1989).
[53] M. D. Smith and D. B. Clewell, *J. Bacteriol.* **160**, 1109 (1984).

Methods of Introducing DNA Using Naturally Competent Strains

In the species or strains of streptococci that are naturally competent (e.g., pneumococci, *S. mutans,* and *S. sanguis* strain Challis), complementation tests can be performed by cloning the wild-type allele of the gene into any of the several streptococcal–*E. coli* shuttle vectors that have been developed[47,54–57] and transforming the chimeric plasmid into the mutant bearing the transposon-inactivated gene. The shuttle vectors contain a streptococcal replication origin in addition to an *E. coli* origin so that the chimeric plasmid will be maintained in both hosts. For example, the gene to be analyzed can be cloned into a suitable site on the shuttle plasmid pVA838.[54] The initial construction can be done in *E. coli* since pVA838 contains an *E. coli* replicon (from pACYC184[58]) in addition to a streptococcal replicon (from pVA380-1[54]). The chimeric plasmid bearing the wild-type form of the potential virulence gene can then be transformed into a naturally competent host. The specific details on how transformation is accomplished will vary depending on the organism of interest; a method for *S. sanguis* strain Challis is presented below.

Transformation of Naturally Competent Streptococcus sanguis

1. The *S. sanguis* Challis strain (V288 or V797; see below) is inoculated into 5 ml of freshly made BHI broth in a screw-top test tube, which is incubated overnight at with the tube tightly capped.

2. The overnight culture is diluted 1 : 500 into 9 ml BHI broth supplemented with 1 ml heat-inactivated (56° for 30 min) horse serum in a side-arm flask and incubated in a water bath without shaking.

3. Growth is monitored at 30-min intervals in a colorimeter (Klett Manufacturing Co., New York, using a red filter). The initial reading is usually 0, and the growth of the culture will lag for approximately 2 hr.

4. When the culture reaches 15 Klett units, a 1-ml aliquot is transferred to a side-arm flask containing 9 ml of fresh BHI and 1 ml of heat-inactivated horse serum.

5. The mixture is incubated at 37° until the culture reaches 5 Klett units. Aliquots of 0.45 ml are transferred to sterile tubes containing 100

[54] F. L. Macrina, J. A. Tobian, K. R. Jones, R. P. Evans, and D. B. Clewell, *Gene* **19,** 345 (1982).

[55] F. L. Macrina, R. P. Evans, J. A. Tobian, D. L. Hartley, D. B. Clewell, and K. R. Jones, *Gene* **25,** 145 (1983).

[56] M. Behnke, S. Gilmore, and J. J. Ferretti, *Mol. Gen. Genet.* **182,** 414 (1981).

[57] H. H. Murchison, J. F. Barrett, G. A. Cardineau, and R. Curtiss III, *Infect. Immun.* **54,** 273 (1986).

[58] A. C. Y. Chang and S. N. Cohen, *J. Bacteriol.* **134,** 1141 (1978).

ng of transforming DNA. The DNA should be suspended in a low-salt buffer (e.g., 10 mM Tris-HCl, 1 mM EDTA, pH 8.00), and the total volume of DNA should not exceed 0.1 ml. A control consists of an aliquot of cells to which no DNA is added.

6. The cell–DNA mixture is incubated for 1 to 2 hr at 37°, then plated on selective media (BHI agar containing 10 μg/ml of erythromycin to select for pVA838). Plates are incubated anaerobically at 37° for 48 hr.

Protoplast Transformation

Protoplast transformation methods have been described for some streptococcal species that are not naturally competent to take up DNA. These methods involve removing all or part of the streptococcal cell wall so that the cells become capable of taking up DNA to which they are exposed. Although protoplast methods have been described for *E. faecalis*[47,48] and *S. lactis*,[59] we have not succeeded in designing a protocol to allow transformation of *S. pyogenes* protoplasts. Thus, at least at present, protoplast methods are not applicable to all streptococcal species.

Transformation of an Intermediate Host

If the strain under study is not transformable, it may be possible to first introduce the chimeric shuttle plasmid into an intermediate host that can be transformed either naturally, after protoplasting, or by electroporation (see below). For example, the shuttle vector pVA838[54] described above can be mobilized into other streptococci using the conjugative plasmid pVA797.[23,53]

Two systems that use an intermediate host have been described. In the first, the intermediate, naturally transformable strain (or protoplasts if it is not naturally competent) contains a conjugative plasmid that mobilizes the chimeric plasmid to the organism of interest, and in the second, the mobilizing plasmid is introduced after the chimeric plasmid. *Streptococcus sanguis* strain V797 contains the conjugative mobilizing plasmid pVA797.[60] Alternatively, the pVA838 chimera can be transformed into a plasmid-free host such as *S. sanguis* V288, and pVA797 can be introduced subsequently by mating with another strain using chloramphenicol (3 μg/ml) to select for pVA797.

The *S. sanguis* strain containing both pVA797 and the pVA838 chimera is then mated with an antibiotic-resistant derivative of the streptococcal strain of interest by the procedure described above for the introduction of

[59] J. K. Kono and L. L. McKay, *Appl. Environ. Microbiol.* **43**, 1213 (1982).
[60] R. P. Evans, Jr., and F. L. Macrina, *J. Bacteriol.* **154**, 1347 (1983).

Tn916. Erythromycin is used to select for the presence of the pVA838 chimera in the transconjugants, and the antibiotic resistance of the recipient is used to select against the donor strain.

The method by which pVA797 mobilizes chimeric derivatives of pVA838 is not clear. The two plasmids have one region in common, so it is possible that homologous recombination between them causes a cointegrate to form and that they are transferred as a single large plasmid. Cointegrates have been observed in *S. sanguis* donor strains.[23] However, in all cases where plasmid DNA has been examined (Ref. 23 and J. R. Scott *et al.,* unpublished observation, 1986), the transconjugant strain contains the two separate plasmids. Either mobilization does not require cointegrate formation, or the cointegrate is immediately resolved in the recipient cell.

Introduction of DNA into Streptococci: Methods Based on Electroporation

Most streptococcal strains are not naturally competent to take up exogenous DNA. Protoplast transformation methods are often difficult and time consuming and have not been described for many species. The use of an intermediate transformable host may also not be possible unless suitable antibiotic resistance markers can be introduced into the recipient strain bearing the mutation in the putative virulence gene. In these instances, complementation analysis with the cloned wild-type virulence gene may still be possible using transformation methods based on electroporation.

Several companies have introduced reasonably priced devices for the transformation of bacterial cells by electroporation. In this technique, a suspension of the cells and DNA is subjected to an electrical field of very high voltage which presumably opens transient "pores" in the cell surface.[61] The DNA traveling in this electrical field then enters the cells through these pores. This method can be used to transform streptococci and is particularly useful for species where other transformation methods are difficult or not available. Electroporation of streptococci generally requires that the cell walls be weakened. We have successfully used the following method adapted from G. Dunny (personal communication, 1989) for *S. pyogenes.*

Transformation of Streptococcus pyogenes by Electroporation

1. The *S. pyogenes* strain to be transformed is cultured overnight at 37° in THY broth supplemented with 20 m*M* glycine. The concentration of

[61] N. M. Calvin and P. C. Hanawalt, *J. Bacteriol.* **170,** 2796 (1988).

glycine may need to be adjusted for other group A strains or streptococcal species.

2. The overnight culture is diluted 1:20 in THY plus glycine (total volume of culture 100 ml), and incubation is continued at 37° with no aeration.

3. Growth is measured spectrophotometrically. When the A_{600} reaches 0.2, which takes approximately 2 hr, incubation is stopped.

4. Cells are harvested by centrifugation (5000 g, 10 min, 10°) and washed twice in 20 ml of ice-cold electroporation medium (EPM: 272 mM glucose, 1 mM MgCl$_2$, with the pH adjusted to 6.5 using NaOH).

5. The final pellet obtained from 100 ml of cells is resuspended in 0.75 ml ice-cold EPM, transferred to a sterile 1.5-ml microcentrifuge tube, and placed on ice.

6. Aliquots (0.2 ml) of the cell suspension are transferred into a chilled 0.2-cm Gene Pulser (Bio-Rad, Richmond, CA) electroporation cuvette, and DNA ($\sim$5 μg per transformation in low-salt buffer) is added. No DNA is added to the aliquot used as the negative control.

7. The Gene Pulser (Bio-Rad) is set with the voltage at 1.75 kV, the capacitance at 25 μF, and the Pulse Controller at 400 ohms resistance.

8. The cuvette containing cells and DNA is placed in the electrode chamber and the current delivered. The time constant should be between 6 and 7 msec.

9. The cell suspension is withdrawn immediately using a sterile Pasteur pipette or a long plastic pipette tip designed for loading gels (e.g., microcap-illary tips, National Scientific, San Rafael, CA), placed directly into 10 ml of THY broth (with no antibiotics or glycine), and incubated 2 hr at 37°.

10. The culture is centrifuged as above (Step 4) and washed once in THY broth. The pellet is suspended in 1 ml THY broth.

11. Aliquots are plated on THY agar supplemented with the appropriate antibiotics and incubated at 37°. Colonies are usually visable in 24–36 hr.

Plasmids and Markers

Electroporation transformation frequencies as high as 1 $\times$ 10^3 to 2 $\times$ 10^4 per microgram have been obtained (M. G. Caparon and J. R. Scott, unpublished observations, 1989) using several approximately 6-kb plasmids based on the pSH71 replicon.[62] Selectable markers that have been used successfully include the pC194-derived chloramphenicol resistance determinant[63] and the *aph*A-3 kanamycin resistance determinant from Tn*1545*.[64]

[62] W. M. De Vos, *FEMS Micribiol. Rev.* **46**, 281 (1987).
[63] S. Horinouchi and B. Weisblum, *J. Bacteriol.* **150**, 815 (1982).
[64] P. Trieu-Cuot and P. Courvalin, *Gene* **23**, 331 (1983).

Introduction of DNA into Streptococci: Methods Using
Shuttle Transposons

Development of Shuttle Transposons

A chimeric shuttle transposon has been designed for the analysis of cloned genes in the streptococci and other gram-positive organisms that can be used for complementation analyses of cloned virulence genes[20] (see below). This method capitalizes on the conjugative ability of Tn916. The advantages of this shuttle transposon include (1) the broad host range of Tn916, (2) its ability to enter organisms which are difficult or impossible to transform, and (3) its low copy number (i.e., 1 per chromosome). This shuttle can also be used for targeted allelic replacements[20] (see below).

Foreign DNA can be inserted into the BstXI site of Tn916 without affecting the conjugative ability of the transposon.[20] The DNA to be analyzed is cloned into this site on a copy of Tn916 present on the E. coli plasmid pAM120, purified from E. coli, and used to transform protoplasts of B. subtilis[65] with selection for the tetracycline resistance determinant of Tn916. The chimeric transposon is then transferred to the gram-positive organism of interest by mating with the B. subtilis transformant.

Limitations to Method

The shuttle transposon method has several drawbacks that can limit its usefulness. First, cloning the DNA fragment of interest is made difficult by the large size of Tn916 and of any plasmid that contains Tn916. Second, the restriction endonuclease BstXI is difficult to use because digestion requires high temperature, and the enzyme generates ambiguous fragment ends. Third, analysis and purification of any potential clone is made more difficult by the fact that Tn916 is unstable and excises at high frequency from plasmids in E. coli. Fourth, when obtained, the chimeric transposon must be introduced into B. subtilis by the tedious protoplast transformation method. Finally, because the ability of a strain to donote Tn916 depends on the location of the transposon in the genome of the donor, several potential donors must be tested to obtain one that will donate the chimeric transposon at an acceptable frequency.

Development of pVIT Vectors

To alleviate all the problems mentioned above for the construction of shuttle transposons, a set of plasmid cloning vectors called pVIT (vector for insertion into Tn916) have been developed.[22a] The simplest version,

---

[65] S. Chang and S. N. Cohen, Mol. Gen. Genet. 168, 111 (1979).

pVIT130 (Fig. 4), is a stable and small (4.6 kb) plasmid based on the *E. coli* vector pUC9.[66] It contains approximately 1.2 kb of DNA from the region immediately to the left and 1.6 kb of DNA from the region immediately to the right of the *Bst*XI site of Tn*916*. The *Bst*XI site itself has been replaced with a *Bam*HI site to simplify subsequent cloning manipulations.

The gene to be analyzed is cloned into the *Bam*HI site of pVIT130 along with a marker selectable in *B. subtilis*. The chimera is then linearized by digestion with a restriction endonuclease that has a recognition site only in the pUC region of the plasmid. This linearized pVIT plasmid is now transformed into competent cells of a *B. subtilis* strain (or any other naturally competent donor of Tn*916*) which has already been characterized as a high-frequency donor of the transposon. By selecting for the marker introduced along with the gene to be analyzed, homologous recombination between the DNA regions present on both the pVIT vector and the transposon will generate the chimeric shuttle transposon (Fig. 4).

Use of Shuttle Transposons for Complementation

The availability of Tn*916* derivatives containing different antibiotic resistance markers (e.g., tetracycline or erythromycin) allows the use of shuttle transposons for complementation tests involving a Tn*916*-inactivated gene since two Tn*916* derivatives with different antibiotic resistance markers can be introduced into the same cell.[67] For example, if the gene was inactivated using Tn*916* (tetracycline resistance), the wild-type allele can be introduced into the mutant on an erythromycin resistance shuttle transposon. Usually the two transposons will insert at different sites of the genome and will be maintained separately,[67] producing a merodiploid strain for complementation analysis.

However, in some of the transconjugants, the incoming transposon can undergo homologous recombination with the resident element so that the tetracycline resistance gene of the resident is replaced by the erythromycin resistance gene of the incoming transposon. This results in transconjugants that are resistant to erythromycin but sensitive to tetracycline.[67] The apparent genotype of any strain containing two transposons should be confirmed by appropriate Southern blots.

Introduction of Mutation into Specific Gene by Targeted Allelic Replacement

The observation that recombination rather than transposition can occur if a homologous region is present in both the incoming transposon and the recipient has served as the basis of a method for targeted allelic

[66] J. Vieira and J. Messing, *Gene* **19**, 259 (1982).
[67] M. Norgren and J. R. Scott, *J. Bacteriol.* **173**, 319 (1991).

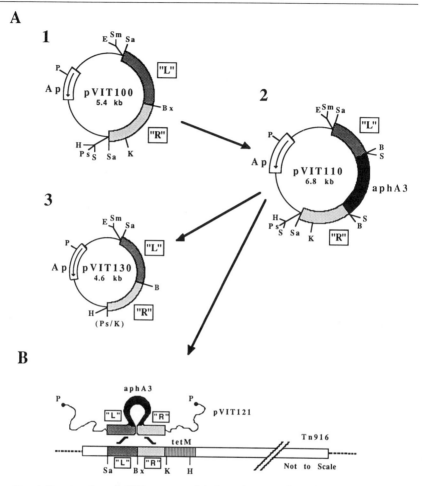

FIG. 4. Construction of pVIT vectors and their use in generating chimeric shuttle transposons. (A) Vector construction (1) pVIT100 was constructed by inserting a 2.8-kb *Sau*3A fragment containing the *Bst*XI site of Tn*916* (see Fig. 1) into the *Bam*HI site of pUC9 (thin line). The *Sau*3A fragment is composed of regions "L," a 1.2-kb segment that consists of the DNA from the *Bst*II site to the leftmost *Sau*3A site, and "R," a 1.6-kb DNA segment from the *Bst*XI site to the closest *Sau*3A site to its right. The direction of transcription of the pUC9 ampicillin resistance gene (Ap) is indicated by the arrow. (2) pVIT110 was constructed by inserting a 1.5-kb *Eco*RI fragment of pUC4-21K [J. Perez-Casal, M. G. Caparon, and J. R. Scott, *J. Bacteriol.* **173**, 2617 (1991)] containing an *aph*A-3 kanamycin resistance gene (originally from Tn*1545*) into the *Bst*XI site of pVIT100. Protruding single-stranded DNA ends were filled in using T4 DNA polymerase prior to ligation. (3) pVIT130 was constructed by digestion of pVIT110 with *Bam*HI followed by ligation to delete a 1.5-kb fragment containing *aph*A-3. To remove the restriction sites in the polylinker, a 0.7-kb fragment was deleted by digestion with *Pst*I and *Kpn*I, and the protruding single-stranded ends were filled in with T4 DNA polymerase and ligated. (B) pVIT110 can be used to construct a chimeric

replacement.[20] The chromosomal regions immediately upstream and downstream of the targeted gene are first introduced into the cloning site of pVIT130 (or the closely related vector pJRS101[20]) so that they flank a selectable marker. As described above, a chimeric shuttle transposon can then be generated by transforming linearized plasmid into a competent *B. subtilis* strain that contains a copy of Tn916 in its chromosome. Recombination between the homologous regions present in both the vector and the resident transposon will introduce the cloned DNA into the *Bst*XI site of Tn916. The *B. subtilis* strain with the chimeric Tn916 is then mated with the organism containing the wild-type allele of the targeted gene. The only transposon marker used for selection is the marker cloned between the upstream and downstream regions of the targeted gene now present in the *Bst*XI site. The marker contained on the copy of Tn916 from which the chimera was derived is not selected for.

In some transconjugants, homologous recombination between the DNA regions upsteam and downstream of the targeted gene present both on the chimeric transposon and on the chromosome will result in the replacement of the selectable marker for the targeted gene (Fig. 5). As a result of this process, the remaining regions of the transposon including its other antibiotic resistance gene are lost from the cell. This then allows organisms in which allelic replacement has occurred to be distinguished from organisms in which transposition has occurred, because the former will be resistant only to the antibiotic whose resistance is encoded by the marker used to replace the targeted gene. If transposition and not replacement has occurred, the organism will be resistant to both antibiotics. This deduction should be confirmed by Southern blot analysis.

Conclusions

Rapid progress has been made in recent years in the development of techniques for genetic analysis in the streptococci. The ability to do sophisticated genetic experiments using transposon mutagenesis, complementation, and allelic replacement will allow detailed molecular analyses of virulence in these organisms. These analyses should allow a rigorous

---

kanamycin-resistant derivative of Tn916. pVIT110 is linearized by digestion with *Pvu*I and transformed into competent cells of a *B. subtilis* strain (MEN202[20]) that bears a copy of Tn916 in its chromosome. Tn916 is represented by the open box, and the *B. subtilis* chromosome is represented by the dashed line. When homologous recombination occurs between the "L" and "R" fragments of the Tn916 in the chromosome and the copy that entered by transformation, *aph*A-3 is inserted at the *Bst*XI site of Tn916. Restriction enzyme sites: P, *Pvu*I; E, *Eco*RI; Sm, *Sma*I; Sa, *Sau*3A; S, *Sal*I; Bx, *Bst*XI; K, *Kpn*I; Ps, *Pst*I; H, *Hin*dIII; B, *Bam*HI.

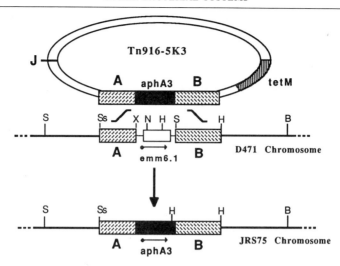

FIG. 5. Targeted alleleic replacement. The chimeric shuttle transposon Tn916-5K3 was used for the targeted allelic replacement of *emm6.1*, the gene for the type 6 M protein in *S. pyogenes* D471 [M. Norgren, M. G. Caparon, and J. R. Scott, *Infect. Immun.* **57**, 3846 (1989).] Tn916-5K3 was constructed using a pVIT vector by a method similar to that shown in Fig. 4; it contains the *aph*A-3 kanamycin resistance gene from Tn*1545* flanked by DNA sequences labeled A and B identical to the segments that flank *emm*6.1 on the D471 chromosome. The incoming transposon is represented as a circular molecule to conform with the current model for its transfer. The "J" shows the location of the transposon circle joint which is involved in transpositional (not homologous) recombination (see Figs. 2 and 3). When Tn916-5K3 was introduced into D471 by conjugation from a *B. subtilis* donor, homologous recombination between fragments A and B present on both transposon and chromosome replaced *emm*6.1 with *aph*A-3 in some of the transconjugants. The rest of Tn916 and its *tet*M tetracycline resistance gene were lost from the cell. The relevant part of the chromosome of JRS75, a strain in which allelic replacement occurred, is shown below the arrow. Restriction enzyme sites: S, *Sal*I; Ss, *Sst*I; X, *Xba*I; N, *Nci*I; H, *Hin*dIII; B, *Bam*HI.

testing of the role of factors previously proposed to be important in pathogenesis as well as permit the identification of new virulence factors. It is hoped that a greater understanding of streptococcal infection will allow the design of new therapeutic agents for the treatment of streptococcal disease.

Acknowledgments

The work in our laboratory was supported by National Institutes of Health Grant AI20723. We appreciate helpful discussions with Bruce Chassy, Garry Dunny, Don Clewell, Vince Fischetti, Susan Hollingshead, Terri Kenney, Charles Moran, Mari Norgren, Jose Perez-Casal, and Kathy Tatti.

[27] Genetic Systems in Staphylococci

By RICHARD P. NOVICK

Staphylococcal Genome

The genus *Staphylococcus* consists of gram-positive, clump-forming, facultatively aerobic cocci and contains some 20 distinct species. On the basis of reassociation kinetics, strains of the same species have 80–100% sequence identity, whereas different species never have more than 20%. Nevertheless, the various species have much in common, can participate broadly in genetic exchange, and probably possess a single common pool of plasmids and transposons. Most of the available molecular and genetic data are for *Staphylococcus aureus;* other species are increasingly being investigated at the molecular genetic level.

Strains of *S. aureus* elaborate a panoply of extracellular proteins, many of which are implicated in pathogenicity. A number of different strains have been used as laboratory prototypes, generally because they are effective producers of one or another of these proteins. Some of the prototype strains are listed in Table I. One strain, NCTC 8325, has been used for most of the genetic studies. This strain carries at least three prophages, ϕ11, ϕ12, and ϕ13. A derivative, 8325-4 (or RN450), that has been UV-cured of all three prophages is the K12 of *S. aureus.*

The staphylococcal genome is fundamentally similar to other prokaryotic genomes, consisting of a single circular chromosome plus an assortment of variable accessory genetic elements (VGE) including prophages, plasmids, transposons, and other uncharacterized types. This section summarizes the current state of information on the various components of the staphylococcal genome.

Chromosome

The 8325 chromosome has been mapped, single-handedly, by Pattee and co-workers using both genetic and physical methods as described below. Some 100 loci have been mapped, including phenotypic markers, silent transposon insertions, and most of the 16 *Sma*I restriction sites.[1] The overall size is about 2.78 megabase pairs (Mbp); two gaps in the

---

[1] P. A. Pattee, H. Lee, and J. P. Bannantine, *in* "Molecular Biology of the Staphylococci," (R. P. Novick, ed.), VCH Publishers, New York, 1990.

METHODS IN ENZYMOLOGY, VOL. 204

TABLE I
Staphylococcus aureus PROTOTYPE STRAINS

Strain	Features
Wood 46	α-Hemolysin producer
TC82	β-Hemolysin producer
TC128	δ-Hemolysin producer
V8	Serine protease producer
Foggi	Nuclease producer
Cowan I	Protein A producer
NCTC 8325	Propagating strain for typing phage 47
NCTC 9789	Propagating strain for typing phage 80
S6	Enterotoxin B producer
COL	Homogeneously methicillin resistant

circular linkage map, at 5 and 9 o'clock, remain (see Fig. 1). Other strains show considerable restriction site polymorphism, and consequently the degree of overall constancy of organization has not been established.

Mapping. Most of the known chromosomal markers were mapped initially by standard multifactor crosses using transduction and competent cell transformation.[2] Linkages representing distances of 100–150 kilobases (kb) can be mapped by transformation using competent cells (see below); evidently the competent recipient can take up a very large molecule, of which several short segments may then be integrated independently by a homology-dependent process. Transduction, limited to the size of the phage genome (45 kb for ϕ11, see below), is useful for analysis of closely linked markers. Subsequently, an overall map of the entire chromosome was obtained by protoplast fusion, using ten widely scattered markers in a single cross, analyzing the results by computer.[3] These results have been corroborated by physical mapping using pulsed-field agarose gel electrophoretic analysis of *Sma*I digests supplemented by Southern blotting.[1] Here, the probes used were mainly transposons Tn*551* and Tn*4001*, for which many genetically mapped insertions were available.[1] At this stage, multiply marked strains exist that can be used to map any new marker by two or at most three crosses (P. Pattee, personal communication, 1986). Refinement of the restriction map is currently in progress (P. Pattee, personal communication, 1990), and in the near future it will be possible to map any cloned fragment by blot hybridization.

Preparation of Whole-Cell DNA. Cultures are grown in 20 ml CY broth at 37° to a Klett value of 250 (540 nm filter), centrifuged, and the pellets

[2] P. A. Pattee and S. Neveln, *J. Bacteriol.* **124,** 201 (1975).
[3] M. L. Stahl and P. A. Pattee, *J. Bacteriol.* **154,** 395 (1983).

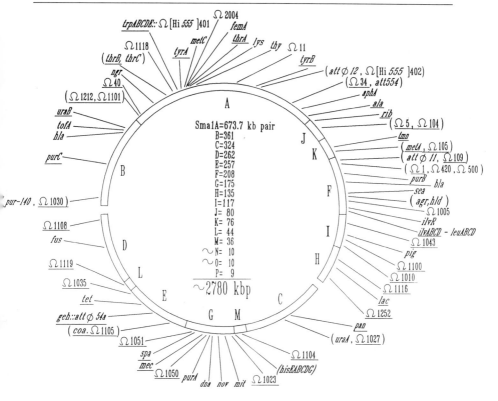

FIG. 1. The staphylococcal chromosome: A physical and genetic map for strain NCTC 8325. The SmaI physical map is drawn to scale, and the estimated sizes of the SmaI fragments are listed in the center. Values for SmaI-N and SmaI-O are conservative estimates, and the value for SmaI-P was provided by R. V. Goering (personal communication). Markers shown are described in greater detail elsewhere [P. A. Pattee, *Genet. Maps* **4**, 148 (1987)]. Markers that are underlined have been used to correlate the physical and genetic maps by DNA hybridization. Ω1 through Ω40 and Ω1005 through Ω1051 represent silent insertions of Tn551. Ω1100 through Ω1119 are silent insertions of Tn916. Ω104, Ω105, Ω109, Ω1212, and Ω1252 are silent insertions of Tn4001, and Ω420 and Ω500 are insertions (perhaps identical) of pI258 [P. A. Pattee, N. E. Thompson, D. Haubrich, and R. P. Novick, *Plasmid* **1**, 38 (1977)]. Ω2004 is one insertion site of Tn551 that impairs the expression of methicillin resistance [B. Berger-Bachi, *J. Bacteriol.* **154**, 479 (1983)], which may be related to *femA* (factor required for expression of methicillin resistance).[133] Ω(Hi555)401 and Ω(Hi555)402 are two sites that can be occupied by the heterologous insertion carrying *tst* [M. C. Chu, B. N. Kreiswith, P. A. Pattee, R. P. Novick, M. E. Melish, and J. J. James, *Infect. Immun.* **56**, 2702 (1988)]. Other determinants of extracellular proteins and virulence factors that have been mapped are one site for the enterotoxin A determinant (*sea*, formerly *entA*); staphylococcal protein A (*spa*); the α-hemolysin determinant (*hla*, formerly *hly*); δ-hemolysin [*hld*; L. Janzon, S. Lofdahl, and S. Arvidson, *Mol. Gen. Genet.* **219**, 480 (1989)]; bound coagulase (*coa*); glycerol ester hydrolase (*geh*), which is subject to negative phage conversion by phage φ54a; and the accessory gene regulator (*agr*). Chromosomal resistance determinants include methicillin

washed with TES buffer and resuspended in 5 ml of the same buffer. (A listing of media and buffer components can be found in the Appendix at the end of this chapter.) The cells are lysed with lysostaphin (Applied Microbiology, Inc., New York, NY) (25 μg/ml, 37°, 1 hr), and the lysate is treated with pronase (500 μg/ml, 37°, 1 hr). The DNA is extracted with an equal volume of buffer-saturated phenol and then with chloroform–isoamyl alcohol; one-tenth volume of 3 M sodium acetate is added, and the DNA is precipitated with ice-cold 70% (v/v) ethanol. The dry pellet is resuspended in 1.0 ml TE buffer and incubated with 2 μl RNase I (100 units/ml) at 37° for 1 hr. The DNA is again precipitated with ethanol, redissolved in 0.5 ml TE buffer, and stored at 4°. The final DNA concentration should be about 200 μg/ml. DNA prepared in this manner is suitable for transformation, digestion with restriction enzymes, cloning, sequencing, etc.

Plasmids

Staphylococcus aureus. Most naturally occurring *S. aureus* strains contain plasmids, ranging in size from approximately 1 to 60 kb and falling into four general classes (recently reviewed by Novick[4]) and listed in Table II. Class I consists of small (1–5 kb) multicopy (15–60 copies/cell) plasmids that are either cryptic or carry a single resistance determinant. Rarely a plasmid carries two markers. Markers include Tc (tetracycline), Em (erythromycin), Cm (chloramphenicol), Sm (streptomycin), Km (kanemycin), Bl (bleomycin), Qa (quaternary amine), and (Cd) cadmium. Twelve of these plasmids have been sequenced and assigned to four families on the basis of homology of replicon functions. Plasmids of this class replicate by an asymmetric rolling circle mechanism similar to that of the filamentous single-stranded coliphages. This mechanism is used by all known

---

[4] R. P. Novick, *Annu. Rev. Microbiol.* **43**, 537 (1989).

---

resistance (*mec*), constitutive tetracycline and minocycline resistance (*tmn*), inducible tetracycline resistance (*tet*), β-lactamase synthesis from Ps53 (*bla*), novobiocin resistance (*nov*), aminoglycoside phosphotransferase (APH3'III; *aphA*), and fusidic acid resistance (*fus*). The remaining markers are L-threonine (*thrA, thrB, thrC*), L-tryptophan (*trpABCDE*), L-tyrosine (*tyrA, tyrB*), L-alanine (*ala*), riboflavin (*rib*), purine (*purA, purB, purC, pur140*), L-isoleucine and L-valine (*ilvABCD–leuABCD*), pantothenate (*pan*), uracil (*uraA, uraB*), and L-histidine (*hisEABCDG*) auxotrophs; prophage and site-specific transposon attachment sites *att*ϕ11, *att*ϕ12, and *att544* (the preferred integration site for Tn554); absence of golden-yellow pigment (*pig*); enhanced sensitivity to mitomycin C (*mit*); regulation of isoleucine and valine biosynthesis (*ilvR*); lactose utilization (*lac*); apurinic endonuclease (*ngr*); temperature-sensitive osmotically remedial peptidoglycan synthesis (*tofA*); and thermosensitive DNA replication (*dna*). (Reprinted from Pattee *et al.* (1990), by kind permission of the publishers.)

TABLE II
STAPHYLOCOCCAL PLASMIDS

Class	Family	Plasmid	Copy number	Size (kb)	Incompatibility group	Phenotype[a]
I[b]	pT181	pT181	22	4.4	3(C)	Tc^r
		pT127	50	4.4	3	Tc^r
		pC221	22	4.6	4(D)	Cm^r
		pC223		4.6	10(J)	Cm^r
		pUB112		4.1	9(I)	Cm^r
		pS194	22	4.4	5(E)	Sm^r
		pCW7		4.2	14(N)	Cm^r
	pC194	pC194	15	2.9	8(H)	Cm^r
		pUB110	10	4.5	13(M)	Km^r Bl^r
		pOX6		3.2		Cd^r
		pRBH1[c]				Km^r
		pBC16[c]		4.5	13(M)	Tc^r
		pWBG32	50	2.4		Qa^r
	pSN2	pSN2	50	1.3		Cryptic
		pTCS1		1.3		Cryptic
		pE12	10	2.2	12(L)	Em^r
		pIM13	10	2.1	12(L)	Em^r
		pE5		2.1		Em^r
		pT48		2.1		Em^r
		pNE131		2.1		Em^r
	pE194	pE194	55	3.7	11(K)	Em^r
II and III	IIα	pI524	5	31.8	1(A)	Pc^r Cd^r Pb^r Hg^r Om^r Asa^r Asi^r Sb^r Bin^+
	IIα	pI258	5	28.2	1(A)	Asa^r Em^r Bin^-
	IIβ	pII147	5	32.6	2(B)	Pc^r Cd^r Pb^r Hg^r Om^r Bin^- Asa^r Bi^hs
	IIα	pI9789		19.7		Cd^r Pb^r Hg^r Om^r Asa^r Asi^r Sb^r
	III	pG01		52		Qa^r

[a] Tc^r, Tetracycline resistance; Cm^r, chloramphenicol resistance; Sm^r, streptomycin resistance; Km^r, kanamycin resistance; Bl^r, bleomycin resistance; Cd^r, cadmium ion resistance; Qa^r, quaternary amine resistance; Em^r, erythromycin resistance; Pc^r, penicillin resistance; Pb^r, lead resistance; Hg^r, mercuric ion resistance; Om^r, organomercurial resistance; Asa^r, arsenate ion resistance; Asi^r, arsenite ion resistance; Sb^r, antimonyl ion resistance; Bin, Tn552 resolvase; Bi^hs, bismuth ion hypersensitivity; Gm^r, gentamicin resistance; Tp^r, trimethoprim resistance; Tra, conjugative transfer.

[b] The list of class I plasmids is essentially complete; only prototypes in classes II and III are listed. A much more extensive list has been presented by B. R. Lyon and R. Skurray [Microbiol. Rev. **51**, 88 (1987)].

[c] *Bacillus subtilis* plasmids listed because of their close relation to the prototype staphylococcal plasmid pUB110 (see text).

small plasmids from gram-positive bacteria.[4,5] All encode initiator (Rep) proteins that act by introducing a site-specific nick in the leading strand replication origin,[6] and all regulate copy numbers at the level of initiator synthesis. In the best studied of these plasmids, the pT181 family, this regulation is accomplished by antisense RNAs, or countertranscripts, that cause attenuation of the Rep protein mRNA.[7] Each plasmid contains a large dyad element that serves as the initiation signal for lagging strand replication[8] and includes a site (RS_B, recombination site B) for sequence-specific interplasmid cointegrate formation mediated by an unknown host function.[9] Some members of this class also contain a site-specific recombination function, *pre* (plasmid recombination gene),[10] that acts at a different recombination site, RS_A. Most of the currently available cloning vectors for gram-positive bacteria are derived from these plasmids (see below). A functional map of the prototype of this class, pT181, is presented in Fig. 2.

Class II plasmids are larger (15–30 kb), have lower copy numbers (4–6/cell), and carry some combination of resistance to β-lactam antibiotics (β-lactamase), macrolides, and a variety of heavy metal ions (arsenic, cadmium, lead, and mercury),[11] some of which are known or predicted to be transposable. Most of the resistance genes are inducible. A vector system based on the pI258 *bla* determinant has been developed. These plasmids also encode initiator proteins[12] but use the θ replication mechanism rather than the rolling circle.[13]

Class III consists of considerably larger (30–60 kb) plasmids that carry a determinant of conjugative transfer (*tra*) plus some combination of resistance markers including Gm (gentamicin), Pc (penicillin) Qa, and Tp, some of which are transposable, and a number of insertion sequence (IS)-like elements. Plasmids that appear to be composites or recombinants between members of these three classes have been isolated as well as others that appear to have resulted from interplasmid transposon movement.[14,15] The

[5] A. D. Gruss and S. D. Ehrlich, *Microbiol. Rev.* **53**, 231 (1989).

[6] R. R. Koepsel, R. W. Murray, W. D. Rosenblum, and S. A. Khan, *Proc. Natl. Acad. Sci. U.S.A.* **82**, 6845 (1985).

[7] S. Iordanescu, S. J. Projan, J. Kornblum, and I. Edelman, *Cell (Cambridge, Mass.)* **59**, 395 (1989).

[8] A. Gruss, H. F. Ross, and R. P. Novick, *Proc. Natl. Acad. Sci. U.S.A.* **84**, 2165 (1987).

[9] R. P. Novick, S. J. Projan, W. Rosenblum, and I. Edelman, *Mol. Gen. Genet.* **195**, 374 (1984).

[10] M. L. Gennaro, J. Kornblum, and R. P. Novick, *J. Bacteriol.* **169**, 2601 (1987).

[11] R. P. Novick and C. Roth, *J. Bacteriol.* **95**, 1335 (1968).

[12] L. Wyman and R. P. Novick, *Mol. Gen. Genet.* **135**, 149 (1974).

[13] R. J. Sheehy and R. P. Novick, *J. Mol. Biol.* **93**, 237 (1975).

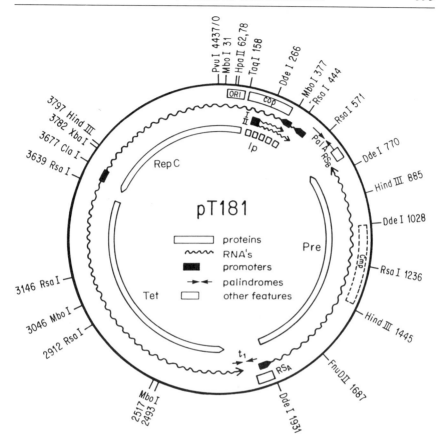

FIG. 2. Functional genetic map of pT181. Important restriction sites are given with nucleotide positions. Wavy lines represent known transcripts, solid blocks represent known promoters, and heavy lines represent reading frames known to encode proteins. Functional elements (in counterclockwise order): lp, putative *repC* leader peptide; *cop*, copy control; *ori*, replication origin; *repC*, initiator protein coding sequence; *tet*, tetracycline resistance determinant; t_1, probable termination signal for *tet*; RS_A, recombination site A, *pre* promoter; *pre* coding sequence; *cmp*, competition determinant; RS_B, recombination site B; *palA*, palindrome A; *repC* and countertranscript promoters. [Reprinted from R. P. Novick, A. D. Gruss, S. K. Highlander, M. L. Gennaro, S. J. Projan, and H. F. Ross, *in* "Antibiotic Resistance Genes: Ecology, Transfer and Expression" (R. P. Novick and S. B. Levy, eds.), p. 225. Cold Spring Harbor Laboratory, Cold Spring Harbor, New York, 1986, by kind permission of the publishers.]

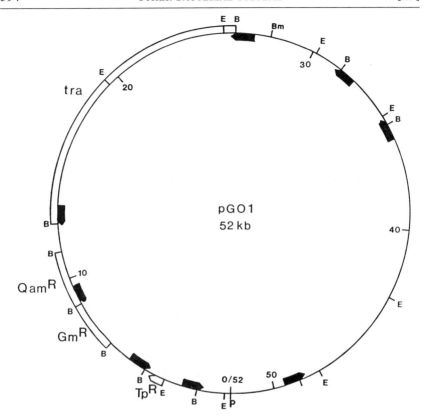

FIG. 3. Physical and functional map of plasmid pG01. Map coordinates are given in kilobases for pG01, assigned using the single *Pst*I site (P) as the origin. Restriction endonuclease cleavage sites for *Eco*RI (E) and *Bgl*II (B) are shown. Phenotypic designations (depicted by open boxes) are as follows: Tp^R, trimethoprim resistance; Gm^R, gentamicin–kanamycin–tobramycin resistance; Qam^R, quaternary ammonium compound resistance; and *tra*, conjugative transfer. Filled boxes represent IS431-like sequences with the arrow indicating the direction of transcription of the single IS431 reading frame. (Reprinted from Thomas and Archer,[16] by kind permission of the publishers.)

map of a typical conjugative class III plasmid, pG01,[16] is presented in Fig. 3.

A few plasmids have been identified that do not seem to belong to any of these three well-defined classes. These have not been studied in any detail and are provisionally placed in a fourth class.[4]

[14] G. S. Gray, *Plasmid* **9**, 159 (1983).
[15] B. R. Lyon and R. Skurray, *Microbiol. Rev.* **51**, 88 (1987).
[16] W. D. Thomas, Jr., and G. L. Archer, *J. Bacteriol.* **171**, 684 (1989).

Other Species. Plasmids similar or identical to *S. aureus* plasmids, of classes I–III, are common in a variety of coagulase-negative staphylococcal species.[17–19] Additionally, plasmids of classes I and III can be readily transferred by protoplast transformation (class I) or by conjugation/mobilization (classes I and III; see below) among various staphylococcal species. Plasmids of class I have been found throughout the gram-positive bacteria, and these can readily be transferred among all gram-positive species tested.

Plasmid Curing. Intercalating agents have been used with minimal success for plasmid curing in *S. aureus;* as these are mutagens, there is always the additional high risk of intercurrent mutations. Prolonged storage on agar slants or cultivation at near-critical temperatures ($\sim$42°–44°) often results in curing of one or more plasmids. Displacement by an incompatible Tsr (thermosensitive for replication) plasmid followed by temperature curing of the latter is a highly reliable curing method. Additionally, we have shown that class I plasmids (but not class II) are cured at high frequency during regeneration of protoplasts.[20]

Preparation and Analysis of Plasmid DNA from Staphylococci. Under their skin, staphylococci are not significantly different, biochemically, from other bacteria, so that standard procedures are effective for preparation of cellular components. Their cell walls, however, are very resistant to mechanical disruption and to lysozyme. Most lytic procedures therefore rely on lysostaphin, a plasmid-determined bacteriocin from *Staphylococcus simulans.*[21] Lysostaphin is an endopeptidase specific for the pentaglycine bridge that connects cell wall peptidoglycan chains and is unique to staphylococci. All *S. aureus* strains are highly sensitive to lysostaphin; coagulase-negative staphylococci, which may contain some serine or alanine residues instead of glycine in the bridge, show great variations in lysostaphin sensitivity. For *Staphylococcus epidermidis,* lysostaphin is often effective at 100 μg/ml at 60° (S. Projan and B. Kreiswirth, personal communication, 1988). Lysozyme at pH 9 has also been found effective (D. Clewell, personal communication, 1982). Pregrowth with a low concentration of penicillin or cycloserine will cause weakening of the cell wall and facilitate enzymatic lysis. In general, when using lysostaphin, sucrose is included in the lysis mixture to prevent immediate lysis of the proto-

[17] F. Gotz, J. Zabielski, L. Philipson, and M. Lindberg, *Plasmid* **9**, 126 (1983).
[18] H. W. Jaffe, H. M. Sweeney, R. A. Weinstein, S. A. Kabins, C. Nathan, and S. Cohen, *Antimicrob. Agents Chemother.* **21**, 773 (1982).
[19] P. Recsei, A. Gruss, and R. Novick, *Proc. Natl. Acad. Sci. U.S.A.* **84**, 1127 (1987).
[20] A. Gruss and R. P. Novick, *J. Bacteriol.* **105**, 878 (1986).
[21] C. Schindler and V. Schuhardt, *Biochim. Biophys. Acta* **97**, 242 (1965).

plasts, as intracellular proteases rapidly degrade the enzyme, preventing complete lysis.

Most staphylococcal plasmids are present as over 90% covalently closed circular (ccc) supercoils with a negative superhelix density σ of approximately 0.04–0.06. Certain class I plasmids, notably pC221, pC223, and pS194, exist partially or largely in the form of a protein–DNA relaxation complex[22] that is relaxed by high salt, proteases, and detergents and is fully strand separated by alkali. With pC221 and pC223 about 50% of the DNA can be recovered in the ccc form; with pS194, less than 10%. They can all be isolated in good yield on preparative sedimentation gradients or by preparative gel electrophoresis. Column chromatography could theoretically also be used.

In using restriction enzymes, it should be remembered that staphylococci do not have GATC methylase activity, so enzymes blocked by GATC methylation are fully active.

Plasmid DNA Replication in Vitro. Replication of supercoiled staphylococcal plasmid DNA in cell-free extracts was first observed by Khan *et al.*[23] and has been used by several laboratories to analyze plasmid replication biochemistry. Thus far, only plasmids of the pT181 family have been shown to replicate in extracts. Supercoiled DNA is required for activity; nicked, linear, and relaxed ccc DNA are all inactive. A single round of replication is observed, producing a double-stranded relaxed circle and a single-stranded monomer corresponding to the displaced leading strand; relaxed ccc DNA, however, is activated in the extracts by endogenous gyrase. Extracts prepared from plasmid-negative bacteria are inactive, and activity can be conferred by addition of the Rep protein specific for the plasmid.

Plasmid Screen Minilysates; Measurement of Copy Numbers. About 5×10^8 colony-forming units (cfu; medium-sized colony) is transferred to 50 μl lysis mixture in a 1.5-ml microcentrifuge tube, resuspended by vortexing gently, and incubated at 37° for 30 min. Fifty microliters 2% (w/v) sodium dodecyl sulfate (SDS) plus 5 μl proteinase K (10 mg/ml) is added, and the mixture incubated at 37° for 30 min. Tracking dye (11 μl) is then added, and the lysate vortexed at top speed for 30 sec, then freeze–thawed 3 times with dry ice–ethanol. The lysate can be stored at 4°. A 5-kb plasmid at 20 copies/cell will give a good agarose gel band with 10 μl of this preparation. If performed with measured samples of cells, this procedure is suitable for the determination of plasmid copy numbers,

[22] R. Novick, *J. Bacteriol.* **127,** 1177 (1976).
[23] S. A. Khan, S. M. Carleton, and R. P. Novick, *Proc. Natl. Acad. Sci. U.S.A.* **78,** 4902 (1981).

based on the ratio of plasmid to chromosomal DNA in the corresponding agarose gel bands, coupled with data on total cellular DNA. This ratio can be determined by fluorescence densitometry[24] or by radioactivity.

Small-Scale Plasmid Preparation for Restriction Analysis. Toothpick about 5×10^8 cfu to 200 μl SET buffer in a 1.5-ml microfuge tube, centrifuge 2 min, wash with 1 ml SET, resuspend in 150 μl lysostaphin mix (100 μl RNase I, 5 mg/ml, plus 50 μl lysostaphin, 1.5 mg/ml in SET), and incubate at 37° for 30 min. Add 350 μl lytic solution (1% SDS (w/v) in 0.2 N NaOH), hold on ice 5 min, add 250 μl of 1.5 M potassium acetate, pH 4.8, hold on ice 10 min, and centrifuge in a microfuge for 5 min at 4°. Place 0.6 ml supernatant (should be clear) in a new tube, add 0.7 ml 2-propanol, hold at room temperature for 10 min, and centrifuge at room temperature for 10 min; wash the pellet with 80% (v/v) ethanol, dry, and resuspend in 30 μl TE. Store at −20°. This DNA can be digested with most restriction enzymes and is suitable for plasmid sequencing, protoplast transformation, and nick translation.

Large-Scale Plasmid Preparation. Cells are grown in 500 ml CY broth by standing overnight at 37° then shaking for 4 hr at 37°. The cells are harvested by centrifugation [10 min at 5000 rpm, Sorvall SS34 rotor (Dupont-Sorvall, Chad Ford, PA)] and resuspended in 20 ml TES. Twenty milliliters of acetone–ethanol (1 : 1) is then added, and the mixture is held on ice for 15 min. Distilled water, 150 ml, is then added, and the suspension is kept on ice for an additional 15 min, collected by centrifugation as above, and washed once with TES. The cell pellet is resuspended in 27 ml of TES, lysostaphin (10 μg/ml, final) is added, and the suspension is incubated at 37° for 30 min. Note that the resulting acetone spheroplasts are not osmotically sensitive. Completion of spheroplasting can be tested by mixing 50 μl with 50 μl 2% SDS (w/v).

The cell suspension is apportioned in 7.5-ml aliquots to 50-ml centrifuge tubes, and 1.5 ml of 5 M NaCl, 0.9 ml of 0.5 M EDTA, and 10 ml of 2% SDS (w/v) plus 0.7 M NaCl solution are added to each. The contents are mixed by inversion, and the clear, viscous SDS–salt lysates are kept on ice for a minimum of 4 hr, then centrifuged for 45 min at 4° at 20,000 rpm in a Sorvall SS34 rotor. The supernatants are pooled into a single 250-ml centrifuge bottle and the volume determined. One-tenth the determined volume of 5 M NaCl plus one-ninth the determined volume in grams of polyethylene glycol (PEG 6000–8000) are then added, and the suspensions are shaken at 37° for 15 min to dissolve the PEG. The PEG solutions are kept on ice for 2 hr, then centrifuged at 7000 rpm for 15 min at 4°. The supernatant is immediately decanted, the DNA pellet resuspended in 2 ml

[24] S. J. Projan and R. P. Novick, *Mol. Gen. Genet.* **204,** 341 (1986).

TES, and the ccc DNA purified by equilibrium density gradient (dye–cesium) centrifugation.

Curing Class I Plasmids by Protoplast Regeneration. Protoplasts are prepared in SMM as described below for transformation and protoplast fusion. These are simply plated on DM3 agar and allowed to regenerate (2–3 days at 37°), selecting for any plasmids whose retention is desired. Regenerants will have lost any unselected plasmids at a frequency of 10–100% depending on the plasmid.[20]

Staphylococcus aureus Plasmid DNA Replication in Cell-Free Extracts[23]

Preparation of Extract. Ten milliliters of a brain–heart infusion (BHI) culture (37°, overnight) is used to inoculate 1 liter BHI (4-liter flask), and the culture is grown with vigorous shaking (250 rpm) to a Klett value of 350 (4 hr). The cells are pelleted (10 min, 5000 rpm at 4°, Sorvall GS3 rotor), resuspended in 75 ml cold TEG buffer (25 mM Tris, pH 8.0, 5 mM EGTA), and recentrifuged. These cells can be stored at $-70°$. In a cold room, resuspend cells in 4 ml TEG buffer (total volume should be 8.5 ml; cell suspension should be even and lack clumps), transfer to SW41 polyallomer tubes, freeze in an ethanol–dry ice bath and thaw at 15°. Add 2 M KCl to final concentration 150 mM, mix, repeat freeze–thaw, add lysostaphin (final concentration 200 μg/ml), mix thoroughly by inversion, and hold on ice for 45 min with periodic mixing by inversion. Repeat freeze–thaw as above. Centrifuge (in prechilled Spinco (Beckman Instruments, Inc., Palo Alto, CA) SW41 rotor) at 33,000 rpm for 60 min, 4°. Transfer the supernatant with Pasteur pipette to a 25-ml beaker. Add one-tenth volume streptomycin sulfate (fresh, 30%, w/v, in water) dropwise while stirring (magnetic stirrer) at 4°. Continue stirring for 30 min at 0°; centrifuge at 13,000 rpm for 10 min (Sorvall SS34 rotor) at 4°. Decant the supernatant, measuring the volume. Slowly add 0.472 g $(NH_4)_2SO_4$ per milliliter of supernatant while stirring slowly at 0°. Continue stirring 30 min, and centrifuge at 13,000 rpm for 10 min, 4° (Sorvall SS34 rotor). Save the protein pellet [$(NH_4)_2SO_4$, 70% saturation fraction]. The 70% $(NH_4)_2SO_4$ pellet is resuspended in TDE buffer [10 mM Tris-HCl, pH 8.0, 1 mM EDTA, 100 mM KCl, 1 mM dithiothreitol (DTT), 10% ethylene glycol], 1 ml/liter of original culture. Dialyze against 500 ml TDE buffer at 4°, 1–2 hr. Determine the protein content on a 1 : 50 dilution; store as 200- to 500-μl aliquots at $-80°$.

Replication assay. This system was developed for pT181/RepC and has been found effective for pC221/RepD. It is not known whether it will be effective for other plasmids. The assay can be used either with extracts prepared from a pT181-containing culture, which will contain endogenous RepC, or with extracts prepared from a plasmid-lacking culture, which will require the addition of RepC. The protein can be used as a crude extract or, better, can be purified by the method of Khan and co-workers.[25] Strains used for the preparation of extracts should be lacking in staphylococcal nuclease. The reaction is carried out in a 30 μl volume assembled as follows, and is incubated at 32° for 1 hr:

0.6 μl of 1 mM dTTP
1.0 μl of [α-^{32}P]dTTP (3 Ci/μmol, 10 μCi/μl)
3.0 μl R-mix*
0.5 μg template DNA (supercoiled pT181 plasmid)
Cell extract containing 0.2–1.0 mg protein†
100–500 ng RepC protein
Deionized, distilled water to 30 μl

The reaction is stopped by adding 30 μl stopping mix (25 μg/ml proteinase K, 50 mM EDTA, 2% SDS) and is incubated an additional 30 min at 37° to allow protein digestion. Forty microliters of carrier tRNA (250 μg/ml) is added, the reaction is extracted 3 times with an equal volume of phenol–chloroform, and the DNA is precipitated with 2-propanol: 40 μl of 7.5 M ammonium acetate plus 160 μl 2-propanol is added; the mixture is placed in a dry ice–ethanol bath for 10 min, then centrifuged in a Beckman microfuge for 10 min. The pellet is washed once with 70% ethanol, then vacuum-dried and resuspended in 10 μl TE buffer plus bromophenol blue–SDS loading dye.

Samples are separated on 1.2% agarose in Tris–acetate–EDTA plus 0.5 μg/ml ethidium bromide, 70 V, 17 hr. The gel is destained for 1 hr in water, then digested with RNase I (10 μg/ml in 10 mM Tris-HCl, pH 8.0) at 37° for 1 hr to remove carrier tRNA. The gel is restained with ethidium bromide, then dried onto Whatman 3MM paper and autoradiographed (Kodak, Rochester, NY, X-ray X-OMAT film).

Bacteriophages

Most *S. aureus* strains are multiply lysogenic, and the temperate phages are usually UV inducible and typically integrate at unique chromosomal sites by the Campbell mechanism.[26] Phages have morphologies

[25] R. R. Koepsel, R. W. Murray, W. D. Rosenblum, and S. A. Khan, *J. Biol. Chem.* **260**, 8571 (1985).
* R-mix: 40 mM Tris-HCl, pH 8.0, 100 mM KCl, 12 mM magnesium acetate, 1 mM DTT, 5% ethylene glycol (v/v), 2 mM ATP, 0.5 mM each of UTP, CTP, and GTP, 50 μM each of dATP, dGTP, and dCTP, 50 μM NAD, and 50 μM cAMP.
† In general, one must titrate RepC versus the extract, for each new set of preparations, to determine the optimal concentrations of each.
[26] C. Y. Lee and J. J. Iandolo, *J. Bacteriol.* **170**, 2409 (1988).

typical of temperate phages from other species and fall into three main serological groups, A, B, and F,[27] of which group B contains most of the known transducing phages, including ϕ11, ϕ147 (R. P. Novick, unpublished data, 1967), and typing phages 53, 79, 80, and 83. This grouping is well correlated with DNA sequence similarities.[28]

ϕ11, a prototypical group B transducing phage,[29] has a latency of around 60 min and a burst size of approximately 250 plaque-forming units, and it requires Ca^{2+} for growth as well as for adsorption. Host protein synthesis is shut down 30–40 min after prophage induction.[30] Temperature- and suppressor-sensitive as well as clear-plaque and virulent mutants have been isolated (R. P. Novick, unpublished data, 1968).[29,31] Lysogenization frequency is low (between 1 and 10%) and can be increased to over 90% by the addition of a growth-inhibitory concentration of chloramphenicol (e.g., 5–10 μg/ml) (R. P. Novick, unpublished observations, 1968). Its 45-kb genome is circularly permuted, terminally redundant, and flush ended.[32] It has been restriction mapped[32,33] and early and late regions plus a number of individual genes identified, including a late switch gene that is required for late protein synthesis and for shutdown of host protein synthesis.[30,31] Its *pac* site has been mapped,[33] and it is assumed that packaging is by sequential headfuls. Available physical-genetic data are summarized in Fig. 4. Virulent T-like phages[34] but no λ-like *cos*-containing phage have been described for staphylococci.

Several examples of lysogenic conversion have been documented; prophages have been found to carry the genes for enterotoxin A[35] and staphylokinase[36] and also transposons[37] and class I plasmids.[38] Additionally, prophages have been observed to cause negative lysogenic conversion, owing to the presence of phage attachment sites within certain structural genes. Two well-studied examples are the lipase[39] and β-hemolysin[40] structural genes.

[27] E. S. Anderson and R. E. O. Williams, *J. Clin. Pathol.* **9**, 94 (1956).

[28] B. Inglis, H. Waldron, and P. R. Stewart, *Arch. Virol.* **93**, 69 (1987).

[29] R. Novick, *Virology* **33**, 155 (1967).

[30] R. Chapple and P. R. Stewart, *J. Gen. Virol.* **68**, 1401 (1987).

[31] P. J. Kretschmer and J. B. Egan, *J. Virol.* **16**, 642 (1975).

[32] S. Lofdahl, J. Zabielski, and L. Philipson, *J. Virol.* **37**, 784 (1981).

[33] B. Bachi, *Mol. Gen. Genet.* **180**, 391 (1980).

[34] P. J. Rees and B. A. Fry, *J. Gen. Virol.* **53**, 293 (1981).

[35] M. J. Betley and J. J. Mekalanos, *Science* **229**, 185 (1985).

[36] I. Kondo, S. Itoh, and Y. Yoshizawa, *Zbl. Bakt. Suppl.* **10**, 357 (1981).

[37] E. Murphy, S. Phillips, I. Edelman, and R. P. Novick, *Plasmid* **5**, 292 (1981).

[38] M. Inoue and S. Mitsuhashi, *Virology* **72**, 322 (1976).

[39] C. Y. Lee and J. J. Iandolo, *J. Bacteriol.* **166**, 385 (1986).

[40] D. C. Coleman, J. P. Arbuthnott, H. M. Pomeroy, and T. H. Birkbeck, *Microb. Pathogen.* **1**, 549 (1986).

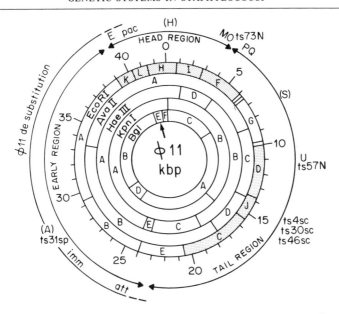

FIG. 4. Phage φ11 map. Restriction sites indicated were mapped by Bachi,[33] Lofdahl *et al.*,[32] and Novick *et al.* [R. P. Novick, I. Edelman, S. Lofdahl, *J. Mol. Biol.* **192**, 209 (1986)]. The arrow pointing to *Bgl*II E indicates the *pac* site[33]; "kbp" refers to the circular scale outside the *Eco*RI map. Early, head, and tail regions were mapped by suppressor-sensitive mutations,[31] and these regions were correlated with the physical map by cloning[32] (the fragments shaded in gray have been shown to cause high-frequency transduction when cloned to a plasmid; see text) and by the analysis of the φ11::pI258 plasmid–phage recombinant, φ11*de*.[29,32] Temperature-sensitive mutations (*ts*) isolated in the author's laboratory (N) (R. P. Novick, unpublished data, 1970), by Sweeney and Cohen (sc) [S. Cohen, H. M. Sweeney, and S. K. Basu, *J. Bacteriol.* **129**, 237 (1977)], or by Sjostrom and Philipson (sp) [J. Sjostrom and L. Philipson, *J. Bacteriol.* **119**, 19 (1974)] have been localized by cloning. Capital letters represent genes mapped by complementation of *sus* mutants[31]; *att* has been mapped by sequencing of the cloned prophage junctions,[26] and *imm* by cloning.[32]

Staphylococcal phages are highly species specific; I am unaware of any *S. aureus* phage that can propagate on any other staphylococcal species.

Phage Typing. Phage typing, widely practiced for epidemiological monitoring, is based on strain variations in susceptibility to a standard set of typing phages, all of which are naturally occurring temperate phages or clear-plaque mutants thereof. The phages all absorb to the same receptor located in the peptidoglycan–teichoic acid complex; differences in susceptibility are thus related to postadsorption phenomena including restriction–modification systems and host-, plasmid-, or prophage-mediated blockage of phage development. Blockage at the adsorption stage is extremely rare in natural isolates (although some strains possess a thick mucopolysaccharide capsule that interferes nonspecifically with phage

TABLE III
TRANSPOSONS AND RELATED ELEMENTS[a]

Element	Size (kbp)	Markers[b]	Terminal repeats (bp)	Target repeats (bp)	Target specificity	Related elements
Tn551	5.3	Em^r	40	5	None	Tn917
Tn554	6.7	Em^r, Sp^r	None	None	High	Tn3853
Tn4001	4.7	Em^r, Tb^r, Km^r	IS256	None	ND[c]	Tn3851
Tn552	6.7	Pc^r	ND	6	ND	Tn3852
						Tn4002
						Tn4201
Tn4003[d]	3.6	Tp^r	IS257	ND	ND	
IS431[d]	0.8	—	17–22	ND	ND	IS257
IS256[d]	1.35	—	ND	ND	ND	
Tn916	15.0	Tc^r	28	None	Low	Tn918
						Tn919
						Tn1545

[a] See text for references.

[b] Em^r, Erythromycin resistance; Sp^r, spectinomycin resistance; Tb^r, tobramycin resistance; Km^r, kanamycin resistance; Pc^r, penicillin resistance; Tp^t, trimethoprim resistance; Tc^r, tetracycline resistance.

[c] ND, Data unavailable.

[d] Transposability inferred from diversity of locations but not thus far demonstrated directly.

penetration), and the vast majority of phage-resistant mutations block phage propagation at a postabsorption stage. As staphylococcal phage typing lacks any firm theoretical basis and the susceptibility patterns are subject to frequent and uncharacterizable variation, its usefulness is rather limited.

Transposons

A variety of transposons have been described, each specifying resistance to one or two antibiotics. These transposons and their important features are listed in Table III (see Murphy[41] for a review). Tn551 from *S. aureus,* its close relative Tn917 from *S. faecalis,* and Tn552 and its relatives are class II transposons in the Tn3 family and transpose to more or less random sites, creating 5- or 6-nucleotide (nt) target duplications. Tn4001 (and presumably Tn3581) are class I transposons, similar in organization to Tn5 or Tn10. A directly repeated 1.35-kb DNA element flanking

[41] E. Murphy, in ''Mobile DNA'' (D. E. Berg, and M. M. Howe, eds.), p. 269. American Society for Microbiology, Washington, D.C., 1989.

the *mer* operon in plasmid pI258 has been designated IS431.[42] Homologous repetitive sequences are widespread in *S. aureus,* and the flanking repeats of Tn*4001*, designated IS256,[43] are very similar or identical to IS431.[42] Both IS431 and IS256 have been assumed to be independently transposable, justifying their designation as IS elements. Both Tn*551* and Tn*917* carry the classic macrolide–lincosamide–streptogramin (MLS) determinant, constitutive in Tn*551*, inducible in Tn*917*. Tn*551* and Tn*917* have strong hotspots, fortunately different, thus tending to favor different target sites.[1] Tn*916* uses a different set of sites.[44] Tn*4001* and Tn*3581*, carrying a Gm^r Km^r Nm^r marker, use target sites not highly preferred by Tn*551* and Tn*917*.[1] Transposon delivery is usually via thermosensitive replication-defective (Tsr) vectors such as pRN3032[45] and pRN5101.[46] Transposition to the chromosome occurs at a frequency of 10^{-3}–10^{-5}/cell; class II and class III plasmids are good targets for many of these transposons. Tn*551* seems unable to transpose to class I plasmids (A. Gruss and R. Novick, unpublished data, 1982).

Tn552 and Related Transposons. Tn*552* is a 6.7-kb β-lactamase transposon of the Tn*3* type that is found intact or as remnants at a variety of plasmid and chromosomal locations.[47] Transposition of intact transposons of this type has been observed in only a very small number of cases.[48] As with Tn*3*, Tn*552* resolvase (*bin*) is encoded in a region flanking the *res* site at which it acts.[47] This *bin–res* complex, present in certain plasmids and on the chromosome of certain *S. aureus* strains without the rest of the transposon, serves as a hotspot for Tn*552* transposition, first described by Asheshov.[49] The insertion of Tn*552* in this region may create an inverted repeat of the *bin–res* element, and in such cases one observes site-specific *rec*-independent inversion at high frequency.[50] Additionally, the *bin–res* complex can mediate high-frequency reversible cointegrate formation between elements that carry it.[51]

Tn554 and Tn3582. Tn*554* is an *S. aureus* transposon that carries Em^r

[42] L. Barberis-Maino, B. Berger-Bachi, H. Weber, W. D. Beck, and F. H. Kayser, *Gene* **59,** 107 (1987).

[43] B. R. Lyon, M. T. Gillespie, and R. A. Skurray, *J. Gen. Microbiol.* **133,** 3031 (1987).

[44] J. Jones, S. Yost, and P. Pattee, *J. Bacteriol.* **169,** 2121 (1987).

[45] R. P. Novick, I. Edelman, M. D. Schwesinger, D. Gruss, E. C. Swanson, and P. A. Pattee, *Proc. Natl. Acad. Sci. U.S.A.* **76,** 400 (1979).

[46] R. P. Novick, G. K. Adler, S. J. Projan, S. Carleton, S. Highlander, A. Gruss, S. A. Khan, and S. Iordanescu, *EMBO J.* **3,** 2399 (1984).

[47] S. J. Rowland and K. G. Dyke, *EMBO J.* **8,** 2761 (1989).

[48] D. A. Weber and R. V. Goering, *Antimicrob. Agents Chemother.* **32,** 1164 (1988).

[49] E. H. Asheshov, *J. Gen. Microbiol.* **59,** 289 (1969).

[50] E. Murphy and R. P. Novick, *Mol. Gen. Genet.* **175,** 19 (1979).

[51] E. Murphy and R. Novick, *J. Bacteriol.* **141,** 316 (1980).

and Sp^r (spectinomycin resistance) markers and has several remarkable features. It transposes with a frequency approaching 100% to a single chromosomal site[37] and neither contains terminal repeats nor generates target duplications.[52] It blocks transposition of a second copy by occupation of its *att* site and also by a specific, trans-acting interference determinant located at one end of the transposon.[53] In naturally occurring strains, it has been found at several additional (secondary) sites, and it can transpose to a secondary site on class II plasmids at a frequency below 10^{-8}.[37] This secondary site is related to one of the chromosomal secondary sites. The closely related Tn*3582* occurs naturally on a conjugative plasmid and behaves as a hitchhiking transposon,[54] by which is meant that, following entry of the carrier plasmid by conjugation, it immediately transposes to its chromosomal *att* site. Tn*554* has been sequenced, and three genes involved in transposition have been identified,[55,56] two of which are related to the λ Int family of recombinases.[57]

Tn*554* provides an ideal vehicle for returning cloned fragments to the chromosome. A vector constructed by cloning the chromosomal *att554* site to a Tsr derivative of pT181 was used as a target for Tn*554* insertion, and we have used the resulting plasmid[55] (see Fig. 5) for cloning to the unique *Pst*I site in the Tn*554* *spc* determinant, which inactivates Sp^r. On transfer of the clone to a Tn*554*$^-$ recipient, Tn*554* transposes to its normal chromosomal site, carrying the cloned insert. The vector is then eliminated by growth at 43°. Such Tn*554* derivatives are stable and express genes in the cloned insert (J. Kornblum and R. Novick, unpublished data, 1989), presumably using the *spc* promoter or any promoters present in the insert. Further refinements of this system are currently in progress in the author's laboratory.

Conjugative Transposons. Conjugative transposons are genetic elements that are capable of intercell transfer by conjugation but are incapable of autonomous replication. They evidently undergo excision and circularization followed by conjugative transfer, then recircularize and insert into

[52] E. Murphy and S. Lofdahl, *Nature (London)* **307**, 292 (1984).

[53] E. Murphy, *Plasmid* **10**, 260 (1983).

[54] D. E. Townsend, S. Bolton, N. Ashdown, D. I. Annear, and W. B. Grubb, *Aust. J. Exp. Biol. Med. Sci.* **64**, 367 (1986).

[55] M. C. F. Bastos and E. Murphy, *EMBO J.* **7**, 2935 (1988).

[56] E. Murphy, L. Huwyler, and M. Bastos, *EMBO J.* **4**, 3357 (1985).

[57] P. Argos, A. Landy, K. Abremski, J. B. Egan, E. Haggard-Ljungquist, R. H. Hoess, M. L. Kahn, B. Kalionis, S. B. L. Narayana, L. S. Pierson III, N. Sternberg, and J. M. Leong, *EMBO J.* **5**, 433 (1986).

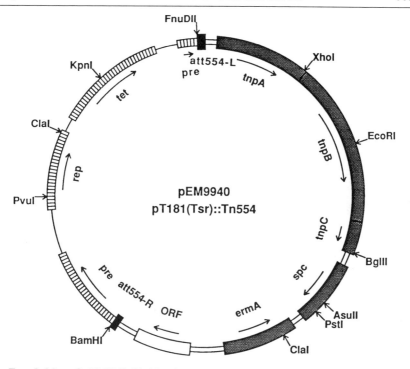

FIG. 5. Map of pEM9940 (E. Murphy, personal communication, 1988). pT181-derived sequences are shown as a thin line and genes as striped boxes (see Fig. 2). Solid boxes represent the chromosomal Tn554 att site cloned to pT181 and subsequently divided in two by the insertion of Tn554 (heavy lines and shaded boxes).

the genome of the target cell.[58] Conjugative transposons Tn916, Tn918, and Tn1545, originally isolated in streptococci,[59–61] can be readily transferred to staphylococci and many other bacterial species, where they insert into the chromosome at a variety of sites.[44] The transfer of Tn918 to S. aureus is a second example of transposon hitchhiking. Here, the transposon inserts into a conjugative plasmid such as pAD373 in the Streptococcus faecalis donor; the complex is transferred to S. aureus where the plasmid cannot replicate, and the transposon then excises and

[58] C. Gawron-Burke and D. B. Clewell, J. Bacteriol. 159, 214 (1984).
[59] P. Courvalin and C. Carlier, Mol. Gen. Genet. 205, 291 (1986).
[60] C. Gawron-Burke and D. B. Clewell, Nature (London) 300, 281 (1982).
[61] J. R. Scott, P. A. Kirschman, and M. G. Caparon, Proc. Natl. Acad. Sci. U.S.A. 85, 4809 (1988).

TABLE IV
TRANSPOSON DELIVERY VECTORS

Plasmid	Replicon	Features	Transposon	Frequency	Selection for jumps
pRN3206[a]	pI258	Tsr, Cd^r	Tn551	10^{-4}–10^{-5}	Em 43°
pLTV1[b]	pE194	Tsr, Tc^r	Tn917	10^{-4}–10^{-5}	Em 43°
pEM9940[c]	pT181	Tsr, Tc^r	Tn554	1.0	Em 43°
pPQ61[d]	pI258	Tsr, Cd^r	Tn4001, Tn551	10^{-4}–10^{-5}	Em 43°
					Gm 43°

[a] R. P. Novick, *in* "Genetic Interaction and Gene Transfer" (C. W. Anderson, ed.), Brookhaven Symposia in Biology, Vol. 29, p. 272. Brookhaven National Laboratory, Upton, New York, 1978.
[b] P. Youngman, personal communication, 1990; see Fig. 6.
[c] E. Murphy, personal communication, 1988, see Fig. 5.
[d] P. Pattee, personal communication, 1990.

reinserts in the *S. aureus* chromosome.[62] Additionally, suggestive evidence exists for conjugative transposons native to *S. aureus*.[63]

In Table IV is a list of plasmids useful as transposon delivery vectors in gram-positive species. Note in particular the Tn917 system. Tn917 has been engineered to a state of high versatility by Youngman.[64] The basic plasmid vector is pE194 or its Tsr mutant, pRN5101.[46] In the Youngman vector set, improved cloning sites have been introduced, a promoterless *lacZ* or *cat* (chloramphenicol acetyltransferase) gene has been inserted at one end, enabling the transposon to serve as a promoter probe, and a derivative has been prepared with the ColE1 replication origin cloned within the transposon. This permits direct cloning in *Escherichia coli* of sequences flanking the inserted transposon. A map of pLTV1, a typical member of this set of vectors, is shown in Fig. 6. These tools have been used extensively in *B. subtilis* but not, thus far, in staphylococci. One problem is that staphylococci are Lac$^+$, and most species use phosphorylated intermediates in the lactose fermentation pathway, so that X-Gal (5-bromo-4-chloro-3-indolyl-β-D-galactoside) would not be useful for screening colonies for promoter-distal insertions. *o*-Nitrophenyl-β-D-galactoside ONPG), however, which is phosphorylated and hydrolyzed, is a useful

[62] D. B. Clewell, F. Y. An, B. A. White, and C. Gawron-Burke, *J. Bacteriol.* **162**, 1212 (1985).
[63] N. El Solh, J. Allignet, R. Bismuth, B. Buret, and J. Fouace, *Antimicrob. Agents Chemother.* **30**, 161 (1986).
[64] P. Youngman, *in* "Plasmids, A Practical Approach" (K. G. Hardy, ed.), p. 79. IRL Press, Oxford, 1987.

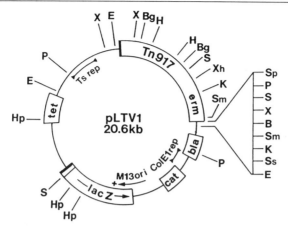

FIG. 6. Map of pLTV1 (P. Youngman, personal communication, 1990). The region from 7 to 12 o'clock represents a pE194 derivative in which the *ermC* gene has been replaced by *tet* and which carries a mutation causing thermosensitive replication. The rest represents a highly engineered derivative of Tn*917*, which contains promoterless *lacZ* at one end, the M13 + o as well as the ColE1 origin, and *cat* and *bla* markers followed by a multiple cloning site and then by the rest of the transposon. Restriction sites: B, *Bam*HI; Bg, *Bgl*II; E, *Eco*RI; H, *Hin*dIII; Hp, *Hpa*II; K, *Kpn*I; P, *Pst*I; S, *Sal*I; Sp, *Sph*I; Sm, SmaI; X, *Xba*I; Xh, *Xho*I.

lac indicator for *S. aureus*. Additionally, some species use the classic *lac* fermentation pathway, and X-Gal can be used with these (G. Stewart, personal communication, 1990).

Transposon Mutagenesis with Tn551 or Tn4001. An exponential culture of the strain to be mutagenized, carrying plasmid pPQ61, which is a Tsr derivative of pI258 and contains both transposons, is plated on media containing either Em (10 μg/ml) or Gm (10 μg/ml) and incubated at 43° for 24–48 hr. Platings should be made over a range of 10^6–10^8 cfu/plate to establish a number that gives 100–300 colonies representing transposon insertions at 43°. Colonies are tested for loss of the plasmid, and the inserted transposon is backcrossed to the target strain (by transduction) to confirm that the observed phenotype is due to the transposon. Note that the entire plasmid can also insert into the chromosome and that this type of insertion is strongly favored in ϕ11 lysogens, in which the plasmid inserts at one end of the prophage.[65] Lysogens of ϕ11 and related phages should clearly be avoided for transposon analysis.

[65] M. D. Schwesinger and R. P. Novick, *J. Bacteriol.* **123**, 724 (1975).

Other Variable Genetic Elements

Variable genetic elements (VGE) occur in some but not other strains of a given species. Aside from plasmids, prophage, IS elements, and transposons, a number of variable chromosomal determinants have been characterized. Typically, these contain one or more identifiable markers flanked by additional unique sequences. Three examples in *S. aureus* are the determinants of toxic shock syndrome toxin 1 (TSST-1), staphylococcal enterotoxin B (SEB), and methicillin resistance (Mc^r). Additionally, a large set of probably linked chromosomal antibiotic resistance genes has been identified in a series of widely dispersed Mc^r hospital isolates sensitive to typing phage 88.[66] Among the variable elements, the TSST-1 element may occupy at least two different chromosomal locations.[67] These elements may turn out to be transposons, prophages, or integrated plasmids; however, presently available data are insufficient to permit any such specific identification, and they have provisionally been designated heterologous insertions (HI).[67] VGE that lack phenotypic markers are provisionally designated IS sequences. Two examples have been identified in staphylococci (see, e.g., Fig. 3). Elements with sequences very similar or identical to those of the flanking repeats of Tn*4001* have been found in various plasmid and chromosomal locations without the rest of the transposon, and are referred to collectively as IS256.[16] A second IS-like element (IS257/IS431) has also been identified at various locations.[16,42] To date, neither of these has been observed either to transpose or to promote plasmid cointegrate formation.

Genetic Exchange

Staphylococci participate in all of the standard types of bacterial genetic exchange. The presence of indistinguishable plasmids and other VGE in various species suggests natural horizontal transfer, and this is presumed to occur by conjugation.[68] Intraspecific plasmid transfer by transduction has been demonstrated in infected tissue[69] and on skin.[70] A poorly characterized phenomenon known as prophage-mediated conjugation[71] is unique to staphylococci.

[66] S. Schaefler, *Antimicrob. Agents Chemother.* **21**, 460 (1982).
[67] M. C. Chu, B. N. Kreiswirth, P. A. Pattee, R. P. Novick, M. E. Melish, and J. J. James, *Infect. Immun.* **56**, 2702 (1988).
[68] R. W. McDonnell, H. M. Sweeney, and S. Cohen, *Antimicrob. Agents Chemother.* **23**, 151 (1983).
[69] R. P. Novick and S. I. Morse, *J. Exp. Med.* **125**, 45 (1967).
[70] M. H. Richmond and R. W. Lacey, *in* "Contributions to Microbiology and Immunology, Volume 1: Staphylococci and Staphylococcal Infections," p. 135. Karger, Basel, 1973.
[71] R. W. Lacey, *J. Gen. Microbiol.* **119**, 423 (1980).

Recombination

On the basis of anecdotal observations, one has the impression that staphylococci are not as proficient in generalized recombination as species such as *Escherichia coli* and *Bacillus subtilis*. For example, differentially marked homologous incompatible multicopy plasmids can be maintained indefinitely in a *rec*⁺ background (by double selection) without generating doubly resistant recombinants.[72] Cloned chromosomal fragments can also be stably maintained in the plasmid state in a *rec*⁺ background, as can directly repeated sequences cloned to a plasmid.

Several *rec*⁻ mutants have been isolated.[73] The most stringent, a nitrosoguanidine-generated *recA*-like 8325-4 derivative, RN981, shows a 10^4-fold reduction in generalized recombination frequency, shows UV-induced DNA degradation, and lacks the SOS response.[74] This strain is a very poor grower and has several other mutations, including a defect in one of the two known restriction–modification systems of NCTC 8325[75]; attempts to transfer the *rec* mutation to a nonmutagenized background have been unsuccessful. A second *recA*-like mutant, *rec-4*, grows better and shows at least a 400-fold reduction in recombination frequency.[73]

Site-specific recombination involving plasmids has been observed, and one group of small plasmids encodes a site-specific *rec* function, *pre*, that acts on a plasmid sequence known as RS_A.[10] Unlike the site-specific *rec* systems of other plasmids, the *pre*/RS_A system is not involved in multimer resolution. As noted above, the Tn552 resolvase,[47] *bin*, is closely related to the *hin* (flagellar antigen invertase), *pin* (P1 invertase), and *gin* (Mu invertase) systems of *E. coli*, and it may be involved in the reversible insertion of plasmids into the staphylococcal chromosome and in interplasmid recombination as well as intraplasmid inversion.[50]

Temperate phage are assumed to encode integrase functions, and phages such as $\phi 11$ are capable of directing generalized recombination.[74]

Markers and Nutritional Factors

Because most staphylococci require several amino acids and vitamins, complex media such as BHI, TSB, and CY, are used normally; various synthetic media have also been designed and used primarily for nutritional analysis or for genetic manipulation using auxotrophic markers.[76,77] The

[72] S. J. Projan and R. P. Novick, *Plasmid* **12,** 52 (1984).
[73] R. V. Goering, *Mutat. Res.* **60,** 279 (1979).
[74] L. Wyman, R. V. Goering, and R. P. Novick, *Genetics* **76,** 681 (1974).
[75] S. Iordanescu, *Arch. Roum. Pathol. Exp. Microbiol.* **35,** 111 (1976).
[76] A. B. Dalen, *J. Gen. Microbiol.* **74,** 53 (1973).
[77] L. Rudin, J. Sjostrom, M. Lindberg, and L. Philipson, *J. Bacteriol.* **118,** 155 (1974).

TABLE V
STAPHYLOCOCCAL RESISTANCE MARKERS

Marker					Resistance level	Selective concentration	Expression lag[b]
Gene[a]	Phenotype[a]	Mechanism	Ind[a]	Source[a]			
Naturally occurring							
cat	Cm^r	Acetylase	+	P	10 μg/ml^c	5 μg/ml	0
tet	Tc^r	Efflux	+	P, C	25 μg/ml	5 μg/ml	1 hr
tmn(tetM)	Tc^r Mn^r		+	P, T, C	25 μg/ml	5 μg/ml	0
erm	Em^r(MLS^r)	Ribosome methylase	+	P, T	>1 mg/ml	5 μg/ml	0
cadA	Cd^r	Efflux	+	P, C	10 mM	40 mM	0
asa	As^r	Efflux	+	P, C	1 mM		0
bla	Pc^r	β-Lactamase	+	P, T, C	0.1 μg/ml^c	0.1 μg/ml	0
mec	Mc^r	PBP2a	+	C	5–1000 μg/ml	5 μg/ml	0
tmp	Tp^r	DHFR		P	>1 mg/ml	—	
aacA-aphD	Gm^r	Acetylation, phosphorylation		P, T, C		5–10 μg/ml	0
aacA-aphD	Km^r	Acetylation, phosphorylation		P		50 μg/ml	0
qac	Qa^r	Exclusion(?)		P	30 μg/ml (Eb^r)		
linA^d	Lm^r	Inactivation		P	64	5 μg/ml	2 hr
Laboratory mutations							
gyrB	Nb^r	Gyrase	−	C		5 μg/ml	
rpo	Rf^r	RNA polymerase	−	C		10 μg/ml	
	Sm^r	Ribosome protein	−	C	>1 mg/ml	1 mg/ml	
	Fa^r	Unknown	−	C			

[a] Ind, Inducibility; P, plasmid; T, transposon; C, chromosome; erm, erythromycin ribosome methylase; lin, lincomycin resistance; qac, quaternary amine resistance; As^r, arsenate resistance; Nb^r, novobiocin resistance. An extensive bibliography for these markers can be found in B. R. Lyon and R. Skurray, Microbiol. Rev 51, 88 (1987).

[b] Time of incubation on nonselective medium for optimal yield of transcipients.

[c] Single-cell resistance levels. Resistance is proportional to inoculum size.

[d] R. Leclercq, C. Carlieur, J. Duval, and P. Courvalin, Antimicrob. Agents Chemother. 28, 421 (1985).

markers used most commonly are resistance markers, either naturally occurring or selected as laboratory mutants. Plasmid- and transposon-linked resistances to a wide variety of antibiotics and other toxic substances have been documented and analyzed, and resistance mechanisms have been determined for a considerable number. These data have been carefully reviewed and tabulated by Lyon and Skurray.[15] Of the naturally occurring resistance markers, Tc^r, Em^r, Cm^r, and Cd^r plus others listed in Table V are the most useful. Most are well expressed whether in single copy on the chromosome or on a high copy plasmid. An exception is that the level of chloramphenicol resistance obtained with the pC221 Cm^r

marker in single copy is barely above background (R. P. Novick, unpublished data, 1985).

The following precautions apply: as most of these drugs are bacteriostatic, sensitive cells may persist for long periods of time in their presence. When using such drugs to aid in the maintenance of an unstable plasmid, it should be remembered that the culture will always contain plasmid-negative cells, and the more unstable the plasmid, the greater will be the proportion of negatives. This situation is particularly problematical with Cm, which is inactivated by the Cm^r gene product. Since a few resistant cells can protect a large population of sensitives, it may be impossible to maintain a highly unstable plasmid using only Cm. This situation is even worse with penicillin, as the staphylococcal β-lactamases are exceptionally effective. Penicillin is therefore an extremely poor selective marker for β-lactamase-mediated resistance in *S. aureus* and is ordinarily used only as a last resort. It is useless for the maintenance of unstable plasmids.

Some aminoglycoside resistance markers, such as Nm^r, Km^r, and Gm^r, are useful for maintenance and for selection, but they have not been successfully used in protoplast transformation. The Tn554-linked Sp^r marker and the plasmid-linked Sm^r can be used for scoring only, in both cases because of the high frequency of spontaneous mutations to resistance. Laboratory-induced chromosomal Rf^r (rifampin resistance) Nb^r, Fa^r (fusidic acid resistance), and Sm^r mutations have also been found useful as selective markers; Fa^r is used in conjunction with another marker as the frequency of spontaneous Fa^r mutants is unacceptably high. Conditions for the use of these and other common markers are given in Table V.

Conjugation and Mobilization

As all early examples of plasmid transfer in mixed cultures of staphylococci could be attributed to spontaneous transduction, it was assumed that conjugation did not occur in this genus. In the early 1980s, however, in parallel with the emergence of gentamicin resistance, interspecies transfer of gentamicin resistance from *S. epidermidis* to *S. aureus* was observed in several laboratories and soon shown to be due to conjugative plasmids carrying this marker.[68,78] These results provided biological relevance for the previously observed plasmid–protein relaxation complexes in *S. aureus*.[22] Similarly to those of ColE1 and related *E. coli* plasmids, these complexes are involved in conjugative mobilization.[79]

Conjugative plasmids in staphylococci belong to class III and always

[78] B. A. Forbes and D. R. Schaberg, *J. Bacteriol.* **153**, 627 (1983).
[79] G. L. Archer, *J. Bacteriol.* **171**, 1841 (1989).

carry gentamicin resistance plus some combination of other resistance markers including Qa[r], Tm[r], Pc[r], and resistances to other aminoglycosides.[15] Most are in the size range of 40–60 kb and maintain low copy numbers. One of the prototypes (pSH6), however, is only 18 kb in size.[68] Conjugative transfer is most efficient in mixed cultures on filters, with frequencies in the range of 10^{-5}–10^{-6}/donor cell. A mutant derivative of the conjugative plasmid pCRG1600 with an elevated transfer frequency has been isolated by Asch et al.[80]

Genetic analysis of the tra systems of two of the plasmids, pG01[16] and pCRG1600,[80] has revealed a 14-kb region flanked by copies of the putative IS sequence, IS431 (IS257), and divisible into at least two complementation units.[16] Neither pili nor conjugative pheromones are involved; however, S. aureus elaborates a conjugative pheromone, cAM373, that activates conjugation by the streptococcal plasmid, pAM373,[62] enabling transfer of the conjugative hitchhiking transposon, Tn918, to S. aureus. Conjugative S. aureus plasmids can efficiently mobilize those small plasmids that can form relaxation complexes.[79] These complexes, described for 4.5-kb Cm[r] and Sm[r] plasmids of the pT181 family, involve two plasmid-coded proteins, MobA and MobB, and a site, MobS, at which they act.[79] MobA and MobS are sufficient for relaxation; all three elements are required for mobilization. Remarkably, mobilization occurs at a 100-fold greater frequency than transfer of the conjugative plasmid itself.

As mentioned, streptococcal conjugative transposons such as Tn916, Tn918, and Tn1545 are functional in S. aureus. Elements of this type do not seem able to mobilize adjacent DNA (D. Clewell, personal communication, 1990); however, this capability offers an attractive potential for development.

In addition to classic mobilization by conjugative plasmids, small S. aureus plasmids participate in a DNase-resistant transfer system that requires cell contact plus the presence in the donor of a transducing phage[71]; this transfer nonetheless occurs at a frequency several orders of magnitude higher than spontaneous transduction. This high frequency in conjunction with the requirement for cell contact has been taken to imply a novel transfer mechanism, which is known as phage-mediated conjugation.[81]

Conjugation and Mobilization. Aliquots (0.3 ml) of overnight 37° standing TSB cultures are mixed and vacuum-filtered onto 47-mm 0.45-μm pore size Millipore (Bedford, MA) membranes which are then transferred to TSA plates (cell side up) and incubated for 4–18 hr at 37°. The bacterial growth is then eluted from the filter and samples plated on media selective

[80] D. K. Asch, R. V. Goering, and E. A. Ruff, *Plasmid* **12,** 197 (1984).
[81] S. Iordanescu and M. Surdeanu, *J. Gen. Microbiol.* **96,** 277 (1976).

for the desired transconjugants. Sm^r and Nb^r are useful as counterselective markers; Rf^r is best used in conjunction with a second marker such as Fa^r to suppress background mutations. Conjugation frequencies with plasmids such as pG01 (Gm^r) are of the order of 10^{-6}/donor cell; mobilization of plasmids such as pC221 occurs at a frequency of about 10^{-4}/donor cell.

Transduction

Transduction was the first genetic exchange mechanism discovered in staphylococci and remains the commonest and simplest routine method. Staphylococcal transducing phages are generalized transducers of the P22 type. The best studied is ϕ11,[29] which is a typical temperate *S. aureus* phage (see above). Transduction frequencies are in the range of 10^{-4}–10^{-6} for plasmids and 10^{-6}–10^{-8} for chromosomal markers; this difference in frequencies is probably due to factors such as abortive transduction, copy number, and packaging efficiency. Because of the low frequency of lysogenization, transduction with phage-sensitive recipients is very difficult or impossible unless phage growth is inhibited; sodium citrate (17 mM in GL agar) is highly effective. Resistant mutants, as well as lysogenic strains or strains containing the cloned immunity determinant, make excellent recipients (it will be recalled that phage resistance in *S. aureus* is at a postabsorption stage). Lysates are sterilized by filtration (0.45 μm Millipore) as chloroform inactivates the phage.

Transduction between natural isolates of *S. aureus* may be difficult or impossible, because of restriction barriers. Conjugation and protoplast fusion (see below) are likely to succeed in such cases.

Staphylococcal transducing phages encapsulate single genome-sized DNA molecules.[82,83] Plasmids larger than the phage genome are truncated; those smaller are induced to form linear multimers during phage growth.[83] It is probable that these multimers are formed as a consequence of inhibition by the phage of termination of the rolling circle plasmid replication. Plasmid transduction frequency by ϕ11 can be increased dramatically by the insertion of a random ϕ11 fragment into the plasmid. Frequencies approaching 50% of the total phage population have been achieved by this means, and phage fragments as small as 300 nt give essentially the full effect.[83,84] As such plasmids are not dramatically amplified during phage growth,[83] it has been suggested that plasmid–phage recombination is responsible for the high transduction frequency. As the transducing particles lack phage DNA other than the cloned fragment and contain only linear

[82] D. W. Dyer, M. I. Rock, C. Y. Lee, and J. J. Iandolo, *J. Bacteriol.* **161**, 91 (1985).

[83] R. P. Novick, I. Edelman, and S. Lofdahl, *J. Mol. Biol.* **192**, 209 (1986).

[84] S. Lofdahl, J. Sjostrom, and L. Philipson, *J. Virol.* **37**, 795 (1981).

multimers of the plasmid, and as the effect of cloned phage fragments is sequence independent, we have proposed that the recombination takes place late in the phage growth cycle, between the vegetative phage and long plasmid multimers. This has the effect of introducing the phage *pac* site into the plasmid multimers.[83]

Preparation of φ11 Transducing Lysates

Infective lysates. A mid-exponential culture in CY broth at approximately 3 × 10^8 cfu/ml (Klett reading 180, 540-nm filter) is diluted 1 : 1 with phage buffer, infected with φ11 at a multiplicity of 0.1–1, and shaken slowly at 30°. Lysis should occur in about 2 hr. The lysate is centrifuged at 10,000 rpm at room temperature for 10 min, then filtered (0.45-μm membrane, Millipore). Plaque titers are determined by dilution and plating in 2.5 ml phage top agar on phage plates, using RN450 as the indicator, 10^8 cfu/plate. Titers are usually in the range of 10^{10}–10^{11}/ml; transduction titers are usually 10^4–10^6 for plasmids, 10^2–10^4 for chromosomal markers.

UV-induced lysates. Ten milliliters of a mid-exponential CY culture at about the same density as above is centrifuged, resuspended in the same volume of phage buffer, and irradiated with a single shortwave 15-W UV bulb at 50 cm for 20 sec with gentle shaking. Ten milliliters of CY broth is added, and the irradiated cells are incubated at 30° with gentle shaking. Lysis should occur within 2 hr. The lysate is centrifuged and filtered as above. Plaque titers are usually 10^9–10^{10}/ml.

Transduction. One-tenth milliliter of the recipient culture, exponential cells in broth at about 5 × 10^8 cfu/ml or cells from a fresh overnight plate, is infected with an equal volume of transducing lysate, either undiluted or diluted 10- or 100-fold depending on the source of the donor marker (plasmid or chromosome), the quality of the lysate, etc. After 5 min at room temperature, GL top agar (at 45°) is added (3 ml), and the mixture is plated on the appropriate medium, depending on the selective marker. For Tc, Cm, or Cd, plating may be directly on media containing the selective agent (Tc and Cm at 5 μg/ml; Cd(NO₃)₂ at 40 μM). For Em, plating is on nonselective medium; plates are incubated 2 hr at 37°, then overlaid with 3 ml GL top agar containing Em, 100 μg/ml. Alternatively, a 10-ml bottom layer containing 300 μg Em may be poured, allowed to harden, and followed by a 20-ml top layer lacking drug. These "underlay" plates should be used immediately on hardening and are particularly useful when the Em^r marker is inducible.

For selection of auxotrophic nucleotide markers, CV medium is satisfactory; if prepared with acid-hydrolyzed casein and if tryptophan is omitted, this medium can be used for tryptophan auxotrophs. For selection of auxotrophic amino acid markers other than tryptophan, synthetic media containing all amino acids except that used as a marker, plus the other ingredients as in CV medium are used. Satisfactory amino acid concentrations can be found in various publications.[2,76] For transduction with phage-sensitive recipients, 0.17 mM sodium citrate should be added to GL medium, including top agar. For other media, the optimal citrate concentration may be different.

Transformation

Competent Cell Transformation. Staphylococcus aureus has a specific competence system that is active very early in exponential growth.[77] Transformation efficiency is increased in nuclease-negative mutants,[85] and these are generally the standard recipients. A remarkable feature of competent cell transformation is its enhancement by lysogeny with group B phages[77] or by their products. Both phage tails[86] and the product of early gene 31[77] have been reported to enhance competence. Typing phage 55 is used most commonly for enhancement of competence.[2] Efficiencies of the order of 10^4 transformants/μg are about the best that one can expect with this system; its main advantage is that it is effective with chromosomal markers whereas protoplast transformation is not.

Protoplast Transformation. Protoplasts prepared with lysostaphin and stabilized with sucrose plus Mg^{2+} have been the primary transformation recipients, especially for plasmids.[87] Transformation is accomplished by shocking the protoplasts with PEG in the presence of DNA, then plating on DM3 regeneration medium,[88] which uses sodium succinate for osmotic stabilization. Tc^r, Em^r, and Cm^r markers work well; heavy metals and aminoglycosides do not perform well in DM3 medium. Regeneration requires at least 2 days, and transformants may continue to appear for as long as 1 week. Transformation frequencies in the range of 10^5-10^6/μg DNA can be achieved, optimally. Note that all equipment coming in contact with protoplasts must be rigorously free of detergents. Glassware is safe if acid-washed; standard disposable commercial glass- and plasticware are satisfactory.

It is widely held that plasmid transformation efficiency is inversely proportional to size. I am unaware of any carefully controlled experiment documenting this and remain unconvinced that the effect is not simply a matter of molecular concentration.

[85] G. S. Omenn and J. Friedman, *J. Bacteriol.* **101**, 921 (1970).
[86] M. P. Jackson, J. DeSena, J. Lednicky, B. McPherson, R. Haile, R. G. Garrison, and M. Rogolsky, *Infect. Immun.* **39**, 939 (1983).
[87] J. Polak and R. P. Novick, *Plasmid* **7**, 152 (1982).
[88] S. Chang and S. N. Cohen, *Mol. Gen. Genet.* **168**, 111 (1979).

Staphylococcus aureus strains have a variety of restriction systems and generally will not accept DNA prepared from *E. coli* or *B. subtilis,* although *S. aureus* DNA can usually be transferred to standard strains of either of these species with little difficulty. We have isolated a mutant of strain 8325-4, RN4220, by selection for the ability to accept DNA from *E. coli,* that is probably defective in one or more key restriction systems that are responsible for this interspecific barrier. Other restriction-defective mutants such as SA113[81] have also proved to be useful as recipients for foreign DNA.

Protoplast Fusion. Protoplast fusion[89] has been used for chromosomal mapping (as noted above) and for plasmid transfer. The usual result of protoplast fusion in *S. aureus* is the segregation of chromosomal haploids containing all plasmids present in both parents (R. P. Novick, unpublished data, 1981). Thus, a common strategy for the selection of fusants is to have a selectable plasmid in each parent and to include both drugs in the regeneration medium. Chromosomal recombinants are infrequent and can ordinarily be isolated only by selection. Protoplast fusion is an effective means of transferring plasmids among different gram-positive species, especially when there are restriction or other barriers to interspecific genetic exchange.

Electroporation. Lack of success in initial attempts at electroporation suggested that this technique might not be useful for staphylococci. However, several laboratories have recently succeeded very well with this technology,[90,91] and it is likely to supplant protoplast transformant in the near future, especially for plasmids. Whether electroporation will be effective for chromosomal markers is not yet clear.

Protoplast Transformation and Fusion. Protoplast transformation is performed according to the method of Chang and Cohen[88] as modified for *S. aureus.*[92] The *S. aureus* recipient is grown in 25 ml of 2× Penassay broth (Difco) to mid-log phase (5 × 10^8 cells/ml), washed once with the same broth, and resuspended in 5 ml SMM. Lysostaphin (50 μg/ml) is added, and the cells are incubated at 37° for 15–20 min. Protoplasts are pelleted as gently as possible by centrifugation (1900 *g* for 10 min at room temperature) and resuspended in SMM at 1/10 the original culture volume. For transformation, DNA, in a volume of 0.25 ml, is mixed with an equal volume of 2× SMM and 0.5 ml of protoplasts added. Then 1.5 ml of 40%

[89] P. Schaeffer, B. Cami, and R. D. Hotchkiss, *Proc. Natl. Acad. Sci. U.S.A.* **73,** 2151 (1976).

[90] J. B. Luchansky, P. M. Muriana, and T. R. Klaenhammer, *Mol. Microbiol.* **2,** 637 (1988).

[91] L. Masson, G. Prefontaine, and R. Brousseau, *FEMS Microbiol. Lett.* **60,** 273 (1989).

[92] S. Carleton, S. J. Projan, S. K. Highlander, S. Moghazeh, and R. P. Novick, *EMBO J.* **3,** 2407 (1984).

(w/v) PEG 6000 is immediately added and mixed by gentle inversion. After 2 min, 5.0 ml of SMMP is added, and the protoplasts are collected by centrifugation at 1900 g for 10 min at room temperature. Protoplasts are resuspended in 1 ml of SMMP and incubated at 32° with gentle shaking for 2 to 3 hr, then plated on DM3 agar containing the appropriate antibiotic (see Table V). Colonies begin to appear after 48 hr and may continue to appear for as long as 1 week. For protoplast fusion, protoplasts of the second strain (0.5 ml) are substituted for DNA, and the procedure is otherwise the same.

Competent Cell Transformation

Preparation of competent cells. Cells from an overnight BHI plate are resuspended in 600 ml oxoid tryptone soya broth plus 1 mM CaCl$_2$ to give an OD reading of 0.01 (540 nm), divided into six 100-ml batches in 300-ml flasks, and incubated with shaking (100 rpm) at 35° for 5 min, then divided in six equal aliquots and washed with Tris–maleate buffer, pH 7.5 (Trizma–maleate 11.9 g/liter, 10 N NaOH 5 ml/liter). The competent cells can be quick-frozen and stored at $-75°$ for at least 1 month.

Transformation procedure. Competent cells from one tube are resuspended in 2 ml Tris–maleate buffer plus 100 mM CaCl$_2$ and transforming DNA added (0.1 ml, 100–500 μg/ml in TE). The tube is vortexed and held on ice 3 min, then incubated for 3 min at 35° with gentle shaking. The cells are centrifuged at 5000 rpm for 5 min at room temperature, resuspended in 3 ml BHI plus 0.4 mM sodium citrate, incubated for 1 hr at 35°, then pelleted, resuspended in 1.0 ml of 0.1 M NaCl, and plated on selective media with dilutions as needed.

Electroporation. Recipient cells are grown in 100 ml TSB at 37° with shaking to an OD$_{540\,nm}$ of 0.2–0.25 and are then washed 4 times with 0.5 M sucrose at 0°, reducing the volume by steps from 100 to 25 to 10 to 5 to 1 ml final. Next, 160 μl of this cell preparation is added directly to an ethanol-precipitated and dried DNA sample (10–100 ng DNA for plasmids such as pC194), and the mixture is held on ice for 15 min. Electroporation is then performed (Bio-Rad, Richmond, CA, Gene Pulser, 2-mm cuvettes) at the following settings: 2.5 kV, 400 Ω, 25 μF, pulse time, less than 5 msec. One milliliter TSB is then added, and aliquots are plated on TSA, GL, BHI, etc., containing selective agents as required (see Table V).

Cloning and Sequence Analysis

Most cloning of chromosomal DNA from *S. aureus* has used *E. coli* as the primary recipient for standard *E. coli* vectors or *E. coli–S. aureus* shuttle vectors (see below). In general, standard procedures have been used successfully, and all genes cloned thus far have been expressed in *E. coli*. For fragments, such as plasmid fragments that can readily be obtained in pure form, direct cloning in *S. aureus* by protoplast transformation or by electroporation is standard.

A series of cloning vectors useful in staphylococci and in several other gram-positive species has been developed from the small multicopy *S. aureus* plasmids, and shuttle vectors using the ColE1 replicon have been constructed using these same plasmids.[93] Promoter–probe vectors using β-lactamase as a reporter gene and an expression vector using the transcriptionally inducible β-lactamase promoter have been developed from the *S. aureus* plasmid pC194 and the classic inducible β-lactamase gene from the class II plasmid, pI258.[94] Similar vectors have been developed using the *Bacillus pumilus cat-86* gene and the *S. aureus* plasmid pUB110.[95] Vectors have also been developed in which a protein of interest can be translationally fused to staphylococcal protein A. Such fusions permit the product to be secreted and purified by affinity chromatography using immunoglobulins.[96] Both the *bla* vectors and protein A vectors are available as *E. coli–S. aureus* shuttle vectors. Table VI is a list of vectors useful for cloning and expression in staphylococci. *Staphylococcus carnosus* has been developed as a nonpathogenic host for cloning and expression of staphylococcal genes, again using these same plasmids as vectors.[97] Two-vector systems suitable for returning cloned fragments to the chromosome are the Tn554-based pEM9940 (see p. 604 and Fig. 5) and a series of vectors which use the cloned phage 42D *att/int* mechanism.[97a] Maps of several of these vectors are shown in Fig. 7.

Preparation of Cell Extracts and Supernatants for Protein Analysis

Cells are harvested by centrifugation, washed and resuspended in TE buffer, lysed with lysostaphin as for plasmid miniscreens, and vortexed to shear DNA. Culture supernatants are concentrated 20-fold by ultrafiltration using Amicon (Danvers, MA) YM10 membranes or precipitated by methanol–chloroform as described by Wessel and Flugge[98] and resuspended in 1/20 volume of water. Cell lysates or concentrated supernatants mixed with an equal volume of $2\times$ sample buffer and boiled for 5 min are suitable for analysis by polyacrylamide gel electrophoresis.

[93] S. D. Ehrlich, *Proc. Natl. Acad. Sci. U.S.A.* **75**, 1433 (1978).

[94] P. Wang, S. J. Projan, K. Leason, and R. P. Novick, *J. Bacteriol.* **169**, 3082 (1987).

[95] P. S. Lovett, D. M. Williams, E. J. Duvall, and S. Mongkolsuk, *in* "Genetics and Biotechnology of Bacilli" (A. T. Ganesan and J. A. Hoch, eds.), p. 275. Academic Press, Orlando, FL, 1984.

[96] B. Nilsson, L. Abrahmsen, and M. Uhlen, *EMBO J.* **4**, 1075 (1985).

[97] G. Keller, K. H. Schliefer, and F. Gotz, *Plasmid* **10**, 279 (1983).

[97a] C. Lee, S. A. Buranen, and Z.-H. Ye, *Gene* (in press) (1991).

[98] D. Wessel and U. I. Flugge, *Anal. Biochem.* **138**, 141 (1984).

TABLE VI
CLONING VECTORS[a]

Plasmid	Original replicon	Selection marker	Size (kb)	Copy number	Useful features	Ref.[b]
pRN5543	pC194	Cm^r	2.9	90–100	pUC19 polylinker in unique HindIII site[a]	(1)
pRN6441	pE194	Em^r	3.7	50–60	pUC18 polylinker in unique PstI site,[c] Tsr	(2)
pRN6725	pC194	Cm^r	3.7	90–100	See Fig. 7B	(3)
pWN1818	pC194, ColE1	Cm^r	10.4	15	See Fig. 7A	(4)
pWN1819	pC194, ColE1	Cm^r	10.4	15	See Fig. 7A	(4)
pHV33	pC194, ColE1	Cm^r			Shuttle vector	(5)
pRIT16	pC194	Cm^r, Ap^r	6.9		See Fig. 7C	(6)
pHV1431	pAMB1, ColE1	Cm^r, Ap^r		2–5	Low copy shuttle vector using θ replication mechanism	(7)
pNZ12		Cm^r, Km^r	4.1	100	Cm^r Km^r single-replicon shuttle vector for gram-positive and gram-negative bacteria	(8)
pPL703	pUB110	Km^r	~5		Promoter–probe vector, see Fig. 7D	(9)

[a] Derived from pSK265 [C. L. Jones, M. B. Johns, G. J. Mussey, and S. A. Khan, *Proc. Natl. Acad. Sci. U.S.A.* **82**, 5850 (1985)] by removal of one of the two HindIII sites in the pSK265 polylinker region (S. Projan and R. Novick, unpublished data, 1988). Ap^r, Ampicillin resistance.

[b] Key to references: (1) C. L. Jones, M. B. Johns, G. J. Mussey, and S. A. Kahn, *Proc. Natl. Acad. Sci. U.S.A.* **82**, 5850 (1985); (2) M. L. Gennaro and R. P. Novick, *J. Bacteriol.* **170**, 5709 (1988); (3) R. P. Novick, H. F. Ross, S. J. Projan, J. Kornblum, B. Kreiswirth, and S. Moghazeh. *EMBO J.* (in press) (1991); (4) P. Wang, S. J. Projan, K. Leason, and R. P. Novick, *J. Bacteriol.* **169**, 3082 (1987); (5) S. D. Ehrlich, *Proc. Natl. Acad. Sci. U.S.A.* **75**, 1433 (1978); (6) L. Abrahmsen, Ph.D. Thesis, Royal Institute of Technology, Stockholm, Sweden (1988); (7) L. Janniere, C. Bruand, and S. D. Ehrlich, *Gene* **87**, 53 (1990); (8) A. F. M. Simons and W. M. de Vos, *Chem. Mag. (Rijswijk, Neth.)* **49**, 813 (1985); (9) E. J. Duvall, D. M. Williams, P. S. Lovett, C. Rudolph, N. Vasantha, and M. Guyer, *Gene* **24**, 171 (1983).

[c] The cloned polylinker is flanked by PstI sites.

Sequence Analysis

Codon Usage. Table VII contains a list of the staphylococcal genes that have been sequenced; the overall codon usage derived from these sequences is essentially as expected for the 35% G + C content typical of staphylococci.

Promoters and Shine–Dalgarno (S–D) Sequences. Table VIII is a list of several staphylococcal promoters that have been mapped to date,

A

GAA GCT TGC CAG TGA ATT CGA GCT CGG TAC CCG GGG ATC CTC TAG AGT CGA CCT GCA GGC ATG CAA GCT TC
Hind III*   EcoRI Sst I*   KpnI*SmaI* BamHI* XbaI   SalI   Pst I* SphI*   Hind III*

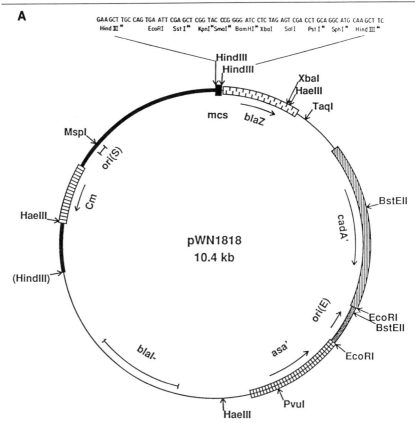

FIG. 7. Maps of cloning vectors useful for *S. aureus*. (A) pWN1818[94] is a promoter–probe vector derived from pA07 [A. Oka, N. Nomura, M. Hiroyuki, and K. Sugimoto, *Mol. Gen. Genet.* 151 (1979)], a fusion of the *cadA, blaZ*-containing 6.9-kb *Eco*RI-B fragment of pI258 [R. P. Novick, E. Murphy, T. J. Gryczan, E. Baron, and I. Edelman, *Plasmid* **2**, 109 (1979)] to a 0.5-kb segment containing the ColE1 origin, oriE (12 to 9 o'clock). pC194 (9 to 12 o'clock) was inserted into the unique *Hin*dIII site of pA07, the *bla* promoter and S–D site were eliminated by *Bal*31 digestion, and finally the pUC18 polylinker (mcs) was inserted as shown.[94] Versions of this vector with stop codons in all three reading frames, with the polylinker in the opposite orientation, or with the *bla* SD sequence intact are available. (*Hin*dIII) represents a *Hin*dIII site that has been filled in and religated; cadA' and asa', truncated *cadA* and *asa* genes; blaI, the *bla* control region with a constitutive mutation. (B) pRN6725 is a promoter vector derived by inserting the *Hin*dIII–*Xba*I fragment containing the *bla* promoter plus two-thirds of the *blaZ* structural gene into the polylinker region of pRN5543, a pC194 derivative. DNA inserted into the polylinker is transcribed from the *bla* promoter, which can be made inducible by providing the *bla* repressor in trans. P-bla, β-Lactamase promoter; blaZ', truncated *blaZ*; palB, dyad region of unknown function; *rep*, *rep* gene of pC194; *ori*, leading strand replication origin; *cat*, Cm[r] gene; palA, lagging strand replication origin. (C) Protein A fusion vectors pRIT16/pRIT21–23 were constructed by cloning to pEMBL9 a 1.1-kb fragment containing a 3' truncated *spa* gene lacking the coding

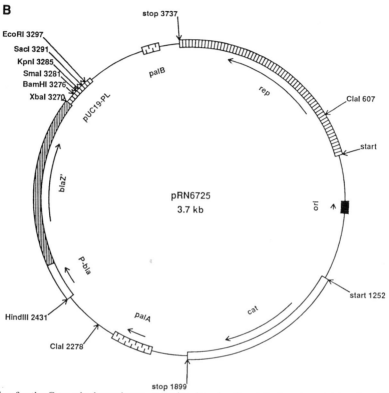

B

stop 3737

EcoRI 3297
SacI 3291
KpnI 3285
SmaI 3281
BamHI 3276
XbaI 3270

pUC19-PL

palB

rep

ClaI 607

start

blaz'

pRN6725
3.7 kb

ori

P-bla

HindIII 2431

ClaI 2278

palA

cat

start 1252

stop 1899

region for the C-terminal membrane-spanning domain of protein A and containing the pUC18
polylinker sites just 3' to the *spa* fragment. Staphylococcal plasmid pC194 was then inserted
between ColE1 and *spa* regions to give pRIT16 and a transcription termination signal, T,
inserted just past the polylinker region to give pRIT21–23. Cloning to the polylinker in this
vector will, if in frame, produce fusion proteins that may be periplasmic in *E. coli,* secreted
in *S. aureus* (depending on the fused protein). The native *spa* promoter is present and is
supplemented by the *E. coli* lac UV5 promoter (arrowhead) [L. Abrahmsen, Ph.D. Thesis,
Royal Institute of Technology, Stockholm, Sweden (1988)]. (D) pPL703 consists of a 1250-bp
*Pst*I–*Bgl*II fragment of *B. pumilus* NCIB 8600 DNA inserted between the *Eco*RI and *Bam*HI
sites of pUB110 by use of a 21-bp *Eco*RI–*Pst*I fragment from M13mp7 [S. Mongkolsuk, Y.
Chiang, R. B. Reynolds, and P. S. Lovett, *J. Bacteriol.* **155,** 1399 (1983); D. M. Williams,
J. Duvall, and P. S. Lovett, *J. Bacteriol.* **146,** 1162 (1981)]. The promoterless gene *cat-86*
resides within the 1250-bp fragment, and the gene is followed by an efficient transcription
termination signal, designated *ter* [S. Mongkolsuk, E. J. Duvall, and P. S. Lovett, *Gene* **37,**
83 (1985)]. The pUB110 portion of pPL703 provides an origin of replication and a neomycin
resistance gene (Neo^r). *cat-86* specifies chloramphenicol acetyltransferase when the gene is
transcriptionally activated by inserting a promoter into any of four unique restriction sites
5' to *cat-86*: *Eco*RI, *Bam*HI, *Sal*I, and *Pst*I. RBS-1, RBS-2, and RBS-3 designate the
approximate locations of ribosome binding sites identified by their complementarity to *B.
subtilis* 16 S rRNA. Since the *cat-86* regulatory sequences (→←) are intact (see Fig. 9), Cm
inducibility is retained. Accordingly, with a promoter-containing derivative of pPL703,
cloning of any gene in frame into the *cat-86* coding sequence will result in a Cm-inducible
fusion protein. (Reprinted from Lovett *et al.,*[95] by kind permission of the publishers.)

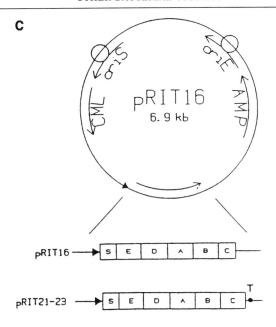

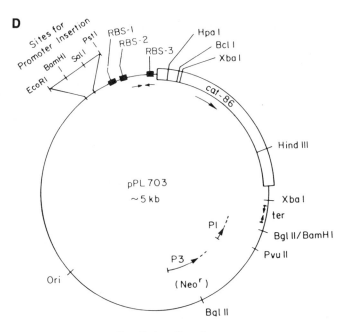

FIG. 7. (*continued*)

TABLE VII

TABLE VII

Staphylococcus aureus Sequences in GenBank[a]

GenBank index	Description	Size (bp)
PB0110CG	Plasmid pUB110 complete genome	4548
PB2CAT	Plasmid pUB112 chloramphenicol acetyltransferase (*cat*) gene	901
PC1CAT	Plasmid pC194 chloramphenicol acetyltransferase (*cat*) gene	1036
PC2CG	Plasmid pC221 complete genome	4555
PI25BLAZA	Plasmid pI258 β-lactamase (*blaZ*) gene	1148
PI25CADA	Plasmid pI258 cadmium resistance (*cadA*) gene	3534
PI25MER	Plasmid pI258 mercury resistance (*mer*) operon	6404
STAAGLSRA	Tn*4001* aacA-aphD aminoglycoside resistance gene	3541
STAAGR	*agr* gene encoding an accessory gene regulator	964
STAATTB	Bacteriophage φ11 attachment site (*attB*)	300
STACATPC	Plasmid pC223 chloramphenicol acetyltransferase (*cat*) gene	3534
STAENTB	Enterotoxin B (*ent*) gene	1712
STAENTE	Enterotoxin type E (*entE*) gene	774
STAETA	Exofoliative toxin A (*eta*)	1391
STAETB	Exofoliative toxin B (*etb*)	1368
STAFEMA	Accessory gene (*femA*) for methicillin resistance	3446
STAFNBP	Fibronectin-binding protein (*fnbA*) gene	3342
STAGEH	Lipase (glycerol esterase) (*geh*) gene	2968
STAHLB	β-Hemolysin (*hlb*) gene	2187
STAL54BOB	Phage L54 *attB* site	320
STAL54BOP	Phage L54 *attL* site	320
STAL54POB	Phage L54 *attR* site	320
STAL54POP	Phage L54 *attP* site	320
STALACS	Enzyme III-lac (*lacF*), enzyme II-lac (*lacE*) genes	3712
STALINA	Lincosamide nucleotidyltransferase (*linA*) gene	1396
STANUCF	Staphylococcal nuclease (*nuc*) gene	966
STAPBP	Gene for PBP2a (methicillin resistance)	2322
STAPE5A	Plasmid pE5, complete genome	2473
STAPT48CG	Plasmid pT48 complete genome	2475
STASCAG	Staphylocoagulase (*coa*) gene	3042
STASP	V8 serine protease (*spr*) gene	1634
STASPA	Protein A (*spa*) gene	1921
STATETM	Tetracycline resistance (*tetM*) gene	2900
STATNIS5	Tn*554* insertion site (strain RN450)	249
STATOXA	α-Hemolysin (*hla*) gene from strain Wood 46	1485
STATOXAA	Enterotoxin A (*entA*) gene	774
STATSST1	Toxic shock syndrome toxin-1 (*tst*) gene	731
TRN4311	IS431, 5′ copy	841
TRN4312	IS431, 3′ copy	853
TRN431MEC	IS431 (from *mec* region)	711
TRN554	Tn*554*, complete genome	6691
STARRASA	5 S ribosomal RNA	115
BPHATTP	Bacteriophage φ11 attachment site (*attP*)	300
L54INTXIS	Bacteriophage L54a *int* and *xis* genes	1626
SPCSAK	Bacteriophage S-φ-C staphylokinase (*sak*) gene	1377
X13290	Plasmid pSK1, complete genome	4884
X14827	*lacC* and *lacD* genes	1995

[a] GenBank.

TABLE VIII
Staphylococcus aureus Promoter Regions

Promoter	Sequence[a]	Ref.[b]
	−35 −10 +1	
pT181*repC*	aaagacttgatctgATTAGAccaaatcttttgatagTGTTATattaataA	1
pT181-*ct*	cacccaaatatataTCTTGAtgtatatttaaatatcgTTTAATatctaaA	1
pT181*pre*	acgtattaacgactTATTAAaaataagtctagtgtgtTAGACTtaaactA	2
pT181*tet*	aaatcgactttaaaAGCGAAcgaaaaacaattgcaaaAACAAAtcgattT	3
agr-P3	cacagagatgtgatGGAAAAtagttgatgagttgtttAATTTTaagaatT	4, 5
agr-P2	acagttaggcaataTAATGAtaaaagattgtactaaaTCGTATaatgacA	4, 5
pI258*blaZ*	ttataataaactaTTGACAccgatattacaattgtaaTATTATtgatttA	6
spa	tatagattttagtaTTGCAAtacataattcgttatatTATGATgactttA	7
hla	taatagttaattaaTTGATTtaattctaagatatttgTATTTTtatattG	8

[a] Nucleotides representing the −35 and −10 consensus elements and the mRNA start are capitalized.

[b] (1) E. C. Kumar and R. P. Novick, *Proc. Natl. Acad. Sci. U.S.A.* **82,** 638 (1985); (2) M. L. Gennaro, J. Kornblum, and R. P. Novick, *J. Bacteriol.* **169,** 2601 (1987); (3) R. P. Novick, G. K. Adler, S. Majumder, S. A. Khan, S. Carleton, W. Rosenblum, and S. Iordanescu, *Proc. Natl. Acad. Sci. U.S.A.* **79,** 4108 (1982); (4) L. Janzon, S. Lofdahl, and S. Arvidson, *Mol. Gen. Genet.* **219,** 480 (1989); (5) R. P. Novick, H. F. Ross, S. J. Projan, J. Kornblum, B. Kreiswirth, and S. Moghazeh, *EMBO J.* In press. (1991); (6) J. R. McLaughlin, C. L. Murray, and J. C. Rabinowitz, *J. Biol. Chem.* **256,** 11283 (1981); (7) D. Sullivan, J. Kornblum, and R. P. Novick, unpublished data (1989); (8) J. Kornblum and R. P. Novick, unpublished data (1990).

consisting of −10 and −35 regions identified by the mapping of transcription start sites. In general, these sequences deviate considerably from the *E. coli* consensus, possibly reflecting the existence of alternative σ factors. In Table IX is a list of Shine–Dalgarno sequences, based on genes whose translational start is known or can be inferred with high probability by inspection of the sequence. As has been suggested,[99] these sequences conform more closely to the canonical complementarity with 16 S ribosomal RNA (although it must be admitted that the ribosomal RNA sequence is unknown for staphylococci and the canonical homology is based on the *B. subtilis* sequence) than do those from *E. coli.*

Methods

Induction and Measurement of β-Lactamase. The prototype staphylococcal β-lactamase is inducible and is encoded by plasmids such as pI524[100] and pI258. The cloned promoter is constitutive and can be regulated in

[99] J. R. McLaughlin, C. L. Murray, and J. C. Rabinowitz, *J. Biol. Chem.* **256,** 11283 (1981).
[100] R. P. Novick, *Biochem. J.* **83,** 229 (1962).

TABLE IX
TRANSLATION START SIGNALS IN *Staphylococcus aureus*[a]

Gene	Sequence
pUB110*kan*	tatctgaaaagggaATGAGAAtagtgaATG
pC194*cat*	tagataaaatttAGGAGGCatatcaaaATG
pC221*cat*	gatttaaaatttAGGAGGAaattatatATG
pI258*blaZ*	tacaactgtaatatCGGAGGGtttattTTG
pT181*repC*	aaattgagattaAGGAGTCgattttttATG
Tn*4001tnp*	caaaaacataccCAGGAGGActttttacATG
Tn*4001aac*	tatgattatgaaaaAGGTGATaaataaATG
entB	gggaatgttggataaAGGAGATaaaaaATG
hla	gttcaaaaaaatAGAAGGAtgatgaaaATG
eta	aaaacgcaaatgttAGGATGAttaataaATG
etb	tataaaagttaaaAGGAGGTtttatatATG
fnbA	attttgcatttaaaGGGAGATattataGTC
hlb	gtaagctatataaaAGGAGTGataatgATG
hld	atcttaattaaggaAGGAGTGatttcaATG
EIII[mtl]	aaaatattaaaaaTGGAGTGatcgattATG
geh	tcaaaaaaatacttTTTGAGGtgattatATG
lacF	aaataagtttaaaAGGGGATtaatgctATG
lacG	tgaaattaagaatAGGAGTTtttcatATG
linA	tcaaattcctaaattAGGAGGGgtaaaATG
pE5*ermC*lp	atactaattttataAGGAGGAaaaaatATG
pE5*ermC*	attataaccaaattaAAGAGGGttataATG
vgh	taatttaaataaaaCGGAGGGgatagaATG
coa	attacatttTGGAGGAAttaaaaaattATG
spr	gtaaataaattttttTGGAGGTttttagATG
spa	ttacaaatacatacAGGGGGTattaatTTG
tetM	ttgggtttttgaaTGGAGGAaaatcacATG
Tn*554tnpA*	ttaaagtatttaAAGAGGTgggaacatGTG
Tn*554tnpB*	atacctcgagagaaAGGAGCAtaagaaATG
Tn*554tnpC*	attggaagtttgACGGGGTaattatcaATG
Tn*554spc*	tatgaacataatcaACGAGGTgaaatcATG
Tn*554ermA*	attataaccagtaAGGAGAAggttataATG

[a] See Table VII for GenBank references. *tnp*, transposase; *aac*, aminoglycoside acetylase; *entB*, enterotoxin B; *eta*, exfoliatin A; *etb*, exfoliatin B; *fnbA*, fibronectin-binding protein A; *EII*[mtl], phosphotransferase enzyme III (mannitol); *geh*, glycerol ester hydrolase (lipase); *lac*, lactase; *ermC*lp, *ermC* leader peptide; *vgh*, virginiamicin; *coa*, coagulase; *spa*, staphylococcal protein A. Nucleotides corresponding to the 16 S ribosomal RNA complementarity and start codons are capitalized.

trans by a coresident copy of the regulatory elements.[100a] In the absence of an inducer, the repressed promoter has a residual activity of 1–3% of the fully induced level; the fully induced level is 2- to 3-fold higher than that obtained with the promoter alone, consistent with the earlier finding that this regulation has both positive and negative components.[101]

The most effective inducer is 2-(2'-carboxyphenyl)benzoyl-6-amino-penicillanic acid (CBAP) (BDH Ltd., Poole, England), a gratuitous inducer that is fully effective at 5 µg/ml. Induction is at the level of transcription (R. P. Novick, unpublished data, 1988), follows classic kinetics, and is therefore essentially complete in 3–4 cell generations (2–3 hr at 37°). There are many methods of measuring β-lactamase, in colonies on agar or in liquid culture.[102,103] Three are described here.

Starch–iodide method for colonies.[103] Plates are prepared with the desired medium plus 0.2% potato starch. After the growth of colonies, the plate is flooded with a 1% fresh aqueous solution of penicillin G and incubated at 37° for 10–30 min. The penicillin is then poured off and the plate flooded with 80 mM I_2 plus 3.2 M KI. β-Lactamase activity is indicated by a clear halo around the colony which increases in size with time. The relative halo size is an indication of relative activity. The concentrations of penicillin and iodine can be varied to change the sensitivity of the method. Inducibility can be tested by streaking a culture on a starch plate and placing a filter disk containing CBAP (10 µg) at one end of the streak.

PNCB method for colonies.[104] Plates with colonies are dried (2 hr, 37°), then flooded with 5 ml of a 0.25% (w/v) solution of the acid–base indicator N-phenyl-1-naphthylamine 4-azo-o-carboxybenzene (BDH Ltd., Poole, England) in N,N-dimethylformamide, 5.8 mM NaOH. The plates are kept at room temperature for 30 min, then the excess dye poured off; the plates are allowed to dry (5–10 min) and then are flooded with a fresh 1% (w/v) aqueous solution of penicillin G. β-Lactamase-positive colonies turn dark red at a rate depending on their activity. Fully induced *S. aureus* containing wild-type pI524 β-lactamase turn in about 5–10 sec, uninduced in 1–2 min. This method is much less sensitive than the starch–iodide method. Its lower limit is about 2 units/mg dry weight, where 1 unit equals 1 µmol/hr. Its advantage is that the acid form of the indicator is insoluble, so that sectoring patterns indicative of plasmid instability may be produced (see Fig. 8). This method requires a medium such as 0.3GL agar that has a

[100a] R. P. Novick, H. F. Ross, S. J. Projan, J. Kornblum, B. Kreiswirth, and S. Moghazeh, *EMBO J.* (in press) 1991.
[101] M. H. Richmond, *J. Mol. Biol.* **26**, 357 (1967).
[102] J. M. T. Hamilton-Miller, J. T. Smith, and R. Knox, *J. Pharm. Pharmacol.* **15**, 81 (1963).
[103] J. Perret, *Nature (London)* **174**, 1012 (1954).
[104] R. P. Novick and M. H. Richmond, *J. Bacteriol.* **90**, 467 (1965).

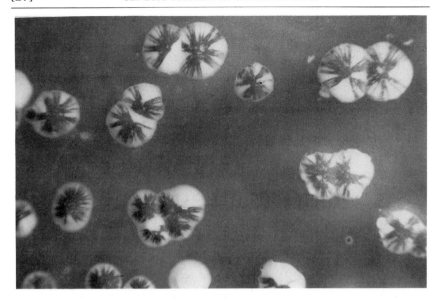

FIG. 8. PNCB test for β-lactamase in colonies.[104] Colonies of a strain containing pRN1053, an unstable derivative of pI524, constitutive for β-lactamase, were grown on GL agar at 32°, stained with PNCB, and developed with penicillin as described. Dark areas represent plasmid-containing sectors; light areas, plasmid-negative segregants.

relatively low buffering capacity and does not undergo major changes in pH during the growth of bacteria.

The indicator may darken during staining owing to exposure to CO_2. This should be corrected by the addition of a few drops of 0.1 N NaOH before pouring off the excess indicator.

Microtiter plate assay with nitrocefin. Nitrocefin is a chromogenic β-lactamase substrate, turning from yellow to pink (absorption maximum 490 nm) on hydrolysis by β-lactamase. We have adapted the published method[105] for use with microtiter plates for convenience; because the maximum attainable concentration of substrate is subsaturating, it is useful to have several different dilutions to choose from. To perform the assay, one dispenses 50 μl of 50 mM potassium phosphate buffer, pH 5.9, into the wells of a microtiter plate and then does serial 2- or 3-fold dilutions of samples to be assayed, starting with a 25-μl sample in well 1. Whole cultures, supernatants, cells, or enzyme preparations at various stages of purification can all be assayed effectively. If viable organisms are present, the phosphate buffer must contain a growth inhibitor such as 0.1 M NaN$_3$.

[105] C. H. O'Callaghan, A. Morris, S. M. Kirby, and A. H. Shingler, *Antimicrob. Agents Chemother.* **1**, 283 (1972).

After the dilutions have been completed, 50 μl of a 0.1 mM solution of nitrocefin (BBL, Baltimore, MD) (5.16 mg/100 ml 50 mM potassium phosphate buffer, pH 7.0, dissolved by stirring overnight with a magnetic stirrer) is added and the plates read (at 490 nm) immediately and at 10-min intervals for 1 hr (or longer if needed). For each dilution series, two or three dilutions are chosen that give a final OD 990 nm below 0.2, the time points are plotted, and slopes are determined for the linear initial portion of the curve. Activities are calculated by the following formula:

$$\text{Units/ml} = \frac{(\text{slope})(Vd)}{(\varepsilon_m)ls}$$

where slope is the slope in absorbance units/hr, V is the volume of reaction (0.1 ml), s the amount of sample in the first well (usually 16.7 μl), d the dilution factor, l the light path in cm (0.3 for a round-bottomed microtiter plate containing 0.1 ml), and ε_m the millimolar extinction coefficient for nitrocefin (15.9). The apparent inconvenience of this assay is offset by the advantage of a simple chromogenic substrate and by the availability of computational software.

Chloramphenicol Acetyltransferase Assay. The chloramphenicol acetyltransferase (CAT) assay measures the CoA sulfhydryl group released by the transfer of acetate from acetyl CoA to Cm. The sulfhydryl group reacts with 5,5'-dithiobis(2-nitrobenzoic acid) (DTNB), releasing 5-thio-2-nitrobenzoate which has a molar extinction coefficient of 13,600 at 412 nm.

The reaction mixture (freshly prepared) contains 4 mg DTNB in 1.0 ml of 1 M Tris-HCl, pH 7.8, 0.2 ml of 5 mM acetyl-CoA (stored frozen), and 8.8 ml water. Two milliliters of this mixture is added to the enzyme sample to be assayed (0.01–1.0 ml), and the reaction is started by adding one-tenth volume of 1.0 mM Cm. The reaction is best run in a temperature-regulated (37°) recording spectrophotometer and the observed reaction rate corrected by subtracting the rate of color change in a blank cuvette (lacking Cm). Enzyme units, defined as micromoles Cm acetylated per minute at 37°, are calculated by dividing the corrected slope (ΔOD/min) by 13.6. To obtain units/ml enzyme, multiply by V/S, where V is the reaction volume and S the volume of the enzyme sample (in ml). Mercaptoethanol or other sulfhydryl-containing compounds interfere with the reaction, as does high thioesterase activity that may be present in crude extracts. Certain varieties of CAT have essential thiol groups that react with DTNB, inactivating the enzyme. If this occurs too rapidly to obtain linear initial reaction rates, some other procedure would have to be used.

RNA Isolation

Guanidium isothiocyanate method.[106] Acid-washed glassware or disposable plasticware should be used once the culture is grown. Diethyl pyrocarbonate 0.1% (w/v) should be added to all solutions coming in contact with RNA. An overnight plate of *S. aureus* is used to inoculate 100 ml of CY broth, and the cells are grown at 37° until late exponential phase, Klett value approximately 400. Cells are pelleted using acid-washed Corex tubes and washed in 1× SMM. The washed pellet is then resuspended in 10 ml of 1× SMM and the cells protoplasted with lysostaphin (100 μg/ml). After 45 min on ice, the protoplasts are centrifuged (1900 g for 10 min at 4°). The pellet is resuspended in 2 ml of 5 M guanidinium isothiocyanate and vortexed for 1 min to shear the DNA. This lysate is layered on a 3-ml CsCl cushion (density 1.76) and centrifuged at 35,000 rpm for 17 hr in a Beckman SW50.1 rotor. Following the gentle removal of the CsCl, the pelleted RNA is resuspended in 1 ml of 8 M guanidine hydrochloride and transferred to a Corex tube. The sides of the centrifuge tubes are rinsed with an additional 1 ml guanidine hydrochloride, and 50 μl of 1 N acetic acid is added to the pooled 2 ml and the RNA precipitated with 1.5 ml of 95% ethanol. After holding at −20° for 1 hr, the precipitate is centrifuged at 10,000 rpm for 15 min and resuspended in 5 ml guanidine hydrochloride and again ethanol precipitated. The pellet is dissolved in 3 ml distilled water, and the RNA is precipitated by the addition of 0.3 ml 3 M sodium acetate, pH 5.2, and 7.5 ml ethanol. This precipitation is repeated and the final pellet rinsed twice with 80% ethanol. The final pellet can be stored in ethanol at −20° or resuspended in 1 ml TE buffer and stored at −75°.

RNA screening lysates. A rapid method of preparing RNA suitable for quantitative blot hybridization is described here.[107] Approximately 5 × 10⁸ fresh exponential phase cells are collected by centrifugation at 4° in a 1.5-ml microcentrifuge tube. The cells are washed with 500 ml cold TES and resuspended in 100 μl SET buffer containing 10 μg/ml lysostaphin. The suspension is incubated on ice for 20 min, then at 37° for 3 min. Lysis is effected by adding 100 μl 2% SDS while vortexing. Ten microliters of 5 mg/ml proteinase K is added, and the lysate is shaken for 15 min at room temperature on an Eppendorf mixer. To relieve viscosity, the lysate is frozen and thawed twice (−70° to 55°). Samples are prepared for electrophoresis by adding 50 μl tracking dye; these can be stored at −75°.

[106] J. Chirgwin, A. Przabyla, R. Macdonald, and W. Rutter, *Biochemistry* **18**, 5294 (1979).
[107] J. Kornblum, S. J. Projan, S. L. Moghazeh, and R. Novick, *Gene* **63**, 75 (1988).

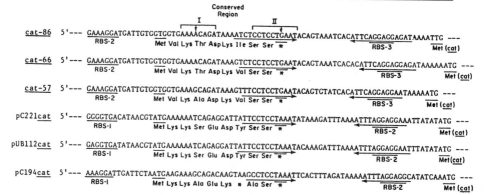

FIG. 9. Leader regions 5′ to the coding sequences for six inducible *cat* genes. Conserved regions I and II are found in all *cat* leaders, and the distance between these regions is 8 nucleotides. The two small vertical arrows designate the bases in conserved regions I and II that vary among the leaders. References for the sequences are as follows: *cat-86*, N. P. Ambulos, Jr., E. J. Duvall, and P. S. Lovett, *J. Bacteriol.* **167**, 842 (1986); *cat-66*, E. J. Duvall, D. M. Williams, P. S. Lovett, C. Rudolph, N. Vasantha, and M. Guyer, *Gene* **24**, 171 (1983); *cat-57*, E. J. Duvall, D. M. Williams, S. Mongkolsuk, and P. S. Lovett, *J. Bacteriol.* **158**, 784 (1984); pC221*cat*, W. V. Shaw, D. G. Brenner, S. F. LeGrice, S. E. Skinner, and A. R. Hawkins, *FEBS Lett.* **179**, 101 (1985); pUB112*cat*, R. Bruckner and H. Matzura, *EMBO J.* **4**, 2295 (1985); pC194*cat*, S. Horinuchi and B. Weisblum, *J. Bacteriol.* **150**, 804 (1982). The leader sequence shown for *cat-86* and related genes (*cat-66* and *cat-57*) is referred to as leader 2 since in *cat-86* another leader sequence (leader 1) is located 5′ to leader 2. Leader 1 is not essential to chloramphenicol induction of *cat-86*. Regulatory regions for *cat* genes on pC221, pC194, and pUB112 contain only the leader shown. The asterisks indicate translational stop codons; the horizontal arrows show inverted repeat sequences. [Reprint from E. J. Duvall, N. P. Ambulos, Jr., and P. S. Lovett, *J. Bacteriol.* **169**, 4235 (1987), by kind permission of the publishers.]

Regulation of Gene Expression

Among the few staphylococcal genes whose regulation has been analyzed are a set of naturally occurring resistance genes that are regulated by transcriptional or translational attenuation. Several of these are genes for resistance to antibiotics that inhibit protein synthesis, namely, *erm*, *cat*, and *tmn*. In each case, a short peptide is encoded in the highly structured leader region; inducing concentrations of the antibiotic interfere with translation of this peptide, causing a configurational change in the secondary structure of the leader. In two cases, *erm*[108] and *cat*,[109] this change engenders a configuration of the leader that permits translation of the resistance gene. The relevant sequences in the *cat* system are shown

---

[108] D. Dubnau, *CRC Crit. Rev. Biochem.* **16**, 103 (1984).
[109] R. Bruckner, T. Dick, and H. Matzura, *Mol. Gen. Genet.* **207**, 486 (1987).

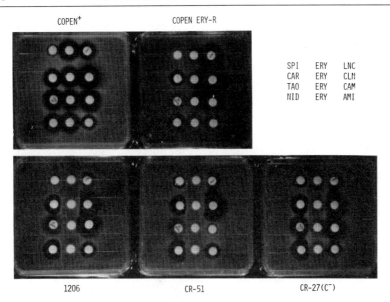

COPEN⁺   COPEN ERY-R

SPI	ERY	LNC
CAR	ERY	CLM
TAO	ERY	CAM
NID	ERY	AMI

1206   CR-51   CR-27(C⁻)

Fig. 10. Erythromycin induction of MLS resistance. The antibiotics used and their disposition on each plate correspond to the pattern given (upper right). SPI, Spiramycin, CAR, carbomycin; TAO, triacetyloleandomycin; NID, niddamycin; ERY, erythromycin; LNC, lincomycin; CLM, clindamycin; CAM, chloramphenicol; AMI, amikacin. Strains: COPEN ERY-R, a laboratory-induced Emʳ mutant; 1206, a naturally occurring MLS-resistant strain containing Tn554; CR-51 and CR-27 (C⁻), other naturally occurring MLS-resistant strains. (Reprinted from Weisblum and Demohn,[111] with kind permission of the publishers).

in Fig. 9. In the *erm* system, the gene product is additionally autoregulated at the level of translation.[110] In the third, *tmn* (*tetM*), in the uninduced state, leader transcripts terminate 5′ to the translational start of the resistance gene; inducing concentrations of the drug prevent this termination (J. Kornblum and R. P. Novick, unpublished data, 1990).

In the case of the erythromycin resistance system, commonly referred to as the MLS system since macrolide, streptogramin B, and lincosamide antibiotics are all affected, erythromycin is an inducer whereas most of the others are not.[111] Uninduced cells are fully sensitive to the noninducer drugs; only cells induced by an inducing drug are resistant (see Fig. 10). Selection for resistance to a noninducer, such as tylosin, yields either constitutive mutants, most of which have mutations or rearrangements within the *erm* leader that prevent formation of the inactive configuration, or mutations increasing plasmid copy number.[112] Plasmids expressing

[110] C. D. Denoya, D. H. Bechhofer, and D. Dubnau, *J. Bacteriol.* **168**, 1133 (1986).
[111] B. Weisblum and V. Demohn, *J. Bacteriol.* **98**, 447 (1969).
[112] B. Weisblum, M. Y. Graham, T. Gryczan, and D. Dubnau, *J. Bacteriol.* **137**, 635 (1979).

MLS resistance constitutively occur naturally, and these also lack sequences in the *erm* leader.[113,114] A different type of transcriptional attenuation regulates synthesis of the replication initiator protein for class I plasmids of the pT181 family. Here, an inhibitory antisense RNA (countertranscript) transcribed from the untranslated 5' leader of the initiator gene complexes with the latter, promoting the formation of a termination hairpin just 5' to the translational start. In the absence of the countertranscript, an upstream sequence pairs with the proximal arm of the terminator stem, preventing termination and permitting transcription of the initiator gene.[7]

The more conventional type of repressor-mediated transcriptional regulation has been documented in two cases, *lac* and *bla*. In the former, there is a regulatory gene, *lacR*, that encodes a single repressor protein that is presumed to function similarly to repressor proteins in other systems.[115] In the case of *bla*, both positive and negative trans-acting regulatory elements have been identified genetically[101] and transcriptional regulation has been demonstrated by gene fusion and Northern blotting analysis.[116] The positively acting elements appear to correspond to a two-component sensory transduction system (P-Z. Wang and R. P. Novick, unpublished data, 1991) as with *Bacillus licheniformis*.[116a] Induction of β-lactamase activity has been demonstrated in cell-free extracts (S. Khan, personal communication, 1981) prepared essentially according to the methods of Lederman and Zubay.[117] Other plasmid-coded inducible genes such as *mer*, *asa*, and *cad* are probably also regulated at the transcriptional level.

A global regulatory system, *agr*, has been identified that controls the postexponential production of exoproteins such as toxins, hemolysins, and exoenzymes.[118] *agr* is a complex polycistronic locus that encodes a two-component signal transduction pathway[119] whose function is to activate transcription of a regulatory RNA molecule that in turn activates

[113] B. C. Lampson and J. T. Parisi, *J. Bacteriol.* **166**, 479 (1986).

[114] M. Monod, C. Denoya, and D. Dubnau, *J. Bacteriol.* **167**, 138 (1986).

[115] B. Oskouian and G. C. Stewart, *J. Bacteriol.* **169**, 5459 (1987).

[116] R. P. Novick, J. Kornblum, B. Kreiswirth, S. J. Projan, and H. Ross, *in* "Molecular Biology of the Staphylococci" (R. P. Novick, ed.), VCH Publishers, Poughkeepsie, New York, 1990.

[116a] T. Kobayashi, Y. F. Zhu, N. J. Nichols, and J. O. Lampen. *J. Bacteriol.* **169**, 3873 (1987).

[117] M. Lederman and G. Zubay, *Biochim. Biophys. Acta* **149**, 253 (1967).

[118] P. Recsei, B. Kreiswirth, M. O'Reilly, P. Schlievert, A. Gruss, and R. Novick, *Mol. Gen. Genet.* **202**, 58 (1985).

[119] B. T. Nixon, C. W. Ronson, and R. M. Ausubel, *Proc. Natl. Acad. Sci. U.S.A.* **83**, 7850 (1986).

transcription of the exoprotein genes.[116] Transcription of genes for surface proteins such as protein A is repressed by this regulatory RNA molecule.[116]

Coupled Transcription–Translation in Cell-Free Extracts

Plasmid DNA. Plasmid DNA is purified by two cycles of equilibrium dye–cesium centrifugation, using $CsCl_2$ at a starting density of 1.54, followed by three cycles of phenol extraction. The final extract is ethanol precipitated and redissolved in 10 mM Tris-HCl with 0.1 mM EDTA, pH 8.0, at a final concentration of 5 mg DNA/ml.

Preparation of S-30 Extract. Two hundred fifty milliliters of a BHI overnight culture is used to inoculate 4 liters of BHI and grown for 4.5–5 hr to a Klett reading of 410–470. The culture is centrifuged and washed with 500 ml of 10 mM Tris–acetate, 14 mM magnesium acetate, 1 mM DTT, 1 M KCl, pH 8.0, and then with 250 ml of the same buffer substituting 50 mM KCl for 1 M KCl. This pellet can be stored at $-80°$. The pellet is resuspended in 30 ml of 10 mM Tris–acetate, 20 mM magnesium acetate 50 mM KCl, 1 mM DTT, pH 8.0; 5 ml lysostaphin, 2 mg/ml in the same buffer, is added and the suspension incubated at 37° for 45 min. Fifty microliters of 0.5 M DTT is added, the lysate is centrifuged at 30,000 g for 30 min, 4°, and the supernatant is removed. The pellet is recentrifuged at 30,000 g for 30 min at 4°, the second supernatant is added to the first, and the combined supernatants are centrifuged at 30,000 g for 30 min at 4°. After this final centrifugation, only the top half of the supernatant is saved. Prior to storage or use, endogenous mRNA is translated and degraded: for each 2 ml of final supernatant is added 0.5 ml of a solution consisting of 670 mM Tris–acetate, pH 8.0, 20 mM magnesium acetate, 7 mM trisodium phosphoenolpyruvate, 7 mM DTT, 5.5 mM ATP, 70 μM each of 20 amino acids, and 75 μg/ml pyruvate kinase. This mixture is incubated at 37° for 30 min, then dialyzed against a buffer consisting of 10 mM Tris–acetate, pH 8, 14 mM magnesium acetate, 60 mM potassium acetate, 1 mM DTT for 18 hr, 4°, with one change of buffer. Protein concentration is determined with Coomassie Brilliant blue[120] and the extract aliquoted and stored under liquid N_2.

Transcription–Translation Reactions. The procedure is essentially that described by Narayanan and Dubnau.[121] The final reaction volume is 20 μl, containing 5 μg plasmid DNA, 45 μM Tris–acetate, pH 8.2, 30 mM ammonium acetate, 45 μM potassium acetate, 1.5 mM DTT, 320 μg PEG 8000, 5 μg *E. coli* ammonium acetate tRNA, 2.3 mM ATP, 0.6 mM each

[120] M. M. Bradford, *Anal. Biochem.* 248 (1976).
[121] C. S. Narayanan and D. Dubnau, *J. Biol. Chem.* **262,** 1756 (1987).

of GTP, CTP, and UTP, 20 mM trisodium phosphoenolpyruvate, 0.6 μg/ml each of pyridoxine hydrochloride, NADP, flavin adenine dinucleotide, and leucovorin, 0.2 μg/ml p-aminobenzoic acid, 11 μCi [^{35}S]methionine (1087 Ci/mmol), 200 μM each of 19 amino acids, 10 mM magnesium acetate, 5 units ribonuclease inhibitor (RNasin, Promega, Madison, WI) and 200 μg S30 extract protein (added last). The reaction mixture is incubated at 60 min for 37° and stopped by the addition of 200 μl acetone to precipitate proteins. After 30 min on ice, the precipitated proteins are pelleted by centrifugation for 30 min 4° in a Beckman microfuge. The pellet is washed with 1.2 ml of 80% (v/v) acetone, then vacuum dried, resuspended in 30 μl SDS–PAGE sample buffer, and boiled for 5 min in a water bath. Samples are analyzed by electrophoresis in 12.5% (w/v) polyacrylamide with 0.1% (w/v) SDS.[122] Prestained protein standards (Diversified Biotech, Newton, MA) are used as molecular weight markers. Gels are dried and exposed to X-ray film (Kodak X-Omat AR) at −80° with an intensifying screen.

Appendix: Media and Buffers

	Component	Concentration
CY broth	Casamino acids (Difco, Detroit, MI)	10.0 g/liter
	Yeast extract (Difco)	10.0 g/liter
	Glucose	5.0 g/liter
	NaCl	5.9 g/liter
	1.5 M β-Glycerophosphate (added after autoclaving)	40.0 ml/liter
0.3GL agar	Casamino acids (Difco)	3.0 g/liter
	Yeast extract (Difco)	3.0 g/liter
	NaCl	5.9 g/liter
	Sodium lactate, 60% syrup	3.3 ml/liter
	25% (v/v) Glycerol	4.0 ml/liter
	Agar (Difco)	15.0 g/liter (bottom agar)
	Agar	7.5 g/liter (top agar)
	Adjust pH to 7.8	
Phage agar	Casamino acids	3.0 g/liter
	Yeast extract	3.0 g/liter
	NaCl	5.9 g/liter
	Agar	15.0 g/liter (bottom agar)
	Agar	5.0 g/liter (top agar)
	Adjust pH to 7.8	
Phage buffer	0.1 M MgSO$_4$	10.0 ml/liter
	0.4 M CaCl$_2$	10.0 ml/liter

[122] U. K. Laemmli, *Nature (London)* **227**, 680 (1970).

	Component	Concentration
	2.5 M Tris, pH 7.8	20.0 ml/liter
	NaCl	5.9 g/liter
	Gelatin	1.0 g/liter
CV agar	Casamino acids (Difco)	10.0 g/liter
	Glucose	50.0 g/liter
	Tryptophan	25.0 mg/liter (optional)
	NaCl	5.9 g/liter
	KPO$_4$	0.1 M
	Agar	15.0 g/liter
	Adjust pH to 7.5.	
	Add after autoclaving:	
	Nicotinic acid	100 μg/liter
	Thiamin	100 μg/liter
DM3	Sodium succinate	135.0 g/liter
	Agar	8.0 g/liter
	H$_2$O	700.0 ml/liter
	Adjust pH of succinate	
	and water to 7.2, then	
	add agar and autoclave.	
	Cool to 65°, then add:	
	5% (w/v) Casamino acids	100.0 ml/liter
	20% (w/v) Glucose	25.0 ml/liter
	10% (w/v) Yeast extract	50.0 ml/liter
	1 M KPO$_4$, pH 7.5	100.0 ml/liter
	1 M MgCl$_2$	20.0 ml/liter
	5% (w/v) BSA (bovine serum albumin)	10.0 ml/liter
	Antibiotics as required	
	Pour plates (30–40 ml) the day before use	
2 × SMM	Sucrose	342.0 g/liter
	Sodium maleate	4.6 g/liter
	MgCl$_2$	8.1 g/liter
SMMP	2 × SMM	500.0 ml/liter
	4 × Penassay broth (Difco)	500.0 ml/liter
TE buffer	Tris-HCl	10 mM
	EDTA	1 mM
	pH 7.5	
TES buffer	Tris-HCl	20 mM
	NaCl	50 mM
	EDTA	10 mM
	pH 7.5	
SET buffer	Tris-HCl	20 mM
	EDTA	50 mM
	Sucrose	200 g/liter
	pH 7.0	

	Component	Concentration
Protein sample	Tris-HCl	30 g/liter
buffer (2×)[122]	SDS	40 g/liter
	Glycerol	200 ml/liter
	2-Mercaptoethanol	100 ml/liter
	Bromphenol blue	40 mg/liter
Lysis mixture	Sucrose	200 g/liter
	Tris-HCl	10 mM
	EDTA	1 mM
	NaCl	100 mM
	RNase I	250 mg/liter
	Lysostaphin	30 mg/liter
	pH 6.0	
Tracking dye	Glycerol	500 ml/liter
	EDTA	0.2 M
	Xylene cyanol	1.0 g/liter
	Bromphenol blue	1.0 g/liter

Author Index

Numbers in parentheses are footnote reference numbers and indicate that an author's work is referred to although the name is not cited in the text.

Asheshov, E. H., 603
Ashwood-Smith, M. J., 251
Askov, S., 239
Asmus, A., 346
Astell, C. R., 110
Astell, P., 102
Atkinson, K., 101
Attfield, P., 102
Atvges, P., 462
Audureau, A., 17
Auerswald, E. A., 361, 362
Aukerman, S. L., 116
Austin, S., 11, 16, 368
Ausubel, F. M., 265, 267, 273(15), 393, 401,
    405, 406, 412, 416(13), 469, 511
Ausubel, F., 400
Ausubel, R. M., 632
Avery, L., 362, 363, 366
Avery, O. T., 558
Ayakawa, G. Y., 577
Ayer, V. N., 405

B

Babykin, M. M., 462, 463(28), 469(28)
Bachhuber, M., 191, 209
Bachi, B., 19, 20(3), 600, 601(33)
Bachmann, B. J., 44, 45(10), 47(10), 61(10,
    17), 100, 101(27), 102(27), 103(27),
    105(27), 106(27), 279
Bachmann, B., 366
Backhaus, H., 35, 37, 38(36)
Backmann, B. F., 48, 57(10), 58(17)
Bade, E. G., 201, 209
Badgasarian, M., 429
Bagdasarian, M. M., 429, 502
Bagdasarian, M., 468, 469, 472(78), 481(78),
    491, 500, 502, 514
Bailey, C. R., 436
Baker, D. J., 286
Baker, H. V., 238, 239, 240(80)
Baker, T. A., 45, 179, 188, 209
Balassa, G., 318
Baldari, C., 290, 292(34)
Baldini, M. M., 523, 529
Baldwin, T. O., 524, 532
Balganesh, M., 334

Ballard, B. T., 465
Ballou, D. P., 507, 514
Baltz, R. H., 431, 433, 435, 454
Bancroft, I., 430
Bang, S. S., 515
Bannantine, J. P., 587, 588(1), 603(1)
Bar-Nir, D., 441
Barany, F., 339
Barat, M., 305, 313(3)
Barbe, J., 464
Barberis-Maino, L., 603, 608(42)
Barcak, G. J., 327, 328(22), 334, 337, 338(22,
    35, 41), 340
Barker, D. F., 141
Barnes, W. M., 115, 121(12), 123
Barnett, L., 3
Barr, P. J., 138
Barrett, J. F., 554, 555(16), 578
Barrett, K. J., 268
Barritt, D. S., 348
Barritt, L. S., 439
Barron, C., 506
Barth, P. T., 368
Barton, J., 486, 487(4)
Bartowsky, E. J., 515
Basche, M., 428
Bassford, P. J., 235, 236(66)
Bastia, D., 214, 219(14), 240
Bastos, M. C. F., 604
Bastos, M., 604
Bateman, C., 286
Batut, J., 412
Bauer, C. E., 466, 480(68, 69), 481(68), 483
Bauer, C., 422, 423(11, 13)
Bauerle, 39
Baumann, L., 515
Baumann, P., 515
Baumberg, S., 433
Bautz, E. K. F., 9
Beattie, K. L., 326   Beatty, J. T., 462,
    463(32), 468(32), 483, 502(32), 507(32)
Bebenek, K., 138
Bechhofer, D. H., 630
Beck, E., 361
Beck, W. D., 603, 608(42)
Beckman, D., 9, 12(21)
Beckwith, J. R., 62, 194, 212, 233, 235,
    244(61)
Beckwith, J., 6, 10, 13, 14, 15, 18, 44, 141,

H

U

Uchida, T., 102, 110
Uhlen, M., 618
Ullman, A., 106
Ullmann, A., 293
Uotila, J., 407

V

Valentine, R. C., 279
van de Putte, P., 116, 181, 182, 183, 210, 211, 212
Van De Rijn, I., 562
Van der Elsacker, S., 9
Van Doren, S., 469, 477(85), 478(85)
Van Gijesgem, F., 188, 204, 212
van Gijsegem, F., 181, 199, 212
Van Sinderen, D., 113
VanMontagu, M., 407
Varga, A. R., 480, 482
Varreiro, V., 278
Vasantha, N., 619
Vats, S., 436
Velloatto, V., 401
Venema, B., 577
Venema, G., 113
Verhoef, C., 52, 54, 55(33)
Vericat, J. A., 464
Viale, A. M., 194, 210
Vieira, J., 109, 128, 278, 327, 328(27), 583
Viera, J., 477
Vieria, J., 393
Vignais, P. M., 462, 463(27), 473, 474(27), 475(97)
Vignais, P., 462
Vijg, J., 286, 287(21)
Vilotte, J. L., 295, 297(51)
Vimr, E. R., 523
Vincent, J. M., 400
Vincze, E., 285
Vining, L. C., 436
Vinopal, R. T., 5, 7(9)
Vinson, C. R., 294
Voeykova, T. A., 437
Vogel, H. J., 119, 120(29)
Vogel, W., 37
Vogelstein, B., 564

W

Vogt, M., 323, 331
Vögtli, M., 457
Volckaert, G., 101, 110
Vollrath, D., 361
von Specht, B.-U., 506
Vonshak, A., 426
Vreemann, J., 62

Waddell, C. S., 195, 212
Waddell, L., 178, 299
Wahl, G., 563, 566(32)
Waldman, A. S., 326
Waldron, H., 600
Walker, E. M., 53, 55(31)
Walker, G. C., 44, 114, 169, 171, 176, 237, 398, 400, 401, 402, 404(6), 406, 407(23), 408(6), 409(2), 410(6), 411(6), 413
Walker, G. W., 387, 390(16)
Walker, G., 96, 400
Walkins, M. M., 405
Wall, J. D., 23, 464, 465, 466
Wallace, J. C., 511, 512(92)
Walsh, R. B., 151
Walter, M. A., 529
Walter, R. B., 323, 327, 328(25), 334, 338
Wandersman, C., 25
Wang, A., 144, 145, 146(21)
Wang, B., 198, 201, 202, 209, 212
Wang, J. C., 267, 268(19)
Wang, M.-D., 202, 209
Wang, P., 618, 619, 620(94)
Wang, Y., 83
Wann, M., 55, 59(42)
Wannamaker, L. W., 568
Wanner, B. L., 44, 186, 211, 229, 236, 237
Wanner, B., 10
Ward, D., 174
Ward, J. E., 392
Ward, J. M., 434, 435(22), 436, 437(42), 441, 444, 456(42), 458
Warren, M., 436
Watkins, M. M., 481
Watkins, M., 473, 475(98)
Watson, J. M., 486
Watson, J., 486
Watt-Tobin, R. J., 3

Subject Index

A

D

N

S